W0051352

CIRCULATING REGULATORY FACTORS AND NEUROENDOCRINE FUNCTION

ADVANCES IN EXPERIMENTAL MEDICINE AND BIOLOGY

Editorial Board:

NATHAN BACK, *State University of New York at Buffalo*

IRUN R. COHEN, *The Weizmann Institute of Science*

DAVID KRITCHEVSKY, *Wistar Institute*

ABEL LAJTHA, *N. S. Kline Institute for Psychiatric Research*

RODOLFO PAOLETTI, *University of Milan*

Recent Volumes in this Series

Volume 266
LIPOFUSCIN AND CEROID PIGMENTS
Edited by Eduardo A. Porta

Volume 267
CONSENSUS ON HYPERTHERMIA FOR THE 1990s: Clinical Practice in
Cancer Treatment
Edited by Haim I. Bicher, John R. McLaren, and Giuseppe M. Pigliucci

Volume 268
EXCITATORY AMINO ACIDS AND NEURONAL PLASTICITY
Edited by Yehezkel Ben-Ari

Volume 269
CALCIUM BINDING PROTEINS IN NORMAL AND TRANSFORMED CELLS
Edited by Roland Pochet, D. Eric M. Lawson, and Claus W. Heizmann

Volume 270
NEW DEVELOPMENTS IN DIETARY FIBER: Physiological, Physicochemical,
and Analytical Aspects
Edited by Ivan Furda and Charles J. Brine

Volume 271
MOLECULAR BIOLOGY OF ERYTHROPOIESIS
Edited by Joao L. Ascensao, Esmail D. Zanjani, Mehdi Tavassoli,
Alan S. Levine, and F. Roy MacKintosh

Volume 272
CIRRHOSIS, HEPATIC ENCEPHALOPATHY, AND AMMONIUM TOXICITY
Edited by Santiago Grisolía, Vicente Felipo, and María-Dolores Miñana

Volume 273
TOBACCO SMOKING AND ATHEROSCLEROSIS: Pathogenesis and Cellular
Mechanisms
Edited by John N. Diana

Volume 274
CIRCULATING REGULATORY FACTORS AND NEUROENDOCRINE FUNCTION
Edited by John C. Porter and Daniela Ježová

A Continuation Order Plan is available for this series. A continuation order will bring delivery of each new volume immediately upon publication. Volumes are billed only upon actual shipment. For further information please contact the publisher.

CIRCULATING REGULATORY FACTORS AND NEUROENDOCRINE FUNCTION

Edited by

John C. Porter

The University of Texas Southwestern Medical Center at Dallas
Dallas, Texas

and

Daniela Ježová

Institute of Experimental Endocrinology
Bratislava, Czechoslovakia

PLENUM PRESS • NEW YORK AND LONDON

Library of Congress Cataloging in Publication Data

Circulating regulatory factors and neuroendocrine function / edited by John C. Porter and Daniela Ježová.

    p.   cm. — (Advances in experimental medicine and biology; v. 274)

    Proceedings of a symposium held June 26–July 1, 1989 in Smolenice Castle, Czechoslovakia, as a satellite symposium to the XXXI International Congress of Physiological Sciences.

    Includes bibliographical references.

    Includes index.

    ISBN-13: 978-1-4684-5801-5   e-ISBN-13: 978-1-4684-5799-5

    DOI: 10.1007/ 978-1-4684-5799-5

    1. Neuroendocrinology — Congresses. 2. Blood-brain barrier — Congresses. 3. Peptide hormones — Physiological effect — Congresses. 4. Interleukin 1 — Physiological effect — Congresses. I. Porter, John C. II. Ježová, Daniela. III. International Congress of Physiological Sciences (31st: 1989: Helsinki, Finland)

    [DNLM: 1. Blood-Brain Barrier — physiology — congresses. 2. Hormones — blood — congresses. 3. Neurosecretory Systems — physiology — congresses. WL 102 C578 1989]

QP356.4.C55   1990

599'.0188 — dc20

DNLM/DLC   90-7365

for Library of Congress   CIP

Proceedings of the symposium on Circulating Regulatory Factors and Neuroendocrine Function, held June 26–July 1, 1989, in Smolenice Castle, Czechoslovakia

© 1990 Plenum Press, New York

Softcover reprint of the hardcover 1st edition 1990

A Division of Plenum Publishing Corporation

233 Spring Street, New York, N.Y. 10013

Contributors

Aguilera, G., *Endocrinology and Reproduction Research Branch, National Institute of Child, Health and Human Development, National Institutes of Health, Bethesda, Maryland, U.S.A.*

Banks, W.A., *Veterans Administration Medical Center, New Orleans, Louisiana, U.S.A.*

Berkenbosch, F., *Department of Pharmacology, Medical Faculty, Free University, Amsterdam, The Netherlands*

de Jong, F.H., *Department of Biochemistry, Division of Chemical Endocrinology, Erasmus University Rotterdam, Rotterdam, The Netherlands*

de Kloet, E.R., *Rudolf Magnus Insitute, Utrecht, The Netherlands*

Ermisch, A., *Department of Cell Biology and Regulation, Section of Biosciences and Interdisciplinary Centre of Neurosciences, Karl Marx University, Leipzig, German Democratic Republic*

Eskay, R.L., *Section of Neurochemistry, Laboratory of Clinical Studies, National Institute on Alcohol Abuse and Alcoholism, Bethesda, Maryland, U.S.A.*

Grossman, A., *Department of Endocrinology, St. Bartholomew's Hospital, West Smithfield, London, Great Britain*

Ježová, D., *Institute of Experimental Endocrinology, Centre of Physiological Sciences, Slovak Academy of Sciences, Bratislava, Czechoslovakia*

Johansson, B.B., *Department of Neurology, University of Lund, Lund, Sweden*

Kozlowski, G.P., *Department of Physiology, University of Texas Southwestern Medical Center, Dallas, Texas, U.S.A.*

Kvetňanský, R., *Institute of Experimental Endocrinology, Centre of Physiological Sciences, Slovak Academy of Sciences, Bratislava, Czechoslovakia*

Lichardus, B., *Institute of Experimental Endocrinology, Centre of Physiological Sciences, Slovak Academy of Sciences, Bratislava, Czechoslovakia*

Linton, E.A., *Department of Biochemistry and Physiology, University of Reading Whiteknights, Reading, Great Britain*

Lorenzi, M., *Eye Research Institute of Retina Foundation and Departments of Ophthalmology and Medicine, Harvard Medical School, Boston, Massachusetts, U.S.A.*

McCann, S.M., *Department of Physiology, Neuropeptide Division, University of Texas, Southwestern Medical Center, Dallas, Texas, U.S.A.*

Meyerhoff, J.L., *Neurochemistry & Neuroendocrinology Branch, Department of Medical, Neurosciences, Division of Neuropsychiatry, Walter Reed Army Institute of Research, Washington, District of Columbia, U.S.A.*

Oliver, C., *Laboratoire de Neuroendocrinologie Expérimentale, INSERM U 297, Faculté de Médecine Nord, Marseille, France*

Reymond, M.J., *Division of Endocrinology, Department of Internal Medicine, C.H.U.V.-EH 19, Lausanne, Switzerland*

Rivier, C., *The Clayton Foundation Laboratories for Peptide Biology, The Salk Institute, San Diego, California, U.S.A.*

Rothwell, N.J., *Department of Physiological Sciences, University of Manchester School of Biological Sciences, Manchester, Great Britain*

Porter, J.C. , *Department of Obstetrics and Gynecology, University of Texas Southwestern Medical Center at Dallas, Dallas, Texas, U.S.A.*

Saavedra, J.M., *Section on Pharmacology, Laboratory of Clinical Science, National Institute of Mental Health, Bethesda, Maryland, U.S.A.*

Samson, W.K., *Department of Anatomy & Neurobiology, University of Missouri, School of Medicine, Columbia, Missouri, U.S.A.*

Tilders, F.J.H., *Department of Pharmacology, Free University, Medical Faculty, Amsterdam, The Netherlands*

Vigaš, M., *Institute of Experimental Endocrinology, Centre of Physiological Sciences, Slovak Academy of Sciences, Bratislava, Czechoslovakia*

Preface

During the past several decades, much research effort has gone into the elucidation of the role of neuroendocrine systems as secretory and metabolic regulators of cells of a variety of organs and structures, including the testes, ovaries, adrenals, thyroid, pituitary gland, and mammary glands. However, the role of cells comprising such organs and structures in the modulation of neuroendocrine processes has received considerably less attention and is generally less well appreciated.

Nonetheless, it is important that we understand the actions on neuroendocrine systems of substances that reach the brain by way of the vasculature, including hormones, cytokines, toxins, amino acids, drugs, and similar agents. In order to analyze the present state of knowledge on this topic, experimental scientists and clinicians, whose shared interests include actions of circulating agents on the brain, met at a satellite symposium of the XXXI International Congress of Physiological Sciences. This symposium, entitled *Circulating Regulatory Factors and Neuroendocrine Function*, was held in Smolenice Castle, Czechoslovakia, June 26-July 1, 1989, and reviews delivered at this symposium as invited presentations are published in this volume. Presentations given as free communications have been published separately and are available in *Endocrinologia Experimentalis* 24: 1-273, 1990.

The success of the meeting was attributable in large part to two factors: first, the participation of scientists working in various fields in disparate countries who would not otherwise have had the opportunity to establish personal collaborations, and second, the excellent organization of every aspect of the program that took place in the charming atmosphere of the 14th century castle, Smolenice. The Castle is situated in a beautiful pastoral setting amidst the Carpathian mountains, not far from the historic cities of Bratislava and Vienna. After World War II, and after a long and colorful history, Smolenice Castle became the property of the Slovak Academy of Sciences who graciously provided the setting for this forum.

We wish to use this opportunity to thank the Slovak Academy of Sciences and the Institute of Experimental Endocrinology, especially the Institute's Director, Professor L. Macho, for providing such excellent circumstances in which to conduct the meeting and for their generous financial support. The importance of the topics discussed at the symposium was also acknowledged by the National Science Foundation, U.S.A., and by the Czechoslovak Medical Society. Their assistance helped bring together several renown scientists. We are indebted to these organizations.

The Chairmen of the symposium (Professor J.C. Porter, Dallas, and Professor M. Vigaš, Bratislava) and the Secretary General (Dr. D. Ježová, Bratislava) appreciate the collaboration of the Executive Secretaries: Professor C. Oliver, Marseille; Professor F.J.H. Tilders, Amsterdam; and Professor C. Rivier, San Diego, whose combined efforts greatly contributed to the high scientific value of the symposium. We also recognize the assistance of the members of the International Organizing Committee, particularly Drs. J.M. Saavedra and J. Farah. The successful evolution of the symposium was due to the excellent work of the Local Organizing Committee, consisting of the non-professional yet enthusiastic staff of the Institute of Experimental Endocrinology. To each contributor we extend a warm thank you.

We also wish to acknowledge the generous contributions of the following organizations: Bachem Company (U.S.A.), Boehringer Mannheim (GmbH, Austria), Duphar, B.V. (The Netherlands), Flamingo Fisch (FRG), Koospol (Czechoslovakia), Nippon Zoki (Japan), Searle (U.S.A.), and Servier (France).

The symposium was made memorable by the congeniality and camaraderie of the participants. Their enthusiasm has stimulated the organizers to consider holding a similar meeting in 1993.

In conclusion, the Editors wish to applaud the extraordinary contributions of Kay Stanley during the planning phases of the symposium and in the preparation of the manuscripts for publication. Her dedication and commitment to these tasks have not deviated. Her contributions from beginning to end were beyond value. To Kay, the Editors express sincere gratitude.

John C. Porter
University of Texas
Southwestern Medical Center at Dallas
Dallas, Texas (U.S.A.)

Daniela Ježová
Institute of Experimental Endocrinology
Bratislava (Czechoslovakia)

January 31, 1990

Contents

Chapter 1
The Tuberoinfundibular Dopaminergic Neurons of the Brain: Hormonal Regulation 1

J.C. Porter, W. Kedzierski, N. Aguila-Mansilla,
B.A. Jorquera, and H.A. González

Introduction 1
Neurotropic Agents 2
Prolactin and Fertility 2
Prolactin and TIDA Neurons 4
Young Adult Female Rats 4
Aged Female Rats 5
Young Adult Male Rats 8
Ovarian Hormones and TIDA Neurons 11
Young Adult Female Rats 11
Aged Female Rats 14
Young Adult Male Rats 17
Discussion 19
Summary 21
Acknowledgements 21
References 21

Chapter 2
The Physiology of the Blood-Brain Barrier

B.B. Johansson

Introduction 25
History 25
Morphology of Brain Endothelial Cells 25
The Role of the Astrocytes 26
Lipid Solubility 26
The BBB: A Modified Tight Epithelium 27
Transport Mechanisms 27
Cell Polarity 29
Enzymatic Barriers 29
Neurogenic Influences 30
Dynamics and Fluctuations of BBB Functions 31
Immunological Aspects 31
Brain Regions Lacking BBB have a CSF Barrier 31

Facilitating Passage of Substances from Blood to Brain 32
   Increasing Lipid Solubility 32
   Cationization 33
   Liposome Entrapment 33
   Coupling to Substances Entering by Receptor-Mediated
   Transport 34
   Experimentally Increasing the BBB Permeability 34
Is it Wise to Alter the BBB? 34
Summary 35
References 35

Chapter 3
Blood-Brain Barrier and Neuroendocrine Regulations   41

   D. Ježová, Z. Opršalová, F. Héry, M. Héry, A. Kiss,
   J. Jurčovičová, J. Chauveau, C. Oliver, B.B. Johansson,
   and M. Vigaš

Catecholamines and ACTH Secretion 49
Serotonin and ACTH Secretion 51
Acknowledgements 56
References 56

Chapter 4
Exchange of Peptides Between the Circulation and the Nervous System: Role of the Blood-Brain Barrier   59

   W.A. Banks and A.J. Kastin

Early Work on Peptides and the BBB 59
Nonsaturable Passage of Peptides 61
   Saturable Transport 63
Modification of Transport 63
   Lighting 64
   Amino Acids 64
   Monoamines 64
   Stress 64
   Aging 65
   Analgesia 66
   Aluminum 66
   Ethanol 66
References 67

Chapter 5
Vasopressin, The Blood-Brain Barrier, and Brain Performance   71

   A. Ermisch and R. Landgraf

Introduction 71

The Peripheral and Central Release of Vasopressin 72
Vasopressin and Information Processing within the Brain 75
The BBB as a Target for Vasopressin 76
AVP and Brain Performance 83
References 85

Chapter 6
Neuroendocrine Responses to Emotional Stress: Possible Interactions Between
Circulating Factors and Anterior Pituitary Hormone Release   91

J.L. Meyerhoff, M.A. Oleshansky, K.T. Kalogeras, E.H. Mougey
G.P. Chrousos, and L.B. Granger

Introduction 91
Methods 92
Results 95
Discussion 97
Summary 104
Acknowledgements 104
References 105

Chapter 7
Regulation of the Sympathetic Nervous System by Circulating Vasopressin   113

R. Kvetňanský, D. Ježová, Z. Opršalová, O. Földes,
N. Michajlovskij, M. Dobrakovová, B. Lichardus, and G.B. Makara

Conclusions 129
References 130

Chapter 8
Antibodies to Neuropeptides: Biological Effects and Mechanisms of Action   135

F.J.H. Tilders, J.W.A.M. van Oers, A. White, F. Menzaghi,
and A. Burlet

Introduction 135
The Humoral Antibody-Peptide Interaction Concept 135
    Epitopes and Biological Activity 137
    Affinity and Biological Activity 139
    Binding Kinetics and Biological Activity 139
Antibody-Cell Interaction Concept 141
    Neuronal Uptake of Antibodies to Enzymes 141
    Antibody-Induced Neuronal Lesions 142
    Antibody to Neuropeptides, Uptake and Biological Effects 142
Conclusions 144
References 144

Chapter 9
Circulating Corticotropin-Releasing Factor in Pregnancy 147

E.A. Linton, C.D.A. Wolfe, D.P. Behan, and P.J. Lowry

References 160

Chapter 10
Therapeutic Effects of Neuroactive Drugs on Hypothalamo-Pituitary in Man 165

P.J. Trainer and A. Grossman

Prolactin 165
Growth Hormone 168
ACTH 172
Conclusions 173
References 173

Chapter 11
Cardiac Hormones and Neuroendocrine Function 177

W.K. Samson

Cardiac Hormones: Isolation and Physiological Effects
  in the Periphery 177
Cardiac Hormones: Presence and Actions in the
  Central Nervous System 178
Cardiac Hormones: Are the Pharmacologic Effects
  Physiologically Significant? 185
Acknowledgements 185
References 186

Chapter 12
Interactions Between the Circulating Hormones Angiotensin and Atrial Natriuretic Peptide and Their Receptors in Brain 191

J.M. Saavedra

Introduction 191
Materials and Methods 192
  Animals 192
  In Vitro Labeling of Peptide Receptors 194
  Receptor Autoradiography 195
Results 195
  Distribution of Peptide Receptors in Brain 195
  Peptide Receptors in Hypertension 197
  Peptide Receptors in Alterations of Water Balance 199
  Peptide Receptors During Stress 200
  Brain Peptide Receptors After Endocrine Manipulations 200

Discussion 201
Conclusions 207
References 207

Chapter 13
The Anterior Third Ventricle Region is a Receptor Site for the Composition Rather than Volume of Body Fluids 211

B. Lichardus, J. Ponec, M.J. McKinley, J. Okoličány,
I. Gabauer, J. Styk, P. Bakoš, C. Oliver, and N. Michajlovskij

Cerebral Origin of a Natriuretic Hormone 211
Sodium Concentration of Cerebrospinal Fluid and Renal Regulation of
the Extracellular Fluid Volume 213
Anterior Wall of the Third Cerebral Ventricle and Renal
Sodium Excretion 214
Atrial Appendectomy Impairs Natriuresis Induced by Isotonic and Hypertonic
Saline Loads 218
The Pituitary—Another Site of a Cerebral Natriuretic System 222
Conclusions 225
References 225

Chapter 14
Peripheral Neurohumoral Factors and Central Control of Homeostasis During Altered Sodium Intake 227

G. Aguilera

Multifactorial Control of Aldosterone Secretion: Role of AII 227
Regulation of the Adrenal Sensitivity to AII 229
Role of Dopamine in the Adrenal Sensitivity to AII 230
Mechanism of Dopaminergic Regulation of Aldosterone Secretion 231
Source of Dopamine During Changes in Sodium Intake 234
Conclusions 239
References 240

Chapter 15
Central Action of Adrenal Steroids During Stress and Adaptation 243

J.M.H.M. Reul, W. Sutanto, J.A.M. van Eekelen, J. Rothuizen,
and E.R. de Kloet

Introduction 243
Differentiation of Corticosteroid Receptor Types: *In Vivo* and
In Vitro Binding Sites 243
Distribution of Corticosteroid Receptor Sites: Radioligand
Binding Studies 244
Distribution of Corticosteroid Receptor Sites: Immunocytochemistry and
Gene Expression 245

Corticosterone- and Aldosterone-Selective Mineralocorticoid Receptors   246
Species-Specificity of Mineralocorticoid Receptors and Glucocorticoid Receptors
    in Rat and Hamster Brain   246
Plasticity of the Corticosteroid Receptor System: Binding Studies   248
Plasticity: Corticosteroid Receptor Gene Expression   249
Corticosteroid Receptors and Senescence   250
Mineralocorticoid Receptor and Glucocorticoid Receptor
    Occupancy Studies   250
Receptor Plasticity *Vs.* Receptor Occupancy   251
Functional Considerations   251
Summary   252
References   253

Chapter 16
Effects of Thyroid Hormones on the Hypothalamic Dopaminergic Neurons   257

M.J. Reymond and T. Lemarchand-Béraud

Effects of Thyroid Hormones on the Hypothalamic DA Neurons During Fetal
    Life or Neonatal Period   257
Effects of Thyroid Hormones on the Hypothalamic DA Neurons During
    Adult Life   258
Activity of the TIDA Neurons in Adult Streptozotocin-Diabetic Rats: Regulation
    by Thyroid Hormones?   264
Effects of Thyroid Hormones on the Neuroendocrine Hypothalamic TRH-Secreting
    Neurons: Analogies with the TIDA Neurons   265
Acknowledgements   266
References   266

Chapter 17
Inhibin and Related Proteins: Localization, Regulation, and Effects   271

F.H. de Jong, A.J. Grootenhuis, I.A. Klaij, and
W.M.O. Van Beurden

Introduction   271
Detection of Inhibin and Related Peptides   271
Structure of Inhibin and Related Peptides   273
Localization of Inhibin and Related Substances   278
Regulation of Inhibin Production   278
    Studies in Male Animals   278
        In Vitro in Male Animals   278
        In Vivo Studies   279
    Studies in Female Animals   281
        Inhibin Production by Granulosa Cells *In Vitro*   281
        In Vivo Studies   281
Extrapituitary Effects of Inhibin and Activin   282
    Effects on Secretion of GnRH and Oxytocin   282
    Intragonadal Effects   283

Effects on Placenta and Conception 283
Effects on Hematopoietic Cell Lines 284
Effects Related to Carbohydrate Metabolism 284
Summary and Conclusions 284
References 285

Chapter 18
Role of Endotoxin and Interleukin-1 in Modulating ACTH, LH, and Sex Steroid Secretion

295

C. Rivier

Introduction 295
Materials and Methods 295
    Experiments 295
    Drugs 296
Results and Discussion 297
    Effect of LPS on the HPA and HPG Axis 297
    Mechanisms Mediating the Effects of LPS or IL-1 on the HPA Axis 299
    Mechanisms Mediating the Effect of LPS or IL-1 on the HPG Axis 300
Summary 300
Acknowledgements 300
References 301

Chapter 19
Neuroendocrinology of Interleukin-1

303

F. Berkenbosch, R. de Rijk, A. Del Rey, and H. Besedovsky

Introduction 303
Pituitary-Adrenal Activity During Immune Challenges 304
Involvement of Monokines? 304
Site of Action 306
Mechanism of Action 307
Other Neuroendocrine Effects of IL-1 309
Functional Significance 310
Acknowledgements 310
References 311

Chapter 20
Role of Monokines in Control of Anterior Pituitary Hormone Release

315

S.M. McCann, V. Rettori, L. Milenkovic, J. Jurčovičová, and M.C. González

Introduction 315
Interleukin-1 316
Tumor Necrosis Factor (Cachectin) 319
Gamma Interferon 325

Summary 327
References 327

Chapter 21
Interleukins, Signal Transduction, and the Immune System-Mediated Stress Response 331

R.L. Eskay, M. Grino, and H.T. Chen

Introduction 331
Early Studies Linking Immune System and Stress Axis Activation 332
CNS-Mediated Cytokine Stimulation of the HPA Axis 332
Pituitary-Mediated Cytokine Stimulation of the HPA Axis 334
Adrenal-Mediated Cytokine Stimulation of the Stress Axis 338
Summary and Conclusions 339
References 340

Chapter 22
Immunoneurology: A Serum Protein Afferent Limb to the CNS 345

G.P. Kozlowski, G. Nilaver, and B.V. Zlokovič

Introduction 345
The Blood-Brain Barrier (BBB) and Blood-Cerebrospinal Fluid (CSF) Barrier . 345
BBB Studies in the Rat 348
BBB Studies in the Rabbit 350
    Introduction 350
    Materials and Methods 350
    Results and Discussion 351
BBB Studies in the Guinea Pig 353
    Introduction 353
    Materials and Methods 354
        Vascular Brain Perfusion Technique 354
        Experimental Allergic Encephalitis (EAE) in the Guinea Pig 355
        Scintillation Counting 355
        Isotopically Labelled IgG 355
        Measurement of IgG and Plasma Albumin Levels in CSF 356
        Immunocytochemical (ICC) Studies 356
        Calculation of the Unidirectional Blood-to-Brain Transfer Constant 356
    Results 357
Autoimmune Diseases 358
    Introduction 358
    Multiple Sclerosis (MS) 359
    Paraneoplastic Cerebellar Degeneration (PCD) 360
    Alzheimer's Disease (AD) 362
Discussion and Conclusions 362
Acknowledgements 363
References 363

Chapter 23
Neuroendocrine Mechanisms in the Thermogenic Responses to Diet, Infection, and Trauma ... 371

    N.J. Rothwell

Thermogenesis—Basic Concepts and Biological Value 371
Effector Mechanisms of Thermogenesis 371
   Central Control of Thermogenesis 371
Hypothalamic-Pituitary-Adrenal Axis 372
Dietary Related Signals for Thermogenesis 374
Fever 375
Thermogenic Responses to Injury 377
Conclusions 377
References 377

Chapter 24
The Blood-Brain Barrier in Diabetes Mellitus ... 381

    M. Lorenzi

Introduction 381
Central Nervous System (CNS) Abnormalities in Diabetes 381
   Cognitive Function 382
   EEG Changes 382
   Nociception and Neuroendocrine Abnormalities 382
   Abnormalities of the Cerebral Microvasculature in Diabetes 383
   Permeability 384
   Transport 386
   Histologic Abnormalities 387
References 388

Chapter 25
Circulating Blood Glucose and Hypothalamic-Pituitary Secretion ... 391

    M. Grino, V. Guillaume, A. Caraty, B. Conte-Devolx,
    P. Joanny, F. Boudouresque, G. Pesce, J. Steinberg,
    G. Peyre, A. Dutour, P. Giraud, and C. Oliver

Introduction 391
Effect of Acute Hyperglycemia on Hypothalamic-Pituitary Secretion 391
Effect of Diabetes Mellitus on Hypothalamus-Pituitary Secretion 392
Effect of Hypoglycemia on Hypothalamic-Pituitary Secretion 399
Conclusions 403
References 403

Chapter 26
Nutritional and Hemodynamic Factors Influencing Adenopituitary Function in Man 407

M. Vigaš, P. Tartár, D. Ježová, J. Jurčovičová,
R. Kvetňanský, J. Malatinsky, and R. Tigranyan

General Conditions of the Investigation 408
Effect of Hyperlipidemia on Growth Hormone Secretion 409
Tryptophan-Induced Release of GH and PRL: Effect of Glucose Administration 412
Effect of Glucose on Exercise-Induced Release of GH, ACTH, and PRL 413
Hypoglycemia-Induced GH and PRL Release: Mediation by Different
    Glucoreceptor Areas 416
Plasma GH, PRL, and Cortisol after Transient Cerebral Ischemia 416
Effects of Open Heart Surgery on Growth Hormone, Cortisol, and Insulin
    Concentrations 421
Conclusions 423
References 424

Chapter 27
Differences in the Effects of Acute and Chronic Administration of Dexfenfluramine on Cortisol and Prolactin Secretion 427

C. Oliver, D. Ježová, M. Grino, V. Guillaume,
F. Boudouresque, B. Conte-Devolx, G. Pesce,
A. Dutour, and D. Becquet

Introduction 427
Effect of Dexfenfluramine on the Hypothalamic-Pituitary-Adrenal Axis
    in Rat 428
Materials and Methods 428
    Animals 428
    Collection of HPB 428
    Cannulation of the Tail Artery 428
    Acute d-F Treatment 428
    Chronic d-F Treatment 429
    Hormone Measurements 430
    Statistical Analysis 430
Results 430
    Acute d-F Treatment 430
    Chronic d-F Treatment 431
Discussion 431
Effects of Dexfenfluramine on the Secretion of Pituitary Hormones
    in Human 434
    Effect in Human 434
Materials and Methods 435
    Subjects 435
    Protocols 435
    Determination of Hormones 435
    Statistical Analysis 436
Results 436

Discussion 437
Conclusion 440
Acknowledgements 441
References 441

Participants 445

Index 449

THE TUBEROINFUNDIBULAR DOPAMINERGIC NEURONS OF THE BRAIN: HORMONAL REGULATION

John C. Porter, Wojciech Kedzierski, Nelson Aguila-Mansilla,
Bernardo A. Jorquera, and Héctor A. González[1]

Departments of Obstetrics and Gynecology and Physiology
University of Texas Southwestern Medical Center at Dallas
Dallas, Texas 75235

Introduction

In the last few decades, our concepts of the mechanisms by which certain brain cells are involved in the regulation of other cells, including other brain cells, have undergone major alterations, and it seems likely that additional changes of equal, if not greater, significance will be forthcoming in the not distant future. Such advances, when they occur, will likely follow the development of new methods or improvement of existing techniques for evaluating various cellular functions, including secretory functions.

In the study of the influence of circulatory factors on neuroendocrine processes, it can be readily appreciated that it is easier to investigate analytically a secretory product of brain cells that is fated for the vasculature than it is to perform a similar feat for a product that does not do so. In addition, other complexities arise. Does a given circulatory factor act directly on a specific set of neuroendocrine cells? Does it act on other brain cells that in turn influence the neuroendocrine cells in question? In the end, a particular result may in a broad sense be the same, but the mechanism by which the result is achieved may vary widely, and a full understanding of a complete process will depend on the correct analysis.

There is a growing appreciation for the concept that a circulatory factor can have a neurotropic or neurotoxic action on secretory neurons of the brain, as exemplified by the occurrence of a symposium, which is dedicated to this issue. And, as a way of introducing this subject, we shall discuss a few substances, *i.e.*, hormones, secreted by cells of the classical endocrine system that affect neuroendocrine cells. Others will discuss different hormones, different concepts, and a variety of circulatory agents.

It is noteworthy that many drugs and substances of abuse influence neurosecretory cells. The effects of some of these agents appear to be more or less transitory, whereas others are known to modify permanently the function of brain cells and may even appear to be specific for a given subset of cells. Thus, through the intermediacy of the circulation we may be victims of our internal as well as external environments to a greater degree than has been appreciated heretofore. In this light, it is interesting to consider the following question: To what extent is aging of the brain a consequence of environmental toxins of internal or external origin? Consider the agent, 1-methyl-4-phenyl-tetrahydropyridine (MPTP)! Exposure to MPTP has been shown to induce a disease in young persons that is usually associated with the aged, *viz.*, Parkinsonism (1-5). Although an

---

[1] Present affiliation: Department of Dermatology, Hospital San Juan de Dios, Santiago, Chile

Circulating Regulatory Factors and Neuroendocrine Function
Edited by J. C. Porter and D. Ježová
Plenum Press, New York, 1990

1

agent such as MPTP may represent an extreme case, neurotropic or neurotoxic substances having subtle actions on mentation, motivation, emotional state, reproduction, *etc.* may be frequently encountered.

Neurotropic Agents

Prolactin and Fertility

The involvement of the pituitary hormone, prolactin, in reproductive function is strikingly dramatic, especially in circumstances where circulating levels of prolactin are abnormally elevated. In women, hyperprolactinemia due to a pituitary adenoma is associated with low levels of circulating gonadotropin, resulting in infertility. Such a phenomenon is exemplified by the findings in a 24-year-old woman who exhibited galactorrhea during the six years preceding her participation in the following study, and she had not experienced a spontaneous menstrual episode during the last four of these six years (6). It was against this background that her 24-hour secretory patterns of prolactin and luteinizing hormone (LH) were studied. As shown in Figure 1, her plasma prolactin concentrations ranged from a low of 175 to a high of 425 ng/ml, and had an average concentration during the 24 hours of observation of 275 ng/ml. (This concentration of circulating prolactin is approximately 20 times that of non-pregnant, normal young women.) With the exception of the first three of the 72 samples, her plasma LH levels were less than the limit of detection (< 5 ng/ml). This young woman was then treated with bromocriptine mesylate, a dopamine agonist. After seven weeks of treatment, her menses recurred spontaneously. Ten days after the onset of menses, her 24-hour secretory patterns of prolactin and LH were again determined. At this time the concentration of prolactin in her plasma was less undetectable (< 5 ng/ml), but LH was readily demonstrable (Figure 2).

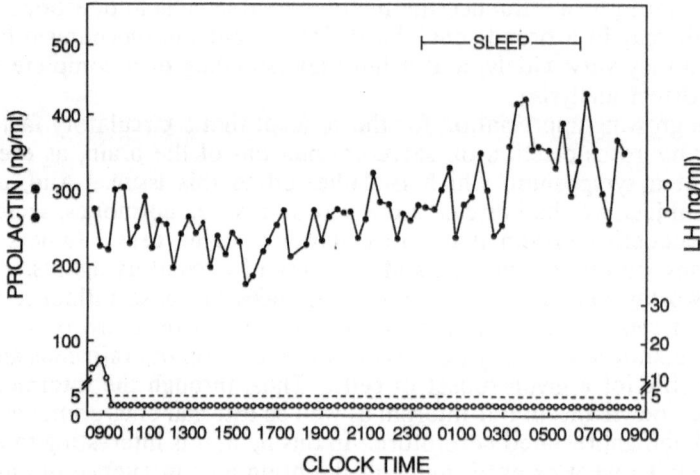

Figure 1.   Twenty-four-hour secretory patterns of prolactin and LH in a 24-year-old woman with a pituitary adenoma. (The study was conducted prior to treatment.) At the times indicated on the abscissa, a blood sample was obtained every 20 min for 24 hr by way of an indwelling venous catheter. Plasma from these samples was analyzed for prolactin *(closed circles)* and LH *(open circles)*. The period when the subject slept is indicated. Reproduced from Porter *et al.* (6) with permission of *FASEB Journal.*

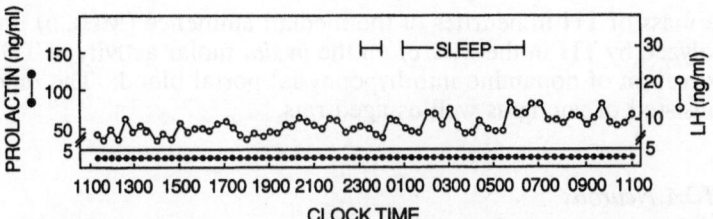

Figure 2. Twenty-four-hour secretory patterns of prolactin and LH in a 24-year-old woman with a pituitary adenoma after treatment for 7 weeks with bromocriptine mesylate (5 mg/day). See Fig. 1 for further details. Reproduced from Porter *et al.* (6) with permission of *FASEB Journal*.

A similar action of prolactin can be demonstrated in rats with experimentally induced hyperprolactinemia. When two pituitary glands or an equivalent amount of muscle tissue were maintained as tissue grafts to the renal cortex of young adult female rats, the average plasma concentration of prolactin was 94 ± 15 (mean and SE) ng/ml in animals with pituitary grafts and 39 ± 7 ng/ml in animals with muscle grafts (Figure 3) (7). These data illustrate the capacity of pituitary grafts to induce hyperprolactinemia as documented earlier by Chen *et al.* (8).

When two pituitaries were grafted to a kidney of each of 52 young female rats, their estrous cycles were promptly interrupted, and most of the animals remained in a state of diestrus during the subsequent four weeks (Figure 4). At the end of this four-week period, the kidney bearing pituitary grafts was ablated in 26 of the 52 animals, and the contralateral kidney, *i.e.*, the kidney not bearing pituitary grafts, was removed from the remaining 26. The animals in which the kidney bearing pituitary grafts was ablated promptly began to exhibit estrous cycles; whereas, most of the animals bearing pituitary grafts continued in diestrus for the remaining four weeks of observation (Figure 4).

Although not conclusive, the data shown in Figures 1, 2, and 4 can be interpreted as evidence that prolactin inhibits gonadotropin secretion by affecting neuroendocrine cells of the hypothalamus. Of the neurosecretory cells of the hypothalamus, those characterized as the tuberoinfundibular dopaminergic (TIDA) neurons have been the most extensively studied, due in part to the ease and accuracy with which various aspects of their secretory activity can be quantitatively evaluated.

Documentation of the existence of dopaminergic neurons in the ventromedial hypothalamus was provided in 1964 by Fuxe (9), in 1966 by Fuxe and Hökfelt (10), and in 1967 by Hökfelt (11). Subsequent work by other investigators has led to the view that these neurons constitute the major catecholaminergic system of the ventral hypothalamus. The perikarya of these cells lie largely in the region of the arcuate and periventricular nuclei, and neurites of these cells lie in the median eminence. Dopamine, a biosynthetic product of hypothalamic neurons (9-11), was first shown by Ben-Jonathan *et al.* in 1977 (12) to be released into hypophysial portal blood. These findings were interpreted as evidence of synthesis and release into portal blood of dopamine by TIDA neurons. Evidence supporting this view has been provided by many subsequent studies. When the intracellular enzymatic activity of tyrosine hydroxylase (TH), an enzyme that is essential for the biosynthesis of dihydroxyphenylalanine (DOPA) (13), the precursor of dopamine, was pharmacologically inhibited, the secretion of dopamine into portal blood was markedly suppressed (14). Indeed, the secretion of dopamine can be suppressed by a variety of agents, including morphine, β-endorphin, enkephalin, serotonin, and 3-hydroxyben-zylhydrazine (NSD 1015) (15-19).

However, it is the purpose of this presentation to discuss the actions of circulatory agents that stimulate, not suppress, TIDA neurons and to analyze the influence of such agents *a)* on the mass of TH in neurites of the median eminence (ME), *b)* on the rate of the reaction catalyzed by TH in the ME, *c)* on the *in situ* molar activity of TH in the ME, and *d)* on the secretion of dopamine into hypophysial portal blood. The models used in these studies consisted of young as well as aged rats.

Prolactin and TIDA Neurons

YOUNG ADULT FEMALE RATS. The mass and *in situ* activity of TH in the ME was evaluated in young female rats bearing two pituitaries grafted to the renal cortex for 5-6 weeks (7). It was found that the quantity of TH in the ME of animals with pituitary grafts was similar to that of animals with muscle grafts (Figure 5), but the total *in situ* activity of TH in the ME (pmol DOPA/ME × hr), in contrast to the amount of TH, was appreciably greater in animals with pituitary grafts than in animals with muscle grafts (Figure 6). In addition, the *in situ* molar activity of TH (mol DOPA/mol TH × hr) in the ME was greater in rats with pituitary grafts than in animals with muscle grafts (Figure 7). These data suggest that the hyperprolactinemia resulting from pituitaries grafted to the renal cortex (Figure 3) may be the causative agent in the stimulation of the *in situ* activity of TH in the ME neurites of TIDA neurons. However, in themselves, these observations can not be taken as conclusive evidence for this interpretation.

For this reason we studied the effect of neutralization of circulating prolactin on TIDA neurons. Neutralization was accomplished by administration of rabbit antiserum against rat prolactin. Two pituitaries were grafted to the renal cortex for two days, and during this time pre-immune rabbit serum or rabbit antiserum against rat prolactin (10 ml/kg body weight) was injected subcutaneously each day (7). (The antiserum contained sufficient antibodies to neutralize five times the amount of prolactin that daily entered the circulation.) At the end of this time, the mass and *in situ* activity of TH in the ME were evaluated.

As shown in Figure 8, *left panel,* there was no significant difference in the quantity of TH in the ME of rats treated with pre-immune serum and that of animals

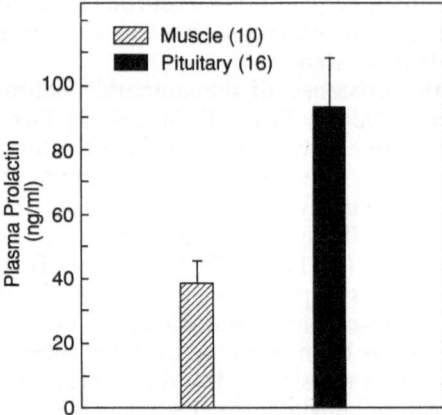

Figure 3.   Hyperprolactinemia in young female rats resulting from two pituitary glands grafted to the renal cortex. Rats with muscle tissue grafted to the renal cortex served as controls. The key is shown in the inset. The number of animals is shown in parentheses. The grafts were of 4 weeks duration. The mean and magnitude of the standard error are indicated by the height of the bar and vertical line, respectively. Redrawn from González and Porter (7) with permission of The Endocrine Society

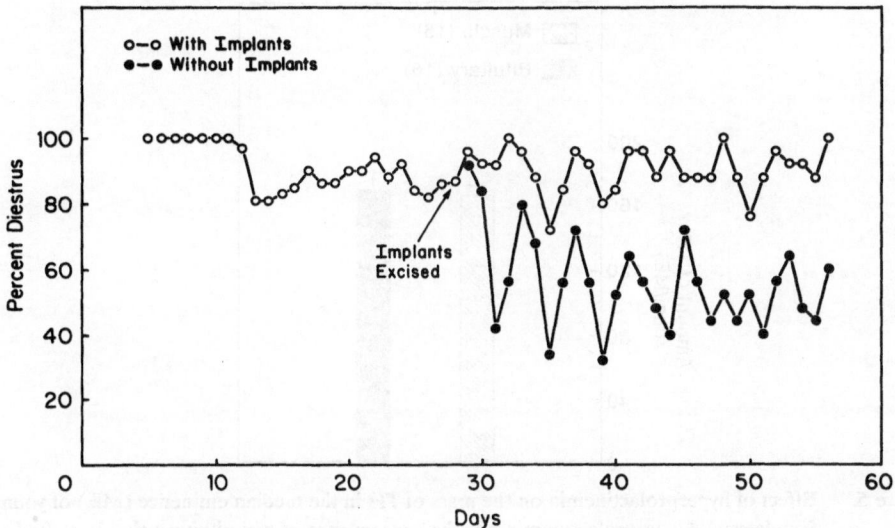

Figure 4. Effect of hyperprolactinemia on the estrous cycles of young female rats. At time 0, two pituitary glands were grafted to the renal cortex of 52 young female rats having regular estrous cycles. Twenty-eight days later (as indicated in the figure), the kidney bearing the pituitary grafts was ablated from 26 of these animals *(closed circles)*, and the kidney not bearing pituitary grafts was ablated from the remaining 26 animals *(open circles)*. Animals that continued to bear pituitary grafts *(open circles)* remained mostly in diestrus. Animals in which the pituitary grafts had been removed by renalectomy promptly began to exhibit estrous cycles *(closed circles)*.

treated with prolactin antiserum. However, neutralization of circulating prolactin greatly reduced the total *in situ* activity of TH in the ME (Figure 8, *middle panel*). In addition, the *in situ* molar activity of TH in the ME of animals receiving prolactin antiserum was much less than that of rats given pre-immune serum (Figure 8, *right panel*).

AGED FEMALE RATS. Aged rats of both sexes have been shown to secrete dopamine into hypophysial portal blood at rates that are appreciably less than those of young animals of the same sex (21,22). The intact aged female rat is a difficult model to study since a population of such animals usually contains rats in constant estrus, rats in constant diestrus, and rats that change back and forth between estrus and diestrus. Moreover, aged female rats with intact ovaries exhibit chronic hyperprolactinemia (23).

For these reasons the following study was performed in aged rats that had been ovariectomized at four months of age. When these animals were 17-24 months old, two freshly excised pituitary glands from young female donors or an equivalent amount of muscle tissue were grafted to the cortex of one kidney. Four or eight weeks after placement of the grafts, the mass of TH and the *in situ* activity of TH in the ME were studied (24). The amount of TH in the ME of these animals is shown in Figure 9. When analyzed by a two-way analysis of variance, it was found that the amount of TH in the ME of aged, ovariectomized rats with pituitary implants was slightly but significantly ($P < 0.05$) greater than that of aged animals with muscle implants. However, the difference was small, and the amount of TH in the ME of animals with pituitary grafts of four weeks duration was the same as those with pituitary grafts of eight weeks duration.

The total *in situ* activity and the *in situ* molar activity of TH in the ME of aged animals are shown in Figures 10 and 11. In both sets of animals, *i.e.*, animals with grafts

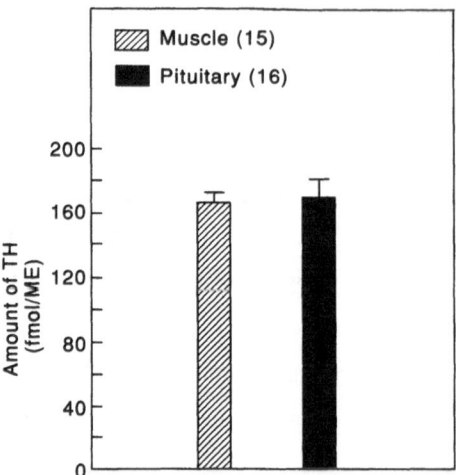

Figure 5. Effect of hyperprolactinemia on the mass of TH in the median eminence (ME) of young female rats. Hyperprolactinemia was the consequence of two pituitary glands grafted to a kidney. Muscle grafts served as controls. Duration of the grafts was 5-6 weeks. In the calculation of the mass of TH, 240,000 was taken as the molecular weight of the enzyme, which is the weight of the tetrameric form that exists in the cell (20). As indicated in the inset, results obtained in rats with muscle grafts is shown by the hatched bar; results obtained in rats with pituitary grafts is shown by the solid bar. The number of animals is given in parentheses. The height of the bar denotes the mean, and the vertical line denotes the standard error. Drawn from González and Porter (7) with permission of The Endocrine Society.

of four weeks duration and of eight weeks duration, the total *in situ* activity and the *in situ* molar activity of TH in the ME were greater in rats with pituitary grafts than in animals with muscle grafts. An evaluation by a two-way analysis of variance revealed that the *in situ* molar activity of TH was significantly ($P < 0.001$) greater in animals with pituitary grafts than in those with muscle grafts (24).

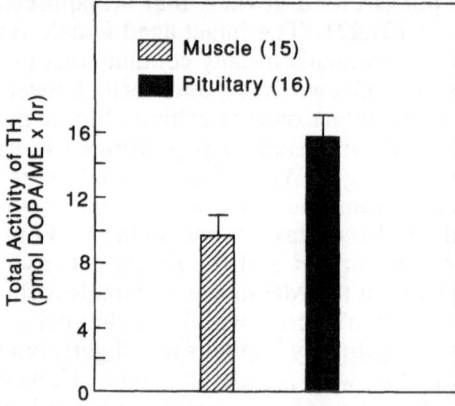

Figure 6. Effect of hyperprolactinemia on the total *in situ* activity of TH in the median eminence. See Fig. 5 for details. Drawn from González and Porter (7) with permission of The Endocrine Society.

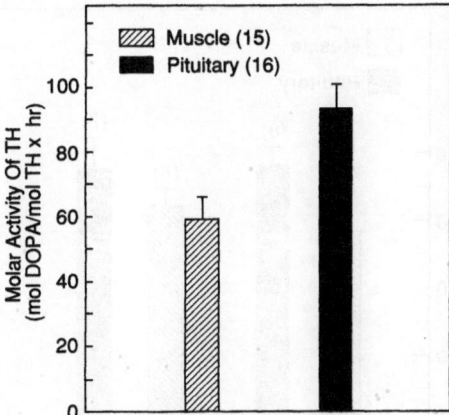

Figure 7. Effect of hyperprolactinemia on the *in situ* molar activity of TH in the median eminence. See Fig. 5 for details. Drawn from González and Porter (7) with permission of The Endocrine Society.

These findings are consistent with the view that hyperprolactinemia induced by pituitary grafts increases the biosynthetic activity of TIDA neurons of the aged brain, and lead to the following question: Does the hyperprolactinemia that occurs spontaneously in aged rats with intact ovaries (23) have a role in maintaining the level of TH activity in the ME? To address this issue, aged intact female rats were treated for two days with pre-immune rabbit serum or antiserum against rat prolactin to neutralize circulating prolactin. As illustrated in Figure 12, the quantity of TH in the ME of animals treated with prolactin antiserum was not significantly different from that of animals treated with pre-immune serum (143 ± 8 *vs.* 160 ± 23 fmol TH per ME). However, the total *in situ* activity and the *in situ* molar activity of TH of the ME of animals treated with prolactin antiserum were significantly ($P < 0.05$) less than that in animals treated with pre-immune

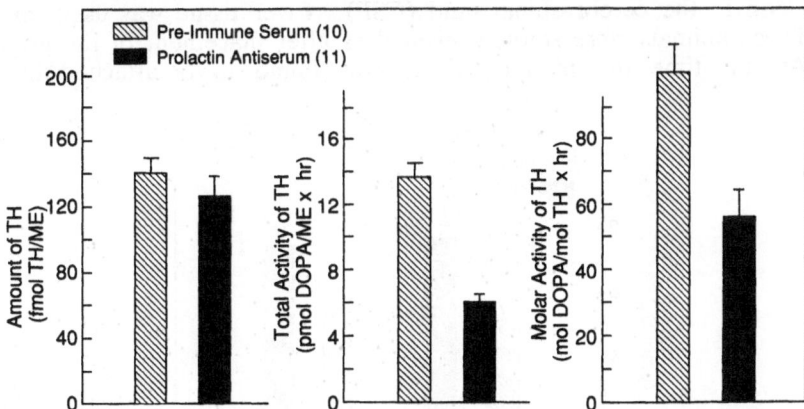

Figure 8. Effect of immunoneutralization of circulating prolactin on the mass and *in situ* activity of TH in the median eminence of young female rats. Hyperprolactinemia was induced by two pituitaries grafted to the renal cortex. The grafts were of 2 days duration, during which the animals were treated with prolactin antiserum or pre-immune serum. Drawn from González and Porter (7) with permission of The Endocrine Society.

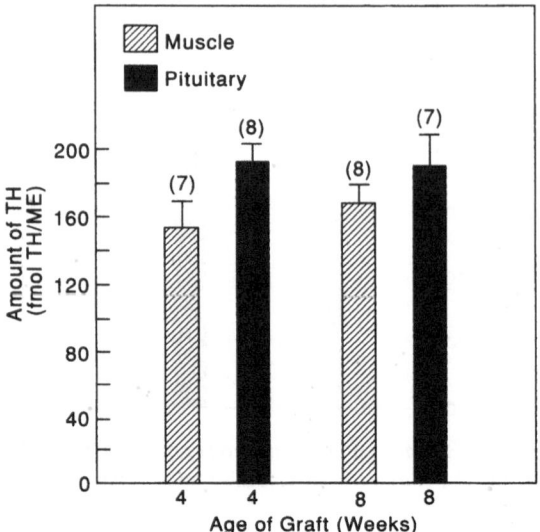

Figure 9.   Effect of hyperprolactinemia on the mass of TH in the median eminence of aged ovariectomized rats. Hyperprolactinemia was induced by two pituitaries grafted to the renal cortex for 4 weeks and 8 weeks. Muscle grafts served as controls. See Fig. 5 for details. Drawn from González *et al.* (24) with permission of S. Karger, AG.

serum (Figure 12). These findings support the view that prolactin has a role in maintaining the synthesis of dopamine in TIDA neurons of the aged brain (24).

YOUNG ADULT MALE RATS. Although it seems unlikely, it can be argued that the alteration in TH activity of TIDA neurons by pituitaries grafted to the renal cortex was dependent on an interaction of the pituitary graft and the kidney or on the sex of the animals. To evaluate these possibilities, an experiment was performed using male rats, and the pituitary graft was placed in a lateral ventricle to induce an elevated concentration of prolactin in the cerebrospinal fluid (CSF). Liver tissue was used for control purposes. These animals were studied seven days after placement of the graft in the ventricle. At this time the pituitary tissue was found to be attached through the

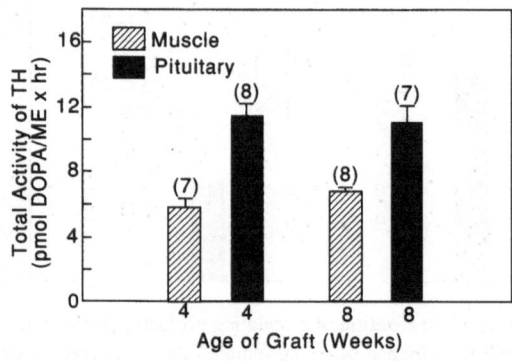

Figure 10.   Effect of hyperprolactinemia on the total *in situ* activity of TH in the median eminence of aged ovariectomized rats. See Fig. 5 and 9 for details. Drawn from González *et al.* (24) with permission of S. Karger, AG.

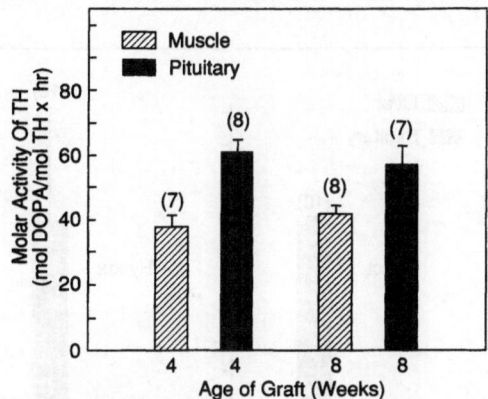

Figure 11. Effect of hyperprolactinemia on the *in situ* molar activity of TH in the median eminence of aged ovariectomized rats. See Fig. 5 and 9 for details. Drawn from González *et al.* (24) with permission of S. Karger, AG.

vasculature to the corpus striatum and nearby structures of the brain (25). When CSF, collected from the cisterna magna, was analyzed for prolactin, it was found that the mean prolactin concentration in the CSF of intact animals bearing intraventricular pituitary grafts was many times that of intact animals bearing liver grafts (Figure 13, *left panel*).

The mass of TH in the ME of intact male animals with intracerebral pituitary grafts was not appreciably different from that of animals with liver grafts (Figure 14, *left panel*). However, the total *in situ* activity of TH was 5.1 ± 0.5 pmol DOPA/ME × hr in intact animals with liver grafts and was 10 ± 1.3 in intact rats with pituitary grafts (Figure 15, *left panel*). The *in situ* molar activity of TH was 46 ± 8 and 109 ± 18 mol DOPA/mol TH × hr in animals with liver grafts and anterior pituitary grafts, respectively (Figure 16, *left panel*). These data demonstrate that anterior pituitary tissue grafted in the brain secretes prolactin into the CSF and is effective in stimulating TH activity in the ME, as were pituitary glands grafted to the kidney that secrete prolactin into blood.

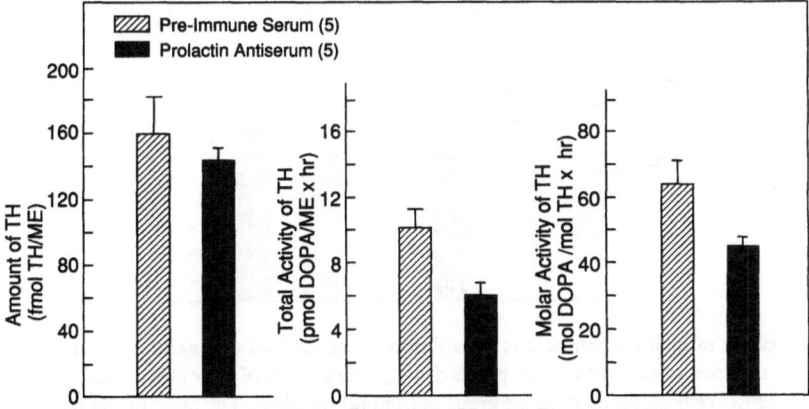

Figure 12. Effect of immunoneutralization of circulating prolactin on the mass and *in situ* activity of TH in the median eminence of aged intact female rats. The animals were treated for 2 days with prolactin antiserum or pre-immune serum. See Fig. 5 for details. Drawn from González *et al.* (24) with permission of S. Karger, AG.

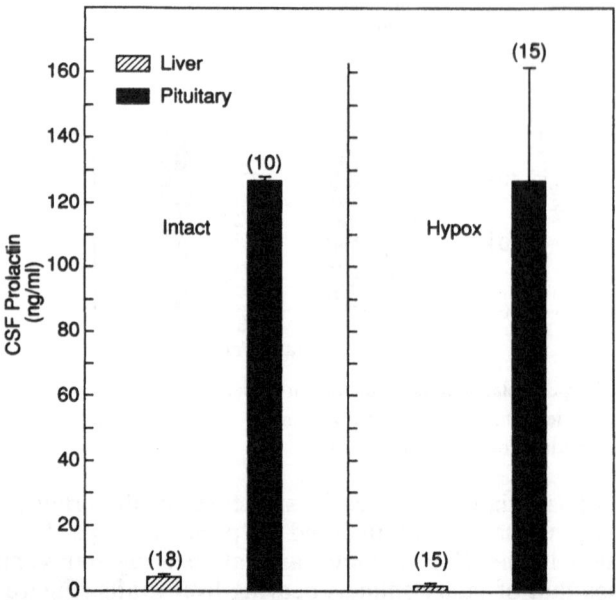

Figure 13. Prolactin in the CSF of male rats bearing pituitary tissue grafts in a lateral ventricle. Liver tissue served as the control. The grafts were of 7 days duration. The recipient animals either had intact pituitary glands *(left panel)* or were hypophysectomized at the time of placement of the grafts *(right panel)*. CSF was obtained from the fourth ventricle 7 days after placement of the tissue in a lateral ventricle. See Fig. 5 for details. Drawn from González *et al.* (25) with permission of The Endocrine Society.

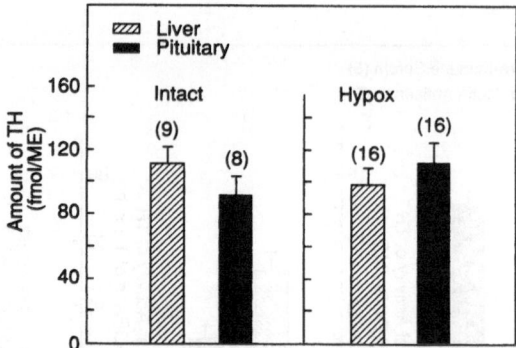

Figure 14. Effect of high concentrations of prolactin in the CSF on the mass of TH in the median eminence of rats with intact pituitaries *(left panel)* and of hypophysectomized rats *(right panel)* bearing pituitary tissue grafts in a lateral ventricle. The mass of TH was measured 7 days after placement on the grafts. See Fig. 5 and 13 for details. Drawn from González *et al.* (25) with permission of The Endocrine Society.

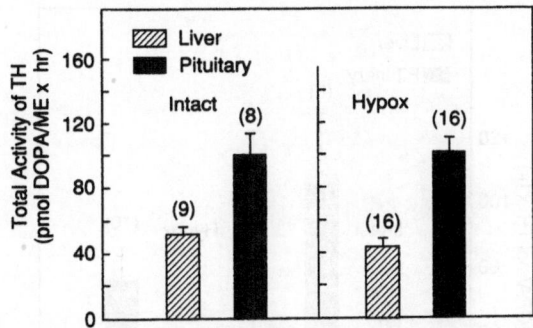

Figure 15. Effect of high concentrations of prolactin in the CSF on the total *in situ* activity of TH in the median eminence of male rats. See Fig. 5 and 13 for details. Drawn from González *et al.* (25) with permission of The Endocrine Society.

There is the possibility that the anterior pituitary grafts in the brain acted by way of the animals own pituitary gland. To evaluate this possibility, the experiment was repeated in hypophysectomized animals. Liver tissue or anterior pituitary tissue was inserted into the lateral ventricle, followed immediately by hypophysectomy. Seven days later, the animals were used for further study. At this time the concentration of prolactin in the CSF of hypophysectomized rats with liver grafts was 1.5 ± 0.2 ng/ml and was 127 ± 35 ng/ml in hypophysectomized animals with anterior pituitary grafts (Figure 13, *right panel*). The mass of TH in the ME in these two paradigms were essentially the same(Figure 14, *right panel*), but the total *in situ* activity of TH was 4.3 ± 0.5 pmol DOPA/ME$\times$ hr in hypophysectomized animals with liver grafts and was 10.1 ± 0.9 pmol DOPA/ME$\times$ hr in similar animals with anterior pituitary grafts (Figure 15, *right panel*). The *in situ* molar activity of TH was 44 ± 4.7 and 76 ± 10 mol DOPA/mol TH $\times$ hr in hypophysectomized rats with liver grafts and anterior pituitary grafts, respectively (Figure 16, *right panel*). These findings demonstrate that anterior pituitary grafts in the lateral ventricles stimulate the enzymatic activity of TH in the neurites of TIDA neurons, and the co-existence of elevated prolactin levels in the CSF and increased *in situ* activity of TH suggest that prolactin is the stimulatory agent. Further studies are needed to establish whether prolactin acts directly on TIDA neurons or other brain cells that in turn act on TIDA neurons.

Ovarian Hormones and TIDA Neurons

YOUNG ADULT FEMALE RATS. The first observations on dopamine in hypophysial portal blood demonstrated that the secretion of dopamine into portal blood of cycling female rats was greater than that of male rats (12). Moreover, the stage of the ovulatory cycle was found to influence the secretion of dopamine, being greatest on estrus. We have subsequently observed on many occasions that the secretion of dopamine is greater in intact female rats than in ovariectomized animals. These findings suggest that a hormone or hormones secreted by the ovaries stimulates TIDA neurons. This suggestion was substantiated by assessment of the *in situ* activity of TH in the ME. As shown in Figure 17, the *in situ* molar activity of TH in the ME of ovariectomized rats was 30 ± 2.1 mol DOPA /mol TH $\times$ hr. In proestrous and estrous animals, the *in situ* molar activity of TH in the ME activity was 40 ± 4 and 87 ± 7, respectively.

From these observations, it cannot, be ascertained whether one ovarian hormone or more than one is involved in the stimulation of TIDA neurons. Moreover, the identity

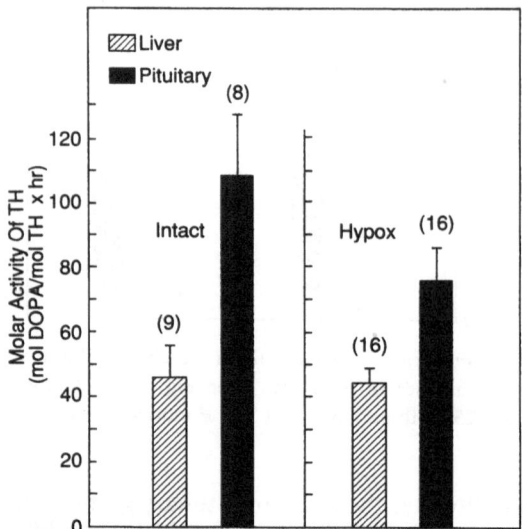

Figure 16. Effect of high concentrations of prolactin in the CSF of male rats on the *in situ* molar activity of TH in the median eminence. See Fig. 5 and 13 for details. Drawn from González *et al.* (25) with permission of The Endocrine Society.

of the ovarian hormone(s) is unclear. On the basis of the study by Butcher *et al.* (26) on the circulating concentrations of various hormones of the female rat throughout the estrous cycle, we hypothesized that the tropic agent(s) of the ovary for the TIDA neurons was estradiol, progesterone, or both. To test this hypothesis, several studies were undertaken. When ovariectomized rats were treated for three days with estradiol or progesterone or with both estradiol and progesterone, a marked increase in the *in situ* molar activity of TH occurred in the ME of animals treated with both estradiol and progesterone (Figure 18). The increase in the activity of TH in the ME of rats treated with estradiol

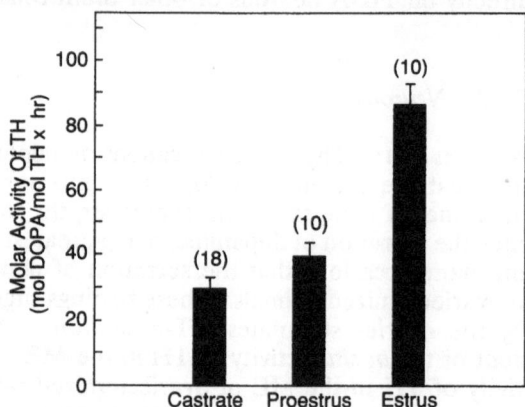

Figure 17. *In situ* molar activity of TH in the median eminence of castrate, proestrous, and estrous rats. See Fig. 5 for details. Drawn from González and Porter (7) with permission of The Endocrine Society.

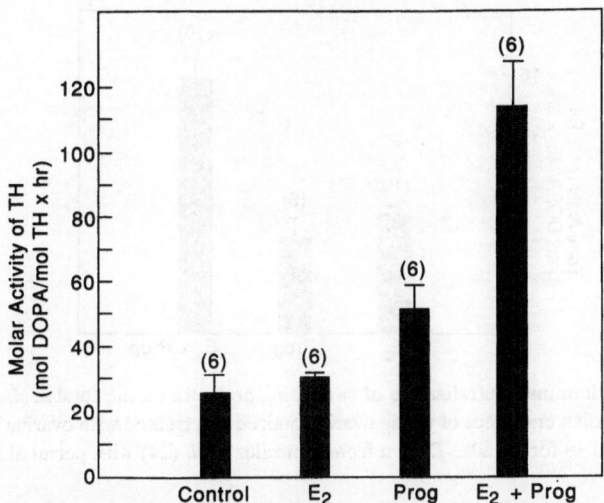

Figure 18. Effect of ovarian hormones on the *in situ* molar activity of TH in the median eminence of young ovariectomized rats. The animals were treated for 3 days with the solvent vehicle, estradiol (E_2), progesterone (Prog), or E_2 and Prog as indicated on the abscissa. See Fig. 5 for details. Drawn from Wang and Porter (27) with permission of the authors.

and progesterone was associated with an increase in the release of dopamine into hypophysial portal blood (27). As shown in Figure 19, the mean concentration of dopamine in portal plasma of ovariectomized rats treated for three days with estradiol and progesterone was 5.7 times that of rats treated with the solvent vehicle.

These data led us to consider the following issue: Was the increase in the biosynthetic and secretory activity of TIDA neurons that was induced by estradiol and

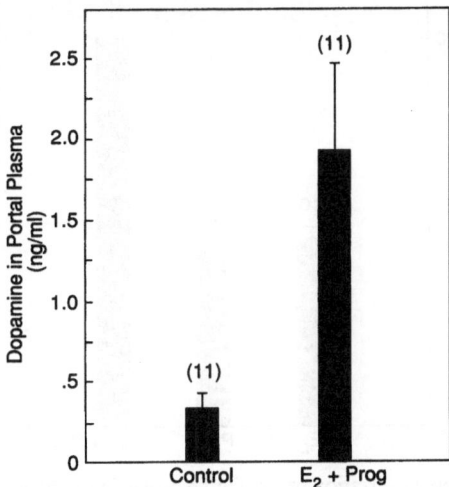

Figure 19. Dopamine in hypophysial portal blood of young ovariectomized rats treated with ovarian hormones. See Fig. 5 and 18 for details. Drawn from Wang and Porter (27) with permission of the authors.

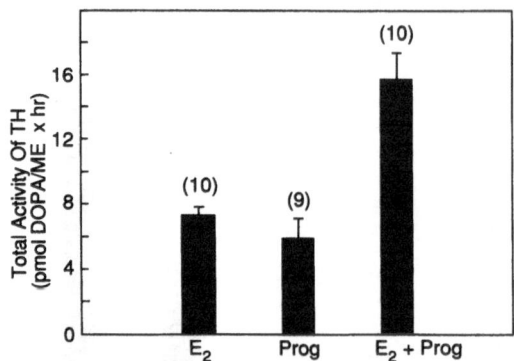

Figure 20. Effect of immunoneutralization of circulating prolactin on the total *in situ* activity of TH in the median eminence of young ovariectomized rats treated with ovarian hormones. See Fig. 5 and 18 for details. Drawn from González *et al.* (24) with permission of S. Karger, AG.

progesterone a consequence of a direct action of these hormones on TIDA neurons or a consequence of the hyperprolactinemia induced by estrogen (28)? This question was addressed in ovariectomized rats that were simultaneously treated for three days with prolactin antiserum and with estradiol, progesterone, or estradiol and progesterone. Antiserum against prolactin did not prevent the increase in the total *in situ* TH activity of the ME (Figure 20) or in the *in situ* molar activity of the ME of estradiol-progesterone-treated rats (Figure 21). The amount of TH in the ME was unaffected by these treatments (Figure 22).

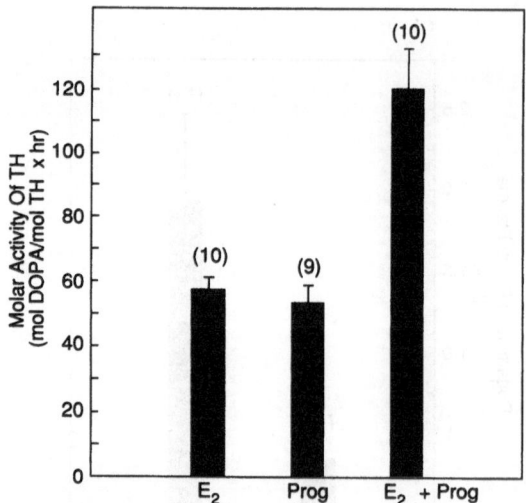

Figure 21. Effect of immunoneutralization of circulating prolactin on the *in situ* molar activity of TH in the median eminence of young ovariectomized rats treated with ovarian hormones. See Fig. 5 and 18 for details. Drawn from González *et al.* (24) with permission of S. Karger, AG.

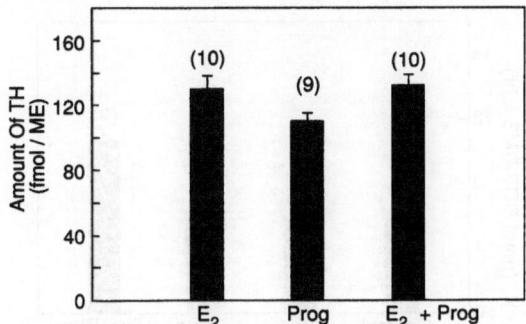

Figure 22. Effect of immunoneutralization of circulating prolactin on the mass of TH in the median eminence of young ovariectomized rats treated with ovarian hormones. See Fig. 5 and 18 for details. Drawn from González *et al.* (24) with permission of S. Karger, AG.

AGED FEMALE RATS. In aged rats of either sex the secretion of dopamine into hypophysial portal blood is appreciably less than that of young rats of the same sex (20,21). This reduction in dopamine secretion is paradoxical in view of the fact that the number of TH-containing neurons in the arcuate nuclei as well as the periventricular region of the hypothalamus of the aged female rat was found to be the same as that of the young female (29). Moreover, the capacity of the ME of the aged rat to form dopamine from exogenous DOPA is high (22), indicating that the reduced ability of the aged rat to secrete dopamine is probably due to an impaired ability to form DOPA from tyrosine.

To obviate the complexity resulting from ovarian function, aged rats ovariectomized at a young age are used in these studies on the response of the TIDA neurons to estradiol and/or progesterone. As shown in Figure 23, after treatment with the solvent vehicle, estradiol, progesterone, or estradiol and progesterone, there was no difference in the amount of TH in the ME, regardless of treatment. However, the total *in situ* activity of TH in the ME of animals treated with estradiol and progesterone was significantly greater than that of animals treated with the solvent vehicle, estradiol, or progesterone (Figure 24). When the *in situ* molar activity of TH of the ME was calculated, it was found that

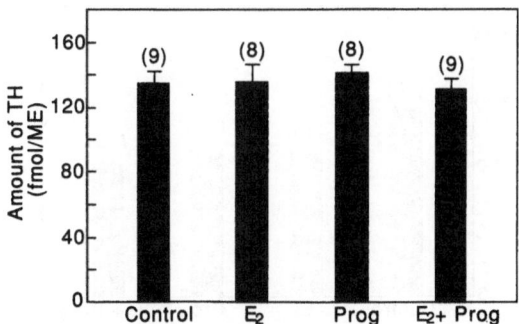

Figure 23. Effect of ovarian hormones on the mass of TH in the median eminence of aged ovariectomized rats. See Fig 5 and 18 for details. Drawn from González *et al.* (24) with permission of S. Karger, AG.

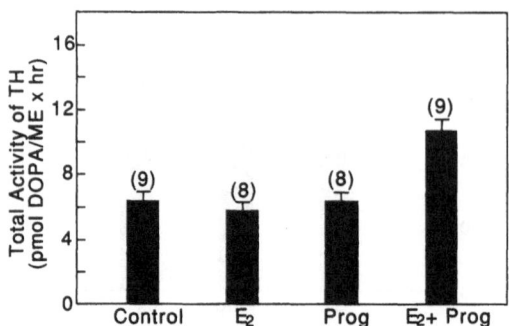

Figure 24.   Effect of ovarian hormones on the total *in situ* activity of TH in the median eminence of aged ovariectomized rats. See Fig. 5 and 18 for details. Drawn from González *et al.* (24) with permission of S. Karger, AG.

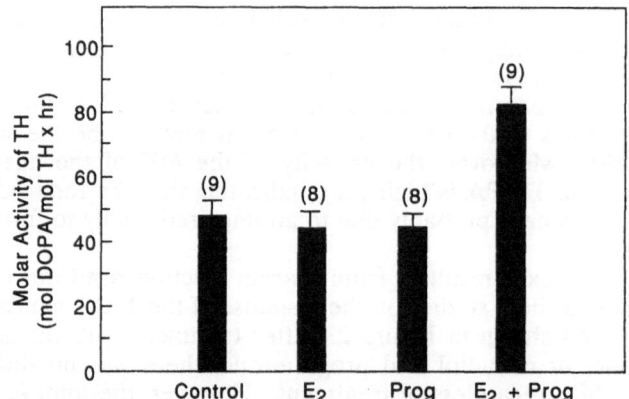

Figure 25.   Effect of ovarian hormones on the *in situ* molar activity of TH in the median eminence of aged ovariectomized rats. See Fig. 5 and 18 for details. Drawn from González *et al.* (24) with permission of S. Karger, AG.

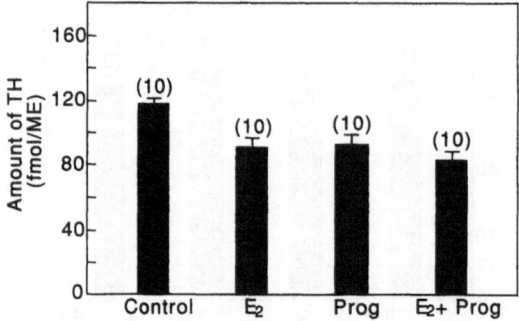

Figure 26.   Effect of ovarian hormones on the mass of TH in the median eminence of young adult male rats. See Fig. 5 and 18 for details. Drawn from González *et al.* (25) with permission of The Endocrine Society.

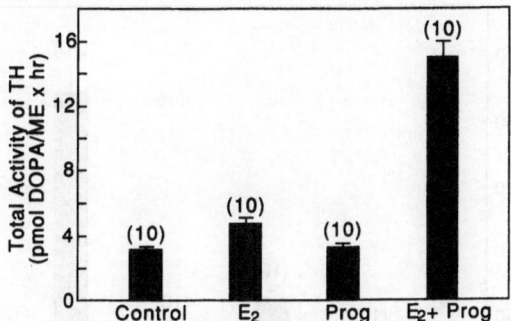

Figure 27. Effect of ovarian hormones on the total *in situ* activity of TH in the median eminence of young male rats. See Fig. 5 and 18 for details. Drawn from González *et al.* (25) with permission of The Endocrine Society.

treatment with estradiol and progesterone resulted in a significant increase in TH activity compared to animals treated with the solvent vehicle, estradiol, or progesterone (Figure 25). These findings show that TIDA neurons of the aged brain as well as the young brain of female rats respond to appropriate ovarian hormone stimulation.

YOUNG ADULT MALE RATS. Although it was shown by Ben-Jonathan *et al.* (12) that the rate of secretion of dopamine into hypophysial portal blood of male rats was less than that of cycling females, it was not clear whether the low rate of secretion of dopamine by TIDA neurons of males was a consequence of low estradiol-progesterone stimulation or was due to a fundamental difference in the TIDA neurons of the two sexes. To address this issue, young male rats were treated for three days with ovarian hormones (25). Then, the biosynthetic activity of the TIDA neurons was evaluated. The amounts

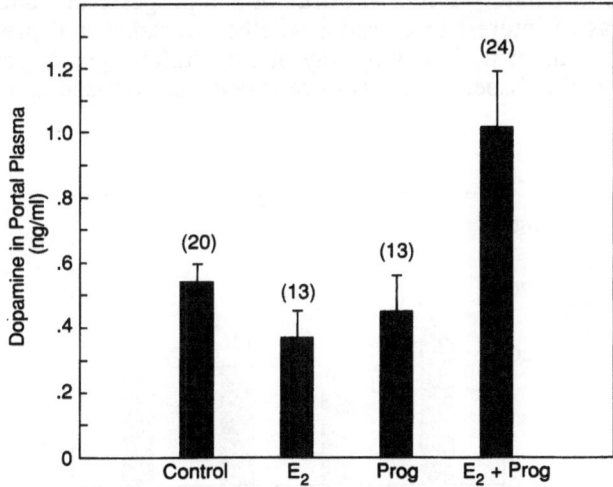

Figure 28. Dopamine in hypophysial portal blood of male rats treated with ovarian hormones. See Fig 5 and 18 for details. Drawn from González *et al.* (25) with permission of The Endocrine Society.

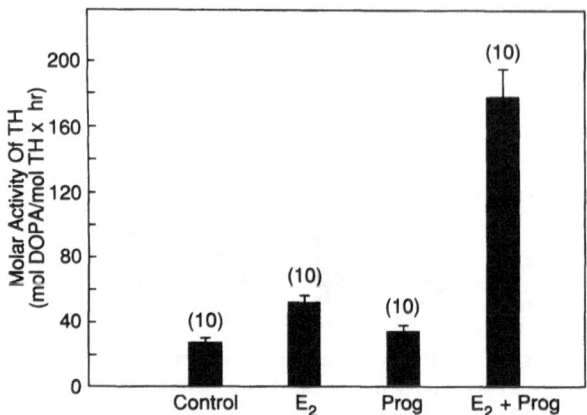

Figure 29. Effect of ovarian hormones on the *in situ* molar activity of TH in the median eminence of male rats. See Fig. 5 and 18 for details. Drawn from González *et al.* (25) with permission of The Endocrine Society.

of TH in the ME of male rats treated with estradiol, progesterone, or both estradiol and progesterone were somewhat less than that of rats treated with the solvent vehicle (Figure 26). However, the total *in situ* activity of TH in the ME of estradiol-progesterone treated rats was 3 to 4.5 times that of the controls or of animals treated with estradiol or progesterone alone (Figure 27). When the secretion of dopamine into portal blood was examined, it was found that estradiol-progesterone treatment induced a significant increase in the concentration of dopamine of portal plasma, compared to controls or to rats treated with estradiol or progesterone (Figure 28). In addition, the *in situ* molar activity of TH was markedly increased in male rats treated with both estradiol and progesterone compared to animals treated with the solvent vehicle or with estradiol or progesterone alone (Figure 29).

In view of the stimulatory action of estradiol and progesterone and of prolactin on TIDA neurons, it was of interest to consider whether estradiol and progesterone acted directly on TIDA neurons or indirectly by way of the pituitary gland, possibly by way of prolactin. To examine this issue, male rats were hypophysectomized and then treated for

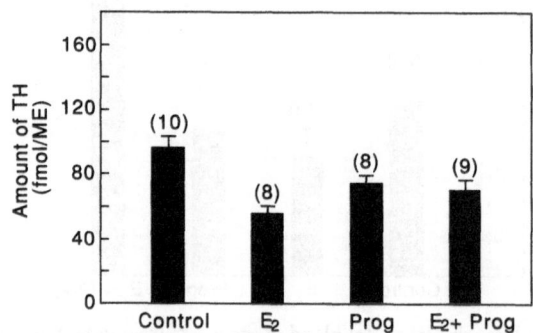

Figure 30. Effect of ovarian hormones on the mass of TH in the median eminence of hypophysectomized male rats. See Fig. 5 and 18 for details.

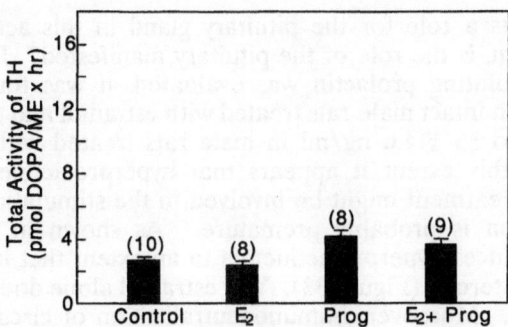

Figure 31. Effect of ovarian hormones on the total *in situ* activity of TH in the median eminence of hypophysectomized male rats. See Fig. 5 and 18 for details. Drawn from González *et al.* (25) with permission of The Endocrine Society.

with estradiol, progesterone, or estradiol and progesterone resulted in a reduction in the three days with ovarian hormones (25). As shown in Figure 30, treatment of male rats amount of TH in the ME compared to hypophysectomized rats treated with the solvent vehicle.

The total *in situ* activity of TH in the ME of hypophysectomized rats treated with estradiol and progesterone was similar to that of animals treated with the solvent vehicle, estradiol, or progesterone (Figure 31). In addition, there was little difference between the *in situ* molar activity of TH in the ME of hypophysectomized rats treated with the solvent vehicle, and those treated with estradiol, progesterone, or both estradiol and progesterone (Figure 32). These findings are in sharp contrast to those seen in similarly treated intact male rats (Figure 27 and 29).

Discussion

The finding that estradiol-progesterone treatment led to increased secretory activity of TIDA neurons in rats with intact pituitary glands but did not do so in hypophysec-

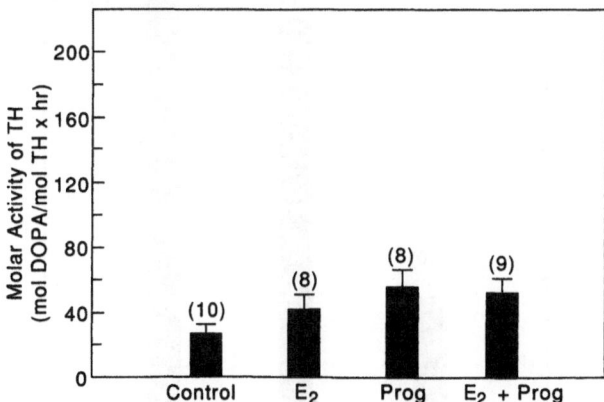

Figure 32. Effect of ovarian hormones on the *in situ* molar activity of TH in the median eminence of hypophysectomized male rats. See Fig. 5 and 18 for details. Drawn from González *et al.* (25) with permission of The Endocrine Society.

tomized animals suggests a role for the pituitary gland in this action of estradiol and progesterone. How, then, is the role of the pituitary manifested? Is prolactin involved? When the level of circulating prolactin was evaluated, it was found that the plasma prolactin concentration in intact male rats treated with estradiol and progesterone was 133 ± 12 ng/ml compared to 35 ± 4.6 ng/ml in male rats treated with the solvent vehicle (Figure 33) (25). To this extent it appears that hyperprolactinemia induced by the estradiol-progesterone treatment might be involved in the stimulation of TIDA neurons. However, this conclusion is probably premature. As shown in Figure 33, estradiol treatment alone also induces hyperprolactinemia to an extent that is similar to that seen with estradiol and progesterone (Figure 33). Yet, estradiol alone does not stimulate TIDA neurons (Figures 31,32). Moreover, immunoneutralization of circulating prolactin with prolactin antiserum did not prevent estradiol and progesterone from stimulating of TIDA neurons (Figures 20,21). Thus, these data are suggestive of the existence of a factor of pituitary origin that is not prolactin, which is necessary for the action of estradiol and progesterone on TIDA neurons. Alternatively, it may be that prolactin is unable to stimulate TIDA neurons in the presence of estradiol and absence of progesterone. Additional research is needed, to clarify this issue.

It is noteworthy that both estradiol and progesterone receptors are localized in perikarya of TH-containing neurons of the hypothalamus (30-32) and that estradiol treatment increases the quantity of progesterone receptors in the hypothalamus (33). These findings support the hypothesis that the synergism of estradiol and progesterone seen here may be due to an increase in progesterone receptors in the perikarya of the TIDA neurons.

The low level of secretory activity of the TIDA neurons in ovariectomized rats speaks to a role for estradiol and progesterone in maintaining an appropriate level of function by these neurons. In this regard it is interesting to speculate that the symptoms of women that are characterized as post-partum depression, pre-menstrual syndrome, and post-menopausal depression may be consequences of reduced dopamine secretion due to

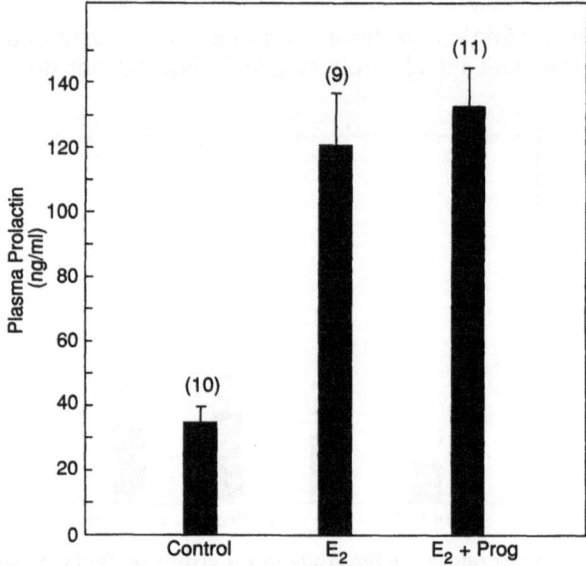

Figure 33. Effect of ovarian hormones on circulating prolactin in male rats. See Fig. 5 and 18 for details. Drawn from González *et al.* (25) with permission of The Endocrine Society.

abrupt cessation of estradiol and progesterone secretion following parturition, regression of the corpus luteum, or ovarian failure at menopause.

These findings on increased secretory activity by TIDA neurons raise the question of whether other dopaminergic neurons of the brain are similarly affected or whether the stimulatory effects of estradiol and progesterone and of prolactin are restricted to the TIDA neurons. If the TIDA neurons are found to respond uniquely to hormonal stimulation, it will be necessary to establish subsets of dopaminergic neurons having different regulatory characteristics.

Summary

The role of prolactin and of estradiol and progesterone in the control of the biosynthetic and secretory activity of TIDA neurons has been investigated in the following animal models: young female rats, aged female rats, and young male rats. The indices of TIDA neuronal function employed were *a)* mass of TH in neurites in the ME, *b)* total *in situ* activity of TH in the ME, *c) in situ* molar activity of TH in the ME, and *d)* secretion of dopamine into hypophysial portal blood.

It was found that prolactin in high concentration in the circulation and in the CSF had little, if any, effect on the mass of TH in the ME. However, a high concentration of prolactin in either the circulation or in the CSF stimulated significantly the *in situ* TH activity in the ME whether expressed in terms of total activity per ME or activity per mole of TH. The stimulation of TH activity with prolactin was prevented by immunoneutralization of circulating prolactin. A high concentration of prolactin in the CSF was as effective in stimulating TH activity in the ME of rats with intact pituitary glands as in hypophysectomized rats.

In addition to prolactin, treatment of animals with intact pituitaries with a combination of estradiol and progesterone markedly stimulated the total *in situ* activity of TH of the ME as well as the *in situ* molar activity of TH of the ME, but neither estradiol nor progesterone alone had an effect on TH activity. Hypophysectomy abolished the stimulatory action of estradiol and progesterone on TH activity of the ME. In addition to the *in situ* activity of TH in the ME, estradiol-progesterone treatment stimulated the secretion of dopamine into hypophysial portal blood. Neither estradiol nor progesterone alone affected dopamine secretion by TIDA neurons.

We conclude that exposure to high concentrations of prolactin or to both estradiol and progesterone stimulate the biosynthetic and secretory activity of TIDA neurons. These hormones are effective in old rats and well as young rats and in males as well as females.

Acknowledgements

The authors thank Kay Stanley for editorial and administrative assistance. We also thank Renon Mical, Robert Lipsey, Jodie Roberts, Nhu-Y Dong, and Sharyn Monroe for excellent technical contributions. This research was supported by grant nos. DK-01237, AG-04344, and AG-08173 from the National Institutes of Health, Bethesda, MD, and by the Chilton Foundation, Dallas, TX.

References

1.   Davis, G.C., A.C. Williams, S.P. Markey, M.H. Ebert, E.D. Caine, C.M. Reichert, and I.J. Kopin,

Parkinsonism secondary to intravenous injection of meperidine analogues, *Psychiatry Res* 1: 249-254, 1979.

2. Bruns, B.S., C.C. Chiuech, S.P. Markey, M.H. Ebert, D.M. Jacobowitz, and I.J. Kopin, A primate model of parkinsonism: selective destruction of dopaminergic neurons in the pars compacta of the substantia nigra by N-methyl-4-phenyl-1,2,3,6-tetrahydropyridine, *Proc Natl Acad Sci USA* 80: 4546-4550, 1983.

3. Langston, J.W., and P. Ballard, Chronic parkinsonism in humans due to a product of meperidine-analog synthesis, *Science* 219: 979-980, 1983.

4. Langston, J.W., and P.A. Ballard, Jr., Parkinson's disease in a chemist working with 1-methyl-4-phenyl-1,2,5,6-tetrahydropyridine, *N Engl J Med* 309: 310, 1983.

5. Langston, J.W., MPTP: insights into the etiology of parkinson's disease, *Eur Neurol* 1: 2-10, 1987.

6. Porter, J.C., D.D. Nansel, G.A. Gudelsky, M.M. Foreman, N.S. Pilotte, C.R. Parker, Jr., G.H. Burrows, G.W. Bates, and J.D. Madden, Neuroendocrine control of gonadotropin secretion, *Fed Proc* 39: 2896-2901, 1980.

7. González, H.A., and J.C. Porter, Mass and *in situ* activity of tyrosine hydroxylase in the median eminence: effect of hyperprolactinemia, *Endocrinology* 122: 2272-2277, 1988.

8. Chen, C.L., Y. Amenomori, K.H. Lu, J.L. Voogt, and J. Meites, Serum prolactin levels in rats with pituitary transplants or hypothalamic lesions, *Neuroendocrinology* 6: 220-227, 1970.

9. Fuxe, K., Cellular localization of monoamines in the median eminence and infundibular stem of some mammals, *Acta Physiol Scand* 58: 383-344, 1963.

10. Fuxe, K., and T. Hökfelt, Further evidence for the existence of tuberoinfundibular dopamine neurons, *Acta Physiol Scand* 66: 243-244, 1966.

11. Hökfelt, T., The possible ultrastructural identification of tuberoinfundibular dopamine-containing nerve endings in the median eminence of the rat, *Brain Res* 5: 121-123, 1967.

12. Ben-Jonathan, N., C. Oliver, H.J. Weiner, R.S. Mical, and J.C. Porter, Dopamine in hypophysial portal plasma of the rat during the estrous cycle and throughout pregnancy, *Endocrinology* 100: 452-458, 1977.

13. Levitt, L., S. Spector, A. Sjöerdsma, and S. Udenfriend, Elucidation of the rate-limiting step in norepinephrine biosynthesis in the perfused guinea-pig heart, *J Pharmacol Exp Ther* 148: 1-8, 1965.

14. Gudelsky, G.A., and J.C. Porter, Release of newly synthesized dopamine into the hypophysial portal vasculature of the rat, *Endocrinology* 104: 583-587, 1979.

15. Gudelsky, G.A., and J.C. Porter, Morphine- and opioid peptide-induced inhibition of the release of dopamine from tuberoinfundibular neurons, *Life Sci* 25: 1697-1702, 1979.

16. Haskins, J.T., G.A. Gudelsky, R.L. Moss, and J.C. Porter, Iontophoresis of morphine into the arcuate nucleus: effects on dopamine concentrations in hypophysial portal plasma and serum prolactin concentrations, *Endocrinology* 108: 767-771, 1981.

17. Reymond, M.J., C. Kaur, and J.C. Porter, An inhibitory role for morphine on the release of dopamine into hypophysial portal blood and on the synthesis of dopamine in tuberoinfundibular neurons, *Brain Res* 262: 253-258, 1983.

18. Reymond, M.J., and J.C. Porter, Hypothalamic secretion of dopamine after inhibition of aromatic L-amino acid decarboxylase activity, *Endocrinology* 111: 1051-1056, 1982.

19. Pilotte, N.S., and J.C. Porter, Dopamine in hypophysial portal plasma and prolactin in systemic plasma of rats treated with 5-hydroxytryptamine, *Endocrinology* 108: 2137-2141, 1981.

20. Okuno, S., and H. Fujisawa, A new mechanism for regulation of tyrosine 3-monooxygenase by end product and cyclic AMP-dependent protein kinase, *J Biol Chem* 260: 2633-2635, 1985.

21. Gudelsky, G.A., D.D. Nansel, and J.C. Porter, Dopaminergic control of prolactin secretion in the aging male rat, *Brain Res* 204: 446-450, 1981.

22. Reymond, M.J., and J.C. Porter, Secretion of hypothalamic dopamine into pituitary stalk blood of aged female rats, *Brain Res Bull* 7: 69-73, 1981.

23. Lu, K.H., B.R. Hopper, T.H. Vargo, and S.S.C. Yen, Chronological changes in sex steroid, gonadotropin and prolactin secretion in aging female rats displaying different reproductive states, *Biol Reprod* 21: 193-203, 1979.

24. González, H.A., W. Kedzierski, and J.C. Porter, Mass and activity of tyrosine hydroxylase in the tuberoinfundibular dopaminergic neurons of the aged brain: control by prolactin and ovarian hormones, *Neuroendocrinology* 48: 663-66, 1988.

25. González, H.A., W. Kedzierski, N. Aguila-Mansilla, and J.C. Porter, Hormonal control of tyrosine hydroxylase in the median eminence: demonstration of a central role for the pituitary gland, *Endocrinology* 124: 2122-2127, 1989.

26. Butcher, R.L., N.W. Fugo, and W.E. Collins, Semicircadian rhythm in plasma levels of prolactin during early gestation in the rat, *Endocrinology* 90: 1125-1127, 1972.

27. Wang, P.S., and J.C. Porter, Hormonal modulation of the quantity and *in situ* activity of tyrosine hydroxylase in neurites of the median eminence, *Proc Natl Acad Sci USA* 83: 9804-9806, 1986.

28. Chen, C.L., and J. Meites, Effects of estrogen and progesterone on serum and pituitary prolactin levels in ovariectomized rats, *Endocrinology* 86: 503-505, 1970.

29. Reymond, M.J., J. Arita, C.A. Dudley, R.L. Moss, and J.C. Porter, Dopaminergic neurons in the mediobasal hypothalamus of old rats: evidence for decreased affinity of tyrosine hydroxylase for substrate and cofactor, *Brain Res* 304: 215-223, 1984.

30. Sar, M., Distribution of progestin-concentrating cells in rat brain: colocalization of 3H-ORG.2058, a synthetic progestin, and antibodies to tyrosine hydroxylase in hypothalamus by combined autoradiography and immunocytochemistry, *Endocrinology* 23: 1110-1118, 1988.

31. Fox, S.R., B.D. Shivers, R.E. Harlan, and D.W. Pfaff, Progesterone receptors in the female rat are localized within dopaminergic neurons of the hypothalamic arcuate nucleus, but not within pituitary lactotrophs, 68th Annual Meeting of the Endocrine Society, (Abstract #434), p. 139, 1986.

32. Sar, M., Estradiol is concentrated in tyrosine hydroxylase containing neurons of the hypothalamus, *Science* 223: 938-940, 1984.

33. MacLusky, N.J., and B.S. McEwen, Oestrogen modulates progestin receptor concentrations in some brain regions but not in others, *Nature* 274: 276-278, 1978.

THE PHYSIOLOGY OF THE BLOOD-BRAIN BARRIER

Barbro B. Johansson

Department of Neurology
Lund University
Lund, Sweden

Introduction

A strict regulation of the neuronal environment is essential for optimal brain function. The blood-brain barrier (BBB), a concept including the morphological and functional mechanisms that restricts or facilitates the passage of substances from blood to brain, enables the brain environment to be regulated relatively independently of fluctuations in plasma concentrations. Although disagreement prevails regarding some specific mechanisms, our knowledge of the BBB physiology has advanced impressively during the last decades (1-13).

History

The historical development of the current BBB concept has recently been reviewed by Davson (14). In short, although the term BBB (Blutgehirnschranke) was first used by Lewandowsky (15), the BBB concept arose from Ehrlich's observation in 1885 that acidic vital dyes while staining the rest of the body left the brain unstained. It was later found that substances injected into the subarachnoidal space could easily enter the brain (16). Spatz (17) and later Broman (18) strongly argued that the BBB must be located in the endothelial cells, a hypothesis that was not generally accepted until the classical study by Reese and Karnowsky (19). In the 1950s and early 1960s, many investigators believed that the BBB function was either due to a glial barrier or could be explained by a lack of extracellular space, an interpretation based on early electronmicroscopical findings that later were shown to be caused by fixation artifacts.

Morphology of Brain Endothelial Cells

The main differences between a brain and systemic capillary are summarized in Figure 1. The endothelial cells in brain capillaries are sealed together by tight junctions and contain few pinocytotic vesicles (19,20). Thus, membrane fusion leaves no gap between the endothelial cells. The tight junctions are extremely complex structures. The importance of the low endocytotic activity for the various BBB functions is controversial. There is evidence that the majority of what appear to be independent vesicles in the endothelial cytoplasm may in fact be part of membrane invaginations that communicate either with the blood or with the perivascular space (21). No single channel seems to open to both blood and interstitial spaces simultaneously. The different aspects of the

Circulating Regulatory Factors and Neuroendocrine Function
Edited by J. C. Porter and D. Ježová
Plenum Press, New York, 1990

anatomical basis of the BBB has been analyzed in an excellent review by Brightman (20).

The mitochondrial content of brain capillary endothelial cells amounts to 8-11% of the cytoplasmic volume, a much higher volume than in brain regions lacking BBB and tissues outside the central nervous system (Figure 2) (22), reflecting the high metabolic activity needed to maintain ion differentials between blood plasma and brain extracellular fluid and to maintain the unique characteristics of central nervous system capillaries.

The Role of the Astrocytes

Although the tight junctions are formed early in the rat fetus, the barrier to molecules like sucrose and inulin tightens markedly between the 4th and 9th day after birth and reaches adult degree of tightness by 2 to 3 weeks (3,23). Ontogenetic studies suggest that the closure of the BBB is related to the maturation of the glial foot processes surrounding the larger part of the endothelial cell surface and the capillary basement membrane. Based on studies with cell cultures, it has been hypothesized that the glial cells surrounding the vessels induce the BBB (24). If purified astrocytes are injected into the anterior chamber of the eye, the vessels from the iris invading the astrocytes develop BBB properties with typical tight junctions and exclude macromolecular tracers (25). It has also been suggested that glial cells induce polarity in brain endothelial cells (26) (see later section).

Lipid Solubility

The similarity between BBB and cell membranes was pointed out by Krogh (27), who suggested the importance of lipid solubility for entrance of drugs into the central nervous system, a concept that has been confirmed in later studies. Oldendorf (28) has documented that the octanol:water partition coefficient can predict the entry of substances into the brain (Figure 3). Restricted uptake depends on polarity of the substance, which is mostly potently decided by $-OH$, $-COOH$, and $-NH_2$ groups. Secondary amines, esters, and ethers are less polar in that order (3,28). Substances that have access to some carrier

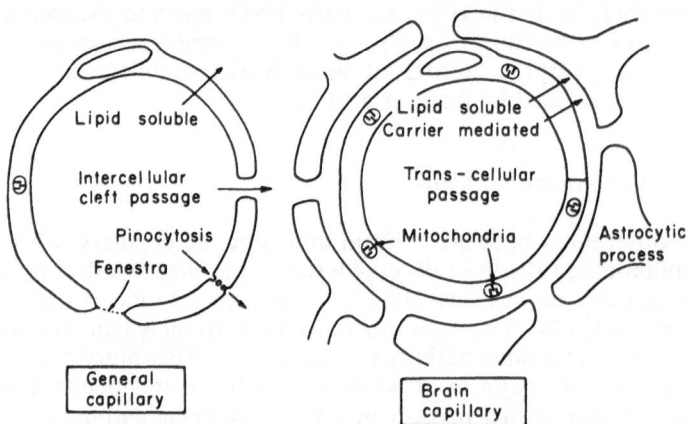

Figure 1. Main differences between a general capillary and brain capillary. Reproduced from Oldendorf (2) with permission of Academic Press, Inc.

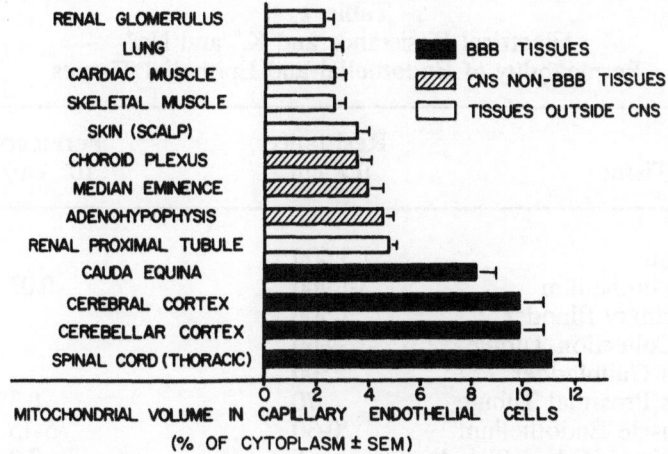

Figure 2. The percentage of cytoplasmic volume occupied by mitochondria is higher in brain capillary endothelial cells than in capillaries from other tissues (rat). Reproduced from Oldendorf *et al.* (22) with permission of Little, Brown and Company.

mechanism (see below) do not follow this rule (Figure 2) and larger molecules will have difficulty in passing regardless of lipophilicity. Thus, cyclosporin, a highly lipid-soluble cyclic peptide with a molecular weight of approximately 12,000, enters the brain poorly (29).

The BBB: A Modified Tight Epithelium

Crone has repeatedly stressed that the brain endothelial cells represent a modified epithelium rather than an endothelium (7,8,80). Like epithelial cells, the brain endothelial cells have a very high electrical resistance reflecting their low ion permeability (Table 1). The transport characteristics and polarity of cerebral endothelial cells are consistent with this concept (see below).

Transport Mechanisms

A large number of carrier-mediated transport systems have been identified in the brain endothelial cells, including carriers for D-glucose, short chain monocarboxylic acids, amino acids, nucleic acid precursors, and choline (2,31-34). D-glucose, the main fuel for the brain, is a water-soluble substance that enters the brain *via* saturable, stereospecific facilitated diffusion (Figure 4) (31). The glucose transporter in the human cerebral microvessels has recently been identified and brain microvessels were found to have a higher density of the glucose transporter than any other tissue except human erythrocytes (35). 2-Deoxy-D-glucose and 3-O-methylglucose use the same carrier and have been extensively used in experimental and clinical research. 2-Deoxy-D-glucose enters the nerve cells and remains there without being phosphorylated, making radiolabelled analogues very useful in experimental studies on the glucose metabolism including *in vivo* studies using positron tomography in man.

Table 1
Electrical Resistance and K^+ and Na^+
Permeability of Endothelial and Epithelial Tissues

Tissue	Resistance $\Omega \times cm^2$	Permeability 10^{-5} cm/sec
Frog Skin	3,600	
Brain Endothelium	1,900	0.03
Toad Urinary Bladder	1,500	
Rabbit Collection Tubule	860	
Necturus Gallbladder	300	
Necturus Proximal Tubule	70	0.3
Frog Muscle Endothelium	20-30	5-15
Rat Proximal Kidney Tubule	5	2.3
Frog Mesenteric Endothelium	1-3	70

Reproduced from Crone and Levitt (6) with permission of The American Physiological Society.

It has been proposed that glucose transport is *down-regulated* in chronic hyperglycemia, *i.e.*, the transporter adapts to the altered serum concentration by reducing the transport which might lead to relative hypoglycemia if blood glucose is lowered too rapidly (36-38). However, some recent findings do not support this concept (39).

The monocarboxylic acid transporter has affinity for acetic, lactic, butyric, pyruvic, and β-hydroxybutyric acids (2). In the neonatal and early postnatal period in the rat, when the glucose transporter is not fully developed, the transport capacity of the monocarboxylic acid carrier is high, and lactate and β-hydroxybutyrate are the main substrates for the brain (40-43). In the adult animal, the monocarboxylic carrier can increase its capacity by adaptation to low glucose levels during starvation when the ketone bodies, β-hydroxybutyrate and acetoacetate, are the main energy substrates for the brain (Figure 5) (37).

Neutral amino acids, basic amino acids, and dicarboxylic amino acids enter the brain *via* separate stereospecific and saturable carriers (33). Amino acids belonging to the same group compete for the carrier. Thus, large doses of L-dihydroxyphenylalanine (DOPA) may influence the passage of other large neutral amino acids, and neurological symptoms seen in patients with phenylketonuria are probably related to insufficient passage of other essential amino acids through the barrier because of the high serum levels of phenylalanine. Amino acids with presumed transmitter function within the brain such as glutamate and aspartate enter the brain poorly. It should also be noted that many transport mechanisms do not, therefore, tell us how much of a substance really enters the brain.

The passage of peptides over the BBB is controversial (10,34,44,45) and is discussed by Dr. Banks in another chapter in this book (46).

The choroid plexuses of the lateral and fourth ventricles have the capacity to transport organic acids such as the monoamine metabolites, 5-hydroxyindoleacetic acid and homovanillic acid, from the cerebrospinal fluid into the blood (47), and a similar transport of organic acids from brain to blood is believed to be present at the brain capillary level.

The permeability to electrolytes is low (Table 1). However, a carrier mediated transport of chloride from blood to brain has been demonstrated by Smith and Rapoport (48). Whereas, the permeability of the BBB for potassium is very low in the blood to

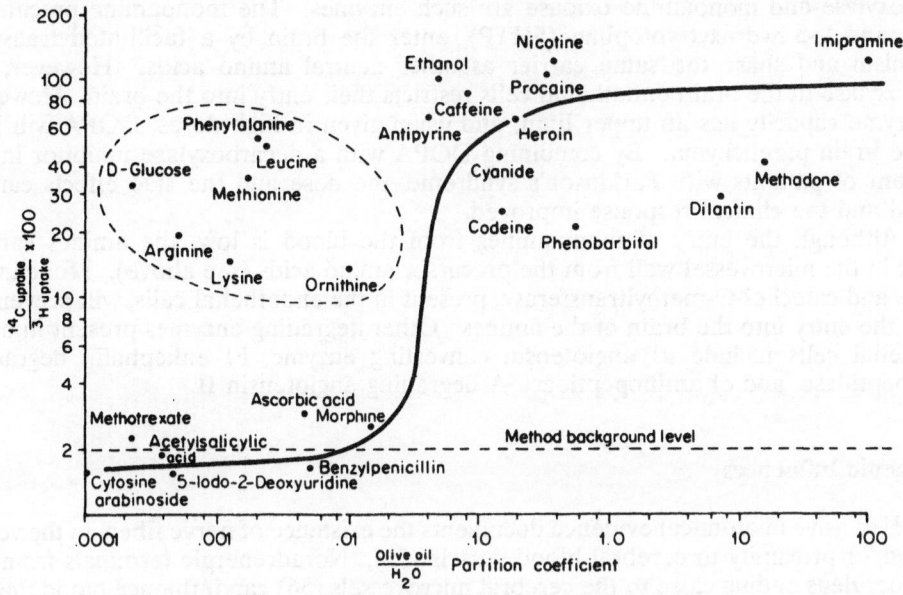

Figure 3. Correlation between brain uptake index and olive oil/water partition coefficient for a number of substances. D-glucose and many amino acids *(within the circle)* have access to transport mechanisms and their penetration is therefore high in spite of low lipid solubility. Reproduced from Oldendorf (28) with permission of Experimental Society of Biology and Medicine.

brain direction. There is a NA,K-ATPase dependent efflux of potassium from the brain (49). For further details on electrolytes and BBB, see Smith and Rapoport (50) and Betz (51).

Cell Polarity

When the two opposing plasma membranes of a cell have different properties, the cell is said to be polar. Epithelial cells have such characteristics. Cellular polarity provides the basis for transcellular transport of many solutes and permits a cell to take up and metabolize a substance from one side while maintaining a barrier preventing the substrate from crossing to the other side. The luminal cell membrane (facing the blood) of the brain capillaries is different from the abluminal membrane (facing the brain). Whereas alkaline phosphatase is equally distributed between the luminal and the abluminal membrane, Na,K-ATPase and 5´-nucleotidase are present primarily on the abluminal side; λ-glutamyl transpeptidase is mainly found on the luminal side (52). Polarity of receptors can allow a cell to respond to an agonist present at one side but not the other.

Enzymatic Barriers

Enzymes in the brain endothelial cells may degrade substances that are not impeded by the luminal endothelial cell membrane (5,53,54). Aromatic L-amino acid decarboxylase and monoamine oxidase are such enzymes. The monoamine precursors, DOPA and L-5-hydroxytryptophan (5HTP), enter the brain by a facilitated transport mechanism and share the same carrier as other neutral amino acids. However, the decarboxylase in the brain endothelial cells restricts their entry into the brain. However, the enzyme capacity has an upper limit, and when given in high doses, DOPA will leak into the brain parenchyma. By combining DOPA with a decarboxylase inhibitor in the treatment of patients with Parkinson's syndrome, the dose and the side effects can be reduced and the clinical response improved.

Although the entry of monoamines from the blood is low, the amines can be formed in the microvessel wall from the precursor amino acids (see above). Monoamine oxidase and catechol-O-methyltransferase, present in the endothelial cells, will prevent or reduce the entry into the brain of the amines. Other degrading enzymes present in brain endothelial cells include *a)* angiotensin converting enzyme, *b)* enkephalin degrading aminopeptidase, and *c)* aminopeptidase A degrading angiotensin II.

Neurogenic Influences

Extensive anatomical evidence documents the existence of nerve fibers in the vessel wall and/or proximity to cerebral blood vessels (55). Noradrenergic terminals from the locus coeruleus ending close to the cerebral microvessels (56) can influence blood flow as well as water permeability (57). In addition to the classical transmitters, a number of peptides including VIP, substance P, neuropeptide Y, and calcitonin gene-related peptide have been localized to peptidergic vascular nerves. Peptidergic nerves seem to have some influence on the cerebrovascular tone and hence blood flow, but little is known about their effects on the BBB.

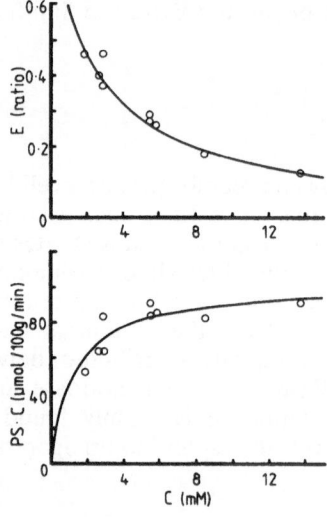

Figure 4.   The *upper curve* shows that the initial unidirectional extraction of D-glucose in the brain during a single passage is higher at low concentrations of glucose (abscissa) in the dog. The *lower part* shows the calculated unidirectional transporter. Reproduced from Crone (7,31) with permission of Ellis Horwood Limited.

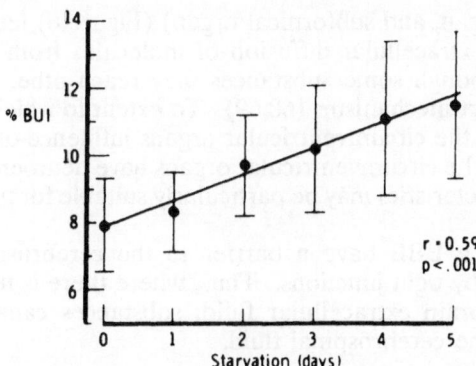

Figure 5. Brain uptake index of β-hydroxybutyric acid during starvation. Reproduced from Gjedde and Crone (37) with permission of The American Physiological Society.

A number of receptors have been identified in isolated brain vessels including α- and β-adrenergic receptors as well as receptors for serotonin, adenosine-2, histamine, angiotensin, and arginine vasopressin. Their influence if any on the BBB is poorly understood.

Dynamics and Fluctuations of BBB Functions

The ontogeny and phylogeny of the BBB has recently been reviewed (58). Diurnal and other rhythmic variations in BBB function have not been widely studied. However, there is a circadian variation in the permeability of the BBB during adrenaline induced hypertension (59), a variation that seems to be under noradrenergic influence from locus coeruleus (60). In view of the fact that most biological systems show rhythmic variations, it is likely that the transport mechanisms and enzymatic activities of the cerebral vessels also show some variation. Furthermore, the possible influence of anesthesia and drugs should be considered (61,62).

Immunological Aspects

The brain has long been considered an *immunologically privileged site*. This has been attributed to such factors as the absence of passage across the BBB of lymphocytes, absence of lymphatic drainage, and absence of antigen-presenting cells. However, activated T-cells seem to be able to cross postcapillary cerebral venules (63), and there is a link between the brain and the lymphatic system (64,65). In addition, intracerebral antigen deposition can lead to an immune-response in deep cervical lymph nodes (66). Thus, the reason for prolonged survival of grafts in the brain compared to other organs is not clear, but is seems unlikely that the BBB plays a major role (67).

Brain Regions Lacking BBB have a CSF Barrier

The choroid plexus and the circumventricular organs (area postrema, median eminence, pituitary neural lobe, organum vasculosum of the lamina terminalis, pineal gland, subcommissural organ, and subfornical organ) (Figure 6) lack a BBB. Functional barriers seem to reduce extracellular diffusion of molecules from the circumventricular organs into the brain although some substances may reach other parts of the brain by retrograde axonal transport mechanisms (68,69). To extent to which substances acting on neuroreceptors or cells in the circumventricular organs influence other parts of the brain is unclear. Since many of the circumventricular organs have neuroendocrine functions, the special permeability characteristics may be particularly suitable for passage in the direction from brain to blood.

Structures lacking a BBB have a barrier to the cerebrospinal fluid formed by tanycytes joined together by tight junctions. Thus, where there is no barrier between the cerebrospinal fluid and brain extracellular fluid, substances cannot enter the circumventricular organs from the cerebrospinal fluid.

Facilitating Passage of Substances from Blood to Brain

The entry of a substance with poor penetration into the brain can be enhanced by

—altering the lipid solubility
—cationization
—liposome entrapment
—coupling to a substance that can enter by receptor-mediated transport
—experimental opening of the BBB.

Increasing Lipid Solubility

Converting water-soluble substances into lipid-soluble pro-drugs has so far been the most commonly used pharmacological principle. Figure 7 illustrates the marked influence

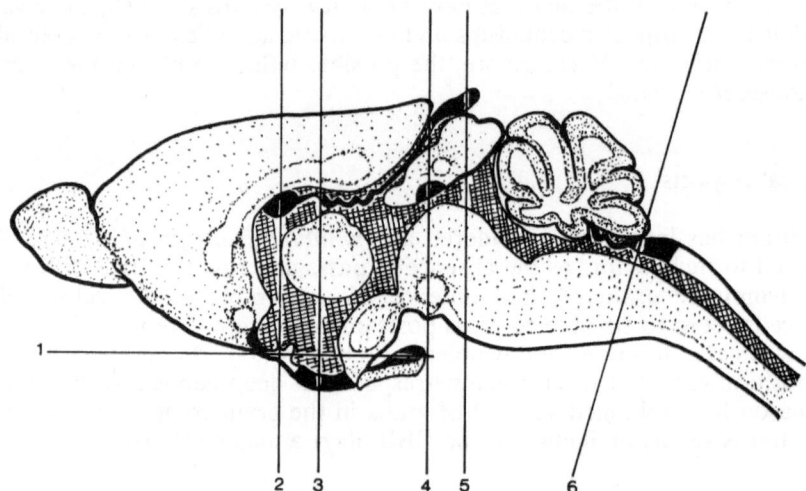

Figure 6. Schematic view of a sagittal section of mouse brain showing the location of the circumventricular organs *(dark areas)*: 1 denotes hypophysis, 2 denotes subfornical organ, 3 denotes medial eminence, 4 denotes subcommissural organ, 5 denotes pineal gland, 6 denotes postremal area. Reproduced from Bigotte *et al.* (88) by permission of Springer-Verlag.

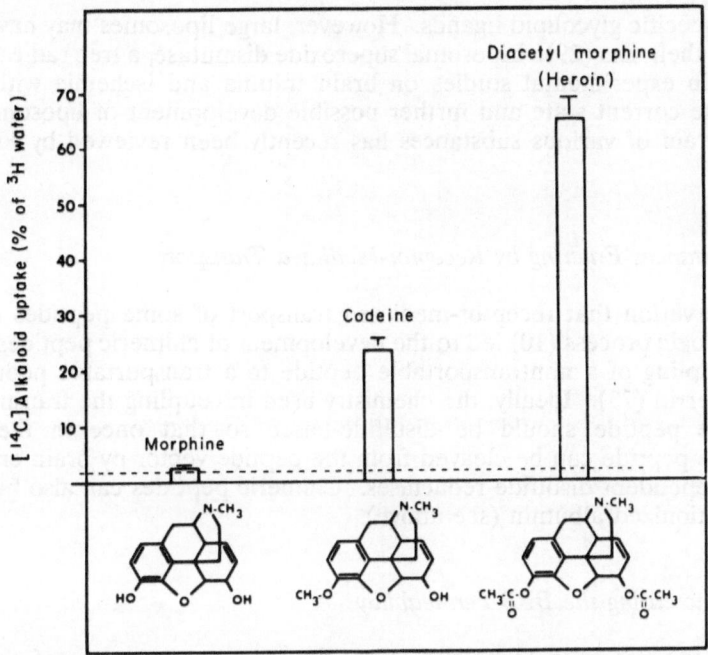

Figure 7. Morphine penetrates the BBB poorly because of the highly polar hydroxyl groups. Codeine and heroin are more lipophilic and can therefore more easily enter the brain. Reproduced from Oldendorf (2) with permission of Academic Press, Inc.

of lipid solubility. The greater uptake of codeine and heroin than morphine reflects their reduced polarities due to substitutions of the highly polar hydroxyl groups in morphine. The two acetyl groups of heroin are rapidly removed in the brain by pericapillary cholinesterase forming morphine.

Cationization

Cationization of proteins enhances cellular uptake in general and has been shown to enhance passage of the BBB (70). The same is true for *glycosylation* of proteins, which could be one reason for the enhanced penetration of proteins in patients with diabetes. Both principles might be used for delivery of monoclonal antibodies through the BBB.

Liposome Entrapment

Liposomes are tiny spheres composed of concentrated lipid bilayers that can be made to entrap various substances such as drugs, hormones, or viruses. Encapsulation of a substance slows the clearance from the circulation and protects it from enzymatic or hydrolytic degradation. Nontoxic and biodegradable liposomes have been developed. A problem of liposomes is that these particulate structures cross the endothelial cells in the lung, liver, and spleen more easily than the brain. Furthermore, liposomes are recognized as foreign particles by the immune system and may be phagocytosed by the reticuloendothelial cells. The permeability of the brain to liposomes can be enhanced by using liposomes with specific glycolipid ligands. However, large liposomes may have restricted entry because of their size (29). Liposomal superoxide dismutase, a free radical scavenger, has been used in experimental studies on brain trauma and ischemia with promising results (71). The current state and further possible development of liposomes to allow entry into the brain of various substances has recently been reviewed by Fishman and Chan (72).

Coupling to Substances Entering by Receptor-Mediated Transport

The observation that receptor-mediated transport of some peptides through the BBB is a physiologic process (10) led to the development of chimeric peptides, formed by the covalent coupling of a nontransportable peptide to a transportable peptide such as insulin or transferrin (73). Ideally, the chemistry used in coupling the transportable and nontransportable peptide should be disulfide-based so that once in the brain the nontransportable peptide can be cleaved from the peptide vector by brain enzymes such as glutathione-dependent disulfide reductases. Chimeric peptides can also be formed by coupling to a cationized albumin (see above).

Experimentally Increasing the BBB Permeability

Through the use of a technique for continuous measurement of the electrical resistance of microvascular endothelium, many substances including serotonin and free radicals, have been shown to increase the permeability of the BBB (74-76).

Current experimental techniques that are used to open unselectively the BBB include acute hypertension, local increase in intravascular pressure, and intracarotid injections of hyperosmolar solutions or protamine sulphate.

An acute rise in the intravascular pressure will lead to patchy protein extravasation in the brain that varies somewhat with the method used to increase the pressure (77,78). BBB dysfunction is a direct consequence of the increased mechanical stress on the endothelial cells and can be modified by factors such as the vascular tone and the abruptness of the pressure increase. When the blood pressure normalizes, the BBB will rapidly normalize. Since blood pressure fluctuations are common in experimental animal research, the potential BBB opening effect of vasoactive drugs or stress-induced hypertension should be considered.

Intracarotid injection of hyperosmolar solutions such as urea, mannitol, and arabinose, has been extensively used in experimental research (79) and also used in patients to deliver drugs or antibodies (80). Protamine sulphate, a polycation that is thought to open the BBB by reduction of the negative endothelial surface charge (81) may have some advantages over osmotic opening (82,83).

Is It Wise to Alter the BBB?

In spite of the importance of an intact BBB for brain homeostasis and hence optimal brain function, little is known about the long-term effects of an altered brain environment induced by a transient dysfunction of the BBB. However, evidence has accumulated during the last few years that permanent neuronal damage can occur after a transient opening of the BBB induced either by hypertension (84) or hyperosmolar solutions (85). This is important because of the introduction of hyperosmotic opening of the BBB in patients for drug or antibody delivery (80), a highly controversial procedure

(79,86). The possible hazards of a transient opening of the BBB were discussed at a recent symposium (87).

Summary

The BBB is a dynamic interface between blood and the central nervous system enabling the brain to keep an optimal internal environment. The endothelial cells of the brain capillaries are unique epithelial-like cells that are fused together by tight junctions and have a low pinocytotic activity. The entry of a specific substance will, therefore, mainly depend on its lipid solubility, and whether or not it has access to any of the carriers in the endothelial cells. Enzymatic degradation in the endothelium can prevent entry into the brain of substances that do enter the endothelial cells. Astrocytes may have an important role by inducing and upholding some barrier functions.

An intact BBB is evidently important for optimal brain function. Manipulation of the BBB to allow entry of therapeutic agents may be justified under certain circumstances but should be done with caution until we know more about the long-term consequences of such manipulation.

References

1. Rapoport, S.I. (ed), *Blood-Brain Barrier in Physiology and Medicine*, Raven Press, New York, pp. 1-316, 1976.
2. Oldendorf, W.H., The blood-brain barrier, *Exp Eye Res, Suppl 1977* 177-190, 1977.
3. Bradbury, M.W.B., *The Concept of a Blood-Brain Barrier*, J. Wiley and Sons, New York, 1979.
4. Bradbury, M.W.B., The structure and function of the blood-brain barrier, *Fed Proc* 43: 186-190, 1984.
5. Hardebo, J.E., and C. Owman, Barrier mechanisms for neurotransmitter monoamines and their precursors at the blood-brain interfact, *Ann Neurol* 8: 1-11, 1980.
6. Crone, C., and D.G. Levitt, Capillary permeability to small solutes, *In* E.M. Renkin and C.C. Michel (eds) *Handbook of Physiology, The Cardiovascular System, Volume 4, Section 2, The Microcirculation*, American Physiology Society, Bethesda, pp. 411-466, 1984.
7. Crone, C., The blood-brain barrier: a modified tight epithelium, *In* M.W.B. Bradbury, M.G. Rumsby, and A.J. Suckling (eds) *The Blood-Brain Barrier in Health and Disease*, Ellis Horwood Limited, Chichester, pp. 17-40, 1986.
8. Crone, C., The blood-brain barrier as a tight epithelium: where is information lacking? *Ann NY Acad Sci* 481: 174-185, 1986.
9. Pardridge, W.M., Brain metabolism: a perspective from the blood-brain barrier, *Physiol Rev* 63: 1481-1535, 1983.
10. Pardridge, W.M., Receptor-mediated peptide transport through the blood-brain barrier, *Endocrine Rev* 7: 314-330, 1986.
11. Goldstein, G.W., and A.L. Betz, The blood-brain barrier, *Sci American* 255: 74-83, 1986.
12. Cserr, H.F. (ed), *In The Neuronal Microenvironment*, New York Academy of Science, New York, 1986.
13. Neuwelt, E.A. (ed) *In Implications of the Blood-Brain Barrier and Its Manipulation, Basic Science Aspects, Volume 1*, Plenum Publishing Corporation, New York, 1989.
14. Davson, H., History of the blood-brain barrier concept, *In* E.A. Neuwelt (ed) *Implications of the Blood-Brain Barrier and Its Manipulation, Volume 1, Basic Science Aspects*, Plenum Publishing Corporation, New York, pp. 27-52, 1989.
15. Lewandowsky, M., Zur lehre der cerebrospinalflussigkeit, *Z Klin Med* 40: 480-494, 1900.
16. Goldmann, E.E., Vitalfärbung am zentralnervsystem, *Abh Preuss Akad Wiss Physmath* 1: 1-60, 1913.
17. Spatz, H., Die bedeutung der vitalen färbung für die lehre vom stoffaustausch zwischen dem zentralnervensystem und dem übrigen körper. Das morphologische substrat der stoffuechselschranken

im zentralorgan, *Arch f Psychiat* 101: 267-358, 1933.

18. Broman, T., The permeability of cerebrospinal vessels in normal and pathological conditions, Munksgaard, Copenhagen, pp. 33-35, 1949.

19. Reese, T.S., and M.J. Karnovsky, Fine structural localization of a blood-brain barrier to exogenous peroxidase, *J Cell Biol* 34: 207-217, 1967.

20. Brightman, M.W., The anatomic basis of the blood-brain barrier, In *Implications of the Blood-Brain Barrier and Its Manipulation, Volume 1, Basic Science Aspects*, Plenum Publishing Corporation, New York, pp. 53-83, 1989.

21. Bundgaard, M., J. Frøkjaer-Jensen, and C. Crone, Endothelial plasmalemmal vesicles as elements in a system of branching invaginations from the cell surface, *Proc Natl Acad Sci USA* 76: 6439-6442, 1979.

22. Oldendorf, W.H., M.E. Cornform, and W.J. Brown, The large apparent work capability of the blood brain barrier: a study of the mitochondrial content of capillary endothelial cells in brain and other tissues of the rat, *Ann Neurol* 1: 409-417, 1977.

23. Ferguson, R.K., and D.M. Woodbury, Penetration of ^{14}C-inulin and ^{14}C-sucrose into brain, cerebrospinal fluid, and skeletal muscle of developing rats, *Exp Brain Res* 7: 181-194, 1969.

24. deBault, L.E., and P.A. Cancilla, γ-glutamyl transpeptidase in isolated brain endothelial cells: induction by glial cells in vitro, *Science* 207: 653-655, 1980.

25. Janzer, R.C., and M.C. Raff, Astrocytes induce blood-brain barrier properties in endothelial cells, *Nature* 325: 253-257, 1987.

26. Beck, D.W, H.V. Vinters, M.N. Hart, and P.A. Cancilla, Glial cells influence polarity of the blood-brain barrier, *J Neuropathol Exp Neurol* 43: 219-224, 1984.

27. Krogh, A., The active and passive exchanges of inorganic ions through the surfaces of living cells and through living membranes generally, *Proc Roy Soc B* 133: 140-200, 1946.

28. Oldendorf, W.H., Lipid solubility and drug penetration of the blood brain barrier, *Proc Soc Exp Biol Med* 147: 813-816, 1974.

29. Cefalu, W.T., and W.M. Pardridge, Restrictive transport of a lipid-soluble peptide (cyclosporin) through the blood-brain barrier, *J Neurochem* 45: 1954-1956, 1985.

30. Crone, C., and S.P. Olesen, Electrical resistance of brain microvascular endothelium, *Brain Res* 241: 49-55, 1982.

31. Crone, C., Facilitated transfer of glucose from blood into brain tissue, *J Physiol* 181: 103-113, 1965.

32. Oldendorf, W.H., Brain uptake of radiolabeled amino acids, amines and hexoses after arterial injection, *Am J Physiol* 221: 1629-1639, 1971.

33. Oldendorf, W.H., and J. Szabo, Amino acid assignment to one of three blood-brain barrier amino acid carriers, *Am J Physiol* 230: 94-98, 1976.

34. Pardridge, W.M., Neuropeptides and the blood-brain barrier, *Ann Rev Physiol* 45: 73-82, 1983.

35. Kalaria, R.N., S.A. Gravina, J.W. Schmidley, G. Perry, and S.I. Harik, The glucose transporter of the human brain and blood-brain barrier, *Ann Neurol* 24: 757-764, 1988.

36. deFronzo, R.A., R. Hendler, and N. Christensen, Stimulation of counterregulatory hormonal responses in diabetic man by a fall in glucose concentration, *Diabetes* 29: 125-131, 1980.

37. Gjedde, A., and C. Crone, Induction processes in blood-brain transfer of ketone bodies during starvation, *Am J Physiol* 229: 1165-1169, 1975.

38. McCall, A.L, W.R. Millington, and R.J. Wurtman, Metabolic fuel and amino acid transport into the brain in experimental diabetes mellitus, *Proc Natl Acad Sci USA* 79: 5406-5410, 1982.

39. Harik, S.I., S.A. Gravina, and R.N. Kalaria, Glucose transporter of the blood-brain barrier and brain in chronic hyperglycemia, *J Neurochem* 51: 1930-1934, 1988.

40. Cremer, J.E., L.D. Braun, and W.H. Oldendorf, Changes during development in transport processes of the blood-brain barrier, *Biochim Biophys Acta* 448: 633-637, 1976.

41. Cremer, J.E., V.J. Cunningham, W.M. Pardridge, L.D. Braun, and W.H. Oldendorf, Kinetics of blood-brain barrier transport of pyruvate, lactate and glucose in suckling, weanling and adult rats, *J Neurochem* 33: 439-445, 1979.

42. Moore, T.J., A.P. Lione, M.C. Sugden, and D.M. Regen, β-Hydroxybutyrate transport in rat brain, developmental and dietary modulation, *Am J Physiol* 230: 619-630, 1976.

43. Cornford, E.M., L.D. Braun, and W.H. Oldendorf, Developmental modulations of blood-brain barrier permeability as an indicator of changing nutritional requirements in the brain, *Pediatr Res* 16: 324-328, 1982.

44. Ermisch, A., H.-J. Rühle, R. Landgraf, and J. Hess, Review: blood-brain barrier and peptides, *J Cereb Blood Flow Metab* 5: 350-357, 1985.

45. Banks, W.A., and A.J. Kastin, Interactions between the blood-brain barrier and endogenous peptides: emerging clinical implications, *Am J Med Sci* 295: 459-465, 1988.

46. Banks, W.A., and A.J. Kastin, Exchange of peptides between the circulation and the nervous system: role of the blood-brain barrier, In J.C. Porter and D. Ježová (eds) *Circulating Regulatory Factors and Neuroendocrine Function*, Plenum Publishing Corporation, New York, pp. 59-69, 1990.

47. Ashcroft, G.W., R.C. Dow, and A.T.B. Moir, The active transport of 5-hydroxyindol-3-ylacetic acid and 3-methoxy-4-hydroxyphenylacetic acid from a recirculatory perfusion system of the cerebral ventricles of the unanaesthetized dog, *J Physiol* 199: 397-425, 1968.

48. Smith, Q.R., and S.I. Rapoport, Carrier-mediated transport of chloride across the blood-brain barrier, *J Neurochem* 42: 754-763, 1984.

49. Bradbury, M.W.B., and B. Stulcová, Efflux mechanism contributing to the stability of the potassium concentration in cerebrospinal fluid, *J Physiol* 208: 415-430, 1970.

50. Smith, Q.R., and S.I. Rapoport, Cerebrovascular permeability coefficients to sodium, potassium and chloride, *J Neurochem* 46: 1732-1742, 1986.

51. Betz, A.L., Transport of ions across the blood-brain barrier, *Fed Proc* 45: 2050-2054, 1986.

52. Betz, A.L., J.A. Firth, and G.W. Goldstein, Polarity of the blood-brain barrier: distribution of enzymes between the luminal and antiluminal membranes of brain capillary endothelial cells, *Brain Res* 192: 17-28, 1980.

53. Bertler, Å., B. Falck, C.H. Owman, and E.B. Rosengrenn, The localization of monoaminergic blood-brain barrier mechanisms, *Pharmacol Rev* 18: 369-385, 1966.

54. Hardebo, J.E., and C. Owman, Enzymatic mechanisms for neurotransmitter monoamines and their precursors at the blood-brain interface, In B.B. Johansson, C. Owman, and H. Widner (eds) *Pathophysiology of the Blood-Brain Barrier and Peptides, The Fernström Symposium*, Series 14, Elsevier, Amsterdam, In Press

55. Owman, C., and J.E. Hardebo (eds) In (eds), *Neural Regulation of Brain Circulation*, Elsevier, Amsterdam, 1986.

56. Hartman, B., The innervation of cerebral blood vessels by central noradrenergic neurons, In E. Usdin and S.H. Snyder (eds) *Frontiers in Catecholamine Research*, Pergamon Press, New York, pp. 91-96, 1973.

57. Raichle, M.E., B.K. Hartman, J.O. Eichling, and L.G. Sharpe, Central noradrenergic regulation of cerebral blood flow and vascular permeability, *Proc Natl Acad Sci USA* 72: 3726-3730, 1975.

58. Johanson, C.E., Ontogeny and Phylogeny of the blood-brain barrier, In E.A. Neuwelt (ed) *Implications of the Blood-Brain Barrier and Its Manipulation, Volume 1, Basic Science Aspects*, Plenum Publishing Corporation, New York, pp. 157-198, 1989.

59. Johansson, B.B., and L. Martinsson, The blood-brain barrier in adrenaline-induced hypertension, Circadian variations and modification by beta-adrenoreceptor antagonists, *Acta Neurol Scan* 62: 96-102, 1980.

60. Johansson, B.B., and O. Isaksson, Circadial variation of cerebral vessel vulnerability during adrenaline-induced hypertension, In D.D. Heistad and M.L. Marcus (eds) *Cerebral Blood Flow: Effect of Nerves and Neurotransmitters*, Elsevier Biomedical, Amsterdam, pp. 367-375, 1982.

61. Johansson, B.B., Effect of an acute increase of the intravascular pressure on the blood-brain barrier. A comparison between conscious and anesthetized rats, *Stroke* 9: 588-590, 1978.

62. Johansson, B.B., Pharmacological modification of hypertensive blood-brain barrier opening, *Acta Pharmacol Toxicol* 48: 242-247, 1981.

63. Wekerle, H., C. Linington, H. Lassmann, and N.R. Meyerman, Cellular immune reactivity within the CNS, *Trends Neurosci* 9: 271-277, 1986.

64. Bradbury, M.W.B., and R.J. Westrop, Lymphatics and the drainage of cerebrospinal fluid, In K. Shapiro, A. Marmarou, and H. Portnoy (eds) *Hydrocephalus*, Raven Press, New York, pp. 69-81, 1984.

65. Widner, H., B.-A. Jönsson, L. Hallstadius, K. Wingårdh, S.-E. Strand, and B.B. Johansson, Scintigraphic method to quantify the passage from brain parenchyma to the deep cervical lymph nodes in rat, *Eur J Nucl Med* 13: 456-461, 1987.

66. Widner H., G. Möller, and B.B. Johansson, Immune response in deep cervical lymph nodes and spleen in the mouse after antigen deposition in different intracerebral sites, *Scan J Immunol* 28: 563-571, 1988.

67. Widner, H., and P. Brundin, Immunological aspects of grafting in the mammalian central nervous system. A review and speculative synthesis, *Brain Res Rev* 13: 287-324, 1988.

68. Krisch, B., H. Leonhardt, and W. Buchheim, The functional and structural border between the CSF- and blood-milieu in the circumventricular organs (organum vasculosum laminae terminalis subfornical organ, area postrema) of the rat, *Cell Tissue Res* 195: 485-497, 1978.

69. Bigotte, L., and Y. Olsson, Cytofluorescence localization of adriamycin in the nervous system III. Distribution of drug in the brain of normal adult mice after intraventricular and arachnoidal injections, *Acta Neuropathol [Berl]* 58: 193-202, 1982.

70. Houthoff, H.J., R.C. Moretz, H.G. Rennke, and H.M. Wisniewski, The role of molecular charge in the extravasation and clearance of protein tracers blood-brain barrier impairment and cerebral edema, In K.G. Go and A. Baethmann (eds) *Recent Progress in the Study and Therapy of Brain Edema*, Plenum Publishing Corporation, New York, pp. 67-79, 1984.

71. Chan, P.H., S. Longar, and R.A. Fishman, Protective effects of liposome-entrapped superoxide dismutase on post traumatic brain edema, *Ann Neurol* 21: 540-547, 1987.

72. Fishman, R.A., and P.H. Chan, Liposome entrapment of drugs and enzymes to enable passage across the blood-brain barrier, In B.B. Johansson, C. Owman, and H. Widner (eds) *The Pathophysiology of the Blood-Brain Barrier. Long Term Consequences for the Brain, The Fernström Symposium, Series 14*, Elsevier, Amsterdam, In Press.

73. Pardridge, W.M., A.K. Kumagai, and J.B. Eisenberg, Chimeric peptides as a vehicle for peptide pharmaceutical delivery through the blood-brain barrier, *Biochem Biophys Res Commun* 146: 307-313, 1987.

74. Olesen, S.-P., A calcium-dependent reversible permeability increase in microvessels in frog brain, induced by serotonin, *J Physiol* 361: 103-113, 1985.

75. Olesen, S.-P, Free oxygen radicals decrease electrical resistance of microvascular endothelium in brain, *Acta Physiol [Scand]* 129: 181-187, 1987.

76. Olesen, S.-P., and C. Crone, Substances that rapidly augment ionic conductance of endothelium in cerebral venules, *Acta Physiol [Scand]* 127: 233-241, 1986.

77. Johansson, B., C.-L. Li, Y. Olsson, and I. Klatzo, The effect of acute arterial hypertension on the blood-brain barrier to protein tracers, *Acta Neuropathol [Berl]* 16: 117-124, 1970.

78. Johansson, B.B., Hypertension and the blood-brain barrier, In E.A. Neuwelt (ed) *Implications of the Blood-Brain Barrier and Its Manipulation, Volume 2, Clinical Aspects*, Plenum Publishing Corporation, New York, pp. 389-410, 1989.

79. Rapoport, S.I., Osmotic opening of the blood-brain barrier, *Ann Neurol* 24: 677-680, 1988.

80. Neuwelt, E.A., and S.A. Dahlborg, Chemotherapy administered in conjunction with osmotic blood-brain barrier modification in patients with brain metastases, *J Neurooncol* 4: 195-207, 1987.

81. Nagy, Z., H. Peters, and I. Huttner, Charge-related alterations of the cerebral endothelium, *Lab Invest* 49: 662-671, 1983.

82. Ježová, D., B.B. Johansson, Z. Opršalová, and M. Vigaš, Changes in blood-brain barrier function modify the neuroendocrine response to circulating substances, *Neuroendocrinology* 49: 428-433, 1989.

83. Westergren, I., and B.B. Johansson, Albumin content in brain and CSF after intracarotid infusion of protamine sulphate. A longitudinal study, *Exp Neurol*, In Press.

84. Sokrab, T.-E.O., B.B. Johansson, H. Kalimo, and Y. Olsson, A transient hypertensive opening of the blood-brain barrier can lead to brain damage, *Aca Neurpathol [Berl]* 75: 557-565, 1988.

85. Salahuddin, T.S., B.B. Johansson, H. Kalimo, and Y. Olsson, Structural changes in the rat brain after carotid infusions of hyperosmolar solutions. An electron microscopic study, *Acta Neuropathol* 77: 5-13, 1988.

86. Fishman, R.A., Is there a therapeutic role or osmotic breaching of the blood-brain barrier (editorial), *Ann*

Neurol 22: 298-299, 1987.

87. Johansson, B.B., C. Owman, and H. Widner (eds), *The Pathophysiology fo the Blood-Brain Barrier. Long Term Consequences for the Brain, The Fernström Symposium, Series 14*, Elsevier, Amsterdam, In Press.

88. Bigotte, L., B. Arvidson, and Y. Olsson, Cytofluorescence localization of adriamycin in the nervous system I. Distribution of the drug in the central nervous system of normal adult mice after intravenous injection, *Acta Neuropathol [Berl]* 57: 121-129, 1982.

BLOOD-BRAIN BARRIER AND NEUROENDOCRINE REGULATIONS

D. Ježová, Z. Opršalová, F. Héry,[1] M. Héry,[1] A. Kiss,
J. Jurčovičová, J. Chauveau,[2] C. Oliver,[1] B.B. Johansson,[3]
and M. Vigaš

Institute of Experimental Endocrinology
CPS, Slovak Academy of Sciences
Bratislava, Czechoslovakia

Neuroendocrine regulatory processes have been intensively studied, and a considerable amount of data has accumulated on the central regulatory mechanisms involved in the control of pituitary hormone release. Special attention has been given to monoamines and peptides of central origin that act as neurotransmitters and neuromodulators, and there are several reviews summarizing the large amount of information on this important topic (1-5). Nevertheless, many results dealing with the effects of various neuroactive substances on the secretion of individual pituitary hormones are controversial and need further elucidation.

One of the frequently used approaches to the study of central neuroendocrine regulations is pharmacological treatment with drugs, hormones, peptides and analogues of these substances. The treatments may be given directly into the central nervous system (CNS) or peripherally, both routes having their advantages and limitations. As to peripheral drug administrations, only substances penetrating the blood-brain barrier (BBB) are believed to directly influence central regulatory mechanisms for pituitary hormone release.

However, some substances having a chemical structure that does not allow their penetration of the BBB have been found to modify the secretion of pituitary hormones. In such cases, several mechanisms may be involved in the action of these substances. In such cases, an action at the level of the pituitary must be considered. Much attention has been given to the direct effects of circulating catecholamines on pituitary adrenocorticotropic hormone (ACTH) release, as will be discussed below.

When the action of a substance at the pituitary level has been excluded or the involvement of hypothalamic regulatory hormones established, an effect within the CNS of structures lacking the BBB is presumed. For example, it has been postulated that the stimulatory effect of systemic angiotensin II on ACTH secretion is mediated through the subfornical organ, a structure with fenestrated capillaries. The neurons of the subfornical organ send angiotensin II-positive efferents to the parvicellular part of the paraventricular nucleus. Activation of this pathway facilitates secretion of corticotropin releasing factor (CRF) and possibly vasopressin into the hypophysial-portal circulation with a consequent increase in ACTH secretion (6,7). Other actions of substances that do not cross the BBB

---

[1]Laboratory of Experimental Neuroendocrinology, Inserm U. 297, Marseille, France

[2]Immunotech, Marseille, France

[3]Department of Neurology, University Hospital, Lund, Sweden

Circulating Regulatory Factors and Neuroendocrine Function
Edited by J. C. Porter and D. Ježová
Plenum Press, New York, 1990

41

may involve an indirect influence on neuroendocrine cell function. Such substances may induce metabolic or hemodynamic changes and thereby modify hormone secretion; they may influence the secretion of other regulatory factors; they may even be transformed to active metabolites.

Another, until now, unconsidered possibility is that the function of the BBB itself may undergo changes and under certain conditions become more permeable. This type of alteration would result in access to brain centers of otherwise non-permeating substances with subsequent modification of brain functions including the neuroendocrine

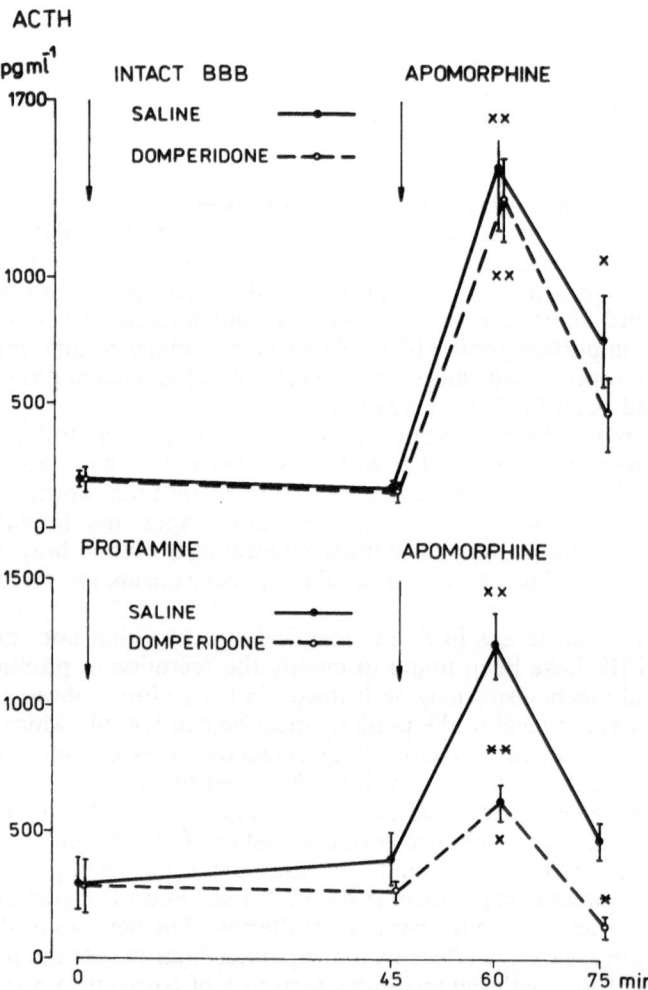

Figure 1. Inhibition of apomorphine-induced (200 μg/kg BW, iv) rise in plasma ACTH by pretreatment with domperidone (1 mg/kg BW, iv) in rats infused with protamine sulfate (15 mg/kg *via* a carotid artery) immediately after saline or domperidone injection to increase BBB permeability. Statistical significance *vs.* values before apomorphine injection: × denotes $P < 0.05$, ×× denotes $P < 0.01$; domperidone *vs.* saline-treated groups: * denotes $P < 0.05$; ** denotes $P < 0.01$. Reproduced from Ježová *et al.* (17) with permission from *Neuroendocrinology*.

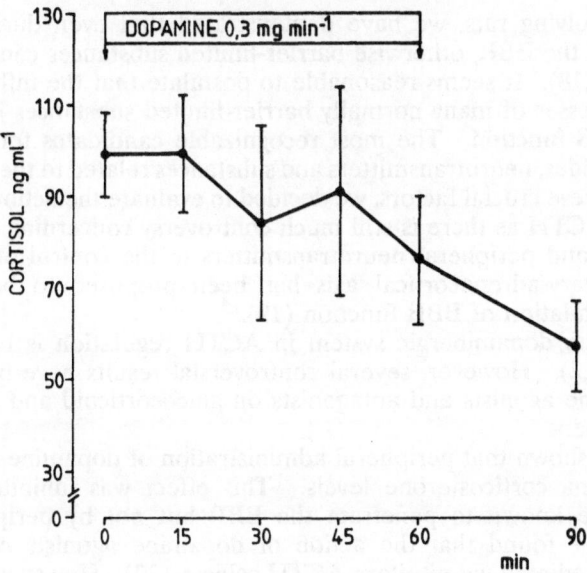

Figure 2. Intravenous infusion of dopamine (0.3 mg/min for 60 min) failed to influence concentrations of cortisol in plasma of healthy men.

system. In this respect, a large irreversible rupture of the BBB is of little interest as it would disturb the homeostasis of the organism in general. Much more interesting are transient and/or selective changes in BBB function that may remain undetected but modify the results of some experimental treatments.

Among neuroendocrinologists, it is a generally accepted notion that the BBB is a stable phenomenon, always effective in the brain regions in which the presence of non-fenestrated capillaries has been described. In fact, the BBB is a dynamic interface (8), and a number of experimental procedures, pathological conditions and even physiologically occurring events, may modify its permeability.

Several conditions, which might play a role during experiments involving neuroendocrine processes, can be mentioned. The permeability of the BBB shows circadian variation, being lower during the night than during the day (9,10). Cerebral vessel permeability is affected by anesthesia; *e.g.*, it is lower in conscious rats than in rats under nitrous oxide anesthesia (11).

Another important factor to be considered is an acute increase in arterial blood pressure. The administration of various agents inducing acute elevations of arterial blood pressure result in transient enhancement of BBB permeability in both anesthetized and conscious rats (11-13). The extent of vascular leakage is dependent on the pre-existing vascular tone. Thus, an increase in BBB permeability may occur following a relatively moderate rise in blood pressure if vasodilatation or hypercapnia are simultaneously present. Drugs affecting vascular tone can make the BBB more or less vulnerable to other treatments, as found in the case of β-blockers (14). An interesting finding is the observation that the blood pressure elevation, induced by intravenous 6-hydroxydopamine, evoked an increase in BBB permeability and thus facilitated the entry of the drug into the brain parenchyma (15). Hypertension-induced opening of the BBB can be influenced by agents that do not have an important effect on cerebrovascular tone, *e.g.*, some psychoactive drugs (16).

In studies involving rats, we have demonstrated that even during a short time reversible opening of the BBB, otherwise barrier-limited substances can reach the active sites in the brain (17,18). It seems reasonable to postulate that the influence on central neuroendocrine processes of many normally barrier-limited substances in the circulation is dependent on BBB function. The most recognizable candidates for such a role are circulating neuropeptides, neurotransmitters and substances related to the immune system.

In a study of these crucial factors, we decided to evaluate the action of monoamines on the secretion of ACTH as there is still much controversy concerning the physiological role of both central and peripheral neurotransmitters in the control of ACTH release. Moreover, the pituitary-adrenocortical axis has been proposed to be physiologically important for the regulation of BBB function (19).

The role of the dopaminergic system in ACTH regulation is believed to be of minor importance (1,2). However, several controversial results have been reported on the effect of dopamine agonists and antagonists on glucocorticoid and ACTH secretion in both rats and humans.

In rats, it was shown that peripheral administration of dopamine agonists resulted in elevation of plasma corticosterone levels. This effect was inhibited by dopamine antagonists that were known to penetrate the BBB but not by peripheral dopamine blockers (20,21). We found that the action of dopamine agonists on corticosterone concentrations was mediated via pituitary ACTH release (22). However, direct evidence demonstrating that the increase in ACTH release is actually mediated by stimulation of central dopamine receptors is not available.

We contributed to the elucidation of this problem by using an unusual approach, viz., experimental opening of the BBB (17). We made use of a conscious rat model in

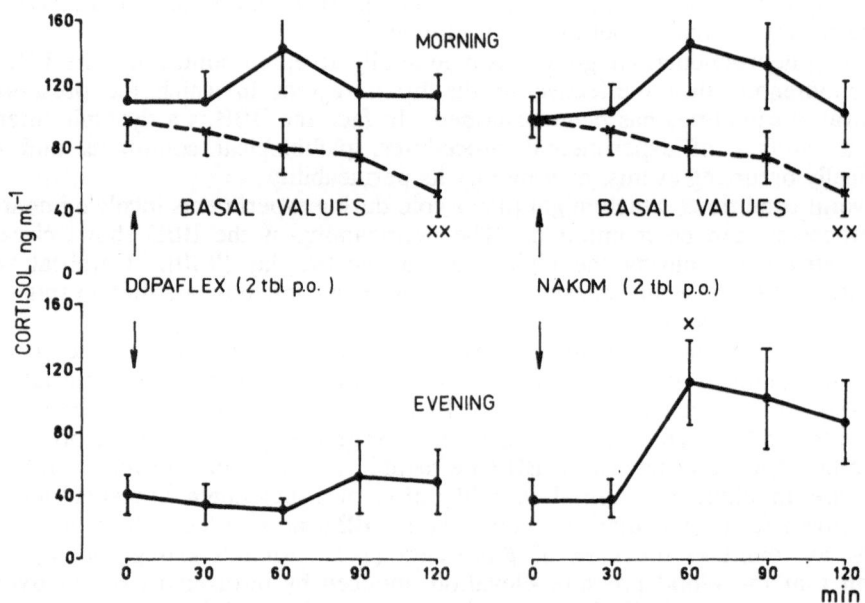

Figure 3. Plasma cortisol levels following administration of DOPA (Dopaflex, 1 g) or DOPA in combination with carbidopa (0.5 g and 50 mg, respectively; Nakom) in healthy men. Without drug treatment the morning values exhibit a decrease in the light of diurnal variation.

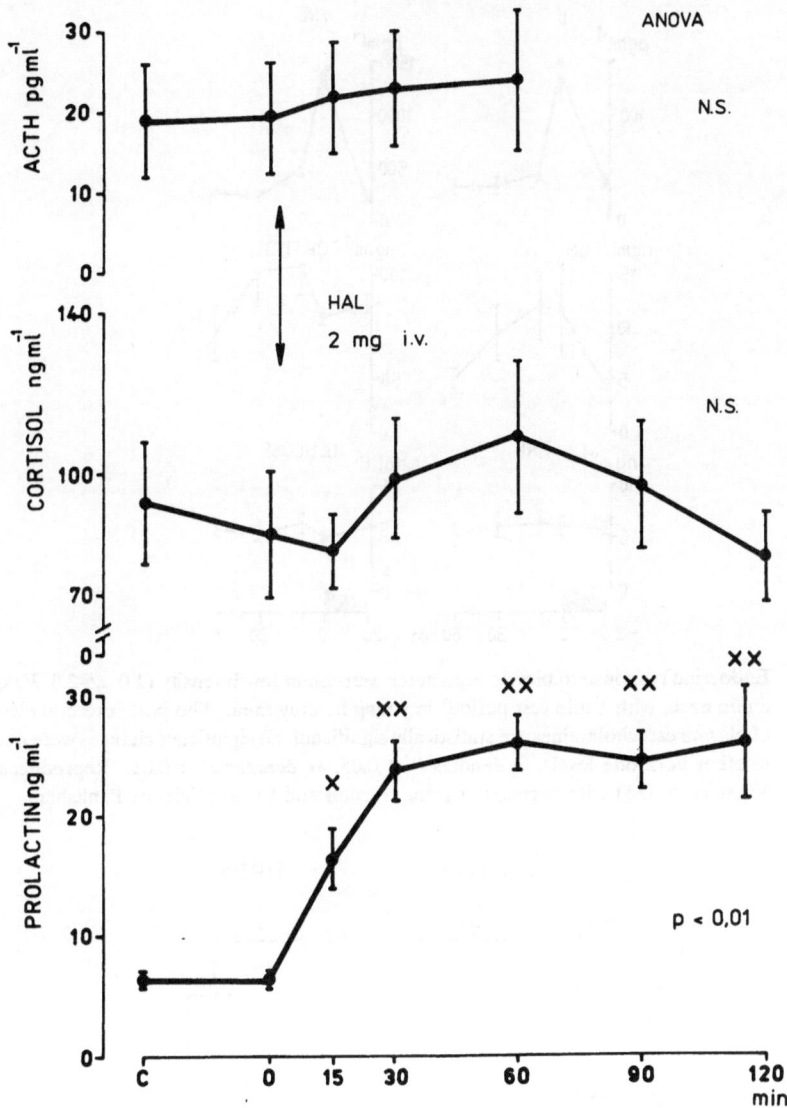

Figure 4. Injection of the dopamine receptor blocker, haloperidol, failed to influence ACTH and cortisol levels though inducing a significant increase in prolactin release in healthy men.

which the permeability of the BBB was increased by administration of protamine sulfate, which did not induce serious non-specific stress effects. The results are demonstrated in Figure 1. In the rats with an intact BBB, intravenous injection of the dopaminergic drug, apomorphine, stimulated ACTH release, and pretreatment with domperidone, a dopamine receptor blocker, which does not cross the BBB, failed to influence initial ACTH levels or the response to apomorphine. This finding is in agreement with the reports mentioned earlier. When pretreatment with domperidone was performed during a transient rise in BBB permeability, the apomorphine-induced ACTH release was found to be significantly inhibited. This result may be considered as

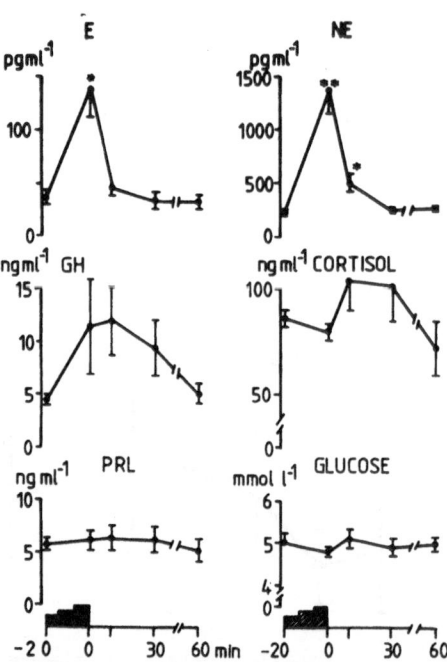

Figure 5.   Endocrine response to bicycle ergometer exercise of low intensity (1.0-1.5-2.0 W/kg BW, 6 min each, with 1 min rest period) in young healthy men. The post- exercise elevations of plasma catecholamines are statistically significant, no significant changes were observed in other hormone levels. ∗ denotes $P < 0.05$; ∗∗ denotes $P < 0.01$. Reproduced from Vigaš *et al.* (35) with permission from Gordon and Breach Science Publishers.

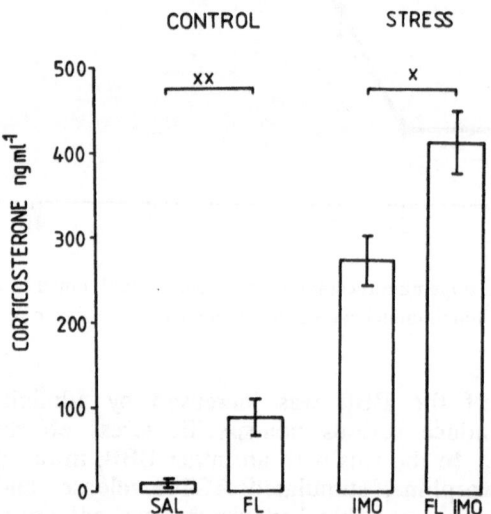

Figure 6.   Stimulatory effect of the serotonin uptake inhibitor fluoxetine on basal and stress-induced corticosterone secretion in rats. SAL - saline; FL - fluoxetine (10 mg/kg BW, sc); IMO - immobilization for 5 min. Statistical significance: × denotes $P < 0.05$; ×× denotes $P < 0.01$.

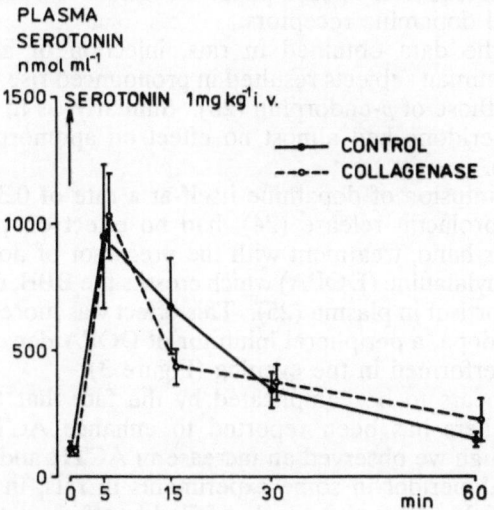

Figure 7. Serotonin in plasma following its iv injection in control rats and rats pretreated with collagenase to induce BBB dysfunction.

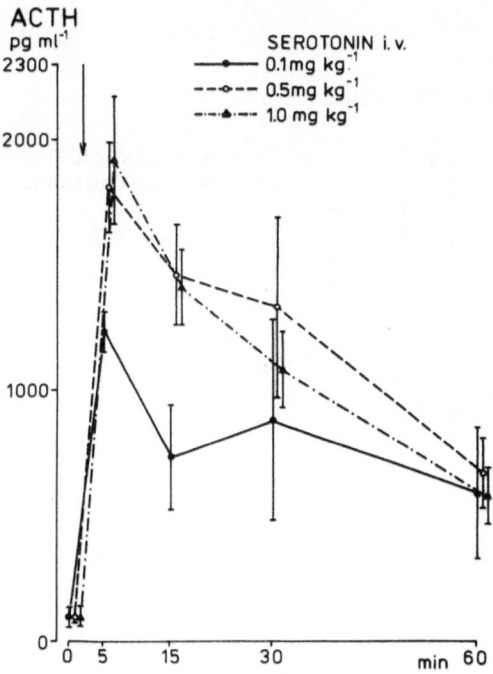

Figure 8. Plasma ACTH response to iv injection of serotonin administered in 3 different doses as creatinine sulfate.

evidence showing that the effect of apomorphine on ACTH release is, at least partially, mediated through central dopamine receptors.

Consistent with the data obtained in rats, injection of a sub-emetic dose of apomorphine to healthy human subjects resulted in pronounced rise in plasma ACTH and cortisol levels as well as those of β-endorphin (23). Similarly, as in rats with intact BBB, pretreatment with domperidone had almost no effect on apomorphine-induced ACTH release (data not shown).

As expected, the infusion of dopamine itself at a rate of 0.3 mg/min for 60 min, which clearly inhibited prolactin release (24), had no effect on plasma cortisol levels (Figure 2). On the other hand, treatment with the precursor of dopamine and norepinephrine, L-dihydroxyphenylalanine (DOPA) which crosses the BBB, resulted in a small rise in the concentration of cortisol in plasma (25). This effect was more evident when DOPA was combined with carbidopa, a peripheral inhibitor of DOPA decarboxylase activity and when the studies were performed in the evening (Figure 3).

This problem appears to be complicated by the fact that the administration of dopamine receptor blockers has been reported to enhance ACTH or glucocorticoid secretion (26-28). Although we observed an increase in ACTH and corticosterone levels following injection of haloperidol in some experiments in rats, in our studies with the human, administration of this dopamine blocker failed to affect cortisol secretion (Figure 4). Altogether, our data in both humans and rats favor a stimulatory role of central but not peripheral dopamine receptors in ACTH release.

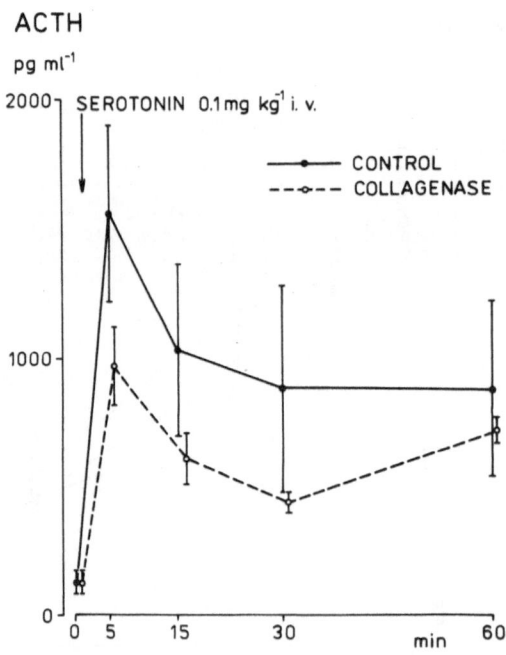

Figure 9.   Plasma ACTH response to the lowest administered dose of serotonin, administered as creatinine sulfate, in control rats and rats pretreated with collagenase to induce BBB dysfunction.

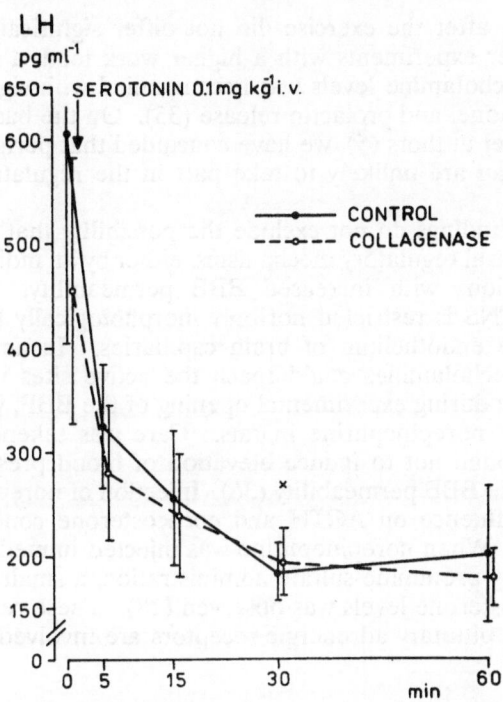

Figure 10.   Decrease in LH concentrations in response to the lowest administered dose of serotonin (creatinine sulfate) in both control rats and rats pretreated with collagenase to induce BBB dysfunction.

Catecholamines and ACTH Secretion

It is generally accepted that central catecholamines play an important role in the regulation of CRF and ACTH secretion; however, it is controversial whether this role of the central adrenergic system is primarily stimulatory or inhibitory. Most of the recent data in man and in the rat suggest that activation of central α-adrenergic receptors stimulates the hypothalamic-pituitary-adrenal axis (29).

Catecholamine receptors have been identified in the adenohypophysis under *in vitro* as well as *in vivo* conditions. In addition, several workers have hypothesized that circulating catecholamines contribute to the control of ACTH release (30).

The main controversy seems to concern the contribution of the stress-induced increase in peripheral catecholamines to the stimulation of ACTH release. In a large series of experiments in rats involving using anterolateral cuts around the mediobasal hypothalamus, knife lesions in the mediobasal hypothalamus, adrenal medullectomy plus sympathectomy, or combinations of these treatments, we have clearly demonstrated that circulating catecholamines have no significant role in the mediation of ACTH response under stress (31-34).

Using different stress models in man, we have been unable to demonstrate a correlation between the incremental increases of norepinephrine, and epinephrine in plasma with those of cortisol or other stress hormones (35). An example is given in Figure 5. A low intensity exercise using a graded bicycle ergometer induced a marked increase in plasma catecholamine levels while plasma concentrations of cortisol, growth hormone, and prolactin after the exercise did not differ significantly from pre-exercise concentrations. In other experiments with a higher work load, a similar or even lower increase in plasma catecholamine levels was accompanied with significant enhancement of cortisol, growth hormone, and prolactin release (35). On the basis of our findings and more recent data of other authors (5), we have concluded that peripheral catecholamines in the rat as well as man are unlikely to take part in the regulation of ACTH release during stress.

However, these findings do not exclude the possibility that circulating catecholamines interfere with central regulatory mechanisms, either by an indirect action, or directly in the brain in situations with increased BBB permeability. The penetration of monoamines into the CNS is restricted not only morphologically but also by degrading enzymes present in the endothelium of brain capillaries. In an attempt to establish whether circulating catecholamines could reach the active sites in the brain and thus influence ACTH release during experimental opening of the BBB, we tested the effect of intravenous injection of norepinephrine in rats. Care was taken to choose a dose of norepinephrine low enough not to induce elevation of blood pressure, as hypertension itself is known to increase BBB permeability (36). Injection of norepinephrine (0.25 μg/kg intravenous) had no influence on ACTH and corticosterone concentrations in plasma under basal conditions. When norepinephrine was injected immediately after increasing the BBB permeability by protamine sulfate administration, a small but significant rise in both ACTH and corticosterone levels was observed (18). These results support the view that central rather than pituitary adrenergic receptors are involved in the stimulation of ACTH secretion.

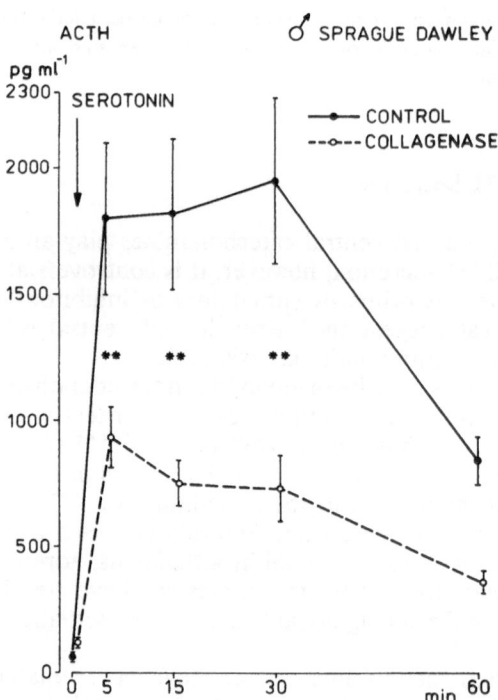

Figure 11. Lower ACTH response to the highest administered dose of serotonin (creatinine sulfate, 1 mg/kg BW, iv) in rats pretreated with collagenase compared to the response in the control group ($P < 0.01$).

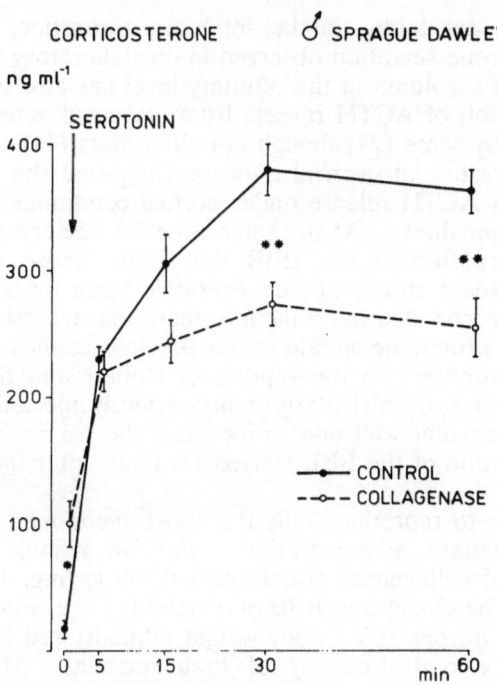

CORTICOSTERONE   ♂ SPRAGUE DAWLEY

Figure 12. Corticosterone response to the highest administered dose of serotonin (creatinine sulfate, 1 mg/kg BW, iv) in rats pretreated with collagenase compared to the response in the control group ($P < 0.01$).

It has to be mentioned that during severe stress conditions, such as immobilization of rats, dysfunction of the BBB was described (18,37,38). Moreover, some other stress situations involve conditions, *e.g.*, acute increase in blood pressure, which might result in a deranged function of the BBB with enhanced permeability, favoring the penetration of catecholamines from the blood into brain tissue. It is reasonable to suspect that changes in BBB function occurring during some experiments, which remained undetected, may have contributed to the controversial literary data concerning the role of peripheral catecholamines in ACTH release.

Serotonin and ACTH Secretion

The serotonergic neuronal system is known to affect ACTH release; however, as in the case of other monoamines, the data on the nature of its action are controversial. As claimed in a review by Tuomisto and Mannisto (1), the data related to the brain serotonergic activity and ACTH secretion are inconclusive, conflicting, and defective. A well documented effect of brain serotonin is its involvement in adrenocorticotropic rhythms (39,40).

Some authors reported an inhibitory action of serotonin particularly on stress-induced secretion of ACTH (41-43). On the other hand, there is considerable pharmacological evidence that is consistent with the hypothesis that central serotonin has a stimulatory influence on pituitary-adrenocortical function (40). An example is the stimulatory effect of a serotonin uptake inhibitor, fluoxetine, on both basal and stress-induced corticosterone secretion observed in our laboratory (Figure 6).

A direct action of serotonin at the pituitary level has also to be considered, as a dose-dependent stimulation of ACTH release from dispersed anterior pituitary cells by serotonin was observed by some (24) though not all authors (44).

In the light of these controversial data we compared the effect of intravenous injection of serotonin on ACTH release under normal conditions and under conditions with increased BBB permeability. At the same time we decided to introduce a model of more prolonged disruption of the BBB that would more closely resemble the disturbances in some disease states, *e.g.*, degenerative brain diseases, and would allow better evaluation of neuroendocrine function than the transient increase of BBB permeability induced by protamine sulfate in our previous studies.

We turned our attention to the reports of Robert and his colleagues (45-47) demonstrating that intracerebroventricularly or intravenously injected collagenase degrades the constituents of the vascular wall and so increases the permeability of the BBB. In their studies, the dysfunction of the BBB started 1-2 hours after the injection and lasted for about 60 hours.

We were not able to reproduce fully the above mentioned data, particularly not with intravenous collagenase administration. Positive results were obtained with intracisternal injection of collagenase (Boehringer 0.199 μg/mg; 16 μg/kg body weight administered in 20 μl). The changes in BBB permeability were tested by extravasation of the dye, trypan blue, (1 ml per 100 g body weight administered intravenously as a 2% solution), measuring the optical density of brain extracts. Macroscopically, many

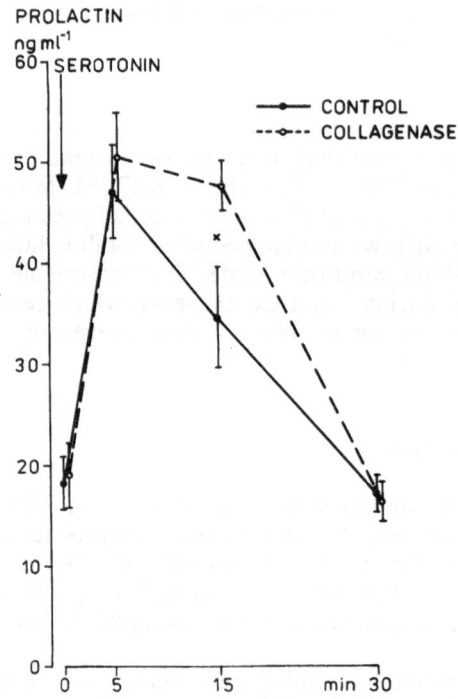

Figure 13. Prolactin response to the highest administered dose of serotonin (creatinine sulfate, 1 mg/kg BW, iv) in rats pretreated with collagenase compared to the response in the control group ($P < 0.05$).

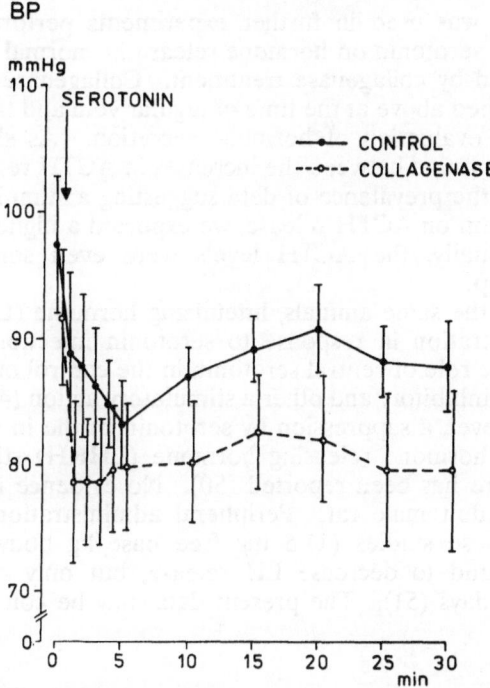

Figure 14. Values of mean arterial blood pressure in response to serotonin (creatinine sulfate, 1 mg/kg BW, iv) injection in control rats and rats pretreated with collagenase to induce BBB dysfunction.

hemorrhagic changes were observed in collagenase treated rats. Because no detailed morphologic studies have been performed as yet, no conclusions as to the extent and nature of BBB lesions can be made from our data Before further evaluation of this model is made, the interpretation of the data described below deserves caution.

In conscious rats bearing permanent catheters in the jugular vein and tail artery, three doses of serotonin were tested. The measurement of serotonin in plasma using a specific enzyme immunoassay (48) showed that the highest dose used (1 mg serotonin creatinine sulfate per kg body weight administered intravenously) resulted in a tenfold increase of its basal levels (Figure 7). The peak in serotonin concentrations was achieved 5 minutes after the injection.

In rats with intact BBB, intravenous administration of serotonin (0.1, 0.5 and 1 mg of serotonin creatinine sulfate per kg body weight) induced a significant rise in plasma ACTH levels already at 5 minutes, the and the ACTH levels were still elevated 30 minutes after the injection (Figure 8). The mechanism by which circulating serotonin triggers ACTH release may involve a direct action on the pituitary, or the increase in ACTH secretion may be a consequence of non-specific stress effects induced by serotonin injection. Indeed, in most of the rats injected with the highest dose of serotonin, some adverse effects were observed. Within the first minutes after injection the rats usually moved around the cage and then remained motionless showing some respiratory problems which were likely to be the result of serotonin-induced vasodilatation. These symptoms were short-lasting, however. The administration of the lowest dose of serotonin (0.1 mg/kg body weight) was free of visible adverse effects.

The lowest dose was used in further experiments performed with the aim of comparing the effects of serotonin on hormone release in normal rats and in rats with BBB dysfunction induced by collagenase treatment. Collagenase was injected into the cisterna magna as described above at the time of jugular vein and tail artery cannulations, i.e., 20-24 hours before evaluation of hormone secretion. As shown in Figure 9, no significant difference was found between the increases in ACTH release in the two groups of animals. Because of the prevalence of data suggesting a stimulatory influence of the central serotonergic system on ACTH release, we expected a higher increase in rats with BBB dysfunction. Actually, the ACTH levels were even somewhat lower in the collagenase-treated group.

In the plasma of the same animals, luteinizing hormone (LH) was measured. A decrease in LH concentration in response to serotonin injection was recorded in all animals (Figure 10). The role of central serotonin in the control of LH is uncertain, with some data suggesting an inhibitory and other a stimulatory action (40,49). As to the direct effects at the pituitary level, a suppression by serotonin of the in vitro release of LH in response to luteinizing hormone releasing hormone (LHRH) stimulation of neonatal pituitary glands in culture has been reported (50). No evidence is available to support such an action in the adult male rat. Peripheral administration of a higher dose of serotonin as used in these studies (11.5 mg free base/kg body weight administered subcutaneously) was found to decrease LH release, but only in rats that had been orchiectomized for two days (51). The present data may be considered to support the

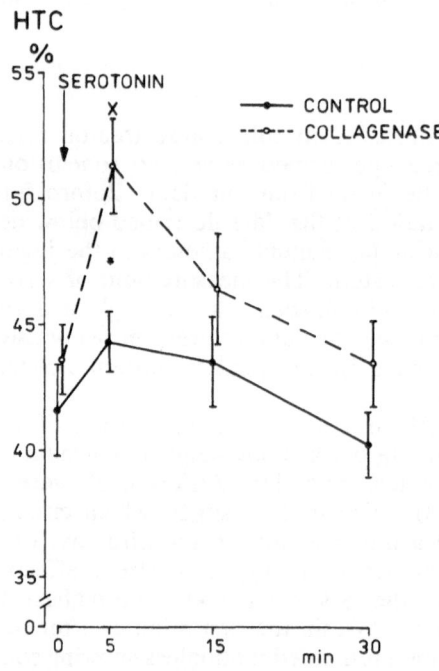

Figure 15. Increase in hematocrit 5 min after serotonin (creatinine sulfate, 1 mg/kg BW) in rats pretreated with collagenase. Statistical significance vs. basal values: × denotes $P < 0.05$; collagenase-treated vs. control group: * denotes $P < 0.05$.

action of serotonin in decreasing LH concentration by mechanisms operating outside the BBB; however, further studies are needed before a final conclusion can be made.

Another explanation for the lack of difference in hormone responses to serotonin between intact rats and rats with BBB dysfunction may be that the level of circulating serotonin following its injection was too low to reach brain regulatory centers in concentrations sufficiently high to influence neuroendocrine function.

Therefore, the experiments were repeated using a higher dose (1 mg serotonin creatinine sulfate/kg body weight). As already mentioned, the changes in animal behavior occurred shortly after the injection, the adverse effects being much more pronounced in collagenase-treated rats. Surprisingly, the increase in ACTH levels was significantly lower in rats injected with collagenase 20-24 hours earlier (Figure 11). The attenuated response of ACTH resulted in a smaller increase of corticosterone concentrations (Figure 12). The experiments were repeated in two strains of rats (Sprague Dawley and Wistar) with similar results, though the serotonin-induced rise in ACTH levels in intact Wistar rats was higher, and the difference between the intact and collagenase-treated groups was less pronounced as in Figure 11. If, in line with our main proposal, circulating serotonin actually reached the brain regulatory centers in our experimental model involving collagenase treatment, then our data demonstrate an inhibitory effect of central serotonin receptors on ACTH release.

This conclusion is supported by the significantly enhanced response of prolactin to serotonin administration in collagenase-treated rats (Figure 13). The stimulatory role of central serotonin in the control of prolactin release seems to be clearly established, and no firm evidence is available showing that serotonin has any effect at the pituitary level (1,40). However, an increase of prolactin secretion after systemic injections of serotonin was observed by several authors, as well as in our experiments. The mechanism of this effect is not known; yet, one explanation could be the interference of serotonin with dopamine secretion (51,52). Enhanced prolactin release in response to systemic serotonin injection in rats with BBB dysfunction as compared to normal rats suggests additional stimulation of prolactin secretion by activation of central serotonin receptors.

It cannot be excluded that serotonin, in combination with collagenase treatment, evokes other changes unrelated to the BBB that influence hormonal responses. We observed that the highest dose of serotonin used in our studies induced only a small insignificant decrease in blood pressure, which was somewhat more pronounced in collagenase treated rats (Figure 14). A finding, which defies explanation, is the increase in hematocrit 5 minutes after serotonin injection in collagenase-treated but not in the control group of rats (Figure 15).

It clearly emerges that further investigations are needed to characterize better the model of prolonged BBB dysfunction and to relate this model to the permeability of the brain vessels and possibly other changes in the CNS as well as other parts of the body. At present, no final conclusion can be made whether the changes in pituitary hormone release are the result of a direct action of serotonin in the brain or whether the hormone responses were modified by other factors which remain to be elucidated.

In summary, the data obtained using an experimental model of transient opening of the BBB by protamine sulfate along with other approaches, suggest a stimulatory role of central dopamine receptor activation in the control of ACTH release and no direct effect of circulating catecholamines on pituitary ACTH secretion. Preliminary data obtained in rats treated with collagenase to induce more prolonged dysfunction of the BBB, show that increase in circulating serotonin may interfere with neuroendocrine regulations. In general, the results of our studies strongly indicate that the changes in BBB permeability, which may occur under different experimental and pathological conditions, have to be considered as a phenomenon involved in neuroendocrine regulatory mechanisms.

Acknowledgements

The generous supply of the materials for ACTH, prolactin, and LH radioimmuno-assays by National Pituitary Agency, NIADDK, is gratefully acknowledged. The authors thank Miss I. Szalayova, Mrs. L. Zilava, Mrs. N. Petrikova, Mrs. M. Masarykova, Mrs. C. Collas, and Mrs. Y. Fahamoe for assistance in individual stages of the experiments.

References

1. Tuomisto, J., and P. Mannisto, Neurotransmitter regulation of anterior pituitary hormones, *Pharmacol Rev* 37: 249-332, 1985.
2. Negro-Vilar, A., C. Johnston, E. Spinedi, M. Valenca, and F. Lopez, Physiological role of peptides and amines on the regulation of ACTH secretion, *Ann NY Acad Sci* 512: 218-236, 1987.
3. Delitala, G., P. Tomasi, and R. Virdis, Neuroendocrine regulation of human growth hormone secretion. Diagnostic and clinical applications, *J Endocrinol Invest* 11: 441-462, 1988.
4. Jones, M.T., and B. Gillham, Factors involved in the regulation of adrenocorticotropic hormone/β-lipotropic hormone, *Physiol Rev* 68: 743-818, 1988.
5. Al-Damluji, S., D. Cunnah, A. Grossman, L. Perry, G. Ross, D. Coy, L.H. Rees, and G.M. Besser, Effect of adrenaline on basal and ovine corticotrophin-releasing factor-stimulated ACTH secretion in man, *J Endocrinol* 112: 145-150, 1987.
6. Ferguson, A.V., Systemic angiotensin acts at the subfornical organ to control the activity of paraventricular nucleus neurons with identified projections to the median eminence, *Neuroendocrinology* 47: 489-497, 1988.
7. Plotsky, P.M., S.W. Sutton, T.O. Bruhn, and A.V. Ferguson, Analysis of the role of angiotensin II in mediation of adrenocorticotropin secretion, *Endocrinology* 122: 538-545, 1988.
8. Cornford, E.M., The blood-brain barrier, a dynamic regulatory interface, *Mol Physiol* 7: 219-260, 1985.
9. Johansson, B.B., and L. Martinsson, The blood brain barrier in adrenaline-induced hypertension. Circadian variations and modification by beta-adrenoreceptor antagonists, *Acta Neurol Scand* 62: 96-102, 1980.
10. Johansson, B.B., O. Isaksson, Circadian variation of cerebral vessel vulnerability during adrenaline-induced hypertension, In D.D. Heistad and M.L. Marcus (eds) *Cerebral Blood Flow: Effects of Nerves and Neurotransmitters*, Elsevier North Holland, Amsterdam, pp. 367-375, 1982.
11. Johansson, B.B., Effect of beta-adrenoreceptor antagonists on the increased cerebrovascular permeability to protein induced by amphetamine, *Prog Neuropsychopharmacol* 2: 529-534, 1978.
12. Johansson, B.B., Blood-brain barrier dysfunction in acute arterial hypertension after papaverine-induced vasodilatation, *Acta Neurolog Scand* 50: 573-580, 1974.
13. Johansson, B.B., The blood-brain barrier in acute and chronic hypertension, In H.M. Eisenber and R.L. Suddith (eds) *The Cerebral Microvasulature*, Plenum Press, New York, pp. 211-226, 1980.
14. Johansson, B.B., and L. Martinsson, β-adrenoreceptor antagonists and the dysfunction of the blood-brain barrier induced by adrenaline, *Brain Res* 181: 219-222, 1980.
15. Johansson, B.B., and M. Henning, 6-Hydroxydopamine and the blood-brain barrier in adult conscious rats, *Acta Physiol Scand* 110: 1-4, 1980.
16. Johansson, B.B., Pharmacological modification of hypertensive blood-brain barrier opening, *Acta Pharmacol Toxicol* 48: 242-247, 1981.
17. Ježová, D., B.B. Johansson, Z. Opršalová, and M. Vigaš, Changes in blood-brain barrier function modify the neuroendocrine response to circulating substances, *Neuroendocrinology* 49: 428-433, 1989.
18. Ježová, D., B.B. Johansson, Y. Olsson, Z. Opršalová, A. Kiss, J. Jurčovičová, J. Grassler, I. Westergren, and M. Vigaš, Can catecholamines and other neurotransmitters cross the blood-brain barrier and modify neuroendocrine function under stress?, In G.R. Van Loon, R. Kvetňanský, and J. Axelrod (eds) *Stress: Neurochemical and Humoral Mechanisms*, Gordon & Breach Science Publishers, New York, In Press.
19. Long, J.B., and J.W. Holaday, Blood-brain barrier: endogenous modulation by adrenal-cortical function, *Science* 227: 1580-1583, 1985.

20. Fuller, R.W., and H.D. Snoddy, Elevation of serum corticosterone concentrations in rats by pergolide and other dopamine agonists, *Endocrinology* 109: 1026-1032, 1981.

21. Fuller, R.W., H.D. Snoddy, N.R. Mason, J.A. Clemens, and K.G. Bemis, Elevation of serum corticosterone in rats by dopamine agonists related in structure to pergolide, *Neuroendocrinology* 36: 285-290, 1983.

22. Ježová, D., J. Jurčovičová, M. Vigaš, K. Murgaš, and F. Labrie, Increase in plasma ACTH after dopaminergic stimulation in rats, *Psychopharmacology* 85: 201-203, 1985.

23. Ježová, D., and M. Vigaš, Apomorphine injection stimulates β-endorphin, adrenocorticotropin, and cortisol release in healthy man, *Psychoneuroendocrinology* 13: 479-485, 1988.

24. Vigaš, M., D. Ježová, J. Jurčovičová, E. Kellerová, V. Strbák, and R. Kvetňanský, Effect of increased plasma catecholamines on adenopituitary secretion, In E. Usdin, R. Kvetňanský, and I.J. Kopin (eds) *Catecholamines and Stress: Recent Advances*, Elsevier North/Holland, New York, pp. 143-148, 1980.

25. Ježová-Repcekovác, D., M. Vigaš, and I. Klimeš, Effect of glucose on plasma cortisol level after L-dopa administration in man, *Horm Metab Res* 12: 280-281, 1980.

26. Balestreri, R., S. Bertolini, and C. Castello, The neural regulation of ACTH secretion in man, In A. Polleri and R.M. MacLeod (eds) *Neuroendocrinology: Biological and Clinical Aspects*, Academic Press, London, pp. 155-185, 1979.

27. Giraud, P., J.C. Lissitzky, B. Conte-Devolx, P. Gillioz, and C. Oliver, Influence of haloperidol on ACTH and β-endorphin secretion in the rat, *Eur J Pharmacol* 62: 215-217, 1980.

28. Murburg, M.M., D. Paly, C.W. Wilkinson, R.C. Veith, K.L. Malas, D.M. Dorsa, Haloperidol increases plasma beta endorphin-like immunoreactivity and cortisol in normal human males, *Life Sci* 39: 373-381, 1986.

29. Al-Damluji, S., Adrenergic mechanisms in the control of corticotrophin secretion, *J Endocrinol* 119: 5-14, 1988.

30. Axelrod, J., and T.D. Reisine, Stress hormones: Their interaction and regulation, *Science* 224: 452-459, 1984.

31. Makara, G.B., R. Kvetňanský, D. Ježová, A. Jindra, I. Kakucska, and Z. Opršalová, Plasma catecholamines do not participate in pituitary-adrenal activation by immobilization stress in rats with transection of nerve fibers to the median eminence, *Endocrinology* 119: 1757-1762, 1986.

32. Ježová, D., R. Kvetňanský, K. Kovács, Z. Opršalová, M. Vigaš, and G.B. Makara, Insulin-induced hypoglycemia activates the release of adrenocorticotropin predominantly via central and propranolol insensitive mechanisms, *Endocrinology* 120: 409-415, 1987.

33. Kvetňanský, R., F.J.H. Tilders, I.D. Van Zoest, M. Dobrakovová, F. Berkenbosch, J. Culman, P. Zeman, and P.G Smelik, Sympathoadrenal activity facilitates beta-endorphin and alpha-MSH secretion but does not potentiate ACTH secretion during immobilization stress, *Neuroendocrinology* 45: 318-324, 1987.

34. Ježová, D., R. Kvetňanský, F.J.H. Tilders, and G.B Makara, Interaction of circulating catecholamines, CRF and AVP in the control of ACTH release during stress, In G.R. Van Loon, R. Kvetňanský, and J. Axelrod (eds) *Stress: Neurochemical and Humoral Mechanisms*, Gordon & Breach Science Publishers, New York, In Press, 1989.

35. Vigaš, M., R. Kvetňanský, J. Jurčovičová, D. Ježová, and P. Tatar, Comparison of catecholamine and adenopituitary hormone responses to various stress stimuli in man, In E. Usdin, R. Kvetňanský, and J. Axelrod (eds) *Stress: The Role of Catecholamines and Other Neurotransmitters*, Gordon and Breach Science Publishers, New York, pp. 865-882, 1984.

36. Johansson, B.B., C.L. Li, Y. Olsson, and I. Klatzo, The effect of acute arterial hypertension on the blood-brain barrier to protein tracers, *Acta Neuropathol* 16: 117-124, 1970.

37. Belova, T.I., and G. Jonsson, Blood-brain barrier permeability and immobilization stress, *Acta Physiol Scand* 116: 21-29, 1982.

38. Sharma, H.S., P.K. and Dey, Influence of long-term immobilization stress on regional blood-brain barrier permeability, cerebral blood flow and 5-HT level in conscious normotensive young rats, *J Neurol Sci* 72: 61-76, 1986.

39. Szafarczyk, A., Le systeme serotoninergique dans les regulations neuroendocriniennes, In: M. Briley (ed) *Le Systeme Serotoninergique et sa Regulation, Les Entretiens du Carla*, Centre de Recherche P. Fabre,

Castres, pp. 47-77, 1986.

40. Montange, M., and A. Calas, Serotonin and endocrinology - the pituitary, In N.N. Osborne and M. Hamon (eds) *Neuronal Serotonin*, John Wiley and Sons, pp. 271-303, 1988.

41. Telegdy, G., I. Vermes, and G.L. Kovács, Effect of drug induced changes in brain monoamines on neuroendocrine and behavioral processes, In K. Magyar (ed) *Second Congress Hungarian Pharmacologic Society: Symposium of Catecholaminergic and Serotonergic Mechanisms*, Akad. Kiado, Budapest, pp. 101-105, 1976.

42. Vernikos-Danellis, J., K.J. Kellar, D. Kent, C. Gonzales, P.A. Berger, and J.D. Barchas, Serotonin involvement in pituitary adrenal function, *Ann NY Acad Sci* 297: 518-526, 1977.

43. Ixart, G., A. Szafarczyk, F. Malaval, J. Nouguier-Soule, and I. Assenmacher, Facteurs de regulations des reponses plasmatiques de l'ACTH et de LH au stress chez le rat, *Ann d'Endocrinol* 41: 10C, 1980 (Abstract #15).

44. Rose, J.C., and W.F Ganong, Neurotransmitter regulation of the pituitary secretion, In W.F. Essman, L. Valzelli, and N.Y. Holliswood (eds) *Current Developments in Psychopharmacology, Volume 3*, Spectrum, pp. 87-123, 1976.

45. Robert, A.M., G. Godeau, M. Miskulin, and F. Moati, Mechanism of action of collagenase on the permeability of the blood-brain barrier, *Neurochem Res* 2: 449-455, 1977.

46. Godeau, G., M. Miskulin, J.M. Tixier, A. Kemeny, and A.M. Robert, Intensité et durée d'action des proteases dan la rupture post-ischémique de la barrière hématoencephalique, *Drugs Dis* 1: 90-101, 1984.

47. Godeau, G., C. Gavignet, N. Groult, and A.M. Robert, Effect of chromocarb diethylamine on the permeability of the blood brain barrier, *Clin Physiol Biochem* 5: 15-26, 1987.

48. Chauveau, J., V. Sert, A.M. Morel, and M.A. Delaage, A new, rapid and specific enzyme immunoassay for serotonin, In *Abstracts of International Symposium on Serotonin from Cell Biology to Pharmacology and Therapeutics*, Florence, March 29-April 1, 1989, In Press.

49. Weiner, R.I., and W.F. Ganong, Role of brain monoamines and histamine in regulation of anterior pituitary secretion, *Physiol Rev* 58: 905-976, 1978.

50. Martin, J.E., J.N. Engel, and D.C. Klein, Inhibition of the in vitro pituitary response to luteinizing hormone-releasing hormone by melatonin, serotonin, and 5-methoxytryptamine, *Endocrinology* 100: 675-680, 1977.

51. Pilotte, N.S., and J.C. Porter, Circulating luteinizing hormone and prolactin concentrations in intact or castrated male rats treated with 5-hydroxytryptamine, *Endocrinology* 105: 875-878, 1979.

52. Pilotte, N.S., and J.C. Porter, Dopamine in hypophysial portal plasma and prolactin in systemic plasma of rats treated with 5-hydroxytryptamine, *Endocrinology* 108: 2137-2141, 1981.

EXCHANGE OF PEPTIDES BETWEEN THE CIRCULATION AND THE NERVOUS SYSTEM: ROLE OF THE BLOOD-BRAIN BARRIER

William A. Banks and Abba J. Kastin

Veterans Administration Medical Center and
Tulane University School of Medicine
New Orleans, LA 70146

Peptides can cross the blood-brain barrier (BBB) bidirectionally, that is, from the central nervous system (CNS) to the blood or from the blood to the CNS. Passage occurs by both saturable and nonsaturable mechanisms that can be modified by factors such as lighting, amino acids, monoamines, aging, and neurotoxins. The role of this regulated exchange may be involved in processes such as analgesia, stress, dementia, and addiction. In this review, we shall emphasize the mechanisms regulating exchange of peptides between the circulation and the CNS and explore some of the implications of this exchange on selected biological functions.

Early Work on Peptides and the BBB

It was observed very early that peptides administered peripherally could affect CNS function (1). It was also noted that the classical gastrointestinal peptide hormones were present in the CNS, raising the possibility of the existence of *gut-brain* axes. Both of these observations were consistent with the passage of peptides across the BBB. The prevalent idea that peptides were like proteins in being too large to cross the BBB ignored the fact that most peptides were 1-2 orders of magnitude smaller than blood proteins. Many peptides, including TRH, Met-enkephalin, Leu-enkephalin, Tyr-MIF-1, kyotorphin, molluscan cardioexcitatory neuropeptide (FMRF), neuromedin N, proctolin, morphiceptin, tuftsin, and numerous biologically active fragments or analogs of large peptides, are smaller than thyroxine, a substance transported across the BBB by one of the classical carrier-mediated systems (2).

Early investigations gave conflicting information about the possibility that peptides might cross the BBB. One approach examined correlations between the concentrations of peptide in the cerebrospinal fluid (CSF) and the plasma (3). Most of the first studies were done with CSF obtained from the lumbosacral region and usually found no relationship between concentrations in the CSF and plasma. This led some investigators to conclude prematurely that peptides did not cross the BBB. This conclusion failed to consider two possibilities. First, some substances, such as potassium, have concentrations in the CNS that are totally derived from the circulation but are so tightly regulated that little or no correlation occurs between CSF and plasma concentrations. Second, very little CSF is produced or reabsorbed in the lumbosacral area so that the exchange rate with cranial CSF is slow. For substances with fluctuating concentrations in the blood, therefore, a better place to obtain CSF would be within the cranium. Later studies showed that

Circulating Regulatory Factors and Neuroendocrine Function
Edited by J. C. Porter and D. Ježová
Plenum Press, New York, 1990

59

when CSF was obtained from the posterior fossa or the lateral ventricle, the concentration of peptides in the CSF often correlated with concentrations in the plasma (3).

However, just as there are mechanisms by which substances derived from the circulation can show dissociated concentrations between CSF and blood, mechanisms might exist for correlations between concentrations in CSF and blood that are not due to passage across the BBB. For example, the release of a peptide into the CNS and into the circulation might be controlled by similar stimuli. Therefore, correlations between concentrations in CSF and plasma would indicate that a CNS-peripheral organ interaction is occurring but would not prove that the interaction is consists of the material across the BBB.

More direct evidence for passage of peptides across the BBB accumulated when investigators began to inject materials peripherally and measure the entry of radioactivity or immunoactivity into the brain. In retrospect, many early studies were flawed and can be viewed as the inevitable process in science of discovering appropriate methodologies. Although the phenomenon of the BBB has been studied about 100 years and methods were well established for studying the passage of amino acids, glucose, and other substances, peptides presented new challenges. In particular, their short half-lives in the circulation with the accumulation of degradation products and a low to modest rate of entry rendered most of the established techniques of little value. Only recently have new methods been established that can quantify transfer rates of peptides across the BBB.

Conceptual as well as technical problems impeded progress. For example, one study found significant entry of methionine enkephalin into the brain (4). When other investigators could not repeat these findings (5), it was concluded by many that methionine enkephalin in particular and peptides in general could not enter the CNS from the circulation. Several years later, a regulated transport of methionine enkephalin was discovered that could reconcile these studies (6,7). Similarly, conflicting studies on the passage of arginine vasopressin (AVP) across the BBB can be resolved, as discussed later (8).

From this early work, several principles have merged that should guide subsequent studies:

1)   Correction for extravascular space: Some early studies mistook the peptide contained within the vascular space of the brain for the amount of peptide that had crossed the BBB. Failure to correct for extravascular content can give erroneous rates of entry and unreliable estimates of peptide degradation.

2)   Identification of transferred material: Material that has presumably crossed the BBB should be rigorously identified. The material should be obtained from the side opposite that of injection (*e.g.*, from the blood if brain to blood transport is being studied). For saturable transport systems, however, competition with fragments can often yield more detailed information about the requirements of the transport system. Degradation of material during the identification process can be a difficult, and often an unrecognized, problem.

3)   CNS to blood as well as blood to CNS transport should be considered. Many peptides that do not appear to enter the CNS are actually crossing only to be immediately transported back into the blood. This can prevent peptide from accumulating to the levels needed for detection and can result in the erroneous impression that the peptide did not enter the CNS.

4)   Sensitive techniques should be used. Newer *in situ* perfusion methods for blood to brain studies (9,10) and an intraventricular injection method for

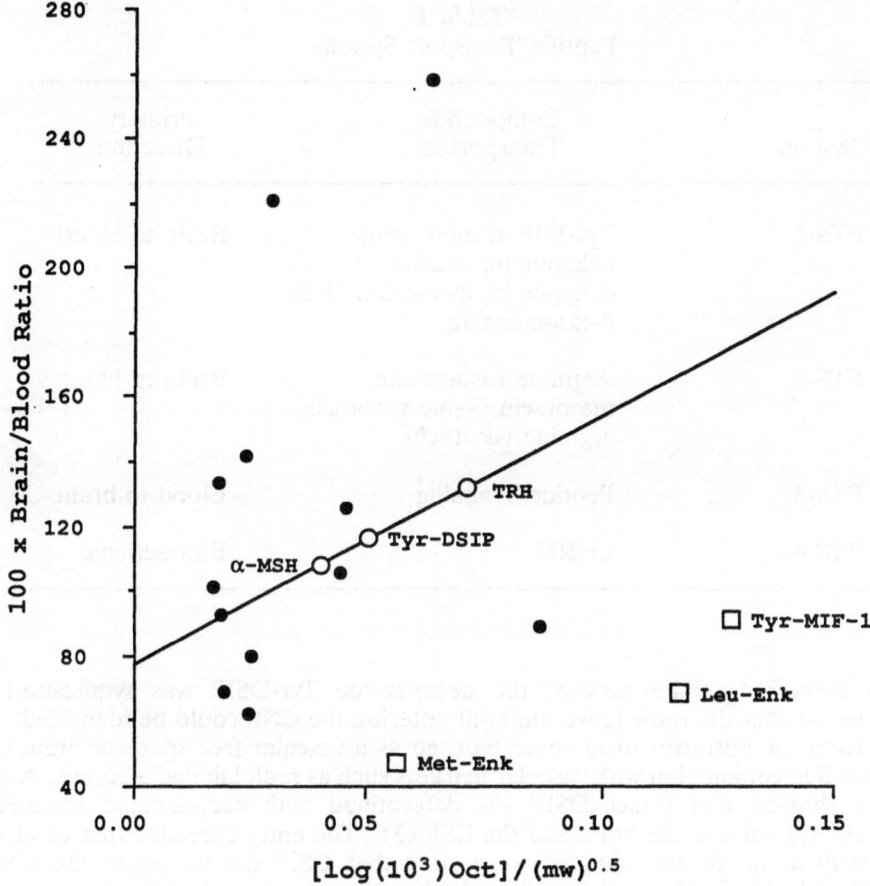

Figure 1. Relationship between the lipid solubility and molecular weight of iodinated peptides and their ability to enter the brain from the blood. *Open squares* indicate peptides for which subsequent studies have shown that a saturable brain to blood transport exists. *Open circles* indicate peptides that appear to enter the brain primarily by nonsaturable mechanisms. Drawn from data of Banks *et al.* (28) with permission of Elsevier Scientific Publishers Ireland Ltd.

brain to blood studies (11) are suitable for peptide studies. For peptides or analogs that are sufficiently resistant to the action of degradative enzymes, a sensitive graphical method can be used (12,13).

5)   CSF should be obtained from cranial, rather than spinal, regions.

Nonsaturable Passage of Peptides

An approach to the study of passage of peptides across the BBB can be illustrated with delta sleep-inducing peptide (DSIP). A radioimmunoassay developed for DSIP requires 8 of its 9 amino acids for detection and the octapeptide and nonapeptide can be readily separated by chromatography. Thus, passage of immunoactive material could

Table 1
Peptide Transport Systems

System	Compounds Transported	Primary Direction
PTS-1	Tyr-MIF-1, methionine enkephalin, leucine enkephalin, dynorphin (1-8), β-casomorphin	Brain to blood
PTS-2	Arginine vasopressin, mesotocin, lysine vasotocin, arginine vasotocin	Brain to blood
PTS-3	Peptide-T analog	Blood to brain
PTS-4	LHRH	Bidirectional

be easily identified. Alternatively, the decapeptide Tyr-DSIP was synthesized and radiolabeled so that the radioactive material entering the CNS could be identified. CSF obtained from the posterior fossa could be used as a vascular free space or brain tissue could be used in conjunction with vascular markers such as radiolabeled albumin. A series of studies showed that intact DSIP (as determined with radiolabeled material or immunoactivity) entered the brain and the CSF (3). The entry exceeded that of vascular markers such as inulin and albumin, suggesting that DSIP did not enter the CNS by leaking through the tightly sealed capillary bed.

Furthermore, analogs of DSIP entered the CNS to varying degrees (14,15). Such differential entry also argued against an unselective mechanism such as leakage. The entry rate of DSIP analogs correlated with their lipid solubilities, suggesting that diffusion through the cell membranes comprising the BBB might be a major mechanism for passage. The nonsaturable nature of such a mechanism was consistent with the inability of unlabeled DSIP or DSIP analogs to inhibit the entry of radiolabeled Tyr-DSIP (16). Lipid solubility of DSIP and its analogs also correlates with this biological activity (17). Taken together, these results suggested that nonsaturable transmembrane diffusion may be a major mechanism by which DSIP and related peptides cross the BBB to enter the CNS, although there may be exceptions to this rule (18).

Semi-quantitative studies with other radiolabeled peptides were consistent with the hypothesis that transmembrane diffusion played a major role in the blood to CNS passage of many peptides (19). Nonsaturable entry into the CNS has been confirmed for α-MSH (20), one of the first peptides clearly shown to enter the CNS from the periphery (21), and TRH (22). For most peptides, lipid solubility correlated only roughly with entry rate, suggesting that other factors were also important (Figure 1). Some substances had entry rates much lower than those predicted by their lipid solubilities. These compounds shared several characteristics, such as similar size (molecular weight: 572-904) and an N-terminal tyrosine. It was subsequently shown that the apparently low entry rate for these peptides, which included Tyr-MIF-1 and the enkephalins, was due to a saturable system that transported peptides from CNS into the blood (6,18). The classification and characterzation of these systems is based on studies of *in vivo* transport of radiolabeled peptides.

Saturable Transport

We have described four Peptide Transport Systems (PTS 1-4), as outlined in Table 1. In addition, evidence exists for a saturable component of the blood to CNS transport for DSIP in the guinea pig (22).

The best studied system is PTS-1. This system transports Tyr-MIF-1 and methionine enkephalin from the brain to the blood (23). It may also transport β-casomorphin and dynorphin 1-8. It certainly transports leucine enkephalin, but preliminary studies suggest that only about half of the saturable brain-to-blood transport may be by the system that transports Tyr-MIF-1/methionine enkephalin (24). Indeed, evidence suggests that there may be a family of transport systems for enkephalin-related peptides differing in their requirements for sodium and energy, sensitivity to modulation by amino acids, structural requirements, modulation by aging, and perhaps location within the CNS. Under some conditions or perhaps with some of the systems, blood-to-brain transport may predominate (4,25,26). It is unclear at this point whether PTS-1 should be subdivided into subsystems or whether new systems will have to be named. It is also possible, given the current data, that many of these apparent differences are due to technique, variation among species, etc. rather than to different transport systems. Unless otherwise noted, the characteristics of PTS-1 described here were elucidated with radiolabeled Tyr-MIF-1 and the intraventricular injection technique.

Arginine vasopressin is transported from the brain to the blood by PTS-2 (8). The existence of this system can explain many anomalies in previous work with AVP transport. For example, the amount of radiolabeled AVP entering the CNS after peripheral injection increases if unlabeled AVP is also given (27). This is consistent with an inhibition of an exiting process in the CNS. This system transports AVP in intact form, and also appears to transport arginine vasotocin, lysine vasotocin, and some fragments with intact ring structures, but not oxytocin, lysine vasopressin, the amino acid tyrosine, Tyr-MIF-1, or the cyclic peptide cycloLeu-Gly (8). Despite this system that transports AVP out of the CNS, peripherally administered AVP can have profound effects on the CNS, and in one study the amount of AVP entering the brain was higher than would be expected from the lipid solubility of this peptide (28). It may be, therefore, that under some circumstances blood to brain transport of AVP may predominate.

PTS-3 was described based on an enzymatically resistant analog of Peptide T (29). Transport is directed from the blood into the CNS. About 1 micromole per mouse of the unlabeled analog inhibited transport of this peptide by almost 25%, while 100 micromoles per mouse of iodotyrosine was without effect. This shows that the entry of the radiolabeled Peptide T analog was inhibitable by peptide but not by the amino acid fragment.

LHRH is transported both into and out of the CNS by a different transport system, termed PTS-4. This is the latest transport system to be described (30), but preliminary evidence suggests that the system is highly sensitive to peptide length, with long fragments being transported more avidly than shorter fragments and the intact peptide transported best of all.

Modification of Transport

Both transmembrane diffusion and saturable transport of peptides can be modified by various factors. These modifications have implications for the normal functioning of organisms and may suggest ways in which peptide BBB interactions are involved in pathological conditions. Some of these are described below (Table 2 and 3).

Lighting

Light is an environmental cue that through the CNS regulates many biological functions. It might be expected that the BBB, as the regulatory interface between the humoral elements of the CNS and the peripheral organs, would also be affected by changes in lighting (31). We have found that entry of radiolabeled Tyr-DSIP was enhanced in rats that were subjected to constant dark or in rats tested during the dark phase of a reversed lighting cycle. By contrast, no change was found in the brain-blood ratios of radioiodinated albumin, ruling out increased leakiness of the BBB as a possible mechanism for the enhanced entry of peptide into the brain. Since radiolabeled Tyr-DSIP appears to cross the BBB primarily by diffusion through cell membranes, this suggests that alterations in the lighting cycle could affect the entry of a host of drugs and other substances that primarily enter the brain by this mechanism.

Amino Acids

Amino acids have many roles, including acting as the building blocks for peptides, monoamines, and structural and regulatory proteins. Amino acids are transported into the brain by a number of saturable systems that may be modulated by some peptides (32). In turn, leucine acts as an allosteric regulator of PTS-1 (33). The site is stereospecific, with D-leucine being about 100 times more potent than L-leucine after intraventricular administration. Evidence now suggests that the regulatory site is found on the brain side of the BBB. Low doses of leucine can stimulate transport, while high doses inhibit transport. Leucine is a neurotoxin in neonates (34) and produces some signs, symptoms, and histologic changes similar to those found with opiate peptide intoxication (35). This raises the possibility that some of the neurotoxic effects of leucine may be mediated through PTS-1.

Monoamines

Most steroid, protein, and peptide hormones are without effect on PTS-1 (36). Some monoamines and related substances, however, do affect transport. Dopamine, acetylcholine, epinephrine, GABA, kainic acid, cAMP, cGMP were without effect when injected intraventricularly (24). By contrast, a low dose of serotonin inhibited and a high dose of histamine stimulated PTS-1. The precursor to serotonin, 5-hydroxytrytophan, was without effect but cyproheptadine, a serotonin antagonist, stimulated transport at a low dose. Antagonists specific for the 5HT2 and 5HT3 receptor subtypes had no effect, but 5HT1 agonists inhibited transport. Concentrations of serotonin are altered in many of the processes that affect PTS-1 and the concentration of opiate peptides. These include aging, ethanol consumption, analgesia, and exposure to opiates. Thus, serotonin may play an important role in the functioning of PTS-1.

Stress

Some stresses selectively affect BBB transport (37). Restraint is associated with an inhibition of PTS-1 but does not affect transport of iodide. Food and water deprivation

Table 2
Modulator of PTS-1

Increased Transport	Decreased Transport
Leucine (low dose)	Leucine (high dose)
Serotonin Antagonist	Serotonin (5HT1) Agonists
Ethanol Withdrawal*	Ethanol Addiction
Histamine (high dose)	Stress (restraint)
	Aging
	Aluminum
	Morphine/Naltrexone

*Relative to addicted mice not actively withdrawing.

do not affect PTS-1. Water loading (5 ml of 0.45% NaCl given intraperitoneally to mice, 30 g body weight) inhibits PTS-2 (transport of AVP), possibly because of the acute stress it produces (8).

Aging

The capillary bed of the brain, which is largely responsible for the functions of the BBB, undergoes changes with aging. These include (a) a decrease in the number and an increase in the length of brain endothelial cells, (b) a decrease in the number of endothelial mitochondria, and (c) a decrease in the luminal diameter and thickness of the wall of the capillary (38). Both the Vmax and Km for the brain-to-blood transport of radiolabeled Tyr-MIF-1 are decreased in aged rats (19,39). The aged choroid plexus takes up—and possibly transports—less Tyr-D-Ala-Gly and more D-Ala-methionine enkephalinamide (40,41).

Table 3
Modulator of the Passage of Other*
Peptides Across the BBB

Modulator	Peptide	Direction of Passage	Effect on Transport
Lighting	DSIP	Blood to brain	Increases nonsaturable
Water loading	AVP	Brain to blood	Decreases saturable
Aluminum	Many peptides	Blood to brain	Increases nonsaturable
Aluminum	Peptide T	Blood to brain	Decreases saturable

*Modulators of PTS-1 are shown in Table 2.

Analgesia

Certain conditions that produce analgesia also result in alterations in the transport rate of PTS-1 (37). Acute and chronic treatment with morphine inhibits transport. Restraint results in analgesia (42,43) that is thought to be mediated through the enkephalins (peptides transported by PTS-1), and is associated with a decrease in the PTS-1 transport rate. However, PTS-1 seems only indirectly related to the level of analgesia. Deprivation of food and water, which results in analgesia (44) possibly mediated through β-endorphin (a peptide not transported by PTS-1), does not affect PTS-1. This indicates that decreases in transport by PTS-1, with a presumed increased retention of enkephalins in the brain, is not necessary for the induction of analgesia. The onset of analgesia induced by morphine precedes the effect on transport rate, indicating that changes in PTS-1 not only do not induce analgesia but could actually be a consequence of such analgesia. Naltrexone affected transport rate but not analgesia under the conditions tested, showing that while similar molecules affect both systems, they do not always result in consistent interrelated effects. Taken together, these findings show that analgesia and PTS-1 are probably not causally related, but may be indirectly linked, possibly by the sharing of some common modulators.

Aluminum

Aluminum is a neurotoxin, causing dialysis dementia (45), and has been implicated in varying degrees with other neurological disorders such as Alzheimer's disease, amyotrophic lateral sclerosis, and Parkinsonism-dementia syndrome of Guam (46). Aluminum increases the entry of many peptides and other substances into the brain by enhancing transmembrane diffusion (47,48,28). It also selectively inhibits some saturable transport systems including PTS-1 (49), PTS-3 (29), and PTS-4 (30). Aluminum does not inhibit the transport of iodide across the BBB nor does it affect the uptake of Tyr-Pro containing peptides by erythrocytes (49). These changes in the function of the BBB could be a mechanism by which aluminum induces neurotoxicity, resulting in an altered internal milieu of the CNS.

Ethanol

Intraventricular or intraperitoneal administration of ethanol does not alter PTS-1 (50). However, in mice physically addicted to ethanol, the transport rate is decreased. This may mirror the decrease in concentrations of enkephalins in the brain that can occur with ethanol addiction (51,52). In mice actively withdrawing from ethanol, however, the transport rate is similar to that in untreated mice (50). Concentrations of enkephalins in the brain are much lower in withdrawing animals (53), raising the possibility that a premature recovery of PTS-1 could contribute to these lower concentrations and possibly even to withdrawal itself. The transport rate of PTS-1 is also lower in mice that have never received alcohol but that have a genetic predisposition to drink ethanol (50). Taken together, these findings suggest a link between ethanol and PTS-1.

These other findings indicate that peptides cross the BBB by both saturable and nonsaturable mechanisms. These mechanisms of passage are altered in a host of conditions, including lighting cycle, amino acids, monoamines, stress, analgesic status, aging, and administration of aluminum and ethanol. This suggests that passage of peptides across the BBB is linked to physiologic processes and may be important in some disease states.

References

1. Kastin, A.J., R.D. Olson, A.V. Schally, and D.H. Coy, CNS effects of peripherally administered brain peptides, *Life Sci* 25: 401-414, 1979.

2. Rapoport, S.I., Transport of sugar, amino acids, and other substances at the blood brain barrier, *Blood-Brain Barrier in Physiology and Medicine*, Raven Press, New York, pp. 177-206, 1976.

3. Banks, W.A., and A.J. Kastin, Permeability of the blood-brain barrier to neuropeptides: the case for penetration, *Psychoneuroendocrinology* 10: 385-399, 1985.

4. Kastin, A.J., C. Nissen, A.V. Schally, and D.H. Coy, Blood-brain barrier, half-time disappearance, and brain distribution for labeled enkephalin and a potent analog, *Brain Res Bull* 1: 583-589, 1976.

5. Conford, E.M., L.D. Braun, P.D. Crane, and W.H. Oldendorf, Blood-brain barrier restriction of peptides and the low uptake of enkephalins, *Endocrinology* 103: 1297-1303, 1978.

6. Banks, W.A., and A.J. Kastin, A brain-to-blood carrier-mediated transport system for small, N-tyrosinated peptides, *Pharmacol Biochem Behav* 21: 943-946, 1984.

7. Sharma, R.R., and R.L.P. Vimal, Theoretical interpretation of extraction (in brain) of peptides including concentration variations, *Brain Res* 308: 201-214, 1984.

8. Banks, W.A., A.J. Kastin, A. Horvath, and E.A. Michals, Carrier-mediated transport of vasopressin across the blood-brain barrier of the mouse, *J Neurosci Res* 18: 326-332, 1987.

9. Takasato, Y., S.I. Rapoport, and Q.R. Smith, An *in situ* brain perfusion technique to study cerebrovascular transport in the rat, *Am J Physiol* 247: H484-H493, 1984.

10. Zlokovič, B.V., D.J. Begley, B.M. Duricic, and D.M. Mitrovič, Measurment of solute transport across the blood-brain barrier in the perfused guinea pig brain: methods and application to N-methyl-a-aminoisobutyric acid, *J Neurochem* 46: 1444-1451, 1986.

11. Banks, W.A., and A.J. Kastin, Quantifying carrier-mediated transport of peptides from the brain to the blood, *Methods Enzymol* 168: 652-660, 1989.

12. Blasberg, R.G., J.D. Fenstermacher, and C.S. Patlak, The transport of a-aminoisobutyric acid across brain capillary and cellular membranes, *J Cereb Blood Flow and Metab* 3: 8-32, 1983.

13. Patlak, C.S., R.G. Blasberg, and J.D. Fenstermacher, Graphical evaluation of blood-to-brain transfer constants from multiple-time uptake data, *J Cereb Blood Flow and Metab* 3: 1-7, 1983.

14. Banks, W.A., A.J. Kastin, D.H. Coy, and E. Angulo, Entry of DSIP peptides into the dog CSF: role of physicochemical and pharmacokinetic parameters, *Brain Res Bull* 17: 155-158, 1986.

15. Kastin, A.J., C. Nissen, and D.H. Coy, Permeability of the blood-brain barrier to DSIP peptides, *Pharmacol Biochem Behav* 15: 955-959, 1981.

16. Banks, W.A., A.J. Kastin, and D.H. Coy, Evidence that [^{125}I]-N-Tyr-delta sleep-inducing peptide crosses the blood-brain barrier by a non-competitive mechanism, *Brain Res* 301: 201-207, 1984.

17. Miller, L.H., B.A. Turnbull, A.J. Kastin, and D.H. Coy, Sleep-wave activity of a delta sleep-inducing peptide analog correlates with its penetrance of the blood-brain barrier, *Sleep* 9: 80-84, 1986.

18. Zlokovič, B.V., M.B. Segal, H. Davson, and R.M. Jankov, Passage of delta sleep-inducing peptide (DSIP) across the blood-cerebrospinal fluid barrier, *Peptides* 9: 533-538, 1988.

19. Banks, W.A., and A.J. Kastin, Peptides and the blood-brain barrier: lipophilicity as a predictor of permeability, *Brain Res Bull* 15: 287-292, 1985.

20. Wilson, J.F., Low permeability of the blood-brain barrier to nanomolar concentrations of immunoreactive alpha-melanotropin, *Psychopharmacology* 96: 262-266, 1988.

21. Kastin, A.J., C. Nissen, K. Nikolics, K. Medzihradszky, D.H. Coy, I. Teplan, and A.V. Schally, Distribution of ^{3}H-α-MSH in rat brain, *Brain Res Bull* 1: 19-26, 1976.

22. Zlokovič, B.V., M.N. Lipovac, D.J. Begley, H. Davson, and L. Rakič, Slow penetration of thyrotropin-releasing hormone across the blood-brain barrier of an *in situ* perfused guinea pig brain, *J Neurochem* 51: 252-257, 1988.

23. Banks, W.A., A.J. Kastin, A.J. Fischman, D.H. Coy, and S.L. Strauss, Carrier-mediated transport of enkephalins and N-Tyr-MIF-1 across blood-brain barrier, *Am J Physiol* 251: E477-E482, 1986.

24. Banks, W.A., and A.J. Kastin, Effect of neurotransmitters on the system that transports Tyr-MIF-1 and the enkephalins across the blood-brain barrier: a dominant role for serotonin, *Psychopharmacology* 98:

380-385, 1989.

25. Zlokovič, B.V., M.N. Lipovac, D.J. Begley, H. Davson, and L. Rakič, Transport of leucine-enkephalin across the blood-brain barrier in the perfused guinea pig brain, *J Neurochem* 49: 310-315, 1987.

26. Zlokovič, B.V., M.B. Segal, H. Davson, and D.M. Mitrovič,, Unidirectional uptake of enkephalins at the blood-tissue interface of the blood-cerebrospinal fluid barrier: a saturable mechanism, *Reg Peptides* 20: 33-44, 1988.

27. Ermisch, A., R. Landgraf, G. Heinold, and G. Sterba, Vasopressin, blood-brain barrier, and memory, In: C.A. Marsan and H. Matthies (eds) *Neuronal Plasticity and Memory Formation*, New York, Raven Press, pp. 147-152, 1982.

28. Banks, W.A., and A.J. Kastin, Aging and the blood-brain barrier: changes in the carrier-mediated transport of peptides in rats, *Neurosci Lett* 61: 171-175, 1985.

29. Barrera, C.M., A.J. Kastin, and W.A. Banks, D-[A1aI]-Peptide T-amide is transported from the blood to the brain by a saturable system, *Brain Res Bull* 19: 629-633, 1987.

30. Barrera, C.M., W.A. Banks, and A.J. Kastin, LHRH crosses the blood-brain barrier by a saturable transport system, *Clinical Res* 37: 31A, 1989.

31. Banks, W.A., A.J. Kastin, and J.K. Selznick, Modulation of immunoactive levels of DSIP and blood-brain barrier permeability by lighting and diurnal rhythm, *J Neurosci Res* 14: 347-355, 1985.

32. Ermisch, A., H.-J. Rühle, R. Landgraf, and J. Hess, Blood-brain barrier and peptides, *J Cereb Blood Flow and Metab* 5: 350-357, 1985.

33. Banks, W.A., and A.J. Kastin, Modulation of the carrier-mediated transport of Tyr-MIF-1 across the blood-brain barrier by essential amino acids, *J Pharmacol Exp Ther* 239: 668-672, 1986.

34. Jeune, M., C. Collombel, M. Michel, M. David, P. Guiboult, G. Guerrier, and J. Albert, Hyper-leucinisoleucinemie par defaut partiel de transamination associee a une hyperprolinemie de type 2. Observation familiale d'une double aminoacidopathie, *Sam Hop Paris (Ann Pediatr)* 17: 85-99, 1970.

35. Brandt, N.J., L. Terenius, B.B. Jacobsen, L. Klinken, A. Nordius, S. Brandt, K. Blegvad, and M. Yssing, Hyper-endorphin syndrome in a child with necrotizing encephalomyelopathy, *N Engl J Med* 303: 914-916, 1980.

36. Banks, W.A., and A.J. Kastin, Twenty-one hormones fail to inhibit the brain to blood transport system for Tyr-MIF-1 and the enkephalins in mice, *J Pharm Pharmacol* 40: 289-291, 1988.

37. Banks, W.A., A.J. Kastin, and B.J. Nager, Analgesia and the blood-brain barrier transport system for Tyr-MIF-1/enkephalins: evidence for a dissociation, *Neuropharmacology* 27: 175-179, 1988.

38. Banks, W.A., and A.J. Kastin, Aging, peptides, and the blood-brain barrier: implications and speculations, In: T. Crook, R. Bartus, S. Ferris, and S. Gerhson (eds), *Treatment Development Strategies for Alzheimer's Disease*, Mark Powley and Associates, Madison, CT, pp. 245-265, 1986.

39. Banks, W.A., and A.J. Kastin, Commentary: Peptides and the senescent blood-brain barrier, *Neurobiol Aging* 9: 48-49, 1988.

40. Huang, J.T., Accumulation of amino acid and peptide by choroid plexus of the aging rat, *Age* 7: 63-65, 1984.

41. Huang, J.T., and A. Lajtha, The accumulation of (^{3}H)enkephalinamide (2-D-alanine-5-methioninamide) in rat brain tissues, *Neuropharmacology* 17: 1075-1079, 1978.

42. Greenberg, R., and E.H. O'Keefe, Thiorphan potentiation of stress-induced analgesia in the mouse, *Life Sci* 31: 1185-1188, 1982.

43. Hersch, L.B., Reaction of opioid peptides with neutral endopeptidase ('enkephalinase'), *J Neurochem* 43: 487-493, 1984.

44. Vaswani, K.K., and G.A. Tejwani, Food deprivation-induced changes in the level of opioid peptides in the pituitary and brain of rat, *Life Sci* 38: 197-201, 1986.

45. Alfrey, A.C., G.R. LeGendre, and W.D. Kaehny, The dialysis encephalopathy syndrome: possible aluminum intoxication, *N Engl J Med* 294: 184-188, 1976.

46. Perl, D.P., D.C. Gajdusek, R.M. Garruto, R.T. Vanagihara, and C.J. Gibbs, Jr., Intraneuronal aluminum accumulation in amyotrophic lateral sclerosis and Parkinsonism-dementia of Guam, *Science* 217: 1053-1055, 1982.

47. Banks, W.A., and A.J. Kastin, Aluminum increases permeability of the blood-brain barrier to labelled

DSIP and β-endorphin: possible implications for senile and dialysis dementia, *Lancet* ii: 1227-1229, 1983.

48. Banks, W.A., and A.J. Kastin, Aluminum alters permeability of the blood-brain barrier to some non-peptides, *Neuropharmacology* 24: 407-412, 1985.
49. Banks, W.A., A.J. Kastin, and M.B. Fasold, Differential effect of aluminum on the blood-brain barrier transport of peptides, technetium, and albumin, *J Pharmacol Exp Ther* 244: 579-585, 1988.
50. Banks, W.A., and A.J. Kastin, Inhibition of the brain to blood transport system for enkephalins and Tyr-MIF-1 in mice addicted or genetically predisposed to drinking ethanol, *Alcohol* 410: 53-57, 1989.
51. Blum, K., S.F.A. Elston, L. DeLallo, A.H. Briggs, and J.E. Wallace, Ethanol acceptance as a function of genotype amounts of brain [Met]-enkephalin, *Proc Natl Acad Sci USA* 80: 6510-6512, 1983.
52. Schulz, R., M. Wuster, T. Duka, and A. Herz, Acute and chronic ethanol treatment changes endorphin levels in brain and pituitary, *Psychopharmacology (Berlin)* 68: 221-227, 1980.
53. Hong, J.S., E. Majchrowicz, W.A. Hunt, and J.C. Gillin, Reduction in cerebral methionine-enkephalin content during the ethanol withdrawal syndrome, *Subst Alcohol Actions Misuse* 2: 233-240, 1981.

5

VASOPRESSIN, THE BLOOD-BRAIN BARRIER, AND BRAIN PERFORMANCE

A. Ermisch and R. Landgraf

Department of Cell Biology and Regulation
Section of Biosciences and Interdisciplinary Centre
of Neurosciences
Karl Marx University, Leipzig, 7010, GDR

Introduction

For about a quarter of a century, nonapeptides of the vasopressin (VP) type, especially arginine-VP (AVP) and oxytocin (OXT), have been studied in relation to the behavioral performance of mammals. The starting point was the observation that, in addition to their *classical roles* in endocrine function, certain peptides modified central neuronal processes. The results of de Wied (1,2) attracted special interest, because they suggested the involvement of nonapeptides in learning and memory processes.

At that time, investigators in our department, lead by Günther Sterba, were engaged in histological, cytochemical, and electronmicroscopical studies of the vertebrate hypothalamic neurons producing nonapeptides. One aim of this research was to visualize extrahypothalamic neuronal fibers that transported nonapeptides from their sites of origin to their targets in the brain. At the same time, this group conducted pharmaco-physiological studies on central nervous reactions after the application of OXT (3,4).

The morphological studies carried out in our department enabled us to identify a large number of target regions for the extrahypothalamic vasopressinergic and oxytocinergic fibers (5-7). Furthermore, histochemical and electronmicroscopical evidence was presented for the localization of the nonapeptides within synaptic vesicles (7,8).

It is now generally accepted that the extrahypothalamic fibers represent the pathways of AVP and OXT that, after their release, are addressed to neurons. The nonapeptides released into the synaptic cleft bind to specific receptors and induce postsynaptic responses. Little is known of the events between *a)* the binding to the receptors, *b)* the cellular events induced by the peptidic signals, and *c)* and the behavioral responses that follow the release of AVP and OXT within the brain, and a major question is whether or not neuronal responses induced by peptides of endogenous origin can be mimicked by the same peptides administered exogenously. Although many studies on learning and memory processes have been made using VP and OXT (see review in Ref. 9), all are confounded by the fact that the blood-brain barrier (BBB) separates a peripherally administered peptide and its receptor sites within the central nervous system. Therefore, while conducting studies on VP and OXT, it was inevitable that we should also investigate the properties of the BBB in regard to peptides.

We have focused on the effects on the central nervous system of AVP and OXT produced within the brain as well as the effects of exogenous synthetic nonapeptides, while bearing in mind the BBB. Consequently, we were able to formulate a concept of synergism induced by different communication sources of the same peptidic signal, *viz.*, AVP. We assume that AVP coming from the same main source, *i.e.*, the hypothalamus,

Circulating Regulatory Factors and Neuroendocrine Function
Edited by J. C. Porter and D. Ježová
Plenum Press, New York, 1990

can be directed *a) via* extrahypothalamic fibers to postsynaptic receptor sites within different brain regions, and *b) via* the blood stream to receptors of distinct cell populations in the periphery, including the endothelial cells that form the BBB. The various events induced by the signal at the different receptor sites can serve a common aim, *viz.*, maintenance of brain performance. Brain performance in this context includes emotional and motivational components, learning, memory, and coping with stress. The model shown in Figure 1 illustrates this synergistic concept.

The Peripheral and Central Release of Vasopressin

AVP is produced in neurons of distinct hypothalamic nuclei, predominantly the supraoptic, paraventricular, and suprachiasmatic nuclei. Areas outside the hypothalamus containing AVP-synthesizing neurons include the bed nucleus of the stria terminalis, amygdala, and others. The axonal processes of the AVP-producing neurons project either to the neurohypophysis—hypothalamic-neurohypophysial pathway, or peripheral axis—or to different brain areas—extrahypothalamic pathways, or central axis (7,10). In response to various stimuli, the hypothalamic-neurohypophysial pathway releases AVP molecules

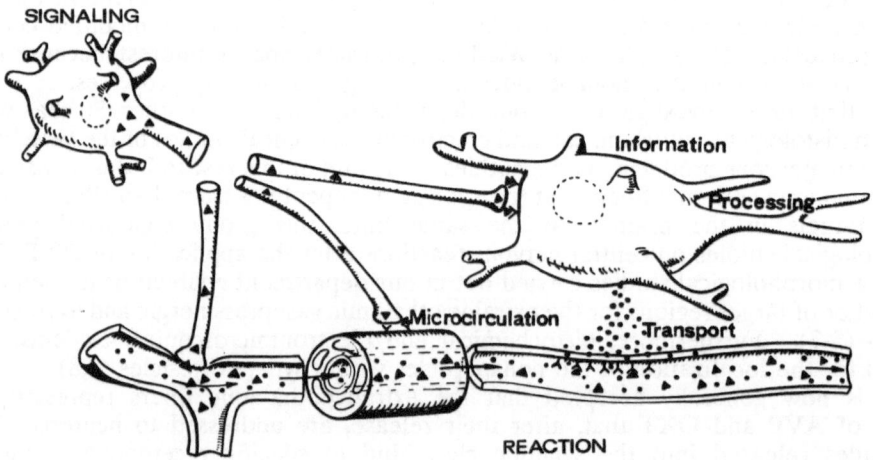

Figure 1.   Vasopressin and neuronal performance. **Signaling:** AVP molecules *(closed triangles)* are produced by distinct hypothalamic neurons and transported through the axons to different terminals. There, the peptidic signal will be released. **Information processing:** The AVP molecules addressed to other neurons of the brain occupy AVP receptors and induce alterations in the electrophysiologically detectable postsynaptic events which lastly alters the information processing of the neuron. **Microcirculation:** There is some evidence that AVP molecules are released from terminals contacting the wall of blood vessels within the brain. The effect of released AVP molecules occupying receptors at, *e.g.*, the smooth muscles might be an alteration of the brain microcirculation. **Transport:** AVP molecules released into the blood stream occupy receptors at the luminal surface of the endothelial cells of the brain vessels and induce an alteration of the transport of essential substances *(closed circles)* from blood to brain. The altered transport of essential substances across the BBB is necessary for the metabolic supply of activated neurons. Lastly, identical peptidic molecules, AVP, induce different reactions at different targets which serve synergistically for neuronal performance.

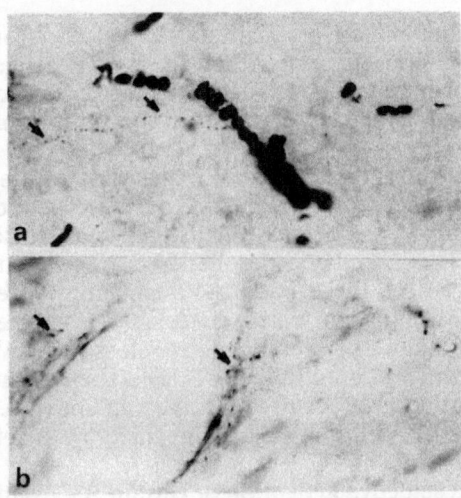

Figure 2. Neurophysin-immunoreactive fibers *(arrows)* contact blood vessels in *a)* the rostral part of the hypothalamus and *b)* the hind brain of the rat (prepared by Dr. H. Petter, Karl Marx University, Leipzig). Magnification: 500X.

into the circulation, whence AVP is transported by the blood and picked up by specific binding sites of a dozen or more target cell populations (11,12).

From extrahypothalamic nerve terminals, AVP is released in different brain regions. In this case, the peptidic signal is addressed to postsynaptic or other neuronal membranes containing AVP receptors (13,14). The presence of AVPergic fibers in contact with blood vessels within the brain (15) does not contradict the assumption of a release of the peptidic signal directed towards the wall of the brain blood vessels (Figure 2). The physiological significance of these contacts, particularly for the cerebral microcirculation, has not, however, been established.

In contrast to the variety of stimuli that result in a peripheral release of AVP, stimuli that produce central release are still poorly understood. Nonetheless, this release within distinct brain areas should be demonstrable with a pharmacological, or better, a physiological stimulus. The release, in turn, could be examined for evidence of some pattern consistent with its hypothesized role. Therefore, we were interested in studying peripheral and central release of AVP including the question of whether these processes are independent of each other or are coordinated.

In a first series of experiments, it could be shown that an extremely strong stimulus, osmotic stimulation induced by intraperitoneal injection of hypertonic saline (1.8 ml of 3.5 M NaCl) results in both a significant rise in plasma AVP and OXT and changes in peptide contents in the septum and hippocampus as detected by radioimmunoassays (16). Changes in peptide content in distinct brain areas, however, may suggest, but are not sufficient evidence, its release from fiber terminals. Therefore, in order to evaluate the dynamic events of septal and hippocampal release of AVP in response to osmotic stimulation, we used the technique of push-pull perfusion of both the medial septum and dorsal hippocampus in conscious, unrestrained rats. This direct approach revealed a significant, stimulus-evoked release of AVP that occurred coincidentally with its peripheral release from the neurohypophysis (16). There are indications that septal and hippocampal release is the result of synaptic or parasynaptic events as suggested by use of agents in the perfusion fluid which either inhibit or facilitate the release from intact fiber terminals (16,17). As shown through different approaches, the rise in plasma AVP following osmotic stimulation does not contribute to peptide levels in the perfusates (16,17). It remains to be seen whether central AVP release occurs in response to more physiological osmotic challenges.

It is of interest that the same osmotic stimulus that resulted in a peripheral as well as septal and hippocampal release of AVP failed to change peptide amounts in perfusates collected from the region of the nucleus tractus solitarius/dorsal motor nucleus of the vagus nerve (NTS/DMV). On the other hand, during electrical stimulation of the paraventricular nucleus (PVN), these basal AVP amounts were found to be increased approximately fivefold indicating an increased release from the PVN-NTS/DMV area projections (18). In this study, the simultaneous release of AVP from the neurohypophysis was suggested by a rise in blood pressure. Electrical stimulation of the PVN also resulted in a significant release of AVP within the septum (19). These results indicate that the peripheral release of AVP may be accompanied by an activation of some, but not necessarily all, centrally projecting AVP pathways.

The possibility of a selective activation of certain AVP pathways is reflected by the next series of experiments. In the urethane-anesthetized rat, prostaglandin E_1-induced fever resulted in a significant release of AVP within the ventral septal area (20), which is the site of the antipyretic action of both exogenous and endogenous AVP (21). In contrast, in the simultaneously perfused dorsal hippocampus, no change in peptide release occurred. This site-specific release indicated selective activation of central AVP pathways and local secretion processes instead of a peptide diffusion from the plasma across a leaky BBB.

In another series of experiments, it was our intention to study central peptide release under more physiological condition, *i.e.*, using naturally occurring stimuli that are known to induce peripheral release of nonapeptides from the neurohypophysis. In the conscious lactating rat, the peripheral release of OXT in response to suckling was accompanied by a septal and hippocampal release of OXT, but not of AVP (22). In contrast, AVP was released within the ventral septal area, the medial septum and the dorsal hippocampus around parturition in the conscious rat (Landgraf, Neumann, and

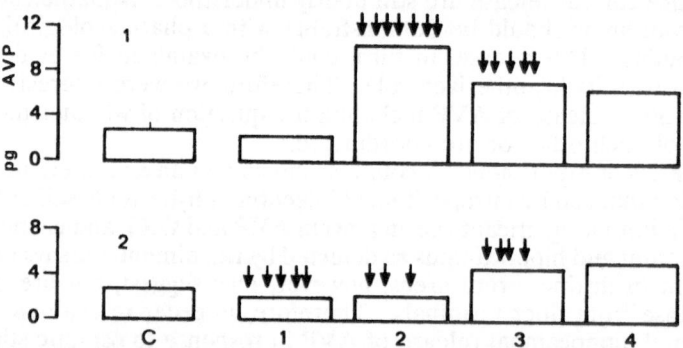

Figure 3.   AVP concentrations in push-pull perfusates sampled from the medial septum of virgin rats (C, n = 6, mean ± SE) and of two animals during parturition (1-4: four consecutive 30-min collection periods). The delivery of a pup is indicated by *arrows*. In contrast to the second animal, the first animal responded with a septal release of AVP. Data from Neumann and Landgraf (unpublished).

Pittman, unpublished). This effect, however, was not a consistent observation. In about half of the animals, central AVP failed to respond to the complex events of parturition (Figure 3). The reason for this variability is not known.

The physiological consequences of the central peptide release are, in most cases, unclear. There is, however, evidence that the septal and hippocampal release of AVP following osmotic stimulation contributes to the antipyretic (23,24) effect in rats that have been treated with hypertonic saline. As revealed by use of the relatively specific V_1 receptor antagonist, d(CH2)5Tyr(Me)AVP, endogenous AVP released within the NTS/DMV area might be involved in autonomic regulation (25), and AVP released within the ventral septal area during fever has been shown to act in this area as an endogenous antipyretic to reduce the febrile body temperature (21,26).

In summary, both neurohypophysial-projecting and centrally-projecting AVP pathways can, in some circumstances, simultaneously release endogenous AVP in a coordinated manner. Furthermore, certain central pathways can be selectively activated by a given stimulus, whereas others containing the same peptide remain unstimulated. These results document the regulatory potency of the vasopressinergic system.

Vasopressin and Information Processing within the Brain

AVP is transported via axonal fibers to numerous target regions of the brain extending from the telencephalon to the medulla spinalis (7,10). Upon release from fiber terminals, the peptide can trigger postsynaptic events involved in information processing of the target cells. The target cell may be integrated in neuronal circuits that, together with other circuits, ensure distinct brain performances. The basic relationships between these circuits are far from being understood. Nevertheless, there are strong indications that AVP is a signal in neuronal circuits that are involved in cardiovascular regulation (18,27,28), defervescence (21,26), feeding behavior (29,30), and analgesia (31). In addition, AVP may be involved in learning and memory processes (4,9). The pioneering work of David deWied and coworkers on the behavioral effects of VP has led to investigation of the pharmacological and physiological actions of VP in the brain. VP has been observed to influence memory formation or retrieval in rodents, primates, and humans (9). Although there has been some variability in these findings (4), the studies mentioned above have, nonetheless, led to extensive investigations exploring this possibility.

In a brightness-discrimination reaction that includes learning and memory mechanisms, our own data (32-34) suggest that endogenous AVP and OXT in certain limbic structures are related directly or indirectly to behavioral performance of animals. Different amounts of the peptides, assayed by radioimmunoassay (35), were found in the unilateral septal area and hippocampus of rats (33) (Figure 4), indicating that both nonapeptides are involved in this kind of behavioral performance. An immunocytochemical demonstration of neurophysin-containing fibers in the contralateral septal area of high and low performance animals revealed that the number of fibers containing neurophysin, the carrier protein of the nonapeptides, is greatest in animals with the best brightness-discrimination performance reaction (34, and unpublished observations). The detailed projection of the neurophysin-containing fibers of the septum is not known. Some fibers may be connected synaptically with septal neurons. Others may reach the hippocampus and form synaptic contacts there. Neuronal connections between the septum and the hippocampus were shown using electrophysiological techniques (36) and cytochemical methods (37,38).

The central pathways containing AVP as a transmitter/modulator and participating, at least partly, in the maintenance of behavioral performance constitute the first component of communication pathways for which we hypothesize a synergistic cooperation.

These central pathways work relatively rapid. The onset of postsynaptic information processing probably occurs within milliseconds following presynaptic release of the nonapeptide. It is not known, however, how long the AVP-induced information processing continues under various circumstances. If there exists any relation to the long term potentiation, now under discussion as a component of learning and memory processes in the hippocampus (39), information processing influenced by AVP may hold for hours or even days. Furthermore, if the peptide directly or indirectly triggers events involved in learning and memory processes, a relationship to metabolic activity, especially to protein and glycoprotein synthetic systems of the respective circuits, should exist (40,41).

The BBB as a Target for Vasopressin

The blood stream is the *classic* distribution system for chemical signals. In the case of AVP, peptide molecules entering the blood from the neurohypophysis are distributed by the circulation within a few seconds to all regions of a vertebrate body. The vessels of the microcirculation allow peptide molecules to cross from the plasma compartment to the interstitial fluid compartment. In this manner, peptides reach their specific receptors localized at about a dozen cell populations. But, How is this done in the brain? Between the blood plasma compartment and the interstitial fluid compartment an interface is localized; this interface has properties of tight epithelial surfaces, *viz.*, the BBB. In the vertebrate brain, there are a few small areas, known as circumventricular organs (CVO), where the BBB does not exist.

The BBB is constituted by endothelial cells of the brain vessels, which are joined by zonulae occludentes. The BBB protects and simultaneously supplies the neuronal information processing. That means, the interface between the blood and brain restricts distinct molecular transport processes and promotes others for information processing in

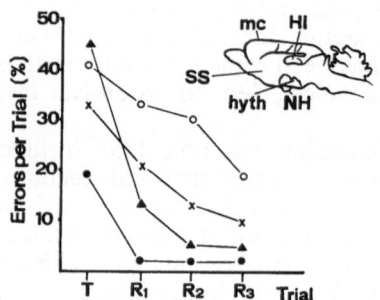

Figure 4. Learning performance and AVP. A population of 55 rats was trained to run into the bright terminal of a Y-chamber to avoid an electric footshock. Each animal completed four sessions of 22 trials each, the *training* (T) and 3 *relearnings* (R_1, R_2, R_3). Of the 55 animals tested (× denotes mean responses), those with extremely high (*closed circles*, n = 5), low (*open circles*, n = 5), and improving performance (*closed triangles*, n = 4), were selected, and the levels of AVP were measured by radioimmunoassay in different regions of the brain. In the septum/striatum (SS), the hippocampus (HI), and the neurohypophysis (NH), we found significant differences of AVP levels among the groups, depending on behavioral performance. In the SS, for instance, the level of AVP of the animals with *high* and with *improving performance* exceeded that of the *low-performance* group. No differences in the AVP levels exist in the hypothalamus (hyth) and motor cortex (mc) of the animals regardless of behavioral performance.

the brain. The restriction and the promotion of transport across the BBB depends on the molecular properties of the plasma membranes of the endothelial cells, including their connections by the tight junctions.

The question arises whether the endothelial cells need information from the environment, the *i.e.*, the outside to implement adequate transport across the thick layer that they constitute. If the endothelial cells require information from *outside*, then they represent receivers, which are targets for signals. How could one demonstrate that the BBB of mammals represents a target for AVP? We have to consider at least three conditions: *a)* the identification of specific binding sites at the BBB, *b)* the characterization of the binding sites as receptors, and *c)* the demonstration of a physiological significance for the receptors and the ligand-receptor interaction.

Experimental approaches to demonstrate specific binding sites are, first of all, orientated towards the question of whether or not they are saturable. Furthermore, specific binding sites are characterized by a specific affinity for a ligand (K_D), and their number as related to distinct biological units, for instance, a single cell. This is expressed by the binding capacity (B_{max}).

For the identification of specific AVP binding sites at the BBB, we used both *in vivo* and *in vitro* techniques. The *in vivo* demonstration of saturable peptide binding at the BBB was based upon the detection of the uptake of radiolabeled peptide by the brain after its injection into the blood stream. The methods for estimation of peptide uptake differ, as do the assumptions underlying the calculations of the data measured (42). Independent of that, saturable binding is lastly demonstrated if the uptake (retention, accumulation, extraction) of a radiolabeled peptide by the brain is depressed by a surplus of the co-injected unlabeled peptide.

Widely used for the detection of the uptake of peptides by the brain and the identification of specific binding at the BBB is the carotid bolus injection technique (43) that allows only a single pass of the tracer through the brain vessels. Depending on the special application, the radiolabeled peptide may be injected with or without a reference substance which is radiolabeled with another isotope.

We have studied the extraction of 10 peptides using the carotid bolus injection technique. Generally, it has been proved that the extraction of peptides is very low, but measurable by brain regions in which the endothelial cells of the brain vessels are connected by tight junctions (tight brain regions) (Table 1). The difference between the lowest extraction, substance P, and the highest, rat atrial natriuretic peptide (ANP), amounts to a factor of about 6.

If there is only a limited number of binding sites at the BBB, the sensitivity of the technique used is of particular interest (42). We have been able to demonstrate a saturable retention of ANP by rat tight regions after intracarotid bolus injection (data not shown). For AVP, a saturable retention by tight brain regions after the peripheral injection of the peptide has been demonstrated by several groups. Through the use of the integral technique, which among others allows longer contact between the injected peptide and the brain endothelial cells than does the carotid bolus injection, Reith *et al.* (44) observed a saturable retention of AVP by the hippocampus. Zlokovič *et al.* (42) were able to detect saturable mechanisms for some peptides, among them AVP, by a minutes-lasting perfusion of the guinea pig brain.

The demonstration of a saturable retention of distinct peptides by tight brain regions *in vivo* does not provide direct information on the localization of the molecules retained. Three locations of the molecules retained are possible: *a)* the luminal membrane of the tight endothelium, *b)* the inside the endothelial cells themselves, and *c)* the inside the fluid compartment of the brain, which would mean behind the BBB.

An accumulation of large amounts of peptide molecules within the endothelial cells has, to the best of our knowledge, never been seriously discussed. In the literature there is a silent agreement that peptide molecules, as far as they penetrate the luminal

Table 1
Extraction of Some Peptides by
Rat Brain Regions

Peptides	Amino Acid Residues	Mean Extraction	References
β-Casomorphine-5	5	1.9	Ermisch *et al.* (65)
Arginine vasopressin	9	2.4	Ermisch *et al.* (66)
Oxytocin	9	1.3	Ermisch *et al.* (66)
Substance P	11	0.5	Landgraf *et al.* (67)
Rat atrial natriuretic peptide	28	2.8	Original data
β-Endorphin	31	1.9	Ermisch *et al.* (66)

Explanations: The extraction was calculated according to the equation $E = [(A_{(x)} - A_{it})/(A_W - A_{it})] \cdot 0.43$. A represents the relative accumulation of radioactivity, and $A = (R_{br} \cdot 100)/(R_{inj} \cdot W_{br})$, where R_{br} represents the radioactivity in the sample, W_{br} its weight, and R_{inj} the intracarotidally injected dose. $A_{(x)}$ represents the accumulation of the peptide tested, A_W that of water, and A_{it} that of inulin in tight regions. It is assumed that the extracted fraction of water is 0.43 (68).

endothelial cell membrane, can be degraded within the cell or leave the cell, passing the abluminal membrane. Recently, a receptor-mediated endocytosis and transcytosis of macromolecules by endothelial cells, especially during BBB-opening, has come under discussion (45). But, even if this possibility really exists, it should concern only a limited number of peptide molecules.

The data available point toward a localization of the peptide molecules retained on the luminal surface of the endothelial cells as well as within the interstitial fluid of the brain. The following evidence can be presented in favor of this conclusion: a saturable retention of peptide molecules, detected by single pass techniques as demonstrated for ANP in our laboratory, indicates, most probably, specific binding sites at the luminal membrane itself. After its intracarotid injection, the bolus containing the labeled peptide molecules passes the whole brain vasculature in a few seconds (46). This means the actual time during which the peptide molecules have a chance to contact the luminal membrane of a single endothelial cell must be only a fraction of a second. The cerebrovascular permeability for some peptides was estimated to be in the range of 10^{-8} m per second (47). An assumed penetration of peptide molecules across the tight junctions of the endothelial cells, that is penetration of a distance in the range of 10^{-6} m, would require some orders of magnitude more time than available. Therefore, the occupation of specific binding sites localized behind the BBB can be largely excluded if we study peptide extraction by single pass techniques.

On the other hand, non-saturable extraction of peptides, demonstrated by both the single pass and the vascular perfusion technique (42), can be explained by passive permeation of the molecules from the plasma compartment of the blood vessels to the

Table 2
Specific Binding Sites on Isolated
Cerebral Microvessels

Peptide	Tracer	Brain Region	Species	Specific Binding	K_D [nmol × liter]	B_{max} [fmol × mg]	References
Angiotensin II	^{3}H	Cortex	Dog	+	1.07	20.6	Speth and Harik (50)
		Cerebellum		+	1.23	23.2	
Arginine-8-vasopressin	^{125}I	Cortex	Rat	-			Kretzschmar et al. (49)
		Striatum		-			
		Hippocampus		+	2.70	390	
			Bovine	+			Kretzschmar and Ermisch (69)
^{3}H		Cortex	Pig	+	2.20	49.0	Pearlmutter et al. (70)
Oxytocin	^{125}I	Cortex	Rat	-			Original data
		Striatum		-			
		Hippocampus		+	18.3	540	
ANP	^{125}I	Cortex	Bovine	+	0.11	58.0	Chabrier et al. (71)
			Rat	+	0.40	52.0	Smith et al. (52)*
				+	0.10	152	Niwa et al. (53)
		Striatum		+	0.23	120	Original data
		Hippocampus		+			
Insulin	^{125}I	Cortex	Bovine	+	0.44	180	Frank and Pardridge (48)
				+	20.0	200	
			Pig	+	0.33	27.0	Pillion et al. (72)
				+	0.34	66.0	Haskell et al. (73)
			Rabbit	+	0.44	140	Frank et al. (74)
					35.7	430	
			Human	+	1.20	170	Pardridge et al. (75)
					48.0	140	

*Data gained from cultured brain capillary endothelial cells.

fluid compartment of the brain. Finally, we have to conclude from the *in vivo* experiments that certain peptides, among them AVP, occupy a limited number of specific binding sites at the luminal membrane of the endothelial cells. Another very small portion of these peptides, and peptides for which a saturable mechanism cannot be detected, are believed to cross the BBB.

Saturable peptide binding on endothelial cells *in vitro* has been demonstrated using isolated brain microvessels. Specific binding sites of four blood-borne peptides, insulin (for review see 48), AVP (for review see 49), angiotensin II (50), and ANP (51-53, Ermisch *et al.*, data not shown) on the tight endothelium of the microvessels have been demonstrated (Table 2). AVP, angiotensin II, and ANP were shown to be peptidic signals in the periphery as well as in neuronal circuits. In this context, it is of interest that other transmitter binding sites have been identified on BBB capillaries (for review see 54).

In contrast to the relatively homogenous distribution of ANP binding sites, AVP binding to rat brain microvessels is regionally heterogenous (49). Measurement of bound iodinated AVP by means of a gamma counter as well as microautoradiography revealed specific binding sites only to hippocampal microvessels. Radioactivity associated with microvessels from the striatum and neocortex represents nonspecific binding. Reith *et al.* (44) corroborated a heterogenous AVP binding, reporting a saturable AVP retention *in vivo* only in hippocampal blood vessels. In an immunocytochemical study, van Zwieten *et al.* (55) visualized AVP binding sites in their highest density on hippocampal endothelial cells. The regionally heterogenous distribution of AVP binding in comparison to the homogenous distribution of the ANP binding indicates differences in the molecular organization of the cells comprising the BBB in various brain areas. On the other hand, the data do not allow us to conclude that the microvessels of the brain regions in which we could not detect AVP binding sites are really without any specific binding sites for the peptide. This depends on the sensitivity of the technique used.

On the basis of the B_{max}, the number of registrable specific binding sites per endothelial cell can be roughly calculated (56). According to data shown in Table 2, the number of AVP binding sites exceeds the number of ANP binding sites by a factor of about three (Table 3). It should be noted, however, that we have no information on the exact localization of the binding sites demonstrated at isolated microvessels. This means binding sites at the abluminal membranes of the endothelial cells may exist in addition to the luminal binding sites demonstrated by *in vivo* techniques (see also Figure 2).

To characterize the AVP binding with a cellular-physiological approach, hippocampal microvessels of rats in different physiological states were studied (Table 3). The isolated hippocampal microvessels of normal Wistar rats, dehydrated animals, as well as heterozygous and homozygous Brattleboro rats are characterized by a single class of high-affinity binding sites only. However, the microvessels differ in their capacity to bind the peptide and in their affinity for the ligand (12). The obvious correlation between the binding capacities and the plasma concentration of endogenous AVP [dehydrated animals reach levels of about 20 pmol × liter^{-1}, whereas in the plasma of homozygous Brattleboros the peptide is not detectable, (35,57)] points to a regulation of the binding sites by the peptidic signal itself. On the other hand, it is difficult to find an adequate explanation for the altered binding affinities seemingly induced by the different concentrations of the endogenous ligand in the environment of the binding sites (12). Until now, however, we cannot definitively decide whether the changed B_{max} and K_D values reflect a regulation of the AVP binding sites (and events associated with this) or merely their occupation by endogenous ligands.

At this stage, looking back to the first condition for establishing the BBB as a target for AVP, we have to conclude that with different approaches specific binding sites for the peptide have been demonstrated, and their parameters, K_D and B_{max}, were estimated. The binding sites could be localized, at least partly, at the luminal membrane of the tight endothelium representing the BBB. There are strong indications of *a)* a

Table 3
Specific AVP and ANP Binding to
Hippocampal Microvessels

Peptides	Animals	B_{max}	R_c	K_D
	Normal Wistar rats	394	4,700*	2.38
	Dehydrated Wistar rats	245	2,000	1.18
AVP	Heterozygous Brattleboro rats	593	7,100	3.67
	Homozygous Brattleboro rats	865	10,300	6.05
ANP	Normal Wistar rats	120	1,400	0.23

*Ermisch (56).
Explanations: In the AVP study normal Wistar rats, animals deprived of water for 48 hrs as well as heterozygous and homozygous Brattleboro rats were used (12). The ANP data are from Ermisch et al. (unpublished observations). The number of binding sites per endothelial cell (R_c) were calculated according to Ermisch (56), assuming the mass of a single endothelial cell to be 2×10^{-11} g. B_{max} (fmol $\times$ mg^{-1}), K_D (nmol $\times$ liter^{-1}).

heterogenous distribution of the AVP binding sites at the tight endothelium of various brain regions with the highest concentrations at the hippocampal vessels, and b) a regulation of the number of binding sites and their affinity caused by the concentration of the ligand in the fluid environment of the endothelial cells.

Effort aimed at characterizing the specific AVP binding sites at the BBB led to the conclusion that they were receptors resembling the V_1 subtype (58,59). Two sets of data point to this conclusion. AVP had no effect on the adenylate cyclase activity in isolated hippocampal microvessels. However, the peptide stimulated intracellular Ca^{2+} storage. This result probably indicates calcium uptake induced by the ligand of the receptor. Recent experiments are focused on further main steps of the proposed V_1 signal cascade.

The search for the physiological significance of the endothelial V_1 receptors was began here a decade ago. At that time our group (60,61) received the first indications that the BBB, as a regulatory interface for the substrate supply to the brain, may itself be regulated by peptidic signals: intracarotidally injected VP enhances the BBB permeability to the RNA precursor, orotic acid, in some brain regions. Furthermore, we have asked the question, What other compounds are affected by peptides in their passage across the BBB? In total, seven substrates and nine peptides were used and up to 18 brain regions and the anterior pituitary were studied (59). The substrates studied with regard to the question whether their passage across the BBB is affected by peptides, include water, orotic acid, glucose, fucose, and three amino acids. The peptides tested with regard to their effect on the passage of the substrates are AVP, five VP analogues, OXT as well as β-casomorphine-5, and ANP. From the results, the following conclusions were drawn: a) there is no general effect of peptides on the passage of compounds across the tight endothelium, b) certain peptides induce alterations of the BBB permeability which are substrate-specific, c) this effect is inducible by physiologically relevant peptide concentrations, especially on the carrier-mediated amino acid transport, and d) there are regional differences in the peptide effects on the substrate transport. Nonetheless, the interpretation of these data remains difficult (59).

The first conclusion implies that none of the peptides studied induces a general opening of the BBB. The extraction of radiolabeled AVP, for instance, was similar following injection of concentrations ranging from 10^{-8} mol $\times$ liter^{-1} to 10^{-4} mol $\times$ liter^{-1} (62) indicating that the cerebral endothelial junctions remained tight. Furthermore, as shown by electronmicroscopical studies, lanthanum ions did not cross the tight junctions even when they were perfused through the vascular bed together with AVP. Also, the accumulation of the space marker, inulin, by the brain is identical after its intracarotid injection alone or together with AVP.

The conclusion that certain peptides induce alterations in the BBB permeability is exemplified by the fact that only some of the eight peptides tested alter the regional transport of leucine. The substrate specificity of the peptide-mediated transport alterations is substantiated, since the effect could be observed using amino acids, but failed to occur using glucose. The signal and the substrate specificity as well as the finding that AVP-induced alterations occur when only 10^{-11} mol $\times$ liter^{-1}, i.e., physiologically relevant doses (35) are applied, point towards the significance of the transport phenomena even under physiological conditions.

To obtain greater insight into the events underlying the transport alterations of the tight endothelium, we studied the effects of co-injection of AVP and three amino acids: leucine, phenylalanine, and lysine. AVP caused significant alterations in the BBB permeability to the two neutral as well as the basic amino acid (Table 4). AVP enhanced the accumulation of the amino acids by the brain if their concentration in the injected solution was far below the physiological level, but diminished the uptake if their concentration in the injected bolus was far above the physiological mean (Figure 5). The reason for the regulatory effect induced by AVP can be explained by the results of kinetic studies. The intracarotid co-injection of AVP and leucine (63) or phenylalanine (Reichel, personal communication) elicited a decrease of the maximum transport capacity (V_{max}) and of the half saturation concentration (K_M) of the transport of both large neutral amino acids (Table 5). The effect of exogenous AVP is not the result of altered cerebral blood flow (44,63). The alteration of V_{max} and K_M induced by AVP concerns two individual amino acids. Under physiological circumstances, however, individual amino acids compete for the blood-brain transfer with other large neutral amino acids. Reith et al. (44) tested the hypothesis of the alteration of the kinetic constants by AVP using the integral technique, being able to detect the effect on the blood-brain transfer of leucine in competition with the other endogenous large neutral amino acids. It was concluded that the findings do not contradict the hypothesis of a role of AVP in the regulation of the large neutral amino acid transfer into brain tissue. So far, we can assume that the effect of AVP on the transfer of amino acids from blood to brain is a regular event in vivo. Studies with Brattleboro rats, i.e., animals that lack plasma AVP if they are homozygous, point in the same direction (64). Homozygous Brattleboro rats show higher V_{max} and K_M values for the transfer of leucine from blood to brain compared to heterozygous animals (64). The plasma AVP concentration of the heterozygous rats is about 70% of that seen in Long-Evans rats (57). Therefore, we can assume that endogenous AVP induces alterations in the brain supply of amino acids. Most probably, the specific effect consists of an alteration of the affinity of the large neutral amino acid-transporter. Enhanced affinity as concluded from the decreased K_M after AVP application might have a selective effect, which means it could lead to a selective preference in the transport of distinct amino acids across the BBB.

In regard to the regional differences of substrate transport as affected by AVP, the studies are not conclusive. Some data indicate an exceptional position of the hippocampal

Table 4
Significant Alterations in the BBB Permeability
to Amino Acids Induced by AVP in Nine Brain Regions

Substrate	Region									Sum
	OL	FC	VC	SE	ST	HI	HY	TH	CO	
Leu*	+	+	+	+	+	+	+	+	+	9
Phe†	+	+	+	-	+	+	+	-	+	7
Lys‡	-	+	+	-	+	-	+	+	+	6
Sum	2	3	3	1	3	2	3	2	3	

*Alterations determined in the kinetic studies by Brust (63) and
†By Reichel (personal communication).
‡Implies different levels of accumulation (59).

Explanations: significant alterations in the regions are indicated by +; - denotes no alteration.

Abbreviations: OL, olfactory lobe; FC, frontal cortex; VC, visual cortex; SE, septum; ST, striatum; HI, hippocampus; HY, hypothalamus; TH, thalamus; CO, colliculi; Leu, leucine; Lys, lysine; Phe, phenylalanine.

BBB (59). This is a remarkable coincidence with the number of AVP receptors of the hippocampal microvessels (49). On the other hand, in the studies using the intracarotid bolus injection technique, regional induction by AVP by heterogenous amino acid transfer from blood to brain is not always observable.

In summary, three main conclusions may be drawn from the data indicating the BBB is a target for peptides: *a)* specific binding sites, presumably receptors, for distinct peptides are localized at the BBB, *b)* the AVP binding sites are now identified as receptors, most likely of the V_1 subtype, and *c)* the AVP receptor is linked to transport processes from blood to brain, at least to those mediating large neutral amino acid transfer. Open questions concerning AVP receptors at the BBB involve their exact regional distribution as well as their subcellular localization, the molecular linkage between the V_1 subreceptor, and the amino acid transporters, including the cascade of cellular reactions induced by the peptide, and the effects of the altered amino acid transfer on brain functions. Proceeding from the molecular level to that of the whole organism, the question arises of the physiological significance of the presented data.

AVP and Brain Performance

Under certain circumstances, there occurs a simultaneous peripheral and central release of AVP, which binds to peripheral targets (*e.g.*, the BBB) and central targets, *viz*, neurons. These *certain circumstances* are poorly understood, but include strong osmotic and hemodynamic challenges, electrical stimulation of hypothalamic nuclei and, in some

Table 5

Kinetic Constants of the Leucine and Phenylalanine
Transport from Blood to Brain by Tight Brain Regions

	Leu			Phe		
	Control	AVP	Ratio	Control	AVP	Ratio
V_{max} (nmol × min^{-1} × g^{-1})	21 ± 8	7.8 ± 2.6	2.7	37.6 ± 3.3	27.4 ± 3.2	1.4
K_M (mmol × liter^{-1})	0.11 ± 0.04	0.029 ± 0.010	3.8	0.11 ± 0.012	0.061 ± 0.008	1.8

Explanations: data according to Brust (61) for Leucine (Leu) and Reichel (personal communication) for phenylalanine (Phe). *Control* means the bolus injected into the animals contained only the labeled amino acid, whereas *AVP* indicates that the bolus also contained unlabeled AVP in a concentration of 10^{-11} mol × liter^{-1}.

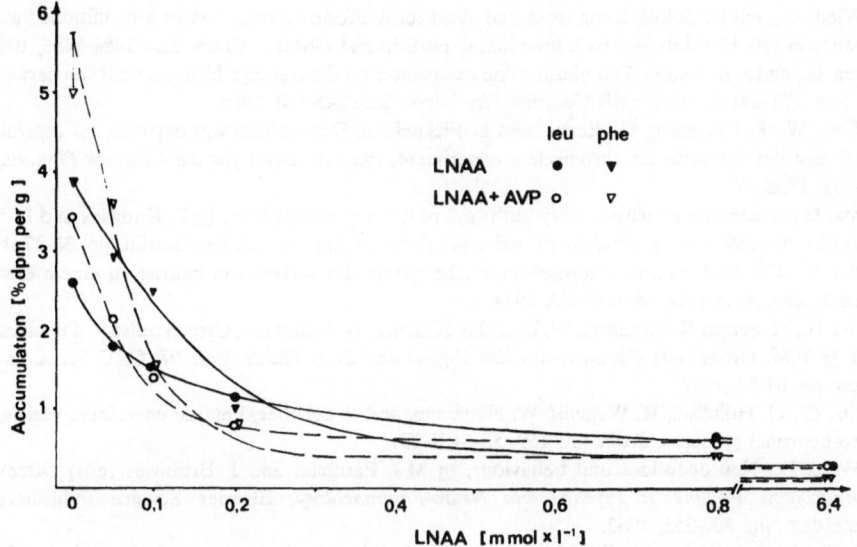

Figure 5. AVP-induced alteration in the accumulation of leucine (leu) and phenylalanine (phe) by the rat brain. The abscissa shows the concentration of the large neutral amino acids (LNAA) injected; the ordinate shows the accumulation of radiolabeled LNAA expressed as percent of the dpm injected per gram of brain. The point 0 on the abscissa indicates that the labeled LNAA was injected in tracer concentration, which was about 10^{-7} mol $\times$ liter^{-1}. The symbols of the curves represent means of 3-5 animals. The standard deviations amount to 15-26%; for clarity only one is marked at point 0. Significant differences ($P < 0.05$) in the accumulation in animals receiving LNAA $vs.$ LNAA + AVP was detected for both, leu and phe, at 10^{-7} mol $\times$ liter^{-1}; for phe at 1×10^{-4} mol $\times 10^{-1}$ liter; and for leu at 2×10^{-4} mol $\times$ liter^{-1}.

animals, parturition. The peripheral release, however, is not necessarily accompanied by an activation of all central pathways containing AVP. Central AVP release within limbic brain areas may result in events underlying information processing that is linked to metabolic alteration in the related neurons.

The metabolic demand for information processing will be ensured by the fine-tuned substrate transport by the tight endothelium of brain capillaries. The occupation of the AVP receptors at this endothelium may lead to a selective preference in the transport of at least distinct large neutral amino acids.

Our hypothesis suggests the possibility of a coordinated peripheral and central release of AVP and a synergistic action following AVP binding to peripheral and central receptors. This synergism seems to be particularly evident in limbic structures and may be of significance for brain performance, including learning and memory processes and coping with stress. It is, however, a long way from the secretion processes and responses of a restricted number of cells to the physiological response of the whole animal.

References

1. de Wied, D., The influence of the posterior and intermediate lobe of the pituitary and pituitary peptides on the maintenance of a conditioned avoidance response in rats, *Int J Neuropharmacol* 4: 157-167, 1965.

2. de Wied, D., and B. Bohus, Long term and short term effects on retention of a conditioned avoidance response in rats by treatment with long acting pitresin and α-MSH, *Nature* 212: 1484-1486, 1966.

3. Sterba, G., and J. Kormann, Der einfluss von oxytocinen auf das ständige hirnpotential von narkotisierten fröschen, *Pflügers Archiv fur die Gesamte Physiologie* 287: 345-350, 1966.

4. Schäker, W., F. Klingberg, G. Sterba, and L. Pickenhain, Der einfluss von oxytocin auf zentralnervöse funktionen bei der ratte im chronischen experiment, *Pflügers Archiv fur die Gesamte Physiologie* 288: 322-331, 1966.

5. Sterba, G., Ascending neurosecretory pathways of the peptidergic type, In F. Knowles and L. Vollrath (eds) *Neurosecretion—The Final Neuroendocrine Pathway*, Springer Verlag, Berlin, pp. 38-47, 1974.

6. Sterba, G., Das oxytocinerge neurosekreterische system der wirbeltiere, beitrag zu einem erweiterten konsept, *Zool Jb Physiol* 78: 409-423, 1974.

7. Sterba, G., H. Petter, R. Landgraf, W. Lösecke, K. Seiler, W. Naumann, Cytochemistry of neurosecretory cells, In P.M. Gross (ed) *Circumventricular Organs and Body Fluids, Vol. III*, CRC Press, Inc., Boca Raton, pp. 63-82, 1987.

8. Sterba, G., G. Hoheisel, R. Wegelin, W. Naumann, and F. Schober, Peptide containing vesicles within neuro-neuronal synapses, *Brain Res* 169: 55-64, 1979.

9. de Wied, D., Neuropeptides and behaviour, In M.J. Parnham and J. Bruinvles (eds) *Discoveries in Pharmacology, Volume 1: Psycho- and Neuro-Pharmacology*, Elsevier Science Publishers, B.V., Amsterdam, pp. 307-353, 1983.

10. Buijs, R.M., Vasopressin localization and putative functions in the brain, In D.M. Gash and G.J. Boer (eds) *Vasopressin, Principles and Properties*, Plenum Press, New York, pp. 91-115, 1987.

11. Jard, S., Vasopressin isoreceptors in mammals: relation to cyclic AMP-dependent and cyclic AMP-independent transduction mechanisms, In A. Kleinzeller and B.R. Martin (eds) *Current Topics in Membranes and Transport, Volume 18 Membrane Receptors*, Academic Press, New York, pp. 255-285, 1983.

12. Kretzschmar, R., and A. Ermisch, Arginine-vasopressin binding to isolated hippocampal microvessels of rats with different endogenous concentrations of the neuropeptide, *Exp Clin Endocrinol* 94: 151-156, 1989.

13. Van Leeuwen, F.W., Vasopressin receptors in the brain and pituitary, In D.M. Gash and G.J. Boer (eds) *Vasopressin, Principles and Properties*, Plenum Press, New York, pp. 477-496, 1987.

14. Poulain, D.A., and D.T. Theodosis, Coupling of electrical activity and hormone release in mammalian neurosecretory neurons, *Curr Top Neuroendocrinol* 9: 73-104, 1988.

15. Jójárt, I., F. Joó, L. Siklós, F.A. Lázló, Immunoelectronhistochemical evidence for innervation of brain microvessels by vasopressin-immunoreactive neurons in the rat, *Neurosci Lett* 51: 259-264, 1984.

16. Landgraf, R., I. Neumann, and H. Schwarzberg, Central and peripheral release of vasopressin and oxytocin in the conscious rat after osmotic stimulation, *Brain Res* 457: 219-225, 1988.

17. Demotes-Mainard, J., J. Chauveau, F. Rodrigues, J.D. Vincent, and D.A. Poulain, Septal release of vasopressin in response to osmotic, hypovolemic and electrical stimulation in rats, *Brain Res* 381: 314-321, 1986.

18. Landgraf, R., T.J. Malkinson, T. Horn, W.L. Veale, K. Lederis, and Q.J. Pittman, Release of vasopressin and oxytocin from nucleus tractus solitarius/dorsal vagal nucleus following PVN stimulation in rats, *Am J Physiol* In Press.

19. Neumann, I., H. Schwarzberg, and R. Landgraf, Measurement of septal release of vasopressin and oxytocin by the push-pull technique following electrical stimulation of the paraventricular nucleus of rats, *Brain Res* 462: 181-184, 1988.

20. Landgraf, R., T.J. Malkinson, W.L. Veale, K. Lederis, and Q.J. Pittman, Vasopressin and oxytocin in the rat brain in response to prostaglandin fever, In Preparation.

21. Pittman, Q.J., A. Naylor, P. Poulin, J. Disturnal, W.L. Veale, S.M. Martin, T.J. Malkinson, and B. Mathieson, The role of vasopressin as an antipyretic in the ventral septal area and its possible involvement in convulsive disorders, *Brain Res Bull* 20: 887-892, 1988.

22. Neumann, I., R. Landgraf, Septal and hippocampal release of oxytocin, but not vasopressin, in the conscious lactating rat during suckling, *J Neuroendocrinol* 1: 305-308, 1989.

23. Kasting, N.W., Potent stimuli for vasopressin release, hypertonic saline and hemorrhage cause antipyresis

in the rat, *Regul Pept* 15: 293-300, 1986.

24.   Koob, G.F., R. Dantzer, F. Rodriguez, F.E. Bloom, and M. Le Moal, Osmotic stress mimics effects of vasopressin on learned behaviour, *Nature* 315: 750-752, 1985.

25.   Pittman, Q.J., and L.G. Franklin, Vasopressin antagonist in nucleus tractus solitarius/vagal area reduces pressor and tachycardia responses to paraventricular nucleus stimulation in rats, *Neurosci Lett* 56: 155-160, 1985.

26.   Kasting, N.W., Criteria for establishing a physiological role for brain peptides. A case in point: the role of vasopressin in thermoregulation during fever and antipyresis, *Brain Res Rev* 14: 143-153, 1989.

27.   Doris, P.A., Central cardiovascular regulation and the role of vasopressin: a review, *Clin Exp Theor Practice* A6: 2197-2217, 1984.

28.   Schmid, P.G., F.M. Sharabi, G.B. Guo, F.M. Abboud, and M.D. Thanes, Vasopressin and oxytocin in the neural control of the circulation, *Fed Proc* 43: 97-102, 1984.

29.   Leibowitz, S.F., Hypothalamic paraventricular nucleus: interaction between α_2-noradrenergic system and circulating hormones and nutrients in relation to energy balance, *Neurosci Biobehav Rev* 12: 101-109, 1988.

30.   Messing, R.B., S.B. Sparker, Greater task difficulty amplifies the facilitatory effect of des-glycinamide arginine vasopressin on appetitivily motivated learning, *Behav Neurosci* 99: 1114-1119, 1985.

31.   Berkowitz, B.A., S. Sherman, Characterization of vasopressin analgesia, *J Pharmacol Exp Ther* 220: 329-334, 1982.

32.   Ermisch, A., M. Koch, and T. Barth, Learning performance of rats after pre- and postnatal application of arginine-vasopressin, In G. Dörner, S.M. McCann, and L. Martini (eds) *Monographs in Neural Sciences, Volume 12, Systemic Hormones, Neurotransmitters and Brain Development*, Karger, Basel, pp. 142-147, 1985.

33.   Ermisch, A., R. Landgraf, and P. Möbius, Vasopressin and oxytocin in brain areas of rats with high or low behavioral performance, *Brain Res* 379: 24-29, 1986.

34.   Ermisch, A., R. Landgraf, P. Möbius, and H. Petter, Behavioral performance of rats and the content of vasopressin and oxytocin in distinct brain areas, In H. Matthies (ed) *Learning and Memory: Mechanisms of Information Storage in the Nervous System, Advances in the Biosciences, Volume 59*, Pergamon Press, Oxford, pp. 369-372, 1986.

35.   Landgraf, R, Simultaneous measurement of arginine vasopressin and oxytocin in plasma and neurohypophyses by radioimmunoassay, *Endokrinologie* 78: 191-204, 1981.

36.   Hasche, W., Grundzüge der neurophysiologie unter dem aspekt der integrativen Tätikeit des ZNS, 3. Aufl., VEB Gustav Fischer Verlag, Jena, 1986.

37.   Melander, T., W.A. Staines, T. Hökfelt, A. Rökaeus, F. Eckenstein, P.M. Salvaterr, and B.H. Wainer, Galanin-like immunoreactivity in cholinergic neurons of the septum-basal forebrain complex projection to the hippocampus of the rat, *Brain Res* 360: 130-138, 1985.

38.   Nyakas, C., P.G.M. Luiten, D.G. Spencer, and J. Traper, Detailed projection patterns of the septal and diagonal band efferents to the hippocampus in the rat with emphasis on innervation of CA1 and dentate gyrus, *Brain Res Bull* 18: 533-545, 1987.

39.   Kennedy, M.B., Synaptic memory molecules, *Nature* 335: 770-772, 1988.

40.   Matthies, H., Plasticity in the nervous system an approach to memory research, In C.A. Marsan and H. Matthies (eds) *Neuronal Plasticity and Memory Formation, IBRO-Monograph Series Volume 9*, Raven Press, New York, pp. 1-15, 1982.

41.   Rose, S.P.R., Obstacles and progress in studying the cell biology of learning and memory, In H. Matthies (ed) *Learning and Memory: Mechanisms of Information Storage in the Nervous System, Advances in the Biosciences Volume 59*, Pergamon Press, Oxford, pp. 165-172, 1986.

42.   Zlokovič, B.V., D.J. Begley, M.B. Segal, H. Davson, L.J. Rakič, M.N. Lipovač, D.M. Mitrovič, and R.M. Jankov, Neuropeptide transport mechanisms in the central nervous system, In L.J. Rakič, D.J. Begley, H. Davson, and B.V. Zlokovič (eds) *Peptide and Amino Acid Transport Mechanisms in the Central Nervous System*, Stockton Press, NY, pp. 3-20, 1988.

43.   Oldendorf, W.H., Measurement of brain uptake of radiolabeled substances using a tritiated water internal standard, *Brain Res* 24: 372-376, 1970.

44.   Reith, J., A. Ermisch, N.H. Diemer, and A. Gjedde, Saturable retention of vasopressin by hippocampus

vessels in vivo, associated with inhibition of blood-brain barrier transfer of large neutral amino acids, *J Neurochem* 49: 1471-1479, 1987.

45. Joó, F., The blood-brain barrier: new aspects to the function of the cerebral endothelium, *Nature* 321: 197-198, 1986.

46. Oldendorf, W.H., and L.D. Braun, [^{3}H]tryptamine and [^{3}H]-water as diffusible internal standards for measuring brain extraction of radio-labeled substances following carotid injection, *Brain Res* 113: 219-224, 1976.

47. Rapoport, S.I., W.A. Klee, K.D. Pettigrew, and K. Ohno, Entry of opioid peptides into the central nervous system, *Science* 207: 84-86, 1980.

48. Frank, H.J.L., and W.M. Pardridge, A direct in vitro demonstration of insulin binding to isolated brain microvessels, *Diabetes* 30: 757-761, 1981.

49. Kretzschmar, R., R. Landgraf, A. Gjedde, and A. Ermisch, Vasopressin binds to microvessels from rat hippocampus, *Brain Res* 380: 325-330, 1986.

50. Speth, R.C., and S.I. Harik, Angiotensin II receptor binding sites in brain microvessels, *Proc Natl Acad Sci USA* 82: 6340-6343, 1985.

51. Chabrier, P.E., P. Roubert, P. Plas, and P. Braquet, Blood-brain barrier and atrial natriuretic factor, *Can J Physiol Pharmacol* 66: 276-279, 1988.

52. Smith, K.R., A. Kato, and R.T. Borchardt, Characterization of specific receptors for atrial natriuretic factor on cultured bovine brain capillary endothelial cells, *Biochem Biophys Res Comm* 157: 308-314, 1988.

53. Niwa, M., M. Ibaragi, K. Tsutsumi, M. Kurihara, A. Himeno, K. Mori, and M. Ozaki, Specific atrial natriuretic peptide binding sites in rat cerebral capillaries, *Neurosci Lett* 91: 89-94, 1988.

54. Cornford, E.M., The blood-brain barrier, a dynamic regulatory interface, *Mol Physiol* 7: 219-260, 1985.

55. Van Zwieten, E.J., R. Ravid, D.F. Swaab, and T.J. Woude, Immunocytochemically stained vasopressin binding sites on blood vessels in the rat brain, *Brain Res* 474: 369-373, 1988.

56. Ermisch, A., Blood-brain barrier and peptides, *Wiss Z Karl-Marx-Univ Leipzig Math-Naturwiss Reihe* 36: 72-77, 1987.

57. Möhring, B., and J. Möhring, Plasma ADH in normal Long-Evans rats and in Long-Evans rats heterozygous and homozygous for hypothalamic diabetes insipidus, *Life Sci* 17: 1307-1314, 1975.

58. Hess, J., A. Gjedde, and H. Jessen, Vasopressin receptors at the blood-brain barrier in rats, *Wizz Z Karl-Marx-Univ Leipzig Math-Naturwiss Reihe* 36: 81-83, 1987.

59. Ermisch, A., R. Landgraf, P. Brust, R. Kretzschmar, and J. Hess, Peptide receptors of the cerebral capillary endothelium and the transport of amino acids across the blood-brain barrier, In L.J. Rakič, D.J. Begley, H. Davson, and B.V. Zlokovič (eds) *Peptide and Amino Acid Transport Mechanisms in the Central Nervous System*, Stockton Press, New York, pp. 41-54, 1988.

60. Landgraf, R., J. Hess, and E. Hartmann, The influence of oxytocin on the regional uptake of [^{3}H] orotic acid by rat brain, *Endokrinologie* 70: 45-52, 1977.

61. Landgraf, R., J. Hess, and A. Ermisch, The influence of vasopressin on the regional uptake of [^{3}H] orotic acid by rat brain, *Acta Biol Med Germ* 37: 655-658, 1978.

62. Ermisch, A., T. Barth, H.J. Rühle, J. Skopková, P. Hrbas, and R. Landgraf, On the blood-brain barrier to peptides: accumulation of labelled vasopressin, desglyHG$_2$-vasopressin and oxytocin by brain regions, *Endocrinologia Experimentalis* 19: 29-37, 1985.

63. Brust, P., Changes in regional blood-brain transfer of L-leucine elicited by arginine-vasopressin, *J Neurochem* 46: 534-541, 1986.

64. Brust, P., and J. Zicha, Kinetics of regional blood-brain barrier transport of L-leucine in Brattleboro rats, *Biomed Biochim Acta* 47: 1013-1021, 1988.

65. Ermisch, A., H.-J. Rühle, K. Neubert, K. Hartrodt, and R. Landgraf, On the blood-brain barrier to peptides: [^{3}H]β-casomorphin-5 uptake by eighteen brain regions in vivo, *J Neurochem* 41: 1229-1233, 1983.

66. Ermisch, A., H.-J. Rühle, R. Landgraf, and J. Hess, Blood-brain barrier and peptides, *J Cereb Blood Flow Metab* 5: 350-357, 1985.

67. Landgraf, R., E. Klauschenz, M. Bienert, A. Ermisch, and P. Oehme, Some observations indicating a low brain uptake of [^{3}H]Nle^{11}-substance P, *Pharmazie* 38: 108-110, 1983.

68. Gjedde, A., and M. Rasmussen, Blood-brain glucose transport in the conscious rat: comparison of the intravenous and intracarotid injection methods, *J Neurochem* 35: 1375-1381, 1980.

69. Kretzschmar, R., and A. Ermisch, Arginine-vasopressin binding to isolated cerebral microvessels, *Wiss Z Karl-Marx-Univ Leipzig Math-Naturwiss Reihe* 36: 78-80, 1987.

70. Pearlmutter, A.F., M. Szkrybalo, Y. Kim, and S.I. Harik, Arginine vasopressin receptors in pig cerebral microvessels, cerebral cortex and hippocampus, *Neurosci Lett* 87: 121-126, 1988.

71. Chabrier, P.E., P. Roubert, and P. Braquet, Specific binding of atrial natriuretic factor in brain microvessels, *Proc Natl Acad Sci USA* 84: 2078-2081, 1987.

72. Pillion, D.J., J.F. Haskell, and E. Meezan, Cerebral cortical microvessels: an insulin-sensitive tissue, *Biochem Biophys Res Comm* 104: 686-692, 1982.

73. Haskell, J.F., E. Meezan, and D.J. Pillion, Identification of the insulin receptor of cerebral microvessels, *Am J Physiol* 248: E115-E125, 1985.

74. Frank, H.J.L, T. Jankovic-Vokes, W.M. Pardridge, and W.L. Morris, Enhanced insulin binding to blood-brain barrier in vivo and to brain microvessels in vitro in newborn rabbits, *Diabetes* 34: 728-733, 1985.

75. Pardridge, W.M., J. Eisenberg, and J. Yang, Human blood-brain barrier insulin receptor, *J Neurochem* 44: 1771-1778, 1985.

NEUROENDOCRINE RESPONSES TO EMOTIONAL STRESS: POSSIBLE INTERACTIONS BETWEEN CIRCULATING FACTORS AND ANTERIOR PITUITARY HORMONE RELEASE[1]

J.L. Meyerhoff, M.A. Oleshansky, K.T. Kalogeras[2], E.H. Mougey, G.P. Chrousos,[2] and L.G. Granger[2]

Neurochemistry & Neuroendocrinology Branch
Department of Medical Neurosciences
Division of Neuropsychiatry
Walter Reed Army Institute of Research
Washington, D.C. 20307-5100

Introduction

Anterior pituitary hormones released during stress have multiple direct and indirect effects in the periphery including mobilization of energy reserves (1) *via* gluconeogenesis and lipolysis (2, 3), effects on the cardiovascular (4) and immune systems (5, 6), and effects on electrolyte balance *via* stimulation of adrenal mineralocorticoid release. It seems logical that the necessity of orchestrating these potent effectors to adapt to specific challenges would require sophisticated modes of signalling between the pituitary cells and numerous circulating factors. We have recently described a stressful social interaction that elicits increases in plasma levels of the anterior pituitary peptide hormones derived from proopiomelanocortin (POMC): *viz.*, adrenocorticotropin (ACTH), β-endorphin (β-EP) and β-lipotrophic hormone (β-LPH) (7). Increases were also seen in plasma levels of the adrenal cortical hormone, cortisol (CS), as well as prolactin (PRL), an anterior pituitary hormone not derived from POMC.

Because of the robustness of these responses, we sought to determine whether circulating regulatory factors arising from sources other than the anterior pituitary would also be affected by this stressful paradigm. Accordingly, we have examined the effect of a stressful interview on plasma levels of arginine vasopressin (AVP), atrial natriuretic peptide (ANP) and plasma renin activity (PRA). Because exertion also produces increases in plasma levels of several humoral factors of peripheral origin (8-11) as well as of anterior pituitary origin (12-20), our experimental design controlled for any effects due to physical exertion.

Individuals in many occupations and professions face evaluative interviews or oral exams as a part of their professional development. Such exams typically entail: *a)* public speaking, *b)* formal presentation before a number of higher-ranking members of a profession, and *c)* submitting to evaluation. As part of their training and preparation for promotion boards, soldiers at the Walter Reed Army Institute of Research (WRAIR) are encouraged to appear before *Soldier of the Month* (SOM) boards—contests held monthly

---

[1]The views of the authors do not purport to reflect the position of the Department of the Army or the Department of Defense (par 4-3, AR 360-5)

[2]Developmental Endocrinology Branch, NICHHD, National Institutes of Health, Bethesda, MD

Circulating Regulatory Factors and Neuroendocrine Function
Edited by J. C. Porter and D. Ježová
Plenum Press, New York, 1990

91

to select and honor an outstanding soldier. These boards are rigorous, competitive, structured interviews conducted by a panel comprised of senior non-commissioned officers. The contest winner is selected on the basis of knowledge of job-related subject matter, personal appearance and quality of oral presentation. Contestants who volunteered to be studied during this competition gave written informed consent for us to monitor their heart rates and draw blood samples during their exam. We have characterized this competitive oral exam as a model for studying endocrine responses to stressful social interactions in an occupational setting. To study stressful social interactions in an occupational environment, it is necessary to satisfy several conditions: the subjects must be protected, yet the paradigm should be stressful; the situation should be occupationally relevant, yet permit rigorous experimental control. The oral exam paradigm we describe below meets these criteria.

Methods

Each candidate was examined individually by the panel for approximately 30 minutes. To minimize effects of diurnal variation, the subjects were always scheduled to appear before the panel at either 0900 or 0930. A nearby room was used as a waiting room for the subjects to provide them with a constant environment for 30 minutes before and after the exam. The typical subject was male, 25 years old, had 14 years of education, was physically very fit, was a corporal or a sergeant in the U.S. Army with 31 months of active duty experience, and was assigned as a research technician at the WRAIR. The examiners were seated at a long table. A soldier would walk to the room, knock on the door, enter, salute and be directed to sit in a chair across the room from the panel. There were usually five candidates (corporals or sergeants) who were interviewed individually during each monthly contest. The winner was awarded a $50.00 savings bond, a weekend pass, a certificate of achievement and an opportunity to compete at higher levels of competition in the future. In addition, the winner was exempt from ancillary duties for one month. Also, a press release was sent to the winner's hometown newspaper and his photograph was displayed in the lobby of the Institute. Candidates who did not win were subjected to no penalties and could enter subsequent competitions as often as they wished. Participation in this study was in no way a requirement or a factor in the SOM contest. Subjects in the study were candidates in the monthly competitions who volunteered to serve as subjects after being fully informed about the experimental procedures. None of the subjects was a supervisee of any of the authors. This study was approved by the Human Subjects Research Review Board of the Office of the Surgeon General of the U.S. Army.

We measured physiological (heart rate) and biochemical responses before, during, and after the exam. Psychological responses were estimated *via* self-report questionnaires administered before and after the exam. Two hours before the exam, electrodes were placed on the subject's chest and connected *via* EKG leads to a Vitalog PMS-8 monitor (Vitalog, Inc, Mountainside, CA) worn on a belt for continuous monitoring of heart rate. A sterile intravenous needle was then inserted into the anterior cubital vein of the left arm and blood was collected through non-thrombogenic tubing using a 3-speed blood withdrawal pump (Cormed Inc., Medina, N.Y., model 6-53R) also worn on the belt. A cylindrically-shaped plastic arm board was worn on the left arm to prevent flexion. Blood was withdrawn continuously at 0.2 ml/min throughout the procedure, except during discrete sampling intervals when pump speed was increased to 5 ml/min. All apparatus was concealed under the uniform jacket, at the back. Blood was sampled at 7 time points on the day of the exam (Table 1). Each blood sample consisted of three sequential 10 ml samples collected in chilled tubes containing EDTA and kept on ice until centrifuged at

Table 1
Blood Sampling Schedule

Sample	Description
REST 22a	After iv insertion, followed by 22 min sitting at rest in medical room.
REST 7	After walking 50 yards in hall, then sitting 7 min in waiting room.
REST 22b	After sitting 22 min in the waiting room, before walking to the exam room.
EXAM 7	7 min after being seated before the panel of examiners.
EXAM 22	22 min after being seated before the panel.
POST 7	7 min after returning to the waiting room and being seated.
POST 22	After sitting 22 min in the waiting room.

4°C. Two tubes were pooled, mixed and 2.5 ml was then aliquoted into storage tubes containing 50 μl (22 T.I.U./ml) of aprotinin (a protease inhibitor) for β-LPH, β-EP and ACTH assays. All storage tubes, as well as tubes of untreated plasma for CS, PRL, PRA, AVP and ANP assays were frozen at -70°C until assayed. The hormone values reported represent 4 minute blood collections beginning at the scheduled sampling point.

An extraction procedure was used to separate and purify β-EP and β-LPH, which were then quantified *via* RIA, using an endorphin kit supplied by DuPont NEN Research Products (7). Plasma CS was measured by radioimmunoassay, using an antibody produced in our laboratory in rabbits against prednisolone-3CMO:BSA (21). Kits obtained from INCSTAR Corp. were used to assay human plasma for ACTH and PRL content (7). After acidification of the plasma and adsorption to a Sep-Pak C_{18} cartridge (Waters Associates), AVP and ANP were eluted with acetonitrile and evaporated. The RIA for AVP used standard from Peninsula Laboratories, AVP antiserum from Arnel Laboratories and ^{125}I-AVP from Amersham Corp. Sensitivity for AVP was 0.9 pg/ml. The RIA for ANP used standard and antisera from Peninsula Laboratories, and ^{125}I-ANP from Amersham Corp. Sensitivity for ANP was 20 pg/ml. PRA was determined by measuring rate of formation of angiotensin I (22). Data were analyzed statistically by analysis of variance (ANOVA) for experiments with repeated measures. When the overall F-score was significant, the ANOVA was followed by a post-hoc test (Tukey's) for multiple comparisons to determine which time points showed significant differences.

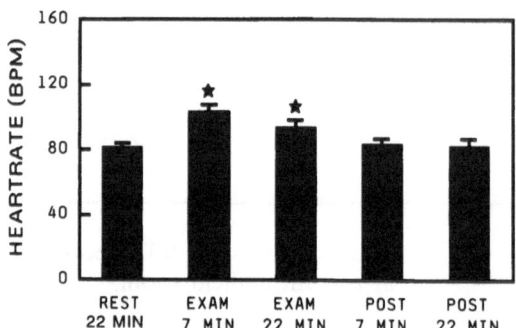

Figure 1. Heart rate response to stressful interview. Data expressed as mean ± SE, n = 11. Data were analyzed by ANOVA for experiments with repeated measures, followed by Tukey's test for multiple comparisons. *Significantly elevated ($P < 0.05$) compared to all values not marked with an asterisk. BPM denotes beats per minute.

Standing for 5 minutes has been reported to increase plasma levels of norepinephrine (NE) (8), and postural changes have been reported to affect AVP (23) and PRA (24). The logistics of the exam situation required that the subjects walk approximately 20 yards from the waiting room to the exam room. Since we planned to take a blood sample 7 minutes after the subject had been seated in the exam, it was apparent that a control for the effects of postural change was required. We therefore collected samples at 3 time points during the hour before the exam to determine if walking 50 yards would affect plasma hormone levels. Following insertion of the intravenous catheter, the subjects were seated 22 minutes before the first blood sample was taken *(REST 22a)*. They then walked 50 yards, were seated in the waiting room and a second blood sample was taken after they had been resting for 7 minutes *(REST 7)*. The subjects remained seated for an additional 15 minutes and the third sample was then taken *(REST 22b)*. There was no apparent effect of the physical activity on any of the

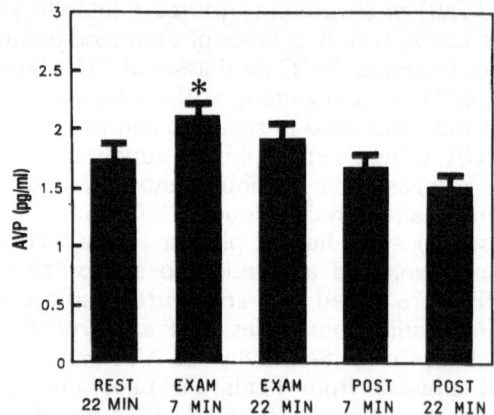

Figure 2. Response of plasma vasopressin to stressful interview. Data expressed as mean ± SE, n = 17. Data were analyzed by ANOVA for experiments with repeated measures, followed by Tukey's test for multiple comparisons. *Significantly elevated ($P < 0.05$) compared to all values not marked with an asterisk. AVP denotes arginine vasopressin.

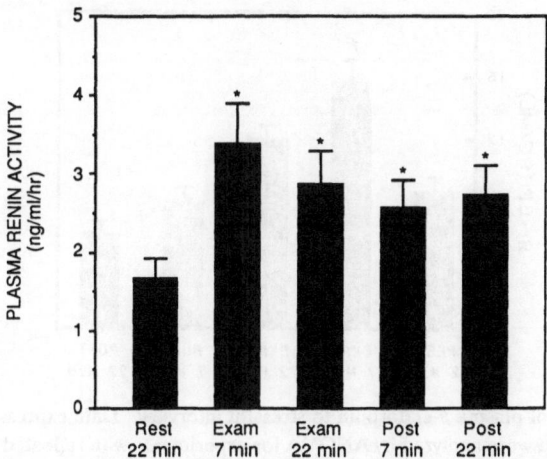

Figure 3. Response of plasma renin activity to stressful interview. Data expressed as mean ± SE, n = 11. Data were analyzed by ANOVA for experiments with repeated measures, followed by Tukey's test for multiple comparisons. *Significantly elevated ($P < 0.05$) compared to all values not marked with an asterisk. PRA denotes plasma renin activity.

physiological or biochemical variables measured. Accordingly, data from *REST 22a* and *REST 7* were omitted from the figures.

Results

As shown in Figure 1, the mean heart rate was 81 beats per minute (bpm) at rest before the exam. This was higher than normally expected in a group of young, physically fit males. This elevated resting heart rate may reflect apprehension about the upcoming exam. In preliminary reports, we have noted increases in self-reported anxiety scores

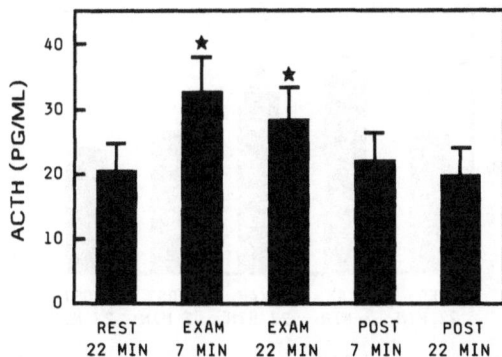

Figure 4. Response of plasma ACTH to stressful interview. Data expressed as mean ± SE, n = 11. Data were analyzed by ANOVA for experiments with repeated measures, followed by Tukey's test for multiple comparisons. *Significantly elevated ($P < 0.05$) compared to all values not marked with an asterisk. ACTH denotes adrenocorticotrophic hormone.

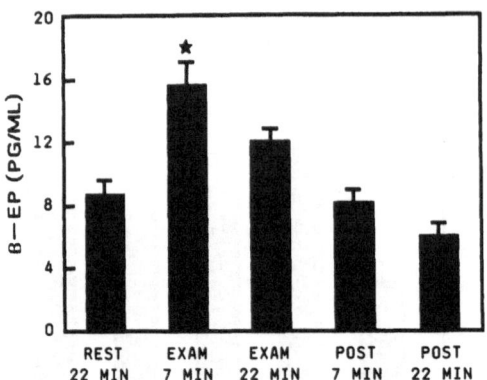

Figure 5.   Response of plasma β-endorphin to stressful interview. Data expressed as mean ± SE, n = 11. Data were analyzed by ANOVA for experiments with repeated measures, followed by Tukey's test for multiple comparisons. *Significantly elevated ($P < 0.05$) compared to all values not marked with an asterisk. β-EP denotes β-endorphin.

(Spielberger State Anxiety Scale) immediately before the exam compared to one week prior to the exam (25, 26). Seven minutes after beginning the exam, the average heart rate increased by 22 bpm, to 103 bpm. Although this represents an increase of 27%, the increase seems even more significant when one considers that 22 bpm represents 3.7 times the standard deviation (5.97) of the average heart rate for the group before the exam. As shown in Figures 2 through 7, the stressful 30 minute interview increased plasma levels of AVP (Figure 2) and PRA (Figure 3), as well as ACTH (Figure 4), β-EP (Figure 5), β-LPH (Figure 6), and PRL (Figure 7). These increases occurred within 7 minutes of beginning the interview and were quite robust: AVP (39%), PRA (123%), ACTH (59%), β-EP (79%), β-LPH (42%), and PRL (46%). As shown in Figure 8, 22 minutes after beginning the interview, cortisol (CS) levels were increased (31%) and all other measures remained significantly elevated, except plasma β-EP and AVP. Plasma levels of ANP were

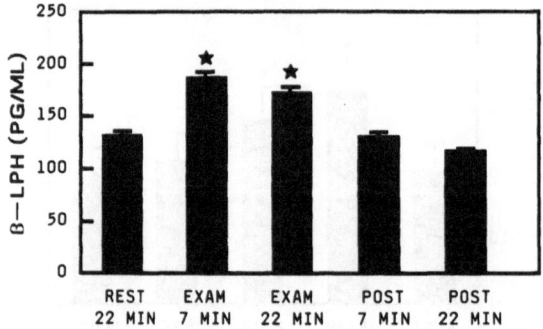

Figure 6.   Response of plasma β-lipotrophic hormone to stressful interview. Data expressed as mean ± SE, n = 11. Data were analyzed by ANOVA for experiments with repeated measures, followed by Tukey's test for multiple comparisons. *Significantly elevated ($P < 0.05$) compared to all values not marked with an asterisk. β-LPH denotes β-lipotrophic hormone.

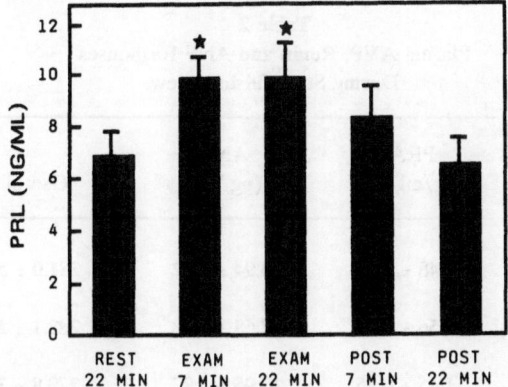

Figure 7. Response of plasma prolactin to stressful interview. Data expressed as mean ± SE, n = 11. Data were analyzed by ANOVA for experiments with repeated measures, followed by Tukey's test for multiple comparisons. *Significantly elevated ($P < 0.05$) compared to all values not marked with an asterisk. PRL denotes prolactin.

unaffected by this particular stressful experience (Figure 9). Seven minutes after leaving the examination room and sitting in the waiting room, heart rate as well as levels of all POMC-derived hormones had declined to levels that were no longer significantly different from those seen before the exam. Plasma CS did not return to *rest* levels until 22 minutes after the exam. PRA remained elevated 22 minutes after subjects returned to the waiting room, the latest time point at which samples were taken.

Table 2 gives actual values for AVP and PRA, which increased. Also shown are values for plasma ANP, osmolality and glucose, which did not change significantly.

Discussion

We believe that this is the first report of increased plasma AVP levels in normal human subjects exposed to a psychological stressor. In the pages that follow, we will

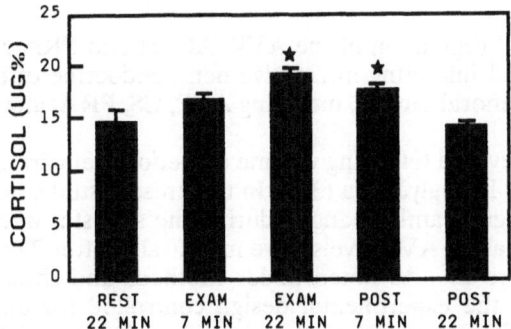

Figure 8. Response of plasma cortisol to stressful interview. Data expressed as mean ± SE, n = 11. Data were analyzed by ANOVA for experiments with repeated measures, followed by Tukey's test for multiple comparisons. *Significantly elevated ($P < 0.05$) compared to all values not marked with an asterisk.

Table 2
Plasma AVP, Renin and ANF Responses
During Stressful Interview

	AVP (pg/ml)	PRA (ng/ml/hr)	ANP (pg/ml)	Osmol	Glucose (mg/dl)
REST 22a	1.46 ± 0.21	1.48 ± 0.29	13.94 ± 1.72	281.0 ± 5.8	97.6 ± 4.5
REST 7	1.47 ± 0.21	1.56 ± 0.29	12.68 ± 1.68	282.1 ± 2.5	95.9 ± 4.0
REST 22b	1.58 ± 0.24	1.45 ± 0.25	12.95 ± 1.87	279.8 ± 3.1	94.2 ± 2.4
EXAM 7	$2.17 \pm 0.24^*$	$3.24 \pm 0.68\dagger$	14.30 ± 3.01	281.7 ± 3.5	95.8 ± 3.0
EXAM 22	1.96 ± 0.22	$2.87 \pm 0.56\dagger$	13.50 ± 1.72	280.3 ± 2.3	99.7 ± 2.3
POST 7	1.51 ± 0.19	$2.66 \pm 0.40\dagger$	13.64 ± 1.85	282.7 ± 2.2	98.9 ± 3.1
POST 22	1.51 ± 0.21	$2.73 \pm 0.42\dagger$	13.32 ± 1.36	282.7 ± 3.3	101.4 ± 2.6
F	3.12	12.15	0.13	0.24	2.0
P	< 0.008	< 0.001	NS	NS	NS
N	21	11	8	19	18

Data expressed as mean $\pm$ SE. Data or log-transformed data were analyzed by ANOVA for experiments with repeated measure. When F score was significant, the ANOVA was followed by Tukey's test for multiple comparisons. AVP denotes arginine vasopressin, PRA denotes plasma renin activity, ANP denotes atrial natuiretic peptide, Osmol denotes plasma osmolality.
*Significantly increased ($P < 0.05$) compared to *REST 22a, REST 7, REST 22b,* and *POST 22.*
†Significantly increased ($P < 0.05$) compared to *REST 22a, REST 7,* and *REST 22b.*

discuss the neurohumoral regulation of the AVP, ACTH and PRA responses to stress, as well as the enhancing and inhibiting interactive neuroendocrine effects of several stress-responsive circulating humoral factors, including AVP, CS, PRA, angiotensin II (AII), NE, and epinephrine (EPI).

Plasma AVP is elevated following volume depletion, dehydration, increased plasma osmolality (23, 27-29), or hypoglycemia (30). In the present study, neither plasma glucose nor plasma osmolality significantly changed during the stressful interview (Table 2), but significant increases in plasma AVP levels were measurable after 7 minutes of a 30 minute stressful interview (Figure 2). As noted under **Methods** and shown in Tables 1 and 2 *(REST 7* and *POST 7)*, the experimental design controlled for any effects of postural change on AVP (23). During the hour that preceded the stressful interview, 90 ml of blood had been withdrawn from our subjects. It is unlikely, however, that volume depletion accounts for the transient rise in AVP at the beginning of the interview because AVP levels returned to baseline despite continued blood withdrawal during the remainder of the experiment (total duration of the experiment was approximately 3 hours).

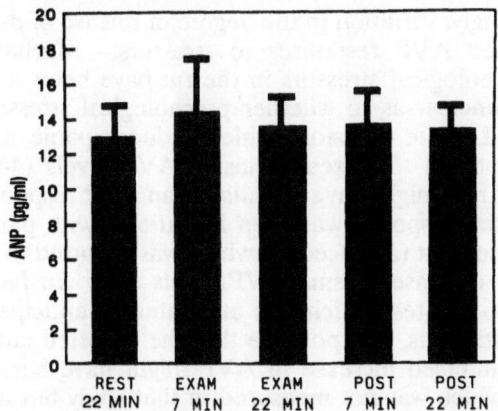

Figure 9. Response of plasma atrial natriuretic peptide to stressful interview. Data expressed as mean ± SE, n = 8. Data were analyzed by ANOVA for experiments with repeated measures, followed by Tukey's test for multiple comparisons. ANP denotes atrial natriuretic peptide.

Moreover, in our study each 30 ml sample was withdrawn over 6-8 minutes, but it has been reported that withdrawal of as much as 485 ml of blood over a 5-7 minute period had no effect on plasma antidiuretic activity (31) and removal of up to 9% of total blood volume can be accomplished without affecting plasma AVP levels (27, 28). Tachycardia can result in decreased left auricular filling during a shortened diastole, which could stimulate release of AVP. But that would only happen at extremely rapid heart rates and seems unlikely at the rates recorded at the beginning of the interview (103 bpm) in the present study.

Reported AVP responses to various stressors in laboratory animals vary widely, but we will cite studies aimed at revealing mechanisms of AVP regulation which might explain some of this variability. An early study reported that diuresis in dogs was inhibited by emotional stress (32). On the other hand, monkeys subjected to a confinement/noise paradigm had increased levels of ACTH, while AVP remained unchanged (33). In the rat, some, but not all stressors increase plasma AVP levels. This literature is well-reviewed by Gibbs (34), who cites reports that electric tail shocks (35), ether exposure, hemorrhage (36), hypoglycemia, and hypoxia increase plasma levels of AVP. Restraint stress has been reported to increase plasma AVP levels (37), but another study found no response using a slightly different paradigm (38). Yet, another study reported increases in AVP after manual restraint, body compression or electric shock to the tail, but no increases due to forced activity, light ether anesthesia, swimming, noise, or restraint in plastic containers (39). During prolonged immobilization stress, plasma AVP was reported to increase acutely, but was significantly decreased after 150 minutes of immobilization (40). There are two reports that footshock increases plasma AVP (41, 42) as well as several negative reports (43, 44).

A series of studies has suggested that endogenous opioid-mediated mechanisms inhibit AVP release. It has been shown that the administration of naloxone (42), intracerebroventricular administration of antibody to β-EP (44) or ablation of the anterior lobe of the hypophysis (43) unmask a strong plasma AVP response to stressors. From these observations, it was inferred that endogenous β-EP may inhibit the stress-induced release of AVP. Accordingly, variation in the degree of release of β-EP might account for the variability in reported AVP responses to stressors. Mechanisms regulating the response of β-EP to psychological stressors in the rat have been recently reviewed (45).

The literature is unclear as to whether psychological stressors elicit increases in plasma AVP in man. Lactate infusion, which induces panic attacks in susceptible individuals, has been reported to increase plasma AVP levels (46). Since the lactate solution was hyperosmolar, it might have stimulated an AVP response *via* altered plasma osmolality. No differential response was seen in patients with panic disorder. On the other hand, the cold pressor test (a procedure which was reported to be painful) increased plasma CS but failed to increase plasma AVP levels (47). In fact, there was a slight decrease in AVP prior to the test which was attributed to anticipation. Since the cold pressor test is a painful stimulus, it is possible that the reported anticipatory decrease in AVP and lack of stress-induced increase in AVP might have been due to increases in plasma β-EP, a peptide which was not measured in that study but might be presumed to have increased prior to the reported increase in CS. In the present study, however, both AVP and β-EP were significantly increased.

To explain the source of the AVP increase in the general circulation, a brief review of the hypothalamic vasopressinergic neurosecretory systems is appropriate. Hypothalamic vasopressin systems include the magnocellular neurosecretory system which projects to the neural lobe of the pituitary, and the parvocellular paraventricular system which projects to the external lamina of the median eminence. Injections of horseradish peroxidase (HRP) into the neural lobe resulted in transport of the HRP to the hypothalamic supraoptic nucleus and the posterior magnocellular division of the paraventricular nucleus of the hypothalamus (48). This magnocellular neurosecretory system includes vasopressinergic neurons which originate in the supraoptic nucleus and the posterior magnocellular paraventricular nucleus and project *via* the internal lamina of the median eminence to nerve terminals in the neural lobe (49-55). Thus, some AVP-containing neurons in the neural lobe appear to be under paraventricular nucleus control. AVP content in the neural lobe is many fold greater than the AVP content of the median eminence. Thus, increases in AVP levels in the general circulation, such as those found in the present study, probably reflect release from the neural lobe.

To give a more complete anatomical picture of the potential bases for influences of AVP on ACTH secretion, a brief description of the parvocellular paraventricular nucleus projection to the median eminence is required. Injection of HRP into the median eminence results in labelling of small neurons in the medial parvocellular division of the paraventricular nucleus (48, 56). Significant numbers of neurons in the medial parvocellular paraventricular nucleus stained immunocytochemically for AVP (55) and 44% of CRF-positive axons and nerve terminals in the median eminence also stained for AVP (57). Lesions of the paraventricular nucleus completely eliminated AVP staining in fibers of the external lamina of the median eminence, suggesting that the origins were in the paraventricular nucleus neurons (58). It is the terminals of these fibers which are generally believed to secrete AVP into the portal system.

AVP was first measured in portal blood by Zimmerman *et al.* (59), where its concentration is reported to increase after hemorrhage, in anesthetized rats (60) and in conscious rams (61). Psychological stressors elicited increases in AVP levels in the hypophysial-portal circulation of conscious sheep (62) and in the pituitary venous effluent of conscious horses (63). Although secretion of AVP into the portal circulation is generally assumed to occur from the external lamina of the median eminence, one group has suggested that AVP might also be released by axons of the magnocellular vasopressinergic neurosecretory system which pass through the internal lamina of the median eminence (64). Also, it has been proposed that AVP from the neural lobe might reach the anterior lobe *via* retrograde flow through the pituitary portal vein (65, 66), but this hypothesis was challenged by the report that removal of the posterior pituitary fails to alter portal venous levels of AVP (67).

Historically, AVP has been prominently considered among factors that were thought to release ACTH or to enhance ACTH secretion to a significant degree (68, 69). Lesions of the paraventricular nucleus, which deplete the median eminence of CRF (70), completely abolish the CRF-mediated stress-induced increase in pituitary cyclic AMP (71), but merely attenuate the stress-induced increase in ACTH (70, 71). Since the CRF/cyclic AMP-dependent component of the ACTH response was eliminated, the residual ACTH response to stress in paraventricular nucleus-lesioned rats may have been mediated *via* AVP, which has been shown to release ACTH by a cyclic AMP-independent mechanism (72, 73). Several elegant studies have documented the ability of AVP to release ACTH from anterior pituitary cells *in vitro* (69, 72, 73), and its ability to enhance CRH-induced release (68, 69, 72-76). Recently, AVP has also been reported to release β-EP from anterior pituitary tissue (77). Several studies have suggested that AVP enhances the ACTH response to stress *in vivo*, by demonstrating that antagonists (78) or antibodies to AVP (75, 76, 78, 79) attenuate the stress-induced increase in ACTH. Such treatments might have blocked AVP released from either the median eminence or the neural lobe. After surgical removal of the neurointermediate pituitary lobe, rats had decreased corticosterone responses to noise and novel stimuli, and the ACTH response to noise was reduced (80), possibly due to elimination of the neural lobe as a source of AVP which might have been enhancing release of ACTH. In the present experiment, the increase in AVP levels in the general circulation might have enhanced ACTH secretion, as the anterior pituitary is outside the blood-brain barrier. In fact, AVP infusion in man has been shown to increase plasma levels of both ACTH and cortisol (81, 82). In the present study, a stressful social interaction elicited percent increases in plasma ACTH equal to those reported during moderate exercise (16). Possibly, the elevated plasma AVP levels augmented the stress-induced, CRF-mediated increase in ACTH secretion and extended its duration.

Plasma levels of CS increased at 22 minutes after beginning the interview (Figure 8). This elevation appears to lag the rise in ACTH by about 15 minutes. A similar lag between increase in ACTH and plasma CS response has been reported to occur during exercise (15). CS responses to psychologically stressful stimuli have been studied in laboratory and clinical situations, as well as in naturalistic settings. Early studies in monkeys reported stress-induced increases in adrenal corticoid secretion (83). In studies with human volunteers, numerous stressors have been reported to elevate plasma CS levels, including mental arithmetic (84). In physicians giving oral presentations at a clinical conference, dexamethasone pre-treatment failed to suppress blood CS levels (85). In clinical studies of patients with phobic symptoms, exposure to phobic stimuli increased anxiety, heart rate and blood pressure as well as plasma levels of CS (86). Increases in urinary excretion of CS were reported in similar studies (87). Anticipation of an expected challenge has also been associated with endocrine changes. Resting heart rate and plasma CS levels were elevated in medical students during final exam week (88) and examinations were also reported to increase urinary excretion of CS (89). In hospital settings, elevated plasma levels of CS were reported in patients awaiting imminent surgical procedures (90). Elevated plasma CS was also found in subjects about to commence vigorous exercise (9) and increases in ACTH and β-EP (91) were seen in trained long distance runners 20 minutes before a race. Plasma CS was also reported to be markedly elevated after parachutists' training jumps—events combining physical activity and psychological arousal (92).

In addition to the well-known inhibitory feedback effect of CS on CRF-producing neurons in the medial parvocellular paraventricular nucleus, the glucocorticoids also strongly inhibit production of AVP by neurons in this division of the paraventricular nucleus. The inhibitory effect of CS on the hypothalamic synthesis of CRF and on the hypothalamic-pituitary-adrenal axis has been extensively reviewed and need not be exhaustively detailed here. Loss of glucocorticoid feedback inhibition after adrenalectomy permits large increases in CRF content and release into the portal system, resulting in marked down regulation of anterior pituitary CRF receptors (93). This decrease in CRF binding might explain the finding that adrenalectomy eliminated the stress-induced increase in pituitary cyclic AMP, *in vivo* (94). As mentioned above, significant numbers of neurons in the medial parvocellular paraventricular nucleus stained immunocytochemically for AVP (55), but only 1% to 2% of parvocellular CRF-positive neurons also stained for AVP (53). By contrast, 44% of CRF-positive axons and nerve terminals in the median eminence stained for AVP (57). Adrenalectomy produced a 14-fold increase in CRF-positive cells and an almost sixfold increase in the total number of AVP-positive cells in the medial parvocellular paraventricular nucleus (55). After adrenalectomy, more than 70% of parvocellular CRF-positive cells (53, 95) and all CRF-positive axons and terminals in the median eminence (57) also stained positively for AVP. Finally, adrenalectomy markedly increased release of AVP from superfused medial basal hypothalamus without affecting release from the neural lobe (96). Thus CS appears to exert a tonic negative inhibitory feedback effect on the medial parvocellular paraventricular nucleus-median eminence neurons secreting CRF and/or AVP, without affecting the vasopressinergic magnocellular neurosecretory system projecting to the neural lobe. Very recently, ACTH also was reported to inhibit AVP immunoreactivity in neurons of the medial parvocellular paraventricular nucleus *via* a short loop negative feedback system (97).

In the present study, the stressful interview elicited a doubling of PRA after 7 minutes of a 30 minute interview (Table 2 and Figure 3). Elevations in PRA have been noted in association with both naturalistic stressors, such as scholastic examinations (98) or experimental stressors such as mental arithmetic (99, 100) and psychosocial stress (101). On the other hand, puzzle-solving (101), watching a disturbing movie (101) and Stroop's color-word conflict test all failed to change PRA (102). Kosunen (100) found that performing a mental arithmetic test under quiet conditions nearly doubled PRA, as well as plasma levels of angiotensin II (AII); both values peaked 15 minutes after termination of the test and had almost returned to baseline by 30 minute post-test. Januszewicz *et al.* (99) reported similar increases in PRA elicited in subjects performing mental arithmetic for 30 minutes during exposure to noise. Moderate exercise in trained athletes, at 70% of maximal oxygen consumption (10), elicited increases only slightly greater than those reported to be elicited by mental arithmetic challenge, and levels had dropped to half of peak values within 10 minutes of completion of exercise. Running to exhaustion produced elevations in PRA which persisted for 190 minutes after termination of exercise (103). In the present study, whereas the levels of most hormones returned to baseline after the subjects returned to the waiting room, PRA remained elevated 22 minutes after termination of the interview (the last time point evaluated), as shown in Figure 3. This represents a more prolonged increase in PRA than those reported after other social stressors, or even moderate exercise. The time course of PRA elevation after the stressful interview was similar to that of NE (104). The extended duration of these increases in plasma PRA and NE may reflect our subjects' anticipation of learning the results of the contest and getting comments from the panel on their interview performance, which was scheduled for a late morning debriefing. The catheters were removed before the debriefing and therefore no samples were taken afterwards.

In studies carried out in rats, PRA was increased by immobilization (105), open field paradigm (106), exposure to a predator (106), and conditioned emotional responses (107). Renal denervation attenuated the stress-induced increase in PRA (105), as did lesions of the paraventricular nucleus (108). Renal nerves have a major adrenergic component. Reserpine-induced depletion of renal nerve catecholamines was accompanied by a concomitant decrease in renin secretion (109), whereas infusions of NE or EPI increased renin release via a β-adrenergic mechanism (109). On the other hand, the stress-induced increase in PRA was attenuated only by very high doses of β-adrenergic blockers, suggesting that additional stimulatory factors might also be involved in the PRA response to stress (105). Renin secretion is inhibited by both AII and AVP (109). Since the PRA increases in the present study were so long lasting, it is hard to imagine that the brief elevation in AVP was an effective inhibitor of renin secretion in this case.

Renin is an enzyme which produces angiotensin I, the precursor of AII, which has been shown to enhance CRH-stimulated release of ACTH from cultured rat or cynomolgus monkey pituitary cells via a cyclic AMP-independent mechanism (72, 110). Although the direct effect of AII on release of ACTH is small (74), the presence of specific receptor sites for AII on the anterior pituitary strengthens the argument that this peptide has a role in the regulation of hormone secretion (111). Theoretically, since the anterior pituitary is outside the blood-brain barrier, increased PRA could enhance anterior pituitary secretion of ACTH by increasing levels of circulating AII. Infusions of renin or AII have been reported to stimulate release of vasopressin as measured by bioassay (112) or RIA (113). The authors of the latter study cautioned, however, that they required supraphysiological amounts of AII to elicit the increases in AVP. Peripheral effects of circulating AII include vasopressor responses, release of aldosterone from the adrenal cortex, and some interesting interactions. The sodium-retaining effects of aldosterone tend to increase serum osmolality, which can stimulate AVP secretion. Increased plasma sodium concentration enhances the vasoactive effects of AII in the periphery, and aldosterone also increases the number of AII binding sites in the brain (114).

A renin-angiotensin system integral to brain has also been described. Its independent neuroendocrine effects have been demonstrated by intracerebroventricular administration of captopril, an inhibitor of angiotensin-converting enzyme. This treatment attenuated the plasma AVP and ACTH responses to hemorrhage, without affecting peripheral levels of AII (115). Although this system was capable of being activated independently of circulating AII, it was also responsive to infusions of AII, which elicited increases in plasma AVP (116-118). These effects were dependent upon an intact subfornical organ (SFO) (116, 117), and were mediated via neurons from the SFO which contain AII and which excite vasopressin-containing neurons in the paraventricular nucleus or supraoptic nucleus (117). Thus, the increases in PRA found in the present study might have had some enhancing effect on ACTH release via direct but slight action of circulating AII on the anterior pituitary, as well as some stimulatory effect on AVP release by an indirect action of circulating AII via the SFO.

ANP is released in response to distention of the right atrium of the heart. This peptide inhibits renin secretion, relaxes blood vessels contracted by AII and inhibits basal and agonist-stimulated aldosterone secretion from adrenal cortex (119). Moreover, ANP inhibits the stimulation of ACTH release by combined AVP and CRF (120). Given the simultaneous secretion of AVP and PRA elicited by the stressful interview, and the fact that ANP elevation can parallel increases in AVP (121), it seemed logical to measure ANP levels during the interview. However, no change in plasma levels of ANP was evident (Table 2 and Figure 9).

The stressful interview elicited increases in PRL which lasted throughout the interview, and returned to pre-exam levels after the subjects returned to the waiting room (Figure 7). Several circulating neurohumors might affect the PRL response. AII has been reported to release PRL (122). It has also been reported that β-EP can overcome dopaminergic inhibition of PRL release from anterior pituitary cells, and it was suggested that this endogenous opioid may partially mediate the PRL response to stressors (123). The marked increase in β-EP in the present study may have contributed to the increases seen in PRL. In conscious rats, infusions of AVP attenuated stress-induced increases in plasma PRL (124), but we have not seen any reports that AVP inhibits PRL secretion in man. While the function of PRL release in response to stressful conditions is not understood, one possible physiological role for PRL might be stimulation of the immune system (6). In this capacity, PRL might serve as a counter-regulatory hormone to adrenocortical hormones, which are increased by many stressors and can suppress immune function (5).

EPI increased significantly in the first 5 minutes of the interview (104), but by 20 minutes into the interview, levels were no longer significantly elevated. This response is similar to that reported by Dimsdale in his studies of public speaking (125). NE also increased in the first 5 minutes of the stress interview, but remained significantly elevated throughout the interview, and for 20 minutes after the subjects returned to the waiting room (104). This pattern is reflected in the PRA response (the facilitative role of catecholamines on renin release is discussed above). It is well established that exposure to various stressors increases circulating EPI and NE (126, 127), but it is unclear whether circulating catecholamines act directly on the normal pituitary gland to release ACTH *in vivo*. Chlorisondamine, which blocks the release of EPI from the adrenal medulla, partially inhibited the stress-induced increase in plasma ACTH (70), and the combination of paraventricular nucleus lesions with chlorisondamine virtually eliminated the ACTH response. Systemic injection of EPI increased plasma ACTH in normal rats and transection of the pituitary stalk attenuated, but did not completely eliminate this increase (128). This is consistent with the report that isoproterenol or salmefalol (a β-2 adrenergic agonist) increased plasma ACTH in stalk-sectioned rats (129). On the other hand, persuasive evidence has been presented to support the assertion that plasma catecholamines do not normally contribute to the ACTH response to forced immobilization (79, 130).

Summary

We have shown that a psychological stressor can elicit increases in plasma AVP levels in normal human subjects. Since AVP can enhance the release of ACTH, and the pituitary gland is outside the blood-brain barrier, AVP present in the general circulation might extend the time course of stress-induced, CRF-mediated release of ACTH from the anterior lobe. Since PRA is involved in the synthesis of angiotensin I, the precursor of AII, and AII is known to enhance CRF-mediated release of ACTH from pituitary cells and to stimulate release of AVP, it is possible that the increase in PRA also contributed to the release of AVP and ACTH in this study. Reports differ as to whether circulating catecholamines can release ACTH *in vivo* by direct action on the pituitary. Finally, it has been reported that β-EP enhances the release of PRL, and inhibits release of AVP. Since the increase in β-EP in the present study was quite robust, it might have extended the PRL release, and truncated the AVP response.

Acknowledgements

The authors gratefully acknowledge the advice and consultation of Sergeant Major Bert J. Mueck, First Sergeant Lawrence D. Olmsted, Lieutenant Colonel Steven R. Hursh, Lieutenant Colonel Frederick J. Manning, and Sergeant First Class Ed Marshall. We are also grateful for the interest and support of the soldiers and non-commissioned officers at the Walter Reed Army Institute of Research. We wish to thank the following individuals for excellent technical support: Lori Wittig, Vinit Jain, Mary Kautz, Michelle

LaPlaca, Michael Bixler, Tim Moore, Lael Lambe, PFC Kevin Walker, SGT Thomasina Maxwell-Irving, SGT Phillip Smigaj, and Mr. Clint Wormley. We thank Dr. H. Ron Smith for consultation and administrative assistance, and John R. Burge and Dr. Sue Hedges for excellent statistical consultation. We also thank Drs. Greti Aguilera, G. Jean Kant, Debra Yourick, Kim Huhman, and Robert Smallridge for helpful comments on the manuscript.

References

1. Mason, J.W., "Over-all" hormonal balance as a key to endocrine organization, *Psychosom Med* 30: 791-808, 1968.
2. Shanker, G., and R.K. Sharma, β-Endorphin stimulates corticosterone synthesis in isolated rat adrenal cells, *Biochem Biophys Res Commun* 86: 1-5, 1979.
3. Richter, W.O., R.J. Naude, W. Oelofsen, and P. Schwandt, In Vitro lypolytic activity of β-endorphin and its partial sequences, *Endocrinology* 120: 1472-1476, 1987.
4. Holaday, J.W., Cardiovascular effects of endogenous opiate systems, In R. George, R. Okun and A.K. Cho (eds) *Ann Rev Pharmacol Toxicol, Volume 23*, Annual Reviews Inc., Palo Alto, pp. 541-594, 1983.
5. Hirschhorn, K., F. Bach, R. Kolodny, I. Firschein, and N. Hashem, Immune response and mitosis of human peripheral blood lymphocytes in vitro, *Science* 142: 1185-1187, 1963.
6. Spangelo, B.L., N.R. Hall, and A.L. Goldstein, Evidence that prolactin is an immunomodulatory hormone, In R.M. MacLeod, M.O. Thorner, and U. Scapagnini (eds) *Prolactin: Basic and Clinical Correlates*, Liviana Press, Padova, pp.343-349, 1985.
7. Meyerhoff, J.L., M.A. Oleshansky, M.A., and E.H. Mougey, Psychologic stress increases plasma levels of prolactin, cortisol and POMC-derived peptides in man, *Psychosom Med* 50: 295-303, 1988.
8. Lake, C.R., M.G. Ziegler, and I.J. Kopin, Use of plasma norepinephrine for evaluation of sympathetic neuronal function in man, *Life Sci* 18: 1315-1326, 1976.
9. Harley, L.H., J.W. Mason, R.P. Hogan, L.G. Jones, T.A. Kotchen, E.H. Mougey, F.E. Wherry, L.L. Pennington, and P.T. Ricketts, Multiple hormonal responses to graded exercise in relation to physical training, *J Appl Physiol* 33: 602-606, 1972.
10. Kotchen, T.A., L.H. Hartley, T.W. Rice, E.H. Mougey, L.G. Jones, and J.W. Mason, Renin norepinephrine, and epinephrine responses to graded exercise, *J Appl Physiol* 31: 178-184, 1971.
11. Grossman, A., P. Boulous, P. Price, P.L. Drury, K.S.L. Lam, T. Turner, J. Thomas, G.M. Besser, and J. Sutton, The role of opioid peptides in the hormonal responses to acute exercise in man, *Clin Sci* 67: 483-491, 1984.
12. Fraioli, F., C. Moretti, D. Paolucci, E. Alicicco, F. Crescenzi, and G. Fortunio, Physical exercise stimulates marked concomitant release of β-endorphin and adrenocorticotropic hormone (ACTH) in peripheral blood in man, *Experientia* 36: 987-989, 1980.
13. Janal, M.N., E.W.D. Colt, W.C. Clark, and M. Glusman, Pain sensitivity, mood and plasma endocrine levels in man following long-distance running: effects of naloxone, *Pain* 19: 13-25, 1984.
14. Gambert S.R., T.L. Garthwaite, C.H. Pontzer, E.E. Cook, F.E. Tristani, E.H. Duthie, D.R. Martinson, T.C. Hagen, and D.J. McCarty, Running elevates plasma β-endorphin immunoreactivity and ACTH in untrained human subjects, *Proc Soc Exp Biol Med* 168: 1-4, 1981.
15. Buono, M.J., J.E. Yeager, and J.A. Hodgdon, Plasma adrenocorticotropin and cortisol responses to brief high-intensity exercise in humans, *J Appl Physiol* 61: 1337-1339, 1986.
16. Luger, A., P.A. Deuster, S.B. Kyle, W.T. Gallucci, L.C. Montgomery, P.W. Gold, D.L. Loriaux, and G.P. Chrousos, Acute hypothalamic-pituitary-adrenal responses to the stress of treadmill exercise: physiologic adaptations to physical training, *N Engl J Med* 316: 1309-1315, 1987.
17. Oleshansky, M., J. Zoltick, R. Herman, E. Mougey, and J. Meyerhoff, Neuroendocrine responses to maximal treadmill exercise (abstract), *Psychiatry Res* 16(4): 72, 1986.
18. Colt, E.W.D., S.L. Wardlaw, and A.G. Frantz, The effect of running on plasma β-endorphin, *Life Sci* 28: 1637-1640, 1981.
19. Elliot, D.L., L. Goldberg, W.J. Watts, and E. Orwoll, Resistance exercise and plasma

beta-endorphin/beta-lipotrophin immunoreactivity, *Life Sci* 34: 515-518, 1984.

20. Bortz, W.M., P. Angwin, I.N. Mefford, M.R. Boarder, N. Noyce, and J.D. Barchas, Catecholamines, dopamine, and endorphin levels during extreme exercise, *N Engl J Med* 305: 466-467, 1981.

21. Mougey, E.H., A radioimmunoassay for tetrahydrocortisol, *Anal Biochem* 91: 566-582, 1978.

22. Menard, J., and K.J. Catt, Measurement of renin activity, concentration and substrate in rat plasma by radioimmunoassay of angiotensin I, *Endocrinology* 90: 422-430, 1972.

23. Robertson, G.L., and S. Athar, The interaction of blood osmolality and blood volume in regulating plasma vasopressin in man, *J Clin Endocrinol Metab* 42: 613-620, 1976.

24. Stella, A., and a. Zanchetti, Control of renal renin release, *Kidney Int [Supple]* 31: S89-S94, 1987.

25. Meyerhoff, J.L., M.A. Oleshansky, E.H. Mougey, H.R. Smith, S.R. Hursh, L.K. Wittig, H.E. Wood, C.R. Glass, and D.B. Arnkoff, Increased heart rat and plasma level of β-lipotropin during a competitive oral examination, *Neurosci Abstr* 12: 1480, 1986.

26. Wood, H.E., C.R. Glass, D.B. Arnkoff, H.R. Smith, J.L. Meyerhoff, and M.A. Oleshansky, An investigation of evaluation anxiety in an employment setting, *Association for Advancement of Behavior Therapy, 20th Annual Convention*, Chicago, IL, November 14-16, 1986.

27. Robertson, G., and E. Mahr, The importance of plasma osmolality in regulation of antidiuretic hormone secretion in man, *J Clin Invest* 51: 79a, 1972 (Abstract 261).

28. Robertson, G.L., The regulation of vasopressin function in health and disease, *Rec Prog Horm Res* 33: 333-385, 1977.

29. Keil, L.C., and W.B. Severs, Reduction in plasma vasopressin levels of dehydrated rats following acute stress, *Endocrinology* 100: 30-38, 1977.

30. Cuneo, R.C., J.H. Livesey, M.G. Nicholls, E.A. Espiner, and R.A. Donald, Effects of alpha-2 adrenoreceptor blockade by yohimbine on the hormonal response to hypoglycemic stress in normal man, *Horm Metab Res* 21: 33-36, 1989.

31. Goetz, K.L., G.C. Bond, and W.E. Smith, Effect of moderate hemorrhage in humans on plasma ADH and renin, *Proc Soc Exp Biol Med* 145: 277-380, 1974.

32. Rydin, H., and E.B. Verney, The inhibition of water diuresis by emotional stress and by muscular exercise, *Quart J Exp Physiol* 27: 343-375, 1938.

33. Kalin, N.H., D.M. Gibbs, C.M. Barksdale, S.E. Shelton, and M. Carnes, Behavioral stress decreases plasma oxytocin concentrations in primates, *Life Sci* 36: 1275-1280, 1985.

34. Gibbs, D.M., Vasopressin and oxytocin: hypothalamic modulators of the stress response: a review, *Psychoneuroendocrinology* 11: 131-140, 1986.

35. Rosella-Dampman, L.M., and J.Y. Summy-Long, Dexamethasone differentially alters naltrexone effects on plasma vasopressin and oxytocin concentrations elevated by tail electroshock in rats, *Neurosci Abstr* 10: 91, 1984.

36. Cameron, V, E.A. Espiner, M.G. Nicholls, R.A. Donald, and M.R. MacFarlane, Stress hormone in blood and cerebrospinal fluid of conscious sheep: effect of hemorrhage, *Endocrinology* 116: 1460-1465, 1985.

37. Kasting, N.W., Simultaneous and independent release of vasopressin and oxytocin in the rat, *Can J Physiol Pharmacol* 66: 22-26, 1988.

38. Gibbs, D.M., Dissociation of oxytocin, vasopressin and corticotropin secretion during different types of stress, *Life Sci* 35: 487-491, 1984.

39. Husain, M.K., W.M. Manger, T.W. Rock, R.J. Weiss, and A.G. Frantz, Vasopressin release due to manual restraint in the rat: role of body compression and comparison with other stressful stimuli, *Endocrinology* 104: 641-644, 1979.

40. Michajlovskij, N., B. Lichardus, R. Kvetňanský, and J. Ponec, Effect of acute and repeated immobilization stress on food and water intake, urine output and vasopressin changes in rats, *Endocrinol Exp* 22: 143-157, 1988.

41. Onaka, T., M. Hamamura, and K. Yagi, Potentiation of vasopressin secretion by footshocks in rats, *Jap J Physiol* 36: 1253-1260, 1986.

42. Knepel, W., D. Nutto, and G. Hertting, Effect of neonatal treatment with monosodium glutamate on vasopressin release during foot shock stress in the rat, *Life Sci* 33: 1703-1709, 1983.

43. Knepel, W., R. Przewlocki, D. Nutto, and A. Herz, Foot shock stress-induced release of vasopressin in

adenohypophysectomized and hypophysectomized rats, *Endocrinology* 117: 292-299, 1985.

44. Knepel, W., D. Nutto, and H. Anhut, β-endorphin controls vasopressin release during foot sock-induced stress in the rat, *Regul Pept* 7: 9-19, 1983.

45. Meyerhoff, J.L., G.J. Kant, B.N. Bunnell, and E.H. Mougey, Regulation of pituitary cyclic AMP, plasma prolactin and POMC-derived responses to stressful conditions, In G.P. Chrousos, D.L. Loriaux, and P.W. Gold (eds) *Mechanisms of Physical and Emotional Stress: Advances in Experimental Medicine and Biology, Volume 245*, Plenum Publishing Corporation, New York, pp. 107-122, 1988.

46. Carr, D.B., S.M. Fishman, N.W. Kasting, and D.V. Sheehan, Vasopressin response to lactate infusion in normals and patients with panic disorder, *Funct Neurol* 1: 123-127, 1986.

47. Edelson, J.T., and G.L. Robertson, The effect of the cold pressor test on vasopressin secretion in man, *Psychoneuroendocrinology* 11: 307-316, 1986.

48. Wiegard, S.J., and J.L. Price, Cells of origin of the afferent fibers to the median eminence in the rat, *J Comp Neurol* 192: 1-19, 1980.

49. Brownstein, M.J., J.T. Russell, and H. Gainer, Synthesis, transport and release of posterior pituitary hormones, *Science* 207: 373-378, 1980.

50. Antoni, F.A., Hypothalamic control of adrenocorticotropin secretion: advances since the discovery of 41-residue corticotropin-releasing factor, *Endocr Rev* 7: 351-378, 1986.

51. Swanson, L.W., P.E. Sawchenko, R.W. Lind, J.-H. Rho, The CRH motor neuron: differential peptide regulation in neurons with possible synaptic, paracrine and endocrine outputs, In W.F. Ganong, M.F. Dallman, and J.L. Roberts (eds) *The Hypothalamic-Pituitary-Adrenal Axis Revisited: Annals of the New York Academy of Sciences* 512: 12-23, 1987.

52. Swanson, L.W., and P.E. Sawchenko, Hypothalamic integration: organization of the paraventricular and supraoptic nuclei, *Ann Rev Neurosci* 6: 269-324, 1983.

53. Sawchenko, P.E., L.W. Swanson, and W.W. Vale, Co-expression of corticotropin-releasing factor and vasopressin immunoreactivity in parvocellular neurosecretory neurons of the adrenalectomized rat, *Proc Natl Acad Sci USA* 81: 1883-1887, 1984.

54. Sawchenko, P.E., L.W. Swanson, and W.W. Vale, Corticotropin-releasing factor: co-expression within distinct subsets of oxytocin-, vasopressin-, and neurotensin-immunoreactive neurons in the hypothalamus of the male rat, *J Neurosci* 4: 1118-1129, 1984.

55. Sawchenko, P.E., Adrenalectomy-induced enhancement of CRF and vasopressin immunoreactivity in parvocellular neurosecretory neurons: anatomic, peptide and steroid specificity, *J Neurosci* 7: 1093-1106, 1987.

56. Lechan, R.M., J.L. Nestler, S. Jacobson, and S. Reichlin, The hypothalamic tuberoinfundibular system of the rat as demonstrated by horseradish peroxidase (HRP) microiontophoresis, *Brain Res* 195: 13-27, 1980.

57. Whitnall, M.H., E. Mezey, and H. Gainer, Co-localization of corticotropin-releasing factor and vasopressin in median eminence neurosecretory vesicles, *Nature* 317: 248-250, 1985.

58. Vandessande, F., K. Dierickx, and J. De Mey, The origin of the vasopressinergic and oxytocinergic fibers of the external region of the median eminence of the rat hypophysis, *Cell Tissue Res* 180: 443-452, 1977.

59. Zimmerman, E.A., P.W. Carmel, M.K. Husain, M. Ferin, M. Tannenbaum, A.G. Frantz, and A.G. Robinson, Vasopressin and neurophysin: high concentrations in monkey hypophyseal portal blood, *Science* 182: 925-927, 1973.

60. Plotsky, P.M., T.O. Bruhn, and W. Vale, Evidence for multifactor regulation of the adrenocorticotropin secretory response to hemodynamic stimuli, *Endocrinology* 116: 633-639, 1985.

61. Caraty, A., M. Grino, A. Locatelli, and C. Oliver, Secretion of corticotropin releasing factor (CRF) and vasopressin (AVP) into the hypophysial portal blood of conscious, unrestrained rams, *Biochem and Biophys Res Commun* 155: 841-849, 1988.

62. Engler, D., T. Pham, M.J. Fullerton, G. Ooi, J.W. Funder, and I.J. Clarke, Studies of the secretion o corticotropin-releasing factor and arginine vasopressin into the hypophysial-portal circulation of the conscious sheep. I. Effect of an audiovisual stimulus and insulin-induced hypoglycemia, *Neuroendocrinology* 49: 367-381, 1989.

63. Alexander, S.L., C.H.G. Irvine, J.H. Livesey, and R.A. Donald, Effect of isolation stress on concentrations

of arginine, α-melanocyte-stimulating hormone and ACTH in the pituitary venous effluent of the normal horse, *J Endocrinology* 116: 325-334, 1988.

64. Holmes, M.C., F.A. Antoni, G. Aguilera, and K.J. Catt, Magnocellular axons in passage through the median eminence release vasopressin, *Nature* 319: 326-329, 1986.

65. Bergland, R.M., and R.B. Page, Can the pituitary secrete directly to the brain? (Affirmative anatomical evidence), *Endocrinology* 102: 1325-1338, 1978.

66. Oliver, C., R.S. Mical, and J.C. Porter, Hypothalamic-pituitary vasculature: evidence for retrograde blood flow in the pituitary stalk, *Endocrinology* 101: 598-604, 1977.

67. Recht, L.D., D.L. Hoffman, J. Haldar, A.J.Silverman, and E.A. Zimmerman, Vasopressin concentrations in hypophysial portal plasma: insignificant reduction following removal of the posterior pituitary gland, *Neuroendocrinology* 33: 88-90, 1981.

68. Gillies, G.E., E.A. Linton, and P.J. Lowry, Corticotropin releasing activity of the new CRF is potentiated several times by vasopressin, *Nature* 299: 355-357, 1982.

69. Rivier, C., and W. Vale, Interaction of corticotropin-releasing factor and arginine vasopressin on adrenocorticotropin secretion *in vivo*, *Endocrinology* 113: 939-942, 1983.

70. Bruhn, T.O., P.M. Plotsky, and W.W. Vale, Effect of paraventricular lesions on corticotropin-releasing factor (CRF)-like immunoreactivity in the stalk-median eminence: studies on the adrenocorticotropin response to ether stress and exogenous CRF, *Endocrinology* 114: 57-62, 1984.

71. Meyerhoff, J.L., E.H. Mougey, and G.J. Kant, Paraventricular lesions abolish the stress-induced rise in pituitary cyclic adenosin monophosphate and attenuate the increases in plasma levels of proopiomelanocortin-derived peptides and prolactin, *Neuroendocrinology* 46: 222-230, 1987.

72. Aguilera, G., J.P. Harwood, J.X. Wilson, J. Morrell, J.H. Brown, and K.J. Catt, Mechanisms of action of corticotropin-releasing factor and other regulators of corticotropin release in rat pituitary cells, *J Biol Chem* 258: 8039-8045, 1983.

73. Bilezikjian, L.M., and W.W. Vale, Regulation of ACTH secretion from corticotrophs: the interaction of vasopressin and CRF, *Ann NY Acad Sci* 512: 85-96, 1987.

74. Abou-Samra, A.-B., K.J. Catt, and G. Aguilera, Involvement of protein kinase C in the regulation of adrenocorticotropin release from rat anterior pituitary cells, *Endocrinology* 118: 212-217, 1986.

75. Ono, N., J.B. de Castro, O, Khorram, and S.M. McCann, Role of arginine vasopressin in control of ACTH and LH release during stress, *Life Sci* 36: 1779-1786, 1985.

76. Linton, E.A., F.J.H. Tilders, S. Hodgkinson, F. Berkenbosch, I. Vermes, and P.J. Lowry, Stress-induced secretion of adrenocorticotropin in rats is inhibited by administration of antisera to ovine corticotropin-releasing factor and vasopressin, *Endocrinology* 116: 966-970, 1985.

77. Sweep, C.G.J., and V.M. Wiegant, Release of β-endorphin-immunoreactivity from rat pituitary and hypothalamus in vitro: effects of isoproterenol, dopamine, corticotropin-releasing factor and arginine-vasopressin, *Biochem Biophys Res Commun* 161: 221-228, 1989.

78. Rivier, C., and W. Vale, Modulation of stress-induced ACTH release by corticotropin-releasing factor, catecholamines and vasopressin, *Nature* 305: 325-327, 1983.

79. Tilders, F.J.H., F. Berkenbosch, I. Vermes, E.A. Linton, and P.G. Smelik, Role of epinephrine and vasopressin in the control of the pituitary-adrenal response to stress, *Fed Proc* 44: 155-160, 1985.

80. Fagin, K.D., S.G. Wiener, and M.F. Dallman, ACTH and corticosterone secretion in rats following removal of the neurointermediate lobe of the pituitary gland, *Neuroendocrinology* 40: 352-362, 1985.

81. DeBold, C.R., W.R. Sheldon, G.S. DeCherney, R.V. Jackson, A.N. Alexander, W. Vale, J. Rivier, and D.N. Orth, Arginine vasopressin potentiates adrenocorticotropin release induced by ovine corticotropin-releasing factor, *J Clin Invest* 73: 533-538, 1984.

82. Liu, J.H., K. Muse, P. Contreras, D. Gibbs, W. Vale, J. Rivier, and S.S.C. Yen, Augmentation of ACTH-releasing activity of synthetic corticotropin releasing factor (CRF) by vasopressin in women, *J Clin Endocrinol Metab* 57: 1087-1089, 1983.

83. Mason, J.W., and J.V. Brady, Plasma 17-hydroxycorticosteroid changes related to reserpine effects on emotional behavior, *Science* 124: 983-984, 1956.

84. Williams, Jr., R.B., J.D. Lane, C.M. Kuhn, W. Melosh, A.D. While, and S.M. Schanberg, Type A behavior and elevated physiological and neuroendocrine responses to cognitive tasks, *Science* 218: 483-485, 1982.

85. Baumgartner, A., K.J. Graf, and I. Kurten, The dexamethasone suppression test in depression, in schizophrenia, and during experimental stress, *Biol Psychiatry* 20: 675-679, 1985.

86. Nesse, R.M., G.C. Curtis, B.A. Thyer, D.S. McCann, M.J. Huber-Smith, and R.F. Knopf, Endocrine and cardiovascular responses during phobic anxiety, *Psychosom Med* 47: 320-332, 1985.

87. Fredrikson, M., O. Sundin, and M. Frankenhaeuser, Cortisol excretion during the defense reaction in humans, *Psychosom Med* 47: 313-319, 1985.

88. Lovallo, W.R., G.A. Pincomb, G.L. Edwards, D.J. Brackett, and M.F. Wilson, Work pressure and the type A behavior pattern exam stress in male medical students, *Psychosom Med* 48: 125-133, 1986.

89. Frankenhaeuser, M., M.R. Von Wright, A. Collins, J. Von Wright, G. Sedvall, and C.-G. Swahn, Sex differences in psychoneuroendocrine reactions to examination stress, *Psychosom Med* 40: 334-343, 1978.

90. Czeisler, C.A., M.C.M. Ede, Q.R. Regestein, E.S. Kisch, V.S. Fang, and E.N. Ehrlich, Episodic 24-hour cortisol secretory patterns in patients awaiting elective cardiac surgery, *J Clin Endocrinol Metab* 42: 273-283, 1976.

91. Oltras, C.M., F. Mora, and F. Vives, Beta-endorphin and ACTH in plasma: effects of physical and psychological stress, *Life Sci* 40: 1683-1686, 1987.

92. Levine, S., Cortisol changes following repeated experiences with parachute training, In H. Ursin, E. Baade, and S. Levine (eds) *Psychobiology of Stress: A Study of Coping Men*, Academic Press, New York, pp. 51-56, 1978.

93. Aguilera, G., P.C. Wynn, J.P. Harwood, R.L. Hauger, M.A. Millan, C. Grewe, and K.J. Catt, Receptor-mediated actions of corticotropin-releasing factor in pituitary gland and nervous system, *Neuroendocrinology* 43: 79-88, 1986.

94. Meyerhoff, J.L., G.J. Kant, C.J. Nielsen, and E.H. Mougey, Adrenalectomy abolishes the stress-induced increase in pituitary cyclic AMP, *Life Sci* 34: 1959-1965, 1984.

95. Sawchenko, P.E., and L.W. Swanson, Localization, and colocalization, and plasticity of corticotropin-releasing factor immunoreactivity in rat brain, *Fed Proc* 44: 221-227, 1985.

96. Knepel, W., D. Nutto, D.K. Meyer, and M. Vlaskovska, Vasopressin release from rat medial basal hypothalamus in vitro after adrenalectomy or lesions of the paraventricular nuclei, *Neurosci Lett* 48: 321-326, 1984.

97. Sawchenko, P.E. Short-loop feedback effects of adrenocorticotrophic hormone on corticotropin-releasing factor and vasopressin-immunoreactivity in the paraventricular nucleus, *Neurosci Abstr* 15: 134, 1989.

98. Tigranian, R.A., L.L. Orloff, N.F. Kalita, N.A. Davydova, and E.A. Pavlova, Changes of blood-levels of several hormones, catecholamines, prostaglandins, electrolytes and cAMP in man during emotional stress, *Endocrinol Exp* 14: 101-112, 1980.

99. Januszewicz, W., M. Sznajderman, B. Wocial, T. Feltynowski, and T. Klonowicz, The effect of mental stress on catecholamines, their metabolites and plasma renin activity, in patients with essential hypertension and in health subjects, *Clin Sci* 57: 229s-231s, 1979.

100. Kosunen, K.J., Plasma renin activity, angiotensin II, and aldosterone after mental arithmetic, *Scand J Clin Lab Invest* 37: 425-429, 1977.

101. Clamage, D.M., A.J. Vander, and D.R. Mouw, Psychosocial stimuli and human plasma renin activity, *Psychosom Med* 39: 393-401, 1977.

102. Hjemdahl, P., and K. Eliasson, Sympatho-adrenal and cardiovascular response to mental stress and orthostatic provocation in latent hypertension, *Clin Sci* 57: 189s-191s, 1979.

103. Adlercreutz, H., K. Kuoppasalmi, S. Narvanen, K. Kosunen, and R. Heikkinen, Use of hypnosis in studies of the effect of stress on cardiovascular function and hormones, *Acta Med Scand [Suppl]* 660: 84-94, 1982.

104. Oleshansky, M.A., J.L. Meyerhoff, E.H. Mougey, and H.R. Smith, Psychological stressor (Oral Exam) increases plasma levels of catecholamines in man, *Neuroendocrinol Lett* 9: 184, 1987 (Abstract).

105. Sigg, E.B, K.L. Keim, and T.D. Sigg, On the mechanism of renin release of restraint stress in rats, *Pharmacol Biochem Behav* 8: 47-50, 1978.

106. Clamage, D.M., C.S. Sanford, A.J. Vander, and D.R. Mouw, Effects of psychosocial stimuli on plasma renin activity in rats, *Am J Physiol* 231: 1290-1294, 1976.

107. Van de Kar, L.D., S.A. Lorens, C.r. McWilliams, K. Kunimoto, J.H. Urban, and L. Bethea, Role of midbrain raphe in stress-induced renin and prolactin secretion, *Brain Res* 311: 333-341, 1984.

108. Morton, K.D.R., L.D. Van de Kar, M.S Brownfield, and C.L. Bethea, Neuronal cell bodies in the hypothalamic paraventricular nucleus mediate stress-induced renin and corticosterone secretion, *Neuroendocrinology* 50: 73-80, 1989.

109. Davis, J.O., and R.H. Freeman, Mechanisms regulating renin release, *Physiol Rev* 56: 1-56, 1976.

110. Millan, M.H., A-B.A. Samra, P.C. Wynn, K.J. Catt, and G. Aguilera, Receptors and actions of corticotropin-releasing hormone in the primate pituitary gland, *J Clin Endocrinol Metab* 64: 1036-1041, 1987.

111. Mukherjee, A., P. Kulkarni, S.M. McCann, and A Negro-Vilar, Evidence for the presence and characterization of angiotensin II receptors in rat anterior pituitary membranes, *Endocrinology* 110: 665-667, 1982.

112. Bonjour, J.P., and R.L. Malvin, Stimulation of ADH release by the renin-angiotensin system, *Am J Physiol* 218: 1555-1559, 1970.

113. Padfield, P.L., and J.J. Morton, Effects of angiotensin II on arginine-vasopressin in physiological and pathological situation in man, *J Endocrinol* 74: 251-259, 1977.

114. Wilson, K.M., C. Sumners, S. Hathaway, and M.J. Fregly, Mineralocorticoids modulate central angiotensin II receptors in rats, *Brain Res* 382: 87-96, 1986.

115. Cameron, V.A., E.A. Espiner, M.G. Nicholls, M.R. MacFarlane, and W.A. Sadler, Intracerebroventricular captopril reduces plasma ACTH and vasopressin responses to hemorrhagic stress, *Life Sci* 38: 553-559, 1986.

116. Iovino, M., and L. Steardo, Vasopressin release to central and peripheral angiotensin II in rats with lesions of the subfornical organ, *Brain Res* 322: 365-368, 1984.

117. Ferguston, A.V., and L.P. Renaud, Systemic angiotensin acts at subfornical organ to facilitate activity of neurohypophysial neurons, *Am J Physiol* R251: R712-R717, 1986.

118. Ferguston, A.V., Systemic angiotensin acts at the subfornical organ to control the activity of paraventricular nucleus neurons with identified projections to the median eminence, *Neuroendocrinology* 47: 489-497, 1988.

119. Atlas, S.A., M. Volpe, R.E. Sosa, J.H. Laragh, M.J.F. Camargo, and T. Maack, Effects of atrial natriuretic factor on blood pressure and the renin-angiotensin-aldosterone system, *Fed Proc* 45: 2115-2121, 1986.

120. Antoni, F.A., and G. Dayanithi, Atriopeptin inhibits pituitary corticotropin release in vitro: involvement of cyclic GMP and potassium channels, *Neurosci Abstr* 15: 1080, 1989.

121. Kalogeras, K.T., M.A. Demitrack, L.G. Granger, E.L. Papioannou, M.J. Hart, G.P. Chrousos, and P.W. Gold, Responses of plasma atrial natriuretic peptide and arginine vasopressin to osmotic and volume stimulation, *Neurosci Abstr* 15: 1078, 1989.

122. Aguilera, G., C.L. Hyde, and K.J. Catt, Angiotensin II receptors and prolactin release in pituitary lactotrophs, *Endocrinology* 111: 1045-1050, 1982.

123. Voigt, K.H., D. Frank, E. Düker, R. Martin, and W. Wuttke, Dopamine-inhibited release of prolactin and intermediate lobe-POMC-peptides: different modulation by opioids, *Life Sci* 33: 507-510, 1983.

124. Mormede, P., J.-D. Vincent, and B. Kerdelhue, Vasopressin and oxytocin reduce plasma prolactin levels of conscious rats in basal and stress conditions. Study of the characteristics of the receptor involved, *Life Sci* 39: 1737-1743, 1986.

125. Dimsdale, J.E., and J. Moss, Short-term catecholamine response to psychological stress, *Psychosom* 42: 493-497, 1980.

126. Natelson, B.H., W.N. Tapp, J.E. Adamus, J.C. Mittler, and B.E. Levin, Humoral indices of stress in rats, *Physiol Behav* 26: 1049-1054, 1981.

127. Kvetňanský, R., C.L. Sun C.R. Lake, N. Thoa, T. Torda, and I.J. Kopin, Effect of handling and forced immobilization of rat plasma levels of epinephrine, norepinephrine, and dopamine-β-hydroxylase, *Endocrinology* 103: 1868-1874, 1978.

128. Kant, G.J., M.A. Oleshansky, D.D. Walczak, E.H. Mougey, and J.L. Meyerhoff, Comparison of the effects of CRF and stress on levels of pituitary cyclic AMP and plasma ACTH *in vivo*, *Peptides* 7: 1153-1158, 1986.

129. Mezey, E., T.D. Reisin, M. Palkovits, M.J. Brownstein, and J. Axelrod, Direct stimulation of β_2-adrenergic receptors in rat anterior pituitary induced the release of adrenocorticotropin *in vivo*, *Proc Natl Acad Sci*

USA 80: 6728-6731, 1983.

130. Makara, G.B., R. Kvetňanský, D. Ježová, A. Jindra, I. Kakucska, and Z. Opršalová, Plasma catecholamines do not participate in pituitary-adrenal activation by immobilization stress in rats with transection of nerve fibers to the median eminence, *Endocrinology* 119: 1757-1762, 1986.

REGULATION OF THE SYMPATHETIC NERVOUS SYSTEM BY CIRCULATING VASOPRESSIN

R. Kvetňanský, D. Ježová, Z. Opršalová, O. Földes,
N. Michajlovskij, M. Dobrakovová, B. Lichardus, and G.B. Makara[1]

*Institute of Experimental Endocrinology, Centre of Physiological
Sciences
Slovak Academy of Sciences
Bratislava, Czechoslovakia*

Neural regulation of the sympathoadrenal system (SAS) is not yet fully understood. It has been shown that various areas within the brain interact with sympathetic preganglionic neurons in the thoracic cord (1,2). Secretion of epinephrine (EPI) or norepinephrine (NE) into the blood of rats was found to be changed after electrical stimulation of certain areas in the medulla oblongata (3), the pons, and the midbrain (4), as well as in different parts of the hypothalamus (5,6). Plasma catecholamines were shown to be modulated also by the action of brain catecholamines (7).

Besides this classical regulation of the SAS, the last decade has witnessed the discovery of many brain neuropeptides, particularly hypothalamic, exerting a regulatory effect on the SAS. This has made the whole situation even more complicated.

Evidence indicating the involvement of neuropeptides in the physiologic control of the SAS activity is scarce. However, recently an excellent review dealing with neuropeptide regulation of the autonomic nervous system (ANS) was given by Brown (8). He evaluated 16 peptides for their actions on the function of the autonomic nervous system. At least five peptides—CRH, substance P, β-endorphin, somatostatin, and TRH—have been demonstrated to exert potentially relevant physiologic effects on ANS function. There is some evidence that CRH increases and somatostatin reduces adrenal EPI secretion, TRH, and substance P enhance sympathetic outflow, and β-endorphin decreases mean arterial pressure. In general, all peptides that increase plasma EPI levels also elevate plasma NE levels. An exception is calcitonin gene related peptide (CGRP) which raises plasma NE without influencing plasma EPI levels. Somatostatin is the only peptide known to decrease plasma EPI levels, and no peptide has been found to reduce plasma NE levels (8).

Brain vasopressin was shown to induce an activation of the sympathoadrenal outflow (9-12). Intracerebroventricular treatment of rats with vasopressin elevated plasma EPI and NE levels, mean arterial pressure (MAP), and heart rate (8). On the other hand, several authors demonstrated that circulating vasopressin could inhibit the activity of sympathetic preganglionic neurons (11,13-16) and diminish plasma catecholamine levels. Already physiologic amounts of circulating vasopressin were reported to have a vasoactive effect (17-22). Brown (8), however, is not inclined to accept the physiologic role of vasopressin in the regulation of ANS activity.

Our aim was to investigate the involvement of hypothalamic peptides, mainly vasopressin and angiotensin II, in mechanisms of regulation of the sympathoadrenal system

---

[1]Institute of Experimental Medicine, Hungarian Academy of Sciences, Budapest, Hungary

Circulating Regulatory Factors and Neuroendocrine Function
Edited by J. C. Porter and D. Ježová
Plenum Press, New York, 1990

113

as well as the involvement of the renin-angiotensin-aldosterone system in rats at resting conditions and especially during stress.

Our previous data have shown that plasma norepinephrine levels were significantly elevated in rats with deafferentation of the medial basal hypothalamus (MBH) by a total or anterolateral cut (ALC), which destroyed corticoliberin (CRH), vasopressin, oxytocin (OXY), and other peptides containing fibers innervating the median eminence (23,24). Plasma epinephrine levels under these conditions, showed only slight but not significant changes. Posterior deafferentation of the MBH did not result in any significant changes in plasma NE and EPI (23,24). These findings suggest that axons innervating the hypothalamus from the anterolateral side are involved in the regulation of sympathetic activity (23).

Since the paraventricular nucleus (PVN), which is the place of synthesis of many neuropeptides that are important for neuroendocrine functions, is localized in the anterior area of the hypothalamus, we concentrated our study on the investigation of sympathoadrenal activity in rats with PVN lesion. The data obtained show that a lesion of PVN, in contrast to anterolateral deafferentation, does not produce elevation of plasma NE levels (24). One of the major differences between animals with a PVN lesion and rats with ALC deafferentation is the presence of vasopressin, as the supraoptic nucleus (NSO), another source of vasopressin, remains intact after the PVN lesion.

In light of the above findings, we investigated vasopressin as a potential factor affecting plasma NE levels in rats with ALC, both in rest conditions and during stress. Male Wistar SPF rats (Velaz, Prague), weighing 280-350 g, were used in the majority of the experiments of this study. The animals were housed under controlled temperature (24°C) with illumination between 0600 and 1700 hours. Standard pellet rat diet and tap water were available *ad libitum*.

In some experiments male Brattleboro rats (Physiological Institute of ČSAV, Prague)—homozygotes with diabetes insipidus based on their minimal ability to synthesize brain vasopressin, and heterozygotes with normal vasopressin synthesis—were used. In all experiments, blood was collected *via* an indwelling cannula inserted under pentobarbital anesthesia into the tail artery 24 hours prior to the experiments (25). After cannulation the rats were caged individually, and between 1600 and 1700 hours, the cannulas were flushed with heparinized saline (150 IU/0.5 ml).

Immobilization (IMO) was used as stressor (26). One ml of blood was collected *via* the tail artery cannula under rest conditions in the morning (0900 hour) and then during stress at the intervals of 5, 20, and 120 minutes of immobilization. Anterolateral deafferentation of the MBH and sham operation were performed according to Makara *et*

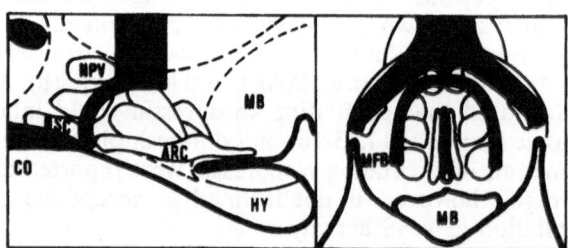

Figure 1. Schematic representation of the sagittal *(left)* and horizontal *(right)* planes of anterolateral deafferentation of the medial basal hypothalamus. Abbreviations: PVN, paraventricular nucleus; NSC, suprachiasmatic nucleus; CO, optic chiasma; ARC, arcuate nucleus; MB, mammillary body; HY, pituitary gland.

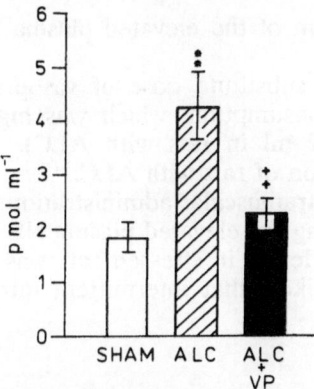

Figure 2. Plasma noradrenaline levels in unstressed rats with anterolateral deafferentation of MBH (ALC) or sham operation after administration of vasopressin (Pitressin Tannat, im 1 IU per rat, seven days). Retransformed geometric means x: RSE (n = 8-18). Statistical significance (Dunn): ** denotes $P < 0.01$ compared to sham group; + + denotes $P < 0.01$ compared to ALC group.

al. (27) seven days before the experiments. The cut is schematically shown in Figure 1. The completeness of the cut was carefully checked histologically (27).

In some experiments double cannulation of the rats was performed. One cannula was inserted into the tail artery for collection of blood and another cannula into the jugular vein for administration of various vasoactive compounds. An infusion pump (UNITA, Braun-Mesungen, FRG) was used at the speed of 50 μl per minute. A 30 minute infusion was started after the control blood collection. Ten minutes later the second blood sample was collected, and the rats were immediately exposed to immobilization stress while the infusion was continuing. Further blood collections were done after 5 and 20 minutes of IMO.

Plasma EPI and NE levels were determined using a radioenzymatic method (28). Plasma ACTH was analyzed by radioimmunoassay without prior extraction, as described earlier (29). Total plasma corticosterone was measured by the competitive protein binding technique (30). Plasma aldosterone was measured using a radioimmunoassay kit (UVVVR, Prague). Plasma renin activity was evaluated by quantifying angiotensin I using a radioimmunoassay kit (UVVVR, Prague).

To study the effect of circulating vasopressin on the activity of the sympathoadrenal system, one group of rats with ALC received daily, 1 IU (intramuscularly) of long-acting vasopressin (Pitressin Tannat, Parke-Davis) per rat (31,32). The administration of vasopressin began immediately after the ALC surgery, and the last dose was given 10 hours prior to blood collection. Olive oil was administered to the control groups.

The effect of vasopressin on plasma NE in unstressed rats with ALC is shown in Figure 2. The ALC-induced elevation of plasma NE was completely abolished in vasopressin-treated rats. Plasma EPI levels in unstressed rats showed no significant change after vasopressin treatment (Figure 3). In stressed animals (immobilization for different intervals), vasopressin administration did not significantly affect the increased plasma NE levels in rats with ALC (Figure 3).

Plasma EPI levels in IMO rats with ALC were not significantly changed compared to those in the sham-operated group (Figure 3). Vasopressin administration did not affect plasma EPI levels in the short IMO intervals, yet after 120 minutes of IMO vasopressin resulted in a significant reduction of the elevated plasma EPI levels in rats with ALC (Figure 3).

The effectiveness of the substitute dose of vasopressin in rats with ALC was checked by measuring water consumption, which was highly elevated (32 ± 1 ml in sham-operated rats *vs.* 152 ± 7 ml in rats with ALC). Vasopressin administration normalized the water consumption of rats with ALC.

These results show that intramuscular administration of vasopressin (1 IU daily for 7 days) was effective in decreasing the elevated plasma NE levels in unstressed rats with ALC; however, the plasma NE levels in stressed rats was not significantly affected by vasopressin substitution. It is likely that intermittent intramuscular administration of

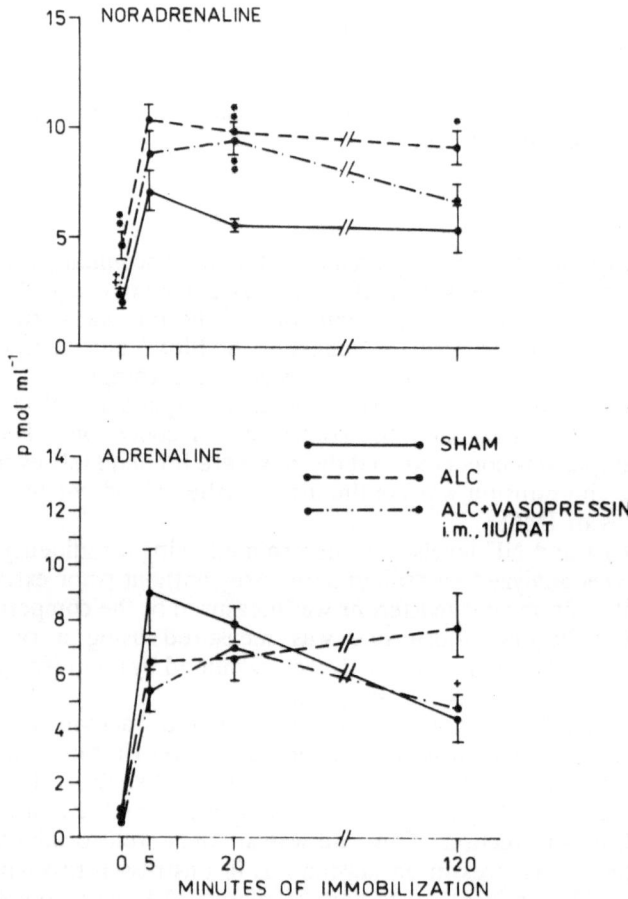

Figure 3.   Plasma catecholamine levels in rats with anterolateral deafferentation of MBH (ALC) and sham operation during immobilization stress. One ALC group was administered im 1 IU of vasopressin per rat for seven days. Levels at all IMO intervals were significantly elevated ($P < 0.01$) compared to the zero interval of the same group. Statistical significance: * denotes $P < 0.05$; ** denotes $P < 0.01$ compared to sham-operated group at the given interval; + denotes $P < 0.05$; ++ denotes $P < 0.01$ compared to ALC group at the given interval.

vasopressin was not the optimal way for its substitution. The last injection of Pitressin was realized 10 hours before blood collection, and it is possible that the vasopressin levels were sufficient for the reduction of potentiated plasma NE levels in animals with ALC under stress.

In the following experiment we administered lysine-vasopressin (Sandoz) continuously using osmotic minipumps (ALZET) after ALC, using 0.01 IU per hour for seven days. The results of this experiment are shown in Figure 4. The elevated plasma NE levels in rats with ALC were completely restored by vasopressin administration in unstressed rats, but in individual IMO intervals the potentiation of plasma NE levels failed to be significantly reduced. On evaluating all the intervals together (ANOVA + Dunn test), the overall course of elevated plasma NE levels in ALC rats was found to be significantly reduced by vasopressin administration ($P < 0.05$) (Figure 4). Plasma EPI levels in rats with ALC were not markedly influenced by continuous administration of vasopressin (Figure 4).

Interactions between circulating vasopressin and sympathetic nerve activity have repeatedly been reported. Landsbert et al. (33) observed increased activity of the heart sympathetic system after hypophysectomy, and Lamprecht and Wooten (34) found after hypophysectomy an increased release of dopamine-β-hydroxylase into plasma, which is another index of increased sympathetic activity. This increase could be prevented by vasopressin administration (34). Similarly, neurophysiological studies showed that an elevation of plasma vasopressin levels resulted in decreased sympathetic nerve activity (14,35) and that vasopressin could inhibit autonomic transmission at the sympathetic ganglion (15). McNeill (20) showed that the vasopressin system acts in an integrated, sequential way with the sympathetic nervous system to maintain blood pressure.

Under stress conditions, however, there are practically no studies on the mechanism of vasopressin-induced inhibition of the sympathetic system. It was shown that during the early phase of IMO, plasma level of vasopressin was increased (36), contributing to both the endocrine and the cardiovascular responses to IMO, which is a severely stressful procedure. Transection of the vasopressin-containing fibers by ALC resulted in a virtually complete disappearance of vasopressin from the posterior lobe of the pituitary during the week after surgery, and prevented the increase of vasopressin secretion during IMO (37).

The two major effects of circulating vasopressin are the regulation of free-water excretion and the modulation of baroreflex function, which regulates sympathetic activity and thus peripheral vascular resistance, the latter in concert with a direct effect of vasopressin and other humoral substances such as angiotensin II. Since vasopressin and dDAVP (31,32), which is virtually without pressor activity, were almost equally potent in decreasing resting NE levels in rats with ALC, it is likely that the lack of vasopressin action was the decisive component in increasing plasma NE levels after ALC.

Attenuation of the activity of the sympathetic system may be brought about by an action of vasopressin at the level of target organs, i.e., peripheral sympathetic endings,, sympathetic ganglia, the adrenal medulla (38-40), and/or by the central regulatory effect on the peripheral sympathetics induced by the baroreceptor reflex (9,12,41). Whether the effect of vasopressin seen in this study was mediated mainly via CNS regulation of the cardiovascular and sympathoadrenal systems or via direct peripheral action on the SAS is not clear.

Immobilization is one of the most powerful stimuli for the secretion of catecholamines, and the secretion of NE during IMO is reliably potentiated by ALC. It is likely that vasopressin replacement sufficient for baseline conditions is not adequate to substitute the increased vasopressin secretion during the early phase of IMO (36), and it is conceivable that the increased NE responses during IMO are caused by the lack of vasopressin in rats with ALC. However, in stressed rats with ALC, some additional stimulatory mechanisms of plasma NE elevation must also be involved.

On balance, the lack of vasopressin accounted for the increased plasma NE levels in unstressed rats with ALC, whereas several further factors, eliminated by anterolateral deafferentation of the MBH, must participate in the regulation of plasma NE levels during stress.

To confirm vasopressin involvement in the physiological regulation of the sympathetic system in animals at rest and under stress, we used Brattleboro diabetes insipidus (DI) rats, which only minimally synthesize brain vasopressin. Both plasma NE and EPI levels in DI homozygote rats at rest were significantly elevated compared to heterozygotes (Table 1). Under IMO stress, plasma NE and EPI levels were significantly

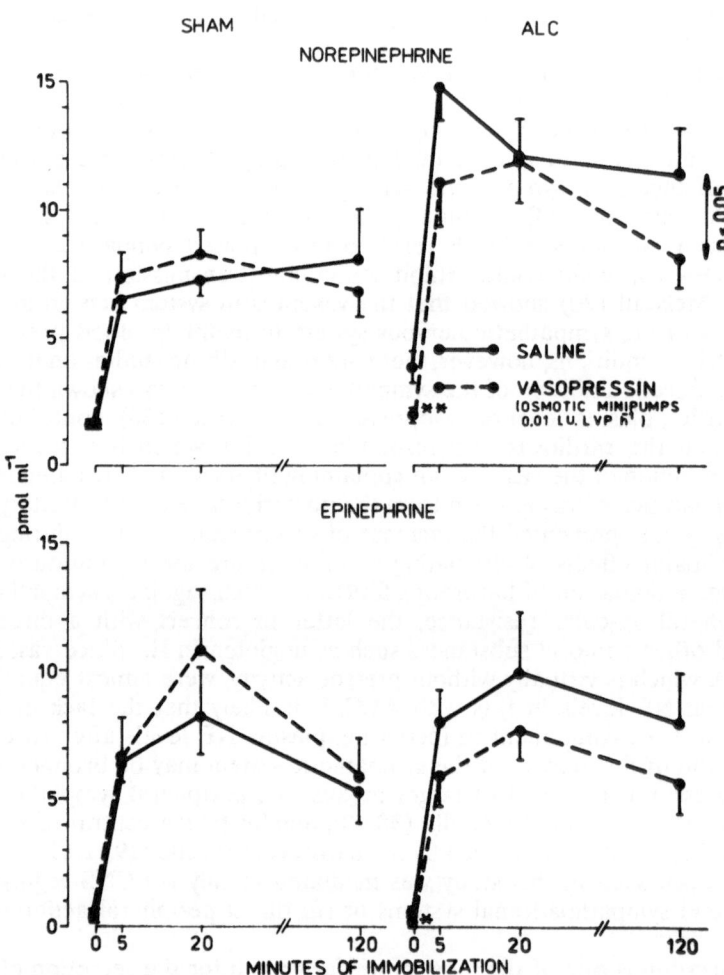

Figure 4. Plasma catecholamine levels in rats with ALC and sham operation during immobilization stress. Some animals were administered vasopressin continuously over seven days by osmotic minipumps (0.01 IU per hr). Values are expressed as geometric means x: RSE (n = 6-9). Statistical significance: * denotes $P < 0.05$; ** denotes $P < 0.01$ compared to ALC group at the given interval. $P < 0.05$ on the right represents statistical evaluation of the pooled intervals (ANOVA + Dunn test).

Table 1

Plasma Catecholamine Levels in Brattleboro DI Homozygote and Heterozygote Rats at Resting Conditions and During Immobilization Stress

Parameter	Group	Control	Immobilization		
			5 Min	20 Min	120 Min
Norepinephrine (pmol ml)	Heterozygote	$1.48 \pm 0.18^*$	5.13 ± 0.27	3.90 ± 0.26	3.48 ± 0.36
	Homozygote DI	2.90 ± 0.66^a	6.73 ± 0.85	6.93 ± 1.10	4.99 ± 1.06
	Homozygote DI + AVP†	1.46 ± 0.21^b	5.03 ± 0.50	4.20 ± 0.39	3.53 ± 0.60
Epinephrine (pmol ml)	Heterozygote	0.33 ± 0.06	5.90 ± 0.58	7.60 ± 1.42	3.21 ± 0.53
	Homozygote DI	1.05 ± 0.27^a	8.12 ± 0.70	10.52 ± 2.48	4.93 ± 0.07
	Homozygote DI	0.41 ± 0.09^b	6.30 ± 0.74	6.60 ± 0.75	4.25 ± 1.67

*Values are mean ± SE; (n = 8-11)

†Vasopressin (AVP, Pitressin 1 IU per day per rat for 7 days, im)

Statistical significance (t test): ^{a} denotes $P < 0.05$ vs. heterozygotes; ^{b} denotes $P < 0.05$ vs. homozygotes.

elevated in homozygotes *vs.* heterozygotes (ANOVA + Dunn test) but the difference in each individual IMO interval was not significant. Repeated vasopressin administration (Pitressin, intramuscular, 1 IU per rat for 7 days) to Brattleboro DI rats resulted in resting and stress-induced plasma NE levels that were comparable to those recorded in heterozygotes (Table 1).

There have been relatively few studies concerned with the sympathetic nervous system in Brattleboro DI rats. Peripheral noradrenergic activity in the kidney was increased in DI rats, and vasopressin replacement produced a sympatho-inhibition in all peripheral tissues examined (42). Both NE and EPI concentrations in the adrenal glands were significantly higher in DI rats (42). Thus, DI rats have the potential to release more catecholamines from the adrenals. In DI rats an enhanced activity of the sympathetic system (42) was also manifested by elevated levels of plasma NE and EPI (43,44), or elevated serum dopamine-β-hydroxylase activity (45), which can be corrected by replacement therapy with vasopressin (42,44,46).

These results suggest that both adrenal glands and sympathetic nerves are more active in the DI rats, probably as a compensatory response to the lack of vasopressin. Actually, baroreflex function was found to be suppressed in Brattleboro DI rats (46).

Our results show that both ALC and the inability of Brattleboro homozygote rats to synthesize normally vasopressin enhance the activity of the sympathetic nervous system at rest and under stress. We have thus contributed to the understanding of the physiological role of vasopressin in the regulation of plasma catecholamine, particularly NE levels not only in resting conditions but also under stress.

In further experiments, we reasoned that if circulating vasopressin is the important factor responsible for the increase in plasma NE levels in Wistar rats with ALC, the ALC

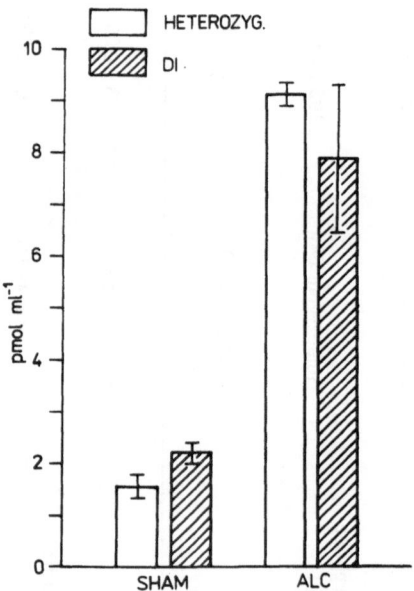

Figure 5. Plasma noradrenaline levels in unstressed Brattleboro heterozygote and diabetes insipidus homozygote (DI) rats with anterolateral deafferentation of MBH (ALC) or sham operation. Plasma NE levels were highly significantly elevated in both ALC group *vs.* sham groups.

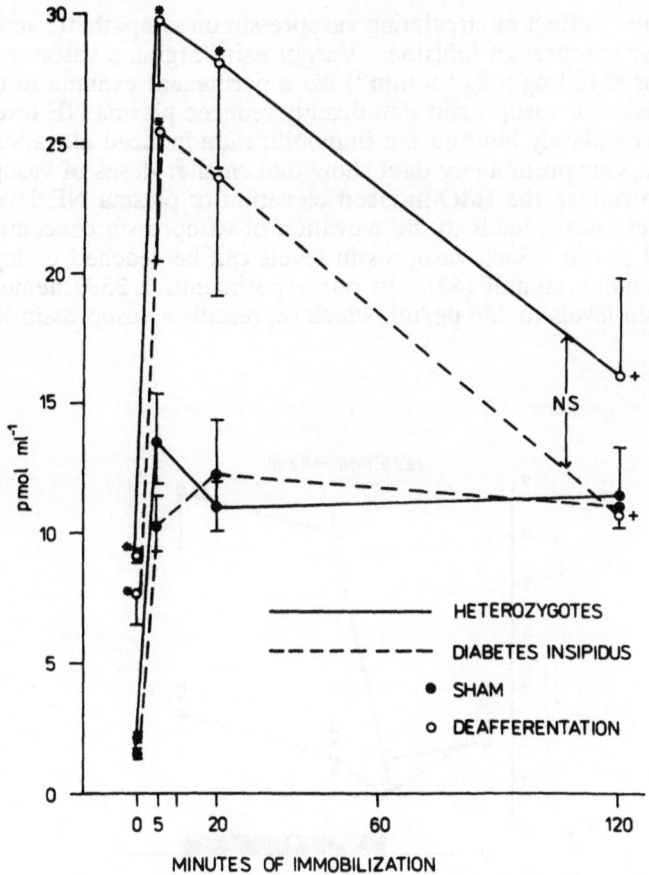

Figure 6. Plasma noradrenaline levels in immobilized Brattleboro heterozygote and diabetes insipidus homozygote rats with ALC or sham operation. Geometric means x: RSE (n = 6-10). Statistical significance (ANOVA + Dunn test): * denotes $P < 0.01$ compared to appropriate sham-operated group; + denotes $P < 0.01$ statistical evaluation of the pooled intervals *vs.* sham group.

in Brattleboro DI rats, which fail to synthesize vasopressin properly, will not result in increased plasma NE levels in these animals. Our results, however, showed that in both heterozygotes and homozygotes ALC significantly elevated plasma NE levels in unstressed (Figure 5) as well as stressed rats (Figure 6).

This finding suggests that in addition to vasopressin another factor(s), eliminated by ALC, participates in the regulation of plasma NE levels. In Brattleboro rats, vasopressin function appears to be compensated by other substances, presumably by peptides with a structure similar to that of vasopressin, possibly oxytocin (47). Diabetes insipidus in Brattleboro homozygote rats represents an osmotic stressor that activates oxytocin-synthesizing neurons (48) resulting in decreased oxytocin levels in the neural lobe of the pituitary and in enhanced plasma oxytocin levels (49). In conclusion, the elevation of the sympathetic activity after ALC observed in Brattleboro diabetes insipidus rats indicates the involvement also of some hypothalamic factor(s) other than vasopressin.

The inhibitory effect of circulating vasopressin on sympathetic activity was studied in intact rats after vasopressin infusion. Vasopressin (arginine vasopressin, FERRING, Malmö) was infused ($80 \text{ ng} \times \text{kg}^{-1} \times \text{min}^{-1}$) *via* a permanent cannula in the jugular vein. A 10 minute infusion of vasopressin significantly reduced plasma NE levels in unstressed rats and almost completely blocked the immobilization-induced elevation of plasma NE levels (Figure 7). Our preliminary data show that smaller doses of vasopressin ($20 \text{ ng} \times \text{kg}^{-1} \times \text{min}^{-1}$) also reduce the IMO-induced elevation of plasma NE levels. Infusion of vasopressin in these doses leads to the elevation of vasopressin concentration in plasma to about 150-200 pg/ml. Such vasopressin levels can be reached endogenously in rats exposed to an osmotic stimuli (43). In our experiments, a 35% hemorrhage elevated plasma vasopressin levels to 250 pg/ml, which represents a vasopressin level higher than

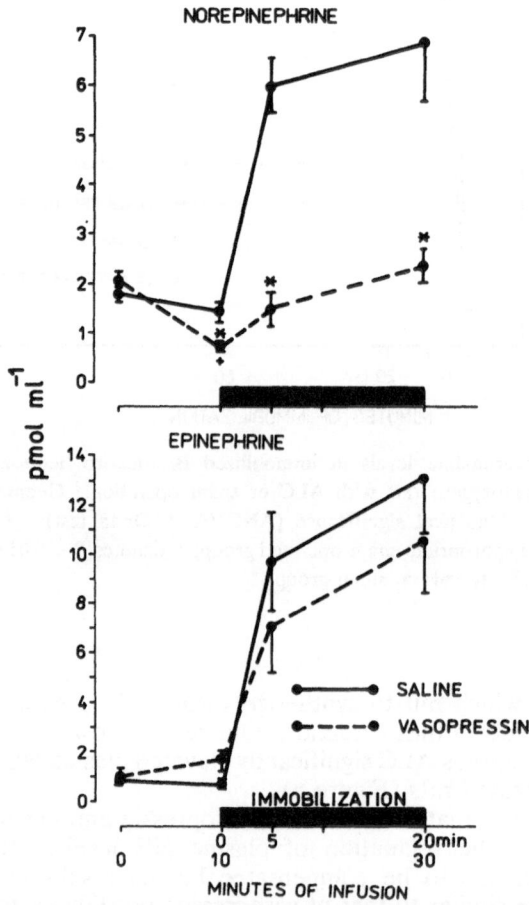

Figure 7. Effect of vasopressin infusion (arginine vasopressin $80 \text{ ng} \times \text{kg}^{-1} \times \text{min}^{-1}$) on plasma catecholamine levels in intact rats at rest conditions and during immobilization stress. Infusion was continuing over the period of 30 min including 20 min IMO. Arithmetic mean ± SE (n = 8-10); statistical significance (Student's *t* test): * denotes $P < 0.01$ *vs.* saline group; + denotes $P < 0.01$ *vs.* control interval.

Table 2

Effect of dDAVP Infusion (80 ng × kg^{-1} × min^{-1}) on Plasma Catecholamine
Levels in Rats under Resting Conditions and During Immobilization Stress

Catecholamine	Group	Resting level	Infusion 10 min	Infusion plus immobilization* 5 min	Infusion plus immobilization* 20 min
Norepinephrine (pmol ml)	Saline	1.66 ± 0.51†	1.24 ± 0.19	5.40 ± 0.63	5.56 ± 0.75
	dDAVP	1.80 ± 0.25	1.18 ± 0.10	5.33 ± 0.54	5.07 ± 0.41
Epinephrine (pmol ml)	Saline	0.49 ± 0.27	0.28 ± 0.10	7.67 ± 0.48	5.74 ± 0.56
	dDAVP	0.59 ± 0.16	0.26 ± 0.10	8.78 ± 1.16	8.64 ± 1.30

*Infusion continued over a 30-min period including 20 min immobilization.
†Values are mean ± SE (n = 5-9).

those found after infusion of 20-40 ng vasopressin $\times$ kg^{-1} $\times$ min^{-1}. Moreover, after 35% hemorrhage the plasma NE levels in rats were not changed in spite of the pronounced rise of plasma levels of EPI, ACTH, etc. (50). These data suggest that endogenous plasma vasopressin could play a significant role in the physiological regulation of the sympathetic activity, at least under certain conditions.

Plasma EPI levels were not significantly influenced by vasopressin infusion (Figure 7). This finding provides further evidence for dissociation of the regulation of the sympathetic and adrenomedullary systems (24).

Vasopressin infusion increased the MAP in a dose-dependent way. Figure 8 shows that AVP infused at 80 ng $\times$ kg^{-1} $\times$ min^{-1} greatly elevated blood pressure.

Vasopressin is known to express both a pressor effect *via* V$_1$-receptors and an antidiuretic effect *via* V$_2$-receptors. Which of these vasopressin effects is involved in the inhibitory mechanism of vasopressin on the sympathetic system? It has been proposed that circulating vasopressin may perform its inhibitory action on the sympathetic system *via* enhancement of the inhibitory influence of baroreflexes (14,16,51).

Our observation based on the effect of vasopressin are in agreement with the data of King *et al.* (9) and Unger *et al.*, showing the attenuating effect of circulating vasopressin on sympathetic activity. A single intravenous injection of vasopressin into normal rats was found to diminish plasma catecholamine levels (9). Numerous investigators have demonstrated that intravenous infusion of vasopressin produces bradycardia and inhibition of sympathetic nerve activity (11,14,16,53). Guo *et al.* (53) have shown that the facilitation of bradycardia and sympatho-inhibition by vasopressin is attenuated when the barorecep-tors are removed. It is clear from this evidence that vasopressin facilitates baroreflex function and is involved in the normal baroreflex regulation of heart rate and peripheral sympathetic activity.

To address the question of the pressor mechanism of vasopressin on sympathetic activity, we measured plasma CA levels after infusion of dDAVP (1-Deamino-8-D-arginine-Vasopressin; Adiuretin-SD, Léčiva, Prague), a vasopressin analogue without pressor activity. dDAVP failed to affect either resting or stress-induced plasma levels of NE and EPI in normal rats (Table 2). These results demonstrate that the inhibitory effect of vasopressin on sympathetic activity is mediated by its pressor activity *via* V$_1$-receptors. This fact allows us to suggest that facilitation of baroreflexes is the mechanism responsible for inhibition of the sympathetic system by vasopressin.

To evaluate the specificity of the inhibitory effect of vasopressin on sympathetic activity, we investigated another vasoactive factor, angiotensin II. Infusion of angiotensin II (Peninsula Laboratories, Belmont, CA) in the same concentration as vasopressin (80 ng $\times$ kg^{-1} $\times$ min^{-1}) produced in rats a pronounced elevation of blood pressure (Table 3), which was similar to that produced by vasopressin infusion (Figure 8). Angiotensin II infusion, however, failed to affect either resting or stress-induced levels of plasma NE and EPI (Table 3). These results show that vasopressin has a specific facilitatory effect on baroreflexes which produce inhibition of plasma NE levels, especially during stress.

Angiotensin II is known to influence baroreflexes, yet its central and peripheral effects are different from those of vasopressin (41,54). Recent studies have demonstrated facilitation by vasopressin and attenuation by angiotensin II of baroreflex-mediated bradycardia and inhibition of the sympathetic system (14,16,41,51). Sander-Jensen *et al.* (55) reported that in man angiotensin II infusion in physiologic doses may augment the sympathetic activity since compensatory decreases in heart rate or in plasma catechola-mines were not observed during increased arterial blood pressure. Furthermore, plasma angiotensin II is responsible for direct activation of the sympathoadrenal system reflected in elevated levels of plasma catecholamines (56,57). This phenomenon is activated under stress (57). These data are in agreement with our results showing no changes in plasma catecholamine levels after angiotensin II infusion.

Table 3

Effect of Angiotensin II Infusion (80 ng × kg^{-1} × min^{-1}) on Plasma Catecholamine Levels and Mean
Arterial Blood Pressure in Rats Under Resting Conditions and During Immobilization Stress

Parameter	Group	Resting level	Infusion 10 min	Immobilization 5 min	20 min
Norepinephrine (pmol ml)	Saline	1.75 ± 0.25*	1.57 ± 0.18	4.64 ± 0.67	4.30 ± 1.01
	Angiotensin II	1.34 ± 0.25	1.53 ± 0.46	4.17 ± 0.57	3.72 ± 0.37
Epinephrine (pmol ml)	Saline	0.56 ± 0.13	0.50 ± 0.11	5.45 ± 1.04	4.61 ± 0.73
	Angiotensin II	0.42 ± 0.03	0.33 ± 0.06	4.51 ± 1.01	4.55 ± 0.90
Mean blood Pressure (mm Hg)	Saline	103.4 ± 4.4	101.5 ± 4.1	103.8 ± 7.0	111.0 ± 6.1
	Angiotensin II	103.6 ± 3.1	145.7 ± 3.4^a	131.3 ± 2.8^a	130.8 ± 4.7^a

*Values are mean ± SE (n = 5-9).
Statistical significance (t test); ^{a} denotes P < 0.01 $vs.$ resting level.

The renin-angiotensin-aldosterone system is, in addition to the sympathoadrenal system and vasopressin, an important system involved in regulation of blood pressure. The activity of plasma renin, which cleaves angiotensin I from its plasma substrate angiotensinogen, is a good indicator of the activity of this system. Plasma renin activity (PRA) is elevated by different stressors (58-61) and shows a severalfold rise during IMO (62). The aim of our work was to investigate the effect of vasopressin on PRA and plasma aldosterone levels in rats exposed to immobilization stress. Vasopressin infusion in doses of 8 to 80 ng $\times$ kg^{-1} $\times$ min^{-1} significantly reduced PRA at rest conditions and blocked the pronounced PRA increase during immobilization stress (Figure 9). Plasma aldosterone levels, which were supposed to exhibit the same changes as PRA since angiotensin II is an effective regulator of aldosterone release, not only failed to do so, but were even significantly elevated after infusion of the highest dose of vasopressin (Figure 9). It is possible that plasma ACTH, which is an important regulator of aldosterone release in the rat, was highly elevated already during the 10-minute vasopressin infusion (Figure 10). This figure also shows plasma corticosterone levels displaying changes comparable to those of plasma ACTH levels. Infusion of dDAVP failed to affect resting or stress-induced levels of PRA and aldosterone (Figure 11).

The changes in PRA during vasopressin or dDAVP infusion are similar to those of plasma NE levels during infusion of these compounds, indicating that activation of the renin-angiotensin system under these conditions is mainly regulated by the sympathetic system. It has been shown that vasopressin inhibits renin secretion in a dose-related manner in a variety of species including rats (63). However, the mechanism of this suppression is not well understood (64). Potential mechanisms of renin inhibition by vasopressin include volume expansion, vasoconstriction, reflex suppression of sympathetic neural activity, altered renal handling of sodium, and a direct action on the jux-

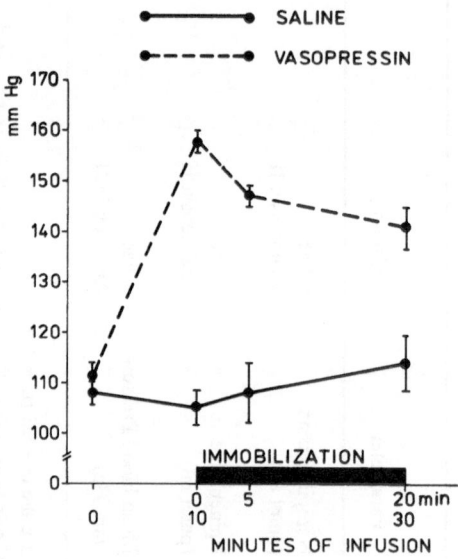

Figure 8. Effect of vasopressin infusion (arginine vasopressin 80 ng $\times$ kg^{-1} $\times$ min^{-1}) on mean arterial blood pressure in intact rats at rest conditions and during immobilization stress. Arithmetic means $\pm$ SE (n = 7-9).

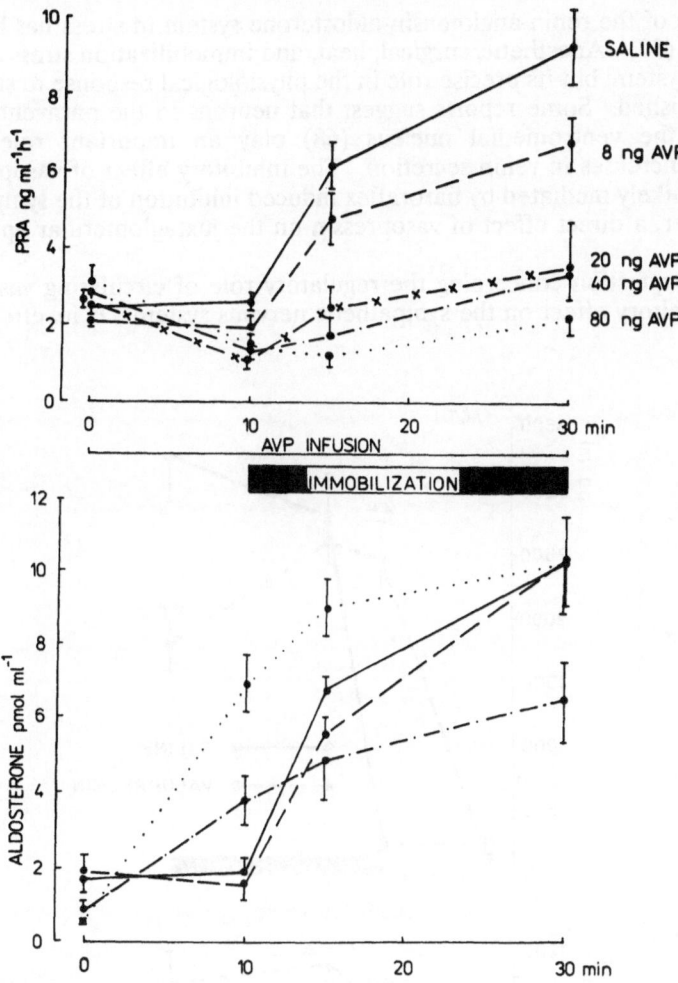

Figure 9. Effect of vasopressin infusion (arginine vasopressin 80 ng $\times$ kg^{-1} $\times$ min^{-1}) on plasma renin activity (PRA) and plasma aldosterone levels in rats at rest conditions and during immobilization stress. Infusion was continuing over the period of 30 min including 20 min IMO. Significant reduction in PRA in all infusion intervals ($P < 0.01$) started at the dose of 20 ng of vasopressin.

taglomerular cells. Vasopressin markedly reduced basal (16) and stimulated (14) renal sympathetic nerve activity, suggesting that the inhibition of renin secretion by vasopressin is a reflex response to vasoconstriction mediated by the renal nerves. This was corroborated by the finding that vasopressin induced decrease of PRA was blocked by sinoaortic cardiac denervation, showing that the reflex response resulted from activation of the high pressure baroreceptors (64). The signal that initiates this reflex suppression of renin release has not been identified. The elevated arterial pressure is not solely responsible for this inhibition since subpressor doses of vasopressin also suppresses renin release (65).

The role of the renin-angiotensin-aldosterone system in stress has been studied in limited fashion (57). Anesthetic, surgical, heat, and immobilization stress all increase the activity of this system, but its precise role in the physiological response to stressors has not yet been established. Some reports suggest that neurons in the paraventricular nucleus (66,67) or in the ventromedial nucleus (68) play an important role in mediating stress-induced increases in renin secretion. The inhibitory effect of vasopressin on PRA during stress is likely mediated by baroreflex induced inhibition of the sympathetic system (57,64); however, a direct effect of vasopressin on the juxtaglomerular apparatus cannot be excluded.

Another question concerning the regulatory role of circulating vasopressin is the level of its inhibitory effect on the sympathetic nervous system. Is its effect mediated by

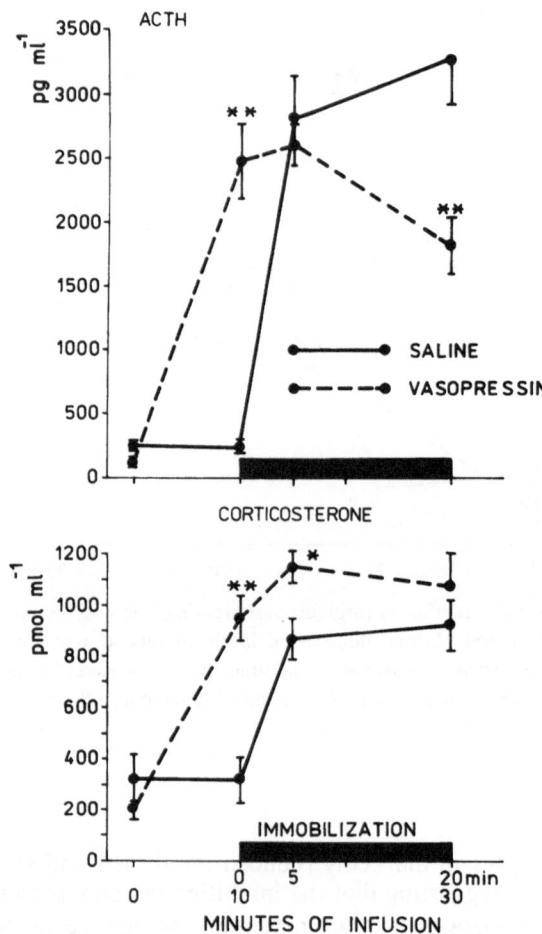

Figure 10. Effect of vasopressin infusion (80 ng × kg^{-1} × min^{-1}) on plasma ACTH and corticosterone levels in rats at rest conditions and during immobilization stress. Arithmetic means ± SE (n = 8-10). Statistical significance (Student's t test): * denotes $P < 0.05$; ** denotes $P < 0.01$ *vs.* appropriate saline treated group.

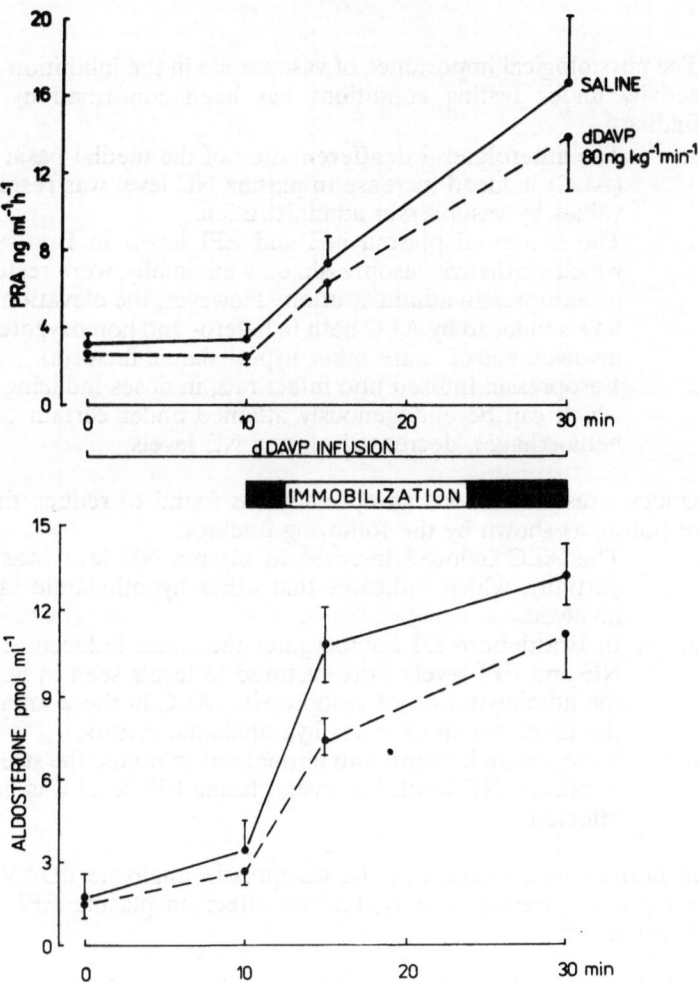

Figure 11. Effect of dDAVP infusion (Adiuretin-SD, 80 ng $\times$ kg^{-1} $\times$ min^{-1}) on plasma renin activity (PRA) and plasma aldosterone levels in rats at rest conditions and during immobilization stress.

peripheral or central mechanisms? Recently it has been reported that circulating vasopressin can pass to the brain at the level of the area postrema and the locus coeruleus to enhance the inhibitory influence of the arterial baroreflex on the sympathetic system (11,51,69). This effect of vasopressin at the level of area postrema appears to be dependent on a catecholaminergic mechanism (51). These data taken together demonstrate that vasopressin given either intracerebroventricularly, subcutaneously, or intravenously in Wistar or Brattleboro diabetes insipidus rats does alter noradrenergic function in various central sites known to be involved in the control of arterial pressure.

Conclusions

1. The physiological importance of vasopressin in the inhibition of sympathetic activity under resting conditions has been confirmed by the following findings:
    a. The anterolateral deafferentation of the medial basal hypothalamus (ALC) induced increase in plasma NE level was restored to normal values by vasopressin administration.
    b. The increased plasma NE and EPI levels in Brattleboro DI rats, which synthetize vasopressin only minimally, were restored to normal by vasopressin administration. However, the elevation of plasma NE levels induced by ALC both in hetero- and homozygotes indicates the involvement of some other hypothalamic factor(s).
    c. Vasopressin infused into intact rats, in doses inducing plasma levels, which can be endogenously attained under certain conditions (*e.g.*, hemorrhage), decreased plasma NE levels.

2. Under stress conditions vasopressin was found to reduce the sympathetic response, as shown by the following findings:
    a. The ALC induced increase in plasma NE level was reduced only partially, which indicates that other hypothalamic factors are also involved.
    b. In Brattleboro DI homozygotes the stress induced elevated plasma NE and EPI levels were restored to levels seen in heterozygotes by the administration of vasopressin. ALC in these animals confirmed the involvement of other hypothalamic factors.
    c. Vasopressin infusion into normal rats inhibited the stress induced rise in plasma NE level; however, plasma EPI level was not significantly affected.

3. In normal rats, infusion of the vasopressin analogue dDAVP, which does not possess pressor activity, had no effect on plasma EPI and NE levels under stress.

4. Infusion of angiotensin II, despite its effect on blood pressure, failed to reduce plasma NE and EPI levels both at rest and under stress.

5. The activity of the renin-angiotensin system exerted changes parallel to those of the sympathetic system under all conditions studied.

In summary, the mechanism involved in the inhibitory action of vasopressin on the sympathetic and renin-angiotensin system in rats at rest and under stress most likely operates by facilitating the baroreflex.

References

1. Tucker, D.C., and C.B. Saper, Specificity of spinal projections from hypothalamic and brainstem areas which innervate sympathetic preganglionic neurons, *Brain Res* 360: 159-164, 1985.
2. Sourkes, T.L., Pathways of stress in the CNS, *Prog Neuro-Psychopharmacol Biol Psychiat* 7: 389-411, 1983.

3. Matsui, H., Adrenal medullary secretory response to stimulation of the medulla oblongata in the rat, *Neuroendocrinology* 29: 385-390, 1979.

4. Matsui, H., Adrenal medullary secretory response to pontine and mesencephalic stimulation in the rat, *Neuroendocrinology* 33: 84-87, 1981.

5. Sun, C.L., and I.J. Kopin, Plasma catecholamines and direct stimulation of rat hypothalamus, In E. Usdin, I.J. Kopin, and J. Barchas (eds) *Catecholamines: Basic and Clinical Frontiers*, Pergamon Press, New York, pp. 1422-1424, 1979.

6. Dobrakovová, M., Z. Opršalová, and R. Kvetňanský, Plasma catecholamines in rats electrostimulated in different brain areas, In E. Usdin, R. Kvetňanský, and J. Axelrod (eds) *Stress: The Role of Catecholamines and Other Neurotransmitters*, Gordon and Breach Science Publishers, New York, pp. 649-659, 1984.

7. Van Loon, G.R., Brain opioid peptide regulation of catecholamine secretion, In E. Usdin, R. Kvetňanský, and J. Axelrod (eds) *Stress: The Role of Catecholamines and Other Neurotransmitters*, Gordon and Breach Science Publishers, New York, pp. 617-635, 1984.

8. Brown, M.R., Neuropeptide regulation of the autonomic nervous system, In Y. Taché, J.E. Morley, and M.R. Brown (eds) *Neuropeptides and Stress*, Springer-Verlag, New York, pp. 107-120, 1989.

9. King, K.A., G. Mackie, C.C.Y. Pang, and R.A. Wall, Central vasopressin in the modulation of catecholamine release in conscious rats, *Can J Physiol Pharmacol* 63: 1501-1505, 1985.

10. Berecek, K.H., Role of central vasopressin in cardiovascular regulation, *J Cardiovasc Pharmacol* 8 [Suppl 7]: S76-S80, 1986.

11. Patel, K.P., and P.G. Schmid, The role of central noradrenergic pathways in the actions of vasopressin on baroreflex control of circulation, In J.B. Buckley and C.M. Ferrario (eds) *Brain Peptides and Catecholamines in Cardiovascular Regulation*, Raven Press, New York, pp. 53-64, 1987.

12. Martin, S.M., T.J. Malkinson, L.G. Bauce, W.L. Veale, and Q.J. Pittman, Plasma catecholamines in conscious rabbits after central administration of vasopressin, *Brain Res* 457: 192-195, 1988.

13. Gilbey, M.P., J.H. Coote, S. Fleetwood-Walker, and D.H. Peterson, The influence of paraventricular - spinal pathway, and oxytocin and vasopressin on sympathetic preganglionic neurons, *Brain Res* 251: 283-290, 1982.

14. Imaizumi, T., and M.D. Thames, Influence of intravenous and intracerebroventricular vasopressin on baroreflex control of renal nerve traffic, *Circ Res* 55: 17-25, 1986.

15. Kiraly, M., M. Maillard, J.J. Dreifuss, and M. Dolivo, Neurohypophysial peptides depress cholinergic transmission in a mammalian sympathetic ganglion, *Neurosci Lett* 62: 89-95, 1985.

16. Undesser, K.P., E.M. Hasser, J.R. Haywood, A.K. Johnson, and V.S. Bishop, Interactions of vasopressin with the area postrema in arterial baroreflex function in conscious rabbits, *Circ Res* 56: 410-417, 1985.

17. Pittman, Q.J., Brain vasopressin and cardiovascular regulation in normotensive and hypertensive animals, In Y. Taché, J.E. Morley, and M.R. Brown (eds) *Neuropeptides and Stress*, Springer-Verlag, New York, pp. 134-145, 1989.

18. Cowley, A.W., E. Monos, and A.C. Guyton, Interaction of vasopressin and baroreceptor reflex system in the regulation of arterial blood pressure in the dog, *Circ Res* 34: 505-514, 1974.

19. Liard, J.F., O. Deriaz, M. Tschopp, and J. Schoun, Cardiovascular effects of vasopressin infused into the vertebral circulation of dogs, *Clin Sci* 61: 345-347, 1981.

20. McNeill, J.R., Role of vasopressin in the control of arterial pressure, *Can J Physiol Pharmacol* 61: 1226-1235, 1983.

21. Bennett, T., and S.M. Gardiner, Involvement of vasopressin in cardiovascular regulation, *Cardiovasc Res* 19: 57-68, 1985.

22. Rascher, W., R.E. Land, and T.H. Unger, Vasopressin, cardiovascular regulation, an hypertension, In D. Ganten and D. Pfaff (eds) *Current Topics in Neuroendocrinology: Neurobiology of Vasopressin*, Springer-Verlag, Berlin, pp. 101-136, 1985.

23. Kvetňanský, R., G.B. Makara, Z. Opršalová, M. Dobrakovová, and D. Ježová, Increased basal and stress-induced sympathetic activity in rats with lesion or deafferentation of the medial basal hypothalamus, *Biogenic Amines* 5: 275-290, 1988.

24. Kvetňanský, R., M. Dobrakovová, D. Ježová, Z. Opršalová, B. Lichardus, and G.B. Makara,

Hypothalamic regulation of plasma catecholamine levels in stress: effect of vasopressin and CRF, In G.R. Van Loon, R. Kvetňanský, R. McCarty, and J. Axelrod (eds) *Stress: Neurochemical and Humoral Mechanisms*, Gordon and Breach Science Publishers, New York, pp. 549-570, 1989.

25. Chiueh, C.C., and I.J. Kopin, Hyperresponsivity of spontaneously hypertensive rat to indirect measurement of blood pressure, *Am J Physiol* 234: H690-H695, 1978.

26. Kvetňanský, R., and L. Mikulaj, Adrenal and urinary catecholamines in rats during adaptation to repeated immobilization stress, *Endocrinology* 87: 738-743, 1970.

27. Makara, G.B., E. Stark, and M. Palkovits, Reevaluation of the pituitary-adrenal response to ether in rats with various cuts around the medial basal hypothalamus, *Neuroendocrinology* 30: 38-44, 1980.

28. Peuler, J.D., and G.A. Johnson, Simultaneous single isotope radioenzymatic assay of plasma norepinephrine, epinephrine and dopamine, *Life Sci* 21: 625-636, 1977.

29. Ježová, D., R. Kvetňanský, K. Kovács, Z. Opršalová, M. Vigaš, and G.B. Makara, Insulin-induced hypoglycemia activates the release of adrenocorticotropin predominantly via central and propranolol insensitive mechanisms, *Endocrinology* 120: 409-415, 1987.

30. Murphy, B.C.P., Some studies on the protein-binding of steroids and their application to routine micro and ultramicro measurement of various steroids in body fluids by competitive protein-binding radioassay, *J Clin Endocrinol Metab* 27: 973-990, 1967.

31. Kvetňanský, R., B. Lichardus, D. Ježová, Z. Opršalová, and G.B. Makara, Vasopressin and 1-deamino-8-D-arginine-vasopressin (DDAVP) reduce elevated plasma catecholamine levels in rats with hypothalamic deafferentation, *Cell Mole Neurobiol* 8: 225-233, 1988.

32. Lichardus, B., R. Kvetňanský, G.B. Makara, Z. Opršalová, N. Michajlovskij, O. Földes, and D. Ježová, Circulating vasopressin attenuates the increased activity of the sympathetic nervous system induced by anterolateral deafferentation of the hypothalamus, In K.D. Döhler and M. Pawlikowski (eds) *Progress in Neuropeptide Research*, Birkhäuser Verlag, Basel, pp. 91-97, 1989.

33. Landsberg, L., J. de Champlain, and J. Axelrod, Increased biosynthesis of cardiac norepinephrine after hypophysectomy, *J Pharmacol Exp Ther* 165: 102-107, 1969.

34. Lamprecht, F., and G.F. Wooten, Effect of hypophysectomy on serum dopamine-β-hydroxylase activity in rat, *Endocrinology* 92: 1543-1546, 1973.

35. Sharabi, F.M., G.B. Guo, F.M. Abboud, M.D. Thames, and P.G. Schmid, Contrasting effects of vasopressin on baroreflex inhibition of lumbar sympathetic nerve activity, *Am J Physiol* 249: H922-H928, 1985.

36. Michajlovskij, N., B. Lichardus, R. Kvetňanský, and J. Ponec, Vasopressin in plasma, median eminence and in anterior pituitary of rats under immobilization stress, In E. Usdin, R. Kvetňanský, and J. Axelrod (eds) *Stress: The Role of Catecholamines and Other Neurotransmitters*, Gordon and Breach Science Publishers, New York, pp. 365-372, 1984.

37. Michajlovskij, N., B. Lichardus, R. Kvetňanský, and J. Ponec, Effect of acute and repeated immobilization stress on food and water intake, urine output and vasopressin changes in rats, *Endocrinol Exp* 22: 143-157, 1988.

38. Hanley, M.R., H.P. Benton, S.L. Lightman, K. Todd, E.A. Bone, P. Fretten, S. Palmer, C.J. Kirk, and R.H. Michell, A vasopressin-like peptide in the mammalian sympathetic nervous system, *Nature* 309: 258, 1984.

39. Antoni, F.A., Characterization of high affinity binding sites for vasopressin in bovine adrenal medulla, *Neuropeptides* 4: 413, 1984.

40. Clements, J.A., and J.W. Funder, Arginine vasopressin and AVP-like immunoreactivity in peripheral tissues, *Endocrine Rev* 7: 449-460, 1986.

41. Patel, K.P., C.A. Shiteis, D.D. Lund, and P.G. Schmid, Effects of intravenous infusions of vasopressin and angiotensin II on central and peripheral noradrenergic function in conscious rabbit, *Can J Physiol Pharmacol* 65: 765-772, 1987.

42. Kline, R.L. K.P. Patel, and P.F. Mercer, Enhance noradrenergic activity in kidney of Brattleboro rats with diabetes insipidus, *Am J Physiol* 250: R567-R572, 1986.

43. Zerbe, R.L., G. Veuerstein, D.K. Meyer, and I.J. Kopin, Cardiovascular sympathetic, and renin-angiotensin system responses to hemorrhage in vasopressin-deficient rats, *Endocrinology* 111: 608-613, 1982.

44. Williams, J.L., and M.D. Johnson, Sympathetic nervous system and blood pressure maintenance in the Brattleboro DI rat, *Am J Physiol* 250: R770-R775, 1986.

45. Wooten, G., T. Hanson, and F. Lamprecht, Elevated serum dopamine-β-hydroxylase activity in rats with inherited diabetes insipidus, *J Neural Transm* 36: 107-112, 1975.

46. Imai, Y., P. Nolan, and C. Johnston, Restoration of suppressed baroreflex sensitivity in rats with hereditary diabetes insipidus (Brattleboro rats) by arginine-vasopressin and dDAVP, *Circ Res* 53: 140-149, 1983.

47. Baertschi, A.J., and J.L. Bény, Central control of ACTH secretion in diabetes insipidus Brattleboro rat, *Ann NY Acad Sci* 394: 591-606, 1982.

48. Morris, J.F., The Brattleboro magnocellular neurosecretory system: a model for the study of peptidergic neurons, *Ann NY Acad Sci* 394: 54-71, 1982.

49. Balment, R.J., M.J. Brimble, and M.L. Forsling, Oxytocin release and renal actions in normal and Brattleboro rats, *Ann NY Acad Sci* 394: 241-253, 1982.

50. Grässler, J., D. Ježová, R. Kvetňanský, and D.W. Scheuch, Hormonal responses to hemorrhage and their relationship to individual shock susceptibility, *Endocrinol Exper* In Press.

51. Bishop, V.S., E.M. Hasser, and K.P. Undesser, Vasopressin and sympathetic nerve activity: involvement of the area postrema, In J.P. Buckley and C.M. Ferrario (eds) *Brain Peptides and Catecholamines in Cardiovascular Regulation*, Raven Press, New York, pp. 373-382, 1987.

52. Unger, T., P. Rohmeiss, G. Demmert, G. Detlev, R.E. Land, and F.C. Luft, Differential modulation of the baroreceptor reflex by brain and plasma vasopressin, *Hypertension* 8 [Suppl II]: 157-162, 1986.

53. Guo, G.B., F.M. Sharabi, F.M. Abboud, and P.G. Schmid, Vasopressin augments baroreflex inhibition of lumbar sympathetic nerve activity in rabbits, *Circulation* 66 [Suppl 2]: 34, 1982.

54. Osborn, J.W., M.M. Skelton, and A.W. Cowley, Hemodynamic effects of vasopressin compared with angiotensin II in conscious rats, *Am J Physiol* 252: H628-H637, 1987.

55. Sander-Jensen, K., N.H. Secher, A. Astrup, N.J. Christensen, M. Damkjaer-Nielsen, J. Giese, J. Warbert, and P. Bie, Angiotensin II attenuates reflex decrease in heart rat and sympathetic activity in man, *Clin Physiol* 8: 31-40, 1988.

56. Peach, M.J., Pharmacology of angiotensin II, In *Kidney Hormones, Volume III*, Academic Press, London, pp. 274-304, 1986.

57. Carey, R.M., C.E. Rose, and M.J. Peach, Role of the renin-angiotensin-aldosterone system in stress, In G.R. Van Loon, R. Kvetňanský. R. McCarty, and J. Axelrod (eds) *Stress: Neurochemical and Humoral Mechanism*, Gordon and Breach Science Publishers, New York, pp. 833-844, 1989.

58. Miller, E.D., J.J. Beckman, J.R. Woodside, J.S. Althaus, and M.J. Peach, Blood pressure control during anesthesia: importance of the peripheral sympathetic nervous system and renin, *Anesthesiology* 58: 32-37, 1983.

59. Kosunen, K.J., A.J. Pakarinen, K. Kuoppasalone, and H. Adlercreutz, Plasma renin activity, angiotensin II and aldosterone during intense heat stress, *J Applied Physiol* 41: 323-327, 1976.

60. Anderson, D.E., C. Gomez-Sanchez, and J.R. Dietz, Suppression of plasma renin and aldosterone in stress-salt hypertension in dogs, *Am J Physiol* 251: R181-186, 1986.

61. Sowers, J.R., M. Tuck, N.D. Asp, and E. Sollars, Plasma aldosterone and corticosterone responses to adrenocorticotropin, angiotensin, potassium and stress in spontaneously hypertensive rats, *Endocrinology* 108: 1216-1221, 1981.

62. Jindra, A., R. Kvetňanský, T.I. Belova, and K.V. Sudakov, Effect of acute and repeated immobilization stress on plasma renin activity, catecholamines and corticosteroids in Wistar and August rats, In E. Usdin, R. Kvetňanský, and I.J. Kopin (eds) *Catecholamines and Stress: Recent Advances*, Elsevier North Holland, New York, pp. 249-254, 1980.

63. Henderson, I.W., R.J. Balment, and J.A. Oliver, Vasopressin effects on plasma renin activity in male and female rats, *Clin Sci Mole Med* 55: 301, 1978.

64. Gregory, L.C., E.W. Quillen, L.C. Keil, and I.A. Reid, Effect of baroreceptor denervation on the inhibition of renin release by vasopressin, *Endocrinology* 123: 319-327, 1988.

65. Malayan, S.A., D.J. Ramsay, L.C. Keil, and I.A. Reid, Effects of increases in plasma vasopressin concentration on plasma renin activity, blood pressure, heart rate and plasma corticosteroid concentration

in conscious dogs, *Endocrinology* 107: 1899-1904, 1980.

66. Richardson-Morton, K.D., L.D. Van de Kar, M.S. Brownfield, and C.L. Bethea, Neuronal cell bodies in the hypothalamic paraventricular nucleus mediate stress-induced renin and corticosterone secretion, *Neuroendocrinology* 50: 73-80, 1989.

67. Porter, J.P., Electrical stimulation of paraventricular nucleus increases plasma renin activity, *Am J Physiol* 254: R325-R330, 1988.

68. Gotoh, E., R.M.A. Golin, and W.F. Ganong, Relation of the ventromedial nuclei of the hypothalamus to the regulation of renin secretion, *Neuroendocrinology* 47: 518-522, 1988.

69. Patel, K.P., and P.G. Schmid, Role of paraventricular nucleus in baroreflex-mediated changes in lumbar sympathetic nerve activity and heart rate, *J Autonom Nerv Sys* 22: 211-219, 1988.

ANTIBODIES TO NEUROPEPTIDES: BIOLOGICAL EFFECTS AND MECHANISMS OF ACTION

F.J.H. Tilders, J.W.A.M. van Oers, A. White,[1] F. Menzaghi,[2] and A. Burlet[2]

Department of Pharmacology
Free University
1081 BT Amsterdam, The Netherlands

Introduction

After the discovery of the conventional neurotransmitters, it became clear that peptides also play an important role as intercellular messengers in the nervous system. Over the past three decades, hundreds of biologically active peptides have been isolated from or identified in brain tissue. These studies were usually directly followed by the identification of the sites of peptide production, storage and release with the aid of radioimmunoassays, immunocytochemistry, and hybridochemistry. In addition to such data, ligand binding studies may generate cues as to the organs or cells that may be influenced by a particular neuropeptide. In contrast to the relative ease with which the biological effects of a peptide are established, studies on the physiological roles of a peptide are complicated and often hampered by the lack of appropriate tools. It is conceivable that a physiological role of a neuropeptide is demonstrated most convincingly by studying the derangements caused by blockade of peptidergic signal transfer by the use of specific peptide receptor antagonists.

In the period between the identification of a peptide and the development of selective receptor antagonists, passive immunization studies with peptide antibodies seem the approach of choice to interfere with peptidergic signal transfer. In contrast to the wealth of papers in which physiological effects are claimed on the basis of positive or negative findings from passive immunization studies, little is known about the mechanism of action of such antibodies and their required characteristics. In the present chapter, we will briefly summarize our present views on this issue.

Although there has been considerable interest in the biological effects of active immunization to hormones and neuropeptides especially in the field of reproduction research (1,2), in this paper we shall focus on the effects of passive immunization with antibodies directed to peptides and other neuron specific substances.

The Humoral Antibody-Peptide Interaction Concept

Although only rarely explicitly stated, the generally accepted concept of the mechanism of action of biologically active peptide-antibodies proposes binding of the

---

[1]Present affiliation: Department of Medicine, University of Manchester, Salford M6 8HD England

[2]Present affiliation: Laboratoire Biologique Cellulaire, INSERM U308 Nancy, France

Circulating Regulatory Factors and Neuroendocrine Function
Edited by J. C. Porter and D. Ježová
Plenum Press, New York, 1990

antibody to the peptide during its intercellular journey from the site of secretion to the site of action as schematically illustrated in Figure 1. This mechanism presupposes that antibodies reach the extracellular compartment involved in the transfer of the peptide signal (e.g., synaptic cleft, plasma) in sufficient concentrations to be effective. After peripheral administration, immunoglobulins easily distribute throughout the vascular and most of the extracellular fluid. The distribution volume can easily be determined by comparing the antibody titers of the injected serum preparation to those in the blood. In rats, the apparent distribution volume of immunoglobulins given intravenously or intraperitoneally is approximately 60 ml per kg body weight (3-5).

In general, circulating immunoglobulins do not have easy access to the brain and interference with central peptidergic mechanisms requires intraventricular or local antiserum administration. However, brain areas that have a poorly developed blood-brain-barrier, such as the ventromedial hypothalamic area (including the median eminence, arcuate nucleus, and ventromedial hypothalamic nucleus) and other circumventricular organs, can be reached rather efficiently by antibodies given peripherally.

If we assume that the liquid phase interaction of the antibody and the peptide is crucial for the observed biological effects, it seems obvious to study the antibody

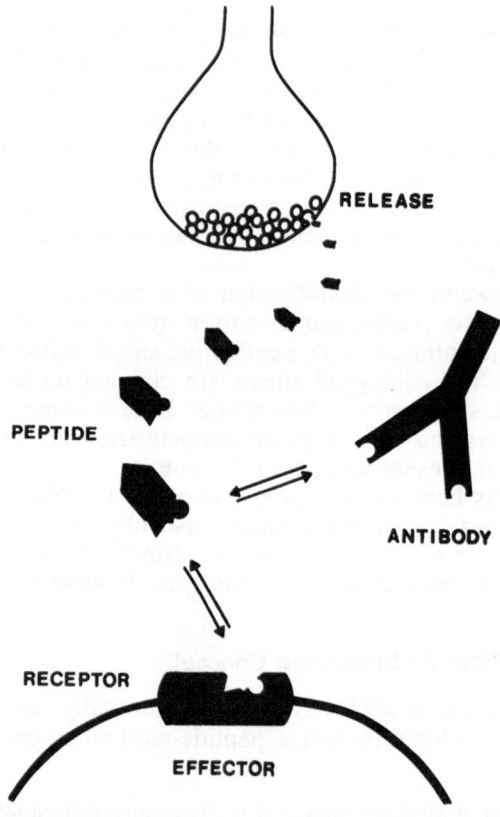

Figure 1.   Schematic representation of the mechanism by which an antibody to a neuropeptide can block peptidergic signal transfer.

characteristics under appropriate conditions. In this respect, surprisingly little attention is paid to the environment in which and the concentrations or titers at which the antibodies are supposed to work.

Antibody-ligand binding characteristics are usually measured in radioimmunoassay protocols involving rather non-physiological media and conditions. It is generally accepted that antibodies may show markedly different binding characteristics in phosphate buffer at 4°C than in plasma at 37°C. We therefore advocate the study of antibody binding characteristics under *in vitro* conditions that mimic as closely as possible the situation *in vivo*.

A most important point in passive immunization studies is the specificity of the interaction. It is frequently stated that *specific* antibodies have been used, a qualification that is based upon competition of substances for binding with the radiolabelled peptide (cross-reactivity). Although such statements are accepted or even required by most referees and are published in all scientific journals, it is not relevant. The interpretation of an antibody effect in terms of a physiological role of a particular peptide, requires that the antiserum does not bind other potentially relevant signals, not even at the high local immunoglobulin concentrations that are usually involved in passive immunization studies. In particular, when conventional (polyclonal) antisera are used, biological effects of serum components other than the immunoglobulins directed to the peptide under study, should be excluded. The problem of antibody specificity in passive immunization studies is reminiscent of the problems concerning the specificity of antibodies for immunocytochemistry, where data from competitive binding studies nowadays are considered as invalid evidence (6,7).

In analogy to immunocytochemistry, an important and necessary control experiment would be to remove the relevant immunoglobulin population (*e.g.*, by affinity chromatography) and demonstrate disappearance of its biological effect. Alternatively, relevant immunoglobulins may be absorbed by incubation with biologically inactive peptide fragments as demonstrated for instance for antibodies to ACTH (8). Although in itself such findings do not demonstrate specificity, they can be considered as bottomline criteria that should be met in all studies with polyclonal sera just as disappearance of staining after preadsorption with the peptide or immunogen should be demonstrated in immunocytochemical studies. It is certainly not sufficient to run pre-immune or control serum as a proof of the specificity. Although only rarely reported (8,9), we know that *control sera* can sometimes be found to exhibit biological effects. In practice, control sera are often selected on the basis of their inability to affect the biological parameter under study.

Also from monoclonal antibodies used in passive immunization studies, ligand binding and specificity characteristics should be tested at relevant concentrations and conditions. With monoclonals, competition studies can give us a hint with regard to the specificity of peptide binding as this will be primarily determined by the occurrence of homologous or similar epitopes on molecules other than the peptide under study.

Epitopes and Biological Activity

It is a general experience that not all antibodies that bind a given peptide do in fact block its biological activity. Thus, before using a particular antiserum to study an unknown role of a peptide, the serum should be demonstrated to be biologically active with respect to an established peptide effect. For instance, in our studies on the involvement of vasopressin in the control of ACTH secretion, we have selected vasopressin antisera according to their capacity to induce polyuria in rats after peripheral administration (3,10).

While conducting these studies, we encountered antisera that efficiently bound radiolabeled CRF but potentiated rather than attenuated CRF or stress-induced ACTH secretion in rats (unpublished observations). Similarly, Aston *et al.* (11) described two growth hormone binding monoclonal antibodies that potentiated the lactogenic and somatotrophic properties of growth hormone. From their studies, potentiation or inhibition of bioactivity seems related to the epitopes involved in antibody binding (12). Little is known on the mechanisms responsible for this phenomenon. Some biopotentiation responses involved the bivalent nature of immunoglobulins (13,14), whereas other responses could also be demonstrated with monovalent antibody fragments (11). As antibody binding to certain epitopes can interfere with receptor binding (see below), it is conceivable that binding to other epitopes may interfere with enzymatic breakdown of the peptide leading to prolongation or potentiation of its biological activity. Finally, when working with conventional antisera, one has to consider the possibility of contamination with anti-idiotype immunoglobulins that may exert peptide-receptor agonistic properties.

It is generally accepted, that biological inactivation requires binding to a particular epitope on the peptide resulting in either a loss of affinity for its receptor or a loss of intrinsic activity, that is to say, receptor binding is preserved but receptor activation is blocked. In fact, studies on the biological activities of antibodies binding to different epitopes of biologically active molecules is used as a strategy to trace molecular domains that are relevant for their biological activity (15). In large protein hormones, domains with different biological activities have been detected by such an approach. For instance, placental lactogen exhibits growth hormone-like and prolactin-like activities that can be blocked selectively by antibodies recognizing different epitopes of the molecule (16). Also in relatively small peptides such as ACTH, the exact binding epitope determines, to a large extent, the biological activity of the antibody. As illustrated in Figure 2, a mouse monoclonal antibody to the sequence 11-18 of the ACTH molecule 1A12 (17) induced a parallel and dose-dependent shift to the right of the ACTH-dose corticosterone-response relationship when tested on rat adrenocortical cells *in vitro* (18). In contrast, the monoclonal antibody 2A3 (17) that bound to the biologically inactive carboxyterminal sequence of ACTH (24-39) with sufficient affinity (see below), has no effect on

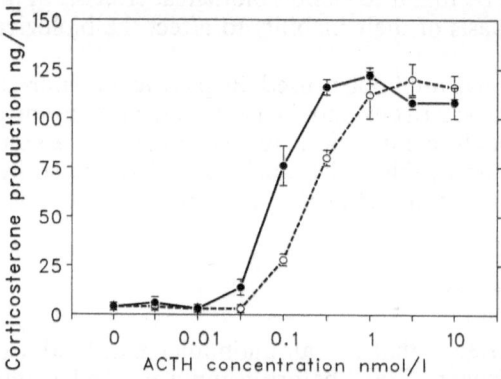

Figure 2.   Biological activity of a monoclonal antibody to ACTH. Antibody 1A12 (directed to the 11-18 sequence of ACTH, 0.2 nmol/liter, *closed symbols*) or control mouse IgG (0.2 nmol/liter *open symbols*) was incubated with different concentrations of ACTH for 1 hr at 37°C. Subsequently, the mixture was added to freshly dispersed adrenocortical cells of rats and corticosterone production was measured after 4 hr of incubation.

ACTH induced corticosterone secretion *in vitro*. Taken together, these observations indicate that the binding epitope, even of small peptides, is an important factor determining the biological activity of a given antibody.

Affinity and Biological Activity

Another antibody characteristic, which is relevant for its biological activity, is the association constant of the antibody-peptide complex. A first approximation of necessary local antibody concentration to bind a given percentage of the peptide can be deduced from the formal description of the law of mass action:

$$Ka = \frac{(Ab\text{-}PEP)}{(Ab)\,(PEP)} \text{ , where}$$

Ab is the concentration of free antibody in the biologically relevant compartment; *PEP* is the concentration of unbound peptide in that compartment; *Ab-PEP* is the local concentration of the antibody-peptide complex; and *Ka* is the association constant of the complex.

When local antibody concentrations are considerably larger than the peptide concentrations, the ratio of the bound to the free peptide equals (AB) × Ka. In other words, if we wish to bind 99% of the peptide, the local antibody concentration should be 100 times higher than the Kd (the reciprocal of Ka).

Let us consider the use of an antiserum in a passive immunization study on a hypophysiotropic peptide in rats and question what its association constant should be. If non-concentrated conventional antisera (*e.g.*, derived from rabbits) are used, the maximal dose to be given peripherally in the rat is in the order of 1 ml. Assuming specific antibody titers in well immunized rabbit of approximately 10 μmol/liter, the circulating antibody titer after administration to a rat will be less than 1 μmol/liter. If we aim to bind 99% of the peptide, such experiments require antibodies with affinity constants of at least 10^8 liter/mol. Obviously, lower affinity antibodies can be used in passive immunization experiments if they can be concentrated and/or properly dosed.

Binding Kinetics and Biological Activity

In the above section, we considered conditions of binding equilibrium, or at least situations that approach equilibrium conditions. However, considering the biology of the systems involved in peptidergic signal transfer, we have to realize that equilibrium conditions are usually not relevant, since binding should occur within a very limited time interval between secretion of the peptide and its interaction with the receptors on the effector cell. Our studies on the rate of dye transport in hypothalamus-pituitary portal vessels indicate that the time interval between secretion of hypophysiotropic factors into the portal blood and distribution in the anterior pituitary is at least 3-5 seconds. When processes of peptidergic neurotransmission are considered, time intervals may be limited to milliseconds. Is it realistic to expect immunoglobulins to act so fast, and if so, what are the relevant parameters?

Thus, it is not sufficient that an antibody can bind a relevant proportion of the peptide, but it should do so within an extremely short time interval. In this respect it is relevant to note that the association constant reflects the ratio between the *on-rate constant* (k^{+1}) and the *off-rate constant* (k^{-1}) of the binding. The higher the on-rate constant, the lower antiserum concentrations are required to bind a certain percentage of the peptide within a given time frame. As illustrated in Figure 3, increasing the local immunoglobulin concentration not only increases the fraction of the peptide that can be bound, but also leads to a concomitant increase of the rate of binding (5). The relevance of this point will be illustrated by two examples from our own work.

Based on actual *in vitro* determinations of the on-rate and association constant of a rat monoclonal antibody (PFU83) to r/hCRF in rat plasma at 37°C, we computed the time-binding characteristics in biologically relevant time frames. From these data (see Figure 3), we can conclude that 98% of CRF in portal plasma can be bound within the time allotted between secretion and action on the corticotrophs when circulating PFU83 concentrations reach 1 μmol/liter. This is in harmony with *in vivo* observations demonstrating that PFU83 at a dose of 10 nmol/rat, which leads to circulating PFU83 concentrations of approximately 1 μmol/liter (4), fully blocks ether and footshock induced ACTH secretion.

The other example refers to monoclonal antibody (1A12) directed to ACTH. In low concentrations this antibody inhibits the ACTH-induced production of corticosterone by rat adrenocortical cells (Figure 2). However, this antibody did not inhibit ether-induced corticosterone secretion in intact rats when given in doses leading to circulating concentrations of 2 nmol/liter, even though such doses were fully effective *in vitro*. In addition, pretreatment of rats with this antibody, reduced by only 50% the plasma corticosterone response in dexamethasone treated rats to an intravenous bolus injection of 100 ng ACTH. In contrast, preincubation of 100 ng ACTH with the same dose of 1A12 *in vitro* for 30-60 minutes and subsequent testing in dexamethasone treated rats, resulted in full prevention of ACTH-induced corticosterone secretion. Detailed analysis showed that the on-rate constant of binding of 1A12 to ACTH in plasma is indeed much too low to result in relevant binding within the time interval between ACTH secretion from the pituitary gland and interaction with adrenal cortex (Tilders *et al.*, data not shown). Based on the characteristics as presented in Figure 3, it is anticipated that further increase of the dose of 1A12 to levels much higher than those needed *in vitro*, may result in an effective blockade of stress-induced corticosterone secretion *in vivo*.

In peripheral endocrine or paracrine signaling, the time scales are in the order of

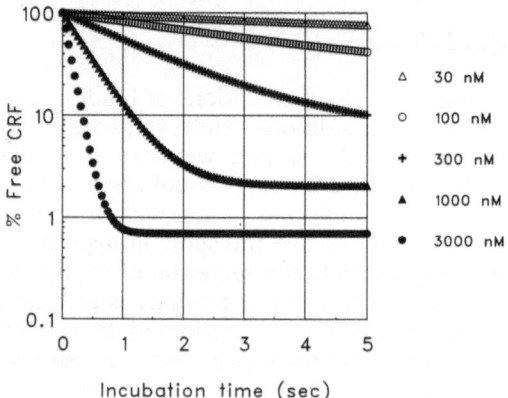

Figure 3. Effects of different concentrations of a monoclonal antibody to rCRF (PFU83) on the time course of rCRF binding in rat plasma at 37°C. Curves are computed according to the following experimentally determined parameters: Ka = 5.107 1/mol, K + 1 = 2.106 1/mol/sec. Note the effect of increasing antibody concentration on the binding rate.

1-10 seconds or more. In peptidergic neurotransmission signal transfer time may be as short as milliseconds. One may therefore wonder whether peptide-antibody interactions can occur to a sufficient extent within such short time intervals (see next section for alternatives). Assuming that liquid phase interactions are responsible for the biological effects, it is most likely that administration of peptide antibodies will primarily interfere with relatively slow (e.g., non-synaptic) transmission processes. When all conditions are optimal (local antibody concentrations, on-rate of binding, etc.), synaptic transmission may become blocked by liquid phase peptide-antibody interactions. It should be noted that in studies on the physiological role of neuropeptides in the brain, antisera are often injected locally in poorly diluted forms leading to high intercellular immunoglobulin concentrations close to the injection site.

Antibody-Cell Interaction Concept

In addition to the most popular concept of antibody-message binding, the possibility of interactions of antibodies with cells should also be considered (19) as a possible explanation antibody action in passive immunization studies. Interactions with cells seem particularly prominent in studies with antibodies to secretory products, enzymes or other constituents of neurons, and over the past 15 years, several authors have reported specific neuronal uptake and biological effects of certain antibodies including neurodegenerative processes.

Theoretically, antibodies may bind to cell membranes because of the presence of the following:

a. *Suitable Fc receptors on certain cells.* Classical examples are Fc receptors on endothelial cells, lymphocytes, etc., which have troubled immunocytochemists so much.

b. *Membrane constituents with epitopes homologous to that of the peptide.* This form can also be considered as a non-specific aspect of the antibody binding, that may nevertheless cause biological effects (see below).

c. *Authentic peptides bound to their receptor(s).* It is obvious that such interactions may lead to disturbances of effector cell function. The observation of Rao et al. (20) that adrenal cells of rats treated with ACTH antisera are insensitive to ACTH *in vitro* may relate to such a mechanism.

d. *Secretory peptides or other intracellular constituents on the cell membrane of neurons and hormone producing cells.* For instance, the membrane of living lactotrophs showed specific binding of a prolactin-directed antibody, a property that has been used to sort lactotrophs from other anterior pituitary cells (21). Similarly, antibodies to LH have been used to select gonadotrophs from pituitary cells (22).

Neuronal Uptake of Antibodies to Enzymes

Before discussing some recent findings with antibodies to neuropeptides, let us briefly consider some of the older observations with antibodies directed to dopamine-β-hydroxylase (DBH), an enzyme that mediates the conversion of dopamine to norepinephrine. DBH occurs in the secretory vesicles of chromaffin cells of the adrenal medulla and in the synaptic vesicles of noradrenergic neurons in the sympathetic and central nervous system. After peripheral administration, DBH antibodies showed up in the chromaffin cells of the adrenal medulla and in sympathetic varicose nerve fibers (23,24). Subsequently, the antibodies were subjected to retrograde transport and the sympathetic ganglia appeared in the form of discrete antibody loaded particles. These findings were confirmed in studies involving injection of radiolabeled DBH antibodies into the anterior eye chamber and measuring the appearance of radioactivity into the superior cervical ganglion (25,26). The observation that DBH antibody uptake was enhanced by treatment of the animals with phentolamine and attenuated after phenylephrine, supports the hypothesis that increased secretory activity leads to increased exposure of DBH on the axonal terminal membrane and thereby to increase antibody binding and uptake (27). Similar conclusions have been drawn from studies on adrenal chromaffin cells (28). Neuron specific uptake and transport is not limited to DBH antibodies. Central administration of antibodies to glutamate decarboxylase have also been found to result in GABA-neuron specific uptake and transport (29).

Antibody-Induced Neuronal Lesions

It is not fully clear by what mechanism antibodies become internalized into specific cells. One possibility is that antibodies that are present in high concentrations in the extracellular fluid, may specifically bind to granular components which are exposed to the exterior of the cell by exocytosis and become internalized by subsequent endocytotic processes.

Alternatively, antibodies may bind to specific components expressed on the cell membrane and induce complement mediated lesions which subsequently allow passive (non-specific) passage of immunoglobulins across the membrane. More than a decade ago, it was observed that administration of DBH antibodies had little effect on sympathetic nerves in mice or rats, but induced marked sympathetic degeneration in guinea pigs. In the latter species, injection of F(ab) fragments instead of intact DBH immunoglobulins, had no neurotoxic effects. Conversely, peripheral administration of DBH antibodies together with complement, also induced in rats sympathetic degenerations (30-32).

Similarly, central intracerebroventricular administration of DBH antibodies in guinea pigs or of DBH antibodies plus complement in rats, was found to induce selective degeneration of central norepinephrine systems as demonstrated both morphologically and biochemically (33-35). Complement mediated antibody interactions may result in a fast increase of cytoplasmic free Ca^{2+}, marked loss of intracellular K^+, membrane depolarization, release of primary secretory products as well as cytoplasmic and particle bound enzymes (28,36,37). In a series of *in vivo* studies, Docherty and colleagues (38-41) demonstrated that the neurotoxic effects of antibodies to DBH were limited to noradrenergic, those to glutamate decarboxylase were limited to GABAergic, and those to tryptophane hydroxylase were limited to serotonergic synaptosomes. Although neurons *in vivo* are possibly less susceptible for complement-mediated membrane attacks, it is worth noting that binding of an antibody to a cell membrane sets the stage for complement mediated interactions that may be associated with short or long lasting interference with neuronal functioning. In a mild form, complement lesions may lead to an immediate and short lasting interference associated with release of neurotransmitters or peptides. Such lesions make the cell accessible for all kinds of substances (37) including immunoglobulins and toxins. In more extensive forms such lesions may lead to degeneration of nerve terminals or neuronal cell death. It is conceivable that these two mechanisms may lead to stimulation and then inhibition of a particular neuronal system and thereby in opposite biological responses.

Antibodies to Neuropeptides, Uptake and Biological Effects

Several studies indicate that antibodies to neuropeptides can become internalized

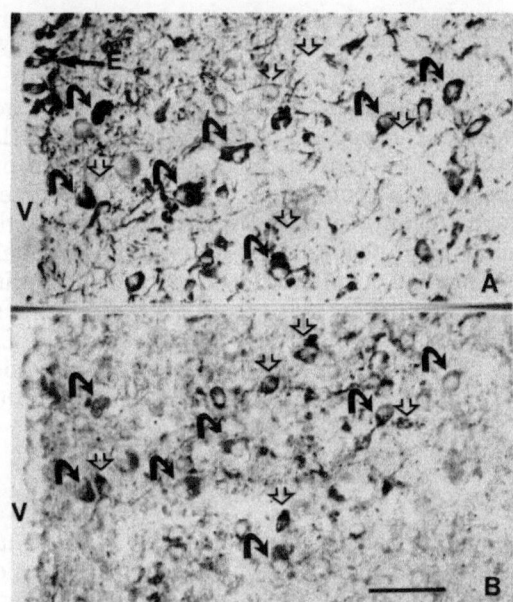

Figure 4. Neuronal uptake of a monoclonal antibody to rCRF in the paraventricular nucleus of an adrenalectomized Long Evans rat after intrahypothalamic microinjection of PFU83. The localization of CRF antibody was visualized by using a peroxidase labelled rabbit anti-rat IgG serum (A). After chemical elution, the same section was immunostained for endogenous CRF (B). *Open arrows:* CRF containing neurons that did not accumulate PFU83. *Closed arrows:* CRF containing neurons that accumulated PFU 83. V: third ventricle. E: ependymal cells show nonspecific uptake of immunoglobulins. *Bar* represents 50 μm.

into specific neurons after local administration in the brain. Chan Palay (42) reported uptake of monoclonal antibodies to substance P in substance P neurons, after local microinjection into certain brain areas and both retrograde and anterograde transport of the antibody. Recently, Burlet *et al.* (43) found that local injection of a monoclonal antibody to vasopressin at a site close to the paraventricular or supraoptic nucleus of rats led to antibody uptake in vasopressin producing neurons and subsequent anterograde transport to the median eminence and posterior lobe. Likewise, intrahypothalamic injection of a monoclonal antibody to pro-vasopressin derived neurophysin resulted in specific internalization into vasopressin synthesizing neurons (44).

In addition to neuronal uptake of injected antibody, local administration of a monoclonal $IgG_{2\alpha}$ to vasopressin close to the supraoptic or paraventricular nucleus has been reported to induce a short lasting polyuria (45). We, therefore, suggest that internalization of antibodies into vasopressinergic cell bodies leads to inhibition of their secretory activity.

In recent experiments on the role of central CRF systems, we have administered our rat monoclonal to rCRF (PFU83) (4,5) into the third ventricle or locally into the hypothalamus following the method described previously (43). Although intraventricular administration of PFU83 as well as of the CRF receptor antagonist (α-helical CRF 9-41) inhibited the metabolic responses to interleukin-1β (46), we were not sure whether this was due to interference of this monoclonal antibody with CRF in liquid phase or with neuronal cells. The reason for our doubts was the finding that microinjection of this antibody leads to uptake into various cells of the paraventricular nucleus as illustrated in Figure 4. Studies on the identification of these neurons revealed uptake of the CRF antibody into CRF producing neurons. In addition, CRF antibodies were also internalized by some vasopressin, oxytocin, neuropeptide Y, and tyrosine hydroxylase immunoreactive neurons (47). As to this point, we can only speculate about the properties that these antibody-accumulating cells have in common. It is unlikely that these various neurons all exhibit CRF-receptors which are occupied by CRF during microinjection of the antibodies in intact animals. In support of this, arousal of pituitary-adrenal activity had been observed under the experimental conditions used in these experiments (unpublished observations). It is of interest, however, that not all CRF producing neurons accumulate CRF antibodies, which may indicate that some exhibit occupied CRF receptors while others do not. Further studies are required to elucidate the significance of neuronal antibody uptake for the biological effects of antibodies to neuropeptides.

Conclusions

Interpretation of the biological effects of antibodies to (neuro)peptides is to a large extent dependent on the underlying mechanism of action. The most popular concept proposes a direct relationship between biological activity and binding of a peptide during its journey from its site of secretion to its site of action. As the time involved in peptide signal transfer may be short, the kinetics (on-rate) of the antibody-peptide binding seem the most crucial parameter. In addition, affinity and binding epitopes are determinants of the biological activity of antibodies. Strategies are suggested to study these parameters as well as the specificity of the interaction under appropriate conditions. We conclude that this mechanism of action may underlie various studies in which antibodies are used to interfere with relatively slow signal transfer processes. In contrast, its role in antibody interference with more rapid peptide signalling processes is less likely. Blockade of central peptidergic transmission requires the administration of antibodies in the brain. This may lead to internalization of antibodies in specific neurons, most likely through endocytotic or complement mediated mechanisms. Assuming that the biological effects of antibodies after central administration are related to neuronal antibody uptake, proper interpretation of central effects requires more detailed knowledge of the parameters that determine antibody uptake in neurons.

References

1. Nieschlag, E. (ed) *Immunization with Hormones in Reproduction Research*, North Holland Publishers, Amsterdam, 1975.
2. Frazer, H.M., Active immunization of stumptailed macaquo monkeys against luteinizing hormone releasing hormone and in effects on menstrual cycles, ovarian steroids and positive feedback, *J Reprod Immunol* 5: 173-183, 1983.
3. Linton, E.A., F.J.H. Tilders, S. Hodgkinson, F. Berkenbosch, I. Vermes, and P.J. Lowry, Stress-induced secretion of adrenocorticotropin in rats is inhibited by administration of antisera to ovine corticotropin-releasing factor and vasopressin, *Endocrinology* 116: 966-970, 1985.
4. van Oers, J.W.A.M., F.J.H. Tilders, and F. Berkenbosch, Characterization and biological activity of a rat monoclonal antibody to rat/human corticotropin-releasing factor, *Endocrinology* 124: 1239-1246, 1989.
5. van Oers, J.W.A.M., F.J.H. Tilders, and F. Berkenbosch, Acute and long-term suppression of stress-induced ACTH secretion by a rat monoclonal antibody to corticotropin releasing factor, In E.E. Müller and R.M. MacLeod (eds) *Neuroendocrine Perspectives, Volume 6*, Springer-Verlag, New York, 287-292, 1989.
6. Van Leeuwen, F., Pitfalls in immunocytochemistry with special reference to the specificity problems in

the localization of neuropeptides, *Am J Anat* 175: 363-377, 1986.

7. Berkenbosch, F., and F.J.H. Tilders, A quantitative approach to cross reaction problems in immunocytochemistry, *Neuroscience* 23: 823-826, 1987.

8. Fleischer, N., J.R. Givens, K. Abe, W.E. Nicholson, and G.W. Liddle, Studies of ACTH antibodies and their reactions with inactive analogues of ACTH, *Endocrinology* 78: 1067-1075, 1966.

9. McGarry, E.E., J.C. Beck, L. Ambe, and R. Nayak, Some studies with antisera to growth hormone ACTH and TSH, *Rec Prog Horm Res* 20: 1-31, 1964.

10. Tilders, F.J.H., F. Berkenbosch, I. Vermes, E.A. Linton, and P.G. Smelik, Role of epinephrine and vasopressin in the control of the pituitary-adrenal response to stress, *Fed Proc* 44: 155-160, 1985.

11. Aston, R., A.T. Holder, M.A. Preece, and J. Ivanyi, Potentiation of the somatogenic and lactogenic activity of human growth hormone with monoclonal antibodies, *J Endocrinol* 110: 381-388, 1986.

12. Aston, R., and J. Ivanyi, Monoclonal antibodies to growth hormone and prolactin, *Pharmacol Therap* 27: 403-424, 1985.

13. Shechter, Y., K.J. Chang, S. Jacobs, and P. Cuatrecasas, Modulation of binding and bioactivity of insulin by anti-insulin antibody: relation to possible role of receptor self aggregation in hormone action, *Proc Natl Acad Sci USA* 76: 2720-2724, 1979.

14. Shechter, Y., L. Hernandez, J. Schlessinger, and P. Cuatrecasas, Local aggregation of hormone-receptor complexes is required for activation by epidermal growth factor, *Nature* 278: 835-838, 1979.

15. Massone, A., C. Baldari, S. Censini, M. Bartalini, D. Nucci, D. Boraschi, and J.L. Telford, Mapping of biologically relevant sites on human 1L-1β using monoclonal antibodies, *J Immunol* 140: 3812-3816, 1988.

16. Chan, J.S.D., Z.R. Nie, N.G. Seidah, and M. Chrétien, Inhibition of binding of ovine placental lactogen to growth hormone and prolactin receptors by monoclonal antibodies, *Endocrinology* 119: 2623-2628, 1986.

17. White, A., C. Gray, and J.G. Ratcliffe, Characterization of monoclonal antibodies to adrenocorticotrophin, *J Immunol Methods* 79: 185-194, 1985.

18. Van Bree, C., M. Rep, A. White, and F.J.H. Tilders, Biological activity of monoclonal antibodies to corticotropin (ACTH): relationship to peptide binding, *Proceedings of the 30th Duth Federative Meeting* Abstract #52, 1989.

19. Butler, J.E., The immune response to protein and protein-conjugated antigens and the possible consequences of resultant *in vivo* reaction, In E. Nieschlag (ed) *Immunization with Hormones in Reproduction Research*, North Holland PBI, Amsterdam, pp. 11-30, 1975.

20. Rao, A.J., J.A. Long, and J. Ramachandran, Effects of antiserum to adrenocorticotropin on adrenal growth and function, *Endocrinology* 102: 371-378, 1978.

21. St. John, P.A., L. Dufy-Barbe, and J.L. Barker, Anti-prolactin cell-surface immunoactivity identifies a subpopulation of lactotrophs from the rat anterior pituitary, *Endocrinology* 119: 2783-2795, 1986.

22. Thorner, M.O., J.C. Borges, M.J. Cronin, D.A. Keefer, P. Hellmann, D. Lewis, L.G. Dabney, and P.J. Quesberry, Fluorescence activated cell sorting of functional anterior pituitary cells, *Endocrinology* 110: 1831-1833, 1982.

23. Jacobowitz, D.M., M.G. Ziegler, and J.A. Thomas, *In vivo* uptake of antibody to dopamine-β-hydroxylase into sympathetic elements, *Brain Res* 91: 165-170, 1975.

24. Ziegler, M.G., J.A. Thomas, and D.M. Jacobowitz, Retrograde axonal transport of antibody to dopamine-β-hydroxylase, *Brain Res* 104: 390-395, 1976.

25. Fillenz, M., G. Gagnon, K. Stoeckel, and H. Thoenens, Selective uptake and retrograde transport of dopamine-β-hydroxylase antibodies in peripheral adrenergic neurons, *Brain Res* 114: 293-304, 1976.

26. Lees, G.R., and R.J. Horsburg, Retrograde transport of dopamine-β-hydroxylase antibodies in sympathetic neurons: effects of drugs modifying noradrenergic transmission, *Brain Res* 301: 281-286, 1984.

27. Lees, G., I. Geffen, and R. Rush, Phentolamine increases neuronal binding and retrograde transport of dopamine-β-hydroxylase antibodies, *Neurosci Lett* 22: 115-118, 1981.

28. Ling, G., R. Fischer-Colbri, W. Schmidt, and H. Winkler, Exposure of an antigen of chromaffin granules on cell surface during exocytosis, *Nature* 301: 610-611, 1983.

29. Chan Palay, V., S.L. Pallay, J.Y. Wu, Gamma-aminobutyric acid pathways in the cerebellum studied by retrograde and anterograde transport of glutamic acid decarboxylase antibody after *in vivo* injections,

Anat Embryol 157: 1-14, 1979.

30. Costa, M., R.A. Rust, J.B. Furness, and L.B. Geffen, Histochemical evidence for the degeneration of peripheral noradrenergic axons following intravenous injection of antibodies to dopamine-β-hydroxylase, *Neurosci Lett* 3: 201-207, 1976.

31. Furness, J.B., S.Y. Lewis, R. Rush, M. Costa, and L.B. Geffen, Involvement of complement of degeneration of sympathetic nerves after administration of antiserum to dopamine-β-hydroxylase, *Brain Res* 136: 67-75, 1977.

32. Lewis, S.Y., R.A. Rush, and L.B. Geffen, Biochemical effects on guinea pig iris of local injection of dopamine-β-hydroxylase antibodies and $F(ab)_2$ fragments, *Brain Res* 134: 173-179, 1977.

33. Blessing, W.W., M. Costa, L.B. Geffen, R.A. Rush, and G. Fink, Immune lesions of noradrenergic neurons in rat central nervous system produced by antibodies to dopamine-β-hydroxylase, *Nature* 267: 368-369, 1977.

34. Costa, M., L.B. Geffen, R.A. Rush, D. Brigges, W.W. Blessing, and J.W. Health, Immuno lesions of central noradrenergic neurons produced by antibodies to dopamine-β-hydroxylase, *Brain Res* 173: 65-78, 1979.

35. Rush, R.A., M. Costa, J.B. Furness, and L.B. Geffen, Changes in tyrosine hydroxylase and dopamine-β-hydroxylase activities during degeneration of noradrenergic axons produced by antibodies to dopamine-β-hydroxylase, *Neurosci Lett* 3: 209-213, 1976.

36. Campbell, A.K., R.A. Daw, and P.J. Luzio, Rapid increase in intracellular free Ca^{++} induced by antibody plus complement, *FEBS Lett* 107: 55-60, 1979.

37. Schweizer, E.S., and M.P. Blaustein, The use of antibody and complement to gain access to the interior of presynaptic nerve terminals, *Exper Brain Res* 38: 443-453, 1980.

38. Docherty, M., H.F. Bradford, B. Anderton, and J.Y. Wu, Specific lysis of GABAergic synaptosomes by an antiserum to glutamate decarboxylase, *FEBS Lett* 152: 57-61, 1983.

39. Docherty, M., H.F. Bradford, C.D. Cash, and M. Maitre, Specific immunolysis of serotonergic nerve terminals using an antiserum against tryptophan hydroxylase, *FEBS Lett* 182: 489-492, 1985.

40. Docherty, M., H.F. Bradford, J.Y. Wu, T.H. Joh, and J.D. Reis, Evidence for specific immunolysis of nerve terminals using antisera against choline acetyltransferase, glutamate decarboxylase and tyrosine hydroxylase, *Brain Res* 339: 105-113, 1985.

41. Docherty, M., H.F. Bradford, and T.H. Joh, Specific lysis of noradrenergic synaptosomes by an antiserum to dopamine-β-hydroxylase, *FEBS Lett* 202: 37-40, 1986.

42. Chan Palay, V., Immunocytochemical detection of substance P neurons their processes and connections by *in vivo* microinjections of monoclonal antibodies, *Anat Embryol* 156: 225-240, 1979.

43. Burlet, A.J., B.P. Leon-Henri, F.R. Robert, A. Abrahmani, B.M.L. Fernette, and C.R. Burlet, Monoclonal anti-vasopressin (VP) antibodies penetrate into VP neurons *in vivo*, *Exper Brain Res* 65: 629-638, 1987.

44. Leon-Henri, B., A. Burlet, J. Chauvet, J. Nicolas, and C. Burlet, Vasopressin neuron is the target of monoclonal antibodies raised against vasopressin-neurophysin injected *in vivo*, *Neuroendocrinology* 49: 125-133, 1989.

45. Haumont-Pellegri, B., A. Burlet, F. Giannangeli, J-P. Nicolas, and C. Burlet, Creation d'un diabete insipide d'origine centrale induit par microinjection intrahypothalamique d'un anticorps anti-vasopressine, *Ann Endocrinol [Paris]* 49: 30 N, 1988.

46. Busbridge, N.J., M.J. Dascombo, F.J.H. Tilders, J.W.A.M. van Oers, E.A. Linton, and N.J. Rothwell, Central activation of thermogenesis and fever by interleukin 1β and interleukin 1α involve different mechanisms, *Biochem Biophys Res Commun* 162: 591-596, 1989.

47. Burlet, A.J., F. Menzaghi, F.J.H. Tilders, J.W.A.M. van Oers, M.P. Nicholas, and C.R. Burlet, Uptake of a monoclonal antibody directed against corticotropin releasing factor in hypothalamic neurons, *Brain Res*, In Press.

CIRCULATING CORTICOTROPIN-RELEASING FACTOR IN PREGNANCY

E.A. Linton, C.D.A. Wolfe,[1] D.P. Behan, and P.J. Lowry

Department of Biochemistry & Physiology
School of Animal and Microbial Sciences
University of Reading, United Kingdom

Normal circulating levels of human corticotropin releasing factor (hCRF), the most potent of the hypothalamic peptides stimulating pituitary ACTH release (1) are low, most workers reporting levels substantially below 100 pg/ml (2-10). This plasma CRF is thought to originate not only from the hypothalamus but also from several tissues outside of the brain such as the spinal cord, alimentary tracts, lung, pancreas, and adrenal gland, which have all been shown to contain CRF-like immunoreactivity (11-14). Release of CRF from these organs is thought to contribute a significant proportion of the CRF in the peripheral circulation, since most studies have been unable to demonstrate modulation of plasma CRF by stress and glucocorticoids (3-5,9,15,16), which would be expected if plasma CRF primarily reflected peptide release from the hypothalamus. Following the demonstration of CRF-like bioactivity (17,18) in extracts of human placenta, several reports appeared describing elevated levels of CRF in the plasma of women in the third trimester of pregnancy (10,18-22). Great variation exists in the absolute levels of the peptide measured and this may reflect, in part, the different assay and extraction systems used.

Our group has used an immunoradiometric assay (IRMA) for the direct measurement of CRF in small volumes of plasma without the need for prior extraction; additionally, this assay does not detect the CRF precursor nor any CRF fragments (9). With this assay, endogenous plasma CRF averaged 16 ± 7 (mean and SD) pg/ml (n = 41) in normal men and 14 ± 7 (mean and SD) pg/ml (n = 27) in normal, non-pregnant women. In the first (n = 20) and second (n = 20) trimester of pregnancy, maternal plasma CRF was within the normal, non-pregnant range (2-28 pg/ml). However, in the third (n = 72) trimester, maternal plasma CRF rose gradually throughout gestation, reaching concentrations of several ng/ml (21). A further small rise was observed in some women throughout labor, followed by a rapid decline postpartum, normal non-pregnant levels being achieved within 24 hours (Figure 1). There is a wide variation in the CRF profiles of pregnant women in the third trimester which is typical of many placental products (Figure 2); this may also explain the great range of maternal plasma CRF reported by different groups, many of whom use small numbers of subjects sampled at random rather than sequentially. This rise in CRF above the non-pregnant range from 28 weeks is unlike most other placental products, which begin to rise early in pregnancy to reach a plateau near term, in keeping with the hypothesis that the release of placental peptides into the maternal circulation is a direct function of the total mass of trophoblast

---

[1]Department of Community Medicine, St. Thomas's Hospital Medical School, London, UK

Circulating Regulatory Factors and Neuroendocrine Function
Edited by J. C. Porter and D. Ježová
Plenum Press, New York, 1990

and its maternal blood supply (23). (It is known that placental weight increases with gestational age but slows in the last 8-10 weeks and this is mirrored in the pattern of plasma estriol and hPL.) The notable exception is pregnancy-associated plasma protein A (PAPP-A), which, like CRF, shows an increase at term and a rise in labor. It has been postulated that the behavior of PAPP-A in late pregnancy is a consequence of uterine activity (24) although others have taken this as an indication of an extra-placental source of PAPP-A (25).

A marked rise in CRF has been reported during labor (10) although a subsequent study could not confirm this (26). In our study, a significant difference ($P < 0.001$) in CRF values was seen between 40 weeks (1320 pg/ml) and labor (1772 pg/ml). However, since there was a strong correlation between weeks of gestation and CRF levels (r = 0.81, $P < 0.001$), it is not clear whether this elevation in labor was due to the process of parturition or the passage of time (27). In 11 subjects in whom the last antenatal value and a labor value were within a few days of each other, a significant difference did exist between them (1117 pg/ml and 1849 pg/ml, respectively; $P < 0.001$).

In keeping with a placental origin for CRF, levels fall dramatically postpartum (10,18,21,26,28). In three patients sampled at 5-10 minute intervals postpartum for a minimum of 2 hours to estimate plasma clearance, CRF was found to have a half-life in the region of 1 hour (Figure 3). This is prolonged in comparison to 25.3 minutes for the slow half-life (representing true metabolic clearance) of synthetic hCRF administered to normal males (29) and may be due to the presence of the CRF binding protein (see later).

Further evidence that the placenta is the source of plasma CRF came with the demonstration that placental CRF was biophysically identical to synthetic hCRF (18,28,30,31), although multiple size classes of immunoreactive CRF have also been demonstrated. The peptide has been localized immunohistochemically in the cytotrophoblast, decidua, and amnion of early pregnancy (32) while indirect immunofluorescent staining of term placenta has shown its presence in the cytotrophoblast (33). It is now known that the CRF gene is expressed in the human placenta (31,34-37). Placental CRF mRNA is similar in size to rCRF mRNA and, like the peptide itself (10,34), message

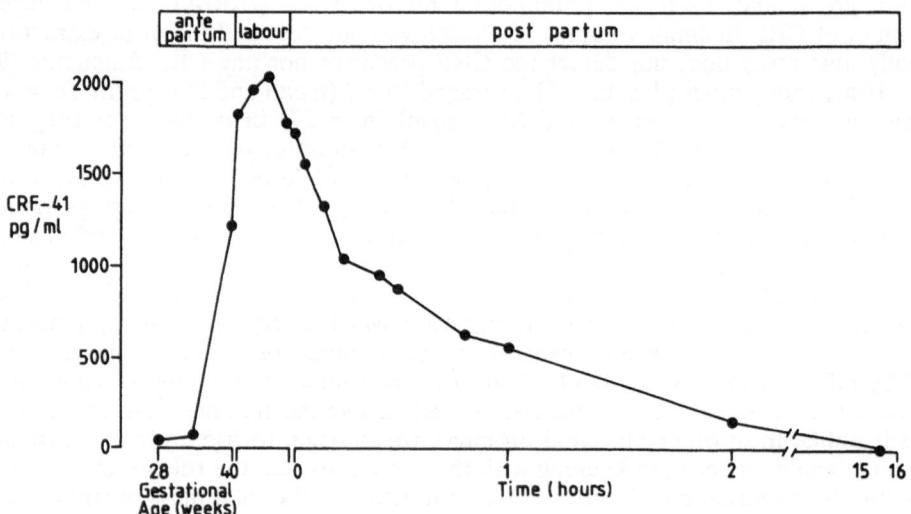

Figure 1. Plasma CRF levels measured sequentially in one woman during gestation, in spontaneous labor and postpartum.

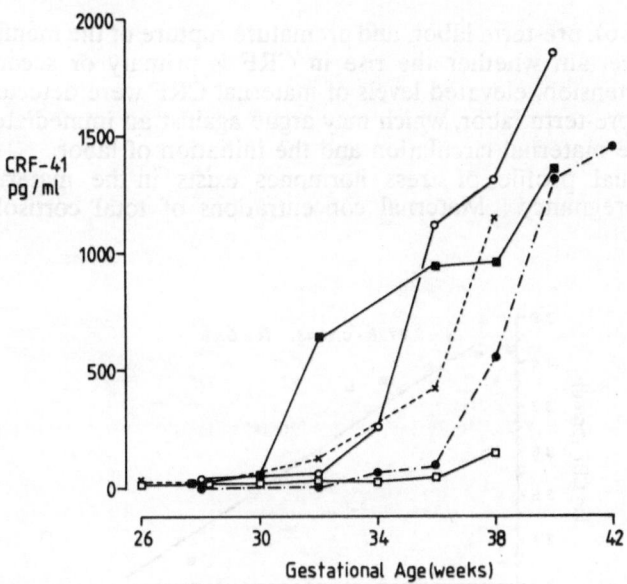

Figure 2. Sequential plasma CRF levels in 5 women in the third trimester of pregnancy, demonstrating the considerable variation in CRF profiles between women.

concentration is low in the first and second trimester, rising approximately 20-fold in placenta of 35-40 weeks gestation. Recent work suggests that hypothalamic and placental CRF are transcribed from the same gene, located on the long arm of chromosome 8 (38). *In vitro* experiments with human placental fragments and cells in culture have shown that the tissue can also secrete this CRF (31-33,37,39-41). We have demonstrated the authenticity of CRF in the maternal circulation, since the immunoaffinity extracted peptide co-elutes with synthetic CRF on HPLC and has equipotent bioactivity (Figure 4).

CRF has been detected in fetal blood collected by fetoscopy at 16-24 weeks (42) and in umbilical cord blood collected at delivery, when there is a significant correlation between umbilical vein and maternal plasma CRF levels (27,28) although fetal levels are only 3-4% of those in the mother. This has been taken as evidence that the placenta is the common source of circulating CRF, and supported by the significant difference between umbilical artery and vein CRF levels (10) although we have not been able to confirm this arterio-venous difference (21). In a very recent study, we obtained cordocentesis plasma samples from 26 abnormal pregnancies between 18 and 36 weeks gestation and found that fetal plasma CRF, ACTH, and cortisol ranged between 94-175 pg/ml, 3-87 pg/ml, and 74-324 nmol/liter, respectively. While no correlation was found between these hormones in the fetus, nor did any relationship exist between the maternal and fetal axes, fetal CRF levels significantly decreased (r_s = 0.88, P < 0.001) with gestational age (Figure 5), which is the opposite situation to that found in the mother.

The function of placental CRF is not yet known and much research effort is now centered on the role that CRF plays in the modulation of the maternal and fetal hypothalamic-pituitary-adrenal axes during gestation. Several authors have speculated that placental CRF may be involved in the initiation of parturition, since maternal plasma CRF has been shown (by some authors at least) to rise in labor when there is a rapid rise in both ACTH (43) and cortisol (44). However, we have found that elevated CRF concentrations outside the normally high range exist in women with pregnancy-induced hypertension (Figure 6), pre-term labor, and premature rupture of the membranes (21,45). Although it is not certain whether the rise in CRF is primary or secondary to pregnancy-induced hypertension, elevated levels of maternal CRF were detected some weeks before the onset of pre-term labor, which may argue against an immediate link between placental CRF in the maternal circulation and the initiation of labor.

A most unusual profile of stress hormones exists in the maternal peripheral circulation during pregnancy. Maternal concentrations of total cortisol are elevated

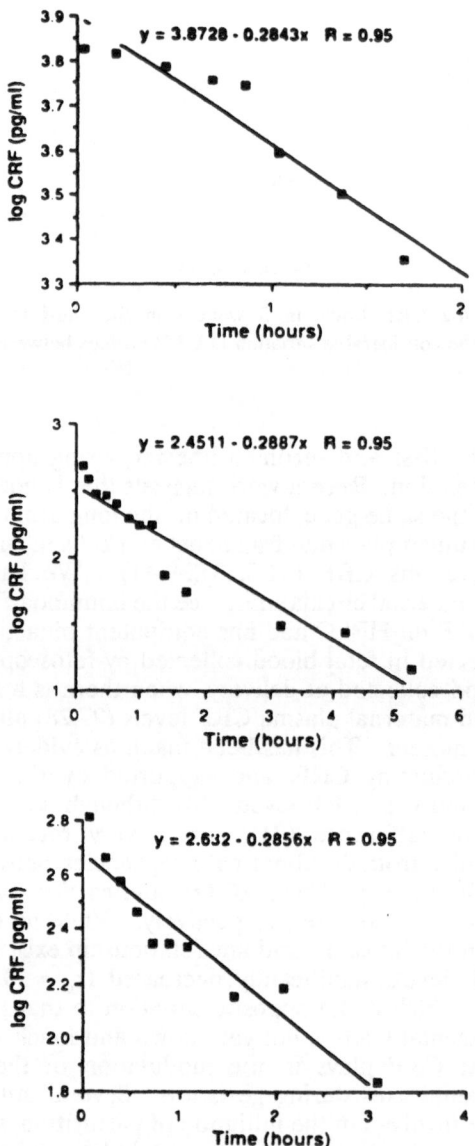

Figure 3. Maternal plasma CRF half-life ($T_{1/2}$) in 3 subjects using linear regression models. $T_{1/2}$ was 63.5 min, 64 min, and 66 min, respectively.

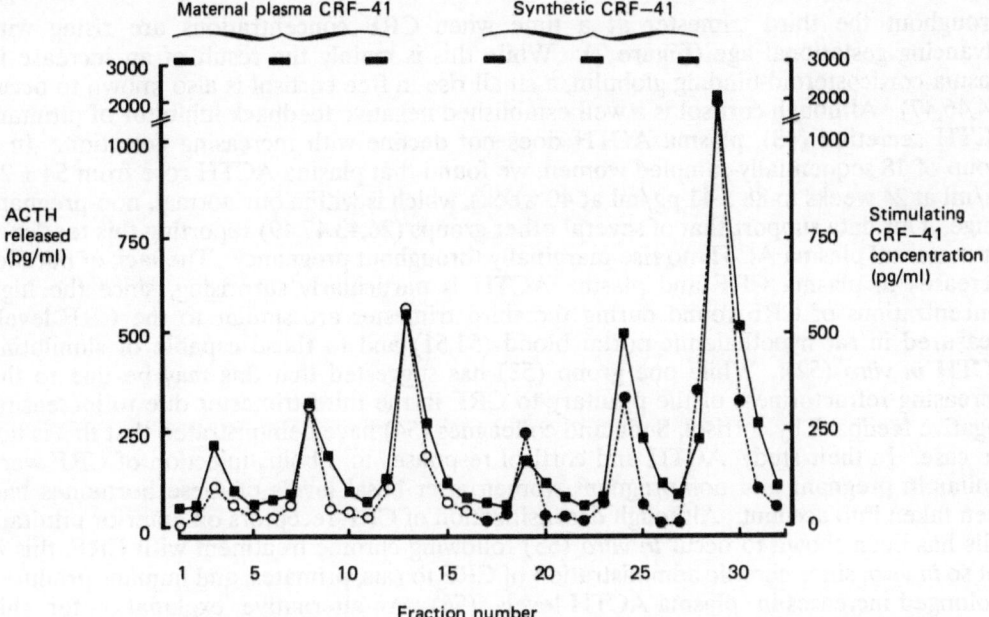

Figure 4. ACTH-releasing activity of pulses of extracted maternal plasma CRF and synthetic hCRF *(solid bars)* in the rat isolated pituitary cell bioassay. *Open circles* denote maternal plasma CRF, *closed circles* denote synthetic CRF and *squares* denote ACTH in the fractions of column effluent.

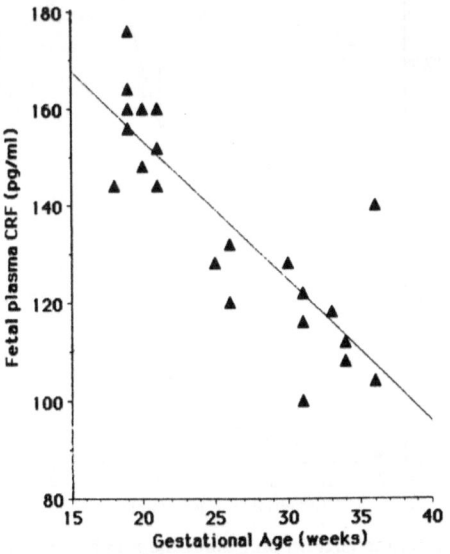

Figure 5. Correlation between fetal CRF levels and gestational age in cordocentesis plasma samples. $R_s = 0.88$.

throughout the third trimester at a time when CRF concentrations are rising with advancing gestational age (Figure 7). While this is mainly the result of an increase in plasma corticosteroid-binding globulin, a small rise in free cortisol is also known to occur (44,46,47). Although cortisol is a well established negative feedback inhibitor of pituitary ACTH secretion (48), plasma ACTH does not decline with increasing gestation. In a group of 18 sequentially-sampled women, we found that plasma ACTH rose from 54 ± 24 pg/ml at 24 weeks to 88 ± 41 pg/ml at 40 weeks, which is within our normal, non-pregnant range. Our data support that of several other groups (26,43,47,49) reporting this tendency for maternal plasma ACTH to rise marginally throughout pregnancy. The lack of parallel increases in plasma CRF and plasma ACTH is particularly surprising, since the high concentrations of CRF found during the third trimester are similar to the CRF levels measured in rat hypothalamic portal blood (51,51) and to those capable of simulating ACTH *in vitro* (52). While one group (53) has suggested that this may be due to the increasing refractoriness of the pituitary to CRF in the third trimester due to increasing negative feedback by cortisol, Suda and colleagues (54) have demonstrated that this is not the case. In their study, ACTH and cortisol responses to a bolus injection of CRF were similar in pregnant and non-pregnant women after basal levels of these hormones had been taken into account. Although desensitization of CRF receptors on anterior pituitary cells has been shown to occur *in vitro* (55) following chronic treatment with CRF, this is not so *in vivo*, since chronic administration of CRF to rats, primates, and humans produces prolonged increases in plasma ACTH levels (56). An alternative explanation for this

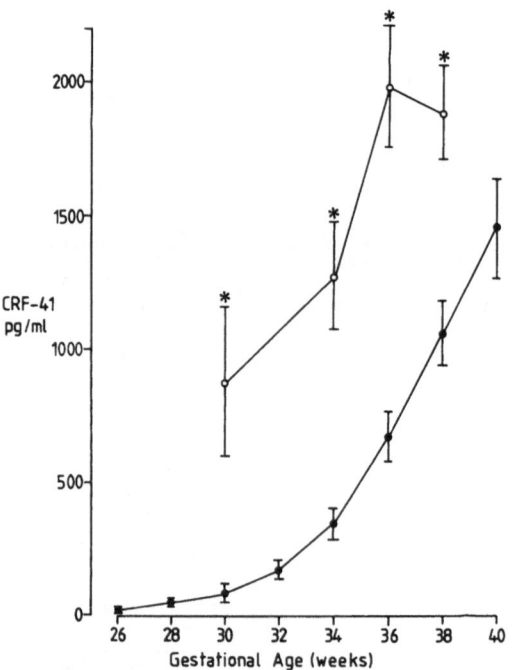

Figure 6.   Elevated levels of plasma CRF in pregnancy-induced hypertension (PIH). *Closed circles* represent the normal range of CRF in sequential samples from 72 women throughout the third trimester. *Open circles* represent CRF values in women with PIH.

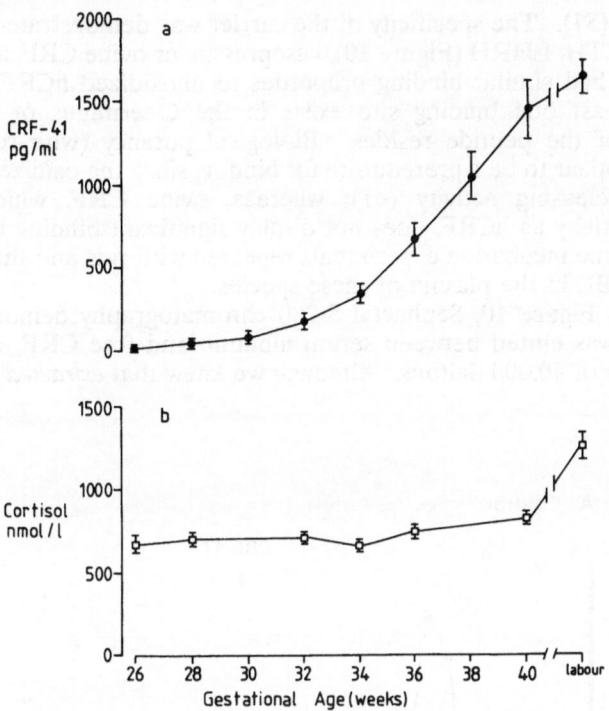

Figure 7. Profiles of *a)* CRF and *b)* total cortisol in maternal plasma with increasing gestational age and at labor.

unusual hormonal profile in pregnancy is now apparent. During our investigations into the nature of the high concentrations of CRF in maternal plasma, we found that most of the endogenous CRF in the maternal circulation is complexed to a binding protein (36,57-60), suggesting that the physiological role of the binding protein in pregnancy may be to attenuate the biological activity of placental CRF.

The existence of the CRF binding protein (CRF BP) was revealed to us following immunoradiometric assay of the chromatographic fractions of a maternal plasma aliquot eluted from Sephadex G50 at neutral pH. Two distinct immunoreactive peaks were found, the major CRF peak (93% of total content) eluted in the void region, with only 7% eluting later in the expected position of free CRF. However, under dissociating conditions (pre-treatment of the plasma with 8 M urea followed by chromatography in a buffer containing 2 M urea), the large molecular weight (LMW) peak was no longer present, and all of the CRF now eluted as the free peptide (Figure 8).

Chromatography of maternal plasma pre-incubated with synthetic human CRF for 5 minutes, 4 hours (data not shown) or 18 hours also resulted in two immunoreactive CRF peaks, with more CRF being incorporated into the LMW peak proportional to the length of the incubation period [38%, 51%, and 71% of total CRF content, respectively (Figure 9)].

The LMW peak in Figures 8 and 9 was unlikely to be due to the CRF precursor, since it dissociated upon urea treatment and could incorporate synthetic CRF in a time-dependent fashion. We concluded that maternal plasma contained a carrier substance for CRF (57). The specificity of the carrier was demonstrated by its failure to bind radioactive ACTH, LHRH (Figure 10), vasopressin, or ovine CRF, although oxidized [Met(0)^{21,38}] hCRF had similar binding properties to unoxidized hCRF. From this, we assumed that at least one binding site exists in the C-terminus of CRF where the species-specificity of the peptide resides. Biological potency (with respect to ACTH release) does not appear to be a prerequisite for binding since the oxidized human peptide has poor ACTH-releasing activity (61); whereas, ovine CRF, which as the same ACTH-releasing activity as hCRF, does not display significant binding to the hCRF BP. Interestingly, the same incubation experiments repeated with rats and sheep indicated the absence of a CRF BP in the plasma of these species.

As shown in Figure 10, Sephacryl S-200 chromatography demonstrated that the CRF BP complex was eluted between serum albumin and free CRF, with a molecular weight in the region of 40,000 daltons. Although we knew that *extracted* maternal plasma

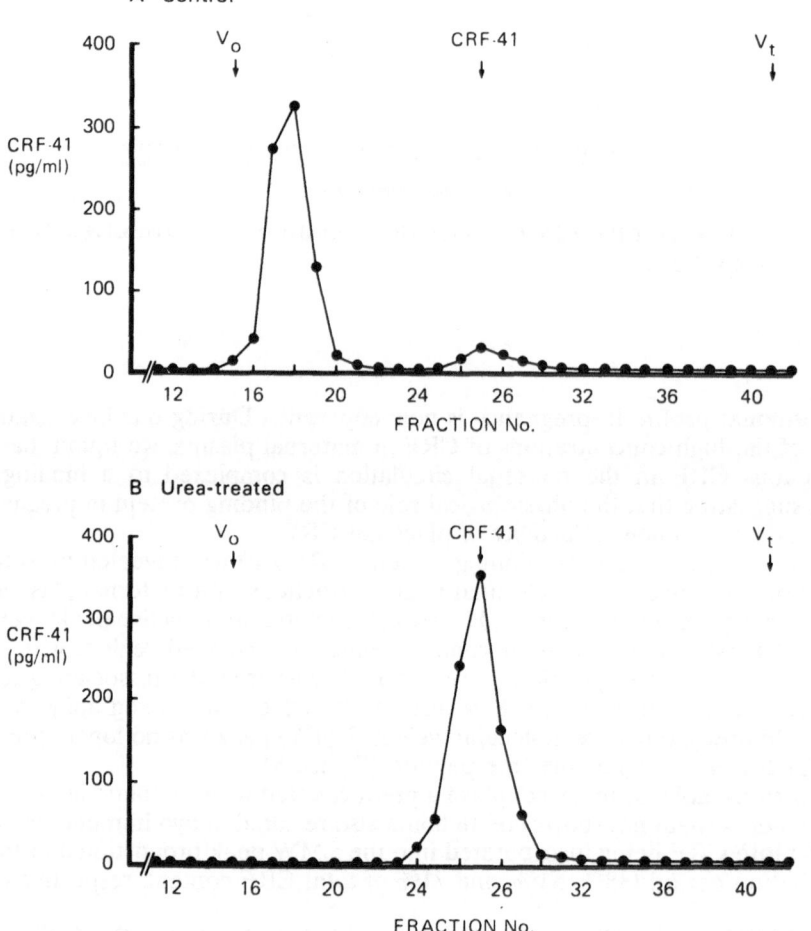

Figure 8. Sephadex G50 chromatogram of maternal plasma in late pregnancy. *A.* Control; *B.* maternal plasma treated with 8 M urea (37˙C, 1 hr) prior to chromatography.

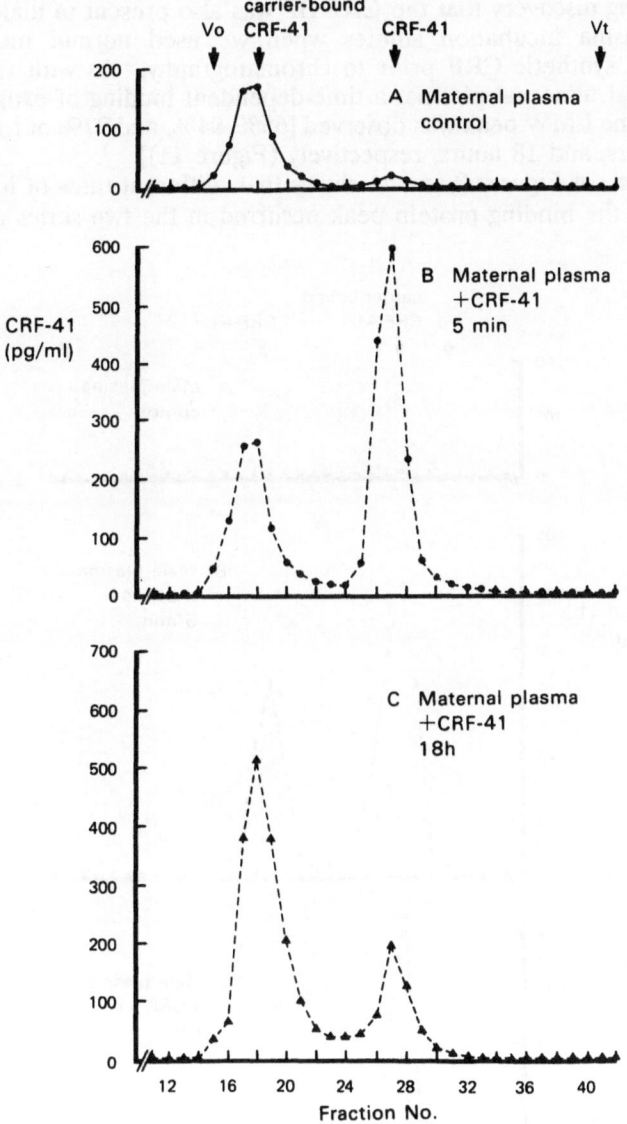

Figure 9. Sephadex G50 chromatogram of late gestation maternal plasma alone *(A)*, or after pre-incubation at 4°C with synthetic hCRF for 5 min *(B)*, or 18 hr *(C)*.

CRF had ACTH-releasing activity (9), we suggested that this need not be the case for the circulating form in pregnancy if the extraction procedure dissociated the inactive CRF BP complex, liberating free CRF. Our attempts to demonstrate inhibition of the ACTH-releasing activity of endogenous CRF in maternal plasma by the CRF BP were confounded by the presence of non-specific factors in dilutions of crude and even partially-purified plasma which induced bioactivity (58). We thus set about purifying the CRF BP to enable definitive bioactivity studies.

The surprising discovery that the CRF BP was also present in male came in light in our control plasma incubation studies when we used normal male plasma for pre-incubation with synthetic CRF prior to chromatography. As with the experiments using late gestational maternal plasma, a time-dependent binding of exogenous CRF to material eluting in the LMW peak was observed [65%, 94%, and 97% of total CRF added at 5 minutes, 4 hours, and 18 hours, respectively (Figure 11)].

A comparison of Figures 9 and 11 shows that different rates of incorporation of synthetic CRF into the binding protein peak occurred in the two series of experiments,

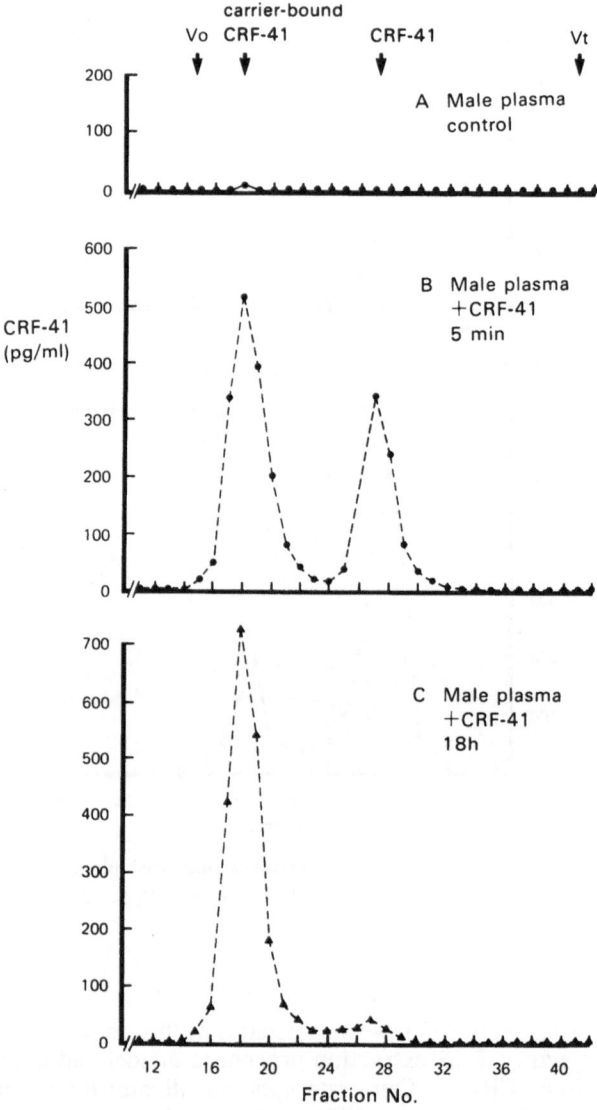

Figure 10. Sephacryl S-200 chromatogram of maternal plasma incubated with ^{125}I-hCRF *(heavy solid line)*, ^{125}I-ACTH *(broken line)*, or ^{125}I-LHRH *(thin solid line)* for 5 min at 4˚C prior to chromatography.

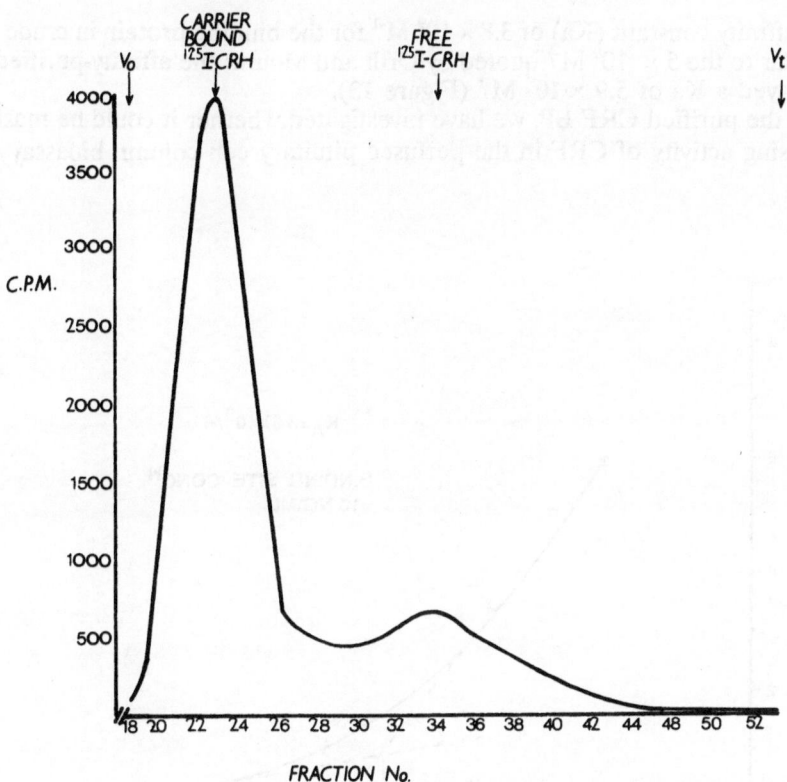

Figure 11. Sephadex G50 chromatogram of normal male plasma alone *(A)*, after pre-incubation at 4·C with synthetic hCRF for 5 min *(B)*, or 18 hr *(C)*.

despite the use of the same quantity of peptide. This could be due either to the presence of endogenous maternal plasma CRF already occupying sites on its binding protein or to different concentrations of binding protein.

Recently, Orth and Mount (59) have demonstrated the existence of a plasma CRF BP by its inhibition of binding of ^{125}I-CRF to CRF antibody, while Ellis *et al.* (62) have also shown that ^{125}I-labelled CRF added to human plasma results in reversible, time-dependent alteration in its molecular size. Carrier-bound CRF in our study is measured as a positive signal in the CRF IRMA, unlike the situation of interference of the signal in radioimmunoassay. Since our incubation studies showed that 1 ml of normal human plasma could bind as much as 10 ng CRF, we chose to use this, rather than maternal plasma, for the purification of the CRF BP. The necessity for dissociating endogenous CRF from the complex is thus removed. Purification has been achieved by repeated affinity chromatography using synthetic hCRF coupled to Sepharose 4B followed by gel filtration (63), tracking the presence of CRF BP throughout by its inhibition of binding of ^{125}I-CRF to CRF antiserum. Two liter batches of human plasma were repeatedly percolated through a CRF solid-phase and the CRF BP desorbed at pH 3.5. Pre-incubation of the affinity-purified material with ^{125}I-CRF and subsequent chromatography resulted in a major bound radioactive peak together with a free ^{125}I-CRF peak, confirming the presence of a high-affinity binding protein which had retained binding ability for CRF following the purification procedure (Figure 12). Scatchard analysis revealed an affinity constant (Ka) of 3.8×10^9 M^{-1} for the binding protein in crude plasma, which is similar to the 5×10^9 M^{-1} quoted by Orth and Mount; the affinity-purified carrier protein displayed a Ka of 5.9×10^9 M^{-1} (Figure 13).

Using the purified CRF BP, we have investigated whether it could be masking the ACTH releasing activity of CRF in the perfused pituitary cell column bioassay (64). A

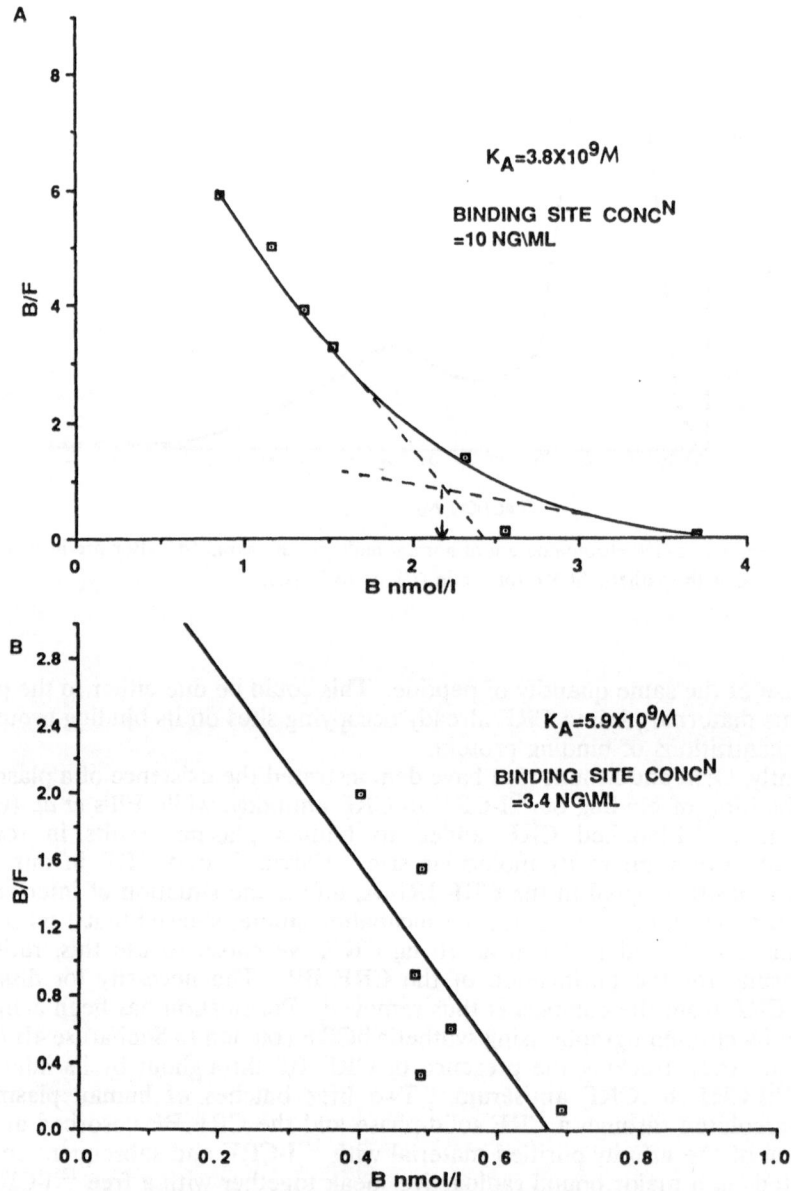

Figure 12. Sephadex G50 chromatogram of affinity-purified CRF BP pre-incubated (5 min at 37°C) with ^{125}I-CRF. V_o, void column volume; V_t, total column volume.

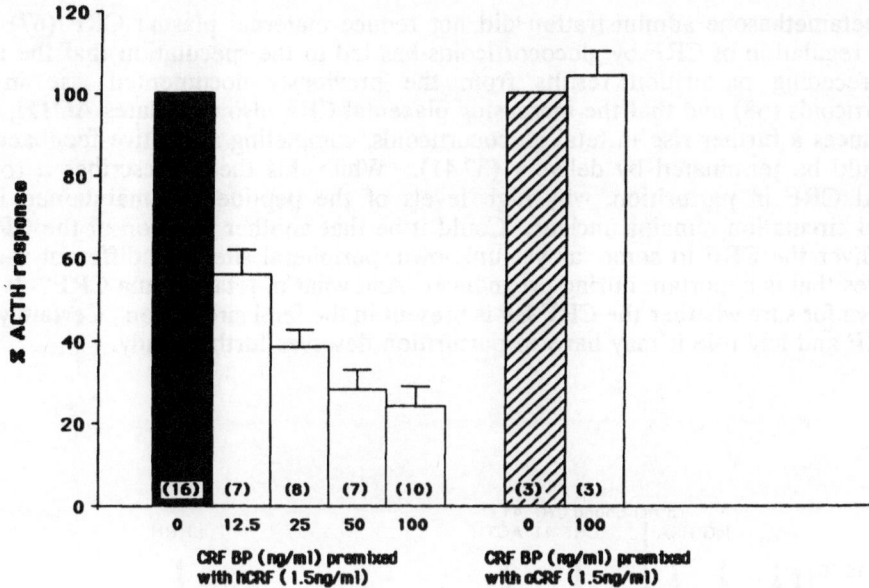

Figure 13. Scatchard plots of *(A)* crude plasma and *(B)* affinity purified CRF BP.

dose-dependent inhibition of ACTH release was obtained with CRF BP pre-incubated (37°C, 15 minutes) with hCRF at 1.5 ng/ml, a concentration commonly found in late gestational maternal plasma. An average of 76% reduction in bioactivity was obtained with binding protein at 100 ng/ml, the dose shown by Scatchard analysis to be equivalent to circulating plasma levels (Figure 14). ACTH release induced by ovine CRF, which is equipotent with the human peptide (61) but which is not bound by the hCRF-specific binding protein (57-59), was not affected by pre-mixing with CRF BP. Ovine CRF thus provided us with an ideal control to demonstrate that the binding protein preparation neither affected the cells directly nor contained any protease activity which could inactivate CRF (Figure 14).

Since we have shown that most of the CRF in maternal plasma circulates in the carrier-bound form (Figure 8), which has greatly reduced ACTH-releasing activity, it is unlikely that the majority of placental CRF released into the maternal circulation is involved in pituitary ACTH secretion. This may well explain why maternal plasma ACTH does not rise in parallel with increasing CRF levels. The very small rise in plasma ACTH which does occur could be due to the small proportion of placental CRF in maternal plasma that remains BP-free, or to ACTH release from the placenta itself. The first of these alternatives is perhaps the least likely, since studies on corticotropic cells have shown that high levels of CRF are necessary to overcome the negative feedback effect of glucocorticoids (65) present during the third trimester. The suggestion that plasma ACTH originates from the placenta rather than from the maternal pituitary, results from the observation that plasma ACTH and urinary cortisol levels in women in late pregnancy are resistant to dexamethasone suppression (43,47,49). Recent studies *in vitro* have suggested that placental ACTH release may be under the paracrine influence of placental CRF (33,66) which is itself subject to positive rather than negative feedback control by cortisol (37,41). Further evidence that the classical suppressive effect of glucocorticoids on hypothalamic CRF is not seen with placental CRF has been provided in an *in vivo* study where betamethasone administration did not reduce maternal plasma CRF (67). The positive regulation of CRF by glucocorticoids has led to the speculation that the rise in CRF preceding parturition results from the previously documented rise in fetal glucocorticoids (68) and that the increasing placental CRF also stimulates ACTH, which then induces a further rise in fetal glucocorticoids, completing a positive feedback loop that would be terminated by delivery (37,41). While this theory describes a role for placental CRF in parturition, why high levels of the peptide are maintained in the maternal circulation remains unclear. Could it be that another function of the CRF BP is to deliver the CRF to some, as yet unknown, peripheral site with different receptor properties that is important during pregnancy? And what of fetal plasma CRF? It is not yet known for sure whether the CRF BP is present in the fetal circulation. Certainly, fetal free CRF and any role it may have in parturition deserves further study.

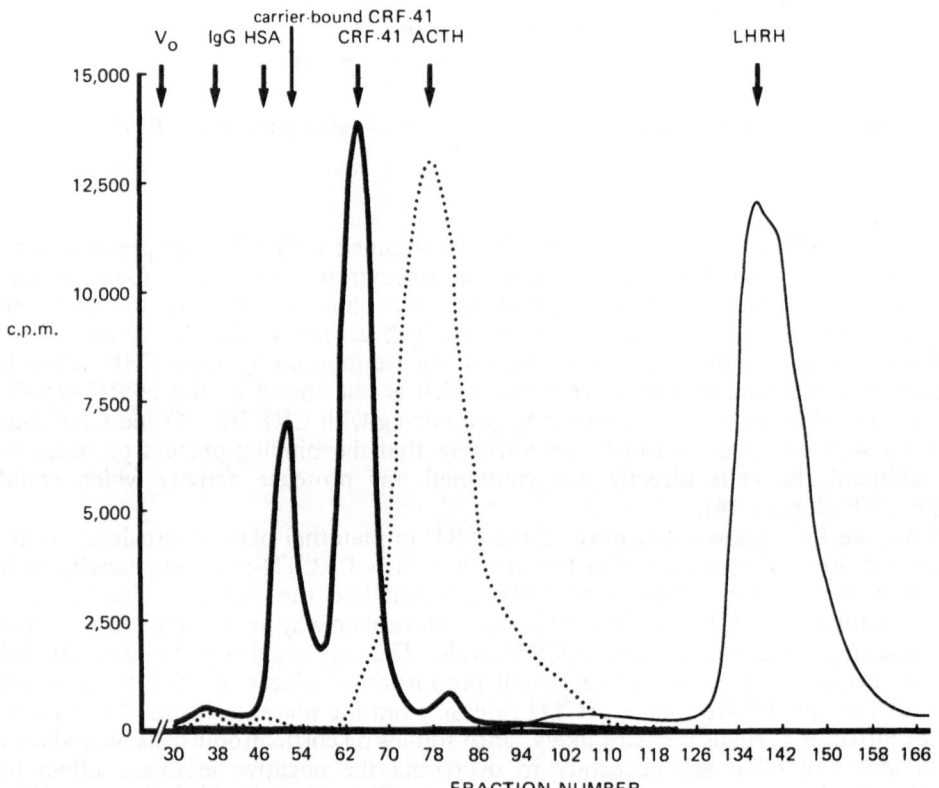

Figure 14.   Effect of the CRF BP on the ACTH-releasing activity of human *(h)* CRF or ovine *(o)* CRF (1.5 ng/ml) after their pre-incubation for 15 min at 37˚C. Mean ± SE of *(n)* stimulations at each dose tested.

References

1. Vale, W., J. Spiess, C. Rivier, and J. Rivier, Characterization of a 41-residue ovine hypothalamic peptide that stimulates secretion of corticotropin and β-endorphin, *Science* 213: 1394-1397, 1981.

2. Suda, T., N. Tomori, F. Yajima, T. Sumitomo, Y. Nakagami, T. Ushiyama, H. Demura, and K. Shizume, Immunoreactive corticotropin-releasing factor in human plasma, *J Clin Invest* 76: 2026-2029, 1985.

3. Cunnah, D., D. Jessop, G.M. Besser, and L.H. Rees, Measurement of plasma levels of human corticotrophin releasing factor, *172nd Meeting of the Society for Endocrinology J Endocrinology [Suppl]* 107, Abstract 31, 1985.

4. Cunnah, D., D. Jessop, L. Perry, F. Afshar, M. Setchell, and L.H. Rees, Measurement of plasma and cerebrospinal fluid levels of human corticotrophin-releasing factor, *68th Annual Meeting of the Endocrine Society*, Anaheim, CA, Abstract 649, p. 193, 1986.

5. Cunnah, D., D.S. Jessop, G.M. Besser, and L.H. Rees, Measurement of circulating corticotropin-releasing factor in man, *J Endocrinology* 113: 123-131, 1987.

6. Stalla, G.K., J. Hartwimmer, J. Schopohl, K. von Werder, and O.A. Muller, Intravenous application of ovine and human corticotrophin releasing factor (CRF): ACTH, cortisol and CRF-levels, *Neuroendocrinology* 42: 1-5, 1986.

7. Stalla, G.K., J. Stalla, J. Schopohl, K. von Werder, and O.A. Müller, Corticotropin-releasing factor in human I. CRF stimulation in normals and CRF radioimmunoassay, *Hormone Res* 24: 229-245, 1986.

8. Hermus, A.R.M.M., G.F.F.M. Pieters, G.J. Pesman, T.J. Benraad, A.G.H. Smals, and P.W.C. Kloppenborg, CRH as a diagnostic and heuristic tool in hypothalamic-pituitary diseases, In E.F. Pfeiffer, G.M. Reaven, W.D. Hetzel, A.R. Hoffman, and O.A. Müller, *Hormone and Metabolism Research Supplement Volume 16*, Theime Medical Publishers, Inc., New York, pp. 68-73, 1987.

9. Linton, E.A., C. McLean, A.C. Nieuwenhuyzen Kruseman, F.J. Tilders, E.A. Van der Veen, and P.J. Lowry, Direct measurement of human plasma corticotropin-releasing hormone by "two-site" immunoradiometric assay, *J Clin Endocrinol Metab* 64: 1047-1053, 1987.

10. Sasaki, A., O. Shinkawa, A.N. Margioris, A.S. Liotta, S. Sato, O. Murakami, M. Go, Y. Shimizu, K. Hanew, and K. Yoshigana, Immunoreactive corticotropin-releasing hormone in human plasma during pregnancy, labour and delivery, *J Clin Endocrinol Metab* 64: 224-229, 1987.

11. Suda, T., N. Tomori, F. Tozawa, T. Mouri, H. Demura, and K. Shizume, Distribution and characterization of immunoreactive corticotropin-releasing factor in human tissues, *J Clin Endocrinol Metab* 59: 861-866, 1984.

12. Nieuwenhuyzen Kruseman, A.C., E.A. Linton, P.J. Lowry, L.H. Rees, and G.M. Besser, Corticotropin releasing factor immunoreactivity in human gastrointestinal tract, *Lancet* II: 1245-1246, 1982.

13. Nieuwenhuyzen Kruseman, A.C., E.A. Linton, J. Ackland, G.M. Besser, and P.J. Lowry, Heterogenous immunocytochemical reactivities of o CRF-41-like material in the human hypothalamus, pituitary and gastrointestinal tract, *Neuroendocrinology* 38: 212-216, 1984.

14. Petrusz, P., I. Merchanthaller, J.L. Maderdrut, S. Vigh, and A.V. Schally, Corticotrophin-releasing factor (CSF)-like immunoreactivity in the vertebrate endocrine pancreas, *Proc Natl Acad Sci USA* 80: 1721-1725, 1983.

15. Charlton, B.G., A. Leake, I.N. Ferrier, E.A. Linton, and P.J. Lowry, Corticotrophin-releasing factor in plasma of depressed patients and controls, *The Lancet* 1: 161-162, 1986.

16. Linton, E.A., A.C. Nieuwenhuyzen Kruseman, C.D.A. Wolfe, E.A. Campbell, and P.J. Lowry, Distribution of immunoreactive CRH in man, *Hormone Metab Res [Suppl]* 16: 38-42, 1987.

17. Shibasaki, T., E. Odagiri, K. Shizume, and N. Ling, Corticotropin-releasing factor-like activity in human placental extract, *J Clin Endocrinol Metab* 55: 384-386, 1982.

18. Sasaki, A., A.S. Liotta, M.M. Luckey, A.N. Margioris, T. Suda, and D.T. Krieger, Immunoreactive corticotropin-releasing factor is present in human maternal plasma during the third trimester of pregnancy, *J Clin Endocrinol Metab* 59: 812-814, 1984.

19.   Stalla, G.K., H. Bost, T. Kaliebe, M. Huber, I. Stalla, D. Pfeiffer, K. von Werder, and O.A. Müller, Human placental corticotropin-releasing factor is identical to hypothalamic corticotropin-releasing factor, *Acta Endocrinol [Suppl 274]* 111: 169-170, 1986.

20.   Campbell, E.A., D.M.G. Halpin, J. Price, A. Dornhorst, B. Gillham, and M.T. Jones, Corticotrophin-releasing factor-like material in maternal plasma and venous and arterial cord plasma, Abstract #98, *J Endocrinol [Suppl]* 104: 65, 1985.

21.   Campbell, E.A., E.A. Linton, C.D.A. Wolfe, P.R. Scraggs, M.T. Jones, and P.J. Lowry, Plasma corticotrophin-releasing hormone concentrations during pregnancy and parturition, *J Clin Endocrinol Metab* 64: 1054-1059, 1987.

22.   Maser-Gluth, C., U. Lorenz, and P. Vecsei, In pregnancy, corticotropin-releasing-factor in maternal blood and amniotic fluid correlates with gestational age, In E.F. Pfeiffer, G.M. Reaven, W.D. Hetzel, A.R. Hoffman, and O.A. Müller, *Hormone and Metabolism Research Supplement Volume 16*, Theime Medical Publishers, Inc., New York, pp. 42-44, 1987.

23.   Gordon, Y.B., and T. Chard, The specific proteins of the human placenta: some new hypotheses, In A. Klopper and T. Chard (eds) *Placental Proteins*, Springer Verlag, Berlin, pp. 1-21, 1979.

24.   Smith, R., M.A.R. Thomson, and W. Cooper, The relationship between changing values of pregnancy-associated plasma protein A in late pregnancy and the onset of labour, *Placenta* 2: 143-148, 1981.

25.   Bishof, P., S. Duberg, W. Herrmann, and P.C. Sizonenko, Amniotic fluid and plasma concentrations of pregnancy-associated plasma protein-A (PAPP-A) throughout pregnancy: comparison with other fetoplacental products, *Br J Obstet Gynaecol* 89: 358-363, 1982.

26.   Laatikainen, T., T. Virtanen, I. Raisanen, and K. Salminen, Immunoreactive corticotrophin-releasing factor and corticotropin in plasma during pregnancy, labor and puerperium, *Neuropeptides* 10: 343-353, 1987.

27.   Wolfe, C.D.A., S.P. Patel, E.A. Campbell, E.A. Linton, J. Anderson, P.J. Lowry, and M.T. Jones, Plasma corticotrophin-releasing factor (CRF) in normal pregnancy, *Br J Obstet Gynaecol* 95: 997-1002, 1988.

28.   Goland, R.S., S.L. Wardlaw, R.I. Stark, L.S. Brown, Jr. and A.G. Frantz, High levels of corticotropin-releasing hormone immunoactivity in maternal and fetal plasma during pregnancy, *J Clin Endocrinol Metab* 63: 1199-1203, 1986.

29.   Schürmeyer, T.H., P.C. Avgerinos, P.W. Gold, W.T. Galluci, T.P. Tomai, G.B. Cutler, Jr., D.L. Loriaux, and G.P. Chrousos, Human corticotropin-releasing factor in man: pharmacokinetic properties and dose-response of plasma adrenocorticotropin and cortisol secretion, *J Clin Endocrinol Metab* 59: 1103-1108, 1984.

30.   Sasaki, A., P. Tempst, A.S. Liotta, A.N. Margioris, L.E. Hood, S.B.H. Kent, S. Sato, O. Shinkawa, K. Yoshinaga, and D.T. Krieger, Isolation and characterization of a corticotropin-releasing hormone-like peptide from human placenta, *J Clin Endocrinol Metab* 67: 768-773, 1988.

31.   Grino, M., G.P. Chrousos, and A.N. Margioris, The corticotropin releasing hormone gene is expressed in human placenta, *Biochem Biophys Res Commun* 148: 1208-1214, 1987.

32.   Saijonmaa, O., T. Laatikainen, and T. Wahlstrom, Corticotrophin-releasing factor in human placenta: localisation, concentration and release *in vitro*, *Placenta* 9: 373-385, 1988.

33.   Petraglia, F., P.E. Sawchenko, J. Rivier, and W. Vale, Evidence for local stimulation of ACTH secretion by corticotropin-releasing factor in human placenta, *Nature* 328: 717-719, 1987.

34.   Frim, D.M., R.L. Emanuel, B.G. Robinson, C.M. Smas, G.K. Adler, and J. Majzoub, Characterization gestational regulation of corticotropin-releasing hormone messenger RNA in human placenta, *J Clin Invest* 82: 287-292, 1988.

35.   Usui, T., Y. Nakai, T. Tsukada, H. Jingami, H. Takahashi, J. Fukata, and H. Imura, Expression of adrenocorticotropin-releasing hormone precursor gene in placenta and other non-hypothalamic tissues in man, *Mole Endocrinol* 2: 871-875, 1988.

36.   Suda, T., M. Iwashita, F. Tozawa, T. Ushiyama, N. Tomori, T. Sumitomo, Y. Nakagami, H. Demura, and K. Shizuma, Characterization of corticotropin-releasing hormone binding protein in human plasma

by chemical cross-linking and its binding during pregnancy, *J Clin Endocrinol Metab* 67: 1278-1283, 1988.

37. Robinson, B.G., R.L. Emanuel, D.M. Frim, and J.A. Majzoub, Glucocorticoid stimulates expression of corticotropin-releasing hormone gene in human placenta, *Proc Natl Acad Sci USA* 85: 5244-5248, 1988.

38. Arbiser, J.L., C.C. Morton, G.A.P. Bruns, and J.A. Majzoub, Human corticotrophin releasing hormone gene is located in the long arm of chromosome 8, *Cytogenet Cell Genet* 47: 113-116, 1988.

39. Liotta, A.S., G.H. Mulder, and D.T. Krieger, Regulation of secretion of human placental corticotropin releasing factor, *67th Annual Meeting of the Endocrine Society*, Baltimore, Abstract 10, 1985.

40. Petraglia, F., S. Sutton, A. Zaccardo, L. Calza, A. Volpe, P. Marrama, L. Giarduno, G. Conkos, P. Sawchenko, A.T.W. Lim, A.R. Genazzani, J. Rivier, and W. Vale, Production of peptide hormones by placental cultures, *8th International Congress of Endocrinology*, Kyoto, Japan, Abstract, p. 24, 1988.

41. Jones, S.A., A.N. Brooks, and J.R.G. Challis, Steroids modulate corticotropin-releasing hormone production in human fetal membranes and placenta, *J Clin Endocrinol Metab* 68: 825-830, 1989.

42. Economides, D., E. Linton, K. Nicolaides, C.H. Rodeck, P.J. Lowry, and T. Chard, Relationship between maternal and fetal corticotrophin-releasing hormone - 41 and ACTH levels in human mid-trimester pregnancy, *J Endocrinology* 114: 497-501, 1987.

43. Carr, B.R., C.R. Parker, J.D. Madden, P.C. MacDonald, and J.C. Porter, Maternal plasma adrenocorticotropin and cortisol relationships throughout human pregnancy, *Am J Obstet Gynecol* 139: 416-422, 1981.

44. Abou-Samra, A.B., M. Pugeat, H. Dechaud, L. Nachury, B. Bouchareb, M. Ferre-Montagne, and J. Tournaire, Increased plasma concentration of N-terminal β-lipotrophin and unbound cortisol during pregnancy, *Clin Endocrinol* 20: 221-228, 1984.

45. Wolfe, C.D.A., S.P. Patel, E.A. Linton, E.A. Campbell, J. Anderson, A. Dornhorst, P.J. Lowry, and M.T. Jones, Plasma corticotrophin releasing factor (CRF) in abnormal pregnancy, *Br J Obstet Gynaecol* 95: 1003-1006, 1988.

46. Burke, C.W., and F. Roulet, Increased exposure of tissues to cortisol in late pregnancy, *Br Med J* 1: 657-659, 1970.

47. Rees, L.H., C.W. Burke, T. Chard, S.W. Evans, and A.T. Letchworth, Possible placental origin of ACTH in normal human pregnancy, *Nature* 254: 620-622, 1975.

48. Reisine, T., and H.-U. Affolter, Hormone receptor regulated proopiomelanocortin gene expression, *Biochem Pharmacol* 36: 191-195, 1987.

49. Genazzani, A.R., F. Fraioli, and J. Hurlimann, Immunoreactive ACTH and cortisol plasma levels during pregnancy: detection and partial purification of corticotropin-like placental hormone, human chorionic corticotropin, *Clin Endocrinol* 4: 1-14, 1975.

50. Plotsky, P.M., and P.E. Sawchenko, Hypophysial-portal plasma levels, median eminence content and immunohistochemical staining of corticotropin-releasing factor, arginine vasopressin and oxytocin after pharmacological adrenalectomy, *Endocrinology* 120: 1361-1369, 1987.

51. Fink, G., I.C.A.F. Robinson, and L.A. Tannahill, Effects of adrenalectomy and glucocorticoids on the peptides CRF-41, AVP and oxytocin in rat hypophysial portal blood, *J Physiol* 401: 329-345, 1988.

52. Gillies, G.E., E.A. Linton, and P.J. Lowry, Corticotropin releasing activity of the new CRF is potentiated several times by vasopressin, *Nature* 299: 355-357, 1982.

53. Schulte, H.M., B. Allolio, D. Weisner, and E.E. Ohnhaus, ACTH and cortisol secretion in response to synthetic human corticotrophin-releasing hormone in late pregnancy, *8th International Congress of Endocrinology*, Kyoto, Japan, Abstract, p. 82, 1988.

54. Suda, T., M. Iwashita, T. Ushiyama, F. Tozawa, T. Sumitomo, Y. Nakagami, H. Demura, and K. Shizume, Responses to corticotropin-releasing hormone and its bound and free forms in pregnant and nonpregnant women, *J Clin Endocrinol Metab* 69: 38-42, 1989.

55. Reisine, T., and A. Hoffman, Desensitization of corticotropin-releasing factor receptors, *Biochem Biophys Res Comm* 111: 919-925, 1983.

56. Schulte, H.M., G.P. Chrousos, D.L. Healy, P.W. Gold, E.H. Odfield, J.D. Booth, G.B. Cutler, Jr., and

D.L. Lorieux, Short- and long-term continuous infusion of corticotropin-releasing factor (CRF) in men and nonhuman primates: physiologic and clinical implications, *The 7th International Congress of Endocrinology*, Quebec City, Canada, Abstract 2268, p. 1394, 1984.

57. Linton, E.A., and P.J. Lowry, A large molecular weight carrier for CRF-41 in human plasma, *J Endocrinol [Suppl]* 111, Abstract 150, 1986.

58. Linton, E.A., C.D.A. Wolfe, D.P. Behan, and P.J. Lowry, A specific carrier substance for human corticotrophin releasing factor in late gestational maternal plasma which could mask the ACTH-releasing activity, *Clin Endocrinol* 28: 315-324, 1988.

59. Orth, D.N., and C.D. Mount, Specific high-affinity binding protein for human corticotrophin-releasing hormone in normal human plasma, *Biochem Biophys Res Commun* 143: 411-417, 1987.

60. Suda, T., M. Iwashita, F. Tozawa, T. Ushiyama, N. Tomori, T. Sumitomo, Y. Nakagami, H. Demura, and K. Shizume, Secretion of corticotropin-releasing factor in pregnancy, In H. Imura, S. Shizume, and K. Yoshid (eds) *Progress in Endocrinology: Proceedings of the 8th International Congress of Endocrinology, Volume 1*, Kyoto, Japan, Elsevier Medica, Amsterdam, pp. 401-405, 1988.

61. Rivier, J., J. Spiess, and W. Vale, Characterization of rat hypothalamic corticotropin-releasing factor, *Proc Natl Acad Sci USA* 80: 4851-4855, 1983.

62. Ellis, M.J., J.H. Livesey, and R.A. Donald, Circulating plasma corticotrophin-releasing factor-like immunoreactivity, *J Endocrinol* 117: 299-307, 1988.

63. Behan, D.P., E.A. Linton, and P.J. Lowry, Isolation of the human plasma corticotropin-releasing factor-binding protein, *J Endocrinol* 122: 23-31, 1989.

64. Gillies, G., and P.J. Lowry, Perfused rat isolated anterior pituitary cell column as bioassay for factor(s) controlling release of adrenocorticotropin: validation of a technique, *Endocrinology* 103: 521-527, 1978.

65. Vale, W., J. Vaughan, M. Smith, G. Yamamoto, J. Rivier, and C. Rivier, Effects of synthetic ovine corticotropin-releasing factor, glucocorticoids, catecholamines, neurohypophyseal peptides and other substances on cultured corticotropic cells, *Endocrinology* 113: 1121-1131, 1983.

66. Margioris, A.N., M. Grino, P. Protos, P.W. Gold, and G.P. Chrousos, Corticotrophin-releasing hormone and oxytocin stimulate the release of placental proopiomelanocortin peptides, *J Clin Endocrinol Metab* 66: 922-926, 1988.

67. Tropper, P.J., R.S. Goland, S.L. Wardlaw, H.E. Fox, and A.G. Frantz, Effects of betamethasone on maternal plasma corticotropin-releasing factor, ACTH and cortisol during pregnancy, *J Perinatal Med* 15: 221-225, 1987.

68. Fencl, M., R.J. Stillman, J. Cohen, and D. Tulchinsky, Direct evidence of sudden rise in fetal corticoids late in human gestation, *Nature* 287: 225-266, 1980.

THERAPEUTIC EFFECTS OF NEUROACTIVE DRUGS ON HYPOTHALAMO-PITUITARY IN MAN

P.J. Trainer and A. Grossman

Department of Endocrinology
St. Bartholomew's Hospital
West Smithfield, London EC1A 7BE   United Kingdom

The treatment of endocrine disorders by neuropharmacological manipulation is a therapeutic field that has only began to be exploited in the last twenty years, and continues to evolve rapidly. In 1947, it was first suggested by Green and Harris (1) that hypothalamic substances regulated hormone secretion from the anterior pituitary. Corticotropin-releasing factor was the first of these substances to be identified in a bioassay system, while in 1982 the most recent hypothalamic peptide, growth hormone releasing hormone (GHRH), was sequenced. Paralleling the identification of the hypothalamic peptides has been progress in understanding the complex interactions of the innervation of the hypothalamus. With this knowledge, and the concomitant availability of biosynthetic hypothalamic peptides, modulation of the secretion of prolactin, growth hormone (GH) and adrenocorticotropin by manipulation of the hypothalamic-pituitary axis has resulted in major therapeutic advances. This chapter will confine itself to the drugs relevant to these hormones, but excludes analogues of the hypothalamic peptides such as octreotide and the GnRH superagonists.

Prolactin

The existence and role of prolactin had been suspected for many years, but only confirmed in 1970 (2). Prolactin secretion, in contrast to the other hormones of the anterior pituitary is principally under tonic inhibitory regulation from the hypothalamus. Dopamine is secreted from nerve terminals in the median eminence of the hypothalamus; it reaches the pituitary *via* the hypothalamic-pituitary portal circulation and binds to specific receptors on the lactotroph, thereby inhibiting prolactin release.

Hyperprolactinemia, and the associated syndrome of reduced libido, galactorrhea, amenorrhea and impotence, can be primary (due to a prolactinoma or to stalk disruption) or secondary to a system disorder. Chronic renal failure, hypothyroidism, polycystic ovaries, and drugs such as chlorpromazine, haloperidol and metoclopramide are among the more common causes of secondary hyperprolactinemia. Clearly, the treatment of secondary hyperprolactinemia is directed towards the underlying cause.

With a single exception so far, the drugs available for the treatment of hyperprolactinemic states are ergot alkaloid derivatives. The side-effects are common to this group of drugs but vary in severity between preparations. Nausea, vomiting, dizziness, postural hypotension and anorexia may be seen, but can almost always be alleviated by initiating therapy with a low dose, 1.25 mg of the case of bromocriptine, taken last thing at night with a snack, and ultimately building up to take it in the middle of a meal. With

Circulating Regulatory Factors and Neuroendocrine Function
Edited by J. C. Porter and D. Ježová
Plenum Press, New York, 1990

165

these precautions, continued intolerance is seen in 5% or less of patients. Nasal congestion, headache, fatigue, constipation and abdominal pain are less frequent symptoms. Psychiatric disturbance has been reported as developing in about 1% of patients on bromocriptine (3), ranging from mood change and depression to psychosis and hallucination, but this is usually reversible on discontinuation of therapy.

Bromocriptine (2-bromo-α-ergocryptine), a semi-synthetic ergoline with a tripeptide sequence, was first reported to lower serum prolactin levels in patients with prolactinomas in 1972 (4), and subsequently it has been the most widely used drug in this field. In prolactinomas, the dose of bromocriptine required to normalize prolactin levels varies between 7.5 and 40 mg daily (5) and can be successfully administered once daily in many patients (6). Shrinkage of prolactinoma size in response to bromocriptine is now well documented (7), occurring in 60-80% of patients. Reported pregnancy rates in patients on bromocriptine with antecedent infertility are as high as 86% (5). Although there is no evidence that bromocriptine is teratogenic or increases abortion rates (8,9), we recommend that patients discontinue therapy as soon as pregnancy is confirmed.

A single intramuscular 50 mg dose of a depot preparation of bromocriptine has recently been shown to be capable of suppressing serum prolactin levels for up to 6 weeks in normal subjects (10), and in patients with hyperprolactinemia. By measuring serum prolactin, a rapid diagnosis of dopamine agonist sensitivity or resistance (11) can be reached. Furthermore, it has generally been found that patients relatively intolerant of bromocriptine by mouth can be initiated easily onto the therapy following a single depot injection (Figure 1). It can also be used pre-operatively in cases of large prolactinomas to reduce tumor bulk and thereby make surgery less hazardous. Computer tomography has been used to demonstrate a decrease in tumor size in 50% of patients with macroadenomas (12), and rapid shrinkage of a macroprolactinoma with associated visual field defects has been reported with depot bromocriptine. A repeatable preparation of

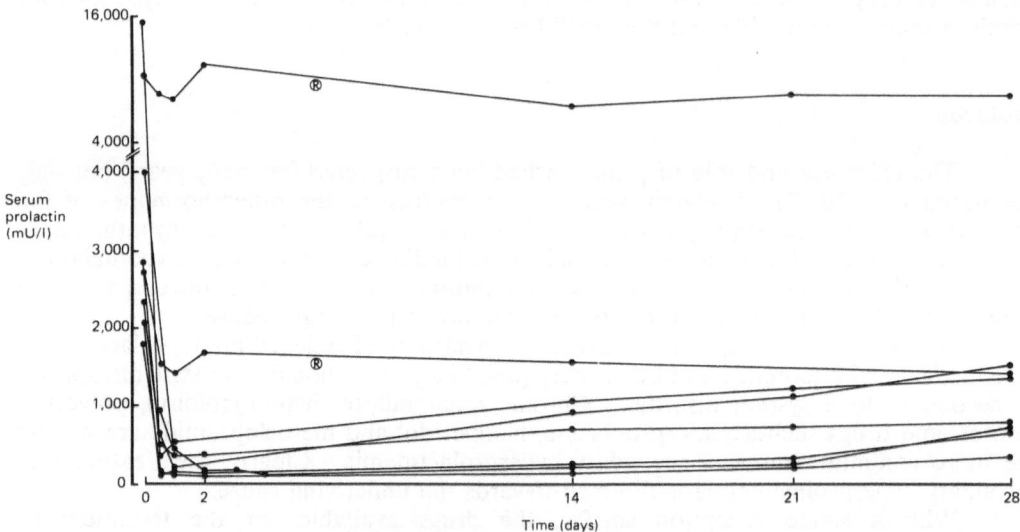

Figure 1. Serum prolactin levels in eight patients with hyperprolactinemia given a single injection of 50 mg depot bromocriptine. Two of the eight patients did not respond to this therapy with a normalization of prolactin (*closed circles*), and were subsequently confirmed to be resistant to oral therapy. Reproduced from A. Grossman *et al.* (11) with permission of the authors.

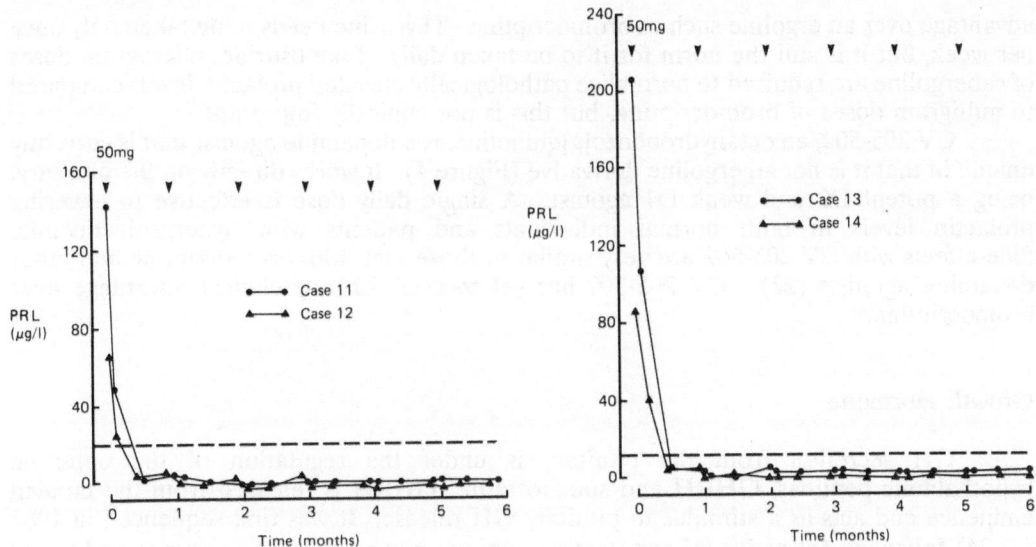

Figure 2. The response of four patients with prolactinomas to repeated injections *(arrows)* of 50 mg Parlodel LAR. These patients all had large pituitary tumors, but showed persistently normal prolactin levels following the first injection (E. Ciccarelli, G.M. Besser, and A. Grossman, unpublished observations).

depot bromocriptine (Parlodel LAR) is now available allowing long-term control of hyperprolactinemia with a once monthly injection in hyperprolactinemic patients. In our experience this is especially useful in patients persistently intolerant of oral bromocriptine, but may also be preferred by patients in terms of convenience (Figure 2).

Several other ergoline dopamine agonists are available but clinical experience with them is limited, and none has yet been shown to possess significant advantages over bromocriptine.

Pergolide and mesulergine are both 8-amino acid ergolines and potent inhibitors of prolactin secretion, pergolide having a longer duration of action than bromocriptine. Three studies have found pergolide to be better tolerated than bromocriptine (13-15), but in general if problems are encountered with one dopamine agonist they are likely to recur with other members of the group. However, this is not always the case, and it is always worth considering these alternatives. Unfortunately, mesulergine has recently been withdrawn from clinical trials.

Lisuride (n-D-6-methyl-8-isoergolenyl-N'-N-diethyl-carbamide hydrogen maleate) is a semi-synthetic isolysergic acid derivative which, in addition to its dopaminergic action (16), has peripheral and central anti-serotonin activity (17). It is effective in lowering prolactin levels, but we have found that is generally less well tolerated than bromocriptine (18). Terguride is a C9-10 dehydrogenated derivative of lisuride which is also an α_2-adrenoceptor antagonist, and therefore is less likely to cause postural hypotension (19). In acute studies we have found it to be better tolerated than bromocriptine (Figure 3), but long-term follow-up of patients is required for a definitive opinion.

Cabergoline is the longest-acting oral ergot-derived dopamine agonist available at present. Maximum suppression of prolactin secretion following a single oral dose of 600 μg being seen at 4 days; at 28 days serum prolactin is still slightly lower than pre-treatment (20). It has been suggested that cabergoline might be administered as a once or possibly twice-weekly oral preparation (10), but it is doubtful if this is a particular advantage over an ergoline such as bromocriptine. Thyroxine needs to be taken only once per week, but it is still the norm for it to be taken daily. Like lisuride, microgram doses of cabergoline are required to normalize pathologically elevated prolactin levels compared to milligram doses of bromocriptine, but this is not clinically important.

CV 205-502, an octahydrobenzo[g]quinoline, is a dopamine agonist that is currently unique in that it is not an ergoline derivative (Figure 4). It works directly on the pituitary, being a potent D2 and weak D1 agonist. A single daily dose is effective in lowering prolactin levels in both normal individuals and patients with hyperprolactinemia. Side-effects with CV 205-502 are very similar to those seen with bromocriptine and other dopamine agonists (22). CV 205-502 has yet to establish any clinical advantage over bromocriptine.

Growth Hormone

GH secretion from the pituitary is under the regulation of the opposing hypothalamic peptides, GHRH and somatostatin. GHRH is released from the median eminence and acts as a stimulus to pituitary GH release. It was first sequenced in 1982 (23,24) following extraction of pancreatic tumors in two acromegalic patients and found to occur as 37, 40, and 44 amino acid sequences; however, structure-activity studies

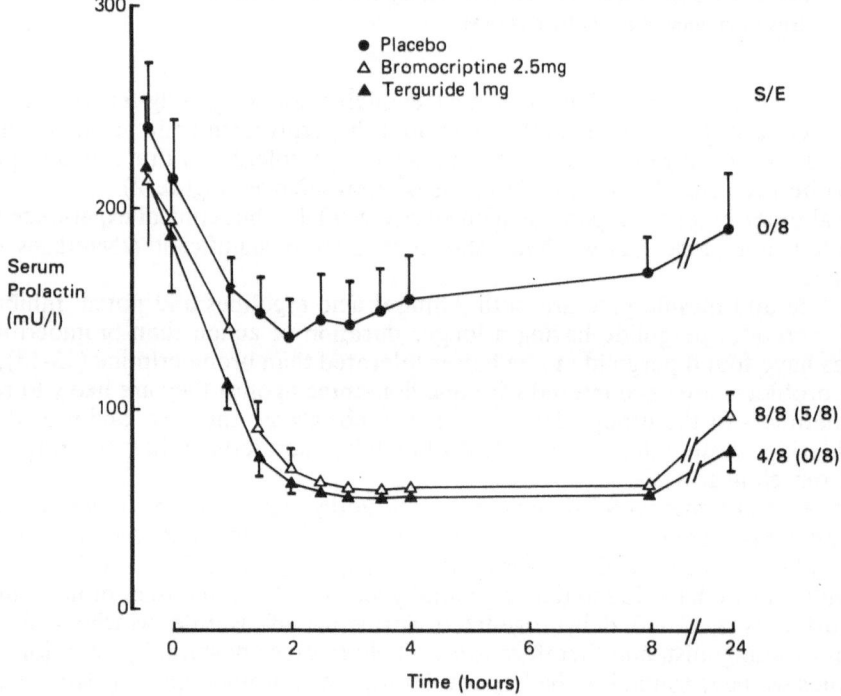

Figure 3.   Change in mean serum prolactin and side effect profiles in eight normal subjects given single doses of bromocriptine (2.5 mg) or terguride (1 mg). Note that the side effects (S/E) were less with terguride compared to bromocriptine, and severe side effects *(shown in parentheses)* did not occur in any subject after terguride. Reproduced from E. Ciccarelli *et al.* (19) with permission of The American Fertility Society.

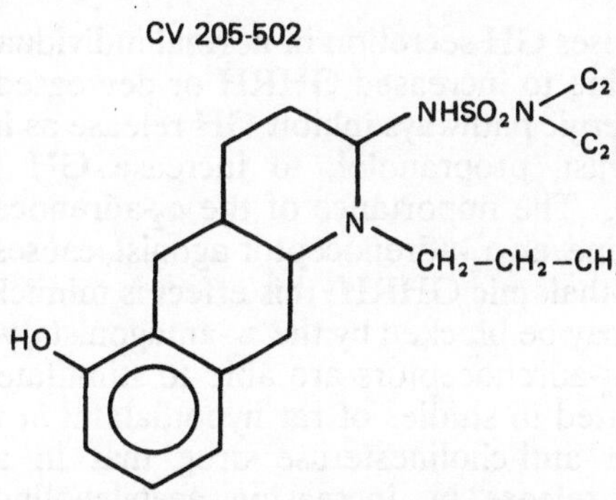

CV 205-502

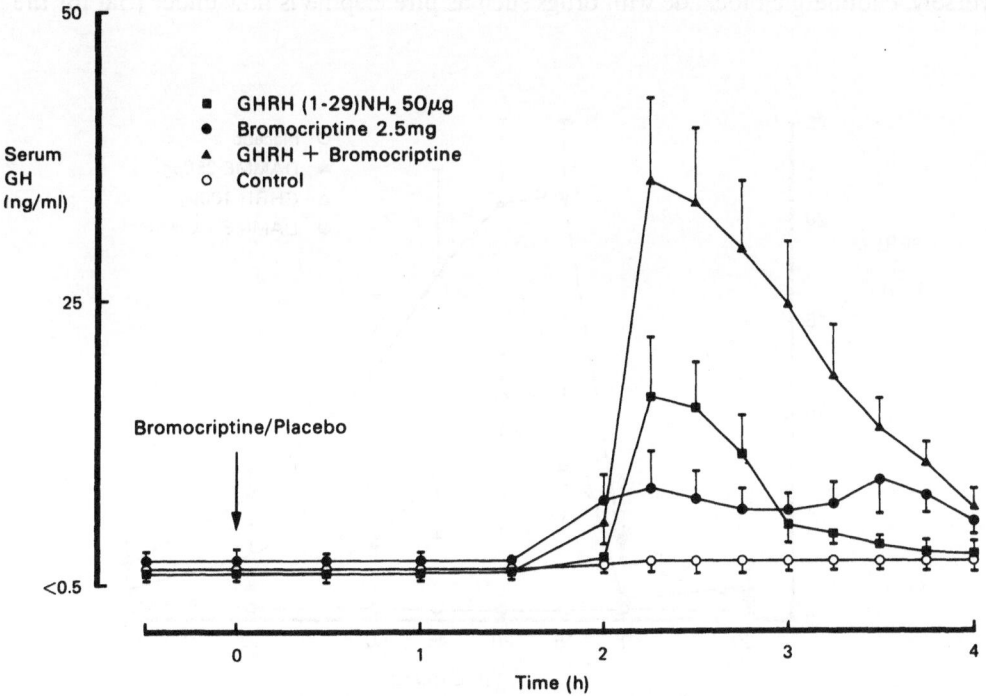

Figure 4. Structure of the new non-ergot dopamine agonist, CV 205-502.

demonstrated that GHRH(1-29)-NH2 had full biological activity. Somatostatin from the median eminence counter-regulates GHRH by inhibiting pituitary GH release, but is itself under tonic inhibitory control by muscarinic cholinergic neurons.

Hypothalamic secretion of GHRH and somatostatin is under complex and sophisticated control. As already mentioned, acetylcholine may decrease somatostatin secretion and hence lead to a rise in GH. Dopaminergic stimulation with drugs such as

- ■ GHRH (1-29)NH₂ 50µg
- ● Bromocriptine 2.5mg
- ▲ GHRH + Bromocriptine
- ○ Control

Serum GH (ng/ml)

Bromocriptine/Placebo

Time (h)

Figure 5. Change in mean serum growth hormone (GH) following bromocriptine or GHRH, or the two in combination, in a group of eight normal subjects. Note that the increase in GH induced by a maximal dose of GHRH was potentiated by bromocriptine, suggesting that dopamine agonists act *via* somatostatin to increase GH in such subjects. Reproduced from G. Delitala *et al.* (26) with permission of S. Karger, A.G.

L-dopa or bromocriptine causes GH secretion in normal individuals, although controversy exists as to whether this is due to increased GHRH or decreased somatostatin secretion (Figure 5) (25,26). β-Adrenergic pathways inhibit GH release as indicated by the abilityof the β-adrenoceptor antagonist, propranolol, to increase GH secretion, probably *via* somatostatin inhibition (27). The importance of the α_2-adrenoceptor on GH regulation is still not clear, while clonidine, an α_2-adrenoceptor agonist, causes GH secretion probably as a result of increased hypothalamic GHRH: this effect is mimicked by the more specific α_2-agonist, S3341 (28), and may be blocked by the α_2-antagonists, yohimbine and idozoxan. This suggests that specific α_2-adrenoceptors are able to stimulate GHRH release, which has been directly demonstrated in studies of rat hypothalami *in vitro* (29).

Pyridostigmine is an anti-cholinesterase drug that in animals clearly inhibits hypothalamic somatostatin release by increasing acetylcholine concentration at the somatostatinergic neurons (30). Oral pyridostigmine in doses of 60 and 120 mg has been shown to increase basal GH secretion and the GH response to GHRH; this synergy is assumed to be secondary to pyridostigmine inhibiting somatostatin secretion. Pyridostigmine potentiates the GH response to GHRH in normal adults and in children with short stature when given in the morning but not when given at night (R. Ross, J. Kirk, and P. Trainer, unpublished data). This suggests that somatostatin tone is already low or absent nocturnally. The attenuated GH responses seen in obesity or acutely after glucose are also counteracted by pyridostigmine, implicating somatostatin in these processes. In theory, decreasing somatostatin tone with drugs such as pyridostigmine should increase growth in children with short stature, but preliminary studies have not been encouraging. Conversely, cholinergic blockade with drugs such as pirenzepine is now under trial for the

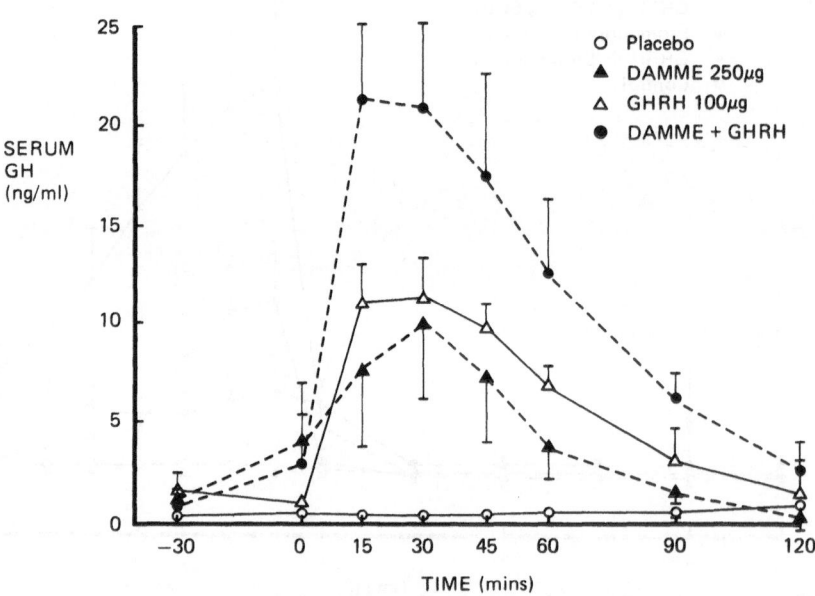

Figure 6. Change in mean serum growth hormone (GH) in normal subjects given the enkephalin analogue, DAMME 250 µg, GHRH 100 µg, or the two in combination. Note that the maximal rise in serum GH induced by GHRH was further potentiated by the enkephalin analogue, suggesting that opiates increased GH *via* somatostatin. (G. Delitala, G.M. Besser, and A. Grossman, unpublished data.)

treatment of tall stature or diabetic retinopathy. In both conditions excess GH release is clearly disadvantageous.

The physiological significance of opiate and serotoninergic pathways in the regulation of GH secretion remains uncertain. However, we have recently found that opiates may also act to increase GH release *via* somatostatin (Figure 6) (G. Delitala, G. Besser, and A. Grossman, unpublished data).

Free fatty acids, glucose, and insulin-like growth factor-1 (IGF-1) all have an inhibitory effect on GH secretion, though it remains unclear whether the main inhibition is at a hypothalamic or pituitary level. Recently it has been shown that 2 IU of intravenous GH, blocks the response of endogenous GH to GHRH administered 2 hours later, this inhibition being reversed by the prior administration of pyridostigmine (31). It is therefore believed that GH autoregulates its own secretion by feeding back on the hypothalamus to increase somatostatin secretion.

Since 1958 exogenous GH, originally derived from human pituitary extracts, has been used with success in the treatment of short stature, however, it has the disadvantages of involving daily injections and being very expensive. It would clearly be attractive if a cheaper and less arduous therapy could be developed. GHRH has been shown to restore pulsatile GH secretion and growth in children with classical GH deficiency (32), indicating that the failure of GH secretion is commonly due to a hypothalamic rather than a pituitary defect. Fifty percent of severely GH-deficient children treated twice-daily with GHRH show significant improvement in growth velocity (33), and up to 70% improved with pulsatile therapy (34). As already mentioned, the bioactivity of GHRH lies in the first 29 amino acids, and it is this shortened molecule that is generally used for therapy. However, at present GHRH therapy, while able to stimulate growth, does not present any advantages over GH as it is as expensive to manufacture and requires at least twice-daily parenteral administration: GH remains currently the treatment of choice. Studies using continuous constant rate infusions of GHRH have demonstrated that there is no attenuation of GH pulsatility with time (35), and this encourages the belief that the therapy of the future may be depot GHRH once weekly or monthly.

Oral medication designed to stimulate endogenous GH secretion is an alternative approach to treating short stature. As we have discussed, while pyridostigmine is a theoretically attractive means of inducing growth, current work suggests that it is probably ineffective. Another candidate is the α_2-adrenoceptor agonist, clonidine, which has been shown in acute studies to stimulate GH release. Three studies (36-38) have been published on the value of clonidine in the treatment of short stature, but their conclusions are discordant. Sixteen prepubertal children treated with 150 $\mu g/m^2$ grew, as did 22 of 34 children with constitutional growth delay treated with 100 $\mu g/m^2$. However, in a further study of children with constitutional growth delay, again using 100 $\mu g/m^2$, no difference was found between six months treatment with placebo or clonidine. It is very difficult to reconcile these results, and the case for clonidine therapy remains unproven. Clonidine causes sleepiness and postural hypotension, but these were apparently not significant problems in the majority of children studied. The concept of an oral preparation, such as clonidine, to treat short stature with simulation of endogenous GH secretion remains very attractive as it overcomes both problems of exogenous GH: clonidine is orally active and at a dose of 100 $\mu g/m^2$ is less than one thirtieth of the price of 2 IU of GH (UK prices).

The opposite end of the clinical spectrum is hypersecretion of GH in acromegaly. Dopamine stimulates GH secretion in normal individuals, probably *via* somatostatin (see above), but paradoxically inhibits GH release in acromegalics, the former effect being mediated at a hypothalamic level and the latter being due to direct binding to tumor dopamine receptors. The ergot derivative, bromocriptine, was the first effective medical treatment for acromegaly. GH levels are improved in 70% with doses ranging between

10-40 mg, rarely 60 mg being required. Tumor shrinkage occurs in acromegalics but is not as impressive as that seen with prolactinomas. As bromocriptine is rarely curative therapy in acromegaly, and only occasionally are GH levels normalized, its role is as adjuvant therapy to definitive treatment such as surgery or radiotherapy. The recent introduction of the somatostatin analogue, octreotide, may further decrease its clinical importance.

ACTH

Adrenocorticotropin (ACTH) is of vital importance to the survival of the organism, and the crucial moment-to-moment fluctuations in its secretion, in response to environmental stimuli, are under complex hypothalamic control. The 41-amino acid peptide, corticotropin-releasing factor (CRF-41), is probably the principal hypothalamic stimulus to ACTH release, although vasopressin is also important. There is good evidence to suggest the existence of another hypothalamic factor, possibly atrial natriuretic factor, which inhibits ACTH release (39). Catecholamine pathways are of central importance, with noradrenaline stimulating ACTH secretion by the combined activation of α_1-adrenoceptors (40), probably mediated by vasopressin, and also a β-adrenoceptor pathway *via* CRF-41 (41). Activation of the serotoninergic and histaminergic pathways also simulates ACTH release, probably *via* CRF-41.

Cushing's disease (hypercortisolemia due to a pituitary ACTH-secreting tumor) is best treated by transsphenoidal microadenectomy (42), if necessary combined with external beam mega-voltage radiotherapy (43). Medical treatment of Cushing's disease is indicated under three circumstances: *a)* in an ill patient during investigation, *b)* in preparation for surgery, or *c)* while waiting for radiotherapy to be curative when surgery has not been successful: this may take between 2 and 10 years. Drugs acting on the adrenal cortex are the most effective in Cushing's disease, but there are in addition several drugs that act on the hypothalamic-pituitary axis.

The dopamine agonist, bromocriptine, has been advocated in the treatment of ACTH-secreting tumors of the pituitary, particularly in those presumed to be of intermediate lobe in origin (44). In our experience, responsiveness to bromocriptine is extremely unusual, and is unrelated to any putative pars intermedia pathogenesis. In one patient, a typical pituitary basophilic adenoma responded both *in vivo* and *in vitro* to bromocriptine with a fall in ACTH; the tumor also secreted α-MSH, which existed in several acetylated and desacetylated forms (45). However, we have not found any other patient to respond similarly.

Cyproheptadine is a anti-serotoninergic drug with additional anti-histamine and anti-cholinergic activity that acts on the hypothalamic-pituitary axis to lower ACTH secretion. Results in the treatment of Cushing's disease have been disappointing (46)—depression and weight gain being the major side effects.

The neurotransmitter γ-amino butyric acid (GABA) has been demonstrated in rats to inhibit hypothalamic CRF-41 release (47; S. Tsagarakis and A. Grossman, unpublished data). Sodium valproate, a GABA aminotransferase inhibitor (48), has been shown to reduce circulating ACTH levels in Nelson's Syndrome (49). Remission of Cushing's disease has also been reported (50), but it has yet to establish a role in long-term management.

It has been suggested that cholinergic tone has a regulatory role in the secretion of ACTH during stress, but studies using pyridostigmine to increase hypothalamic cholinergic tone have failed to confirm this (A. Freeman, R. Ross, and A. Grossman, unpublished data). It appears that ACTH secretion is independent of cholinergic tone, which is compatible with the observation that somatostatin fails to affect the ACTH response to CRF-41 (51) and somatostatin analogues are ineffective in the treatment of Cushing's disease.

If transsphenoidal surgery for Cushing's disease is successful and the entire adenoma is excised, the patient will be rendered hypoadrenal, as the remaining pituitary corticotrophs will be suppressed and may remain so for several years. It is an attractive proposition to believe that ACTH secretion can be stimulated by administration of parenteral CRF-41. Unfortunately, treatment with CRF-41 fails to accelerate recovery of the pituitary-adrenal axis (52), indicating that the suppression is primarily at a supra-hypothalamic level. Thus, the neuropharmacology of ACTH has so far generally failed to be of major therapeutic importance.

Conclusions

In the last twenty years, the introduction of drugs that act on the hypothalamic-pituitary axis to modulate pituitary secretion has led to major advances in the management of several endocrinological disorders. Dopamine agonists are established as having a central role in the therapy of prolactinomas and acromegaly. To date, bromocriptine has been the most widely prescribed of these drugs, but in the future it may be superseded by better tolerated and longer-acting analogues.

GHRH is being investigated for the treatment of short stature. Although it has yet to be shown to have any intrinsic advantage over GH, the advent of a depot preparation of GHRH requiring children to have only a once weekly or possibly monthly injection might make it preferable to GH. In the treatment of Cushing's disease, no drug has been demonstrated to be consistently effective in lowering ACTH secretion long-term. Bromocriptine and cyproheptadine will on occasion induce remission, but drugs acting on the adrenal cortex are of more use therapeutically. Finally, it should be mentioned that occasional tumors secreting TSH may respond to dopamine agonists or somatostatin analogues. Nevertheless, we are likely to see increasing use of drugs affecting the neuroendocrine axis in the next few years.

References

1. Green, J.D., and G.W. Harris, The neurovascular line between the neurohypophysis and adenohypophysis, *J Endocrinol* 5: 136-146, 1947.
2. Frantz, A., and D.L. Kleinberg, Prolactin: evidence that it is separate from growth hormone in human blood, *Science* 170: 745-747, 1972.
3. Turner, T.H., J.C. Cookson, J.A.H. Wass, P.L. Drury, P.A. Price, and G.M. Besser, Psychotic reactions during treatment of pituitary tumours with dopamine agonists, *Br Med J* 289: 1101-1103, 1984.
4. Besser, G.M., L. Parke, C.R.W. Edwards, I.A. Forsyth, and A.S. McNeilly, Galactorrhoea: successful treatment with reduction of plasma prolactin levels by bromergocryptine, *Br Med J* 3: 669-672, 1972.
5. Besser, G., Medical management of prolactinomas, In J.R. Givens (ed) *Hormone-Secreting Pituitary Tumours*, Year Book Medical Publisher, Chicago, pp. 255-273 1982.
6. Ciccarelli, E., E. Mazza, E. Ghigo, F. Guidoni, A. Barbaeris, F. Massara, and F. Camanni, Longterm treatment with oral single administration of bromocriptine in patients with hyperprolactinaemia, *J Endocrinol Invest* 10: 51-53, 1987.
7. Wass, J.A.H., P.J.A. Moult, M.O. Thorner, J.E. Dacie, M. Charlesworth, A.E. Jones, and G.M. Besser, Reduction in pituitary tumour size in patients with prolactinomas and acromegaly treated with bromocriptine with or without radiotherapy, *Lancet* 2: 66-69, 1979.
8. Thorner, M.O., G.R.W. Edwards, M. Charlesworth, J.E. Dacie, P.J.A. Moult, L.H. Rees, A.E. Jones and G.M. Besser, Pregnancy in patients presenting with hyperprolactinaemia, *Br Med J* 2: 771-774, 1979.
9. Griffith, R.W., I. Turkalji, and P. Braun, Outcome of pregnancy in mothers taking bromocriptine, *Br J Clin Pharmacol* 5: 227-231, 1978.

10. Del Pozo, E., K. Schulter, E. Nuesch, J. Rosenthaler, and L. Kerp, Pharmokinetics of a long-acting bromocriptine preparation (Parlodel LAR) and its effect on release of prolactin and growth hormone, *Eur J Clin Pharmacol* 29: 615-618, 1986.

11. Grossman, A., J.A.H. Wass, and M. Besser, The rapid diagnosis of sensitivity or resistance to dopamine agonists with depot bromocriptine, *Acta Endocrinol* 116: 275-281, 1987.

12. Grossman, A., R. Ross, J.A.H. Wass, and G.M. Besser, Depot-bromocriptine treatment for prolactinomas and acromegaly, *Clin Endocrinol* 24: 231-238, 1986.

13. Franks, S., P.M. Horrocks, S.S. Lynch, W.W. Butt, and D.R. London, Effectiveness of pergolide mesylate in long term treatment of hyperprolactinaemia, *Br Med J* 286: 1177-1179, 1983.

14. Kleinberg D.L., A.E. Boyd, S. Wardlaw, A.G. Frantz, A. George, N. Bryan, S. Hilal, J. Greising, D. Hamilton, T. Seltzer, and C.J. Sommers, Pergolide for the treatment of pituitary tumours secreting prolactin or growth hormone, *N Engl J Med* 309: 704-709, 1983.

15. Grossman, A., P.M.G. Bouloux, R. Loneragan, L.H. Rees, J.A.H. Wass, and G.M. Besser, Comparison of the clinical activity of mesulergine and pergolide in the treatment of hyperprolactinaemia, *Clin Endocrinol* 22: 611-616, 1985.

16. Horoski, R., and H. Wachtel, Direct dopamine action of lisuride hydrogen maleate, an ergot derivative, in mice, *Eur J Pharmacol* 36: 373-383, 1976.

17. Votava, Z., and I. Lamplova, Antiserotonin activity of some ergolenyn and isoergolenyn derivatives in comparison with LSD and the influence of monoamine inhibiton of the serotonin effect, In E. Roshlin (ed) *Neuropsychopharmacology II* Elsevier/North, Amsterdam, pp. 68-73, 1961.

18. Bouloux, P.M.G., P.J.A. Moult, G.M. Besser, and A. Grossman, Clinical evaluation of lysuride in the management of hyperprolactinaemia, *Br Med J* 294: 1323-1324, 1987.

19. Ciccarelli, E., R. Touzel, M. Besser, and A. Grossman, Terguride - a new dopamine agonist drug: a comparison of its neuroendocrine and side effect profile with bromocriptine, *Fertil Steril* 49(4): 589-594, 1988.

20. Potirolli, A.E., L. Cammelli, P. Baroldi, and G. Pozza, Inhibition of basal and metoclopramide-induced prolactin release by cabergoline, and extremely long-acting dopaminergic drug, *J Clin Endocrinol Metab* 65: 1057-1059, 1987.

21. Ferrari, C., C. Barberi, R. Caldara, M. Mucci, F. Codecasa, A. Paracchi, C. Romano, M. Boghen, and A. Dubini, Long-lasting prolactin lowering effect of cabergoline, a new dopamine agonist, in hyperprolactinemic patients, *J Clin Endocrinol Metab* 63: 941-945, 1986.

22. Gaillard, R.C., K. Abeywickrama, J. Brownell, and A.F. Muller, Specific effect of CV 205-502, a potent nonergot dopamine agonist, during a combined anterior pituitary function test, *J Clin Endocrinol Metab* 68: 329-335, 1989.

23. Guillemin, R., P. Brazeau, P. Boehlen, F. Esch, N. Ling, and W.B. Wehrenberg, Growth hormone-releasing factor from a human pancreas that caused acromegaly, *Science* 218: 585-587, 1982.

24. Rivier, J., J. Spiess, M. Thorner, and W. Vale, Characterization of a growth hormone-releasing factor from a human pancreatic islet tumour, *Nature* 300: 276-278, 1982.

25. Vance, M.L., D.L. Kaiser, L.A. Frohman, J. Rivier, W.W. Vale, and M.O. Thorner, Role of dopamine in the regulation of growth hormone secretion: dopamine and bromocriptine augment growth hormone (GH)-releasing hormone stimulated GH secretion in normal men, *J Clin Endocrinol Metab* 64: 1136-1141, 1987.

26. Delitala, G., M. Palermo, R. Ross, D. Coy, M. Besser, and A. Grossman, Dopaminergic and cholinergic influences on the growth hormone response to growth hormone-releasing hormone in man, *Neuroendocrinology* 45: 243-247, 1987.

27. Chihara, K., H. Kodama, H. Kaji, T. Kita, Y. Kashio, Y. Okimura, H. Abe, and T. Fujita, Augmentation by propranolol of growth hormone-releasing hormone-(1-44)-NH_2-induced growth hormone release in normal short and normal children, *J Clin Endocrinol Metab* 61: 229-233, 1985.

28. Grossman, A., K. Weerasuriya, S. Al-Damluju, P. Turner, and G.M. Besser, Alpha 2-adrenoceptor agonists stimulate growth hormone secretion but have no effects on plasma cortisol under basal conditions, *Horm Res* 25: 65-71, 1987.

29. Tsagarakis, S., G. Feng, L.H. Rees, G.M. Besser, and A. Grossman, Stimulation of alpha-adrenoceptors facilitates the release of growth hormone-releasing hormone from rat hypothalamus in vitro, *J Neuroendocrinology* 1: 129-133, 1989.

30. Locatelli, V., A. Torsello, M. Redaelli, E. Ghigo, F. Massara, and E.E. Müller, Cholinergic agonists and antagonist drugs modulate the growth hormone response to growth hormone-releasing hormone in the rat: evidence for mediation by somatostatin, *J Endocrinol* 111: 271-278, 1986.

31. Ross, R.J.M., S. Tsagarakis, A. Grossman, L. Nhagafoong, R.J. Touzel, L.H. Rees, and G.M. Besser, GH feedback occurs through modulation of hypothalmaic somatostatin under cholinergic control: studies with pyridostigmine and GHRH, *Clin Endocrinol* 27: 727-733, 1987.

32. Thorner, M.O., J. Reschke, J. Chitwood, A.D. Rogol, R. Furlanetto, J. Rivier, W. Vale, and R.M. Blizzard, Acceleration of growth in two children treated with human growth hormone-releasing factor, *N Engl J Med* 312: 4-9, 1985.

33. Ross, R.J.M., S. Tsagarakis, A. Grossman, M.A. Preece, C. Rodda, P.S.W. Davies, L.H. Rees, M.O. Savage, and G.M. Besser, Treatment of growth-hormone deficiency with growth hormone-releasing hormone, *Lancet* 1: 5-8, 1987.

34. Low, L.C.A., C. Wang, P.T. Cheung, P. Ho, K.S.L. Lam, R.T.T. Young, C. Yeung, and N. Ling, Long term pulsatile growth hormone (GH)-releasing hormone therapy in children with GH deficiency, *J Clin Endocrinol Metab* 66: 611-617, 1988.

35. Brain C., P.C. Hindmarsh, C.G.D. Brooks, and D.R. Matthews, Continuous subcutaneous growth hormone releasing factor analogue augments growth hormone secretion in normal male subjects with no desensitization of the somatotroph, *Clin Endocrinol* 28: 543-549, 1988.

36. Pintor, C., S.G. Cella, R. Corda, V. Locatelli, R. Puggioni, S. Loche, and E.E. Müller, Clonidine accelerates growth in children with impaired growth hormone secretion, *Lancet* 1: 1482-1484, 1985.

37. Castro-Magnana, M., M. Angulo, B. Fuentes, M.E. Catelar, A. Canas, and B. Espinoza, Effect of prolonged clonidine administration on growth hormone concentrations and rate of linear growth in children with constitutional growth delay, *J Pediatr* 109: 784-787, 1986.

38. Pescovitz, O.H., and E. Tan, Lack of benefit of clonidine treatment for short stature in a double-blind, placebo-controlled trial, *Lancet* 2: 874-877, 1988.

39. Dayanithi, G., and F. Antoni, Atriopeptins are potent inhibitors of corticotropin secretion by anterior pituitary cells in vitro: involvement of the β-ANF receptor domain of membrane-bound cuanylate cyclase, *Biochem Biophys Res Commun*, In Press.

40. Al-Damluji, S., S. Tomlin, L. Perry, P. Bouloux, A. Grossman, L.H. Rees and G. Besser, Alpha adrenoceptor stimulation of ACTH secretion by a specific central mechanism in man, *J Endocrinol* 104: 40, Abstract #48, 1985.

41. Tsagarakis, S., J.M.P. Holly, L.H. Rees, G.M. Besser, and A. Grossman, Acetylcholine and norepinephrine stimulate the release of corticotrophin-releasing factor-41 from the rat hypothalamus in vivo, *Endocrinology* 123: 1962-1969, 1988.

42. Mamplan, T., J. Blake Tyrell, and C.B. Wilson, Transsphenoidal microsurgery for Cushing's disease, *Ann Int Med* 109: 487-493, 1988.

43. Howlett, T., P. Plowman, J. Wass, L. Rees, A. Jones, and G. Besser, Megavoltage pituitary irradiation in the management of Cushing's disease and Nelson's syndrome, long-term follow-up, *Clin Endocrinol*, In Press.

44. Lamberts, S.W.J., S.A. DeLange, and S.Z. Stefanko, Adrenocorticotrophin-secreting pituitary adenomas originate from the anterior or the intermediate lobe in Cushing's disease: differences in the regulation of hormone secretion, *J Clin Endocrinol Metab* 54: 286-291, 1982.

45. Hale, A.C., P.J. Coates, I. Doniach, T.A. Howlett, A. Grossman, L.H. Rees, and G.M. Besser, A bromocriptine-responsive corticotroph adenoma secreting α-MSH in a patient with Cushing's disease, *Clin Endo* 28: 215-223, 1988.

46. Scott, R., E.A. Espinger, and R.A. Donald, Cyproheptadine for Cushing's disease, *N Engl J Med* 296: 57-58, 1977.

47. Jones, M.T., E.W. Hillhouse, and J. Burden, Effect of various putative neurotransmitters on the release

of corticotrophin-releasing hormone from the rat hypothalamus in vitro—a model of the neurotransmitters involved, *J Endocrinol* 69: 1-10, 1976.

48. Godin, Y., L. Heiner, J. Mark, and P. Madel, Effect of di-n-propyl-acetate anticonvulsant compound on GABA metabolism, *J Neurochem* 16: 869-873, 1969.

49. Dornhurst, A., J.S. Jenkins, S.W.J. Lamberts, R.R. Abraham, A.V. Wynn, U. Beckford, B. Gillham, and M.T. Jones, The evaluation of sodium valproate in the treatment of Nelson's syndrome, *J Clin Endocrinol Metab* 56: 985-991, 1983.

50. Koppeschaar, H.P.F., R.J.M. Croughs, J.H.H. Thijssen, and F. Schwarz, Sodium valproate and cyproheptadine may independently induce a remission in the patient with Cushing's disease, *Acta Endocrinol* 104: 160-163, 1983.

51. Stafford, P., P. Kopelman, K. Davidson, T. Loughlin, A. White, L. Reese, G. Besser, D. Coy, and A. Grossman, The pituitary-adrenal response to CRF-41 is unaltered by intravenous somatostatin in normal subjects, *Clin Endocrinol* 30: 651-666, 1989.

52. Avgerinos, P.C., L.K. Nieman, E.H. Oldfield, T. Loughlin, K.M. Barnes, D.L. Loriaux, and G.B. Cutler, Jr., The effect of pulsatile corticotrophin releasing hormone administration on the adrenal insufficiency that follows cure of Cushing's disease, *J Clin Endocrinol Metab* 68: 912-916, 1989.

11

CARDIAC HORMONES AND NEUROENDOCRINE FUNCTION

Willis K. Samson

University of Missouri, School of Medicine
Department of Anatomy & Neurobiology
Columbia, Missouri 65212

Cardiac Hormones: Isolation and Physiological Effects in the Periphery

There is abundant evidence that the heart, in addition to its function as a muscular pump, is an endocrine tissue which secretes a potent family of peptides with diverse biologic actions (1,2). First described by Flynn and colleagues (3), the atrial natriuretic peptides (ANPs) are produced by conventional protein synthesis, yet unconventional post-translational processing (4-7), in atrial and ventricular myocytes. The primary stimulus for secretion of these peptides is atrial stretch (8,9), induced by increased venous return, although interactive effects of hormones (10-12) and neural agents (13,14) have been described. The major secreted form of ANP is the 28 amino acid peptide possessing a 17 membered ring structure formed by an internal disulfide bond which appears necessary for the expression of this hormone's bioactivity (15). In the rat ANP is initially synthesized as a larger molecular weight precursor, the 152 amino acid prepro-ANP, and is stored in secretory granules as the 126 amino acid prohormone, proANP (4-7). The final step in postranslational processing, cleavage of the Arg-Ser bond between positions 98 and 99, which occurs at the time of secretion (6,7), generates the mature 28 amino acid hormone and a small percentage of N- and C- terminally shortened fragments. The peptides circulate apparently unbound to plasma proteins and display an extremely short half-life, approximately 30 seconds (16). ANP is removed from plasma primarily in the kidney (17,18), although sequestration from plasma by clearance receptors (19) plays an important role as well. At least two classes of ANP been identified, one recently cloned and structurally sequenced (20) is called the B receptor. This receptor is thought to be the biologic receptor through which ANP exerts its major actions, *via* activation of particulate guanylate cyclase (21). In fact, the B receptor has now been shown to be an integral part of the particulate guanylate cyclase G-protein. A second ANP receptor, the C (clearance) receptor has been cloned and sequenced (19). While little is known of the biologic significance of this receptor, activation of which does not affect cGMP formation, Murad and colleagues (22) have demonstrated it to be linked to the phosphoinositide second messenger system. Presently, its function is thought to be solely one of sequestration and degradation; however, its link to inositol triphosphate formation suggests possible significant, yet unidentified, biologic actions as well.

Since the stimulus for ANP release is volume overload, as sensed by the atria and ventricles, it is not surprising that the major actions of the hormone are related to fluid and electrolyte homeostasis (1,2,23). ANP was originally named due to its natriuretic effects in the kidney, where diuretic actions also exist. These effects are expressed *via* a combination of direct tubular actions (24-27), effects on interstitial osmotic pressures (28), and its observed action to increase glomerular filtration rate (29,30) due to relaxant and

Circulating Regulatory Factors and Neuroendocrine Function
Edited by J. C. Porter and D. Ježová
Plenum Press, New York, 1990

177

constrictive effects on the pre-glomerular and post-glomerular arterioles, respectively. ANP inhibits renin secretion by not only altering the load of sodium delivered to, but also by directly affecting, the juxtaglomerular cells (31,32). Adrenal effects of ANP include an inhibition of both basal and stimulated aldosterone production and secretion (33-36) and possible inhibitory effects on corticosteroid synthesis (37-39). Selective vascular effects have been described which result in redistribution of blood flow, changes in vascular resistances and increases in capillary permeability (40-47). All of these actions of ANP are coordinated to remove the stimulus for its secretion (Figure 1).

Cardiac Hormones: Presence and Actions in the Central Nervous System

Central nervous system actions of the peptides seem to mirror in part the peripheral actions regarding fluid and electrolyte homeostasis. ANP acts centrally to reduce water intake (48,49) and salt preference (50,51), actions presumably expressed at one or more of the circumventricular organs (CVOs). Additionally, elevations in circulating ANP can significantly inhibit basal and stimulated vasopressin (AVP) secretion (52-56), by a hypothalamic site of action (56), an action that has also been demonstrated in humans (Figure 2). Finally, ANP antagonizes angiotensin II-stimulated adrenocorticotropin (ACTH) secretion, a central effect (57,58) which complements the renal and adrenal actions of the peptide. ANP also antagonizes the central effects of

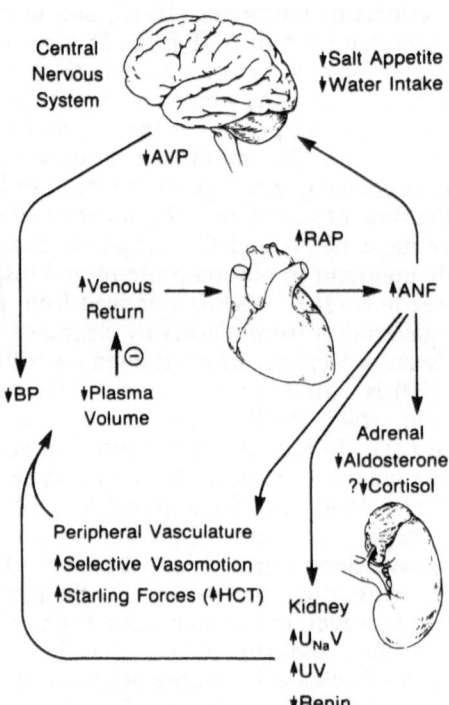

Figure 1.   Central and peripheral actions of atrial natriuretic factor (ANF) related to hydromineral balance. (RAP, right atrial pressure; BP, blood pressure; HCT, hematocrit; UnaV, urinary sodium excretion; UV, urine volume; AVP, vasopressin). Reproduced from Samson (75) with permission of Pergamon Press, Inc.

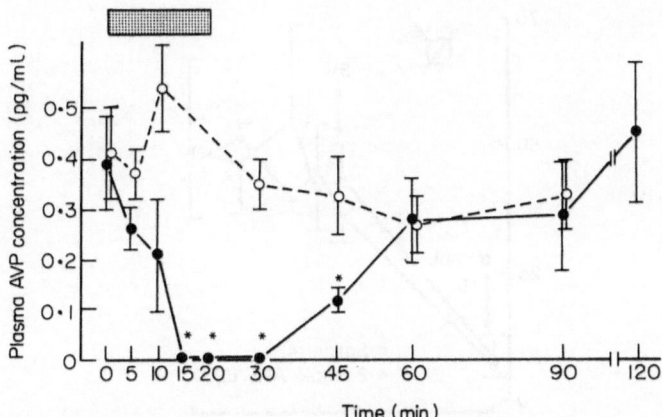

Figure 2. Effect of 20 min iv infusion of 0.1 μg ANP/kg BW/min on vasopressin levels in plasma of healthy human subjects (solid circles, experimentals; open circles, saline infused controls). * denotes $P < 0.05$ Reproduced from Fujio *et al.* (54) with permission of the authors and Blackwell Scientific Publications, Ltd.

angiotensin II on blood pressure (59). ANP positive neurons (60-62) have been identified in a variety of brain regions as have receptors (63-65) for the peptide. The ANP gene message can be detected in brain (66) suggesting its local production and potential neuromodulatory effects in brain. Similarly, in the anterior pituitary gland cellular localization of the peptide (67) and receptors for it have been described. While ANP exposure effects cGMP accumulation in pituitary cells (68), a number of experienced

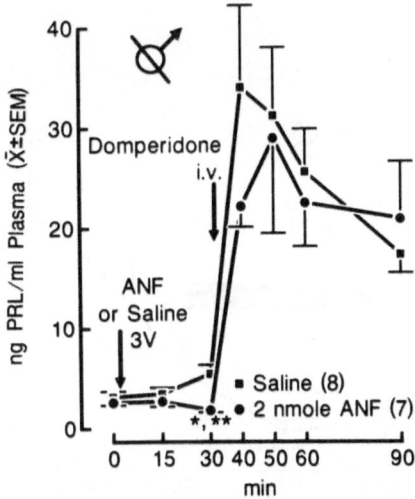

Figure 3. The prolactin inhibitory effect of centrally administered (3V, third ventricle) atrial natriuretic factor (ANF) is reversed by subsequent peripheral administration of the dopamine receptor blocker, domperidone. Reproduced from Samson *et al.* (79) with permisssion of S. Karger, AG.

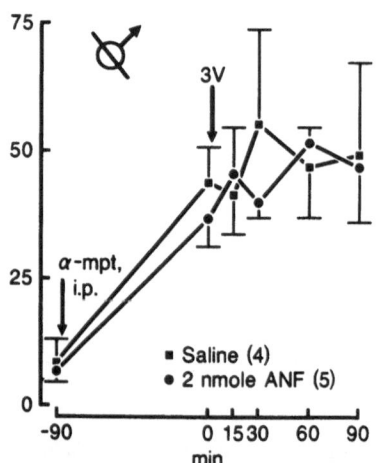

Figure 4. Prior depletion of central dopamine concentrations using the tyrosine hydroxylase inhibitor, α-methyl-p-tyrosine, prevents the ability of centrally administered atrial natriuretic factor to inhibit prolactin secretion. Reproduced from Samson *et al.* (79) with permission of S. Karger, AG.

laboratories have failed to detect any effect of ANP on pituitary hormone synthesis or secretion (68-71) with the possible exception of occasional inhibitory effects on corticotropin-releasing factor (CRF)- stimulated ACTH release (72,73). Although a pituitary site of action of circulating ANP seems unlikely, the presence of ANP receptors in the CVOs including the median eminence (ME), as well as the innervation of the hypophysiotropic diencephalon by endogenous, brain-derived ANP neurons, predicted some, but not all, of the neuroendocrine actions of the hormone we (71,74,75) and Imura (76) have described.

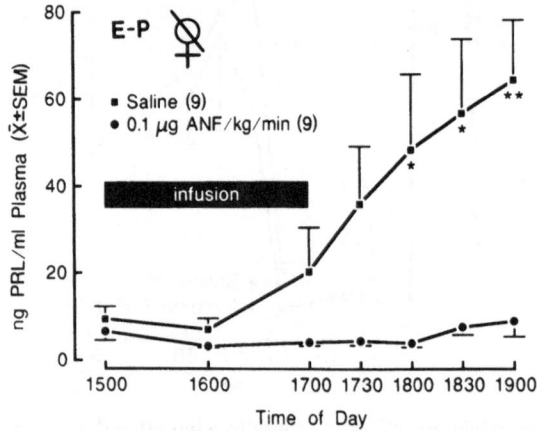

Figure 5. Intravenous infusion of atrial natriuretic factor prevents the steroid-induced (estrogen and progesterone) surge of prolactin in ovariectomized rats. * denotes $P < 0.05$, ** denotes $P < 0.025$.

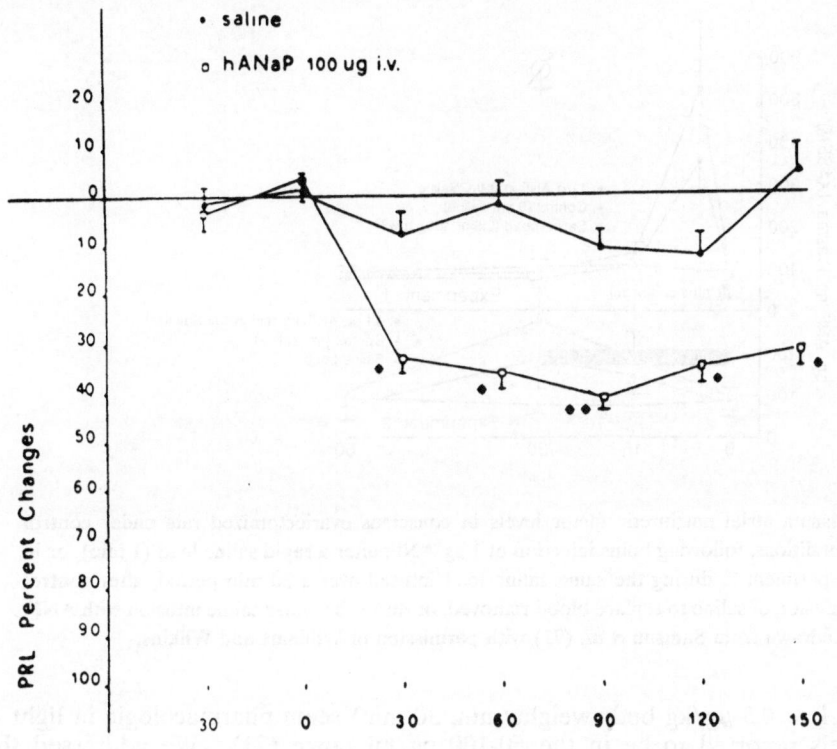

Figure 6. α-Human atrial natriuretic peptide inhibits basal prolactin secretion in healthy men. *denotes $P < 0.01$, ** denotes $P < 005$. Reproduced from Foresta *et al.* (80) with permission of the authors and Thieme Medical Publishers, Inc.

Since prolactin (PRL) in birds and fishes is a renotropic agent (77), it was not entirely surprising that central or peripheral administration of ANP in the rat, a species thought by some to still utilize PRL as an antinatriuretic compound (78), resulted in profound inhibition of PRL. As in other systems, ANP appears not only to express its tissue specific action (in the kidney, diuresis and natriuresis) but also to act at other sites (adrenal gland, brain) to inhibit the release of counterregulatory agents (aldosterone or in this instance, prolactin). In castrated male rats, intracerebroventricular infusion of ANP significantly lowered PRL levels (74), an effect which could be antagonized by pretreatment of the animals with domperidone, a dopaminergic D-2 receptor blocker (Figure 3), and prevented by prior depletion of endogenous, brain-derived dopamine (Figure 4) with the synthesis inhibitor, α-methyl-p-tyrosine (79). This indicated an effect of ANP on tuberoinfundibular dopamine (DA) neurons, which by increasing their activity and release of DA into the hypophysial portal vessels, results in the inhibition of PRL secretion (79). Peripheral infusion of ANP also inhibits PRL secretion in estrogen-primed, ovariectomized rats (Figure 5), suggesting that ANP of cardiac origin can potentially alter neuroendocrine control of this hormone. Foresta and colleagues (80) have demonstrated a PRL inhibiting effect of ANP in normal human volunteers (Figure 6). But can these effects be considered physiologic? The doses required when given

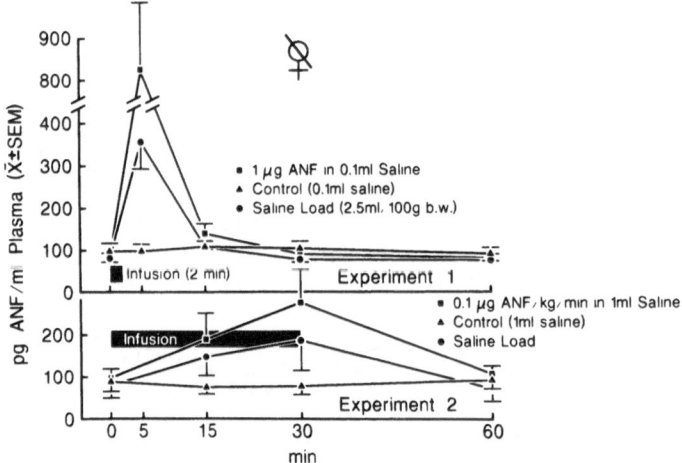

Figure 7.   Plasma atrial natriuretic factor levels in conscious ovariectomized rats under control conditions, following bolus injection of 1 μg ANF, after a rapid saline load (1 min), or in experiment 2, during the same saline load infused over a 30 min period, after control infusion of saline to replace blood removed, or during a similar saline infusion with ANF. Redrawn from Samson *et al.* (71) with permission of Williams and Wilkins.

systemically (0.1 or 0.5 μg/kg body weight/min, 30 min) seem pharmacologic in light of circulating levels reported to be in the 50-100 pg/ml range (71). We addressed this question by quantitating plasma levels of the hormone before, during, and after infusion of ANP and during volume expansion (71), a known physiologic manipulation for hormone release (9,81,82). Additionally, we compared our values to those reported in physiologic and pathologic states of ANP secretion (83-85). Our results indicate (Figure 7) that the elevations in plasma ANP we induced by intravenous infusion of the peptide resulted in circulating levels which approximated those seen in other conditions. Although these 2

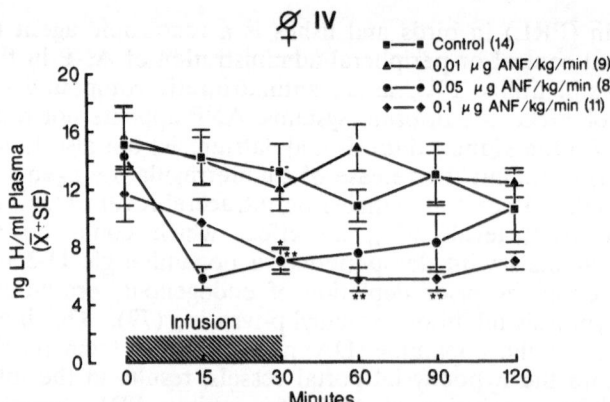

Figure 8.   Plasma luteinizing hormone levels in conscious, ovariectomized rats prior to, during and after iv infusion of 1 ml saline alone or containing atrial natriuretic factor. * denotes $P < 0.05$, ** denotes $P < 0.025$. Redrawn from Samson *et al.* (71) with permission of Williams and Wilkins.

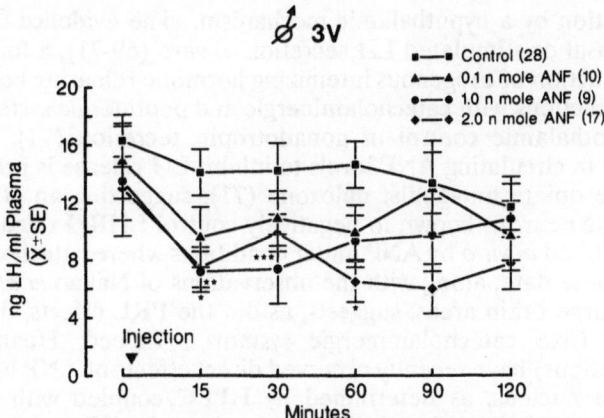

Figure 9. Plasma luteinizing hormone levels in conscious, castrated rats prior to and following third cerebroventricular injection of 2 μl saline alone or containing atrial natriuretic factor. * denotes $P < 0.05$, ** denotes $P < 0.025$. Redrawn from Samson *et al.* (71) with permission of Williams and Wilkins.

to 3-fold elevations exceed those necessary for the expression of ANP's renal effects, they are well within the range required for some of the peptide's vascular and adrenal actions, and therefore are potentially physiologic. Whether the observed effects of ANP are related to the reproductive role of PRL or to its potential renotropic actions is unclear at present; however either effect could be considered part of a coordinated system of positive and negative elements linking more than one endocrine tissue in the homeostatic regulation of several physiologic systems.

The reproductive effects of ANP, known now to include direct actions on testicular steroidogenesis (86) and perhaps even ovarian function (87), are expressed by an additional neuroendocrine mechanism. Peripheral (Figure 8) or central (Figure 9) administration of ANP significantly and in a dose-related fashion inhibits luteinizing

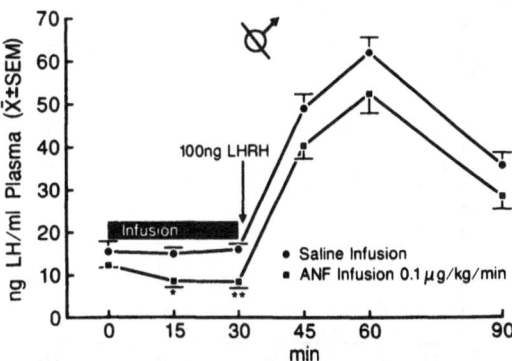

Figure 10. Plasma luteinizing hormone levels in castrated rats prior to, during and following iv infusion of saline or saline containing atrial natriuretic factor, and subsequent to an iv challenge with luteinizing hormone releasing hormone. * denotes $P < 0.5$, ** denotes $P < 0.025$. Reproduced from Samson (23) with permission of Alan R. Liss, Inc.

hormone (LH) secretion by a hypothalamic mechanism. The evidence for this includes: a lack of effect on basal or stimulated LH secretion *in vitro* (69-71), a failure *in vivo* (71) to block the pituitary action of exogenous luteinizing hormone releasing hormone (LHRH) (Figure 10), and interactions with catecholaminergic and peptidergic systems known to be involved in the hypothalamic control of gonadotropin secretion (71). The ability of peripheral elevations in circulating ANP levels to inhibit LH release is prevented by prior administration of the opiate antagonist naloxone (71), suggesting an effect of ANP on endogenous opiatergic neurons known to negatively control LHRH release (88). LHRH release itself was inhibited *in vitro* by ANP under conditions where catecholaminergic drive was present (71). These data, along with the observations of Nakao *et al.* (89) that ANP alters DA levels in large brain areas, suggests, as did the PRL effects, neuromodulatory effects of ANP on CNS catecholaminergic systems. Indeed, Huang and Samson (unpublished observations) have recently observed direct effects of ANP on catecholamine levels in the arcuate nucleus, as determined by HPLC coupled with electrochemical detection, further substantiating this possible mechanism of ANP action.

One final modulatory effect of ANP on neuroendocrine function was described by Imura's group (76). When infused centrally in the early morning when growth hormone (GH) levels are at their diurnal lowest, the peptide stimulates GH release (Figure 11). This effect was demonstrated not to be *via* an action on GH releasing factor secretion into the portal blood, but instead *via* an interaction with hypothalamic somatostatin release. This effect was not seen by us (71) when ANP was infused centrally later in the day when GH levels are elevated and less stable, nor did intravenous infusion of the peptide significantly alter GH release. These observations together with the fact that experimental

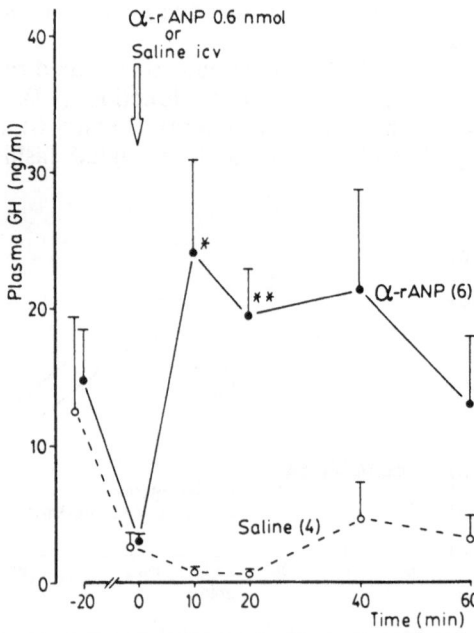

Figure 11. Intracerebroventricular infusion of α-rat atrial natriuretic peptide stimulates the secretion of GH in conscious male rats. Reproduced from Murakami *et al.* (76) with permission of the authors and Williams and Wilkins.

elevations of plasma ANP levels by volume expansion had no effect on GH release, cast doubt on the physiologic significance of ANP as a circulating regulator of GH secretion (71). However, the central effects (76) might reflect a significant role for endogenous brain ANP since we have observed that antiserum to ANP when infused into the cerebroventricle lowered GH levels in conscious rats (90).

Selective brain actions of ANP, not directly mirrored by peripheral effects, are certainly possible; yet, most CNS actions do relate to fluid and electrolyte homeostasis. These ANP actions are even repeated by the most recently discovered (91,92) member of the family of natriuretic hormones, brain natriuretic peptide (BNP). BNP is a structural homolog of 26 amino acids, sharing 13 of 17 identities within the internal disulfide ring, a shortened N-terminus with three homologies, and a C-terminus homologous to that of ANP that is required for bioactivity. Although its initial discovery in the CNS created great enthusiasm and potential, it has now been found to be produced and released as is ANP from the heart, and to display the same spectrum of bioactivity (93). Thus, it does not appear to be necessary to hypothesize BNP to be a distinct CNS neuromodulator, but instead another structural and biologic homolog of ANP. Just the same it does, like ANP, act centrally to inhibit AVP release (94), water intake (95) and A-II stimulated increased in blood pressure (96).

Cardiac Hormones: Are the Pharmacologic Effects Physiologically Significant?

Are then circulating natriuretic peptides physiologically significant regulators of neuroendocrine function? To date this can be only be inferred but not directly proven. Passive immunoneutralization by peripheral administration of large volumes of anti-ANP serum has not, for financial reasons, been conducted. Central administration of smaller volumes of serum, however, have revealed potentially significant physiologic roles in the control of fluid intake (49) and LH and GH (90) release. These results probably reflect effects of centrally derived peptides. Failing the existence of selective, potent ANP antagonists, the role of cardiac-derived ANP in neuroendocrine function is uncertain. One rather drastic, and certainly difficult, approach would be to study neuroendocrine function in experimental animals or man when the heart has been replaced by a mechanical pump. Another, perhaps more promising approach, failing the existence of antagonists, would be the use of genetic mutants lacking the ability to transcribe or translate the ANP gene. Alternatively, methodologies could be developed to selectively destroy only those myocytes or brain elements which produce the peptide.

We have developed novel technology for such studies. Our approach has been to target the plant cytotoxin ricin (97) to select antigen-producing cells by conjugation with immunoglobulin molecules specific for that antigen (98). We have now succeeded in selectively destroying AVP, OT or CRF neurons in hypothalamus and are developing this approach to attack both the central and peripheral sources of ANP. Our goal is to refine this methodology such that the physiology of ANP deficiency can be examined, by inference providing insights into its actions in the normal subject.

Acknowledgements

The author acknowledges the collaborative assistance of Dr. R.J. Fulton, Dr. M.C. Aquila, Dr. R.J. Mogg, Renee Bianchi, Debra Sherman, Lee Martin and Michele Norris. Funding for these studies was provided to the author by the American Heart Association and the National Institutes of Health (HD09988 and HD25373).

References

1. Cantin, M., and J. Genest, The heart and the atrial natriuretic factor, *Endocrine Rev* 6: 107-127, 1985.
2. Needleman, P., S.P. Adams, B.R. Cole, M.G. Currie, D.M. Geller, M.L. Michener, C.B. Saper, D. Schwartz, and D.G. Standaert, Atriopeptins as cardiac hormones, *Hypertension* 7: 469-482, 1985.
3. Flynn, T.G., M.L. deBold, and A.I. deBold, The amino acid sequence of an atrial peptide with potent diuretic and natriuretic properties, *Biochem Biophys Res Commun* 117: 859-865, 1983.
4. Glembotski, C.C., G.M. Wildey, and T.R. Gibson, Molecular forms of immunoreactive atrial natriuretic peptide in the rat hypothalamus and atrium, *Biochem Biophys Res Commun* 129: 671-678, 1985.
5. Sugawara, A., K. Nakao, N. Morii, M. Sakamoto, M. Suda, M. Shimokura, Y. Kiso, M. Kihara, Y. Yamori, K. Nishimura, J. Soneda, T. Ban, and H. Imura, α-human atrial natriuretic polypeptide is released from the heart and circulates in the body, *Biochem Biophys Res Commun* 129: 439-446, 1985.
6. Michener, M.L., J.K. Gierse, R. Seetharam, K.F. Fok, P.O. Olins, M.S. Mai and P. Needleman, Proteolytic processing of atriopeptin prohormone, *Mol Pharmacol* 30: 552-557, 1986.
7. Imada, T., R. Takayanagi, and T. Inagami T, Identification of a peptidase which processes atrial natriuretic factor precursor to its active form with 28 amino acid residues in particulate fractions of rat atrial homogenate, *Biochem Biophys Res Commun* 143: 587-592, 1987.
8. Katsube, N., D. Schwartz, and P. Needleman, Release of atriopeptide in the rat by vasoconstrictors or water immersion correlates with changes in right atrial pressure, *Biochem Biophys Res Commun* 133: 937-944, 1985.
9. Lang, R.E., H. Tholken, D. Ganten, F.C. Luft, H. Ruskoaho, and T. Unger, Atrial natriuretic factor-a circulating hormone stimulated by volume loading, *Nature* 314: 264-266, 1985.
10. Sonnenberg, H., and A.T. Veress, Cellular mechanism of release of atrial natriuretic factor, *Biochem Biophys Res Commun* 124: 443-449, 1984.
11. Tang, J., C.W. Xie, X.M. Gao, and J.K. Chang, Dynorphin A (1-10) amide stimulates the release of atrial natriuretic polypeptide (ANP) from rat atrium, *Eur J Pharmacol* 136: 449-450, 1987.
12. Zamir, N., M. Haass, J.R. Dave, and Z. Zudowska-Grojec, Anterior pituitary gland modulates the release of atrial natriuretic peptides from cardiac atria, *Proc Natl Acad Sci USA* 84: 541-545, 1987.
13. Kaczmarczyk, G., A. Drake, R. Eisele, R. Mognhaupt, M.I.M. Noble, B. Simgen, J. Stubbs, and H.W. Reinhardt, The role of the cardiac nerves in the regulation of sodium excretion in conscious dogs, *Pfleugers Arch* 390: 125-130, 1981.
14. Rankin, A.J., N. Wilson, and J.R. Lesome, Effects of autonomic stimulation on plasma immunoreactive atrial natriuretic peptide in the anesthetized rabbit, *Can J Physiol Pharmacol* 65: 532-537, 1987.
15. Misono, K.S., H. Fukumi, R.T. Grammar, and T. Inagami, Rat atrial natriuretic factor: complete amino acid sequence and disulfide linkage essential for biological activity, *Biochem Biophys Res Commun* 119: 524-529, 1984.
16. Thibault, G., K.K. Murthy, J. Gutkowska, N.G. Seidah, C. Layure, M. Chretien, and M. Cantin, NH$_2$-terminal fragment of rat pro-atrial natriuretic factor in the circulation: identification, radioimmunoassay and half-life, *Peptides* 9: 47-53, 1988.
17. Rascher, W., T. Tulassey, and R.E. Lang, Atrial natriuretic peptide in plasma of volume-overloaded children with chronic renal failure, *Lancet* 2: 303-305, 1985.
18. Larochelle, P., V. Beroniade, J. Gutkowska, J.R. Cusson, A. Lecrivain, P. DuSouich, M. Cantin, and J. Genest, Influence of hemodialysis on the plasma levels of the atrial natriuretic factor in chronic renal failure, *Clin Invest Med* 10: 350-354, 1987.
19. Maack, T., M. Suzuki, F.A. Almeida, D. Nussengveig, R.M. Scarborough, G.A. McEnroe, and J.A. Lewicki, Physiological role of silent receptors for atrial natriuretic factor, *Science* 238: 675-678, 1987.
20. Chinkers, M., D.L. Garbers, M-S. Chang, D.G. Lowe, H. Chin, D.V. Goeddel, and S. Schultz, A membrane form of guanylate cyclase is an atrial natriuretic peptide receptor, *Nature* 338: 78-83, 1989.
21. Leitman, D.C., J.W. Andersen, R.M. Catalano, S.A. Waldman, J.J. Tuan, and F. Murad, Atrial natriuretic peptide binding, cross linking and stimulation of cyclic GMP accumulation and particulate guanylate cyclase activity in cultured cells, *J Biol Chem* 263: 3720-3728, 1988.
22. Murad, F., The guanylate cyclase cGMP system mediates the effects of ANP and other vasodilators,

Peptides, In Press, 1989.

23. Samson, W.K., Hypothalamic actions of the atrial factors to alter hormone secretion from both the anterior and posterior pituitary, In P. Needleman (ed) *Biological and Molecular Aspects of Atrial Factors*, Alan R. Liss, New York, pp. 217-230, 1988.

24. Ichikawa, S., T. Saito, K. Okada, T. Kuzuya, K. Kangawa, and H. Matsuo, Atrial natriuretic factor increases GMP and inhibits cyclic AMP in rat renal papillary collecting tubule cells in culture, *Biochem Biophys Res Commun* 130: 1147-1153, 1985.

25. Burnett, J.C., Jr, T.J. Opgenorth, and J.P. Granger, The renal action of atrial natriuretic peptide during control of glomerular filtration, *Kidney Int* 30: 16-19, 1986.

26. Dillingham, M.A., and R.J. Anderson, Inhibition of vasopressin action by atrial natriuretic factor, *Science* 231: 1572-1573, 1986.

27. Zeidel, M.L., J.L. Seifter, S. Lear, B.M. Brenner, and P. Silva, Atrial peptides inhibit oxygen consumption in kidney medullary collecting duct cells, *Am J Physiol* 251: F379-F383, 1986.

28. Schwab, T.R., B.S. Edwards, R.S. Zimmerman, D.M. Heublein, and J.C. Burnett, Jr., Renal interstitial pressure increases during atrial natriuretic peptide-induced natriuresis, In B.M. Brenner and J.H. Laragh (eds) *Biologically Active Atrial Peptides* Volume 1, pp. 413-420, 1987.

29. Burnett, J.C., Jr, J.P. Granger, T.J. Opgenorth, Effects of synthetic atrial natriuretic factor on renal function and renin release, *Am J Physiol* 247: F863-F866, 1984.

30. Maack, T., D.N. Marion, M.J. Camargo, H.D. Kleinert, J.H. Laragh, E.D. Vaughn, Jr., and S.A. Atlas, Effect of auriculin (atrial natriuretic factor) on blood pressure, renal function, and the renin-aldosterone system in dogs, *Am J Med* 77: 1069-1075, 1984.

31. Obana, K., M. Naruse, K. Naruse, H. Sakurai, H. Demura, T. Inagami, and K. Shizume, Synthetic rat atrial natriuretic factor inhibits *in vitro* and *in vivo* renin secretion in rat, *Endocrinology* 117: 1282-1284, 1985.

32. Opgenorth, T.J., J.C. Burnett, Jr., J.P. Granger, and T.A. Scriven, Effects of atrial natriuretic peptide on renin secretion in nonfiltering kidney, *Am J Physiol* 250: F798-F801, 1986.

33. Biollaz, J., J. Nussberger, M. Porchet, F. Brunner-Feber, E.S. Otterbein, H. Gomez, B. Waeber, and H.R. Brunner, Four-hour infusions of synthetic atrial natriuretic peptide in normal volunteers, *Hypertension* 8(2): 96-105, 1986.

34. Cody, R.J., S.A. Atlas, J.H. Laragh, S.H. Kobo, A.B. Covit, K.S. Ryman, A. Shaknovich, K. Pondolfino, M. Clark, M.J.F. Camargo, R.M. Scarborough, and J.A. Lewicki, Atrial natriuretic factor in normal subjects and heart failure patients, *J Clin Invest* 78: 1362-1374, 1986.

35. Cuneo, R.C., E.A. Espiner, M.G. Nicholls, T.G. Yandle, S.L. Joyce, and N.L. Gilchrist, Renal, hemodynamic and hormonal responses to atrial natriuretic peptide infusions in normal man, and effect of sodium intake, *J Clin Endocrinol Metab* 63: 946-953, 1986.

36. Richards, A.M., G. Tonolo, P. Montorsi, J. Finlayson, R. Fraser, G. Inglis, A. Towrie, and J.J. Morton, Low dose infusion of 26- and 28-amino acid human atrial natriuretic peptides in normal men, *J Clin Endocrinol Metab* 66: 465-472, 1988.

37. Ohashi, M., N. Fujio, K. Kato, H. Nawata, H. Ibayashi, and H. Matsuo, Effect of human α-atrial natriuretic polypeptide on adrenocortical function in man, *J Endocrinol* 110: 287-292, 1986.

38. Weidmann, P., L. Hasler, M.P. Gnadinger, R.E. Lang, D.E. Uehlinger, S. Shaw, W. Rascher, and F.C. Reubi, Blood levels and renal effects of atrial natriuretic peptide in normal man, *J Clin Invest* 77: 734-742, 1986.

39. Cuneo, R.C., E.A. Espiner, M.G. Nicholls, T.G. Yandle, and J.H. Livesey, Effect of physiological levels of atrial natriuretic peptide on hormone secretion: inhibition of angiotensin-induced aldosterone secretion and renin release in normal man, *J Clin Endocrinol Metab* 65: 765-772, 1987.

40. Wakitani, K., B.R. Cole, D.M. Geller, M.G. Currie, S.P. Adams, K.F. Fok, and P. Needleman, Atriopeptins: correlation between renal vasodilation and natriuresis, *Am J Physiol* 249: F49-F53, 1985.

41. Dunn, B.R., I. Ichikawa, J.M. Pfeffer, J.L. Troy, and B.M. Brenner, Renal and systemic hemodynamic effects of synthetic atrial natriuretic peptide in the anesthetized rat, *Circ Res* 59: 237-246, 1986.

42. Fried, T.A., R.N. McCoy, R.W. Osgood, and J.H. Stein, Effect of atriopeptin II on determinants of glomerular filtration rate in the in vitro perfused dog glomerulus, *Am J Physiol* 250: F1119-F1122, 1986.

43. Lappe, R.W., J.A. Todt, and R.L. Wendt, Hemodynamic effects of infusion versus bolus administration of atrial natriuretic factor, *Hypertension* 8:866-873, 1986.

44. Trippodo, N.C., F.E. Cole, E.D. Frohlich, and A.A. MacPhee, Atrial natriuretic peptide decreases circulatory capacitance in areflexic rats, *Circ Res* 59: 291-296, 1986.

45. Weidmann, P., B. Hellmeuller, D.E. Uehlinger, R.E. Lang, M.P. Gnadinger, L. Hasler, S. Shaw, and C. Bachmann, Plasma levels and cardiovascular, endocrine, and excretory effects of atrial natriuretic peptide during different sodium intakes in man, *J Clin Endocrinol Metab* 62: 1027-1036, 1986.

46. Bolli, P., F.B. Muller, L. Linder, A.E.G. Raine, T.J. Resink, P. Erne, W. Kiowski, R. Ritz, and F.R. Buhler, The vasodilator potency of atrial natriuretic peptide in man, *Circulation* 75: 221-228, 1987.

47. Trippodo, N.C., and R.W. Barbee, Atrial natriuretic factor decreases whole-body capillary absorption in rats, *Am J Physiol* 252: R915-R920, 1987.

48. Antunes-Rodrigues, J., S.M. McCann, L.C. Rogers, and W.K. Samson, Atrial natriuretic factor inhibits dehydration and angiotensin II-induced water intake in the conscious, unrestrained rat, *Proc Natl Acad Sci USA* 82: 8720-8723, 1985.

49. Katsuura, G., M. Nakamura, K. Inouye, M. Kono, K. Nakao, and H. Imura, Regulatory role of atrial natriuretic polypeptide in water drinking in rats, *Eur J Pharmacol* 121: 285-287, 1986.

50. Fitts, D.A., R.L. Thunhorst, and J.B. Simpson, Diuresis and reduction of salt appetite by lateral ventricular infusion of atriopeptin II, *Brain Res* 348: 118-124, 1985.

51. Antunes-Rodrigues, J., S.M. McCann, W.K. Samson, Central administration of atrial natriuretic factor inhibits saline preference in the rat, *Endocrinology* 118: 1726-1729, 1986.

52. Samson, W.K., Atrial natriuretic factor inhibits dehydration and hemorrhage-induced vasopressin release, *Neuroendocrinology* 40: 277-279, 1985.

53. Crandall, M.E., and C.M. Gregg, In vitro evidence for an inhibitory effect of atrial natriuretic peptide on vasopressin release, *Neuroendocrinology* 44: 439-445, 1986.

54. Fujio, N., M. Ohashi, H. Nawata, K. Kato, H. Ibayashi, K. Kangawa, and H. Matsuo, α-human atrial natriuretic polypeptide reduces the plasma arginine vasopressin concentration in human subjects, *Clin Endocrinol* 25: 181-187, 1986.

55. Yamada, T., K. Nakao, N. Morii, H. Itoh, S. Shiono, M. Sakamoto, A. Sugawara, Y. Saito, H. Ohno, A. Kanai, G. Katsuura, M. Eigyo, A. Matsushita, and H. Imura, Central effect of atrial natriuretic polypeptide on angiotensin II-stimulated vasopressin secretion in conscious rat, *Eur J Pharmacol* 125: 453-456, 1986.

56. Samson, W.K., M.C. Aguila, J. Martinovic, J. Antunes-Rodrigues, and M. Norris, Hypothalamic action of atrial natriuretic factor to inhibit vasopressin secretion, *Peptides* 8(3): 449-454, 1987.

57. Hattori, T., K. Hashimoto, H. Inoue, M. Sugawara, S. Suemaru, J. Kageyama, and Z. Ota, Effect of synthetic atrial natriuretic polypeptide on hemorrhage-induced adrenocorticotropin secretion of the rat, *Endocrinol Jpn* 33: 533-539, 1986.

58. Itoh, H., K. Nakao, G. Katsuura, N. Morii, T. Yamada, S. Shiono, M. Sakamoto, A. Sugawara, Y. Saito, M. Eigyo, A. Matsushita, and H. Imura, Possible involvement of central atrial natriuretic polypeptide in regulation of hypothalamo-pituitary-adrenal axis in conscious rat, *Neurosci Lett* 69: 254-258, 1986.

59. Itoh, H., K. Nakao, N. Morii, T. Yamada, S. Shiono, M. Sakamoto, A. Sugawara, Y. Saito, G. Katsuura, T. Shiomi, M. Eigyo, A. Matsushita, and H. Imura, Central action of atrial natriuretic polypeptide on blood pressure in conscious rats, *Brain Res Bull* 16: 745-749, 1986.

60. Kawata, M., K. Nakao, N. Morii, Y. Kiso, H. Yamashita, H. Imura, and Y. Sano, Atrial natriuretic polypeptide: topographical distribution in the rat brain by radioimmunoassay and immunohistochemistry, *Neuroscience* 16: 521-546, 1985.

61. Skofitsch, G., D.M. Jacobowitz, R.L. Eskay, and N. Zamir, Distribution of atrial natriuretic factor-like immunoreactive neurons in the rat brain, *Neuroscience* 16: 917-948, 1985.

62. Standaert, D.G., P. Needleman, and C.B. Saper, Organization of atriopeptin-like immunoreactive neurons in the central nervous system of the rat, *J Comp Neurol* 253: 315-341, 1986.

63. Quirion, R., M. Dalpe, A. DeLeon, J. Gutkowska, M. Cantin, and J. Genest, Atrial natriuretic factor (ANF) binding sites in brain and related structures, *Peptides* 5: 1167-1172, 1984.

64. Kurihara, M., J.M. Saavedra, and K. Shigematsu, Localization and characterization of atrial natriuretic

peptide binding sites in discrete areas of rat brain and pituitary gland by quantitative autoradiography, *Brain Res* 408: 31-39, 1987.

65. Mantyh, C.R., L. Kruger, N.C. Brecha, and P.W. Mantyh, Localization of specific binding sites for atrial natriuretic factor in the central nervous system of rat, guinea pig, cat and human, *Brain Res* 412: 329-342, 1987.

66. Gardner, D.G., G.P. Vlasuk, J.D. Baxter, J.C. Fiddes, J.A. Lewicki, Identification of atrial natriuretic factor gene transcripts in the central nervous system, *Proc Natl Acad Sci USA* 84: 2175-2179, 1987.

67. McKenzie, J.C., I. Tanaka, K.S. Misono, and T. Inagami, Immunocytochemical localization of atrial natriuretic factor in the kidney, adrenal medulla, pituitary and atrium of rat, *J Histochem Cytochem* 33: 828-832, 1985.

68. Heisler, S., J. Simard, E. Assayag, Y. Mehri, and F. Labrie, Atrial natriuretic factor does not affect basal, forskolin- and CRF-stimulated adenylate cyclase activity, cAMP formation or ACTH secretion, but does stimulate cGMP synthesis in anterior pituitary, *Mol Cell Endocrinol* 44: 125-131, 1986.

69. Simard, J., F.F. Hubert, F. Labrie, E. Assayag, and S. Heisler, Atrial natriuretic factor-induced cGMP accumulation in rat anterior pituitary cells in culture is not coupled to hormonal secretion, *Regul Pept* 15: 269-278, 1986.

70. Abou-Samra, A., K.J. Catt, G. G. Aguilera, Synthetic atrial natriuretic factors (ANFs) stimulate guanine 3′,5′-monophosphate production but not hormone release in rat pituitary cells: peptide contamination with a gonadotropin-releasing hormone agonist explains luteinizing hormone-releasing activity of certain ANFs, *Endocrinology* 120: 18-24, 1987.

71. Samson, W.K., M.C. Aguila, and R. Bianchi, Atrial natriuretic factor inhibits luteinizing hormone secretion in the rat: evidence for a hypothalamic site of action, *Endocrinology* 122: 1573-1582, 1988.

72. Samson, W.K., and R.L. Eskay, Endocrine and neuroendocrine action of cardiac peptides, In TW Moody (ed) *Neural and Endocrine Peptides and Receptors*, Plenum Publishing Corp., New York, pp. 521-540, 1986.

73. Shibasaki, T., M. Naruse, N. Yamauchi, A. Masuda, T. Imaki, K. Naruse, H. Demura, N. Ling, T. Inagami, and K. Shizume, Rat atrial natriuretic factor suppresses proopiomelanocortin-derived peptides secretion from both anterior and intermediate lobe cells and growth hormone release from anterior lobe cells of rat pituitary in vitro, *Biochem Biophys Res Commun* 135: 1035-1041, 1986.

74. Samson, W.K., and R. Bianchi, Further evidence for a hypothalamic site of action of atrial natriuretic factor: inhibition of prolactin secretion in the conscious rat, *Can J Physiol Pharmacol* 66: 301-305, 1988.

75. Samson, W.K., Central nervous system actions of atrial natriuretic factor, *Brain Res Bull* 20: 831-837, 1988.

76. Murakami, Y., Y. Kato, K. Tojo, T. Inoue, N. Yanaihara, and H. Imura, Stimulation of growth hormone secretion by central administration of atrial natriuretic polypeptide in the rat, *Endocrinology* 122: 2103-2108, 1988.

77. Loretz, C.A., and H.A. Bern, Prolactin and osmoregulation in vertebrates, *Neuroendocrinology* 35: 292-304, 1982.

78. Nicoll, C.S., Role of prolactin in water and electrolyte balance in vertebrates, In R.B. Jaffe (ed) *Current Endocrinology, Prolactin*, Elsevier, New York, pp. 127-166, 1981.

79. Samson, W.K., R. Bianchi, and R. Mogg, Evidence for a dopaminergic mechanism for the prolactin inhibitory effect of atrial natriuretic factor, *Neuroendocrinology* 47: 268-271, 1988.

80. Foresta, C., R. Mioni, G. Scanelli, R. Vettor, B. Busnardo, and C. Scandellari, Alpha human atrial natriuretic peptide and anterior pituitary hormone secretion in man, *Horm Metab Res* 20: 313-315, 1988.

81. Eskay, R., Z. Zukowska-Grojec, M. Haass, R.J. Dave, N. Zamir, Circulating atrial natriuretic peptides in conscious rats: regulation of release by multiple factors, *Science* 232: 636-639, 1986.

82. Ledsome, J.R., N. Wilson, C.A. Courneya, and A.J. Rankin, Release of atrial natriuretic peptide by atrial distension, *Can J Physiol Pharmacol* 63: 739-742, 1985.

83. Lang, R.E., T. Unger, D. Ganter, J. Weil, F. Bidlingmaier, and D. Dohlemann, α-atrial natriuretic peptide concentrations in plasma of children with congenital heart and pulmonary diseases, *Br Med J* 291: 1241, 1985.

84. Tikkanen, I., F. Fyhrquist, K. Metsarinne, and R. Leidenius, Plasma atrial natriuretic peptide in cardiac disease and during infusion in healthy volunteers, *Lancet* 2: 66-69, 1985.

85. Lang, R.E., R. Dietz, A. Merkel, T. Unger, R. Ruskoaho, and D. Ganter, Plasma atrial natriuretic peptide values in cardiac disease, *J Hypertens* 4: S119-S123, 1986.

86. Pandey, K.N., S.N. Pavlou, W.J. Kovaco, and T. Inagami, Atrial natriuretic factor regulates steroidogenic responsiveness and cyclic nucleotide levels in mouse leydig cells in vitro, *Biochem Biophys Res Commun* 138: 399-404, 1986.

87. Pandey, K.N., K.G. Osteen, and T. Inagami, Specific receptor-mediated stimulation of progesterone secretion and cGMP accumulation by rat atrial natriuretic factor in cultured human granulosa-lutein (G-L) cell, *Endocrinology* 121: 1195-1197, 1987.

88. Kalra, P.S., and S.P. Kalra, Control of gonadotropin secretion, In H. Imura (ed) *The Pituitary Gland*, Raven Press, New York, pp. 189-220, 1985.

89. Nakao, K., G. Katsuura, N. Morii, H. Itoh, S. Shiono, T. Yamada, A. Sugawara, M. Sakamoto, Y. Saito, M. Eigyo, A. Matsushita, and H. Imura, Inhibitory effect of centrally administered atrial natriuretic polypeptide on brain of dopaminergic system in rat, *Eur J Pharmacol* 131: 171-177, 1986.

90. Samson, W.K., R. Bianchi, and R. Mogg, Atrial natriuretic peptide modulates hypothalamic control of pituitary function, *Peptides*, In Press, 1989.

91. Sudoh, T., K. Kangawa, N. Minamino, and H. Matsuo, A new natriuretic peptide in porcine brain, *Nature* 332: 78-81, 1988.

92. Itoh, H., K. Nakao, Y. Saito, T. Yamada, G. Shirakami, M. Mukoyama, H. Arai, K. Hosoda, S. Suga, N. Minamino, K. Kangawa, H. Matsuo, and H. Imura, Radioimmunoassay for brain natriuretic peptide (BNP) detection of BNP in canine brain, *Biochem Biophys Res Commun* 158: 120-128, 1989.

93. Song, D.L., K.P. Kohse, and F. Murad, Brain natriuretic factor. Augmentation of cellular cyclic GmP, activation of particulate guanylate cyclase and receptor binding, *FEBS Lett* 232: 125-129, 1988.

94. Yamada, T., K. Nakao, H. Itoh, G. Shirakami, K. Kangawa, N. Minamino, H. Matsuo, and H. Imura, Introcerebroventricular injection of brain natriuretic peptide inhibits vasopressin secretion in conscious rats, *Neurosci Lett* 95: 223-228, 1988.

95. Itoh, H., K. Nakao, T. Yamada, G. Shirakami, K. Kangawa, N. Minamino, H. Matsuo, and H. Imura, Antidipsogenic action of a novel peptide, brain natriuretic peptide, in rats, *Eur J Pharmacol* 150: 193-196, 1988.

96. Shirakami, G., K. Nakao, T. Yamada, H. Itoh, K. Mori, K. Kangawa, N. Minamino, H. Matsuo, and H. Imura, Inhibitory effect of brain natriuretic peptide on central angiotensin II-stimulated pressor response in conscious rats, *Neurosci Lett* 91: 77-83, 1988.

97. Vitteta, E.S., R.J. Fulton, R.D. May, M. Till, and J.W. Uhr, Redesigning nature's poisons to create anti-tumor reagents, *Science* 238: 1098-1104, 1987.

98. Fulton, R.J., R.J. Mogg, D. Sherman, W.K. Paull, and W.K. Samson, Use of immunotoxins to create novel neuropeptide deficient rats, *FASEB J* 3: 480, 1989.

12

INTERACTIONS BETWEEN THE CIRCULATING HORMONES ANGIOTENSIN AND ATRIAL NATRIURETIC PEPTIDE AND THEIR RECEPTORS IN BRAIN

J.M. Saavedra

Section on Pharmacology, Laboratory of Clinical Science
National Institute of Mental Health
Bethesda, MD 20892

Introduction

Angiotensin II (ANG II) and atrial natriuretic peptide (ANP) play important roles in the regulation of cardiovascular function and fluid balance (1,2). Circulating ANG II stimulates specific receptors to induce vasoconstriction, aldosterone production, vasopressin release, and sodium retention (2). There are alterations in the peripheral renin-angiotensin system in genetic and experimental hypertension (3). The renin-angiotensin system is also important in human hypertension, and blockade of the last step of ANG II formation by inhibition of the angiotensin converting enzyme (ACE) is one of the standard therapies. ANP, produced in the heart, is released to the circulation and is involved in the control of fluid volume and cardiovascular function (1). ANP metabolism is altered in hypertension (4). The peripheral effects of ANP, increased sodium and water excretion by the kidneys, decreased aldosterone production, vasodilation and antihypertensive actions are antithetical to those of the water conservation peptides, vasopressin, and ANG II (1). For these reasons peripheral ANP and ANG II are considered to be part of physiologically antagonistic regulatory systems.

Most of the ANG II and ANP effects were first described and extensively studied in the periphery, *i.e.*, in organs other than the brain. It was later recognized that some effects of these peptides could be centrally mediated by way of a localization in the brain. The central ANG II actions include stimulation of drinking, increase in systemic blood pressure, stimulation of vasopressin release, and an increase in salt appetite (5). Proposed central ANP actions include antagonism of ANG II-induced drinking and salt appetite, decrease in systemic blood pressure, inhibition or blockade of vasopressin release, diuresis, and direct effects on neuronal activity (6-8).

Central actions should be produced by stimulation of peptide receptors in brain. Localization and characterization of brain receptors for neuropeptides, therefore, was crucial for the understanding of central effects of peripherally originated peptides. It was first determined that receptors for both ANG II and ANP was localized in high concentration the in circumventricular organs, which are highly vascularized structures lacking a blood-brain barrier and, hence, are accessible to circulating peptides (9-12).

In addition to receptors outside the blood-brain barrier, specific brain regions inside the blood-brain barrier contain ANG II and/or ANP receptors (9-13). Since peripherally formed peptides have no access to receptors in areas protected by the blood-brain barrier, it was considered that these receptors were part of distinct, central peptidergic systems (5,14).

Circulating Regulatory Factors and Neuroendocrine Function
Edited by J. C. Porter and D. Ježová
Plenum Press, New York, 1990

Receptors are essential for the regulation of hormone effects. Changes in receptor number or affinity modulate the physiological response to hormonal stimulation. In the brain, hormone actions are regulated in specific, anatomically selected systems, and should be studied with consideration of the anatomical localization. Quantitative autoradiographic methods provide the opportunity to combine sensitivity and anatomical precision to the study of receptors in the brain. For this reason we chose to focus our study on the characterization and regulation of brain receptors.

We were able to determine that specific alterations in brain ANG II and ANP receptors occur when blood pressure, hormone levels, or fluid consumption are modified, and that at least some of these central alterations are directly or indirectly related to changes in the peripheral peptide systems. Our results indicate a role for the central ANG II and ANP systems in fluid, endocrine and cardiovascular control. The results also suggest that peripheral and central peptidergic systems are associated, perhaps *via* peptidergic receptors that are outside the blood-brain barrier and in contact with the general circulation.

Materials and Methods

Animals

For ANP and ANG II receptor mapping studies, male Sprague-Dawley rats, 250. to 300 g body weight, were purchased from Zivic Miller, Allison Park, PA. The rats were

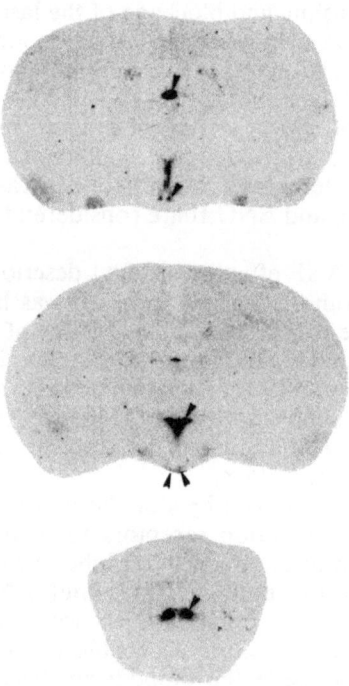

Figure 1. Angiotensin II receptor distribution in the rat brain. Arrows correspond to subfornical organ and suprachiasmatic nucleus (*top figure*), paraventricular nucleus, pars parvo-cellularies, and median eminence (*middle figure*) and nucleus of the solitary tract (*lower figure*).

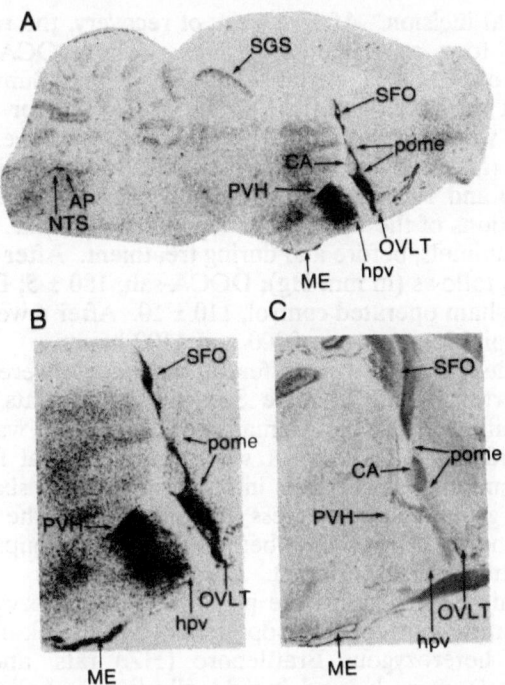

Figure 2. Angiotensin II receptors in forebrain sagittal sections. *A*: midsagittal section. *B*: close up of *A*. *C*: adjacent section stained with Luxol Fast blue. SFO: subfornical organ; pome: median preoptic nucleus; CA: anterior commissure; hpv: periventricular nucleus: PVH: paraventricular nucleus; ME: median eminence; OVLT: organon vasculosum laminae terminalis; SGS: superficial grey layer superio colliculus; AP: area postrema; NTS: nucleus of the solitary tract. Reproduced from Sigematsu *et al.* (12) with permission of Elsevier Science Publishers.

housed under a constant temperature, 24°C, with lights on from 0600 to 1800 hours, and were given free access to food and water. For studies on genetic hypertension, 14-week-old male, spontaneously hypertensive (SH) rats and their age-matched normotensive controls, Wistar-Kyoto (WKY) rats, were obtained from Taconic Farms, Germantown, NY. In a separate experiment we analyzed the effect of a peripherally administered ACE inhibitor, enalapril, in SH rats and WKY rats. These animals were separated into groups of eight animals. One group of each strain was treated daily for 14 days with enalapril (25 mg/kg body weight) given orally as a solution containing 2.5 mg/ml. After 2 weeks of treatment, systolic blood pressures were 190 ± 8, 145 ± 6, 135 ± 5, and 130 ± 8 mm Hg for untreated SH, treated SH, untreated WKY, and treated WKY rats, respectively (15). Twenty-four hours after the last enalapril dose, all animals were killed by decapitation between 0900 and 1100 hours. Data were subjected to a two-way analysis of variance according to the experimental design followed by *post hoc* analysis with Newman-Keuls' test.

To prepare the deoxycorticosterone acetate (DOCA)-salt model of experimental hypertension, 14-weeks-old, male WKY rats were anesthetized with sodium pentobarbital (40 mg/kg body weight, intraperitoneally), and unilateral nephrectomy was performed through a retroperitoneal incision. After 1 week of recovery, the rats were weighed and randomly divided into four experimental groups. The DOCA-salt group received subcutaneous injections of DOCA (Percoten, Ciba Geigy Corp., Summit, NJ) twice weekly (25 mg/kg body weight per week) and 1% solution of NaCl for drinking water. The DOCA-treated animals were treated similarly with DOCA but were given tap water to drink. The salt-treated rats received twice-weekly subcutaneous injections of the vehicle (0.5 ml/kg body weight) and 1% solution of NaCl for drinking water. The control group received the same injections of the vehicle and tap water to drink. Blood pressures were measured weekly in all animals, before and during treatment. After 4 weeks of treatment, blood pressures were as follows (in mm Hg): DOCA-salt, 180 ± 5; DOCA alone, 122 ± 8; salt alone, 115 ± 4; and sham operated control, 110 ± 10. After 4 weeks of treatment, rats were sacrificed by decapitation between 0900 and 1100 hours.

For studies on dehydration, two different rat models were used. Water-deprivation studies were performed in 12 male Sprague-Dawley rats, 250 g body weight, obtained from Zivic Miller. First, all animals were given tap water and rat chow *ad libitum* and were individually housed for 1 week under normal laboratory conditions. Then, the rats were randomly distributed into two groups of six animals each. The water-satiated (control) group had free access to tap water, and the water-deprived group with no access to drinking water for 5 days before use. Both groups were given access to chow *ad libitum* throughout the experiment.

Studies on chronic dehydration were performed in homozygous Brattleboro rats, which lack the antidiuretic hormone, vasopressin. Nine-week-old male homozygous Brattleboro (DI) rats, heterozygous Brattleboro (HZ) rats, and normally hydrated Long-Evans (LE) controls, were housed individually in metabolic cages under normal laboratory conditions for 1 week after being purchased from Blue Spruce Farms, Altamont, NY. The mean daily water intake was 8 ± 2, 9 ± 5, and 75 ± 5 ml/100 g body weight for LE, HZ, and DI rats, respectively. Nine additional LE rats were water deprived for 5 days.

For studies on stress, seven male Sprague-Dawley rats (Zivic Miller), weighing 217 ± 6 g at the beginning of the experiment, were subjected to immobilization for 2 hours per day between 1000 and 1200 hours for 10 consecutive days (16). Control rats were killed by decapitation between 1000 and 1200 hours while avoiding stressful handling. Stressed rats were decapitated immediately after the last stress period.

For endocrine studies, hypophysectomized, adrenalectomized and sham-operated male Wistar rats (8 weeks old) were purchased from Charles River Farms (Wilmington, MA). Hypophysectomized rats were given 5% glucose to drink and killed 7 days after the surgery. Adrenalectomized rats were given 0.9% NaCl solution to drink, and they were killed 11 days after the surgery. Corticosterone replacement was given to eight adrenalectomized rats by adding corticosterone (100 mg/liter) for one week to their drinking water. This route of administration maintains almost normal circadian variation in plasma corticosterone levels. To replace selectively mineralocorticoid deficiency, DOCA (12.5 mg/kg body weight) was injected subcutaneously into eight rats twice weekly for one week. All rats were killed by decapitation between 0930 and 1100 hours. Their brains (in some cases the pituitaries as well) were immediately removed and frozen by immersion in isopentane at -30°C.

In Vitro *Labeling of Peptide Receptors*

ANG II receptors were labeled *in vitro* by incubation of sections with [^{125}I]-[SAR^1]-ANG II, Peninsula Laboratories, iodinated by New England Nuclear, Boston, MA as described earlier (17).

Table 1

Apparent ANG II Binding Density in Discrete Brain Nuclei of Control and
Enalapril-Treated Wistar-Kyoto (WKY) and Spontaneously Hypertensive (SH) Rats

Brain Nuclei	----------Apparent Binding Density---------- (fmol/mg protein)			
	WKY Control	SH Control	WKY Treated	SH Treated
Subfornical Organ	167 ± 7	$247 \pm 8^*$	177 ± 13	$127 \pm 13\dagger$
Median Preoptic Nucleus	77 ± 9	99 ± 11	78 ± 12	86 ± 9
Paraventricular Nucleus	96 ± 6	92 ± 14	83 ± 14	80 ± 11
Suprachiasmatic Nucleus	162 ± 13	143 ± 12	143 ± 14	133 ± 16
Area Postrema	42 ± 4	$85 \pm 5^*$	56 ± 5	82 ± 9
Nucleus of the Solitary Tract	62 ± 4	$85 \pm 5^*$	73 ± 2	82 ± 4
Inferior Olive	20 ± 3	$34 \pm 7^*$	19 ± 1	29 ± 4

Values are means ± SE of 16 animals.
$^*P < 0.05$ compared to WKY control group.
$\dagger P < 0.05$ compared to WKY and SH control groups. Reproduced from Nazarali *et al.* (15) with permission of The American Physiological Society.

Rat ANP receptors were labeled *in vitro* by incubation with $(3-[^{125}I]iodotry-rosyl^{28})$-ANP (specific activity 1750 to 2050 Ci/mmol; Amersham Corp., Arlington Heights, IL) (10,18).

Receptor Autoradiography

Autoradiographic images of ANP and ANG II receptors were obtained by placing the previously incubated sections in X-ray cassettes (CGR Medical Corp., Baltimore, MD) and apposing them against $[^3H]$-Ultrofilm (LKB Industries, Rockville, MD) at room temperature for 2 to 5 days. Film development and computerized microdensitometry were performed as described previously (17,19).

Results

Distribution of Peptide Receptors in Brain

ANG II receptors were highly localized. The density of receptors was high in the circumventricular organs and in structures inside the blood-brain barrier, such as the hypothalamic paraventricular and suprachiasmatic nuclei (Figures 1 and 2). Within the paraventricular nucleus, most of the ANG II receptors were located in the parvicellular region (Figure 1). A number of brain stem areas contained high levels of ANG II receptors with the highest relative concentrations being present in the nucleus of the solitary tract (Figure 1).

Sagittal sections of the brain revealed ANG II binding in a continuous band situated along the subependymal space connecting the subfornical organ (SFO) and the organum vasculosum laminae terminalis (OVLT), and involving the median preoptic nucleus and the lamina terminalis (Figure 2). Caudally, the ANG II receptor band was in direct continuity with ANG II receptors highly localized to the periventricular and paraventricular nuclei and to the median eminence (Figure 2).

The highest concentrations of ANP receptors in brain were present in the

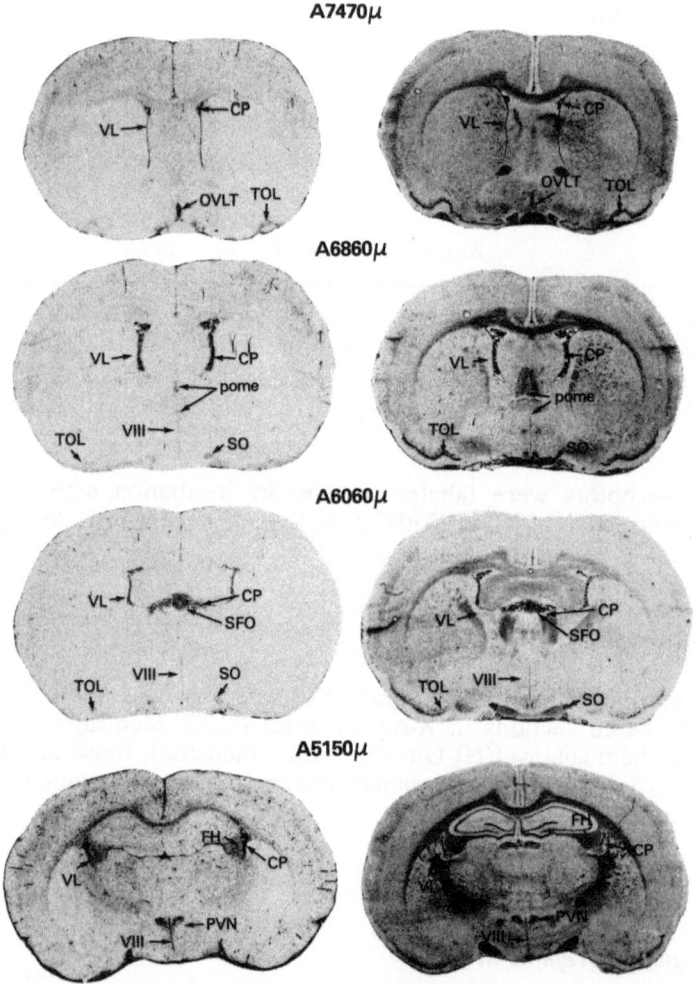

Figure 3. Atrial natriuretic peptide receptors in rat forebrain. LMIO: olfactory bulb, lamina medullar interna; LPIB: olfactory bulb, limina plexiforme interna; VL: lateral ventricle; CP: choroid plexus; OVLT: organon vasculosum laminae terminalis; TOL: lateral olfactory tract; pome: median preoptic nucleus; V III: third ventricle; SFO: subfornical organ. Reproduced from Kurihara *et al.* (13) with permission of Elsevier Biomedical Press.

Table 2

Apparent ANP Binding Density in Discrete Brain Nuclei
of Wistar-Kyoto (WKY) and Spontaneously Hypertensive (SH) Rats

Brain Nuclei	----------------Apparent Binding Density---------------- (fmol/mg protein)	
	WKY	SH
Subfornical Organ	135 ± 10	38 ± 3*
Choroid Plexus	152 ± 8	42 ± 4*
Paraventricular Nucleus	41 ± 4	8 ± 1*
Area Postrema	97 ± 6	23 ± 4*
Nucleus of the Solitary Tract	33 ± 3	n.d.

Values are means ± SE of 8 animals.

$*P < 0.05$; n.d. = not detected. Reproduced from Saavedra *et al.* (18) with permission of Macmillan Magazines Ltd.

circumventricular organs (Figure 3). In addition, ANP receptors were localized in selected areas inside the blood-brain barrier, such as the supraoptic and the paraventricular nuclei. ANP receptors were highest in the magnocellular area (Figure 3). In the brainstem, in addition to the area postrema, a small number of ANP receptors were located in the nucleus of the solitary tract (13).

The choroid plexus contained high numbers of ANP receptors (Figure 3). ANP receptors were also detected in the ependymal layer of the ventricles, vasculature, and pial membranes (Figure 3).

Peptide Receptors in Hypertension

Compared to WKY rats, SH rats had higher ANG II receptor concentrations in the median preoptic nucleus, SFO, paraventricular nucleus and nucleus of the solitary tracts (Table 1). Enalapril treatment reduced the systolic blood pressure only in SH rats (see Methods) and altered ANG II receptors in only one brain area: the SFO of SH rats (Table 1). Conversely, SH animals had much fewer ANP receptors in the SFO, the choroid plexus, the area postrema, and the nucleus of solitary tract than did their normotensive controls, *viz.*, WKY rats (Table 2).

When the complete DOCA-salt experimental hypertension model was studied, it was found that treatment with salt alone significantly increased ANG II binding only in median preoptic nucleus and SFO (Table 3). DOCA administration alone significantly increased ANG II receptors only in the median preoptic nucleus, SFO, and paraventricular nucleus. In DOCA-salt hypertensive rats, ANG II receptor number was high in the median preoptic nucleus, SFO, and in the paraventricular nucleus. DOCA-salt treatment increased ANG II binding in the median preoptic nucleus to the same extent as DOCA or salt treatment alone. In the SFO and paraventricular nucleus, however, the increase in ANG II binding in DOCA-salt hypertensive rats was significantly greater than that seen

Table 3
Apparent ANG II Receptor Concentration in Brain
Nuclei of DOCA-Salt Hypertensive Rats

Area	Group	Apparent Binding Density (fmol/mg protein)
Olfactory Bulb	Control	45 ± 5
	DOCA	56 ± 8
	Salt	61 ± 8
	DOCA-salt	61 ± 6
Median Preoptic Nucleus	Control	131 ± 9
	DOCA	$166 \pm 10^*$
	Salt	$175 \pm 9^*$
	DOCA-salt	$182 \pm 6^*$
Subfornical Organ	Control	175 ± 8
	DOCA	$229 \pm 12^*$
	Salt	$232 \pm 11^*$
	DOCA-salt	$271 \pm 11^*\dagger$
Suprachiasmatic	Control	124 ± 14
	DOCA	142 ± 20
	Salt	113 ± 9
	DOCA-salt	128 ± 18
Paraventricular Nucleus	Control	78 ± 8
	DOCA	$118 \pm 6^*$
	Salt	105 ± 9
	DOCA-salt	$149 \pm 6^*\dagger$
Nucleus of Solitary Tract	Control	81 ± 8
	DOCA	77 ± 6
	Salt	109 ± 9
	DOCA-salt	$151 \pm 10^*\dagger$
Area Postrema	Control	73 ± 7
	DOCA	62 ± 5
	Salt	91 ± 6
	DOCA-salt	$125 \pm 10^*\dagger$

Values are means and SE of 8 animals assayed individually.
$^*P < 0.05$ when compared with the control group.
$\dagger P < 0.05$ when compared with DOCA only and salt only groups. Reproduced from Gutkind *et al.* (25) with permission of The American Physiological Society.

Table 4

Apparent ANG II Binding Density in Discrete Brain
Nuclei of Control and Water-Deprived Sprague-Dawley Rats

Brain Nucleus	-------Apparent Binding Density------- (fmol/mg protein)		% Change
	Control	Dehydrated	
Subfornical Organ	171 ± 12 (8)	373 ± 10 (8)*	118.1
Organum Vasculosum of the Lamina Terminalis	185 ± 15 (8)	210 ± 20 (8)	13.4
Median Preoptic Nucleus	142 ± 14 (4)	185 ± 7 (8)	30.2
Paraventricular Nucleus	116 ± 12 (6)	135 ± 10 (8)	17.2
Area Postrema	140 ± 9 (8)	136 ± 10 (6)	-3.0
Nucleus of the Solitary Tract			
Anterior Portion Bregma, -13.8 mm	136 ± 7 (6)	126 ± 9 (8)	-7.6
Medial Portion Bregma, -14.3 mm	166 ± 8 (8)	171 ± 9 (7)	2.8
Inferior Olive	118 ± 10 (8)	110 ± 16 (7)	-6.4

Values are means ± SE. Number of animals is indicated in parentheses.
*$P < 0.01$ compared to controls. Reproduced from Nazarali *et al.* (40) with permission of Plenum Publishing Corporation.

with either salt or DOCA treatment alone. None of the treatments produced significant alterations in the concentrations of ANG II receptors in other forebrain areas, the olfactory bulb or in the suprachiasmatic nucleus. In the brainstem, DOCA-salt hypertensive rats had increased ANG II binding in the nucleus of the solitary tract and in the area postrema. Neither salt alone nor DOCA alone produced any significant modification of ANG II binding in brainstem nuclei (Table 3).

Peptide Receptors in Alterations of Water Balance

After dehydration, only the SFO showed a significant increase in ANG II receptor density (Table 4). ANG II receptor density was also altered in Brattleboro rats, but the changes were in the opposite direction. DI rats had lower density of SFO ANG II receptors than LE rats (Table 5). ANG II receptors density was unaffected in both magnocellular and parvocellular paraventricular nucleus of DI rats.

DI rats had a much higher number of ANP receptors in the SFO than their LE controls (Table 6), and HZ animals had receptor concentrations that were between those of LE controls and DI rats. When water-deprived for 4 days, LE controls showed up-regulation of ANP receptors in the SFO, and the maximum concentration attained

Table 5
ANG II Receptors in Brain of Brattleboro and Long-Evans Rats

Strain	N	------------Apparent Binding Density------------ (fmol/mg protein)		
		SFO	PVN (Parvocellular)	PVN (Magnocellular)
Long-Evans	19	494 ± 25	190 ± 13	53 ± 6
Brattleboro (homozygous)	19	327 ± 18*	212 ± 10	54 ± 9
Brattleboro (heterozygous)	8	451 ± 35	202 ± 7	55 ± 5

Values are means ± SE. SFO, subfornical organ; PVN, paraventricular nucleus.
*$P < 0.05$ against LE controls. Reproduced from Castrén and Saavedra (42) with permission of the authors.

after water deprivation in LE animals was similar to that present in DI rats (Table 6). In contrast, the binding affinities of the various groups were similar (Table 6).

Peptide Receptors During Stress

The concentration of ANG II receptors was higher in parvocellular paraventricular nucleus (Figure 4 and Table 7) and SFO (Table 7) of chronically stressed rats than in the controls. There was no apparent alteration in the binding affinity (Table 7). ANG II binding in anterior pituitary was not altered by repeated stress (118 ± 8 and 117 ± 14 fmol/mg protein in control and stressed rats, respectively).

Brain Peptide Receptors After Endocrine Manipulations

Hypophysectomized rats had a lower density of ANG II receptors in both magnocellular and parvocellular paraventricular nucleus than did the sham-operated animals (Table 8). Hypophysectomy did not influence ANG II receptors in the SFO.

Adrenalectomy decreased ANG II receptors in the SFO as well as in the parvocellular and magnocellular paraventricular nucleus (Table 8). Corticosterone reversed the effects of adrenalectomy in all three nuclei, but mineralocorticoid replacement was effective only in the paraventricular nucleus. Conversely, ANG II receptors in pituitary gland were not altered by adrenalectomy.

Compared to controls, the density of ANP binding sites of hypophysectomized rats was significantly increased in the supraoptic nucleus (SON) as well as the magnocellular and parvocellular areas of the PVN (Table 9). Adrenalectomy and corticoid treatments did not alter the ANP binding in either the magnocellular or parvocellular paraventricular nucleus. In addition, neither hypophysectomy nor adrenalectomy had any effect on ANP binding sites in the SFO (Table 9).

Table 6
Brain ANP Binding Sites in Long-Evans, Brattleboro and
Water-Deprived Rats

Rat Strain	Binding Affinity (K_a) ($\times 10^9$ M^{-1})	Maximum Binding Capacity (B_{max}) (fmol/mg protein)
LE Control (8)	16.3 ± 2.8	29.0 ± 2.0
LE Water-Deprived (8)	14.1 ± 3.0	74.4 ± 5.3*†
HZ (5)	15.3 ± 1.5	43.7 ± 1.0*
DI (5)	12.4 ± 2.9	67.4 ± 6.9*†

Values are means $\pm$ SE. Number of animals per group is shown in parentheses; they were assayed individually.

*$P < 0.005$ *vs.* control LE.

†$P < 0.005$ *vs.* HZ. Reproduced from Saavedra *et al.* (50) with permission of The Society of Experimental Biology and Medicine.

Discussion

Circulating ANG II and ANP influence brain function by stimulation of their circumventricular organ receptors. Receptors inside the blood-brain barrier indicate the presence of brain (endogenous) peptide systems (5,10,13,14,20,21). Peripheral and central ANG II systems are anatomically linked as indicated by the ANG II forebrain receptor band (11,12). There is an integrated forebrain circuit, responding to both peripheral and central ANG II. The brainstem contains another important site for peripheral and central ANG II interactions, *viz.*, the area postrema-nucleus of the solitary tract complex.

The brain ANG II receptor distribution indicates that the peptide is involved in the central regulation of cardiovascular function and fluid metabolism (5). Peptide receptors are highly concentrated in the parvocellular zone of the paraventricular nucleus (22), the site of formation of the corticotropin releasing factor (CRF) (23), and in the median eminence, where they could regulate CRF release. In the periphery, ANG II regulates the formation and release of aldosterone (2). It is thus possible that the central regulation of CRF, and therefore ACTH, is integrated with the peripheral regulation of mineralocorticoid secretion.

Many of the ANP receptors and neurons are located in areas related to cardiovascular and fluid regulation, and brain ANP may be involved in the modulation of similar functions as brain ANG II. The presence of ANP receptors in the choroid plexus (10,24) suggests a role for ANP in the control of fluid regulation in the central nervous system.

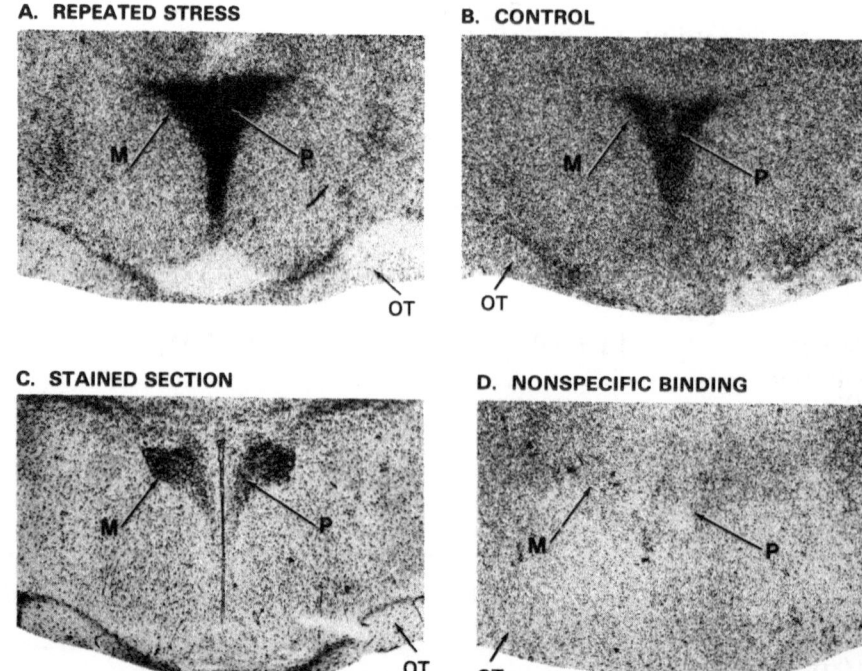

A. REPEATED STRESS

B. CONTROL

C. STAINED SECTION

D. NONSPECIFIC BINDING

Figure 4. Angiotensin II receptor in subfornical organ of stressed rats. *A* and *B*: typical sections from stressed and control rats, respectively. *C*: adjacent section stained with toluidine blue. *D*: non-specific binding. P: parvocellular paraventricular nucleus. M: magnocellular paraventricular nucleus. OT: optic tract. Reproduced from Castrén and Saavedra (22) with permission of The Endocrine Society.

Table 7
ANG II Receptor Concentration in Brain
of Chronically Stressed Rats

Area	Control		Stressed	
	B_{max} (fmol/mg protein)	K_d (nM)	B_{max} (fmol/mg protein)	K_d (nM)
Subfornical Organ	230 ± 8	0.71 ± 0.008	311 ± 13*	0.78 ± 0.07
Paraventricular Nucleus (Parvocellular Subdivision)	63 ± 7	0.54 ± 0.09	105 ± 11*	0.61 ± 0.06

Values are means ± SE.

Kinetic parameters were calculated from 7 different Scatchard plots per group, measured individually in different rats.

*$P < 0.05$, stressed *vs.* control groups. Reproduced from Castrén and Saavedra (22) with permission of The Endocrine Society.

Alterations in ANG II and ANP receptors occur in a model of genetic hypertension, the SH rat. These changes are in opposite directions, ANG II receptors increase in number in SH rats, whereas ANP receptors decrease in number (18).

Increased ANG II receptor concentrations occur in a model of mineralocorticoid-salt excess hypertension, the DOCA-salt rat (25), and in the same areas as those observed in SH rats. The changes are more widespread and of a higher magnitude than those in rats treated with salt or DOCA only, which do not develop hypertension. The alterations in the median preoptic nucleus, part of the forebrain ANG II receptor band, are of special interest, since lesions of this area prevent the development of DOCA-salt hypertension (26). Salt-treated rats show increased ANG II receptor concentrations into the median preoptic nucleus and the SFO. In DOCA-treated rats, increased ANG II binding also occurs in the paraventricular nucleus. Increases in brainstem ANG II receptors occur only in DOCA-salt hypertensive rats. Despite suppression of the peripheral angiotensin system by chronic administration of mineralocorticoids, DOCA-salt hypertensive rats also show evidence of increased central ANG II system activity (27-29). Our results support these findings.

Thus, in two models of hypertension, genetic and mineralocorticoid-salt induced, receptors for a pro-hypertensive, fluid conserving peptide such as ANG II is increased in key areas of the brain related to cardiovascular control and fluid regulation. Conversely, receptors for an antihypertensive, diuretic and natriuretic peptide, ANP, are greatly decreased in similar areas of genetically hypertensive rats, and do not increase after a salt load in DOCA-salt hypertensive animals. Such imbalance in receptor numbers of mutually antagonistic peptides may reflect an increased central response to even normal concentrations of ANG II in the blood of SH rats, a model presenting normal or even lower plasma renin concentrations. Our findings could be interpreted as an up-regulation of central ANG II receptors and a down-regulation of central ANP receptors, subsequent to primary alterations in peripheral peptide systems. The physiological significance of the decreased brain ANP receptor number in SH rats, however, is an open question. ANP receptors are also decreased in sympathetic ganglia from SH animals (30), but in these tissues the increase of the second messenger, cyclic GMP, after ANP was unchanged. This indicated a discrepancy between the number of ANP receptors and the ANP mediated biochemical response in SH rats.

Increased activity of the brain ANG II system in hypertension could partially explain the therapeutic effects of ACE inhibitors. By decreasing the number of central ANG II receptors in the SFO, ACE inhibitors may reduce the central response to circulating ANG II in SH animals. Our results indicate that systemic enalapril treatment does not modify ANG II receptors in brain areas located inside the blood-brain barrier. However, systemic enalapril could indirectly affect the central ANG II system through decreased ANG II binding in the SFO and inhibition of the ANG II forebrain receptor band.

Our results could explain certain autonomic and endocrine abnormalities in hypertension. Changes in receptor concentrations in the SFO and the paraventricular nucleus may be associated with alterations in vasopressin formation and release in hypertension (31). Circulating ANG II plays a role in the release of vasopressin, and ANG II-induced vasopressin release is enhanced by mineralocorticoid treatment (32). Since the posterior pituitary is devoid of ANG II receptors (17,33), it is likely that this is a central effect. The first step in this chain of events is probably the stimulation of ANG II receptors in the SFO, followed by stimulation of the forebrain ANG II band and stimulation of the posterior pituitary with the end result of increased vasopressin release.

The changes in ANG II receptors in the paraventricular nucleus may be related to those in the SFO. When administered by iontophoresis into the paraventricular

Table 8
Effects of Endocrine Manipulations on Angiotensin II
Receptor Concentrations

Treatment	N	SFO	PVN (Parvocellular)	PVN (Magnocellular)	Anterior Pituitary
			Apparent Binding Density (fmol/mg protein)		
Control	12	308 ± 13	223 ± 10	56 ± 6	
Hypophysectomy	12	326 ± 17	175 ± 13*	30 ± 4*	
Control	12	365 ± 33	182 ± 7	43 ± 3	345 ± 30
Adrenalectomy	12	175 ± 22*	106 ± 8*	19 ± 1*	325 ± 35
ADX + Corticosterone	8	379 ± 34	206 ± 12	46 ± 4	358 ± 59
ADX + DOCA	8	215 ± 30*	182 ± 19	40 ± 3	348 ± 60

Values are means ± SE.
Tissue sections were incubated with 2.5 nM ^{125}I-labelled [Sar^1]angiotensin II. SFO, subfornical organ; PVN, paraventricular nucleus; ADX, adrenalectomized.
*P < 0.05 against controls. Reproduced from Castrén and Saavedra (42) with permission of the authors.

Table 9
Density of ANP Binding Sites in Subfornical Organ (SFO),
Supraoptic Nucleus (SON), and Paraventricular Nucleus (PVN)
After Endocrine Manipulations

Treatment	N	SFO	SON	PVN (Magnocellular)	PVN (Parvocellular)
			Apparent Binding Density (fmol/mg protein)		
Control	12	63 ± 5	20 ± 2	16 ± 2	12 ± 1*
Hypophysectomy	12	62 ± 4	37 ± 2†	35 ± 2†	19 ± 1
Control	12	90 ± 10	20 ± 5	20 ± 3	9 ± 2
Adrenalectomy	12	83 ± 4	19 ± 1	14 ± 2	7 ± 3
ADX + Corticosterone	8	84 ± 4	14 ± 2	17 ± 1	5 ± 1
ADX + DOCA	8	82 ± 12	15 ± 2	15 ± 2	8 ± 3

Values are means ± SE.
*P < 0.05.
†P < 0.001 $vs.$ controls. Reproduced from Castrén and Saavedra (59) with permission of The American Physiological Society.

nucleus, ANG II stimulates neurosecretory cells (34). Lesions of the paraventricular nucleus decrease the hypertension produced by SFO stimulation (35). Endogenous ANG II is co-localized with vasopressin in paraventricular magnocellular neurons (36). However, most, if not all, ANG II receptors are located in the parvocellular zone, site of formation of CRF. Thus, the observed increase in ANG II receptors in the parvocellular paraventricular nucleus may be related to increased ACTH release in SH rats. We have also found increased numbers of ANG II receptors in the paraventricular nucleus of DOCA-salt treated rats. Since these receptors are high during administration of DOCA alone, it is conceivable that part of this increase could be related to a local effect of mineralocorticoids. Increased number of ANG II receptors in the median preoptic nucleus of SH rats can be explained if we consider the stimulation of the complete forebrain ANG II receptor band necessary for increased vasopressin release and increased blood pressure in SH animals. The median preoptic nucleus may play an important role in the integration of fluid and cardiovascular regulation, as shown by the deficits in drinking and vasopressin secretion which follow its lesion (37). In DOCA-salt rats, however, increased ANG II receptors in this nucleus occurs to the same extent as that seen in DOCA-treated or salt-treated normotensive animals. Thus, the changes may be related to alterations in fluid homeostasis and to effects secondary to mineralocorticoid administration.

In the nucleus of the solitary tract, ANG II receptors were increased in both SH and DOCA-salt rats. Both models have impaired baroreflex function, and that of DOCA-salt rats returns to normal after treatment of the hypertension with ACE inhibitors (38). This area may represent still another site for a central interaction between ANG II and ANP systems, since ANP receptors are decreased in the nucleus of the solitary tract and area postrema of SH animals. These alterations may be related to the increased peripheral sympathetic drive in SH rats and to changes in catecholamines, specially epinephrine (39).

In addition to increased number of ANG II receptors in the SFO of salt-treated rats, a model of extracellular dehydration, similar changes in ANG II receptors are present in another model of dehydration, the acutely dehydrated, water-deprived rat. In dehydration due to water deprivation, vasopressin formation and release is significantly enhanced and the peripheral ANG II system is greatly stimulated. The increase in ANG II receptors (40) is probably the result of up-regulation triggered by increased circulating ANG II concentrations (41) and may represent an amplification of the central response to angiotensin. Vasopressin may be needed for receptor up-regulation in the SFO, since in DI, vasopressin-deficient, dehydrated rats, SFO ANG II receptors are down-regulated (42).

Of particular interest is the lack of change in the paraventricular nucleus during dehydration, which indicates that peptide receptors in this area are probably related to regulation of anterior pituitary function rather than vasopressin release. Our investigations on ANG II receptors during stress lend further support to this hypothesis. During repeated immobilization stress, the number of ANG II receptors in the paraventricular nucleus and in the SFO are greatly increased. This kind of stress increases plasma renin activity (43), ACTH release, and peripheral sympathetic activity (16). In addition to its role in the regulation of ACTH release, the paraventricular nucleus is involved in the control of peripheral sympathetic activity (23). Our results may indicate the presence of an enhanced central effect of circulating ANG II, probably through sensitization of neuronal connections between the SFO and the paraventricular nucleus. It is also possible that corticoids, highly elevated in blood during stress, could play a direct role in the regulation of ANG II receptors at the SFO-paraventricular nucleus axis. Steroid hormones may alter ANG II receptor properties in peripheral organs (44). Alterations in ANG II receptors in the paraventricular nucleus of stress and DOCA-treated rats, may indicate an influence of steroid hormones in the local regulatory mechanisms for ANG II receptors in paraventricular parvocellular neurons.

If circulating ANG II is involved, directly or indirectly, in the increased ACTH secretion during stress, this is probably through central, rather than pituitary, mechanisms, since the ANG II receptors in anterior pituitary are not different in number in stressed animals (22).

Hypophysectomized and adrenalectomized rats that have greatly reduced corticoid levels have a lower number of ANG II receptors in the parvocellular portion of the paraventricular nucleus. Both corticosterone and DOCA restore the receptor densities. These data suggest that the activity of the SFO-paraventricular nucleus connection may be increased in stress and decreased in the absence of corticoids. CRF release is, however, increased in both conditions. The secretion of CRF is thus regulated by different factors in stress and after adrenalectomy. ANG II may be one of the factors that maintains elevated CRF secretion during stress, when circulating corticoids are high. When the corticoid production is abnormally low, other factors such as catecholamines and the lack of negative feedback by corticoids may control the stimulation of CRF secretion (45). The density of ANG II receptors is decreased in the SFO of adrenalectomized and DI, vasopressin-deficient rats. ANG II receptors in SFO are increased after chronic stress and dehydration (22,40). Plasma ANG II levels are high in all these conditions (41,43). The level of corticoids in the peripheral circulation is increased after dehydration and stress (43), but absent or normal in adrenalectomized and DI rats, respectively (46). Plasma renin activity is within normal range after hypophysectomy (47), and ANG II receptors in the SFO are unaltered. These data suggest that the response of ANG II receptors in the SFO to increased plasma ANG II may be dependent on the plasma levels of corticoids and/or vasopressin.

Replacement with both corticosterone and DOCA reverse the effect of adrenalectomy on ANG II receptor binding in the paraventricular nucleus, but only corticosterone is effective in the SFO. It is possible that corticosterone acts through mineralocorticoid receptors in the paraventricular nucleus, since corticosterone has significant mineralocorticoid activity. This is consistent with reports that mineralocorticoid excess increases brain ANG II receptor binding (48).

Recent observations demonstrate that adrenalectomized rats are less sensitive than sham-operated rats to blood pressure elevating effects of peripherally administered ANG II (49). The subsensitivity can be reversed by dexamethasone, but not by aldosterone (49). This is consistent with our results showing that lower number of ANG II receptors in the SFO of adrenalectomized rats could be reversed by glucocorticoid, but not by mineralocorticoid replacement. We suggest that the low number of ANG II receptors in the SFO and the subsensitivity to peripheral ANG II in adrenalectomized rats could be directly related. The regulatory effect of glucocorticoids on central ANG II receptors might therefore be of physiological importance in the regulation of blood pressure.

Both water-deprived rats and DI rats have increased ANP receptors in the SFO (50,51), a finding that could be interpreted as a compensatory increase secondary to decreased ANP release during dehydration (52). Hypophysectomy increases the density of ANP binding sites in the supraoptic and magnocellular paraventricular nuclei. We studied ANP binding to hypothalamic nuclei of adrenalectomized rats and multiple hormone deficiencies, and a gradual recovery of vasopressin secretion occurs within several weeks after hypophysectomy (53). Adrenalectomy with or without corticosterone replacement therapy does not influence ANP binding. Conversely, ANP binding is significantly increased in the supraoptic and magnocellular paraventricular nuclei of DI rats, suggesting that those binding sites may be associated with vasopressin neurons. Increased binding in these nuclei in hypophysectomized rats supports this hypothesis.

Central administration of ANP inhibits vasopressin release (54), and electrophysiological data suggests that the likely sites for this action are SON and PVN (55). Since plasma vasopressin is absent in DI rats, there is no obvious physiological need to release inhibitory regulators. Thus, the increase in the density of ANP binding sites may reflect an up-regulation process due to the decreased release of ANP from nerve terminals in the SON and PVN. We have shown that ANP binding sites in the nuclei where vasopressin is produced are increased when plasma vasopressin levels are reduced. Our findings suggest that ANP binding sites in these nuclei may represent physiologically active receptors that participate in the feedback control of vasopressin secretion.

In the SFO, ANP binding sites are significantly increased in DI rats, but no alteration in ANP binding is seen after hypophysectomy. Plasma ANP levels are normal in DI rats (52) and decreased after hypophysectomy (56). In water-deprived rats, plasma ANP levels are decreased (57), and ANP binding sites in SFO are increased (58). Thus, there does not appear to be a direct correlation between plasma ANP levels and the density of ANP binding sites in the SFO in these conditions.

Conclusions

Both ANG II and ANP receptors are located in selected brain areas; some are accessible—*via* the circumventricular organs—and some are not accessible to circulating peptides. The ANG II and ANP receptor distribution overlaps in many areas related to cardiovascular and fluid regulation, and to the central control of sympathetic activity and pituitary function.

Specific alterations in brain peptide receptor number occur in several animal models of disease. The imbalance between brain ANG II and ANP receptors in hypertension, with a relative increase in receptors for ANG II, should favor central pro-hypertensive mechanisms, and may have therapeutic implications.

During dehydration, the brain up-regulates its circumventricular ANG II receptors in the presence of increased peripheral ANG II concentrations. This is probably a mechanism to secure the necessary biochemical and behavioral responses to the loss of body fluids that are of fundamental importance to survival.

The alterations in brain ANG II receptors during stress provide the first concrete evidence for a role of the peptide in the central control of the pituitary-adrenal axis.

Alterations in peripheral and central peptide systems, and of ANG II and ANP systems, are associated. There is a close balance between peripheral and central ANG II and ANP systems, which is of importance for homeostasis and for hormonal and sympathetic regulation.

References

1. DeBold, A.J., Atrial natriuretic factor: a hormone produced by the heart, *Science* 230: 767-770, 1985.
2. Vallotton, M.B., The renin-angiotensin system, *Trends Pharmacol Sci* 8: 69-74, 1987.
3. Niwa, M., A. Israel, and J.M. Saavedra, Pindolol decreases plasma angiotensin-converting enzyme activity in young spontaneously hypertensive rats, *Eur J Pharmacol* 110: 133-136, 1985.
4. Morii, N., K. Nakao, M. Kihara, A. Sugawara, M. Sakamoto, Y. Yamori, and H. Imura, Decreased content in left atrium and increased plasma concentration of atrial natriuretic polypeptide in spontaneously hypertensive rats (SHR) and SHR stroke-prone, *Biochem Biophys Res Commun* 135: 74-81, 1986.
5. Phillips, M.I., Functions of angiotensin in the central nervous system, *Ann Rev Physiol* 49: 413-435, 1987.
6. Antunes-Rodrigues, J., S.M. McCann, and W.K. Samson, Central administration of atrial natriuretic factor inhibits saline preference in the rat, *Endocrinology* 118: 1726-1728, 1986.

7. Haskins, J.T., G.J. Zingara, and R.W. Lappe, Rat atriopeptin III alters hypothalamic neuronal activity, *Neurosci Lett* 67: 279-284, 1986.

8. Obana, K., M. Naruse, T. Inagami, A.B. Brown, K. Naruse, F. Kurimoto, H. Sakurai, H. Demura, and K. Shizune, Atrial natriuretic factor inhibits vasopressin secretion from rat posterior pituitary, *Biochem Biophys Res Commun* 132: 1088-1094, 1985.

9. Mendelsohn, F.A.O., R. Quirion, J.M. Saavedra, G. Aguilera, and K. Catt, Autoradiographic localization of angiotensin II receptors in rat brain, *Proc Natl Acad Sci USA* 81: 1575-1579, 1984.

10. Quirion R., M. Dalpe, and T.V. Dam, Characterization and distribution of receptors for atrial natriuretic peptides in mammalian brain, *Proc Natl Acad Sci USA* 83: 174-178, 1986.

11. Saavedra, J.M., A. Israel, L.M. Plunkett, M. Kurihara, K. Shigematsu, and F.M.A. Correa, Quantative distribution of angiotensin II binding sites in rat brain by autoradiography, *Peptides* 7: 679-687, 1986.

12. Shigematsu, K., J.M. Saavedra, L.M. Plunkett, M. Kurihara, and F.M.A. Correa, Angiotensin II binding sites in the anteroventral-third ventricle (AV3V) area and related structures of the rat brain, *Neurosci Lett* 67: 37-41, 1986.

13. Kurihara, M., J.M. Saavedra, and K. Shigematsu, Localization and characterization of atrial natriuretic peptide binding sites in discrete areas of rat brain and pituitary gland by quantitative autoradiography, *Brain Res* 408: 31-39, 1987.

14. Jacobowitz, D.M., G. Skofitsch, H.R. Keiser, R.L. Eskay, and N. Zamir, Evidence for the existence of atrial natriuretic factor-containing neurons in the rat brain, *Neuroendocrinology* 40: 92-94, 1985.

15. Nazarali, A.J., J.S. Gutkind, Fernando M.A. Correa, and J.M. Saavedra, Enalapril decreases angiotensin II receptors in subfornical organ of SHR, *Am J Physiol* 256: H1609-H1614, 1989.

16. Kvetñanský, R., V.K. Weise, and I.J. Kopin, Elevation of adrenal tyrosine hydroxylase and phenylethanol-amine-N-methyl transferase by repeated immobilization of rats, *Endocrinology* 87: 744-749, 1970.

17. Israel, A., L.M. Plunkett, and J.M. Saavedra, Quantitative autoradiographic characterization of receptors for angiotensin II and other neuropeptides in individual brain nuclei and peripheral tissues from single rats, *Cell Mol Neurobiol* 5: 211-222, 1985.

18. Saavedra, J.M., F.M.A. Correa, L.M. Plunkett, A. Israel, M. Kurihara, and K. Shigematsu, Binding of angiotensin and atrial natriuretic peptide in brain hypertensive rats, *Nature* 320: 758-760, 1986.

19. Nazarali, A.J., J.S. Gutkind, and J.M. Saavedra, Calibration of [^{125}I]-polymer standards with [^{125}I]-brain paste standards for use in quantitative receptor autoradiography, *J Neurosci Methods*, In Press.

20. Ganten, D., K. Hermann, C. Bayer, T. Unger, and R.E. Lang, Angiotensin synthesis in the brain and increased turnover in hypertensive rats, *Science* 221: 869-871, 1983.

21. Wong, M., W. K. Samson, C.A. Dudley, and R.L. Moss, Direct, neuronal action of atrial natriuretic factor in the rat brain, *Neuroendocrinology* 44: 49-53, 1986.

22. Castrén, E., and J.M. Saavedra, Repeated stress increases the density of angiotensin II binding sites in rat paraventricular nucleus and subfornical organ, *Endocrinology* 122: 370-372, 1988.

23. Swanson, L.W., and P.E. Sawchenco, Paraventricular nucleus: a site for the integration of neuroendocrine and anatomic mechanisms, *Neuroendocrinology* 31: 410-417, 1980.

24. Steardo, L., and J.A. Nathanson, Brain barrier tissues: end organs for atriopeptins, *Science* 235: 470-473, 1987.

25. Gutkind, J.S., M. Kurihara, and J.M. Saavedra, Increased angiotensin II receptors in brain nuclei of DOCA-salt hypertensive rats, *Am J Physiol* 255: H646-H650, 1988.

26. Brody, M.J., and A.K. Johnson, Role of the anteroventral third ventricle region in fluid and electrolyte balance, arterial pressure regulation and hypertension, In W.F. Ganong and L. Martini (eds) *Frontiers in Neuroendocrinology*, Raven Press, New York, Volume 6, pp. 249-292, 1980.

27. Chen, Y.F., M.D. Lindheimer, and S. Oparil, Increased vasopressinergic activity following DOCA administration in the rat, *Brain Res Bull* 16: 93-98, 1986.

28. Itaya, Y., H. Suzuki, S. Matsukawa, K. Kondo, and T. Saruta, Central renin-angiotensin system and the pathogenesis of DOCA-salt hypertension in rats, *Am J Physiol* 251: H261-H268, 1986.

29. Weyhenmeyer, J.A., and J.M. Meyer, Angiotensin II in the brain and brainsten of the DOCA salt hypertensive rat, *Clin Exp Hyperten* A7: 73-92, 1985.

30. Gutkind, J.S. M. Kurihara, E. Castrén, and J.M. Saavedra, Atrial natriuretic peptide receptors in

sympathetic ganglia: biochemical response and alterations in genetically hypertensive rats, *Biochem Biophys Res Commun* 149: 65-72, 1987.

31. Crofton, J.T., L. Share, R.E. Shade, C. Allen, and D. Tanowski, Vasopressin in the rat with spontaneous hypertension, *Am J Physiol* 235: H361-H366, 1978.

32. Wilson, K.M., C. Sumners, S. Hathaway, and M.J. Fregly, Mineralo-corticoids modulate central angiotensin II receptors in rats, *Brain Res* 382: 87-96, 1986.

33. Israel, A., F.M.A. Correa, M. Niwa, and J.M. Saavedra, Quantitative determination of angiotensin II binding sites in rat brain and pituitary gland by autoradiography, *Brain Res* 322: 341-345, 1984.

34. Akaishi, T., H. Negoro, and S. Kobayasi, Electrophysiological evidence for multiple sites of actions of angiotensin II for stimulating paraventricular neurosecretory cells in the rat, *Brain Res* 220: 386-390, 1981.

35. Ferguson, A.V., and L.P. Renaud, Hypothalamic paraventricular neuclus lesions decrease pressor responses to subfornical organ stimulation, *Brain Res* 305: 361-364, 1984.

36. Killcoyne, M.M., D. L. Hoffman, and E.H. Zimmerman, Immunocytochemical localization of angiotensin II and vasopressin in rat hypothalamus: evidence for production in the same neuron, *Clin Sci* 59: 57S-60S, 1980.

37. Mangiapane, M.L., T.N. Thrasher, L.C. Keil, J.B. Simpson, and W.F. Ganong, Deficits in drinking and vasopressin secretion after lesions of the nucleus medianus, *Neuroendocrinology* 37: 73-77, 1983.

38. Matsuguchi, H., and P.G. Schmid, Pressor response to vasopressin and impaired baroreflex funcion in DOCA-salt hypertension, *Am J Physiol* 242: H44-H49, 1982.

40. Nazarail, A.J., J.S. Gutkind, and J.M. Saavedra, Regulation of angiotensin II binding sites in subfornical organ and other rat brain nuclei after water deprivation, *Cell Mol Neurobiol* 7: 447-455, 1987.

41. Mann, J.F.E., A.K. Johnson, and D. Ganten, Plasma angiotensin II: dipsogenic levels and angiotensin-generating capacity of renin, *Am J Physiol* 238: R372-R377, 1980.

42. Castrén, E., and J.M. Saavedra, Angiotensin II receptors in paraventricular nucleus, subfornical organ, and pituitary gland of hypophysectomized, adrenalectomized, and vasopressin-deficient rats, *Proc Natl Acad Sci USA* 86: 725-729, 1989.

43. Jindra, A., Jr., R. Kvetňanský, T.I. Belova, and K.V. Sudakov, Effect of acute and repeated immobilization stress on plasma renin activity, catecholamines and corticosteroids in Wistar and August rats, In E. Usdin, R. Kvetňanský, and I.J. Kopin (eds) *Catecholamines and Stress: Recent Advances*, Elsevier/North Holland, Amsterdam, pp. 249-254, 1980.

44. Douglas, J.G., Corticosteroids decrease glomerular angiotensin receptors, *Am J Physiol* 252: F453-457, 1987.

45. Spinedi, E., C.A. Johnston, A. Chisari, and A. Negro-Vilar, Role of central epinephrine on the regulation of corticotropin-releasing factor and adrenocorticotropin secretion, *Endocrinology* 122: 1977-1983, 1988.

46. Sokol, H.W., and E.A. Zimmerman, The hormonal status of the Brattleboro rat, In H.W. Sokol and H. Valtin (eds) *The Brattleboro Rats*, The New York Academy of Sciences, New York, 394: 535-548, 1982.

47. Hasegawa, H., A. Nasjletti, K. Rice, and G.M.C. Masson, Role of pituitary and adrenals in the regulation of plasma angiotensinogen, *Am J Physiol* 225: 1-6, 1973.

48. King, S.J., J.W. Harding, and K.E. Moe, Elevated salt appetite and brain binding of angiotensin II in mineralocorticoid-treated rats, *Brain Res* 448: 140-149, 1988.

49. Yagil, Y., and L.R. Karkoff, The differential effect of aldosterone and dexamethasone on pressor responses in adrenalectomized rats, *Hypertension* 11: 174-178, 1988.

50. Saavedra, J.M., A. Israel, F.M.A. Correa, and M. Kurihara, Increased atrial natriuretic peptide (6-33) binding sites in the subfornical organ of water deprived and Brattleboro rats, *Proc Soc Exp Biol Med* 182: 559-563, 1986.

51. Saavedra, J.M., A. Israel, and M. Kurihara, Increased atrial natriuretic peptide binding sites in the rat subfornical organ after water deprivation, *Endocrinology* 120: 426-428, 1987.

52. Ogawa, K., M.A. Henry, J. Tange, E.A. Woodcock, and C.I. Johnston, Atrial natriuretic peptide in dehydrated Long-Evans rats and Brattleboro rats, *Kidney Int* 31: 760-765, 1987.

53. Lloyd, C.W., E. Loewy, S. Pierog, K. Bradwick, and R. Sostheim, Presence of antidiuretic material in blood of hypophysectomized rats, *Proc Soc Exp Biol Med* 85: 333-336, 1954.

54. Samson, W.K., M.C. Aguila, J. Martinovic, J. Antunes-Rodrigues, and M. Norris, Hypothalamic action

of atrial natriuretic factor to inhibit vasopressin secretion, *Peptides* 8: 449-454, 1987.

55.   Sandaert, D.G., D.F. Cechetto, P. Needleman, and C.P. Saper, Inhibition of the firing of vasopressin neurons by atriopeptin, *Nature* 329: 151-153, 1987.

56.   Zamir, N., M. Haass, J.R. Dave, and Z. Zukowska-Grojec, Anterior pituitary gland modulates the release of atrial natriuretic peptides from cardiac atria, *Proc Natl Acad Sci USA* 84: 541-545, 1987.

57.   Januszewicz, P., G. Thibault, J. Gutkowska, R. Garcia, C. Mercure, F. Jolicoeur, J. Genest, and M. Cantin, Atrial natriuretic factor and vasopressin during dehydration and rehydration in the rat, *Am J Physiol* 251: E497-E501, 1986.

58.   Saavedra, J.M., Regulation of atrial natriuretic peptide receptors in the rat brain, *Cell Mol Neurobiol* 7: 151-173, 1987.

59.   Castrén, E., and J.M. Saavedra, Lack of vasopressin increases hypothalamic atrial natriuretic peptide binding sites, *Am J Physiol* 257: R168-R173, 1989.

THE ANTERIOR THIRD VENTRICLE REGION IS A RECEPTOR SITE FOR COMPOSITION RATHER THAN VOLUME OF BODY FLUIDS

B. Lichardus, J. Ponec, M.J. McKinley,[1] J. Okoličány,[2]
I. Gabauer,[2] J. Styk,[2] P. Bakoš, C. Oliver,[3] and
N. Michajlovskij

Institute of Experimental Endocrinology, Centre of Physiological Sciences
Slovak Academy of Sciences
Bratislava, Czechoslovakia

The first International Symposium on natriuretic hormones took place in 1969 at Smolenice Castle. At that time, the prevailing evidence pointed to the existence of only one specific natriuretic hormone: a small peptide. This hormone was postulated to inhibit the activity of the transporting enzyme, Na,K-ATPase, and was later classified as an endogenous inhibitor of the sodium pump—a digoxin-like substance. At a subsequent meeting in 1980 in Bratislava, it was decided to change the title to *natriuretic hormones*. The reason for the change was that in addition to the suggested, but still unidentified, natriuretic, hormone a number of well known hormones had been described that promoted renal sodium excretion (certain prostaglandins, kinins, dopamine, vasopressin, oxytocin, substance P, *etc.*) (1).

A year later, the discovery of the most important hormones with natriuretic properties, the atrial natriuretic peptides, was announced by de Bold and his colleagues (2). However, atrial natriuretic peptides do not inhibit the activity of the sodium pump. It is, therefore, reasonable to believe that the biological evidence for an endogenous inhibitor of the sodium pump points to a discrete substance. Substances having *endogenous digoxin-like activity* (EDLA) are numerous and may not always include the particular natriuretic hormone that probably plays a role in the regulation of extracellular fluid volume by promoting renal sodium excretion. This uncertainty is due in part to the various cross-reactivities of digoxin antiserum used in the characterization of EDLA. One may hope that a circulating natriuretic hormone, identified by a digoxin antiserum, has a physiological correlate such as natriuresis, blood pressure increase, and/or expansion of the extracellular fluid volume (ECFV). Thus, references to atrial natriuretic peptides and the natriuretic hormone co-exist in the contemporary literature as they probably work together in all mammals and possibly in many, if not most, vertebrates.

Cerebral Origin of a Natriuretic Hormone

There is evidence that the brain is the site of synthesis and secretion of a

---

[1]Howard Florey Institute, University of Melbourne, Parkville, Victoria, Australia

[2]Institute of Experimental Surgery, Center of Physiological Sciences, Bratislava, Czechoslovakia

[3]INSERM, U 297, Marseille, France

Circulating Regulatory Factors and Neuroendocrine Function
Edited by J. C. Porter and D. Ježová
Plenum Press, New York, 1990

211

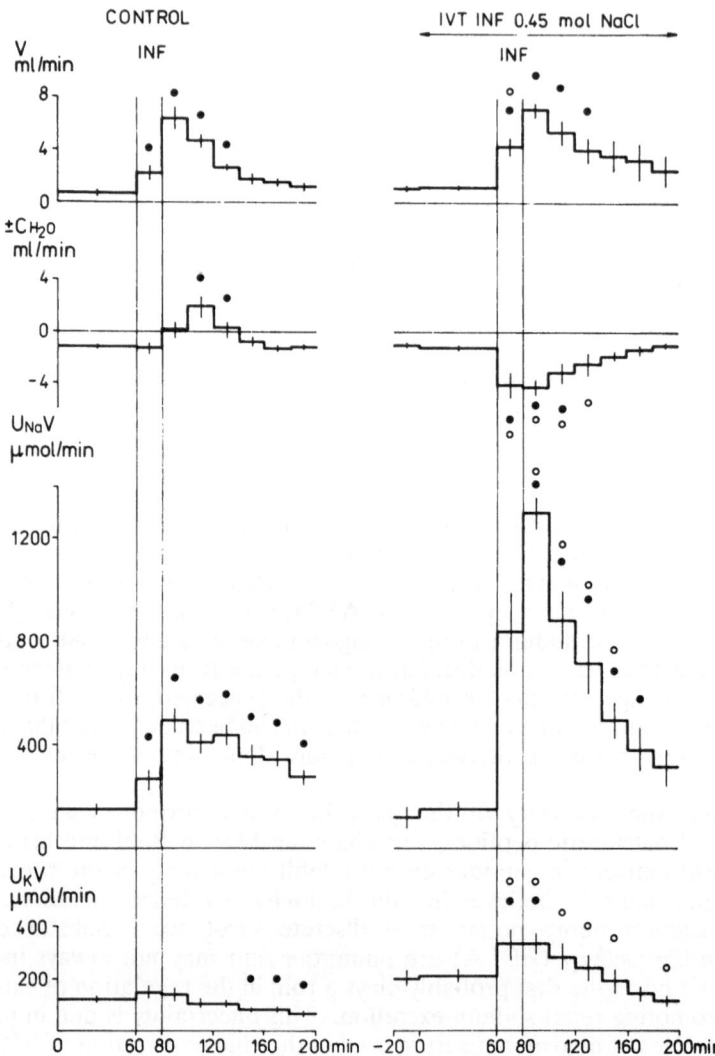

Figure 1. Effect of ECFV expansion by iv infusion of isotonic saline (INF; 3% BW in 20 min) on urine (V) and sodium ($U_{Na}V$) excretion in conscious sheep (mean ± SE) with normal (Control) and increased sodium concentration in the cerebrospinal fluid (increased CSF_{Na}; intraventricular infusion of hypertonic artificial CSF, 450 mmol NaCl/liter, infused into the lateral ventricle at 20 µl/min). The increase in CSF_{Na} was followed by a large increase in renal sodium excretion compared to the effect of ECFV expansion alone in the control animals, or to the essentially non-existent natriuretic action of increased CSF sodium concentration alone during 60 min period before ECFV expansion. IVT signifies intraventricular; C_{H_2O} denotes clearance of osmotically free water; U_KV denotes renal potassium excretion; a closed circle signifies that the mean is statistically different from the preinfusion value; an open circle denotes that the mean of the experimental group (n = 6) is statistically different from the mean of the control group (n = 7) when analyzed by ANOVA.

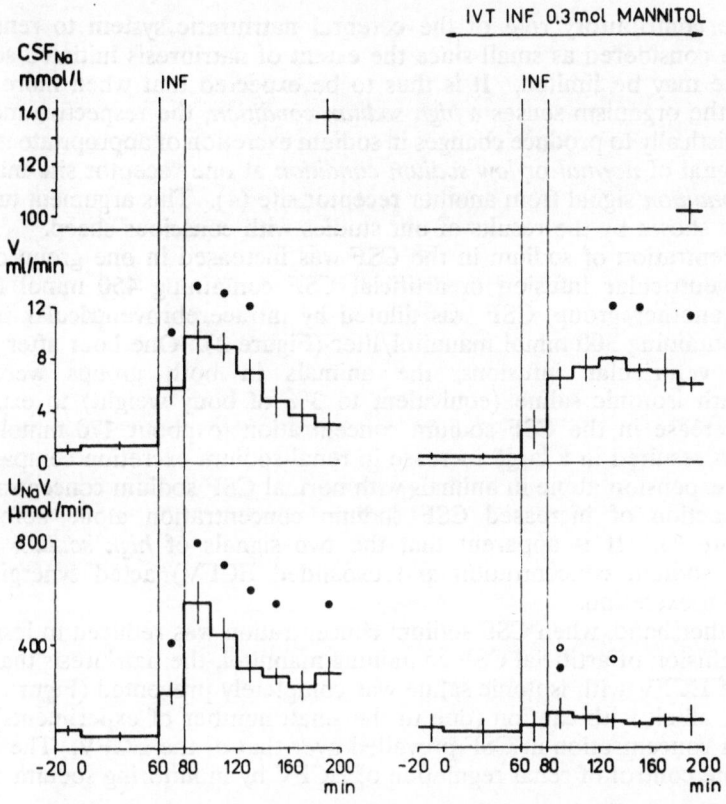

Figure 2. Effect of ECFV expansion by iv infusion of isotonic saline (INF; 3% BW in 20 min) on urine (V) and sodium ($U_{Na}V$) excretion in conscious sheep (mean ± SE) with a normal CSF concentration of sodium (CSF_{NA}) (*left panel*) or decreased CSF_{Na} induced by intraventricular infusion (20 µl/min) of mannitol in artificial CSF (*right panel*) (300 mmol/liter). The lowered CSF_{Na} in two sheep prevented natriuresis following ECFV expansion. See Figure 1 for further details.

natriuretic hormone that is different from the atrial and certain brain natriuretic peptides. Various approaches to studies in this field have used a variety of interventions of brain function to study renal sodium excretion: *e.g.*, intracarotic and intracerebroventricular infusions of hypertonic solutions or dilute solutions of neuroactive drugs, surgical ablations of various parts of the brain, and interruption of central or peripheral neural pathways. Whether all of these experimental procedures affect sodium excretion in the same way is unclear (3).

Sodium Concentration of Cerebrospinal Fluid and Renal Regulation of the Extracellular Fluid Volume

It has been repeatedly shown in various species of anesthetized or conscious experimental animals that increased concentrations or dilutions of sodium salts in the cerebrospinal fluid (CSF) enhance or impede, respectively, renal sodium excretion. But, in this case, the contributory role of the cerebral natriuretic system to renal sodium excretion can be considered as small since the extent of natriuresis initiated solely from one receptor site may be limited. It is thus to be expected that when more than one receptor site in the organism senses a *high sodium condition*, the respective mechanisms would act synergistically to produce changes in sodium excretion of appropriate magnitude. Conversely, a signal of *normal* or *low sodium condition* at one receptor site may weaken a *high sodium condition* signal from another receptor site (4). This argument turns out to be acceptable as shown by the results of our studies with conscious sheep.

The concentration of sodium in the CSF was increased in one group of animals by intracerebroventricular infusion of artificial CSF containing 450 mmol NaCl/liter (Figure 1). In another group, CSF was diluted by intracerebroventricular infusion of artificial CSF containing 300 mmol mannitol/liter (Figure 2). One hour after beginning the intracerebroventricular infusions, the animals in both groups were infused intravenously with isotonic saline (equivalent to 3% of body weight) to expand their ECFV. The increase in the CSF sodium concentration to about 170 mmol/liter and ECFV expansion resulted in a large increase in renal sodium excretion compared to the effect of ECFV expansion alone in animals with normal CSF sodium concentration or to the natriuretic action of increased CSF sodium concentration alone before ECFV expansion (Figure 1). It is apparent that the two signals of *high sodium conditions* (increased CSF sodium concentration and expanded ECFV) acted synergistically in enhancing sodium excretion.

On the other hand, when CSF sodium concentration was reduced to less than 120 mmol/liter by infusion of artificial CSF containing mannitol, the natriuresis that followed the expansion of ECFV with isotonic saline was completely prevented (Figure 2). Thus, it can be stated, albeit with caution (due to the small number of experiments), that the signal of sodium concentration in CSF prevailed over that of the ECFV. The brain may be involved in the control of renal regulation of ECFV by monitoring sodium concentration in CSF (3).

Anterior Wall of the Third Cerebral Ventricle and Renal Sodium Excretion

A critical brain area where the sodium concentration or osmolality of CSF is monitored is probably the anterior wall of the third cerebral ventricle (A3V). It was found by the push-pull technique, which enables one to vary the sodium concentration of CSF within a region adjacent to the ventricles, that sodium (osmotic?) sensors are located in the anterodorsal part of the A3V (AD3V) (5,6).

The periventricular organs, which lack the blood-brain barrier, seem to be critical not only for the central stimuli represented by the CSF sodium concentration but also for changes in the systemic ECF sodium concentration. This conclusion is supported by experiments in conscious sheep in which the A3V was ablated. [The lesion involved the organum vasculosum of the lamina terminalis, both dorsal and ventral divisions of the preoptic medianus nucleus, preoptic periventricular tissue, the anterior commissurae, and parts of the medial septum and diagonal band (7)].

The non-lesioned, control sheep were in a spontaneous water balance, whereas water balance in the animals with the chronically ablated A3V region was re-established 1 to 3 days before the experiment by forced application of drinking water through an intrarumenal tube. One group of animals was infused intravenously with hypertonic saline (0.23 mmol NaCl/minute for 40 minutes), and the other was infused with an equal amount of sodium as isotonic saline in 20 minutes in a volume equivalent to 3% of body weight. It was found that the integrity of the anterior wall of the third ventricle was necessary for the natriuresis induced by intravenous infusion of hypertonic saline. In comparison to

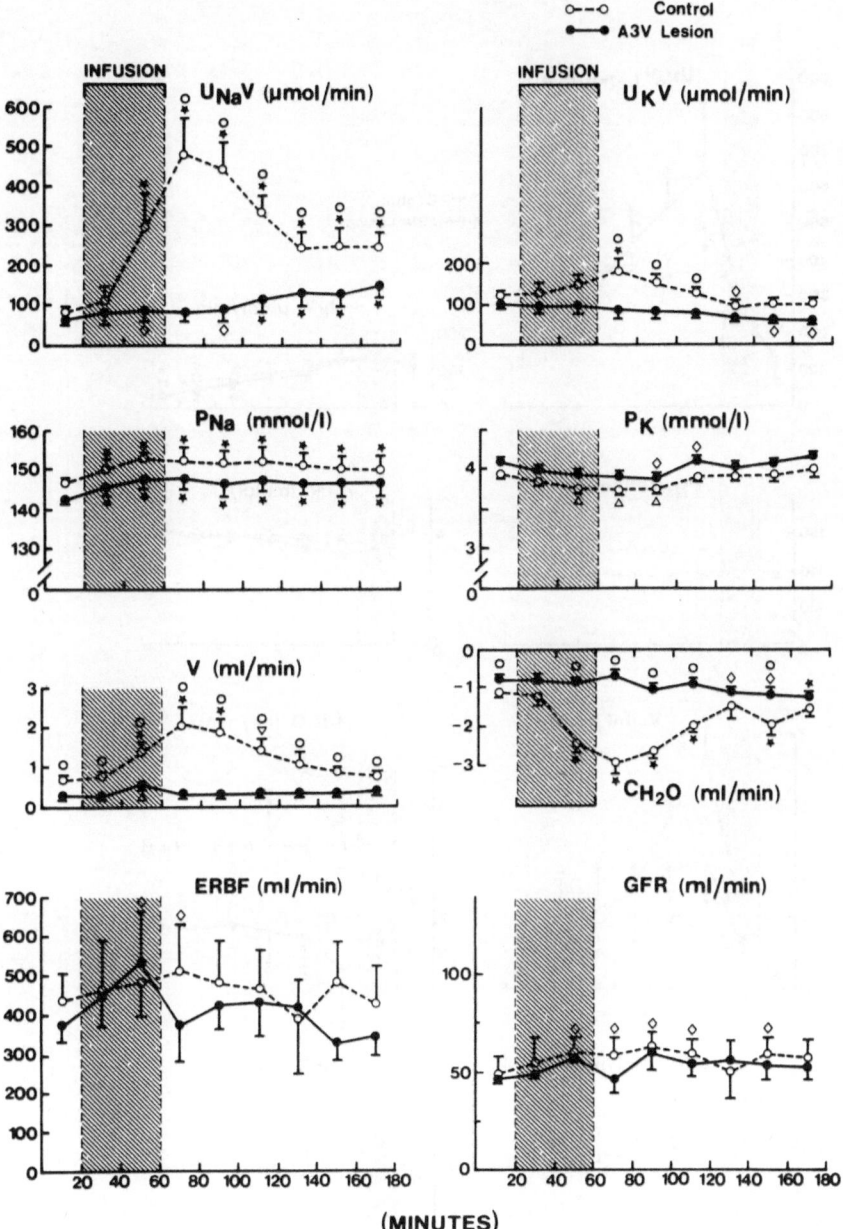

Figure 3. Effect of hypertonic saline load (0.23 mmol NaCl/min) on renal sodium excretion ($U_{Na}V$), potassium excretion (U_KV), plasma sodium (P_{Na}) and potassium (P_K) concentrations, urine excretion (V), clearance of osmotically free water (C_{H_2O}), effective renal blood flow (ERBF), and glomerular filtration rate (GFR). The *open circles* and *vertical lines* denote mean ± SE in the control conscious sheep (n = 8); the *closed circles* pertain to sheep with an ablated A3V region (n = 6). *Asterisks*, *diamonds* or *triangles* indicate statistical difference between the respective parameter and the preinfusion value; *open circles* indicate a statistical difference between the control and experimental groups (ANOVA).

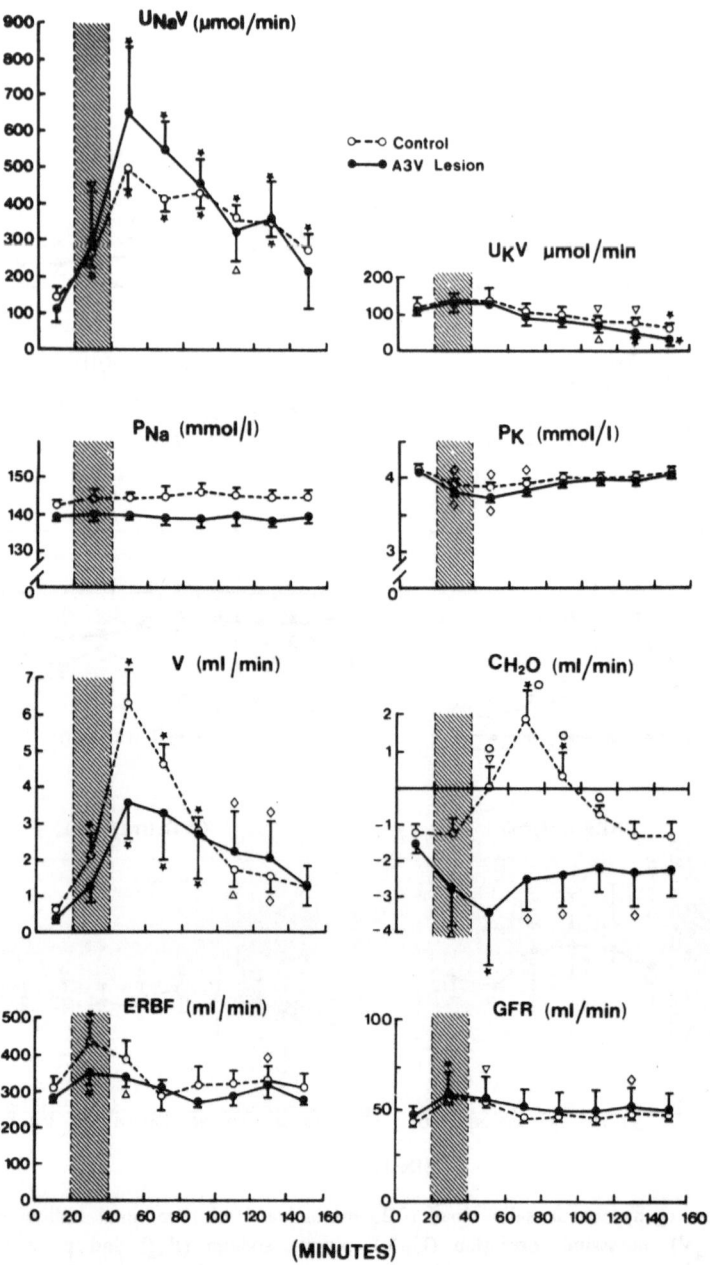

Figure 4. Effect of isotonic saline load (3% BW in 20 min) in the conscious sheep (n = 7) and in sheep with an ablated A3V region (n = 4). See Figure 3 for further details.

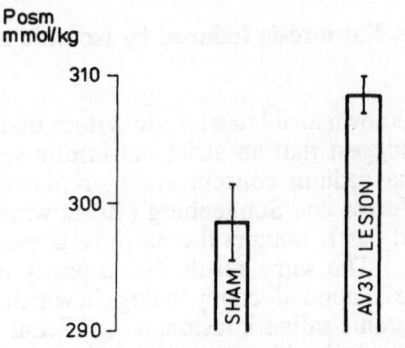

Figure 5. Effect of an AV3V lesion on plasma osmolality in rats [Posm (mmol/kg H$_2$O)]. Figure is produced from data of Brody and Johnson (9) with permission of Raven Press.

controls, the lesioned animals did not increase renal sodium excretion or urine output (Figure 3), which is in agreement with the findings of Thornborough *et al.* (8).

On the other hand, the anterior wall of the third ventricle does not seem to be critical for the mechanism of natriuresis induced by ECFV expansion with isotonic saline (Figure 4). In both normal and lesioned animals, homeostatic natriuresis and urine output were in the same range. In other words, the A3V does not, at least in the sheep, have a direct role in the mechanism of ECFV regulation. This result contrasts with the distinct reduction of natriuresis induced by ECFV expansion with isotonic solutions seen in other mammalian species (*e.g.*, rats or rabbits) with an ablated anteroventral region (9). Inasmuch as the clinical picture of the A3V lesion was identical with that of AV3V lesion (adipsia, hypernatremia, inadequate vasopressin secretion), we do not believe that differences in the extent of the lesioned brain regions in the sheep and rat are responsible for the different reactions to ECFV expansion.

Based on the fact that animals with A3V or AV3V lesions are chronically hypernatremic and hyperosmotic due mainly to inadequate water intake, vasopressin secretion, and probably impaired sodium excretion (Figure 5) (9), species specific differences could provide an explanation for the ineffectiveness of the A3V lesions. However, the ECFV was found to be normal in the lesioned animals despite the deficit in the plasma water. In contrast to the situation in which non-responsive rats were obviously dehydrated, water balance in the sheep in our studies was re-established before the experiment by forced water application. Therefore, we propose that normalization of water balance is a prerequisite for natriuresis following ECFV expansion in animals with an A3V lesion (3).

The integrity of the anterior third ventricle region seems to be important in the mechanism of natriuresis following a hypertonic saline load. Natriuresis following an isotonic saline load can be prevented by such a lesion, but the effect seems to be secondary to the water deficit that develops following ablation of the anterior third ventricle.

Consequently, it is tempting to conclude *a)* that the natriuretic mechanism of the brain is modulated by the sodium concentration in body fluids and *b)* that the ECF either modulates some natriuretic function in the brain that is not located in the anterior third ventricle. Alternatively, a non-cerebral, *i.e.*, peripheral, natriuretic system may be involved.

Atrial Appendectomy Impairs Natriuresis Induced by Isotonic and Hypertonic Saline Loads

In addition to the well-known atrial natriuretic system that is activated by increased ECFV, experiments in dogs suggest that an atrial natriuretic system of the heart is also activated by increased plasma sodium concentration or plasma osmolality. We have repeated the experiment of Veress and Sonnenberg (10) in which they showed that right atrial appendectomy in the rat partly impairs the natriuresis evoked by ECFV expansion with Ringer-albumin solution. The same result, *i.e.*, a partly impaired natriuresis, was obtained following bilateral atrial appendectomy in dogs in which the ECFV was expanded with isotonic saline. After isotonic saline infusion, a significant negative correlation was found between the weight of the excised heart auricular tissue and renal sodium excretion

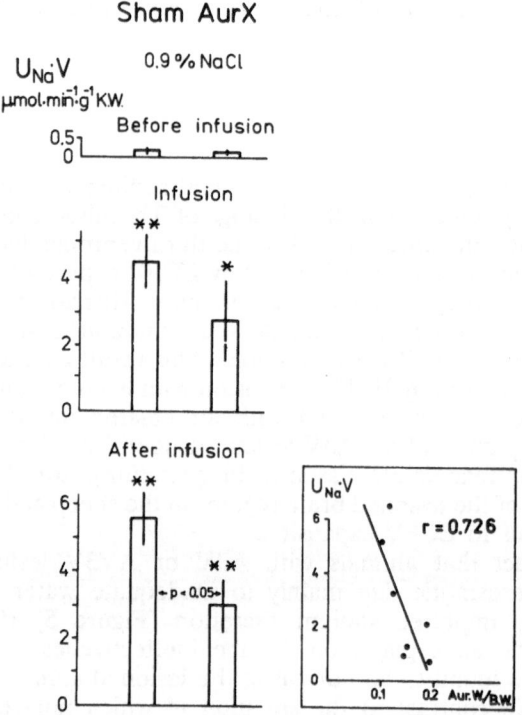

Figure 6.   Renal sodium excretion ($U_{Na}V$) in two groups of anesthetized dogs in 20 min periods *before, during* and *after* iv infusion of isotonic saline (3% BW). In the sham operated control animals (*left bars*; mean ± SE; n = 6), a purse-string was inserted in the wall of the right and left heart atrial appendage one hour before beginning of urine collections to mimic the trauma caused to the dogs in which the right and left appendages were acutely removed (AurX; *right bars*; n = 7). ECFV expansion by isotonic saline infusion evoked significant natriuresis in both groups. *Asterisks* signify statistically significant differences in $U_{Na}V$ during ECFV expansion as compared to the respective pre-expansion values. Bilateral auriectomy attenuated the homeostatic natriuresis. A significant negative correlation was found between the weight of the excised heart auricular tissue (Aur. W/BW)) and renal sodium excretion ($U_{Na}V$). Sodium pentobarbital (30 mg/kg BW) was used for anesthesia of the dogs (8-14 kg) of either sex. Reproduced from Lichardus *et al.* (3) with permission of Springer-Verlag.

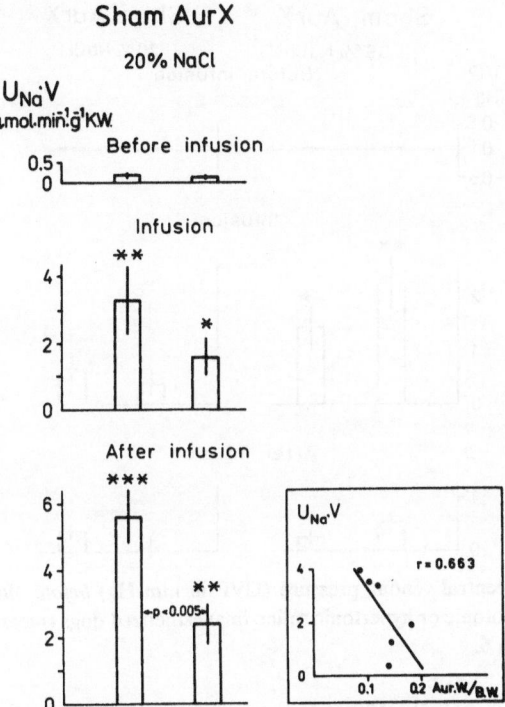

Figure 7. Renal sodium excretion in two groups of dogs treated in the manner described in Figure 6. However, instead of isotonic saline infusion, the animals were infused with an equivalent amount of sodium in a 20% solution of NaCl; the infusion produced an increase in $U_{Na}V$ in both groups. The hypertonic saline-induced natriuresis in dogs with bilateral auriectomy was attenuated to the same degree as seen in the experiment in which isotonic saline was used to expand ECFV (Figure 6). A significant correlation was found between the weight of the heart auricular tissue excised and sodium excretion. For details see Figure 6. Reproduced from Lichardus et al. (3) with permission of Springer-Verlag.

(Figure 6). This finding points to a causal relationship between the heart atrial natriuretic system and the ensuing natriuresis.

A similar attenuation of natriuresis was found when dogs with bilateral atrial appendectomy were infused hypertonic saline. The infusion contained the same amount of sodium in a 20% solution of NaCl as used in the preceding experiment in which isotonic saline was infused). Again, a significant negative correlation was found between this type of natriuresis and the weight of the excised heart auricular tissue (Figure 7) (11).

These findings suggest that the heart natriuretic system is also under the control of the sodium concentration (or osmolality) of the extracellular fluid. Activation of the atrial natriuretic system by infusion of hypertonic saline has been described earlier by Eskay et al. (12). These authors, however, concluded that even after a hypertonic saline load the appropriate stimulus for the atrial system is ECFV expansion as hypertonic saline caused transfer of water from the extravascular to the intravascular compartment. A similar water transfer did not appear to be effective in our experiments since the plasma osmolality remained high throughout the experiment. It is noteworthy that compared with the isotonic saline infusion in dogs, the increase in the central venous pressure following hypertonic saline load was trivial (Figure 8). Thus, it is likely that the critical stimulus for

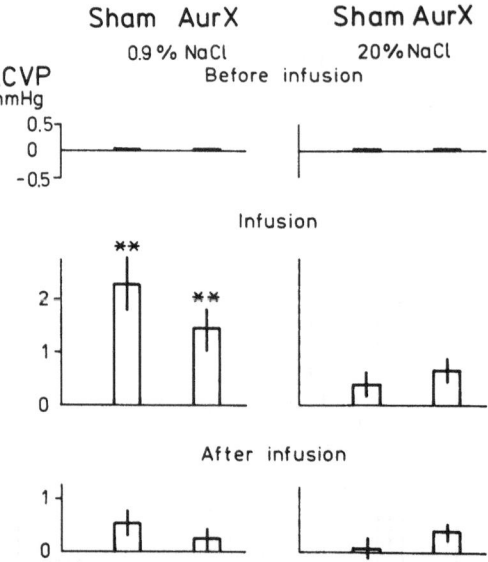

Figure 8. Changes in central venous pressure (CVP in mm Hg) *before, during* and *20 min after* the infusion of isotonic or hypertonic saline in anesthetized dogs (mean ± SE; n = 6). For details see to Figure 6.

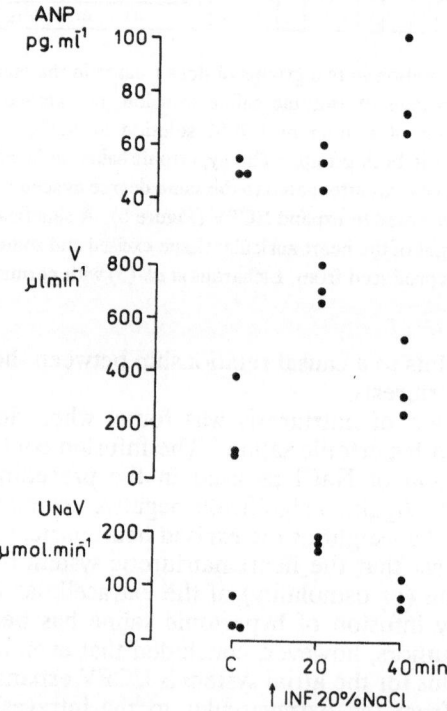

Figure 9. Effect of iv infusion of hypertonic saline (20% solution; 0.13% of BW iv in 20 min) in dogs under sodium pentobarbital anesthesia on plasma level of atrial natriuretic peptide (ANP), urine (V) and sodium ($U_{Na}V$) excretion. Closed circles represent individual values of the respective measures.

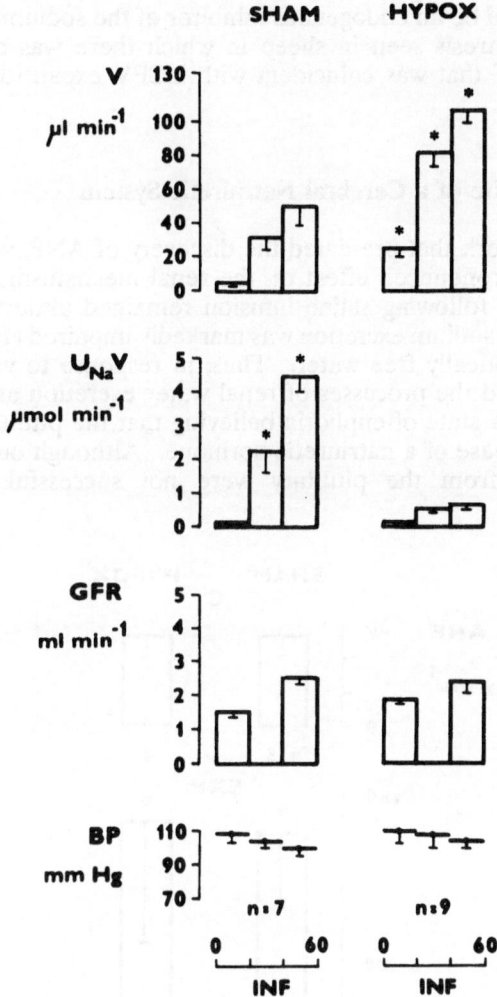

Figure 10. Effect of iv infusion of isotonic saline (INF) (4% BW in 20 min) in acutely hypophysectomized (HYPOX) and SHAM operated rats under Inactin anesthesia on urine (V) and sodium ($U_{Na}V$) excretion, glomerular filtration rate (GFR), and arterial blood pressure (BP) (mean ± SE). *Asterisks* indicate statistically significant intergroup differences of the respective measures (*t*-test).

the activation of the atrial natriuretic system is an increased plasma osmolality or sodium concentration or both (11). Moreover, modestly elevated plasma levels of atrial natriuretic peptide occur in dogs following infusion of hypertonic saline (Figure 9).

The results of our experiments with auricular ablations suggest that the increase of atrial natriuretic peptide (ANP) in plasma following intravenous infusion of hypertonic saline may be due to the release of this hormone from the heart. On the other hand, natriuresis following an increase in CSF sodium concentration was not accompanied by a release of ANP, and it was suggested that this effect was mediated by an unidentified factor (13,14), which could be an endogenous inhibitor of the sodium pump (15). It might also potentiate the natriuresis seen in sheep in which there was an increased sodium concentration in the CSF that was coincident with ECFV expansion due to infusion of isotonic saline (Figure 1).

The Pituitary–Another Site of a Cerebral Natriuretic System

In experimental work that pre-dated the discovery of ANP, we showed that acute hypophysectomy had a pronounced effect on the renal mechanism of ECFV regulation (16,17). Urine excretion following saline infusion remained almost the same as in the control animals, but renal sodium excretion was markedly impaired (Figure 10). The urine consisted mostly of osmotically free water. Thus, in response to volume loading acute hypophysectomy separated the processes of renal water excretion and sodium excretion. At that time we lived in a state of euphoria believing that the pituitary could be the site of production and/or release of a natriuretic hormone. Although our attempts to isolate a natriuretic hormone from the pituitary were not successful, the model of acute

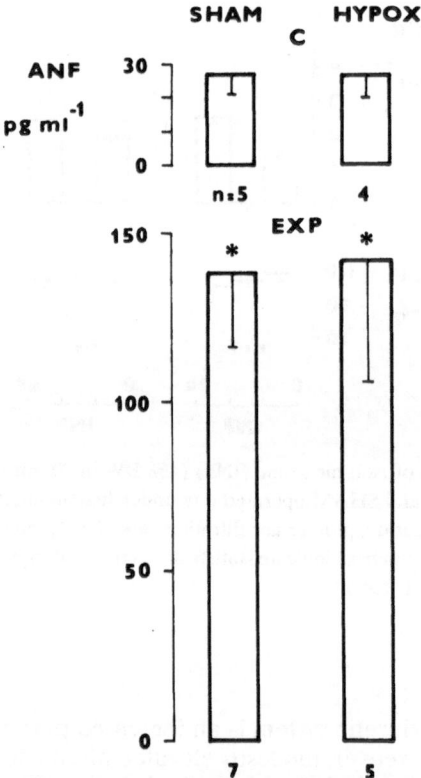

Figure 11. Atrial natriuretic factor (ANF) (mean ± SE) in plasma in acutely hypophysectomized (HYPOX) and SHAM operated rats under Inactin anesthesia before (C) and immediately after (EXP) iv infusion of isotonic saline (4% BW in 20 min). Asterisks indicate the statistically significant differences between the control values of ANF and those found during the ECFV expansion with saline (t-test).

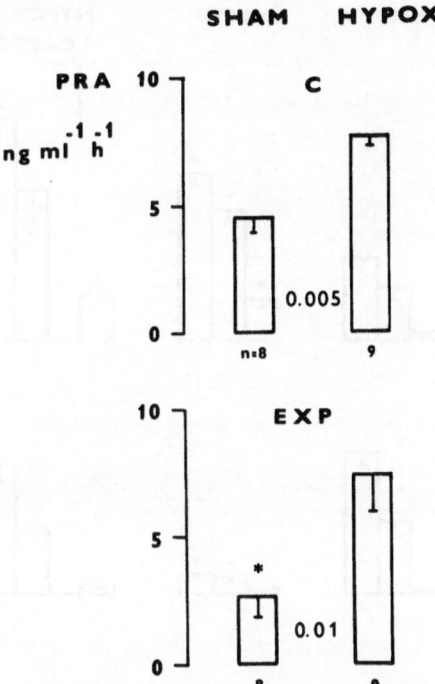

Figure 12. Plasma renin activity (PRA) (mean ± SE) in acutely hypophysectomized (HYPOX) and SHAM operated rats under Inactin anesthesia before (C) and immediately after (EXP) the iv infusion of isotonic saline (4% BW in 20 min). *Asterisks* indicate the statistically significant difference between the control value and that obtained during the ECFV expansion with saline. Intergroup differences are expressed by *P* values.

hypophysectomy remains useful for studying ANP secretion and its effect on renal natriuresis. In regard to the control of ANP secretion, it has been shown by others that chronic hypophysectomy impairs basal as well as stimulated secretion of ANP. Thus, the question arises: Is ANP release regulated directly by the pituitary?

Our experiments revealed that in contrast to chronic hypophysectomy, which is accompanied by thyroid and adrenocortical hypofunction, the acutely hypophysectomized rat (HYPOX) does not exhibit decreased basal levels of circulating ANP. Upon volume loading with isotonic saline, HYPOX animals retain the ability to increase ANP release (Figure 11).

For these reasons, it seems that there is no direct pituitary control of ANP release. If there had been direct control, ANP release should have been impaired by acute hypophysectomy. Moreover, since volume loading did not increase renal sodium excretion despite the elevated plasma levels of ANP, the natriuretic activity of ANP must have been blunted.

It is known that the threshold for the natriuretic response to ANP is altered by decreased renal perfusion pressure, dehydration, or increased sympathetic and/or renin-angiotensin activity. Along this line, it is of interest that acute hypophysectomy increases plasma renin activity, implying an increase of plasma angiotensin II levels, and this increased renin activity is not suppressed by the ECFV expansion (Figure 12). We

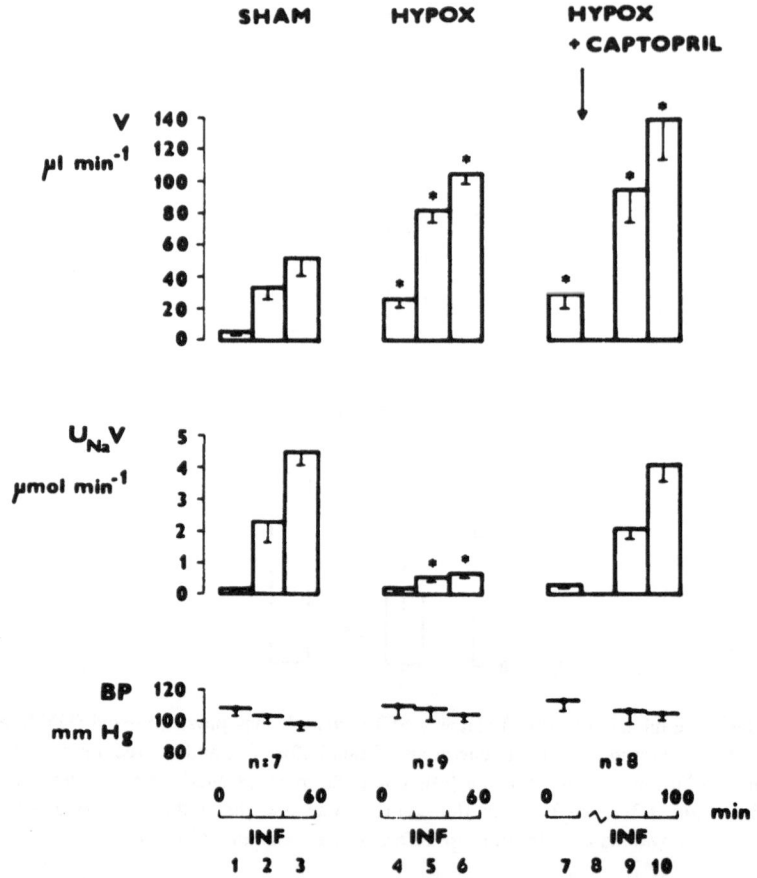

Figure 13. Effect of isotonic saline infusion (INF) (4% BW in 20 min) in acutely hypophysectomized (HYPOX), SHAM, and HYPOX rats given captopril (1.6 mg/kg BW ip) immediately before the saline infusion. For further details see Figure 10.

believe that this finding is the result of increased activity of the sympathetic nervous system (unpublished observations).

It is thus concluded that the increase of both renin-angiotensin and sympathetic nervous activity could play a role in blunting natriuresis in acutely hypophysectomized rats. The reason for the increased activity of these vasoconstrictory systems was probably a consequence of the surgical procedure and the absence of vasopressin in the circulation.

Other evidence of a role for the renin-angiotensin system in the ineffectiveness of ANP in acutely hypophysectomized rats is provided the effect of captopril on natriuresis in volume-expanded, acutely hypophysectomized rats. This converting enzyme inhibitor of angiotensin I to angiotensin II was injected intraperitoneally (1.6 mg/kg body weight) to rats immediately before ECFV expansion with isotonic saline. Captopril restored completely the volume natriuresis in acutely hypophysectomized rats (Figure 13). Angiotensin II may thus be the major ANP antagonist in the mechanism of renal sodium excretion in volume-expanded, acutely hypophysectomized rats. Angiotensin II may promote tubular sodium reabsorption and decrease glomerular filtration rate (18). Acute hypophysectomy may be considered as a model of natriuretic ineffectiveness of ANP in which natriuresis can be reestablished by attenuating the vasoconstrictory activity at the renal level.

Conclusions

1. The sodium concentration in the CSF seems to be important in the control of renal regulation of ECF volume. Increased concentrations of sodium in the CSF potentiates the natriuresis evoked by an expanded ECF volume, induced by isotonic saline infusion (volume natriuresis). Conversely, a decreased concentration of sodium in the CSF blunts natriuresis.

2. The anterior region of the third ventricle is a receptor site for the sodium concentration or osmolality of body fluids.

3. The atrial natriuretic system may be stimulated by hyperosmolality of plasma or possibly by hypernatremia.

4. The pituitary does not directly control the activity of the atrial natriuretic system.

5. Acute hypophysectomy in rats is a model of natriuretic ineffectiveness of ANP. In this model natriuresis can be restored by attenuating vasoconstrictory activity at the renal level.

References

1. Lichardus, B., Natriuretic hormones, In B. Lichardus, R.W. Schrier, J. Ponec (eds), *Hormonal Regulation of Sodium Excretion*, Elsevier North/Holland Biomedical Press, Amsterdam, pp. 1-7, 1980.
2. de Bold, A.J., H.B. Borenstein, A.T. Veress, and H. Sonnenberg, A rapid and potent natriuretic response to intravenous injection of atrial myocardial extracts in rats, *Life Sciences* 28: 89-94, 1981.
3. Lichardus, B., M.J. McKinley, D.A. Denton, and J. Ponec, Brain involvement in the regulation of renal sodium excretion, *Klinische Wochenschrift* 65 (Suppl VIII): 33-39, 1987.
4. Mouw, D.R., A.J. Vander, C. Landis, S. Kutschinski, N. Mathias, D. Zimmerman, Dose-response relation of CSF sodium and renal sodium excretion, and its absence in homozygous Brattleboro rats, *Neuroendocrinology* 30: 206-212, 1980.
5. Tarjan, E., D.A. Denton, B. Lichardus, M.J. McKinley, and R.S. Weisinger, The effect of local increase in CSF (Na) in the third ventricle on salt intake and excretion, Abstract, *Annual Meeting of the Australian Society for Medical Research*, 1984.
6. Cox, P.S., D.A. Denton, D.R. Mouw, and E. Tarjan, Natriuresis induced by localized perfusion within the third cerebral ventricle of sheep, *Am J Physiol* 252: R1-R6, 1987.
7. McKinley, M.J., M. Congiu, D.A. Denton, B. Lichardus, R.G. Park, E. Tarjan, and R.S. Weisinger, Cerebrospinal fluid composition and homeostatic responses to dehydration, In R.W. Schrier (ed), *Vasopressin*, Raven Press, New York, pp. 299-309, 1985.
8. Thornborough, J.R., S.S. Passo, and A.B. Rothballer, Forebrain lesion blockade of the natriuretic response to elevated carotid blood sodium, *Brain Res* 58: 355-363, 1973.
9. Brody, M.J., and A.K. Johnson, Role of anteroventral third ventricle region in fluid and electrolyte balance, arterial pressure regulation and hypertension, In L. Martini, W.F. Ganong (eds), *Frontiers in Neuroendocrinology*, Raven Press, New York, pp. 249-292, 1980.
10. Veress, A.T., and H. Sonnenberg, Right atrial appendectomy reduces the renal response to acute hypervolemia in the rat, *Am J Physiol* 247: R610-R613, 1984.
11. Okoličány, J., B. Lichardus, I. Gabauer, and J. Ponec, Bilateral acute heart atrial auriectomy reduced

diuresis and natriuresis following hypertonic sodium load in anaesthetized dogs, *Physiol Bohemoslov* 38: 179-187, 1989.

12.   Eskay, R., Z. Zukowska-Grojec, M. Haas, J.R. Dave, and N. Zamir, Circulating atrial natriuretic peptide in conscious rats: regulation of release by multiple factors, *Science* 232: 636-639, 1986.

13.   Ulfendahl, H.R., A. Goransson, P. Hansell, M. Karlsson, and M. Sjoquist, Natriuresis obtained by stimulation of the cerebroventricular system with sodium ions indicates a blood-borne natriuretic factor, *Acta Physiol Scand* 127: 269-271, 1986.

14.   Hansell, P., A. Goransson, Leppaluoto, O. Arjamaa, O. Vakkuri, and H.R. Ulfendahl, CNS-induced natriuresis is not mediated by the atrial natriuretic factor, *Acta Physiol Scand* 129: 221-227, 1987.

15.   Jandhyala, B.S., and A.F. Ansari, Elevation of sodium levels in the cerebral ventricles of anaesthetized dogs triggers the release of an inhibitor of ouabain-sensitive sodium-potassium-ATPase into the circulation, *Clin Sci* 70: 103-110, 1986.

16.   Lichardus, B., and J. Ponec, Effect of hypophysectomy on sodium excretion in rats without blood dilution during blood volume expansion, *Experientia* 28: 471-472, 1972.

17.   Szalay, L., P. Bencsath, L. Takacs, J. Ponec, and B. Lichardus, Acute effect of hypophysectomy on the natriuresis following saline infusion in dogs, *Experientia* 31: 1298-1299, 1975.

18.   Liu, F-Y, and M.G. Cogan, Atrial natriuretic factor does not inhibit basal or angiotensin II-stimulated proximal transport, *Am J Physiol* 255: F434-F437, 1988.

PERIPHERAL NEUROHUMORAL FACTORS AND CENTRAL CONTROL OF HOMEOSTASIS DURING ALTERED SODIUM INTAKE

Greti Aguilera

Endocrinology and Reproduction Research Branch
National Institute of Child Health and Human Development
National Institute of Health
Bethesda, MD

Altered sodium intake leads to activation of compensatory mechanisms to restore sodium balance and circulatory homeostasis through modulation of renal sodium and water transport, vascular tone and sodium appetite (1). The humoral, neural, and local factors involved and the sites at which regulation occurs are shown in Table 1. In addition, coordinated interaction between peripheral and central regulators is critical to obtain the appropriate level of response in the target tissue. Examples of this type of interaction are *a)* the modulatory effect of plasma potassium, atrial natriuretic peptide, and somatostatin on the adrenal sensitivity to angiotensin II (AII) (2), *b)* the regulatory effect of vasopressin on baroreflex sensitivity (3), and *c)* the modulatory effect of circulating factors such as AII and steroids on the central regulation of sympathetic activity (4,5). Mineralocorticoids from the zona glomerulosa of the adrenal have a major role in the control of sodium and water metabolism through its effects on sodium reabsorption at the distal tubule of the nephron. The main mineralocorticoid in man and other mammalian species is aldosterone. The secretion of aldosterone by the adrenal glomerulosa is controlled by a complex set of regulators which include peripheral and central components. In this discussion I shall focus on the regulation of aldosterone secretion and in particular the manner in which humoral and neural elements are coordinated.

Multifactorial Control of Aldosterone Secretion: Role of AII

The secretion of aldosterone by the adrenal glomerulosa is influenced by a number of stimulatory and inhibitory factors (Table 2). The major physiological conditions that influence aldosterone secretion are changes in sodium and potassium intake. During changes in potassium intake, very small increments in plasma potassium can increase aldosterone secretion by direct stimulation of the adrenal glomerulosa cells and by increasing the sensitivity of the cells to other regulators such as AII (6). In contrast, the regulation of aldosterone secretion during changes in sodium intake has been more difficult to elucidate. Although AII appears to be the main regulator of zona glomerulosa function, there are several clinical and experimental conditions in which AII cannot explain the changes in aldosterone secretion (7-9). Of particular relevance is the inability of AII to reproduce the levels of aldosterone secretion observed during sodium restriction (8-10). Furthermore, increases in aldosterone secretion during progressive sodium loss

Circulating Regulatory Factors and Neuroendocrine Function
Edited by J. C. Porter and D. Ježová
Plenum Press, New York, 1990

227

Table 1
Adaptation to Change in Sodium Intake

I. Levels of Regulation

1. Kidney: sodium and water transport
2. Blood vessels: vascular tone
3. Water and sodium intake: water and sodium appetite

II. Mechanisms of Regulation

1. Hormonal regulators

   —Mineralocorticoids
   —Peptide hormones: AII, ANF*, VP†, others
   —Adrenal catecholamines

2. Neural regulation

   —Baroreflex
   —Renal afferent signals
   —Sympathetic tone

3. Local regulators

   —Renal dopamine
   —Paracrine factors: renal AII, SRIF‡, others

*Atrial natriuretic factor.
†Vasopressin.
‡Somatostatin.

have been observed under conditions in which renin production is inhibited by infusion of AII in the renal artery and when plasma AII levels are maintained at a constant level by systemic infusion of the peptide (11).

Comparison of the changes in adrenal glomerulosa cell function during sodium restriction with those elicited by administration of other stimuli of aldosterone secretion has clarified the roles of individual regulators in the physiological control of aldosterone secretion. Increases in aldosterone secretion following sodium restriction are accompanied by increases in adrenal AII receptors and 18-hydroxylase activity and enhanced responses to AII and other stimuli in isolated adrenal glomerulosa cells. Sodium loading, on the other hand, has the opposite effects on glomerulosa-cell function, with a decrease in AII receptors and reduced aldosterone responses to the octapeptide (2).

The anterior pituitary gland is necessary for maintenance of the basal activity of those enzymes involved in the early portion of the aldosterone biosynthetic pathway (9). In hypophysectomized animals and humans, the decreased basal activity of these enzymes contributes to the slower and reduced aldosterone responses to sodium restriction (12,13). However, sodium restriction for 6 days in hypophysectomized rats increases adrenal glomerulosa AII receptors and enzymes of the early and late aldosterone biosynthetic

Table 2
Factors Known to Influence Aldosterone Secretion

Stimulatory	Inhibitory
AII	Somatostatin
ACTH	Atrial Natriuretic Factor
Extracellular Potassium	Dopamine
Serotonin	
α-MSH	
β-MSH	
Aldosterone Stimulating Hormone?	
Endothelin?	
Vasopressin	

pathways (9), indicating that the pituitary gland is not essential for the adrenal effects of sodium restriction. Of the pituitary factors, adrenocorticotropic hormone (ACTH) is a potent stimulator of aldosterone secretion *in vivo* and *in vitro* but plasma levels of ACTH are not elevated during sodium deficiency. Furthermore, prolonged stimulation with ACTH *in vivo* leads to a *decrease* in aldosterone secretion, accompanied by marked reduction in adrenal glomerulosa AII receptors and 18-hydroxylase activity (3).

Increases in extracellular potassium mimic the increases in plasma aldosterone, AII receptors (14), and 18-hydroxylase activity (15) observed during sodium restriction and potentiate the effects of AII (6,14). However, plasma potassium levels are often unaltered during sodium deficiency and cannot solely explain the changes in aldosterone secretion during altered sodium intake.

Of the known regulators of aldosterone secretion, AII has been shown to have a primary role during sodium deficiency. Circulating levels of the peptide are elevated during sodium restriction, and its effects on the adrenal glomerulosa cell are largely responsible for increases in aldosterone secretion and the enhanced sensitivity of the adrenal during sodium restriction (2,16). The increases in AII receptors and 18-hydroxylase activity that occur during sodium restriction can be reproduced by prolonged infusion of AII and are abolished by the simultaneous administration of the converting enzyme inhibitor, captopril (Table 3). Such findings demonstrate that AII is essential for the aldosterone responses to sodium restriction and that the peptide has major regulatory actions in the adrenal glomerulosa cell. However, during high sodium intake the prolonged infusion of AII does not increase plasma aldosterone, adrenal AII receptors, or 18-hydroxylase activity, as shown in Figure 1. This finding indicates that factors other than AII must be involved in the altered sensitivity of the adrenal glomerulosa during changes in sodium intake. Despite this discrepancy, the demonstration that AII is essential for the aldosterone responses to sodium restriction suggests that such factors probably interact with AII to modify the action of the octapeptide in the adrenal glomerulosa cell rather than acting as independent regulators.

Regulation of the Adrenal Sensitivity to AII

The sensitivity of the adrenal glomerulosa cell to AII varies according to the level of sodium intake, being decreased during sodium restriction and increased during sodium

Table 3
Dependence of the Adrenal Responses to Sodium Restriction
Upon AII, as Shown by the Effects of Sodium Restriction,
AII Infusion and Converting Enzyme Inhibition (CEI) on Adrenal
Glomerulosa Function

	Control	Sodium Restriction	AII Infusion	Sodium Restriction Plus CEI
Plasma Aldosterone	+	+ + +	+ +	+
Adrenal AII Receptors	+	+ + +	+ +	+
Side Chain Cleavage Enzyme	+	+ + +	+ +	+
18-Hydroxylase	+	+ + +	+ +	+

loading. As discussed above, the trophic effects of AII on adrenal AII receptors and on enzymes of the aldosterone biosynthetic pathway are important determinants of adrenal responsiveness. However, it is also clear that other factors modulate adrenal effects of the peptide. Several of the factors shown in Table 2 are known to interact with the action of AII in the adrenal glomerulosa cell, but their precise physiological role is unclear. Possible mechanisms of interaction between AII and other aldosterone regulators at the level of the adrenal glomerulosa cell are shown in Figure 2. In addition to atrial natriuretic factor and somatostatin, peptides that exert preferential inhibitory effect on AII stimulated aldosterone secretion, the dopaminergic system has been shown to have a role in the modulation of adrenal sensitivity to AII (2,17). The sections below will focus on the role of the dopaminergic system in the control of aldosterone secretion.

Role of Dopamine in the Adrenal Sensitivity to AII

The participation of the dopaminergic system in the regulation of aldosterone secretion was first suggested by the demonstration that the dopamine antagonist, metoclopramide, stimulates aldosterone secretion *in vivo* (17). Most initial studies seem to indicate that the stimulatory action of metoclopramide on aldosterone production is independent of the renin-angiotensin system (18,19). In humans and in rats, increases in plasma aldosterone by metoclopramide are not modified by administration of AII antagonists. Also, the aldosterone responses to metoclopramide in man are decreased during high sodium intake, and metoclopramide does not modify the magnitude or sensitivity of the aldosterone responses to AII. In these studies the interaction between dopaminergic mechanisms and AII was examined during normal or low sodium intake, or without AII replacement during high sodium intake when maximum dopaminergic inhibition can be expected. However, other studies clearly show that in some experimental conditions a dopaminergic mechanism can modulate the adrenal actions of AII. In these studies, the possibility that dopamine may be responsible for the reduced adrenal responses to AII during sodium loading was examined by analysis of the adrenal effects of AII during high sodium intake and metoclopramide infusion (2,17,20). In the rat,

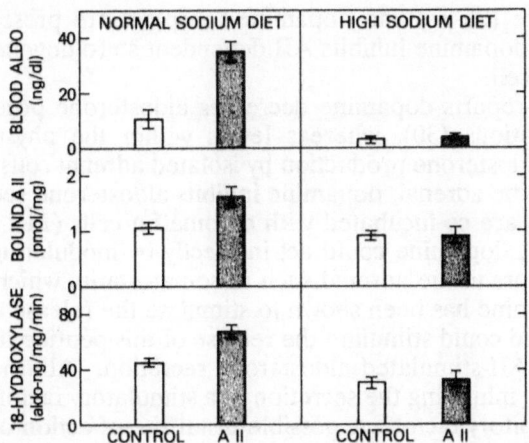

Figure 1. Comparison of the effects of AII infusion on blood aldosterone, adrenal AII receptors, and 18-hydroxylase activity in rats on low and high sodium intake.

infusion of AII or metoclopramide alone for 48 hours had no effect on plasma aldosterone, AII receptors and 18-hydroxylase activity. However, simultaneous infusion of metoclopramide and AII elevated plasma aldosterone, adrenal AII receptors, and 18-hydroxylase activity to the levels observed after AII infusion in animals on normal sodium diet (Figure 3). The sensitization of the adrenal to AII in these animals on high sodium intake was attributable to the dopamine antagonist properties of metoclopramide, since these effects of metoclopramide were blocked by simultaneous infusion of dopamine. Comparable sensitization of the adrenal to AII during high sodium intake has been observed in humans during experiments in which metoclopramide restored the sensitivity of the aldosterone responses to graded AII infusion to that observed during normal sodium intake (20).

Further evidence supporting a modulatory action of dopamine on the actions of AII is provided by experiments performed to analyze the effects of dopamine agonists during low sodium intake, a condition in which the sensitivity of AII is enhanced. The observation that bromocriptine administration decreased the aldosterone response to furosemide, could be attributed to dopaminergic inhibition of endogenous AII (21). However, against this proposal are reports indicating that bromocriptine is ineffective in reducing the plasma aldosterone responses to sodium restriction and posture (22,23). More consistent results have been obtained with dopamine agonists and AII during sodium deficiency. In sodium-restricted humans, treatment with bromocriptine for 6 days decreased aldosterone responses to AII infusion but not to ACTH infusion. In another study, dopamine infusion decreased the aldosterone responses to graded infusions of AII in sodium-restricted humans (24). Similar results have been obtained in sodium-restricted rats (2), in which dopamine infusion with osmotic minipumps (1 μg/min) caused no change in basal plasma aldosterone levels, but significantly decreased the sensitivity to AII infusion.

Mechanism of Dopaminergic Regulation of Aldosterone Secretion

Possible mechanisms by which dopamine modulate the responsiveness of the adrenal to AII are shown in Table 3. Since circulating dopamine levels are increased during high sodium intake (25-27) and dopamine receptors are present in the adrenal (28,29), it is possible that dopamine inhibits AII dependent steroidogenesis directly at the level of the glomerulosa cell.

However, in most reports dopamine decreases aldosterone production *in vitro* at only very high concentrations (30), whereas levels within the physiologic range are ineffective in modifying aldosterone production by isolated adrenal cells (17). Also, it has been reported that in bovine adrenal, dopamine inhibits aldosterone secretion only when adrenal glomerulosa cells are co-incubated with chromaffin cells (31).

On the other hand, dopamine could act indirectly by modulating the secretion of local or paracrine regulators in the adrenal such as somatostatin, which is present in the zona glomerulosa. Dopamine has been shown to stimulate the release of somatostatin in the hypothalamus (32), and could stimulate the release of this peptide in the adrenal with consequent inhibition of AII-stimulated aldosterone secretion. Alternatively, dopamine could act at a distal site by inhibiting the secretion of a stimulatory factor or by stimulating the production of an inhibitory factor. A possible distal site of action of dopamine is the pituitary gland. The secretion of hormones from the anterior and intermediate pituitary are under dopaminergic control. Although the pituitary is not essential for maintenance of adrenal glomerulosa function, several reports show attenuated aldosterone responses following hypophysectomy that cannot be restored by ACTH replacement (12,13). Therefore, a pituitary factor other than ACTH may be involved in the dopaminergic regulation of the adrenal sensitivity to AII. In this regard, dopamine has been shown to decrease the urinary excretion, and presumably the pituitary secretion, of the still uncharacterized *aldosterone stimulating factor*, and consequently decrease aldosterone secretion (33). Non-ACTH POMC peptides have long been implicated in aldosterone regulation and circulatory homeostasis (Table 4). These are mainly of intermediate pituitary origin, and are presumably under dopaminergic control (34-44).

To examine directly the role of the pituitary on the permissive effect of metoclopramide on AII stimulated aldosterone during high sodium intake, the effects of AII and metoclopramide infusion were studied in sodium-loaded intact and 7-day hypophysectomized rats. As in previous reports, in rats receiving a diet containing 2% sodium, infusion of AII (25 ng/min) or metoclopramide (1.5 μg/min) for 2 days had no effect on plasma aldosterone levels or adrenal capsular AII receptors, whereas simultaneous infusion of both agents caused an 11.4-fold increase in plasma aldosterone and a 1.9-fold increase of AII receptors (Figure 4). In contrast, in hypophysectomized rats significantly smaller ($P < 0.01$) responses to the simultaneous infusion of AII and metoclopramide were

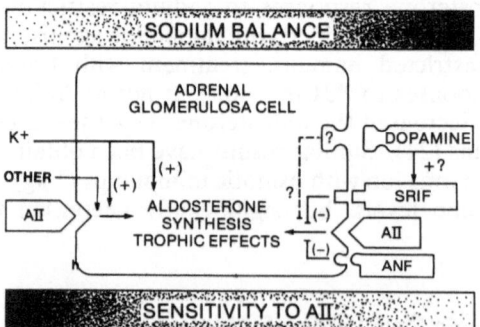

Figure 2.   Stimulatory and inhibitory factors modulating the sensitivity of the adrenal glomerulosa cell to AII (K denotes potassium, SIRF denotes somatostatin, ANF denotes atrial natriuretic factor).

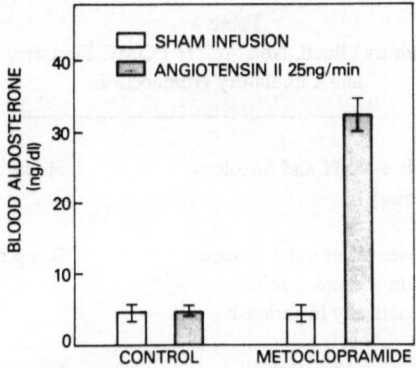

Figure 3. Effect of infusion of metoclopramide and/or AII on plasma aldosterone in rats on high sodium intake. Bars are the mean and SE of the values of 8 rats.

observed with a 4.1-fold increase in plasma aldosterone and a 0.33-fold increase in adrenal AII receptors. As observed in intact rats, in hypophysectomized animals infusion of AII or metoclopramide alone had no effect on plasma aldosterone or AII receptors. Basal plasma aldosterone was reduced in hypophysectomized rats on normal sodium diet to 5.6 ± 0.65 ng/dl, (n = 15), compared with 10.05 ± 1.2 ng/dl, (n = 9) in the intact rats. During high sodium diet there was no difference between intact (2.9 ± 0.55 ng/dl) and hypophysectomized (2.2 ± 0.23 ng/dl) rats respectively (Figure 4).

The plasma levels of renin activity, aldosterone, corticosterone, ACTH, and prolactin in control and metoclopramide-treated rats are shown in Table 5. Plasma renin activity and aldosterone were reduced on high sodium diet, but no effect of metoclopramide infusion was observed with either sodium intake conditions. Plasma prolactin was unaffected by sodium intake, but metoclopramide infusion caused the expected increase in prolactin in both groups. Plasma ACTH and corticosterone levels were unchanged by either sodium diet or metoclopramide infusion. As expected, in hypophysectomized rats plasma corticosterone, ACTH, and prolactin were also undetectable, and these levels were unchanged by sodium intake or metoclopramide infusion.

Since the intermediate lobe of the pituitary is under dopaminergic influence and contains high affinity AII receptors, it was of interest to study how these two systems interacted during changes in sodium intake. The effect of sodium intake on pituitary AII receptor content is shown in Figure 5. As previously reported (45), sodium intake had no effect on anterior pituitary AII receptors, whereas high sodium diet resulted in a 50% decrease. Receptor affinity estimated by Scatchard analysis was unchanged.

To determine whether regulation of AII receptors in the intermediate pituitary is under the influence of dopamine, AII receptor content was measured in neurointermediate pituitary lobes from rats on high sodium diet receiving an infusion of AII, or metoclopramide, or in combination. Figure 6 shows the decrease in AII receptors in the neurointermediate pituitary following 4 days on high sodium diet. Infusions of AII or metoclopramide alone during the 4 days of sodium loading caused a small and not significant increase in AII receptors. However, similar to the effects in the adrenal glomerulosa, combined infusion of AII and the dopamine antagonist markedly increased AII receptors in the neurointermediate lobe to levels similar to those in the control rats on normal sodium diet.

The parallel changes in intermediate pituitary AII receptors and adrenal function during manipulations of sodium diet, angiotensin and dopamine levels suggest a contribution of the intermediate pituitary in the regulation of aldosterone secretion by AII.

Table 4
Pituitary Gland, Non-ACTH POMC Peptides,
and Circulatory Homeostasis

1.	Hypertonic saline modify α-MSH and histology of the intermediate pituitary	Howe and Thody (43)
2.	High sodium intake causes differential changes in pituitary cleft colloid in sodium sensitive and soidum resistant genetically hyperleusive rats	Rapp and Dahl (44)
3.	α-MSH increases aldosterone secretion	Vinson *et al.* (34)
4.	β-MSH increases aldosterone secretion	Matsouta *et al.* (35)
5.	γ-MSH stimulates adrenal steroidogenesis	Pederson *et al.* (36)
6.	β-endorphin stimulates aldosterone secretion	Gullner and Gill (37)
7.	N-terminal fragment mediates reflex natriuresis after unilateral nephrectomy	Lin *et al.* (39)
8.	Hypothalamic CRF mRNA is differentially regulated by sodium intake and dehydration	Young (40)
9.	Sodium loading affect POMC mRNA, biosynthesis and products secretion in mouse pituitary	Elkabes and Loh (41)
10.	γ-MSH mediates natriuresis following increased ureteral pressure	Humphrey and Wiedeman (42)

Source of Dopamine During Changes in Sodium Intake

In addition to its location in specific neurons in the central nervous system, where it acts as a neurotransmitter, dopamine is formed in autonomic ganglia, chromaffin tissue, and peripheral nerves, as well as in certain extraneural tissues such as the kidney tubules (46) and possibly the adrenal cortex. The latter is suggested by the persistence of high dopamine levels in the cortex after adrenal medullectomy or ganglionic blockade (47). Dopamine levels in plasma and urine increased during sodium loading and are reduced during sodium restriction (25-27,46). Dopamine is believed to be synthesized in tubular cells of the nephron independently of the renal innervation. This hypothesis is supported by the data in Figure 7, which shows equal increases in dopamine levels in renal tissue of intact or kidney denervated rats on high sodium diet. The mechanisms of the changes in dopamine in plasma are not clear, but may reflect changes in activity of dopaminergic neurons or non-neuronal dopaminergic secreting cells.

The contribution of circulating dopamine to pituitary regulation, especially of the intermediate pituitary, is uncertain. The blood supply of the intermediate pituitary is poor, and it is more likely to receive dopaminergic input from the hypothalamus or the posterior pituitary. In fact, preliminary experiments showed that changes in dietary sodium result

Table 5

Effect of Metoclopramide (MCP) Infusion on PRA, Plasma Aldosterone,
Corticosterone, ACTH, and Prolactin in Rats on
Normal and High Sodium Diet

| | ----Normal Sodium Diet---- | | ----High Sodium Diet---- | |
	Control	MCP	Control	MCP
PRA (AI ng/ml hr)	4.12 ± 0.5	3.96 ± 0.5	$0.99 \pm 0.1^*$	$1.01 \pm 0.16^*$
Aldosterone (ng/dl)	10.1 ± 1.2	12.9 ± 1.5	$2.9 \pm 0.6^*$	$1.9 \pm 0.1^*$
Corticosterone (μg/dl)	15.9 ± 2.6	17.6 ± 2.7	16.7 ± 2.5	17.2 ± 2.7
ACTH (pg/ml)	16.2 ± 1.7	15.6 ± 1.8	17.6 ± 2.3	17.3 ± 2.0
Prolactin (ng/ml)	7.8 ± 1.5	$25.9 \pm 2.5†$	8.2 ± 1.4	$24.5 \pm 2.5†$

Values are mean $\pm$ SE (n = 9-12).
$^*P < 0.01$ compared with normal sodium diet.
$†P < 0.01$ compared with the respective control.

in changes in dopamine turnover in the neurointermediate pituitary lobe. Rats were maintained on high or low sodium diet for 4 days and then injected intraperitoneally with the DOPA decarboxylase inhibitor, α-methyltyrosine (400 μg/kg). Rats were killed 30 or 60 minutes after the injections, and neurointermediate pituitary lobes rapidly removed and collected individually into microcentrifuge tubes in liquid nitrogen. Tissue was sonicated in perchloric acid and dopamine content measured by HPLC. Unexpectedly,

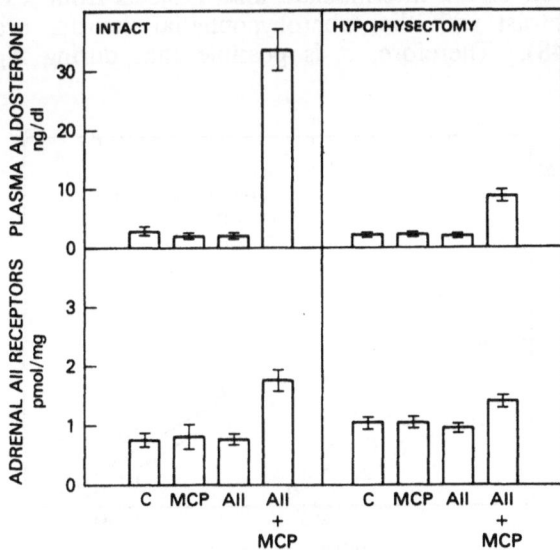

Figure 4. Effects of the infusion of metoclopramide (MCP) (1.2 μg/min iv) and/or AII (50 μg/min, sc) on plasma aldosterone and AII receptors in intact and hypophysectomized rats on a high sodium diet.

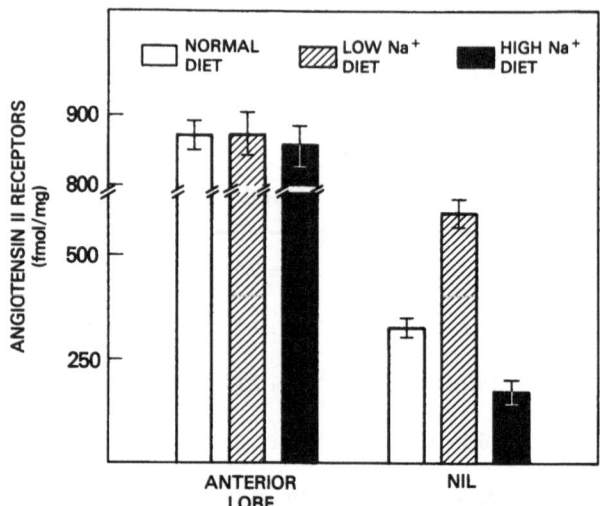

Figure 5. Effect of sodium intake on AII receptors in the anterior and neurointermediate (NIL) pituitary of the rat.

dopamine turnover in the neurointermediate pituitary was decreased instead of increased in those rats on high sodium intake (Figure 8). The relevance of this finding to the dopaminergic activity of the different pituitary lobes and in the regulation of mineralocorticoid secretion is not clear. As shown in Figure 9, the dopaminergic innervation of the pituitary is complex. Fibers originating in the arcuate and periventricular nucleus project to the external layer of the median eminence (tuberoinfundibular system) and to the neural lobe of the pituitary (tuberohypophysial system). The dopaminergic innervation of the intermediate lobe projects from a different group of neurons, which in contrast with the tuberohypophysial neurons are under negative dopamine feedback (48). Therefore, it is possible that during high sodium diet, a

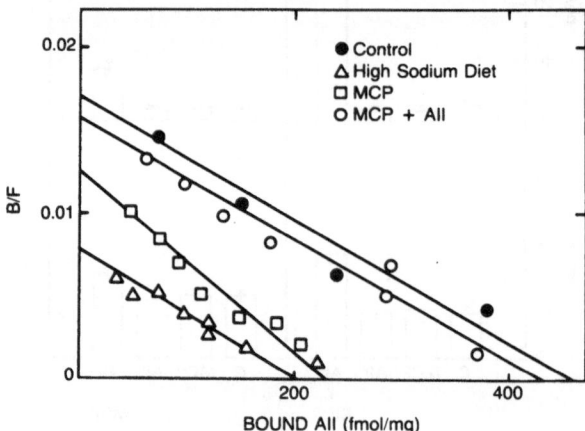

Figure 6. Scatchard analysis of the binding of $^{125}I[Sar^1,Ile^8]AII$ to neurointermediate pituitary membranes from rats on high sodium diet receiving infusions of metoclopramide (MCP) and/or AII. Data was obtained from the pooled tissues of 15 rats in each group.

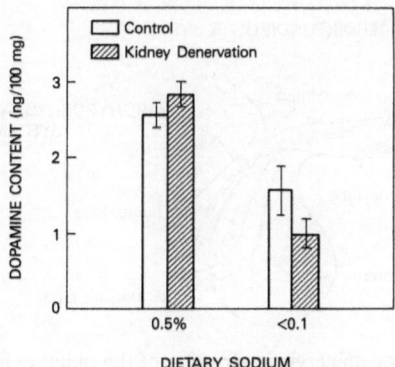

Figure 7. Dietary sodium-induced changes in dopamine content in renal tissue in intact and kidney denervated rats.

decreased negative feedback by dopamine from the neural lobe may result in increased dopaminergic activity in the intermediate lobe. Although further experiments measuring dopamine turnover in separated neural and intermediate lobes will be necessary to elucidate this problem, two important conclusions arise from the latter experiments. First, changes in sodium intake can affect hypothalamic dopaminergic activity, and second, these changes in central dopaminergic activity during altered sodium intake are opposite to those in the periphery.

Another important question is the mechanism by which altered sodium intake results in changes in hypothalamic dopaminergic tone. As depicted in the diagram in Figure 10, the control mechanisms must include a sensor and a humoral or neural signalling system to relay the information to the brain. Possible sensor systems are the kidney and baroreceptor. Experiments involving kidney and baroreceptor denervation will be necessary to determine whether or not afferent innervation from these sites is responsible for changes in hypothalamic dopaminergic activity.

Humoral factors likely to convey information from the periphery to the brain during changes in sodium intake are the renin-angiotensin system and mineralocorticoids. Plasma

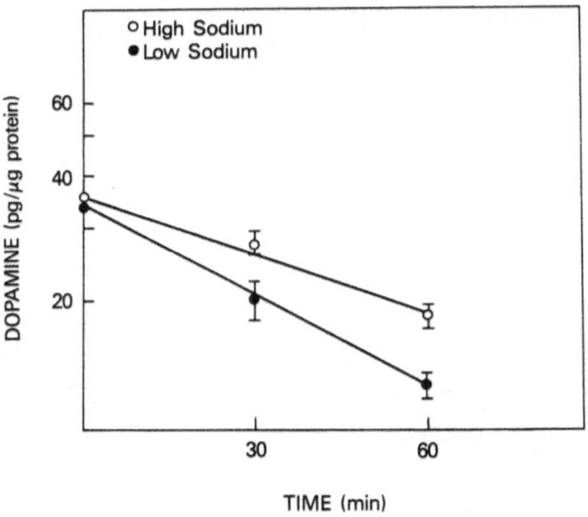

Figure 8. Dopamine turnover in neurointermediate pituitary lobe from rats maintained on high or low sodium diet for 5 days. Data points are the mean and SE of the values in 5 rats.

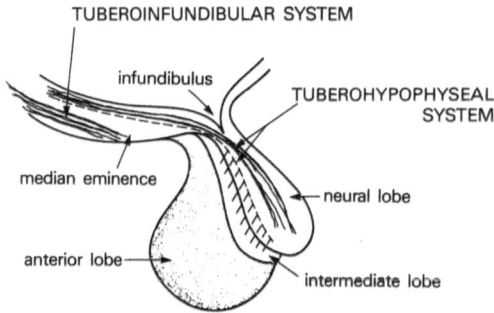

Figure 9. Dopaminergic innervation of the pituitary in the rat.

levels of both AII and aldosterone undergo marked changes during variations in sodium intake and can influence central activity (45). Steroid hormones freely cross the blood-brain barrier to interact with central receptors (49). Preliminary experiments indicate that administration of the mineralocorticoid antagonist, spironolactone, to rats during the period of sodium restriction reduced the increase in dopamine turnover observed in the neurointermediate pituitary. The results of this experiment suggest that the increase in circulating mineralocorticoids during sodium restriction may contribute to the observed increase in hypothalamic dopaminergic tone. In addition to mineralocorticoids, intracerebroventricular administration of AII has been shown to increase dopamine turnover in the hypothalamus (50,51). Circulating levels of AII are elevated during sodium restriction and the peptide could influence central activity by acting through specific receptors located in the circumventricular organs, structures located outside the blood-brain barrier (45) (Figure 11). Whether such a mechanism is involved in the dopaminergic control of mineralocorticoid secretion remains to be elucidated.

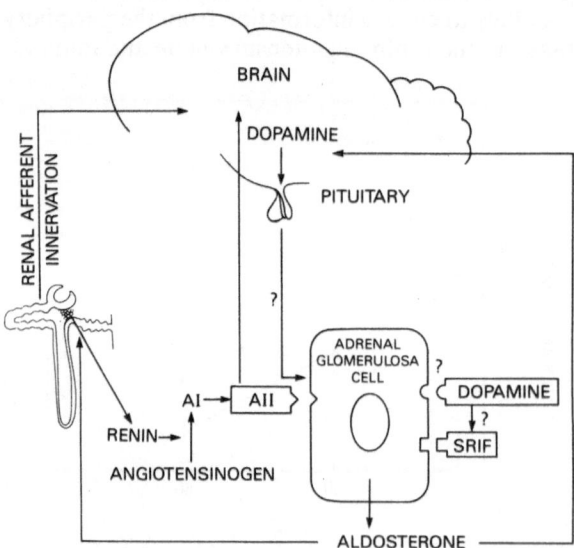

Figure 10. Possible levels of interaction between peripheral and central mechanisms in the control of the adrenal sensitivity to AII.

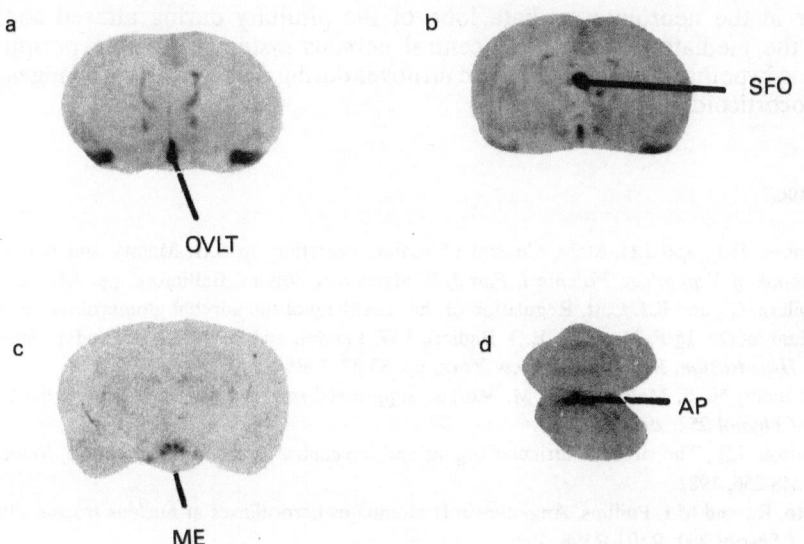

Figure 11. AII receptors in the circumventricular organs (SFO denotes subfornical organ, OVLT denotes organum vasculosum of the lamina terminalis, ME denotes median eminence, AP denotes area postrema). In addition, the figure shows binding in the piriform cortex (a and b), lateral septum and choroidal plexus (b), dorsal thalamic nuclei (c), and nucleus of the solitary tract (d).

Conclusions

The presence of rapid and efficient mechanisms for the regulation of aldosterone secretion is essential for the maintenance of sodium balance and circulatory homeostasis. Although AII is recognized as the main regulator of aldosterone secretion during altered sodium intake, the responsiveness of the adrenal to the peptide varies according to the level of sodium intake. The regulation of adrenal sensitivity to AII involves circulating and neural factors interacting in the adrenal glomerulosa cell and central nervous system.

At the adrenal level, the effect of AII is modulated by stimulatory and inhibitory regulators including extracellular potassium, somatostatin, atrial natriuretic factor, dopamine and others not yet defined. Although these factors clearly increase or decrease adrenal responsiveness to AII, their precise physiological role and mode of participation in the regulation of the adrenal sensitivity to the peptide are still undefined.

There is evidence for the involvement of the dopaminergic system in the decreased adrenal responsiveness to AII during high sodium intake. Despite initial observations indicating that dopaminergic regulation of aldosterone secretion might be independent of the known regulators, more recent studies in which the effects of dopamine agonists and antagonists have been examined in relation with sodium balance have demonstrated that the sensitivity of the adrenal glomerulosa cell to AII can be modulated by dopamine. Thus, during high sodium intake under presumably maximum dopaminergic inhibition, the dopamine agonist metoclopramide restores adrenal sensitivity to AII, while the converse occur upon administration of dopamine during low sodium diet.

Although it cannot be ruled out that dopamine has a direct effect in the adrenal, it is now evident that the pituitary gland is necessary for the full expression of the sensitizing effect of dopamine antagonists. The demonstration of changes in dopaminergic turnover in the neurointermediate lobe of the pituitary during altered sodium intake, suggest the mediatory role of the central nervous system. Possible peripheral factors regulating hypothalamic dopaminergic turnover during dietary sodium changes are AII and mineralocorticoids.

References

1. Reineck, H.J., and J.H. Stein, Control of sodium excretion, In S.G. Massry and R.J. Glassock (eds) *Textbook of Nephrology, Volume 1, Part 2*, Williams and Wilkins, Baltimore, pp. 3.12-3.21, 1983.

2. Aguilera, G., and K.J. Catt, Regulation of the sensitivity of the adrenal glomerulosa cell during altered sodium intake, In F. Mantero, E.G. Biglieri, J.W. Funder, and B.A. Scoggins (eds) *The Adrenal Gland and Hypertension*, Raven Press, New York, pp. 33-53, 1985.

3. Alexander, N., S. Melmed, and M. Morris, Suppressed serum prolactin in sinoaortic-denervated rats, *Am J Physiol* 252: 290-293, 1987.

4. Simpson, J.B., The circumventricular organs and the central actions of angiotensin, *Neuroendocrinology* 32: 248-256, 1981.

5. Casto, R., and M.I. Phillips, Angiotensin II attenuates baroreflexes at nucleus tractus solitarius of rats, *Am J Physiol* 250: R193-R198, 1986.

6. Aguilera, G., K. Fujita, A. Schirar, and K.J. Catt, Role of angiotensin II on the regulation of aldosterone secretion, In I.A. Cumming, J.W. Funder, and F.A.O. Mendelsohn (eds) *Role of Endocrinology*, Australian Academy of Science, Canberra, pp. 389-392, 1980.

7. Bojesen, E., Concentrations of aldosterone and corticosterone in peripheral plasma of rats. The effects of salt depletion, salt repletion, potassium loading and intravenous injections of renin and angiotensin II, *Eur J Steroids* 1: 145-169, 1966.

8. Blair-West, J.R., J.P. Coghlan, D.A. Denton, J.R. Goding, M. Wintour, and R.D. Wright, The control of aldosterone secretion, *Rec Prog Horm Res* 19: 311-383, 1963.

9. Aguilera, G., and K.J. Catt, Regulation of aldosterone secretion during altered sodium intake, *J Steroid Biochem* 19: 525-530, 1983.

10. Boyd, G.W., A.R. Adamson, M. Arnold, V.H.T. James, and W.S. Peart, The Role of angiotensin II in the control of aldosterone in man, *Clin Sci* 42: 91-104, 1972.

11. Blair-West, J.R., J.P. Coghlan, D.A. Denton, and B.A. Scoggins, Aldosterone regulation in sodium deficiency: role of ionic factors and angiotensin II, In I.H. Page and F.M. Bumpus (eds) *Angiotensin*, Springer-Verlag, New York, pp. 337-368, 1974.

12. Lee, T.C., B. van der Val, and D. deWied, Influence of the anterior pituitary on aldosterone secretory response to dietary sodium restriction in the rat, *J Endocrinol* 42: 465-475, 1968.

13. Williams, G.H., L.I. Rose, R.G. Dluhy, J.F. Dingman, and D.P. Lauler, Aldosterone response to sodium restriction and ACTH stimulation in panhypopituitarism, *J Clin Endocrinol Metab* 32: 27-35, 1975.

14. Douglas, J.G., Effects of high potassium diet on angioensin II receptors and angiotensin-induced aldosterone production in rat adrenal glomerulosa cells, *Endocrinology* 106: 983-990, 1980.

15. Boyd, J., L. Manuelidis, and P.J. Mulrow, The importance of potassium int he regulation of aldosterone biosynthesis, In *Proceedings IV International Congress of Endocrinology*, Excerpta Medica, Washington, D.C., pp. 785-789, 1972.

16. Oelkers, W., M. Schöneshöfer, G. Shultze, J.J. Brown, R. Fraser, J.J. Morton, A.F. Lever, and J.I.S. Robertson, Effect of prolonged low-dose angiotensin II infusion on the sensitivity of adrenal cortex in man, *Circ Res* 36: 149-156, 1975.

17. Aguilera, G., F.A.O. Mendelsohn, and K.J. Catt, Dopaminergic regulation of aldosterone secretion, In W.F. Ganong and L. Martini (eds) *Frontiers in Neuroendocrinology, Volume 8*, Raven Press, New York, pp. 265-291, 1984.

18. Sowers, J.R., M.L. Tuck, M.S. Golub, and E.G. Sollars, Dopaminergic modulation of aldosterone secretion is independent of alterations in renin secretion, *Endocrinology* 107: 937-941, 1980.

19. Carey, R.M., Acute dopaminergic inhibition of aldosterone secretion is independent of angtiotensin II and

adrenocorticotrophin, *J Clin Endocrinol Metab* 54: 463-469, 1982.

20. Gordon, M.B., T.J. Moore, R.G. Dluhy, and G.H. Williams, Dopaminergic modulation of aldosterone responsiveness to angiotensin II with changes in sodium intake, *J Clin Endocrinol Metab* 56: 340-345, 1983.

21. Edwards, C.R.W., P.A. Miall, J.P. Hanker, M.O. Thorner, E.A.s. Al-Dujaili, and G.M. Besser, Inhibition of plasma aldosterone response to frusemide by bromocryptine, *Lancet* II: 903-905, 1975.

22. Carey, R.M, and G.R. Van Loon, Bromocriptine does not inhibit the aldosterone response to sodium depletion, *J Clin Endocrinol Metab* 55: 162-165, 1982.

23. Sower, J.R., R. Catania, J. Paris, and M. Tuck, Effects of bromocriptine on renin, aldosterone, and prolactin responses to posture and metoclopramide in idiopathic edema: possible therapeutic approach, *J Clin Endocrinol Metab* 54: 510-516, 1982.

24. Drake, C.R., Jr., N.V. Ragsdale D.L. Kaiser, and R.M. Carey, Dopaminergic suppression of angiotensin II-induced aldosterone secretion in man: differential responses during sodium loading and depletion, *Metabolism Clin Exper* 33: 696-702, 1984.

25. Alexander, R.W., J.R. Gill, Jr., H. Yamabe, W. Lovenberg, and H.R. Haister, Effects of dietary sodium and of acute saline infusion on the interrelationship between dopamine excretion and adrenergic activity in man, *J Clin Invest* 54: 194-200, 1974.

26. Faucheux, B., N.T. Buu, and O. Kuchel, Effects of saline and albumin on plasma and urinary catecholamines in dogs, *Am J Physiol* 232: F123-F127, 1977.

27. Ball, S.G., N.S. Oates, and M.R. Lee, Urinary dopamine in man and rat: effects of inorganic salts on dopamine excretion, *Clin Sci Mol Med* 55: 167-173, 1978.

28. Dunn, M.G., and H.B. Bosmann, Peripheral dopamine receptor identification: properties of a specific dopamine receptor in the rat adrenal zona glomerulosa, *Biochem Biophys Res Commun* 99: 1081-1087, 1981.

29. Bevilacqua, M., T. Vago, D. Scorza, and G. Norbiato, Characterization of dopamine receptors by ^{3}H-ADTN binding in calf adrenal zona glomerulosa, *Biochem Biophys Res Comm* 108: 1661-1669, 1982.

30. McKenna, T.J., D.P. Island, W.E. Nicholson, and G.W. Liddle, Dopamine inhibits angiotensin stimulated aldosterone biosynthesis in bovine adrenal cells, *J Clin Invest* 64: 287-291, 1979.

31. Racz, K., A. Deléan, O. Kuchel, and N.T. Buu, Adrenomedullary mechanisms in aldosterone regulation, In F. Mantero and P. Vecsei (eds) *Serono Symposium Publication (title)*, Raven Press, Volume 39, pp. 77-90, 1987.

32. Wakabayashi, I., Y. Miyazawa, M. Kanda, N. Miki, R. Demuro H. Demura and K. Shizume, Stimulation of immunoreactive somatostatin release from hypothalamic synaptosomes by high (K^+) and dopamine, *Endocrinol Jpn* 24: 601-604, 1977.

33. Carey, R.M., and S. Sen, Recent progress in the control of aldosterone secretion, In R.O. Greep (ed) *Rec Prog Horm Res* 42: 251-296, 1986.

34. Vinson, G.P., B.J. Whitehouse, A. Dell, T. Etienne, and H.R. Morris, Characterisation of an adrenal zona glomerulosa-stimulating component of posterior pituitary extracts as α-MSH, *Nature* 284: 464-467, 1980.

35. Matsuoka, H., P.J. Mulrow, and R. Franco-Saenz, Effects of β-lipotropin and β-lipotropin-derived peptides on aldosterone production in the rat adrenal gland, *J Clin Invest* 68: 752-759, 1981.

36. Pedersen, R.C., A.C. Brownie, and N. Ling, Pro-adrenocorticotropin/endorphin-derived peptides: coordinated action on adrenal steroidogenesis, *Science* 208: 1044-1046, 1980.

37. Güllner, H.-G., and J.R. Gill, Beta endorphin selectively stimulates aldosterone secretion in hypophysectomized, nephrectomized dogs, *J Clin Invest* 71: 124-128, 1983.

38. Rabinowe, S.L., T. Taylor, R.G. Dluhy, and G.H. Williams, β-Endorphin stimulates plasma renin and aldosterone release in normal human subjects, *J Clin Endocrinol Metab* 60: 485-492, 1985.

39. Lin, S.-Y., E. Wiedmann, and M.H. Humphreys, Role of the pituitary in reflex natriuresis following acute unilateral nephrectomy, *Am J Physiol* 18: F282-290, 1985.

40. Young, III, W.S., Corticotropin-releasing factor mRNA in the hypothalamus is affected differently by drinking saline and by dehydration, *FEBS Lett* 208: 158-162, 1986.

41. Elkabes, S., and Y.-P. Loh, Effect of salt loading on proopiomelanocortin (POMC) messenger ribonucleic

acid levels, POMC biosynthesis, and secretion of POMC products in the mouse pituitary gland, *Endocrinology* 123: 1754-1760, 1988.

42. Humphrey, M.H., and E. Wiedemann, γ-Melanocyte stimulating hormone like peptide (γ-MSH) mediates natriuresis after acute unilateral ureteral pressure elevation (UPE), *Clin Res* 37: 395A, 1989.

43. Howe, A., and A.J. Thody, The effect of ingestion of hypertonic saline on the melanocyte-stimulating hormone content and histology of the pars intermedia of the rat pituitary gland, *J Endocrinol* 46: 201-208, 1970.

44. Rapp, J.P., and L.K. Dahl, Anatomical and protein electrophoretic observations on pituitary cleft colloid in rats genetically susceptible or resistant to salt hypertension, *Lab Invest* 30: 417-426, 1974.

45. Catt, K.J., F.A.C. Mendelsohn, M.A. Millan, and G. Aguilera, The role of angiotensin II receptors in vascular regulation, *J Cardiovasc Pharmacol* 6: S575-S586, 1984.

46. Lee, M.R., Dopamine and the kidney (editorial review), *Clin Sci* 62: 439-448, 1982.

47. Kvetňanský, R., V.K. Weise, N.B. Thoa, and I.J. Kopin, Effect of chronic guanethidine treatment and adrenal medullectomy on plasma levels of catecholamines and corticosterone in forcibly immobilized rats, *J Pharmacol Exp Ther* 209: 287-291, 1979.

48. Lookingland, K.J., J.M. Farah, Jr., K.L. Lovell, and K.E. Moore, Differential regulation of tuberohypophysial dopaminergic neurons terminating in the intermediate lobe and in the neural lobe of the rat pituitary gland, *Neuroendocrinology* 40: 145-151, 1985.

49. McEwen, B.S., E.R. de Kloet, and W. Rostene, Adrenal steroid receptors and actions in the nervous system, *Physiol Rev* 66: 1121-1188, 1986.

50. Anderson, K., K. Fuxe, L.F. Agnati, D. Ganten, I. Zini, P. Eneroth, F. Mascagni, and F. Infantilina, Intraventricular injections of renin increase amine turnover in the tuberoinfundibular dopamine neurones and reduce the secretion of prolactin in the male rat, *Acta Physiol Scand* 116: 317-320, 1982.

51. Steele, M.K., S.M. McCann, and A. Negro-Vilar, Modulation by dopamine and estradiol of the central effects of angiotensin II on anterior pituitary hormone release, *Endocrinology* 111: 722-729, 1982.

CENTRAL ACTION OF ADRENAL STEROIDS DURING STRESS AND ADAPTATION

J.M.H.M. Reul, W. Sutanto, J.A.M. van Eekelen, J. Rothuizen,[1]
and E.R. de Kloet

Rudolf Magnus Institute
Vondellaan 6
NL-3521 GD Utrecht
The Netherlands

Introduction

Adrenal glucocorticoid hormones have a potent influence on disparate aspects of brain function. These steroids control circadian and stress-induced aspects of neuroendocrine regulation, affect different aspects of adaptive behavior, and modulate the activity of the neurotransmitter circuitry underlying these processes. During the past twenty years, extensive research has been conducted to elucidate the mechanism of glucocorticoid action in the brain. Great progress was accomplished when it was discovered that adrenal steroids interact with two receptor systems, mineralocorticoid and glucocorticoid receptors, that display different affinities, capacities, specificities, and neuroanatomical localizations in the central nervous system (1-4).

Our aim is to review recent developments in the chemistry, molecular biology, and functional implications of receptor diversity with special emphasis on the topographical localization and regulatory aspects of the brain corticosteroid receptors and the *in vivo* ligand specificity of the mineralocorticoid receptors.

Differentiation of Corticosteroid Receptor Types: *In Vivo* and *In Vitro* Binding Studies

Intracellular binding sites for corticosteroids were first observed by McEwen *et al.* (5) after they injected a tracer dose (0.5-1 μg) of [^{3}H]-corticosterone into adrenalectomized (ADX) rats. Autoradiograms of the *in vivo* labelled brain sections showed an abundant retention of radioactivity in cell nuclei of extrahypothalamic limbic brain regions, particularly in the hippocampal formation. A sixfold increase in the dose of [^{3}H]-corticosterone only slightly increased cell nuclear uptake in the hippocampus, suggesting that the uptake system already became saturated at this low concentration (6). Further neuroanatomical inspection of the autoradiograms revealed that [^{3}H]-corticosterone was selectively retained by hippocampal neurons of the CA_1, CA_2, and CA_3 cell field and by granular cells in the dentate gyrus (7-8). Profound labelling was also found in the dorso-lateral part of the septum and motor neurons of the spinal cord (8). When corticosterone was determined by a radioimmunoassay procedure in brain tissue and cell

---

[1]Department of Clinical Sciences of Companion Animals, Faculty of Veterinary Medicine, University of Utrecht, The Netherlands

Circulating Regulatory Factors and Neuroendocrine Function
Edited by J. C. Porter and D. Ježová
Plenum Press, New York, 1990

243

nuclear extract of non-ADX rats, killed in the morning under resting conditions, the same regional distribution pattern was found as observed after administration of [^{3}H]-corticosterone to ADX animals (9). Apparently, in the presence of other adrenocortical steroids, the receptor system also prefers to bind corticosterone.

Administration of tracer amounts of [^{3}H]-adlosterone to ADX rats resulted in a similar distribution of cell nuclear uptake as observed after [^{3}H]-corticosterone (10). The notion that corticosterone and aldosterone interact with the same receptor system, which consequently is of mineralocorticoid-like nature, was substantiated by the observation that a tracer dose of unlabelled aldosterone could effectively antagonize the cell nuclear retention of [^{3}H]-corticosterone given 30 minutes after aldosterone (11). Other workers, employing a different approach, reported that the physicochemical properties of the corticosterone- and aldosterone-binding receptor in the brain were similar to the kidney mineralocorticoid receptor (Type 1) (12,13). Krozowski and Funder (14) showed that the hippocampal and kidney receptor displayed the same intrinsic steroid binding specificity. The similarity of the brain and kidney mineralocorticoid receptor system was eventually confirmed by the cloning and the *in vitro* expression of the human mineralocorticoid receptors (15).

The first evidence for heterogeneity of brain corticosteroid receptors appeared from experiments in which [^{3}H]-dexamethasone was administered to ADX rats (16). The retention of tracer doses of [^{3}H]-dexamethasone was strikingly different from that observed after [^{3}H]-corticosterone administration, *e.g.*, the synthetic glucocorticoid was retained by corticotrophs in the anterior pituitary but not by hippocampal neurons (11,16,17). Evidently, dexamethasone revealed the presence of a glucocorticoid-binding receptor that was not recognized by corticosterone or aldosterone. Resolution of the issue of corticosteroid receptor heterogeneity became possible after the development of synthetic glucocorticoid steroids, RU 26988 and RU 28362, which display exclusive glucocorticoid action and receptor binding in peripheral glucocorticoid target tissues (18,19). These selective glucocorticoid steroids were employed to discriminate between [^{3}H]-corticosterone binding to mineralocorticoid receptors and glucocorticoid receptors (Type 2). Scatchard analysis of [^{3}H]-corticosterone binding in absence and presence of RU 28362 in hippocampal cytosol of ADX rats revealed that mineralocorticoid receptors display a tenfold higher affinity for corticosterone than do glucocorticoid receptors (1).

Distribution of Corticosteroid Receptor Sites: Radioligand Binding Studies

The distribution of mineralocorticoid receptors as determined by *in vitro* labelling in frozen brain tissue sections with [^{3}H]-corticosterone in the presence of excess RU 26988 or RU 28362 and subsequent autoradiography, showed that mineralocorticoid receptors were predominantly present in neurons of the septo-hippocampal complex (20,21). Mineralocorticoid receptors were particularly abundant in neurons of the CA_1 and CA_2 cell field and the granular cell field of the dentate gyrus, and to a lesser extent in the dorso-lateral septum and the central amygdala. In a parallel study involving measurement of mineralocorticoid receptors in cytosol of micro-dissected brain regions, a regional distribution of mineralocorticoid receptor concentration was found that was similar to that observed with *in vitro* autoradiography (1). Furthermore, the *in vitro* localization of mineralocorticoid receptors agreed with the identification of mineralocorticoid receptors by *in vivo* [^{3}H]-corticosterone retention in hippocampal neurons.

Incubation of brain tissue sections with [^{3}H]-RU 28362 in order to label glucocorticoid receptors and subsequent autoradiography confirmed and extended previous data from cytosol binding studies (1,21,22), and showed that glucocorticoid receptors occurred in both neuronal and glial cells. Glucocorticoid receptors were present in high quantities throughout the forebrain, particularly in cortical, thalamic, hypothalamic (paraventricular, arcuate, and supraoptic nucleus) and septo-hippocampal (dorso-lateral septum, dentate gyrus) regions (1,21,22).

Distribution of Corticosteroid Receptor Sites: Immunocytochemistry and Gene Expression

Immunocytochemical studies using a monoclonal antibody against rat liver glucocorticoid receptor clearly showed the antigen to be localized in neuronal and glial cells, with a distribution pattern similar to that found in the studies with radiolabelled glucocorticoids (23-26).

The recent cloning and *in vitro* expression of the mineralocorticoid receptor and glucocorticoid receptor cDNAs (15,27-29) has opened up new avenues for establishing the neuroanatomical localizations of mineralocorticoid receptor and glucocorticoid receptor gene expression sites and for research in the field of glucocorticoid physiology.

Through the use of Northern blot and/or solution hybridization/S_1 nuclease analysis, the mineralocorticoid receptor and glucocorticoid receptor mRNA concentrations were determined in a number of central and peripheral tissues (15,30). Among the nervous tissues, mRNA coding for mineralocorticoid receptors was highest in the hippocampus. Moderate to low levels were found in the hypothalamus and cerebellum, respectively. *In situ* hybridization histochemistry revealed that in the brain mineralocorticoid receptors appear to be expressed exclusively in neurons (31,32). Moreover, this technique also identified the hippocampus as the nervous tissue with the highest mineralocorticoid receptor mRNA concentration. Heavy labelling occurred in the pyramidal neurons of Ammon's horn with the CA_3 cell field exhibiting the highest mineralocorticoid receptor mRNA levels. The density of labelling was also high in the CA_1 and CA_2 cell field, in the granular layer of the dentate gyrus, and in neurons of the subiculum. Moderate labelling occurred in the dorso-lateral septum and in subdivisions of the amygdala (31,32).

The distribution of glucocorticoid receptor mRNA, in contrast to the topography of mineralocorticoid receptor mRNA, was rather evenly distributed over the brain. Nevertheless, hypothalamus tissue contained more glucocorticoid receptor mRNA than did the hippocampus and cerebellum, as measured by Northern blot and solution hybridization/S_1 nuclease analysis (30). *In situ* hybridization showed glucocorticoid receptor mRNA within the hippocampus to be primarily present in neurons of the CA_1 and CA_2 field (32,33). Very low concentrations of glucocorticoid receptor mRNA were found in the CA_3 and CA_4 cell field. Apart from the hippocampal formation, high labelling of glucocorticoid receptor mRNA was present over the granular layer of the cerebellum, in subdivision of the amygdala, and in the arcuate nucleus and parvocellular region of the paraventricular nucleus of the hypothalamus (33). Further, profound labelling was found over the aminergic cell groups of the locus coeruleus and the raphe nuclei. Moderate glucocorticoid receptor mRNA concentrations were present in well-defined layers of the cortex, central nucleus of the amygdala, thalamic region, and most of the preoptic and hypothalamic nuclei. Glucocorticoid receptor mRNA was low in the caudate nucleus, putamen, accumbens, septal region, and substantia nigra. Surprisingly, no labelling was evident of the supraoptic nucleus (33), although both *in vitro* autoradiography and immunocytochemistry have shown the glucocorticoid receptor protein to be present in this nucleus (34). However, when comparing data from *in vivo* and *in vitro* labelling of receptor protein, immunocytochemistry and molecular studies, it is fair to state that the distribution of mineralocorticoid receptor and glucocorticoid receptor protein usually matches with the localization of the mineralocorticoid receptor and glucocorticoid receptor mRNA and antigen.

Corticosterone- and Aldosterone-Selective Mineralocorticoid Receptors

Recent findings from two laboratories have shed some light on the issue of mineralocorticoid- *vs.* glucocorticoid-selectiveness of the mineralocorticoid receptors in brain and peripheral target tissues. The key to the solution of this problem came when it was found that mineralocorticoid target tissues contain the enzyme, 11β-hydroxysteroid dehydrogenase, which converts the physiological glucocorticoids, corticosterone and cortisol, but not the mineralocorticoid aldosterone, to their 11-keto analogs. The analogs have a very low affinity for mineralocorticoid receptors. Although glucocorticoids circulate in much higher quantities than does aldosterone, in mineralocorticoid target tissues the enzyme converts these glucocorticoids and thereby enables aldosterone to reach the mineralocorticoid receptors (35,36). The enzyme has been localized in kidney, parotid and colon, but seems to be absent in hippocampus and heart. For the future it is intriguing to speculate on the presence of 11β-hydroxysteroid dehydrogenase in mineralocorticoid target sites in the brain.

Other mechanisms conferring specificity to the mineralocorticoid receptors may also be implicated in tissues, such as liver, anterior pituitary, and kidney. These tissues express the transcortin (CBG, corticosterone binding globulin) gene and contain high quantities of extravascular transcortin, that selectively binds physiological glucocorticoids, but exhibits only very low affinity for aldosterone and dexamethasone. Transcortin, thus by binding corticosterone, lowers the amount of free corticosterone able to compete with aldosterone for mineralocorticoid receptor occupancy (4,14,37).

Employing a more functional approach, McEwen *et al.* (38) re-examined the uptake of corticosterone and aldosterone thereby using the effectiveness of both steroids in affecting salt appetite as a functional marker. Infusion of corticosterone (50 μg/hr), producing hormone levels found after stress, but not affecting the ability of aldosterone to suppress salt appetite in ADX rats, resulted in a reduction by 70% of [^{3}H]-aldosterone uptake in forebrain regions such as the hippocampus. However, uptake in the circumventricular region was only slightly affected. Administration of the potent antimineralocorticoid, RU 28318 (50 μg/hr), not only reduced the uptake of [^{3}H]-aldosterone by 95%, but also the effect of aldosterone on salt appetite. Thus, these *in vivo* studies indicate the existence of a population of mineralocorticoid receptors, which appear to be aldosterone-selective, even in the presence of excess corticosterone (38). In the periphery, these aldosterone-selective mineralocorticoid receptors occur in classical mineralocorticoid target tissues, such as kidney, parotid, and colon (14,39-41). In the brain, the aldosterone-selective mineralocorticoid receptor sites seem to be localized in the circumventricular organs, but recent experiments also point to the antero-ventral part of the third ventricle (AV3V) of the hypothalamus as a site of aldosterone-selective mineralocorticoid receptors (42). A conflicting finding in this respect is the observation that mineralocorticoid receptor mRNA was found to be absent in circumventricular organs such as the subfornical organ and vascular organ of the lamina terminalis (31). In addition to effects on salt appetite, specific effects of intracerebroventricularly injected aldosterone on blood pressure have been reported (43,44).

Species-Specificity of Mineralocorticoid Receptors and Glucocorticoid Receptors in Rat and Hamster Brain

Differentiation of corticosteroid receptor sites into mineralocorticoid receptors and glucocorticoid receptors has not only been conducted in rat brain, but also in the central nervous system of other animals, *viz.*, hamster (45,46), and dog (47). These species secrete both cortisol and corticosterone with cortisol being the predominantly circulating

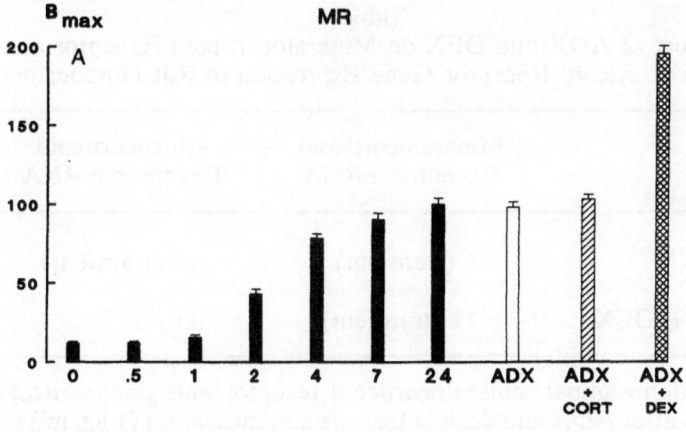

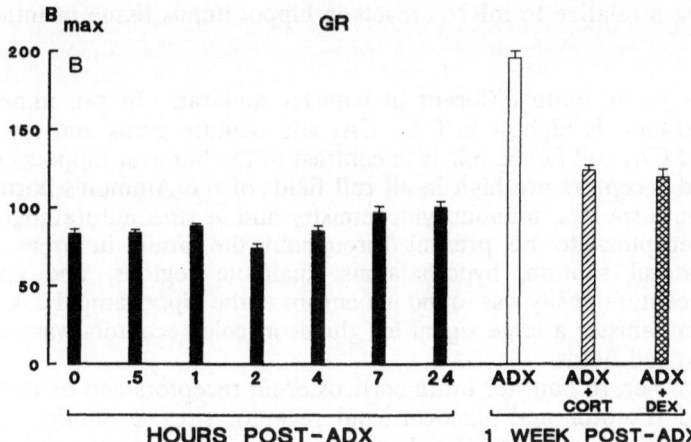

Figure 1. Effect of ADX, corticosterone and dexamethasone administration on mineralocorticoid receptor (1A) and glucocorticoid receptor (1B) concentration in rat hippocampus. Male rats were adrenalectomized and killed at indicated hrs post-ADX. ADX rats that were maintained for 1 week were equipped with a pellet containing cholesterol (ADX control), corticosterone (ADX + CORT) or dexamethasone (ADX + DEX) immediately after surgery. Data are expressed as relative maximum binding capacity (B_{max}) with the mineralocorticoid receptor and glucocorticoid receptor level at 24 hrs post-ADX set at 100%.

glucocorticoid (48-51). Using RU 28362 to mask glucocorticoid receptors, it was found that mineralocorticoid receptors in ADX hamster hippocampus binds both cortisol and corticosterone with similar affinity (1 nM), while aldosterone displayed a considerably lower affinity (40 nM). Surprisingly, concerning glucocorticoid receptors in hamster brain, this receptor site displayed the highest affinity for corticosterone (4 nM), slightly lower affinity (5-6 nM) for cortisol and dexamethasone, and lowest affinity (85 nM) for aldosterone (45).

Like the rat, the hamster mineralocorticoid receptors also prevail in the hippocampal formation, and to a lesser extent are present in the lateral septum and central amygdala. However, within the hippocampus the topography of mineralocorticoid

Table 1
Effect of ADX and DEX on Mineralocorticoid Receptor and Glucocorticoid Receptor Gene Expression in Rat Hippocampus

	Mineralocorticoid Receptor mRNA	Glucocorticoid Receptor mRNA
ADX	↑ (transient)	↑↑ (transient)
ADX + DEX	↑↑ (transient)	↓↓

Changes in hippocampal mineralocorticoid receptor and glucocorticoid receptor mRNA levels after ADX and dexamethasone administration (1 μg/ml) *via* drinking water as determined by solution hybridization and S_1 nuclease analysis. *Arrows* indicate changes relative to mRNA levels in hippocampus tissue of intact rats.

receptors appear to be quite different in hamster and rat. In rat, mineralocorticoid receptor concentration is highest in CA_1, CA_2 and dentate gyrus and lower levels are found in CA_3 and CA_4 cell fields, this is in contrast to the hamster hippocampus in which mineralocorticoid receptors are high in all cell fields of the Ammon's horn and dentate gyrus (46). In both species, immunocytochemistry and *in vitro* autoradiography showed glucocorticoid receptors to be present throughout the brain in areas such as the hippocampus, lateral septum, hypothalamus, thalamic regions, and cortex. High glucocorticoid receptor density was found in neurons of the hippocampal CA_1 and CA_2 cell fields, but only in hamster a large signal for glucocorticoid receptors was also present in the CA_3 and CA_4 cell fields.

Taken together, in hamster brain corticosteroid receptors can be differentiated in mineralocorticoid receptor and glucocorticoid receptor sites, exhibiting an alternative ligand binding specificity and a parallel, though not identical, neuroanatomical distribution compared to the receptor sites in rat brain.

Plasticity of the Corticosteroid Receptor System: Binding Studies

When studying the function of glucocorticoid action in the central nervous system it is important to know the conditions leading to an altered steroid signal, and, thus to delineate the effect of such circumstances and manipulations on the corticosteroid receptor capacity. This is in order to define endogenous factors and environmental conditions operating to adjust the receptor concentration to physiological requirements and thereby serving homeostasis.

After ADX, the mineralocorticoid receptor concentration in the rat hippocampus acutely increases and reaches its maximum within 24 hours (Figure 1A). During the next two weeks the mineralocorticoid receptor concentration remains constant (22,52). After ADX, glucocorticoid receptors remain virtually unchanged during the first 24 hours (Figure 1B), but thereafter their concentration increases reaching maximum levels around 5 to 7 days. Chronic exposure of ADX rats to corticosterone levels normally (*i.e.*, in the intact animal) observed after moderate stress, results in a decreased number of glucocorticoid receptors, but no effect is noted on mineralocorticoid receptors. Interestingly, giving dexamethasone to ADX rats produces a dose-dependent up-regulation

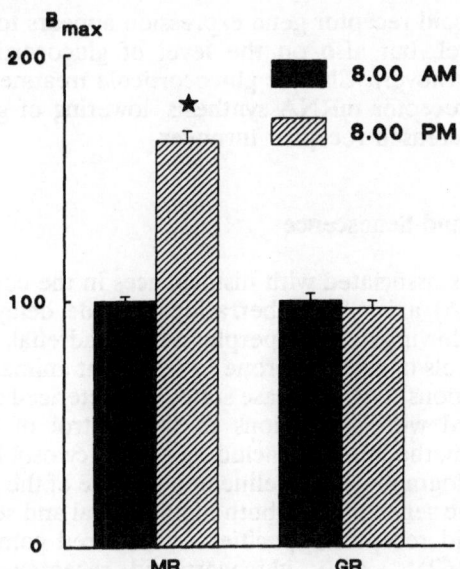

Figure 2. Diurnal variation in mineralocorticoid receptor and glucocorticoid receptor concentration in rat hippocampus. Male rats were adrenalectomized 0800 or 2000 and killed 24 hrs later. Mineralocorticoid receptor and glucocorticoid receptor concentration are indicated as relative B_{max}. The AM receptor level is set at 100%. The data show clearly that hippocampal mineralocorticoid receptors displays circadian variation ($P < 0.05$, Student's t test); whereas, glucocorticoid receptors do not.

of mineralocorticoid receptors and down-regulation of glucocorticoid receptor number (Figure 1) (22). For the moment the underlying mechanisms involved in the different effects of corticosterone and dexamethasone (both being glucocorticoids) on the hippocampal mineralocorticoid receptors remain unclear. An answer may lie in the different affinities of dexamethasone (low) and corticosterone (high) for the mineralocorticoid receptor, and thus, in light of the effects of ligand occupancy on receptor turnover (53,54), these glucocorticoids may affect the half-life of the receptor in a different manner.

Plasticity: Corticosteroid Receptor Gene Expression

In a recent study a remarkably different response in mineralocorticoid receptor and glucocorticoid receptor mRNA was observed following ADX and chronic dexamethasone replacement (Table 1) (30). ADX resulted in transient increments in both mineralocorticoid receptor and glucocorticoid receptor mRNA, suggesting negative regulation of both genes by corticosterone. Dexamethasone administration *via* the drinking water (1 μg/ml) augmented the transient ADX-induced increase in mineralocorticoid receptor mRNA, but blocked the rise in glucocorticoid receptor mRNA (Table 1) (30). These findings suggest that glucocorticoids affect mineralocorticoid receptor and glucocorticoid receptor gene expression through different mechanisms. In particular, it seems that glucocorticoids regulate mineralocorticoid receptor gene expression through post-transcriptional mechanisms involving an improvement of mineralocorticoid receptor mRNA translation and/or receptor stability after chronic glucocorticoid treatment. On the other hand, glucocorticoid receptor gene expression appears to be regulated not only on the transcriptional level, but also on the level of glucocorticoid receptor mRNA translation and receptor turnover. Chronic glucocorticoid treatment seems to result in a decreased glucocorticoid receptor mRNA synthesis, lowering of glucocorticoid receptor mRNA translation, and increased receptor turnover.

Corticosteroid Receptors and Senescence

The aging process is associated with disturbances in the control of the hypothalamic-pituitary-adrenal (HPA) axis. The aberrations include delayed attenuation of the corticosterone response following stress, hyperplasia of the adrenal, and some authors have reported elevated basal levels of corticosterone in senescent animals (55-57; de Kloet and Reul, unpublished observations). Also disease states characterized by depressed mood and affect are often associated with aberrations in the control of HPA activity (58-61). Various receptor binding methodologies, including *in vitro* cytosol binding and autoradiography and *in vivo* autoradiography, have delineated the fate of the corticosteroid receptor systems during aging. In the senescent rat, both hippocampal and septal mineralocorticoid receptor and glucocorticoid receptor capacities are reduced compared to the levels in young mature animals (62). Also, glucocorticoid receptor concentration in the hypothalamus and amygdala were diminished in the aged animals (unpublished observation); thus the impact of the aging process on the brain corticosteroid receptor systems is not restricted to the hippocampal receptors. However, the age-related reduction in septo-hippocampal mineralocorticoid receptor number appear to be reversible. Treatment of aged rats with the $ACTH_{4-9}$ analog, ORG 2766, a peptide to which neurotrophic properties have been ascribed (63,64), reversed the reduced mineralocorticoid receptor concentration to the level seen in young mature animals. In contrast, glucocorticoid receptors were not affected by the peptide therapy (62). Interestingly, treatment of senescent rats with the sapogenin, RG_1, also reversed the age effects on mineralocorticoid receptors, but was ineffective with respect to glucocorticoid receptors (65).

Concerning the factors involved in the regulation of the two corticosteroid receptors, there is a dichotomy in the responsiveness of mineralocorticoid receptors and glucocorticoid receptors to certain conditions and treatments. In short-term, the rat hippocampal mineralocorticoid receptor concentration is responsive to ADX, but on a longer time scale ADX is ineffective and chronic glucocorticoid treatment does not seem to down-regulate mineralocorticoid receptors. Dexamethasone even up-regulates mineralocorticoid receptors. Mineralocorticoid receptor capacity is also subject to regulation by other factors, such as peptides and neurotransmitters. Glucocorticoid receptor concentration is responsive to withdrawal of endogenous corticosteroids (ADX) and high levels of glucocorticoids, and hence this receptor system is under homologous control.

Mineralocorticoid Receptor and Glucocorticoid Receptor Occupancy Studies

In the exploration of the function of brain corticosteroid receptors, it is not only important to define factors controlling receptor plasticity, but also to know the physiological conditions leading to receptor occupancy. Given the tenfold difference in affinity between mineralocorticoid receptors and glucocorticoid receptors for corticosterone (K_d 0.2-0.5 nM for mineralocorticoid receptors; K_d 3-5 nM for glucocorticoid receptors), a difference in the extent of occupancy of each receptor type by corticosteroids may be expected. Indeed, it was found that when intact animals were sacrificed under basal morning conditions or after stress, mineralocorticoid receptor occupancy by endogenous hormones equaled 80% or more (1). Since corticosterone circulates in a 100 to 1000-fold higher concentration than aldosterone, the principal occupant will be corticosterone, even under basal morning resting levels (10 ng corticosterone per ml plasma). Occupancy of glucocorticoid receptors developed parallel with increasing plasma corticosterone due to stress or circadian rhythm (1,22).

Receptor Plasticity Vs. Receptor Occupancy

Occupancy studies led to the notion that corticosterone might exert via the mineralocorticoid receptors and glucocorticoid receptors a dual action on the brain. Mineralocorticoid receptors are thought to mediate tonic influences of corticosterone on hippocampus-associated processes. Since the occupancy of this receptor system is under any condition 80% or more, the mineralocorticoid receptor concentration itself seems to be the rate limiting factor, and not the extent of receptor occupancy by corticosterone. One important consequence of this apparent property of the mineralocorticoid receptors is that modulation of the magnitude of the corticosterone signal via mineralocorticoid receptors appear to occur by adjusting the concentration of this receptor in the hippocampal neurons. The magnitude of the glucocorticoid signal via glucocorticoid receptors is primarily dependent on the extent of occupancy with the steroid being the rate limiting factor (3).

Functional Considerations

It is interesting to note that the nocturnal rise in plasma glucocorticoids stimulates exploratory behavior (66,67) and food-seeking behavior (68), which are associated with the beginning of the active period. In addition, glucocorticoids are implicated in the timing of diurnal patterns such as the diurnal variation in synaptic efficacy in the hippocampus (69,70) and the timing of sleep/wake cycles (71). After ADX, these patterns in daily activity are disrupted or shifted, and re-installed after replacement with a low dose of corticosterone (67,71,72). These findings point to a role of the mineralocorticoid receptors as a mediator of these steroid effects associated with the circadian rhythm in pituitary-adrenal activity. In this respect, it is of interest that the hippocampal mineralocorticoid receptor concentration displays a circadian variation in parallel with the circadian rise in pituitary-adrenal activity (Figure 2) (52).

Occupancy of glucocorticoid receptors occurs in parallel with the stress-induced rise, subsequent fall, and the circadian variation of plasma corticosterone. This observation illustrates that glucocorticoid receptors mediate the feedback action of corticosterone on stress-activated brain processes (3,73). This view is supported by the presence of glucocorticoid receptors in neurons involved in the regulation of the stress response, i.e., neurons of the paraventricular and supraoptic nuclei, the ascending aminergic neurons, and cortical and limbic neurons (1,21,23,24,26,32-34). Dallman and co-workers have postulated that limbic mineralocorticoid receptors are involved in the basal control of ACTH secretion from the anterior pituitary (74). These findings endorse the view that corticosterone exerts via mineralocorticoid receptors a tonic influence on basal activities (diurnal rhythm) of the HPA axis, and a feedback action via glucocorticoid receptors on stress-induced disturbances (65).

This view is supported by recent experiments involving the effect of intracerebroventricularly administered selective mineralocorticoid receptor and glucocorticoid receptor antagonists on basal and stress-induced HPA activity. Administration of 100 ng intracerebroventricularly of the mineralocorticoid antagonist, RU 28318, was followed by a small transient increase in the basal concentration of circulating corticosterone, whereas the glucocorticoid antagonist, RU 38486, was ineffective (75). The mineralocorticoid antagonist appeared to enhance stress-induced HPA-activity, induced by exposing the rats to a novel environment. In a parallel experiment, after exposure of rats to stress, the glucocorticoid antagonist attenuated the negative feedback exerted by endogenous corticosterone. These experiments support our hypothesis that mineralocorticoid receptors are associated with the subtle adjustment of HPA activity under basal circumstances, as well as the response to stress. Glucocorticoid receptors control the termination of the stress response.

In this manner, mineralocorticoid receptors and glucocorticoid receptors appear to control neuronal networks involved in organizing the response to stress. They do so with a different range of control and direct different (temporal) aspects of the stress response. We postulate that mineralocorticoid receptors control the threshold (activational component) and that glucocorticoid receptors control the termination of the stress response, together being responsible for the stress responsiveness of the system.

Summary

Corticosteroids interact with receptors in the central nervous system. These receptors display heterogeneity and can be distinguished as corticosterone- and aldosterone-binding mineralocorticoid receptors and dexamethasone-binding glucocorticoid receptors. Ligand specificity of mineralocorticoid receptors for either corticosterone or aldosterone seems to be determined by co-localized transcortin and the enzyme, 11β-hydroxysteroid dehydrogenase. Aldosterone-selective mineralocorticoid receptors appear to be present in the circumventricular organs and the AV3V region of the hypothalamus and mediate behavior that is driven by salt appetite. Highest concentrations of mineralocorticoid receptors are found in neurons of the hippocampus. These limbic mineralocorticoid receptor sites mediate tonic influences of corticosterone on brain processes.

Glucocorticoid receptors bind corticosterone with a tenfold lower affinity than do mineralocorticoid receptors, and are widely distributed in neuronal and glial cells of the brain. Glucocorticoid receptors are involved in the termination of the stress response (negative feedback). Studies involving measurement of glucocorticoid receptor mRNA and binding sites have revealed that glucocorticoid receptors are subject to autoregulation. After ADX, glucocorticoid receptor concentration increases, but is reduced after chronic stress, chronic administration of glucocorticoids, and at senescence. A diminished glucocorticoid receptor concentration may compromise the negative feedback action exerted by glucocorticoids after stress. After ADX, mineralocorticoid receptor binding is acutely up-regulated and reaches its maximum between 7 and 24 hours post-ADX. Mineralocorticoid receptor mRNA level shows a transient increase following ADX. Long-term ADX has no effect on the mineralocorticoid receptor concentration, but, interestingly, chronic dexamethasone treatment results in an up-regulation of mineralocorticoid receptors. Mineralocorticoid receptor level is decreased at senescence, but this age-related decrement can be reversed by chronic treatment with the $ACTH_{4.9}$ analog, ORG 2766. Functionally, mineralocorticoid receptors and glucocorticoid receptors are involved in different aspects of the organization of the stress response, and in conjunction they control the stress responsiveness of the animal.

References

1. Reul, J.M.H.M., and E.R. de Kloet, Two receptor systems for corticosterone in rat brain: microdistribution and differential occupation, *Endocrinology* 117: 2505-2511, 1985.
2. McEwen, B.S., E.R. de Kloet, and W. Rostene, Adrenal steroid receptors and actions in the nervous system, *Physiol Rev* 66: 1121-1188, 1986.
3. de Kloet, E.R., and J.M.H.M. Reul, Feedback action and tonic influence of corticosteroids on brain function: a concept arising from the heterogeneity of brain receptor systems, *Psychoneuroendocrinology* 12: 83-105, 1987.
4. Funder J.W., and K. Sheppard, Adrenocortico steroids and the brain, *Ann Rev Physiol* 49: 397-411, 1987.
5. McEwen, B.S., J.M. Weiss, and L.S. Schwartz, Selective retention of corticosterone by limbic structures in rat brain, *Nature* 220: 911-912, 1968.
6. McEwen, B.S., E.R. de Kloet, and G. Wallach, Interaction *in vivo* and *in vitro* of corticoids and progesterone with cell nuclei and soluble macromolecules from rat brain region and pituitary, *Brain Res* 105: 129-136, 1976.
7. Gerlach, J.L., and B.S. McEwen, Rat brain binds adrenal steroid hormone: autoradioautography of hippocampus with corticosterone, *Science* 175: 1133-1136, 1972.
8. Stumpf, W.E., and W. Sar, Anatomical distribution of corticosterone concentrating neurons in rat brain, In W.E. Stumpf and L.A. Grant (eds) *Anatomical Neuroendocrinology*, Karger, Basel, pp. 254-261, 1975.
9. McEwen, B.S., B.S. Stephenson, and L.C. Krey, Radioimmunoassay of brain tissue and cell nuclear corticosterone, *J Neurosci Meth* 3: 57-65, 1980.
10. Ermisch, A., and H.J. Rühle, Autoradiographic demonstration of aldosterone-concentrating neuron populations in rat brain, *Brain Res* 147: 154-158, 1978.
11. Veldhuis, H.D., C. van Koppen, M. van Ittersum, and E.R. de Kloet, Specificity of the adrenal steroid receptor system in rat hippocampus, *Endocrinology* 110: 2044-2051, 1982.
12. Wrånge, O., and Z.-Y. Yu, Mineralocorticoid receptor in rat kidney and hippocampus: characterization and quantitation by isoelectric focusing, *Endocrinology* 113: 243-250, 1983.
13. Coirini, H., E.T. Marusic, A.F. DeNicola, T.C. Rainbow, and B.S. McEwen, Identification of minerocorticoid binding sites in rat brain by competition studies and density gradient centrifugation, *Neuroendocrinology* 37: 354-360, 1983.
14. Krozowski, Z.S., and J.W. Funder, Renal minerocorticoid receptors and hippocampal corticosterone binding species have identical intrinsic steroid specificity, *Proc Natl Acad Sci USA* 80: 6056-6060;, 1983.
15. Arriza, J.L., C. Weinberger, G. Cerelli, T.M. Glaser, B.L. Handelin, D.E. Housman, and R.M. Evans, Cloning of human mineralocorticoid receptor complementary DNA: structural and functional kinship with the glucocorticoid receptor, *Science* 237: 268-275, 1987.
16. de Kloet, E.R., G. Wallach, and B.S. McEwen, Differences in corticosterone and dexamethasone binding in rat brain and pituitary, *Endocrinology* 96: 598-609, 1975.
17. Rees, H.D., W.E. Stumpf, and M. Sar, Autoradiographic studies with [^{3}H]-dexamethasone in the rat brain and pituitary, In W.E. Stumpf and L.D. Grant (eds) *Anatomical Neuroendocrinology*, Karger, Basel, pp. 262-269, 1975.
18. Moguilewsky, M., and J.P. Raynaud, Evidence for a specific mineralocorticoid receptor in rat pituitary and brain, *J Ster Biochem* 12: 309-314, 1980.
19. Philibert, D., and M. Moguilevsky, RU 28362, a useful tool for the characterization of glucocorticoid and mineralocorticoid receptors, *Proceeding of the 65th Annual Meeting of The Endocrine Society USA*, San Antonio, Texas, pp. 335, Abstract 1018, 1983.
20. Sarrieau, A., M. Vial, D. Philibert, and W. Rostene, *In vitro* autoradiographic localization of [^{3}H]-corticosterone binding sites in rat hippocampus, *Eur J Pharmacol* 98: 151-152, 1984.
21. Reul, J.M.H.M., and E.R. de Kloet, Anatomical resolution of two types of corticosterone receptor sites in rat brain with *in vitro* autoradiography and computerized image analysis, *J Ster Biochem* 24: 269-272, 1986.
22. Reul, J.M.H.M., F.R. van den Bosch, and E.R. de Kloet, Relative occupation of type-I and type-II corticosteroid receptors in rat brain following stress and dexamethasone treatment: functional implication,

J Endocrinol 115: 459-467, 1987.

23. Fuxe, K., A.C. Wikström, S. Ökret, Z-Y. Yu, L. Granholm, M. Zoli, W. Vale, and J-Å Gustafsson, Mapping of glucocorticoid receptor immunoreactive neurons in the rat tel- and diencephalon using a monoclonal antibody against rat liver glucocorticoid receptors, *Endocrinology* 117: 1803-1812, 1985.

24. Fuxe, K., A. Häfstrand, L.F. Agnati, Z-Y. Yu, A. Cintra, A-C. Wikström, S. Ökret, E. Cantoni, and J-Å Gustafsson, Immunocytochemical studies on the localization of glucocorticoid receptor immunoreactive nerve cells in the lower brain stem and spinal cord of the male rat using a monoclonal antibody against rat liver glucocorticoid receptor, *Neurosci Lett* 60: 1-6, 1985.

25. Agnati, L.F., K. Fuxe, Z.-Y. Yu, A. Häfstrand, S. Ökret, A.C. Wikström, M. Goldstein, M. Zoli, W. Vale, and J-Å Gustafsson, Morphometrical analysis of the distribution of corticotrophin releasing factor, glucocorticoid receptor and phenylethanolamine-N-methyltransferase immunoreactive structures in the paraventricular hypothalamic nucleus of the rat, *Neurosci Lett* 54: 147-152, 1985.

26. van Eekelen, J.A.M., J.Z. Kiss, H.M. Westphal, and E.R. de Kloet, Immunocytochemical study on the intracellular localization of the type 2 glucocorticoid receptor in rat brain, *Brain Res* 436: 120-128, 1987.

27. Hoolenberg, S.W., C. Weinberger, E.S. Ong, G. Cerelli, A. Oro, R. Lebo, E.B. Thompson;, M.G. Rosenfeld, and R.M. Evans, Primary structure and expression of a functional human glucocorticoid receptor cDNA, *Nature* 318: 635-641, 1985.

28. Meisfeld, R., S. Rusconi, P.J. Godowski, B.A. Maler, S. Ökret, A-C. Wikström, J-Å Gustafsson, and K.R. Yamamoto, Genetic complementation of a glucocorticoid receptor deficiency by expression of cloned receptor cDNA, *Cell* 46: 389-399, 1985.

29. Evans, R.M., The steroid and thyroid hormone receptor superfamily, *Science* 240: 889-895, 1988.

30. Reul, J.M.H.M., P.T. Pearce, J.W. Funder, and Z.S. Krozowski, Type-I and type-II corticosteroid receptor gene expression in the rat: effect of adrenalectomy and dexamethasone administration, *Mol Endocrinol* 3: 1674-1680, 1989.

31. Arriza, J.L., R.B. Simerly, L.W. Swanson, and R.M. Evans, The neuronal mineralocorticoid receptor as a mediator of glucocorticoid response, *Neuron* 1: 887-900, 1988.

32. van Eekelen, J.A.M., W. Jiang, E.R. de Kloet, and M.C. Bohn, Distribution of the mineralocorticoid and the glucocorticoid receptor mRNAs in the rat hippocampus, *J Neurosci Res* 21: 88-94, 1988.

33. Aronsson, M., K. Fuxe, Y. Dong, L.F. Agnati, S. Ökret, and J-Å Gustafsson, Localization of glucocorticoid receptor mRNA in the male rat brain by *in situ* hybridization, *Proc Natl Acad Sci USA* 85: 9331-9335, 1988.

34. Kiss, J.Z., J.A.M. Eekelen, J.M.H.M. Reul, H.M. Westphal, and E.R. de Kloet, Glucocorticoid receptor in magnocellular neurosecretory cells, *Endocrinology* 122: 444-449, 1988.

35. Funder, J.W., P.T. Pearce, R. Smith, and A.I. Smith, Mineralocorticoid action: target tissue specificity is enzyme, not receptor, mediated, *Science* 242: 583-585, 1988.

36. Edwards, C.R.W., P.M. Stewart, D. Burt, L. Brett, M.A. McIntyre, W.S. Sutanto, E.R. de Kloet, and C. Monder, Localization of 11β-hydroxysteroid dehydrogenase—tissue specific protector of the mineralocorticoid receptor, *The Lancet II* October 29, 1988, pp. 986-989.

37. de Kloet, E.R., Th.A.M. Voorhuis, J.L.M. Leunissen, and B. Koch, Intracellular CBG-like molecules in the rat pituitary, *J Ster Biochem* 20: 367-371, 1984.

38. McEwen, B.S., L.T. Lambdin, T.C. Rainbow, and A.F. de Nicola, Aldosterone effects on salt appetite in adrenalectomized rats, *Life Sci* 43: 38-43, 1986.

39. Rousseau, G., J.D. Baxter, J.W. Funder, I.S. Edelman, and G.M. Tomkins, Glucocorticoid and mineralocorticoid receptors for aldosterone, *J Ster Biochem* 3: 219-227, 1972.

40. Funder, J.W., D. Feldman, and I.S. Edelman, Specific aldosterone binding in rat kidney and parotid, *J Ster Biochem* 3: 209-218, 1972.

41. Beaumont, K., and D.D. Fanestil, Characterization of the rat brain aldosterone receptors reveals high affinity for corticosterone, *Endocrinology* 113: 2043-2051, 1983.

42. de Kloet, E.R., D. van den Berg, H. van Dijken, E. van der Peet, W. Sutanto, and W. de Jong, Brain corticosteroid receptors and central cardiovascular control, *Proceedings of the Satellite Symposium of the 8th International Congress of Endocrinology*, Serono Symposia Obligations, In Press.

43. Gomez-Sanchez, E.P., Intracerebroventricular infusion of aldosterone induces hypertension in rats,

Endocrinology 118: 819-823, 1986.

44. van den Berg, D.T.W.M., E.R. de Kloet, H.H. van Dijken, and W. de Jong, Differential central effects of mineralo- and glucocorticoid agonists and antagonists on blood pressure, *Endocrinology*, In Press.

45. Sutanto, W., and E.R. de Kloet, Species-specificity of corticosteroid receptors in hamster and rat brains, *Endocrinology* 121: 1405-1411, 1987.

46. Sutanto, W., J.A.M. van Eekelen, J.M.H.M. Reul, and E.R. de Kloet, Species-species topography of corticosteroid receptors types in rat and hamster brain, *Neuroendocrinology* 47: 398-404, 1988.

47. Reul, J.M.H.M., E.R. de Kloet, and J. Rothuizen, Mineralocorticoid and glucocorticoid receptors in dog brain and pituitary, In Preparation.

48. Schindler, W.J., and K.M. Knigge, Adrenal secretion by the Golden hamster, *Endocrinology* 65: 739-747, 1959.

49. Schindler, W.J., and K.M. Knigge, *In vitro* studies and adrenal steroidogenesis by the Golden hamster, *Endocrinology* 65: 748-765, 1959.51.

50. Whitehouse, B.J., and G.P. Vinson, Specific variation in steroid biosynthetic pathways: the formation of cortisol in hamster adrenal tissue *in vitro*, *J Ster Biochem* 2: 307-312, 1971.

51. Dunlap, N.E., and W.E. Grizzle, Golden Syrian Hamsters: a new experimental model for adrenal compensatory hypertrophy, *Endocrinology* 114: 1490-1495, 1984.

52. Reul, J.M.H.M., F.R. van den Bosch, and E.R. de Kloet, Differential response of type-I and type-II corticosteroid receptors to changes in plasma steroid levels and circadian rhythmicity, *Neuroendocrinology* 45: 407-412, 1987.

53. McIntyre, W.R., and H.H. Samuels, Triamcinolone acetonide regulates glucocorticoid receptor levels by decreasing the half-life of the activated nuclear receptor form, *J Biol Chem* 260: 418-427, 1985.

54. Dong, Y., L. Poellinger, J-Å Gustafsson, and S. Ökret, Regulation of glucocorticoid receptor expression: evidence for transcriptional and posttranslational mechanisms, *Mol Endocrinol* 2: 1256-1264, 1988.

55. Sapolsky, R.M., L.C. Krey, and B.S. McEwen, Glucocorticoid-sensitive hippocampal neurons are involved in terminating the adrenocortical stress response, *Proc Natl Acad Sci USA* 81: 6174-6177, 1984.

56. Sapolsky, R.M., L.C. Krey, and B.S. McEwen, The neuroendocrinology of stress and aging: the glucocorticoid cascade hypothesis, *Endocr Rev* 7: 284-301, 1986.

57. Aus der Muhlen, K.A., and H. Ockenfels, Morphologische veränderungen im diencephalon und telencephalon: störungen des regelkreises adenohypophyse-nebennierenrinde, *Zeitung der Zellforschung* 93: 126-141, 1969.

58. Sachar, E.J., L. Hellman, H.P. Roffwarg, F.S. Halpern, D.K. Fukushima, and T.F. Gallagher, Disrupted 24-hour patterns of cortisol secretion in Psychotic depression, *Arch Gen Psychiatry* 28: 19-24, 1973.

59. Von Zerssen, D., Mood and behavioural changes under corticosteroid therapy, <u>In</u> T.M. Itil, G. Laudahn, and W.M. Herrmann (eds) *Psychotropic Action of Hormones* Spectrum, New York, pp. 195-222, 1976.

60. Carroll, B.J., M. Feinberg, J.F. Greden, J. Tarika, A.A. Albala, R.F. Haskett, N.M. James, Z. Kronfol, N. Lohr, M. Steiner, J.P. de Vigne, and E. Young, A specific laboratory test for the diagnosis of melancholia: standardization, validation and clinical utility, *Arch Gen Psychiatry* 38: 15-22, 1981.

61. Carroll B.J., The dexamethasone suppression test for melancholia, *Br J Psychiatry* 140: 292-304, 1982.

62. Reul, J.M.H.M., J.A.D.M. Tonnaer, and E.R. de Kloet, Neurotrophic ACTH analogue promotes plasticity of type-I corticosteroid receptor in brain of senescent male rats, *Neurobiol Aging* 9: 253-260, 1988.

63. de Wied, D., and J. Jolles, Neuropeptides derived from pro-opiocortin: behavioural, psychological, and neurochemical effects, *Physiol Rev* 62: 976-1059, 1982.

64. de Koning, P., and W.H. Gispen, A rationale for the use of melanocortins neural injury, <u>In</u> D.G. Stein and B.A. Sabel (eds) *Pharmacological Approaches to the Treatment of Brain and Spinal Cord Injuries*, Plenum Press, New York, pp. 233-258, 1988.

65. de Kloet, E.R., J.M.H.M. Reul, F.R. van den Bosch, J.A.D.M. Tonnaer, and H. Saito, Ginsengoide RG_1 and corticosteroid receptors in rat brain, *Endocrinol Jpn* 34: 213-220, 1987.

66. McIntyre, D.C., Adrenalectomy: protection from kindled emulsion induced amnesia in rats, *Physiol Behav* 17: 789-795, 1976.

67. Veldhuis, H.D., E.R. de Kloet, I. van Zoest, and B. Bohus, Adrenalectomy reduces exploratory activity in the rat: a specific role of corticosterone, *Horm Behav* 16: 191-198, 1982.

68. Jhanwar-Uniyal, M., C.R. Roland, and S.F. Leibowitz, Diurnal rhythm of α_2-noradrenergic receptors in the paraventricular nucleus and other brain areas: relation to circulating corticosterone and feeding behavior, *Life Sci* 38: 473-482, 1986.

69. Dana, R.C., and J.L. Martinez, Effect of adrenalectomy on the circadian rhythm of LTP, *Brain Res* 308: 392-396, 1984.

70. Barnes, C.A., B.L. McNaughton, G.V. Goddard, R.M. Douglas, and R. Adamec, Circadian rhythm of synaptic excitability in rat and monkey central nervous system, *Science* 197: 91-92, 1977.

71. Micco D.J., J.S. Meyer, and B.S. McEwen, Effects of corticosterone replacement on the temporal patterning of activity and sleep in adrenalectomized rats, *Brain Res* 200: 206-212, 1980.

72. Bohus, B., and E.R. de Kloet, Adrenal steroids and extinction behaviour: antagonism by progesterone, deoxycorticosterone and dexamethasone of a specific effect of corticosterone, *Life Sci* 28: 433-440, 1981.

73. Munck, A., P.M. Guyre, and N.J. Holbrook, Physiological function of glucocorticoids in stress and their relation to pharmacological action, *Endocr Rev* 5: 25-44, 1984.

74. Dallman, M.F., S.F. Akana, C.S. Cascio, D.N. Darlington, L. Jacobson, and N. Levin, Regulation of ACTH secretion: variations on a theme of B, In J.H. Clark (ed) *Recent Progress in Hormone Research*, Academic Press, Orlando, Volume 43: 113-173, 1987.

75. Ratka, A., W. Sutanto, M. Bloemers, and E.R. de Kloet, On the role of brain mineralocorticoid (type-I) and glucocorticoid (type-II) receptors in neuroendocrine regulation, *Neuroendocrinology* 50: 117-123, 1989.

EFFECTS OF THYROID HORMONES ON THE HYPOTHALAMIC DOPAMINERGIC NEURONS

M.J. Reymond and T. Lemarchand-Béraud

Division of Endocrinology, Department of Internal Medicine
C.H.U.V. 1011 Lausanne, Switzerland

On the basis of numerous studies on structural, biochemical, and functional impairments of the central nervous system in instances of fetal or neonatal thyroid deficiency (1-4), it is now acknowledged that thyroid hormones play an important role in brain development. Whereas, trophic effects of the thyroid hormones on brain morphogenesis are generally recognized during fetal and neonatal periods of life, the influence of thyroid hormones on the adult brain is less documented. Yet, functional disturbances—and to a lesser extent morphological impairments—occur in the brain of hypothyroid adult animals and humans. Most of these alterations are reversible through substitutive treatments with thyroid hormones, supportive of the view that thyroid hormones have subtle modulatory effects in the adult central nervous system: effects that may greatly differ from one brain region to another one.

In the present paper, we shall focus on the actions of thyroid hormones in the hypothalamus, and more particularly on the hypothalamic-dopaminergic (DA) neurons, during fetal/neonatal life as well as adult life in the rodent. A review of the data reported in the literature regarding the effects of the thyroid hormones on the hypothalamic DA neurons during the fetal and neonatal periods will be presented as well as our data concerning thyroid hormone regulation of the tuberoinfundibular dopaminergic (TIDA) neurons in adult rats. Finally, a short review will be presented of recent data published on the influence of thyroid hormones on the hypothalamic TRH-secreting neurons that present notable analogies with the hypothalamic DA neurons in their regulation by thyroid hormones, both in fetal and neonatal life as well as adult life.

Effects of Thyroid Hormones on the Hypothalamic DA Neurons During Fetal Life or Neonatal Period

The ontogenesis of the hypothalamic DA neurons in the rodent has been studied by several authors using histofluorescent (5-7) and lately immunohistochemical techniques (8-10). In the rat hypothalamus, a few tyrosine hydroxylase (TH)-containing cells, presumably DA neurons, are detected as early as fetal day 13 (9). The morphological differentiation of these neurons from the 13th fetal day until the 9th postnatal day has been beautifully described by Ugrumov *et al.* (9,10). Ramifications of the nerve processes and increases in the size of the TH-positive perikarya are particularly noted from the 18th fetal day, which corresponds to the day of onset of thyroid function in the fetus rat (11,12). The fetal thyroid gland is the main source of thyroid hormones in the fetus, although evidence has been presented for a maternal contribution before onset of fetal thyroid function (13-15).

Circulating Regulatory Factors and Neuroendocrine Function
Edited by J. C. Porter and D. Ježová
Plenum Press, New York, 1990

The later appearance in the fetus of thyroid function than that of the hypothalamic neurons—particularly the TH- as well as the TRH-containing neurons as will be seen subsequently—is suggestive of a role of thyroid hormones on the differentiation or maturation of the hypothalamic neurons rather than on their proliferation. Indeed, the effects of thyroid hormones, mainly triiodothyronine (T3), on the morphological maturation of the hypothalamic DA neurons have been extensively investigated by Puymirat and collaborators under *in vitro* conditions, using cultures of fetal mouse hypothalamic cells (16). T3 enhances cell size, neurite length, and arborization of DA neurons without affecting the cell number (17). The morphological effects of T3 are accompanied by functional effects, *viz.*, increased DA uptake and release by the neurons in the presence of T3 (17,18). The number of nuclear receptors for T3 also increases with age (19), suggestive of an enhanced responsiveness of the fetal DA perikarya to the trophic effects of thyroid hormones.

When the data obtained *in vitro* by Puymirat and collaborators are related to those reported by Ugrumov *et al.* (9,10) on the morphological ontogenesis of the DA neurons, it can be envisioned that the marked morphological differentiation of the DA nerve processes observed by the latter authors from the 18th fetal day—with prolongation of the processes to the internal zone of the median eminence, superficial layer, and later primary portal plexus—is influenced by thyroid hormones. It is worth mentioning here that T3 has not been found to influence synaptogenesis of fetal hypothalamic neurons whereas other factors, in particular polyunsaturated fatty acids, appear to play a key role in the maturation process (20). Thus, trophic effects of thyroid hormones, mainly T3, have been clearly established on the morphological maturation of the hypothalamic DA neurons, with associated effects on the functional properties of these neurons.

Effects of Thyroid Hormones on the Hypothalamic DA Neurons During Adult Life

Regulatory effects of thyroid hormones on the activity of hypothalamic neurons in adult life have long been suspected but no definite data were obtained. Yet, the presence in the hypothalamus of adult rats of a large number of T3 receptors (21-23) as well as 5'-deiodinase—the specific enzyme for thyroid hormone metabolism—which is modulated by the thyroid state (24-27), is supportive of the existence of such effects.

In regard to the TRH-secreting neurons, numerous investigations have been conducted to detect possible regulatory effects exerted by thyroid hormones, but methodological difficulties in the evaluation of TRH neuronal activity, including TRH biosynthesis in the neurons and TRH secretion into hypophysial portal circulation, prevented until recently conclusive studies of the regulation by thyroid hormones of the TRH-secreting neurons in the adult rat. For the hypothalamic DA neurons, scarce data were available until lately in favor of an influence of the thyroid state on these neurons in the adult rat. Although increased activity of TH and enhanced turnover of DA were observed in the median eminence of hypothyroid rats by some (28-30), contradictory data were, however, reported by others (31-33). Yet, reduced plasma levels of prolactin (PRL) were often reported in hypothyroid adult rats (30,34,35).

To address the issue of a possible regulation by thyroid hormones on the activity of TIDA neurons, we used methodologies allowing *in vivo* evaluation of the secretory activity of the TIDA neurons: *viz.*, the biosynthetic activity of the neurons, which was estimated by DOPA accumulation in the median eminence after inhibition of DOPA decarboxylase activity (36,37), and the *in vivo* releasing activity of the neurons, which was estimated by DA concentration in hypophysial portal plasma (38,39). These two parameters, as well as DA accumulation in the adenohypophysis and PRL concentration in the adenohypophysis and in arterial plasma, were determined in adult male rats that

Table 1
Plasma T3, T4, and TSH Concentrations in Adult Male
Rats in Various Thyroid States

Treatment	N	T3 (ng/ml)	T4 (ng/ml)	TSH (ng/ml)
Sham TX	13	0.70 ± 0.04	29.5 ± 2.0	1.5 ± 0.3
TX	13	$0.33 \pm 0.03^*$	$4.0 \pm 1.1^*$	$9.4 \pm 0.7^*$
TX + T4	11	$0.91 \pm 0.05\dagger$	$54.1 \pm 3.8\dagger$	$0.1 \pm 0.02\dagger$
Intact T4	13	$0.80 \pm 0.03^*$	$55.4 \pm 2.9^*$	$0.2 \pm 0.03^*$

Values shown are mean ± SE.
$^*P < 0.001$ vs. sham TX.
$\dagger P < 0.001$ vs. TX. Reproduced from Reymond et al. (40) with permission of S. Karger, AG.

were either sham operated (sham TX), thyroidectomized (TX), TX and treated with T4 (20 µg/kg body weight × day for 7 days), or in intact rats treated with T4 (20 µg/kg body weight × day for 7 days). Additional control groups were included in the original study (40), but here we will present only the data obtained in the four representative groups of animals.

The plasma levels of T3, T4, and TSH determined in these animals are shown in

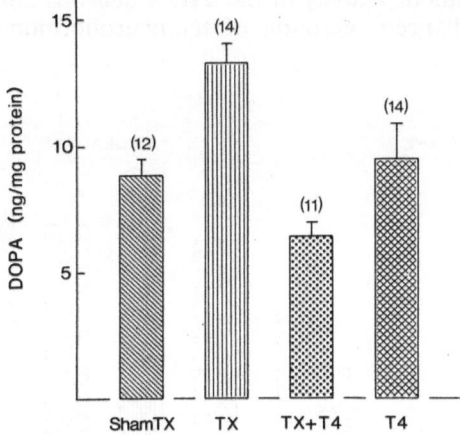

Figure 1. Dopamine synthesis in the TIDA neurons of adult male rats in various thyroid states. Rats were thyroidectomized (TX) or sham operated (sham TX) 15 days before the experiment. A group of thyroidectomized rats was treated with T4 (20 µg/kg BW × day, sc) for 7 days (TX + T4). Intact rats (T4) received a similar T4 treatment. Dopamine synthesis is estimated by DOPA concentration in the median eminence of the rats treated for 30 min with an inhibitor of DOPA decarboxylase, NSD 1015 (100 mg/kg BW, ip). The bars and vertical lines denote mean ± SE. The number of animals is shown in parentheses. $P < 0.001$, TX vs. sham TX, TX + T4 vs. TX; $P < 0.02$, TX + T4 vs. sham TX. Drawn from Reymond et al. (40) with permission of S. Karger AG.

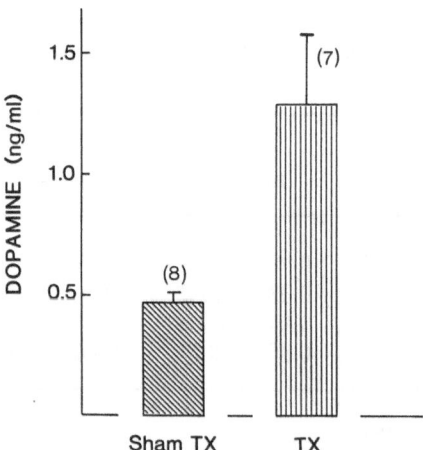

Figure 2. Dopamine concentration in hypophysial portal plasma of sham TX and TX rats. See Figure 1 for further details. $P < 0.02$, TX *vs.* sham TX.

Table 1; these values verify the various treatments of the animals. The biosynthetic activity of the TIDA neurons was found to be clearly increased in the hypothyroid (TX) rats, the increase being abolished by treatment of the TX rats with T4 (Figure 1). No effect of T4 treatment alone was observed on DA synthesis in the TIDA neurons, but it is noteworthy that the dose of T4 used (20 µg/kg body weight × day) was a substitutive dose, and not a dose inducing a chronic hyperthyroid state.

The increased biosynthetic activity of the TIDA neurons observed in the TX rats was associated with an enhanced secretion of the neurohormone into the hypophysial

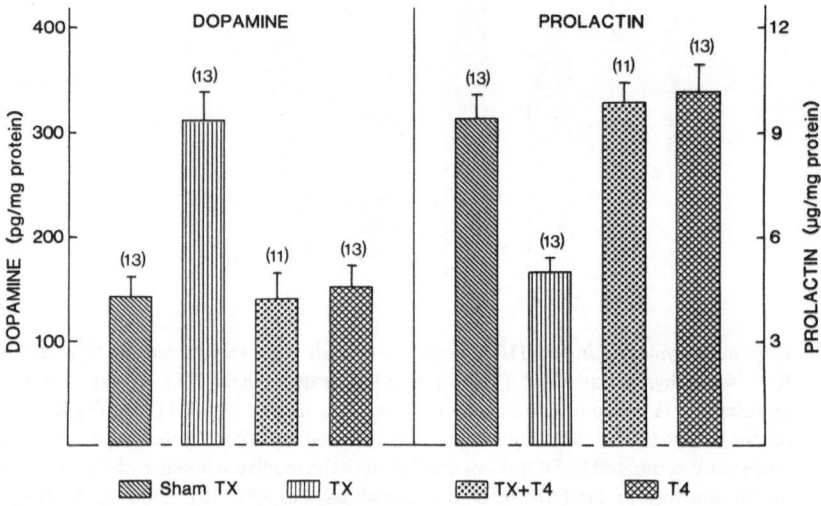

Figure 3. Dopamine (*left panel*) and prolactin (*right panel*) concentrations in the adenohypophysis of adult male rats in various thyroid states. See Figure 1 for further details. $P < 0.001$, TX *vs.* sham TX; $P < 0.005$, TX + T4 *vs.* TX.

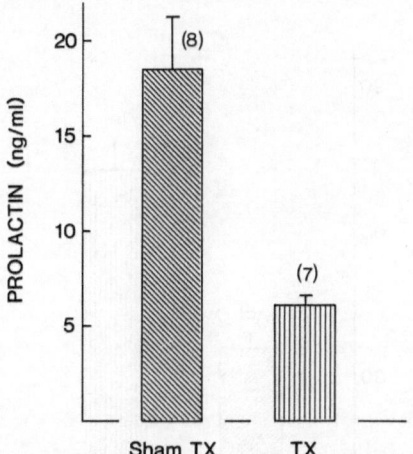

Figure 4. Prolactin concentration in plasma of blood collected from a femoral artery in sham TX and TX rats. See Figure 1 for further details. $P < 0.002$, TX vs. sham TX.

portal circulation, as illustrated in Figure 2. The latter finding has been recently confirmed by two groups of invesigators (41,42). In addition, DA was found in greater concentration in the adenohypophysis of TX than of sham TX rats, which is in accord with the greater secretion of DA into the hypophysial portal circulation, and the pituitary accumulation of DA observed in TX rats was corrected by treatment with T4 (Figure 3). As regards PRL secretion, a marked reduction in the adenohypophysial production and secretion was observed in the TX rats (Figures 3 and 4), which was also corrected by T4 treatment.

The influence of thyroid hormones on the biosynthetic activity of TIDA neurons has been further investigated by Wang *et al.* (43), by measurement of TH mass and *in situ*

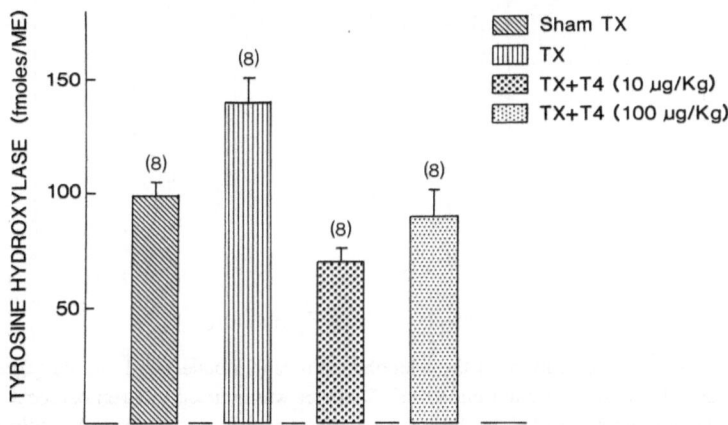

Figure 5. Mass of tyrosine hydroxylase in the median eminence of adult male rats in various thyroid states. $P < 0.01$, TX vs. sham TX, TX + T4 (10 μg/kg BW × day) vs. TX. Drawn from Wang *et al.* (43) with permission of S. Karger AG.

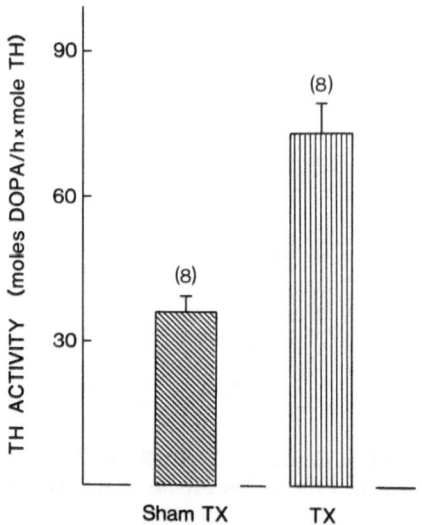

Figure 6. *In situ* molar activity of tyrosine hydroxylase in the median eminence of sham TX and TX rats. $P < 0.01$, TX *vs.* sham TX rats. Drawn from Wang *et al.* (43) with permission of S. Karger AG.

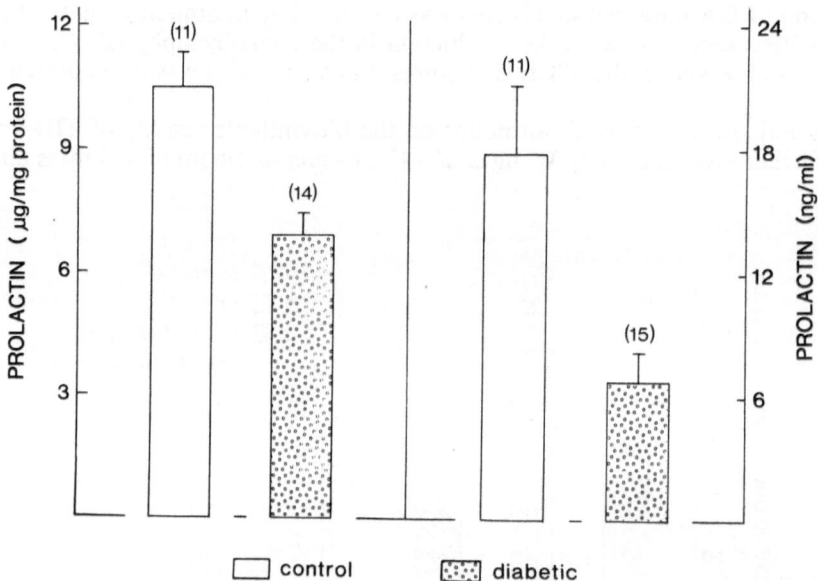

Figure 7. Prolactin concentrations in the adenohypophysis (*left panel*) and plasma (*right panel*) of control and diabetic adult male rats. Diabetes was induced by streptozotocin (40 mg/kg BW, iv) one month before the measurements; control rats received the solvent vehicle. Prolactin was reduced in the adenohypophysis ($P < 0.02$) and in the plasma ($P < 0.005$) of diabetic compared to control rats.

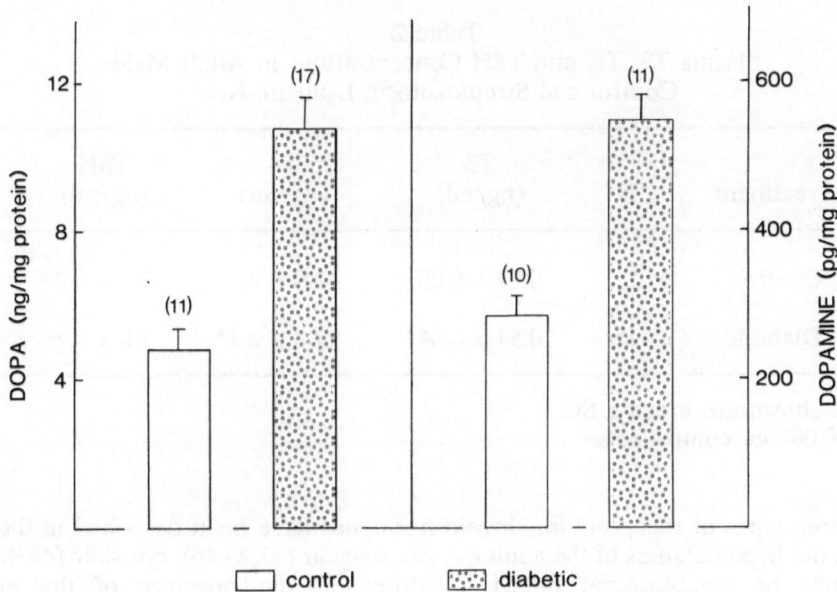

Figure 8. Dopamine synthesis in the TIDA neurons (*left panel*) and dopamine accumulation in the adenohypophysis (*right panel*) of control diabetic rats. See Figure 7 for further details. $P < 0.001$, diabetic *vs.* control rats.

molar activity in the median eminence. A net increase in the mass of TH was observed in the median eminence of the TX rats compared to the sham-operated controls (Figure 5). The enhancement was prevented by substitutive treatment with T4, indicating an inhibitory effect of thyroid hormones on the synthesis of TH in the TIDA neurons. Morever, upon evaluation of the *in situ* molar activity of TH in the median eminence, a 100% increase in the activity of TH was found in the TX rats (Figure 6), suggestive of more active form(s) of TH in the TIDA neurites of the hypothyroid than euthyroid rats. The possibility that increased phosphorylation of TH was responsible for the greater TH activity in the TX rats was investigated, using a procedure previously described (44); however, under the conditions of the analysis, no evidence was found for an increased phosphorylation of TH in TX compared to sham TX rats (unpublished observations).

These data are therefore illustrative of an inhibitory effect by thyroid hormones on the synthesis of TH in the TIDA neurons of adult rats, with an additional effect on the catalytic activity of TH in these neurons (43). Further studies are needed to elucidate the mechanisms underlying the effects of thyroid hormones on the synthesis and activation of TH. Such effects are expected to occur at transcriptional as well as at translational level of TH gene expression, but to our knowledge, no data are as yet available pertaining to such regulations. A transcriptional regulation of TH gene expression in the rat TIDA neurons has been demonstrated after estrogen treatment (45); it could be enviewed that thyroid hormones exert analogous regulatory effects on the TIDA neurons.

At this point it is tempting to speculate that in adult hypothalamic DA neurons the regulatory effects of thyroid hormones on TH synthesis are exerted via nuclear receptors, thyroid hormones being considered as genomic factors; whereas, the effects of thyroid hormones on the enzymatic activity of TH would be non-genomic, *i.e.*, exerted outside the nucleus, possibly after binding of thyroid hormones to membrane or cytosolic receptors.

Table 2
Plasma T3, T4, and TSH Concentrations in Adult Male
Control and Streptozotocin-Diabetic Rats

Treatment	N	T3 (ng/ml)	T4 (ng/ml)	TSH (ng/ml)
Control	11	0.89 ± 0.05	49.0 ± 2.4	2.8 ± 0.2
Diabetic	15	$0.54 \pm 0.04^*$	$27.8 \pm 2.1^*$	$1.6 \pm 0.2^*$

Values shown are mean $\pm$ SE.
$^*P < 0.001$ *vs.* control rats.

Indeed, three types of receptors for thyroid hormones have been described in the brain or even in the hypothalamus of the adult rat, *viz.*, nuclear (21,23,46), cytosolic (46-48), and membranous or synaptosomal (49-51). Moreover, the presence of the enzyme, $5'$-deiodinase, in the hypothalamus and particularly in the median eminence of adult rats may be suggestive of direct modulatory effects exerted by thyroid hormones at the synaptic or nerve terminal level, especially in the median eminence.

Activity of the TIDA Neurons in Adult Streptozotocin-Diabetic Rats: Regulation by Thyroid Hormones?

In previous studies on the neuroendocrine morphological and functional disturbances observed in experimental diabetes induced by streptozotocin (SZ) in adult rats, we have documented a severe hypothyroidism in these diabetic rats, with markedly reduced secretions of T3, T4, and TSH (Table 2), as well as a blunted TSH response to TRH (52). These parameters are indicative of a hypothyroidism of pituitary or central origin in the SZ-diabetic rats, contrary to the hypothyroidism observed in thyroidectomized (TX) rats, characterized by an enhanced TSH secretion. SZ-diabetic rats disclosed reduced secretion of PRL, as found in the TX rats, with decreased levels of PRL in the adenohypophysis and in plasma (Figure 7). We, therefore, evaluated the *in vivo* activity of the TIDA neurons in the SZ-diabetic rats and demonstrated an enhanced activity of these neurons, as revealed by an increased synthesis of DOPA in the neurons of the diabetic compared to control rats (Figure 8, *left panel*); moreover, a greater concentration of DA was found in the adenohypophysis of the diabetic rats (Figure 8, *right panel*).

In view of these findings, we asked the following question: Is the increased activity of the TIDA neurons observed in the SZ-diabetic rats due to the severe hypothyroidism observed in these animals, and, it so, can it be corrected by a substitutive treatment with thyroid hormones? To address these issues, we treated SZ-diabetic rats for 7 days with a substitutive dose of T4 (20 μg/kg body weight × day) as previously done with the TX rats (see above) or with T3 (4 μg/kg body weight × day). Treatment with T3 was tested in addition to T4, since impairment in the deiodination of T4 has been demonstrated in SZ-diabetic rats (53). We observed that the increased activity of the TIDA neurons detected in the SZ-diabetic rats was not simply caused by hypothyroidism (54). We presently have no definite explanation for the mechanisms underlying the hyperactivity of the TIDA neurons in the SZ-diabetic rats. Indeed, we have previously obtained evidence for a reduction—and not an increase—of the neuronal activity of two other hypothalamic neuronal systems, *viz.*, the LHRH- and TRH-secreting neurons, in the SZ-diabetic rats (55-57). The isolated finding of an increased activity of a hypothalamic neuronal system—the TIDA neurons—in the SZ-diabetic rats, therefore, led us to suggest that it was an indirect consequence of diabetes, via the associated hypothyroidism, rather than a direct sequel, but this hypothesis does not seem to be valid. It is noteworthy that Joanny and collaborators (58) recently obtained evidence for an increased hypothalamic secretion of somatostatin in adult SZ-diabetic rats, as observed *in vitro* by an enhanced release of somatostatin by fragments of mediobasal hypothalamus as well as *in vivo* by greater levels of somatostatin in hypophysial portal plasma. These findings pertaining to somatostatin are, therefore, when considered together with our data on the TIDA neurons, supportive of the view that the neuroendocrine changes observed in the SZ-diabetic rats are complex and diverse, including both impairments and enhancements of neuronal activities in the hypothalamus. In this respect, factors related to metabolism or nutrition are expected to play an important role in the regulation of hypothalamic neuroendocrine functions; in particular, hypoglycemia and hyperglycemia are now recognized as influencing hypothalamic somatostatin as well as CRH secretion (58).

Effects of Thyroid Hormones on the Neuroendocrine Hypothalamic TRH-Secreting Neurons: Analogies with the TIDA Neurons

Whereas a trophic role for thyroid hormones on the morphological maturation of the hypothalamic DA neurons during fetal/neonatal life has been well documented (see above), less is known regarding the hypothalamic TRH neurons. Indeed, to our knowledge, no data are available regarding a determinant role of thyroid hormones on the morphological differentiation or maturation of the hypothalamic TRH neurons, except for the report of a stimulatory role of T3, in combination with corticosterone and polyunsaturated fatty acids, on the *in vitro* release of TRH by mouse fetal hypothalamic cells (59). Studies of the ontogenesis of the TRH neurons in the rat hypothalamus have demonstrated the presence of TRH in the fetus (60-62), as early as on the 12th fetal day (63). A role for the thyroid hormones on the maturation of the TRH neurons can therefore be considered as analogous to what has been reported for the DA neurons.

As for effects of thyroid hormones on the TRH neurons during adult life, an inhibitory feedback regulation—long suspected but hampered by methodological difficulties—has recently been demonstrated, using molecular biology technologies. Evidence for a direct cell-specific regulation of TRH gene expression in the hypothalamic paraventricular nucleus has been obtained, hypothyroidism being associated with an increase in proTRH mRNA as well as in proTRH, and substitutive treatment with T3 abolishing these effects (64-67). The inhibitory effect of thyroid hormones on the mRNA of TRH has been clearly demonstrated in intact rats treated with thyroid hormones, and it has been shown that the modulatory effects of T3 or T4 on TRH mRNA are independent of the pituitary gland and of TSH (67).

Moreover, TRH secretion into the hypophysial portal circulation has been lately shown to be regulated by thyroid hormones, an increase and a reduction in TRH levels in portal plasma being observed in hypothyroid and in hyperthyroid rats, respectively (42), in accordance with the regulation of TRH synthesis observed in the paraventricular nucleus. Previous studies on TRH in hypophysial portal blood had failed to disclose any significant variations of TRH secretion with the thyroid state (68,69).

In conclusion, it appears that thyroid hormones exert trophic effects on the maturation of hypothalamic neuroendocrine neurons—at least the DA and TRH neurons—during fetal/neonatal life, and that thyroid hormones disclose regulatory effects on the secretory activity of these neurons during adult life. It can, therefore, be inferred that analogous regulatory effects of thyroid hormones are exerted on other hypothalamic secretory neurons. To our knowledge no direct evidence for such effects has been obtained regarding the CRH-, the GHRH- or the somatostatin-secreting neurons; indirect evidence for an influence of thyroid hormones on GHRH neurons in adult rats has been proposed on the basis of the observation of an abolished pulsatile secretion of GH in adult thyroidectomized rats (70), with a reduction in GHRH but not in somatostatin hypothalamic content (71). An influence of thyroid hormones on the hypothalamic secretion of somatostatin is still debated, contrasting data being reported under *in vivo* (72) and *in vitro* conditions (73). Further studies will certainly provide new insights into the nature of the complex trophic and regulatory effects played by thyroid hormones in the neuroendocrine systems.

Acknowledgements

The authors thank Wilma Benotto and Marie-Claude Evraere for excellent technical and editorial assistance. Funding to support the studies reported in this paper was provided by grants 3.550-0.83 and 3.394-0.86 from the Swiss National Science Foundation.

References

1. Legrand, J., Hormones thyroidiennes et maturation du systeme nerveux, *J Physiologie (Paris)* 78: 603-652, 1982/1983.
2. Nemeskeri, A., D. Grouselle, A. Faivre-Bauman, and A. Tixier-Vidal, Developmental changes of thyroliberin (TRH) in the rat brain, *Neurosci Lett* 53: 279-284, 1985.
3. Nunez, J., Thyroid hormones, In A. Lajtha (ed) *Handbook of Neurochemistry*, Volume 8, Plenum Press, New York, pp. 1-28, 1984.
4. Timiras, P.S., Thyroid hormones and the developing brain, In E. Meisami and P.S. Timiras (eds) *Handbook of Human Growth and Developmental Biology, Volume 1: Neural, Sensory and Integrative Development, Part C: Factors Influencing Brain Development*, CRC Press, Boca Raton, pp. 59-82, 1988.
5. Hyyppä M., A histochemical study of the primary catecholamines in the hypothalamic neurons of the rat in relation to the ontogenetic and sexual differentiation, *Z Zellforsch* 98: 550-560, 1969.
6. Smith, G.C., and R.W. Simpson, Monoamine fluorescence in the median eminence of foetal, neonatal and adult rats, *Z Zellforsch* 104: 541-556, 1970.
7. Loizou, L.A., The postnatal development of monoamine-containing structures in the hypothalamo-hypophyseal system of the albino rat, *Z Zellforsch* 114: 234-252, 1971.
8. Daikoku, S., H. Kawano, Y. Okamura, M. Tokuzen, and I. Nagatsu, Ontogenesis of immunoreactive tyrosine hydroxylase-containing neurons in rat hypothalamus, *Develop Brain Res* 28: 85-98, 1986.
9. Ugrumov, M.V., J. Taxi, A. Tixier-Vidal, J. Thibault, and M.S. Mitskevich, Ontogenesis of tyrosine hydroxylase-immunopositive structures in the rat hypothalamus. An atlas of neuronal cell bodies, *Neuroscience* 29: 135-156, 1989.
10. Ugrumov, M.V., A. Tixier-Vidal, J. Taxi, J. Thibault, and M.S. Mitskevich, Ontogenesis of tyrosine hydroxylase-immunopositive structures in the rat hypothalamus. Fiber pathways and terminal fields, *Neuroscience* 29: 157-166, 1989.
11. Geloso, J.P., and G. Bernard, Effets de l'ablation de la thyroide maternelle ou foetale sur le taux des hormones circulantes chez le foetus de rat, *Acta Endocrinol* 56: 561-566, 1967.
12. Fisher, D.A., J.H. Dussault, J. Sack, and I.J. Chopra, Ontogenesis of hypothalamic-pituitary thyroid function and metabolism in man, sheep, and rat, *Recent Prog Horm Res* 33: 59-107, 1977.
13. Obregon, M.J., J. Mallol, R. Pastor, G. Morreale de Escobar, and F. Escobar del Rey, L-thyroxine and

3,5,3′-triiodo-L-thyronine in rat embryos before onset of fetal thyroid function, *Endocrinology* 114: 305-307, 1984.

14. Morreale de Escobar, G., R. Pastor, M.J. Obregon, and F. Escobar del Rey, Effects of maternal hypothyroidism on the weight and thyroid hormone content of rat embryonic tissues, before and after onset of fetal thyroid function, *Endocrinology* 117: 1890-1900, 1985.

15. Morreale de Escobar, G., M.J. Obregon, and F. Escobar del Rey, Fetal and maternal thyroid hormones, *Hormone Res* 26: 12-27, 1987.

16. Puymirat, J., Effects of dysthyroidism on central catecholaminergic neurons, *Neurochem Intl* 7: 969-977, 1985.

17. Puymirat, J., A. Barret, R. Picart, A. Vigny, C. Loudes, A. Faivre-Bauman, and A. Tixier-Vidal, Triiodothyronine enhances the morphological maturation of dopaminergic neurons from fetal mouse hypothalamus cultured in serum-free medium, *Neuroscience* 10: 801-810, 1983.

18. Puymirat, J., A. Barret, A. Faivre-Bauman, and A. Tixier-Vidal, Biochemical characterization of the uptake and release of [^{3}H] dopamine by dopaminergic hypothalamic neurons: a developmental study using serum-free medium cultures, *Dev Biol* 119: 75-84, 1987.

19. Puymirat, J., and A. Faivre-Bauman, Evolution of triiodothyronine nuclear binding sites in hypothalamic serum-free cultures: evidence for their presence in neurons and astrocytes, *Neurosci Lett* 68: 299-304, 1986.

20. Tixier-Vidal, A., R. Picart, C. Loudes, and A. Faivre-Bauman, Effects of polyunsaturated fatty acids and hormones on synaptogenesis in serum-free medium cultures of mouse fetal hypothalamic cells, *Neuroscience* 17: 115-132, 1986.

21. Schwartz, H.L., and J.H. Oppenheimer, Nuclear Triiodothyronine receptor sites in brain: probable identity with hepatic receptors and regional distribution, *Endocrinology* 103: 267-273, 1978.

22. Dozin, B., and P. De Nayer, Triiodothyronine receptors in adult rat brain: topographical distribution and effect of hypothyroidism, *Neuroendocrinology* 39: 261-266, 1984.

23. Ruel, J., R. Faure, and J.H. Dussault, Regional distribution of nuclear T3 receptors in rat brain and evidence for preferential localization in neurons, *J Endocrinol Invest* 8: 343-348, 1985.

24. Kaplan, M.M., U.D. McCann, K.A. Yaskoski, P.R. Larsen, and J.L. Leonard, Anatomical distribution of phenolic and tyrosyl ring iodothyronine deiodinases in the nervous system of normal and hypothyroid rats, *Endocrinology* 109: 397-402, 1981.

25. Kaplan, M.M., The role of thyroid hormone deiodination in the regulation of hypothalamo-pituitary function, *Neuroendocrinology* 38: 254-260, 1984.

26. Ködding, R., H. Fuhrmann, and A. von zur Mühlen, Investigations of iodothyronine deiodinase activity in the maturing rat brain, *Endocrinology* 118: 1347-1352, 1986.

27. Riskind, P.N., J.M. Kolodny, and P.R. Larsen, The regional hypothalamic distribution of type II 5′-monodeiodinase in euthyroid and hypothyroid rats, *Brain Res* 420: 194-198, 1987.

28. Kizer, J.S., J. Humm, G. Nicholson, G. Greeley, and W. Youngblood, The effect of castration thyroidectomy and haloperidol upon the turnover rates of dopamine and norepinephrine and the kinetic properties of tyrosine hydroxylase in discrete hypothalamic nuclei of the male rat, *Brain Res* 146: 95-107, 1978.

29. Nakahara, T., H. Uchimura, M. Hirano, M. Saito, J.S. Kim, and T. Matsumoto, Effects of gonadectomy and thyroidectomy on tyrosine hydroxylase in discrete areas of the rat median eminence, *Brain Res* 179: 396-400, 1979.

30. Jahnke, G., G. Nicholson, G.H. Greeley, W.W. Youngblood, A.J. Prange, and J.S. Kizer, Studies of the neural mechanisms by which hypothyroidism decreases prolactin secretion in the rat, *Brain Res* 191: 429-441, 1980.

31. Andersson, K., K. Fuxe, P. Eneroth, L. Agnati, and V. Locatelli, Hypothalamic dopamine and noradrenaline nerve terminal systems and their reactivity to changes in pituitary-thyroid and pituitary-adrenal activity and to prolactin, In F. Brambilla, G. Racagni, and D. de Wied (eds) *Progress in Psychoneuroendocrinology*, Elsevier/North Holland, Amsterdam, pp. 395-406, 1980.

32. Andersson, K., and P. Eneroth, Regression analysis of catecholamine utilization in discrete hypothalamic and forebrain regions of the male rat: effects of thyroidectomy, *Acta Physiol Scand* 123: 105-119, 1985.

33. Anderson, K., and P. Eneroth, The effects of acute and chronic treatment with triiodothyronine and

thyroxine on the hypothalamic and telencephalic catecholamine nerve terminal systems of the hypophysectomized male rat, *Neuroendocrinology* 40: 398-408, 1985.

34. Ottenweller, J.E., and G.A. Hedge, Thyroid hormones are required for daily rhythms of plasma corticosterone and prolactin concentration, *Life Sci* 28: 1033-1040, 1981.

35. Tang, T.K., S. W. Wang, and P.S. Wang, Effects of thyroidectomy and thyroxine replacement on the responsiveness of the anterior pitiutaries from male rats to thyrotropin releasing hormone in vitro, *Experientia* 42: 1031-1034, 1986.

36. Demarest, K.T., and K.E. Moore, Accumulation of L-dopa in the median eminence: an index of tuberoinfundibular dopaminergic nerve activity, *Endocrinology* 106: 463-468, 1980.

37. Reymond, M.J., and J.C. Porter, Hypothalamic secretion of dopamine after inhibition of aromatic L-amino acid decarboxylase activity, *Endocrinology* 111: 1051-1056, 1982.

38. Porter, J.C., and K.R. Smith, Collection of hypophysial stalk blood in rats, *Endocrinology* 81: 1182-1185, 1967.

39. Porter, J.C., Methods for studying pituitary-hypothalamic axis in situ, *Methods Enzymol* 39: 166-183, 1975.

40. Reymond, M.J., W. Benotto, and T. Lemarchand-Béraud, The secretory activity of the tuberoinfundibular dopaminergic neurons is modulated by the thyroid status in the adult rat: consequence on prolactin secretion, *Neuroendocrinology* 46: 62-68, 1987.

41. Eckland, D.J.A., S. Biswas, and S.L. Lightman, Hypothalamo-hypophyseal portal blood sampling from laboratory rats. The effects of endocrine manipulations on portal blood catecholamine concentrations, *Exper Brain Res* 72: 640-644, 1988.

42. Rondeel, J.M.M., W.J. De Greef, P. van der Schoot, B. Karels, W. Klootwijk, and T.J. Visser, Effects of thyroid status and paraventricular area lesions on the release of thyrotropin-releasing hormone and catecholamines into hypophysial portal blood, *Endocrinology* 123: 523-527, 1988.

43. Wang, P.S., H.A. González, M.J. Reymond, and J.C. Porter, Mass and *in situ* molar activity of tyrosine hydroxylase in the median eminence. Effect of thyroidectomy and thyroid hormone replacement, *Neuroendocrinology* 49: 659-663, 1989.

44. Porter, J.C., In situ activity and phosphorylation of tyrosine hydroxylase in the median eminence, *Mol Cell Endocrinol* 46: 21-27, 1986.

45. Blum, M., B.S. McEwen, and J.L. Roberts, Transcriptional analysis of tyrosine hydroxylase gene expression in the tuberoinfundibular dopaminergic neurons of the rat arcuate nucleus after estrogen treatment, *J Biol Chem* 262: 817-821, 1987.

46. Dozin-van Roy, B., and P. De Nayer, Triiodothyronine binding to brain cytosol receptors during maturation, *FEBS Lett* 96: 152-154, 1978.

47. Geel, S.E., Development-related change of triiodothyronine binding to brain cytosol receptors, *Nature* 269: 428-430, 1977.

48. Lennon, A.M., J. Osty, and J. Nunez, Cytosolic thyroxine-binding protein and brain development, *Mol Cell Endocrinol* 18: 201-214, 1980.

49. Dratman, M.B., F.L. Crutchfield, J. Axelrod, R.W. Colburn, and N. Thoa, Localization of triiodothyronine in nerve ending fractions of rat brain, *Proc Natl Acad Sci USA* 73: 941-944, 1976.

50. Dratman, M.B., Y. Futaesaku, F.L. Crutchfield, N. Berman, B. Payne, M. Sar, and W.E. Stumph, Iodine-125-labeled triiodothyronine in rat brain: evidence for localization in discrete neural systems, *Science* 215: 309-312, 1982.

51. Mashio, Y., M. Inada, K. Tanaka, H. Ishii, K. Naito, M. Nishikawa, K. Takahashi, and H. Imura, High affinity 3,5,3´-triiodothyronine binding to synaptosomes in rat cerebral cortex, *Endocrinology* 110: 1257-1261, 1982.

52. Mashio, Y., M. Inada, K. Tanaka, H. Ishii, K. Naito, M. Nishikawa, K. Takahashi, and H. Imura, Synaptosomal T3 binding sites in rat brain: their localization on synaptic membrane and regional distribution, *Acta Endocrinol* 104: 134-138, 1983.

53. Bestetti, G.E., M.J. Reymond, I.V. Perrin, P.C. Kniel, T. Lemarchand-Béraud, and G.L. Rossi, Thyroid and pituitary secretory disorders in streptozotocin-diabetic rats are associated with severe structural changes of these glands, *Virchows Arch-Cell Pathology [B]* 53: 69-78, 1987.

54. Zaninovich, A.A., T.J. Brown, R. Boado, N.R. Bromage, and A.J. Matty, Thyroxine metabolism in diabetic rats, *Acta Endocrinol* 86: 336-343, 1977.

55. Reymond, M.J., and T. Lemarchand-Béraud, Hyperactivity of the hypothalamic dopaminergic neurons and hyposecretion of prolactin in diabetic rats: influence of the thyroid status? *8th International Congress of Endocrinology*, Kyoto, Japan, (Abstract) July 17-23, 1988.

56. Bestetti, G.E., C.E. Boujon, M.J. Reymond, and G.L. Rossi, Functional morphological changes in mediobasal hypothalamus of streptozotocin-induced diabetic rats, *Diabetes* 38: 471-476, 1989.

57. Bestetti, G.E., M.J. Reymond, C.E. Boujon, T. Lemarchand-Béraud, and G.L. Rossi, Impaired release of TRH by the mediobasal hypothalamus of streptozotocin-diabetic rats: functional and morphological aspects, *Diabetes,* Submitted.

58. Joanny, P., G. Peyre, J. Steinberg, B. Conte-Devolx, and C. Oliver, Secretion hypothalamique de somatostatine chez les rats diabetiques, *Diabete & Metabolisme* 14: XXIV (Abstract), 1988.

59. Grino, M., V. Guillaume, A. Caraty, B. Conte-Devolx, P. Joanny, F. Boudouresque, G. Pesce, J. Steinberg, G. Peyre, A. Dutour, P. Giraud, and C. Oliver, Circulating blood glucose and hypothalamic-pituitary secretion, In J.C. Porter and D. Ježová (eds) *Circulating Regulatory Factors and Neuroendocrine Function*, Plenum Press, New York, pp. 391-406, 1990.

60. Loudes, C., A. Faivre-Bauman, A. Barret, D. Gouselle, J. Puymirat, and A. Tixier-Vidal, Release of immunoreactive TRH in serum-free cultures of mouse hypothalamic cells, *Develop Brain Res* 9: 231-234, 1983.

61. Oliver, C., R.L. Eskay, and J.C. Porter, Developmental changes in brain TRH and in plasma and pituitary TSH and prolactin levels in the rat, *Biol Neonate* 37: 145-152, 1980.

62. Schaeffer, J.M., and M.J. Brownstein, Ontogeny of TRH-like material in several regions of the rat brain, *Brain Res* 182: 207-210, 1980.

63. Burgunder, J.M., and T. Taylor, Ontogeny of thyrotropin-releasing hormone gene expression in the rat diencephalon, *Neuroendocrinology* 49: 631-640, 1989.

64. Nemeskeri, A., D. Grouselle, A. Faivre-Bauman, and A. Tixier-Vidal, Developmental changes of thyroliberin (TRH) in the rat brain, *Neurosci Lett* 53: 279-284, 1985.

65. Segerson, T.P., J. Kauer, H.C. Wolfe, H. Mobtaker, P. Wu, I.M.D. Jackson, and R.M. Lechan, Thyroid hormone regulates TRH biosynthesis in the paraventricular nucleus of the rat hypothalamus, *Science* 238: 78-80, 1987.

66. Koller, K.J., R.S. Wolff, M.K. Warden, and R.T. Zoeller, Thyroid hormones regulate levels of thyrotropin-releasing-hormone mRNA in the paraventricular nucleus, *Proc Natl Acad Sci USA* 84: 7329-7333 1987.

67. Dyess, E.M., T.P. Segerson, Z. Liposits, W.K. Paull, M.M. Kaplan, P. Wu, I.M.D. Jackson, and R.M. Lechan, Triiodothyronine exerts direct cell-specific regulation of thyrotropin-releasing hormone gene expression in the hypothalamic paraventricular nucleus, *Endocrinology* 123: 2291-2297, 1988.

68. Zoeller, R.T., R.S. Wolff, and K.J. Koller, Thyroid hormone regulation of messenger ribonucleic acid encoding thyrotropin (TSH)-releasing hormone is independent of the pituitary gland and TSH, *Mol Endocrinol* 2: 248-252, 1988.

69. Eskay, R.L., C. Oliver, N. Ben-Jonathan, and J.C. Porter, Hypothalamic hormones in portal and systemic blood, In M. Motta, P.G. Crosignani, and L. Martini (eds) *Hypothalamic Hormones: Chemistry, Physiology, Pharmacology and Clinical Uses*, Academic Press, London, pp. 125-137, 1975.

70. Ching, M.C.H., and R.D. Utiger, Hypothalamic portal blood immunoreactive TRH in the rat: lack of effect of hypothyroidism and thyroid hormone treatment, *J Endocrinol Invest* 6: 347-352, 1983.

71. Martin, D., J. Epelbaum, M.T. Bluet-Pajot, M. Prelot, C. Kordon, and D. Durand, Thyroidectomy abolishes pulsatile growth hormone secretion without affecting hypothalamic somatostatin, *Neuroendocrinology* 41: 476-481, 1985.

72. Katakami, H., T.R. Down, and L.A. Frohman, Decreased hypothalamic growth hormone-releasing hormone content and pituitary responsiveness in hypothyroidism, *J Clin Invest* 77: 1704-1711, 1986.

73. Gillioz, P., P. Giraud, B. Conte-Devolx, P. Jaquet, J.L. Codaccioni, and C. Oliver, Immunoreactive somatostatin in rat hypophysial portal blood, *Endocrinology* 104: 1407-1410, 1979.

74. Berelowitz, M., K. Maeda, S. Harris, and L.A. Frohman, The effect of alterations in the pituitary-thyroid

axis on hypothalamic content and in vitro release of somatostatin-like immunoreactivity, *Endocrinology* 107: 24-29, 1980.

INHIBIN AND RELATED PROTEINS: LOCALIZATION, REGULATION, AND EFFECTS

F.H. de Jong, A.J. Grootenhuis, I.A. Klaij, and W.M.O. Van Beurden

Department of Biochemistry
Division of Chemical Endocrinology
Erasmus University Rotterdam
The Netherlands

Introduction

The anterior lobe of the pituitary gland secretes two gonadotropic hormones, lutropin (luteinizing hormone, LH) and follitropin (follicle-stimulating hormone, FSH) (1), which stimulate specific cells of the testis and ovary. The biosynthesis and secretion of these hormones is influenced by a number of stimulating and inhibiting factors. The hypothalamic gonadotropin-releasing hormone (GnRH) stimulates the secretion of both LH and FSH (2), whereas steroid hormones suppress peripheral concentrations of gonadotropins by inhibition of GnRH release into the hypothalamic-hypophysial portal circulation (3) and by direct effects on the pituitary gonadotrophs (4,5).

These factors affect pituitary secretion of LH and FSH in a parallel fashion. However, in a number of physiological and experimental conditions discrepancies are observed between the changes of peripheral levels of LH and FSH. A number of explanations for these discrepancies have been suggested. First, the existence of a separate hypothalamic neurohormone for the secretion of FSH (FSH-releasing hormone, FSHRH) has been postulated (6,7). The possibility of the existence of such a hormone is emphasized by the observation of a dissynchrony between the pulses of LH and FSH (8,9) and the maintenance of FSH pulsatility after administration of antiserum against GnRH (10,11). Second, the differences in the metabolic clearance rates of LH and FSH and in the responses of peripheral levels of gonadotropins to LHRH pulses may influence the ratio of FSH to LH in peripheral plasma (12). Third, a special gonadal steroid or combination of gonadal steroids may suppress LH or FSH selectively (13). Fourth, differential changes of peripheral levels of LH and FSH may be explained by a existence of the gonadal protein hormone(s), inhibin, which has a selective suppressive action on the secretion of FSH (14). A role for inhibin in the regulation of the pituitary-testicular axis was postulated in 1923 (15). However, it was only recently that the hormone was purified to homogeneity, characterized as a glycoprotein, and its amino acid sequence derived from the nucleotide sequence of its cDNA.

The aim of this chapter is to review recent developments in our understanding of the localization, regulation, and actions of inhibin. In addition, a closely related protein, activin, will be discussed. For more detailed information on other aspects of inhibin the reader is referred to recently published reviews (16-20).

Detection of Inhibin and Related Peptides

Inhibin activity can be detected using a number of *in vivo* and *in vitro* bioassay

Circulating Regulatory Factors and Neuroendocrine Function
Edited by J. C. Porter and D. Ježová
Plenum Press, New York, 1990

271

systems, and radioimmunological methods for the measurement of inhibin has been reported. Methods for the assay of inhibin activity have been discussed in earlier reviews (17,21,22), and will be discussed here only briefly.

The first methods used to detect inhibin activity in a quantitative way were *in vivo* methods, based on suppression of the human chorionic gonadotropin (HCG)-induced increase of reproductive organ weights in immature female rats (23) or mice (24). These methods are based on suppression of the endogenous secretion of FSH by inhibin. However, interference between FSH and its ovarian receptor will yield similar results, thus causing aspecificity of this type of assay. The combination of such aspecificity and the dependence of the results of the assay on the age of the animals, timing, and dose of HCG (21) renders it difficult to interpret data obtained using this type of assay. Furthermore, the sensitivity of this type of assay is relatively low.

More specific results can be obtained using direct *in vivo* methods for the detection of inhibin activity in which suppression of peripheral levels of FSH in experimental animals is measured after the administration of inhibin. In general, blood samples are collected within 3-24 hours after the injection of the inhibin-containing preparations, since the decrease of peripheral FSH levels begins a few hours after the injection and disappears between 6 and 16 hours after the injection (25-27). The best *in vivo* model for detection of inhibin activity appears to be the short-term (2-12 days) ovariectomized adult rat. However, relatively large amounts of inhibin have to be used, and the precision of this type of assay is low (21).

These problems can be obviated by using *in vitro* methods for measurement of inhibin activity. Such methods have better precision than the *in vivo* methods (22). The *in vitro* bioassays, which are most widely used for the measuring inhibin activity, make use of the suppression of secretion of FSH from non-stimulated or GnRH-stimulated pituitary cells or of the suppression of the intracellular levels of FSH in these cells. FSH released from non-GnRH-stimulated pituitary cells is related in a dose-dependent manner to inhibin. In general, the release of LH is not affected (28). However, when GnRH-stimulated release of FSH and LH from rat pituitary cells is measured in the presence of inhibin-containing materials, the release of both gonadotropins is suppressed in a dose-dependent way (28-30). Suppression of GnRH-induced release of LH in rat pituitary cells appears to be an intrinsic effect of inhibin. van Dijk *et al.* (31) observed similar immunoneutralization curves for suppression of GnRH-stimulated FSH and LH release from pituitary cells, using an antiserum against a partly purified bovine inhibin. Farnworth *et al.* (32) reached a similar conclusion on basis of experiments in which a comparison of intra- and extracellular LH levels was made in a pituitary cell system to which purified inhibin was added.

The same authors also observed suppression of the secretion of both FSH and LH under basal conditions when long-term effects of purified inhibin were studied (33). This suppression of gonadotropin release by inhibin may be partly due to increased intracellular metabolism of gonadotropins. Wang *et al.* (34) observed suppression of the number of GnRH receptors in pituitary cells by inhibin. This decreased number of GnRH receptors may account for the suppression of GnRH-stimulated release of LH and FSH.

The sensitivity of these *in vitro* bioassays using rat pituitary cells is sufficient to detect inhibin in gonadal fluids, but not in peripheral blood under basal conditions. However, recently it was reported that the sensitivity of an inhibin bioassay system based on the suppression of spontaneous release of FSH from ovine pituitary cells is 30- to 40-fold greater than that of the rat pituitary cell bioassay, thus making it possible to measure peripheral levels of inhibin (35).

A number of radioimmunoassay systems for the estimation of inhibin have been described using antibodies against native inhibin (36-38) or against synthetic peptides derived from amino acid sequences in inhibin (39-42). McLachlan *et al.* (36) found good correlation between bio- and immunopotencies of a number of inhibin preparations of bovine and human origin, and they indicated a high cross-reactivity between the various molecular forms of inhibin (see below).

The antisera, raised against the N-terminal amino acids of the α-subunit of porcine, bovine, or human inhibin, were used to investigate the regulation of inhibin secretion in *in vitro* systems. In some of these studies the specificity of this type of assay, which may be susceptible to cross-reactivity of loose α-subunits (see below), was investigated by showing co-elution of inhibin bioactivity and immunological activity after gel filtration of rat granulosa cell culture medium (39), Sertoli cell culture medium (43), or semi-purified inhibin from bovine follicular fluid (42). The latter authors found greatly varying bioactivity/immunoactivity in the semi-purified preparations, indicating that results, obtained using this type of assay, should be interpreted with caution.

Structure of Inhibin and Related Peptides

Most of the earlier work on the purification and characterization of inhibin was performed using rete testis fluid or seminal plasma as a source. Since it became apparent that ovarian follicular fluid is also a rich source of inhibin (44), this fluid has been used extensively.

Bovine and porcine follicular fluid inhibin has been found by some to have an apparent molecular weight of 32,000 (45-49). However, Robertson *et al.* (50) and Leversha *et al.* (51) found molecular weights of 58,000 and 65,000 for bovine and ovine follicular fluid inhibin, respectively. Robertson *et al.* (50) found that the 58 kD protein could be transformed to a 32 kD protein by acid precipitation or after incubation with peripheral plasma. The 32 and 58 kD inhibins consists of two dissimilar subunits (α and β), that are linked by disulfide bonds as becomes apparent after reduction with mercaptoethanol. The subunits have no known biological activity. Amino acid sequences of the N-terminal ends of the subunits of the inhibin molecules are very similar for rat, human, porcine, ovine, and bovine (Table 1).

On the basis of these amino acid sequences, oligonucleotide probes were synthesized and used to isolate cDNAs coding for the inhibin subunits from porcine (52,53), bovine (54), rat (55,56), and human ovaries (57,58). The following data can be derived from these nucleotide sequences (Figure 1). The subunits are separately synthesized in a pre-pro form. The α-subunit, of which the pre-pro form consists of about 360 amino acids, contains pairs of basic amino acids that precede the positions of the amino acids that were present in the N-terminal sequence of the α-subunits of the 58 kD and 32 kD inhibins. Similarly, the pro-form of the β-subunit contains 5 basic amino acids, preceding the amino acids found in the N-terminal sequence of the isolated β-subunit for which two isoforms (βA and βB) were found (52,43,55). There is a high degree of homology between the α-subunits of human, porcine, rat, and bovine; the β-subunits appear to be almost identical (Table 1). Mason *et al.* (52) reported that Northern blot analysis revealed a 10-times higher abundance of α-precursor mRNA when compared with the amount of mRNA for the βA-precursor, which itself was twice as abundant as the mRNA for the β-chain in porcine ovarian tissue. The α-subunits of porcine, bovine, and human inhibin contain two, two, and three glycosylation sites, respectively. This is consistent with reports on the glycoprotein character of gonadal inhibin (59,60).

On the basis of these results, the confusion in the literature about the physicochemical characteristics of inhibin can be explained: cleavage of the complete pre-pro-hormone, which should have a molecular weight of about 100 kD leads to a molecular weight of around 60 kD as suggested by a number of authors. This 60 kD molecule might be split up peripherally into the 32 kD protein (36), which apparently has biological activity. The

Table 1

N-Terminal Amino Acid Sequences of the Subunits of 32 kDA Inhibin,
Derived from Nucleotide Sequences of cDNA (52-58) or From
Direct Amino Acid Sequencing (62) in Various Species

α-Subunit	1									10										20										30
Porcine[52,53]	S	T	A	P	L	P	W	P	W	S	P	A	A	L	R	L	L	Q	R	P	P	E	E	P	A	V	H	A	D	C
Bovine[54]	-	-	P	-	-	-	-	-	-	-	-	-	-	-	-	-	-	-	-	-	-	-	-	-	A	-	-	-	-	-
Human[53,57,58]	-	-	P	L	M	S	-	-	-	-	S	-	-	-	-	-	-	-	-	-	-	-	-	A	-	-	N	-
Rat[55,56]	-	A	P	S	M	-	-	-	-	-	-	-	-	-	-	-	-	-	-	-	-	-	-	S	A	-	-	F	-
Ovine[62]	-	-	P	-	-	-	-	-	-	-	-	-	-	-	-	-	-	-	-	-	-	-	-	-	A	-	-	-	-	-

β-A-Subunit	1									10										20										30
Porcine[52]	G	L	E	C	D	G	K	V	N	I	C	C	K	K	Q	F	F	V	S	F	K	D	I	G	W	N	D	W	I	I
Bovine[54]	-	-	-	-	-	-	-	-	-	-	-	-	-	-	-	-	-	-	-	-	-	-	-	-	-	-	-	-	-	-
Human[57,58]	-	-	-	-	-	-	-	-	-	-	-	-	-	-	-	-	-	-	-	-	-	-	-	-	-	-	-	-	-	-
Rat[55,56]	-	-	-	-	-	-	-	-	-	-	-	-	-	-	-	-	-	-	-	-	-	-	-	-	-	-	-	-	-	-
Ovine[62]	-	-	-	-	-	-	-	-	-	-	-	-	-	-	-	-	-	Y	-	-	-	-	-	-	-	-

β-B-Subunit	1									10										20										30
Porcine[52]	G	L	E	C	D	G	R	T	N	L	C	C	R	Q	Q	F	F	I	D	F	R	L	I	G	W	S	D	W	I	I
Human[57]	-	-	-	-	-	-	-	-	-	-	-	-	-	-	-	-	-	-	-	-	-	-	-	-	-	N	-	-	-	-
Rat[55]	-	-	-	-	-	-	-	S	-	-	-	-	-	-	-	-	-	-	-	-	-	-	-	-	-	N	-	-	-	-

Dashes indicate the presence of the same amino acid as found in the porcine sequence.

various sizes of the inhibin molecules, which may be predicted on the basis of this model, have been shown to be present in bovine follicular fluid by Miyamoto *et al.* (61), who detected 32 kD, 55 kD, and 65 kD molecules using monoclonal antibodies against the α- and β-subunits after polyacrylamide gel electrophoresis and subsequent blotting experiments. These authors also described 120, 108, and 88 kD molecules, detected by this technique, which they propose to be trimers of one α-subunits and two β-subunits of different length. The observation of multiple molecular weight forms of follicular fluid inhibin was confirmed by Grootenhuis *et al.* (42) (Figure 2a).

In contrast with this multiplicity of inhibin forms found in follicular fluid, the testis appears to contain predominantly 32 kD inhibin. Bardin *et al.* (62) purified inhibin from ovine rete testis fluid and detected two forms of 32 kD $\alpha\beta$A dimers, which are homologous with the ovine ovarian inhibin. These two isoforms differ at the N-terminal part of the α-subunit in that one had 15 amino acids less than the other. Grootenhuis *et al.* (42) found that the inhibin in rat testis homogenates and in rat Sertoli cell conditioned medium only consists of a 32 kD molecule (Figure 2b). Sequencing of this material revealed the presence of the $\alpha\beta$B dimer (63). The difference in composition of ovarian and testicular inhibin may be the cause of the differential immunoneutralization of biological activities of these materials by an antiserum against bovine follicular fluid inhibin (31).

After the structure of porcine inhibin was elucidated, a significant homology between the β-subunits of inhibin and the subunits of transforming growth factor-β

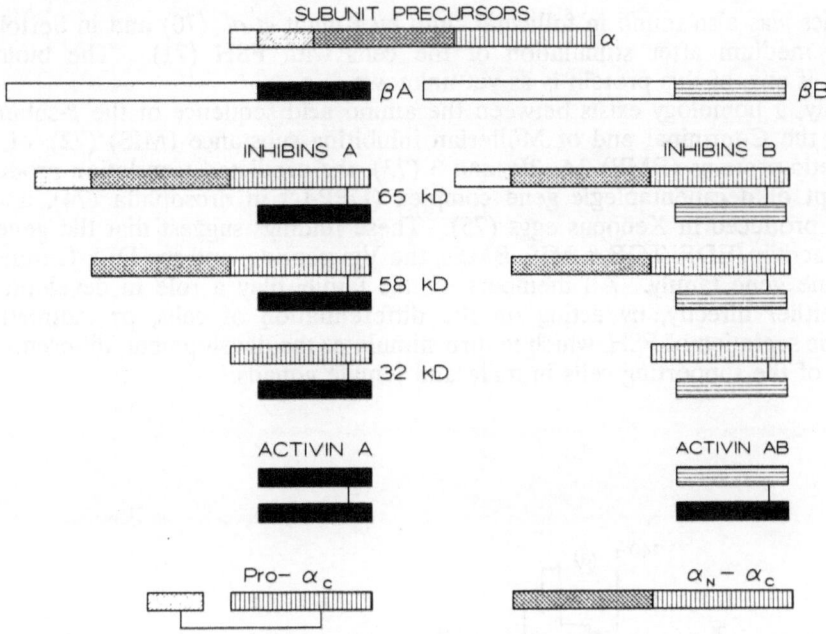

Figure 1. Schematic representation of inhibin subunits and their possible combinations.

(TGF-β) was noticed (52). Subsequent experiments indicated that the secretion of FSH from pituitary cells can be stimulated 1.5-fold by addition of TGF-β (64). This effect was specific; the secretion of other pituitary hormones was not affected. These authors also found that a fraction of porcine follicular fluid could exert a similar stimulatory effect on FSH secretion and postulated the presence of a TGF-β-like molecule in the follicular fluid. The molecule with TGF-β-like activity was subsequently identified as a homo- or heterodimer of the inhibin-β-subunits: $\beta A \beta A$ or $\beta A \beta B$ (47,65,66). The name *activin* was proposed for the active agent. The issue of whether the dimers of the inhibin β-subunit produced in the ovary can reach the circulation in sufficient quantity to affect pituitary FSH secretion, has not yet been resolved. However, inhibin and activin might also have antagonizing effects at the level of the gonads (see below). The possibility of inhibin-antagonizing effects of activin after the addition of gonadal preparations to the culture medium of pituitary cells cannot be excluded. This indicates the necessity to separate inhibin from activin in order to estimate inhibin bioactivity in gonadal preparations.

Recently, Eto *et al.* (67) described the isolation of erythroid differentiation factor (EDF) from the culture medium of human leukemia cell line. The protein appeared to be a homodimer composed of two subunits that have an identical molecular weights and 42 identical N-terminal amino acids, when compared with the $\beta A \beta A$ activin dimer. Later, these authors found that this material specifically stimulated the release of FSH from cultured pituitary cells, indicating that EDF not only shares physicochemical properties but also biological activity with activin (68).

The third inhibin-like protein, which as been purified from ovarian follicular fluid, is a dimer of the C-terminal part of the α-subunit, α_C, covalently linked to the pro-sequence of the N-terminal part of the same subunit (69). Material with similar characteristics was also found in follicular fluid by Knight *et al.* (70) and in Sertoli cell conditioned medium after stimulation of the cells with FSH (71). The biological significance, if any, of this protein is as yet unknown.

Finally, a homology exists between the amino acid sequence of the β-subunit of inhibin and the C-terminal end of Müllerian inhibiting substance (MIS) (72), of bone morphogenetic proteins (BMP) 2A, 2B, and 3 (73), the predicted translation product of the transcript of decapentaplegic gene complex (DPP-C) in drosophila (74), and the Vgl-peptide produced in Xenopus eggs (75). These findings suggest that the genes for inhibin and activin/EDF, TGF-β, MIS, BMPs, the Vgl-peptide, and the DPP-C transcript belong to one gene family. All members of this family play a role in developmental processes, either directly, by acting on the differentiation of cells, or indirectly, by regulating the secretion of FSH, which in turn stimulates the development, differentiation, and activity of the supporting cells in male and female gonads.

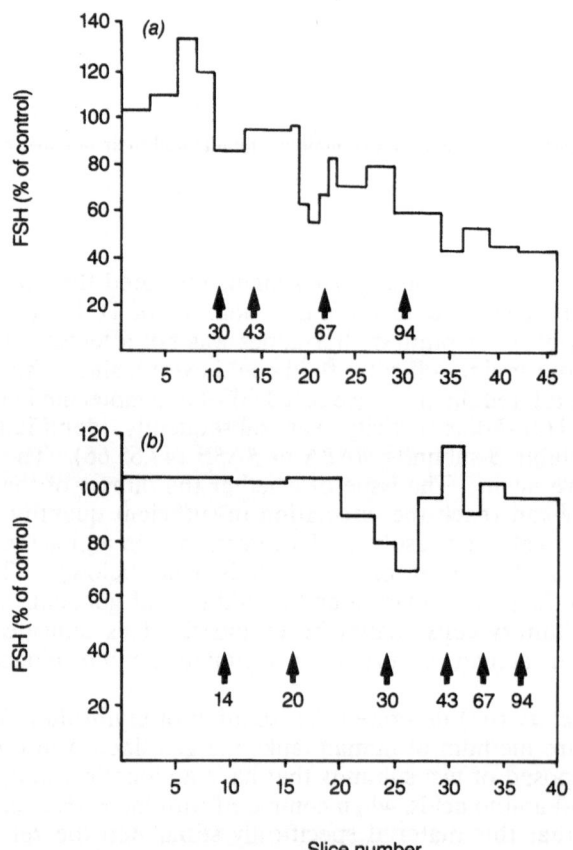

Figure 2. Suppression of FSH in an *in vitro* bioassay system of inhibin activity to which eluates of slices from a sodium dodecylsulphate-polyacrylamide gel were added after electrophoresis of *a)* bovine follicular fluid and *b)* rat testis homogenate. Reproduced from Grootenhuis *et al.* (42) with permission of the Journal of Endocrinology, Ltd.

Table 2
Localization of Inhibin-Subunits

Organ Cell Type	Subunit α	β_A	β_B	Technique Used	Reference
Testis	+	+	+	S_1 Nuclease	Meunier et al. (147)
Sertoli Cells	+	-	+	Northern Blot	Toebosch et al. (84)
	+			Immunocytochemistry	Merchenthaler et al. (152)
	+				Cuevas et al. (153)
	+				Rivier et al. (100)
Leydig Cells	+			Northern Blot	Risbridger et al. (80)
(tumor like)	-	+	+		Lee et al. (79)
Ovary	+	+	+	S1 Nuclease	Meunier et al. (147)
Granulosa Cells	+	+	+	In situ hybridization	Meunier et al. (154)
	+	+			Woodruff et al. (56)
	+	+	+		Rivier et al. (155)
	+			Immunocytochemistry	Merchenthaler et al. (152)
	+	+			Meunier et al. (154)
	+				Cuevas et al. (153)
Theca Cells	+	-	-	In situ Hybridization	Meunier et al. (154)
	+				Torney et al. (82)
	+				Rivier et al. (155)
Corpus Luteum	+	+		Northern Blot	Davis et al. (113,156)
	-	-		In situ Hybridization	Woodruff et al. (157,158)
	+	-	-		Meunier et al. (154)
	+			Immunocytochemistry	Cuevas et al. (153)
Adrenal Gland	+	+		S_1 Nuclease	Meunier et al. (147)
Cortex	+			In situ Hybridization	Crawford et al. (159)
Placenta	+	+	+	Northern Blot	Meunier et al. (147)
Cytotrophoblast	+			Immunocytochemistry	Petraglia et al. (144)
	+				Merchenthaler et al. (152)
Pituitary Gland	+	-	+	S_1 Nuclease	Meunier et al. (147)
Gonadotrophs	+	-	+	Immunocytochemistry	Roberts et al. (160)
Central Nervous System	+	+	+	S_1 Nuclease	Meunier et al. (147)
	-	+	-	Immunocytochemistry	Sawchenko et al. (134)
Bone Marrow	-	+	-	S_1 Nuclease	Meunier et al. (147)
Spleen	+	+	-	S_1 Nuclease	Meunier et al. (147)
Kidney	+	-	-	S_1 Nuclease	Meunier et al. (147)

Localization of Inhibin and Related Substances

Inhibin was originally defined as a testicular hormone, which was responsible for negative regulation of pituitary FSH secretion (76). It was found that inhibin is synthesized and secreted by Sertoli cells (77,78), both *in vitro* and *in vivo*. These observations have been confirmed by the detection of the mRNA for the inhibin α- and β-subunits in Sertoli cells, and by results of immunocytochemistry. Very recently, the production of bioactive activin and inhibin in Leydig cells *in vitro* was described (79,80). Data on localization of mRNA, coding for the inhibin subunits and their immunocytochemical localization are summarized in Table 2.

It has been known since 1976 that inhibin activity is present in ovarian follicular fluid (44). The cell type responsible for the production and secretion of inhibin is the granulosa cell. Here, too, mRNAs for the inhibin subunits were detected, whereas immunocytochemical data confirm the presence of inhibin subunits (Table 2). Finally, a number of authors suggest the presence of mRNA of inhibin-subunits in theca and corpus luteum cells. However, other authors could not confirm the presence of inhibin-subunit mRNAs in corpora lutea from cows and ewes (81,82).

Apart from the data for the gonads, inhibin subunit-mRNA was detected in a large number of other tissues, including the placenta, hypothalamus, pituitary gland, and spleen (Table 2). Bioactive inhibin was detected in the placenta (83); however, for the other tissues no data are available at present. Possible physiological roles of inhibin or activin in these tissues will be discussed in the last section of this chapter.

Regulation of Inhibin Production

Studies in Male Animals

IN VITRO EXPERIMENTS. Most of the studies on regulation of inhibin production by

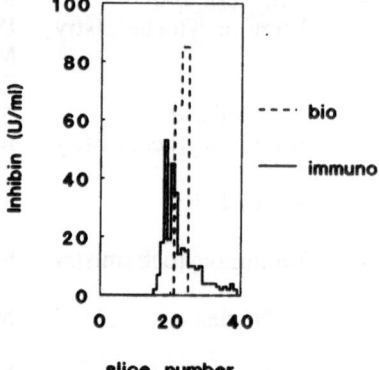

Figure 3.   Biological and immunological inhibin activity in conditioned media from FSH-stimulated cultured Sertoli cells. The media were concentrated and electrophoesed on a sodium-dodecylsulphate polyacrylamide gel. Subsequently, gels were sliced, proteins were eluted and assay by bioassay *(open bars)* and immunoassay *(broken line bars)*. Reproduced from de Jong *et al.* (71) with permission of Raven Press.

Sertoli cells *in vitro* used cells from immature (20-22 days old) rats. In these *in vitro* systems, FSH causes increased expression of the mRNA for the inhibin α-subunit, while the expression of the β-subunit is not affected (84-86). This increased amount of α-subunit may be the cause of the increase in inhibin secretion, as measured by radioimmunoassay using antisera against N-terminal peptides of the α-subunits (41,43,62), against native inhibin (87,88) or by bioassay (85,88,89). However, other investigators found no change in the amount of bioactivity secreted (90), whereas an increase of immunoactivity, which was not related to bioactivity, was reported by de Jong *et al.* (71). The latter authors also found that the increased immunoactivity was due to the presence of a 28 kD molecule (Figure 3), which is presumably the Pro-α_C dimer described above. This indicates that the authors who used antisera raised against the N-terminal peptide from the α-subunit may have mistakenly interpreted their data as an increase in the production of the inhibin $\alpha\beta$ dimer.

Increased secretion of inhibin has been reported after addition of testosterone or 5α-dihydrotestosterone to the medium. This effect could be blocked by addition of the anti-androgen cyproterone acetate (90). Others found that in a system where hormones were added sooner after the isolation of the Sertoli cells, testosterone suppressed the release of inhibin (89) whereas absence of effects of dihydrotestosterone and estradiol was also described (43). The addition of testosterone together with FSH to Sertoli cells suppressed the secretion of inhibin, when compared with the effect of addition of FSH alone (89). Addition of agents, which stimulate intracellular cAMP levels, such as a phosphodiesterase inhibitor, choleratoxin, forskolin, or dibutyryl cAMP, stimulated the expression of the mRNA for the α-subunit (86) and immunoreactive inhibin production (43), which is in keeping with the stimulatory action of FSH.

Little certainty exists on the influence of spermatogenic cells on the production of inhibin by Sertoli cells. The experiments of Steinberger (91) suggest that the presence of spermatogenic cells (spermatocytes or spermatids) does not affect the secretion of inhibin by Sertoli cells *in vitro*. However, in these studies the amount of inhibin added to the pituitary cells caused maximal suppression of the secretion of FSH, thus making it impossible to judge if the addition of the spermatogenic cells increased, or even decreased the amount of inhibin secreted. In similar experiments it was found that at 32°C the recombination of spermatogenic cells with Sertoli cells suppressed the amount of inhibin in the medium of these cells; the addition of thymocytes had no such negative influence (89). Similar inhibitory effects of germ cells on RNA (92) and estradiol (93) synthesis were reported. Recent studies indicate that the amount of inhibin produced may depend on the stage of spermatogenesis in the seminiferous tubules (94).

In vivo STUDIES. The production of inhibin is decreased after hypophysectomy of adult rats but stimulated after administration of FSH (95). In contrast, injection of human menopausal gonadotropins (HMG), which contains FSH activity, into intact adult male rats, caused no increase in the testicular level of inhibin, as measured by bioassay. Injections of FSH or HMG into immature male rats did, however, cause increased testicular inhibin levels (96). These observations indicate that in male rats, exogenous FSH can stimulate inhibin production during the prepubertal period or after hypophysectomy of adult animals, but no effect in intact adult rats.

In the human male, McLachlan *et al.* (97) showed that the administration of FSH after suppression of endogenous gonadotropins by administration of androgens, increased peripheral radioimmunoassayable inhibin levels. The administration of HCG in men or male rats led to similar increases of peripheral inhibin levels (97,98). This observation led the authors to the conclusion that HCG acted directly on Leydig cells, which, in turn, stimulated inhibin production in the Sertoli cells. More recently, it became clear that Leydig cells may also produce inhibin (80) or activin (79).

Table 3
Effects of Inhibin and Activin on
Non-Pituitary Cell Systems

| Organ *Parameter* | --------Effects of------- | | Reference |
	Inhibin	Activin	
Hypothalamus			
Oxytocin Secretion	n.d.	+	Sawchenko *et al.* (134)
Testis			
Testosterone Production	+	-	Hsueh *et al.* (141)
	=	n.d.	Grootenhuis *et al.* (142)
Spermatogonia	-	n.d.	van Dissel-Emiliani *et al.* (136)
Ovary			
Granulosa Cells			
LH-Receptor	n.d.	+	Sugino *et al.* (161)
FSH-Receptor	n.d.	+	Hasegawa *et al.* (162)
Estradiol	-	+	Ying *et al.* (138)
	=	+	Hutchinson *et al.* (139)
Progesterone	=	-	Hutchinson *et al.* (139)
	n.d.	+	Sugino *et al.* (161)
Inhibin	n.d.	+	Sugino *et al.* (161)
Oocytes			
Maturation Division	-	=	O *et al.* (163)
Placenta			
GnRH; HCG	-	+	Petraglia *et al.* (145)
Hematopoietic Cells			
Colony Formation	-	+	Yu *et al.* (164)
	-	+	Broxmeyer *et al.* (165)
Cell Differentiation	-	+	Yu *et al.* (164)
	n.d.	+	Eto *et al.* (67)
	n.d.	+	Murato *et al.* (166)
Cell Proliferation	+	-	Hedger *et al.* (167)
Fibroblasts			
Proliferation	n.d.	+	Kojima & Ogata (168)
Pancreas, Liver			
Insulin Secretion	n.d.	+	Totsuka *et al.* (150)
Glycogenolysis	n.d.	+	Mine *et al.* (151)

+ denotes stimulatory effect; - denotes inhibitory effect; = denotes no effect detected; n.d. denotes no data available.

The effect of age on inhibin production in male rats was studied in an indirect way by Hermans *et al.* (99), who showed that removal of inhibin by castration causes an acute five-fold increase of FSH levels in immature male rats, whereas castration of adult rats causes only a small increase of FSH levels. This observation was confirmed by Rivier *et al.* (100), who immunoneutralized inhibin and found an acute rise of FSH levels in immature, but not in adult male rats. In the human, levels of inhibin in boys going through puberty increase in parallel with increasing FSH levels (101). This suggests that FSH may regulate inhibin secretion rather than being dependent on the amount of inhibin present in the peripheral circulation.

Studies in Female Animals

INHIBIN PRODUCTION BY GRANULOSA CELLS *IN VITRO*. Granulosa cells are the source of inhibin from the ovary as shown first by Erickson and Hsueh (102). They cultured rat granulosa cells and added the medium from these cells to cultured pituitary cells. This observation has been amply confirmed. From these studies it becomes clear that FSH as well as androgens and LH may be factors in the regulation of inhibin secretion by granulosa cells from rat (39,41,103), bovine (104,105), and human infant ovaries (106); insulin-like growth Factor-I (IGF-I) synergizes with the effect of FSH (39,103). Henderson *et al.* (107) indicated that FSH could only stimulate inhibin production by granulosa cells from healthy bovine follicles, and testosterone caused an increase of inhibin production by granulosa cells from both healthy and atretic follicles. This action of testosterone could also be provoked by synthetic androgens, and could be blocked by addition of anti-androgens (105). Stimulatory effects of androstenedione and estradiol were also reported, but only in the presence of FSH (41). From these results it appears that androgens should be aromatized before they may stimulate inhibin production.

The stimulatory action of FSH on inhibin production is apparently mediated by cyclic AMP, since it could be mimicked by the addition of forskolin or a cAMP analogue to the culture medium of rat granulosa cells (39). The production of inhibin by rat granulosa cells also depends on the stage of the estrous cycle in which granulosa cells are harvested (108).

Finally, Tsonis *et al.* (109) found that human granulosa-lutein cells, obtained from ovarian follicles after induction of ovulation by administration of HCG, produce inhibin activity. The production of this material is stimulated after addition of LH or testosterone, but not after the addition of FSH. These results indicate that the human corpus luteum may secrete inhibin and that the regulation of inhibin production after the LH peak differs from that before the peak.

IN VIVO STUDIES. Lee *et al.* (110,111) showed that the injection of PMSG into immature female rats caused a spectacular increase in peripheral levels of inhibin. The concentration of inhibin activity depended on the dose of PMSG injected, and was significantly correlated with the number of ovulations per ovary (112). The increased secretion of inhibin results from an increase of the amount of mRNA for inhibin in the ovaries of these PMSG-treated animals (53,113).

Estimations of inhibin in follicular fluid also show an increase of inhibin concentrations with development of the follicle; highest levels were found on the day of proestrus in rats (114). The highest levels of inhibin in ovarian venous plasma were also detected at this time of the cycle (115). Similar data were found in women (116,117), ewes (118), cows (107), and pigs (119).

Peripheral levels of inhibin throughout the cycle have been reported for women (120) and pigs (121). Immunoreactive inhibin concentrations increase during the follicular phase of the cycle. However, even higher levels are found during the luteal phase of the cycle in women (120-123). These concentrations are directly correlated with progesterone concentrations, indicating the corpus luteum as the likely source of this inhibin. Finally, Robertson *et al.* (123) showed that the bioactivity-immunoactivity ratio of inhibin in peripheral blood during the luteal phase of the cycle is lower than during the follicular phase. This changing ratio may be due to the presence in peripheral plasma of substances, that interfere with either of the two assays. Further studies are needed to resolve this point.

McLachlan *et al.* (122,124) showed that ovarian stimulation for *in vitro* fertilization in women causes an increase in the peripheral levels of inhibin. This increase was related to the number of developing follicles in the ovaries and to peripheral estradiol concentrations. After ovulation, a further increase of peripheral inhibin levels was found, which culminated in even higher levels during early pregnancy (122) The placenta is the most likely source of this circulating inhibin.

The physiological significance of inhibin in female animals has been studied by a number of authors by actively or passively immunizing animals against inhibin or inhibin subunits. Increased peripheral levels of FSH and numbers of ovulations have been reported in rats (125) and sheep (126,127). This result indicates that inhibin plays a role in the regulation of peripheral levels of FSH and, through this suppressive action, of the number of developing follicles in the ovaries.

Extrapituitary Effects of Inhibin and Activin

Inhibin was originally defined as a biological substance that suppresses the release of FSH from the pituitary gland. However, results of various experiments in which more or less purified preparations, containing inhibin activity, were injected *in vivo* or added to non-pituitary cells *in vitro*, suggest that inhibin may also have extra-pituitary actions, both at the level of the hypothalamus and at the gonadal level. Furthermore, effects in non-reproduction related systems have been reported. The interpretation of these results is difficult in those experiments in which non- or partially purified inhibin preparations were used, because it is not clear if inhibin or one or more other factors present in the preparations are the cause of these effects. With the availability of completely pure inhibin preparations, a number of points have been re-investigated. The actions of inhibin at the level of the pituitary gland have already been described in the section on detection of inhibin. Therefore, the following sections describe the data on extra-pituitary actions of inhibin- and activin-containing preparations. The effects have been summarized in Table 3 for easy reference.

Effects on Secretion of GnRH and Oxytocin

Apart from the effects on the secretion of FSH from the pituitary gland, effects of inhibin at the level of the hypothalamic production or secretion of GnRH have been postulated. The concentration of GnRH in pituitary stalk blood was suppressed after intraperitoneal injection of charcoal-treated bovine follicular fluid (128). However, when more purified fractions of inhibin were administered, no such effects on GnRH-secretion were observed, indicating that inhibin has no effect on GnRH secretion. The same conclusion was reached by other investigators, who studied effects of peripheral injections of ovine follicular fluid in ovariectomized ewes on the pulsatile release of LH and FSH (129), did similar experiments in ovariectomized rats (130), or found no changes of LH pulsatility after injection of bovine follicular fluid into ovariectomized heifers (131). After intraventricular administration of steroid-free bovine follicular fluid, no effects on peripheral concentrations of FSH or LH were found (9). This result contrasts with that of earlier studies in which suppressed peripheral FSH concentrations were detected after intraventricular or intrahypothalamic administration of inhibin-containing material (132,133). Although the reason for this difference is not clear, it seems likely that purified inhibin preparations do not have direct effects on hypothalamic GnRH production.

Sawchenko et al. (134) reported that injection of activin into the hypothalamus caused a very rapid release of oxytocin from the posterior lobe of the pituitary gland. This is a very interesting observation, because the authors also described the immunocytochemical presence of the inhibin βA-subunit in the nucleus of the tractus solitarius, which might, therefore, produce activin in the same nucleus and thereby regulate oxytocin secretion.

Intragonadal Effects

A possible direct effect of inhibin-containing preparations from ovine rete testis fluid on the incorporation of radiolabelled thymidine in the DNA of differentiated spermatogonia was reported by Franchimont et al. (135). Intratesticular injection of crude and purified inhibin in mice or hamsters caused a decrease in the number of differentiating spermatogonia (136). Direct effects of inhibin-containing fractions on the ovary have also been described. Injection of human follicular fluid fractions diminished follicular development in immature rats. Similar fractions had inhibitory effects on aromatization in porcine granulosa cells (137). This effect was confirmed by Ying et al. (138), who showed that purified porcine inhibin suppressed FSH-induced aromatase activity of rat granulosa cells in vitro, while TGF-β and activin had the opposite effect. In contrast, Hutchinson et al. (139) showed that purified bovine 32K inhibin did not affect estrogen or progesterone production in FSH-stimulated cultured rat granulosa cells. Finally, Sugino et al. (140) described the presence of a specific binding protein for activin on rat follicular granulosa cells.

For the androgen producing cells in both testis and ovary, the Leydig and theca cells, respectively, antagonizing actions of inhibin and activin on LH-stimulated androgen production have been described (141). Inhibin appears to activate secretion of steroids in long-term cultures, whereas activin suppresses this activity in Leydig and theca cells. Short-term stimulatory effects of rat Sertoli cell conditioned medium on Leydig cell steroidogenesis cannot be due to the presence of inhibin, since purified inhibin preparations do not cause this effect (142).

Effects on Placenta and Conception

Injection of inhibin-containing preparations from ovine testes showed an anti-implantation effect in pregnant hamsters (143), probably by affecting the production of progesterone by the corpora lutea of these animals. However, an effect through suppression of chorionic gonadotropin production cannot be ruled out. In fact, Petraglia et al. (144,145) showed that inhibin can indeed suppress the placental inhibin production through a cAMP mediated mechanism (144), suggesting a negative feedback system (Figure 4).

Finally, Findlay et al. (146) indicated that immunization of ewes against the N-terminal part of the 45 kD α-subunit, α_N, of inhibin (Figure 1) may interfere with normal gestation. Although a normal number of corpora lutea of pregnancy was observed, the number of lambs born from the immunized ewes was significantly suppressed.

Effects on Hematopoietic Cell Lines

The observation of Eto *et al* (67) that the erythropoiesis differentiation factor (EDF) produced by human leukemia cells is equivalent to activin A along with the detection of inhibin subunit mRNA in bone marrow (147), led to investigation of the action of inhibin and activin on hematopoiesis. In most of the systems studied, inhibin and activin appear to have opposite effects (Table 3). This make it of extreme importance to control the ratio of the production of the subunits and the post-translational modification of these subunits. Detailed information on this point is lacking, however, and it is not clear if hematopoietic cells can produce the inhibin α-subunit.

Two types of specific receptors for activin were found on Friend leukemia and embryonal carcinoma cells (148). Campen and Vale (149) described binding sites for activin on the human leukemia cells, cell line K562.

Effects Related to Carbohydrate Metabolism

Totsuka *et al.* (150) showed that activin can stimulate the secretion of insulin from pancreatic islands. As a sequel to this observation, Mine *et al.* (151) studied the effect of activin on glycogenolysis in liver cells, and observed stimulation of this activity. The latter authors also studied the mechanism of action upon which this effect was based, and found that activin acts through protein kinase C and inositol phosphatidyl metabolism.

Summary and Conclusions

Inhibin has originally been defined as a gonadal hormone that exerts a specific negative feedback action on the secretion of FSH from the gonadotropic cells of the pituitary gland. The existence of inhibin was postulated by Mottram and Cramer (15) as early as 1923. However, only after reliable and sensitive bioassay systems had been

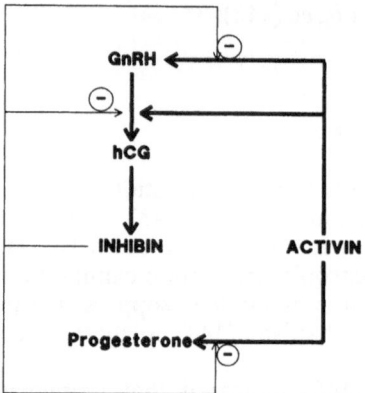

Figure 4. Schematic representation of the possible roles of inhibin and activin in the regulation of the production of placental GnRH, hCG, and progesterone. *Thick arrows* indicate stimulatory actions while inhibin has suppressive actions at the indicated levels. From data of Petraglia *et al.* (145) with permission of the authors.

developed for detection and estimation of inhibin and an ample source of inhibin was found in the form of ovarian follicular fluid, was progress made on the isolation and characterization of the hormone. It is apparent now that inhibin, which itself consists of a dimer of two different subunits, α and β, is a member of a much larger family of (glyco)protein hormones and growth factors that includes Müllerian inhibiting substance, transforming growth factor-β, activin/erythroid differentiation factor, bone morphogenetic proteins, and an insect and a Xenopus protein. All play important roles in cell differentiation.

Gonadal inhibin is produced in the Sertoli cells in the testis and in the granulosa cells in the ovary. The production of inhibin is stimulated by FSH, but controversy exists about other factors that might play a role in the regulation of the production of inhibin. It appears likely that inhibin plays an important role in the feedback regulation of peripheral concentrations of FSH during the period in which Sertoli cells and granulosa cells—the target cells for FSH—divide, $i.e.$, during puberty in male animals and during the development of ovarian follicles in female animals. In this way, inhibin may be an important regulator of the number of developing Sertoli cells and of the length of the seminiferous tubuli in the testis and of the number of developing follicles in the ovary.

Apart from its function in the pituitary-gonadal axis, inhibin and activin may be produced and act in a number of other organs such as the placenta, hypothalamus, adrenal, and bone marrow. Investigation of the role of the members of the inhibin family in these systems has only begun, but will certainly be a field of major interest in the near future.

References

1. Pierce, J.G., and T.F. Parsons, Glycoprotein proteins: structure and function, *Ann Rev Biochem* 50: 465-495, 1981.

2. Schally, A.V., A. Arimura, A.J. Kastin, H. Matsuo, Y. Baba, T.W. Redding, R.M.G. Nair, L. Debeljuk, and W.F. White, Gonadotropin-releasing hormone: one polypeptide regulates secretion of luteinizing and follicle-stimulating hormones, *Science* 173: 1036-1038, 1971.

3. Sarkar, D.K., and G. Fink, Luteinizing hormone releasing factor in pituitary stalk plasma from long-term ovariectomized rats: effects of steroids, *J Endocrinol* 86: 511-524, 1980.

4. Drouin, J., and F. Labrie, Selective effect of androgens on LH and FSH release in anterior pituitary cells in culture, *Endocrinology* 98: 1528-1534, 1976.

5. Drouin, J., and F. Labrie, Interactions between 17β-estradiol and progesterone in the control of luteinizing hormone and follicle-stimulating hormone release in rat anterior pituitary cells in culture, *Endocrinology* 108: 52-57, 1981.

6. Fuchs, S., E. Lundanes, J. Leban, K. Folkers, and C. Bowers, On the existence and separation of the follicle stimulating hormone releasing hormone from the luteinizing hormone releasing hormone, *Biochem Biophys Res Commun* 88: 92-96, 1979.

7. Mizunuma, H., W.K. Samson, M.D. Lumpkin, and S.M. McCann, Evidence for an FSH-releasing factor in the posterior portion of the rat median eminence, *Life Sci* 33: 2003-2009, 1983.

8. DePaolo, L.V., Differential regulation of pulsatile luteinizing hormone (LH) and follicle-stimulating hormone secretion in ovariectomized rats disclosed by treatment with a LH-releasing hormone antagonist and phenobarbital, *Endocrinology* 117: 1826-1833, 1985.

9. de Greef, W.J., G.A.M. Eilers, J. de Koning, B. Karels, and F.H. de Jong, Effects of ovarian inhibin on pulsatile release of gonadotrophins and secretion of LHRH in ovariectomized rats: evidence against a central action of inhibin, *J Endocrinology* 113: 449-455, 1987.

10. Culler, M.D., and A. Negro-Vilar, Evidence that pulsatile follicle-stimulating hormone secretion is independent of endogenous luteinizing hormone-releasing hormone, *Endocrinology* 118: 609-612, 1986.

11. Culler, M.D., and A. Negro-Vilar, Pulsatile follicle-stimulating hormone secretion is independent of

luteinizing hormone-releasing hormone (LHRH): pulsatile replacement of LHRH bioactivity in LHRH-immunoneutralized rats, *Endocrinology* 120: 2011-2021, 1987.

12. Lincoln, G.A., Differential control of luteinizing hormone and follicle-stimulating hormone by luteinizing hormone releasing hormone in the ram, *J Endocrinol* 80: 133-140, 1979.

13. Wiebe, J.P., and P.H. Wood, Selective suppression of follicle-stimulating hormone by 3α-hydroxy-4-pregnen-20-one, a steroid found in Sertoli cells, *Endocrinology* 120: 2259-2264, 1987.

14. Burger, H.G., and M. Igarashi, Inhibin: definition and nomenclature, including related substances, *Endocrinology* 122: 1701-1702, 1988.

15. Mottram, J.C., and W. Cramer, On the general effects of exposure to radium on metabolism and tumour growth in the rat and the special effects on testis and pituitary, *Quart J Exp Physiol* 13: 209-229, 1923.

16. Findlay, J., The nature of inhibin and its use in the regulation of fertility and diagnosis of infertility, *Fertil Steril* 46: 770-783, 1986.

17. de Jong, F.H., Inhibin, *Physiol Rev* 68: 555-607, 1988.

18. de Kretser, D.M., and D.M. Robertson, The isolation and physiology of inhibin and related proteins, *Biol Reprod* 40: 33-47, 1989.

19. Vale, W., C. Rivier, A. Hsueh, C. Campen, H. Meunier, T. Bicsak, J. Vaughan, A. Corrigan, W. Bardin, P. Sawchenko, F. Petraglia, J. Yu, P. Plotsky, J. Spiess, and J. Rivier, Chemical and biological characterization of the inhibin family of protein hormones, *Rec Prog Horm Res* 44: 1-34, 1988.

20. Ying, S.-Y., Inhibins, activins, and follistatins: gonadal proteins modulating the secretion of follicle-stimulating hormone, *Endocr Rev* 9: 267-293, 1988.

21. Hudson, B., H.W.G. Baker, L.W. Eddie, R.E. Higginson, H.G. Burger, D.M. de Kretser, M. Dobos, and V.W.K. Lee, Bioassays for inhibin: a critical review, *J Reprod Fertil Suppl* 26: 17-29, 1979.

22. Baker, H.W.G., L.W. Eddie, R.E. Higginson, B. Hudson, E.J. Keogh, and K.D. Niall, Assays of inhibin, In P. Franchimont and C.P. Channing (eds) *Intragonadal Regulation of Reproduction*, Academic Press, London, pp. 193-228, 1981.

23. Chari, S., S. Duraiswami, and P. Franchimont, A convenient and rapid bioassay for inhibin, *Horm Res* 7: 129-137, 1976.

24. Ramasharma, K., H.M. Shashidara Murthy, and N.R. Moudgal, A rapid bioassay for measuring inhibin activity, *Biol Reprod* 20: 831-835, 1979.

25. Nandini, S.G., H. Lipner, and N.R. Moudgal, A model system for studying inhibin, *Endocrinology* 98: 1460-1465, 1976.

26. Hermans, W.P., M.H.M. Debets, E.C.M. van Leeuwen, and F.H. de Jong, Time-related secretion of gonadotrophins after a single injection of steroid-free bovine follicular fluid in prepubertal and adult female rats, *J Endocrinol* 90: 69-76, 1981.

27. Findlay, J.K., T.W. Gill, and B.W. Doughton, Influence of season and sex on the inhibitory effect of ovine follicular fluid on plasma gonadotrophins in gonadectomized sheep, *J Reprod Fertil* 73: 329-335, 1985.

28. de Jong, F.H., S.D. Smith, and H.J. van der Molen, Bioassay of inhibin-like activity using pituitary cells *in vitro*, *J Endocrinol* 80: 91-102, 1979.

29. Eddie, L.W., H.W.G. Baker, R.E. Higginson, and B. Hudson, A bioassay for inhibin using pituitary cell cultures, *J Endocrinol* 81: 49-60, 1979.

30. Franchimont, P., J. Verstraelen-Proyard, M.T. Hazee-Hagelstein, C. Renard, A. Demoulin, J.P. Bourguignon, and J. Hustin, Inhibin: from concept to reality, *Vitam Horm* 37: 243-302, 1979.

31. van Dijk, S., J. Steenbergen, J.Th. Gielen, and F.H. de Jong, Sexual dimorphism in immunoneutralization of bioactivity of rat and ovine inhibin, *J Endocrinol* 111: 255-261, 1986.

32. Farnworth, P.G., D.M. Robertson, D.M. de Kretser, and H.G. Burger, Effects of 31 kilodalton bovine inhibin on follicle-stimulating hormone and luteinizing hormone in rat pituitary cells in vitro: actions under basal conditions, *Endocrinology* 122: 207-213, 1988.

33. Farnworth, P.G., D.M. Robertson, D.M. de Kretser, and H.G. Burger, Effects of 31 kDa bovine inhibin on FSH and LH in rat pituitary cells *in vitro*: antagonism of gonadotrophin-releasing hormone agonists, *J Endocrinol* 119: 233-241, 1988.

34. Wang, Q.F., P.G. Farnworth, J.K. Findlay, and H.G. Burger, Inhibitory effect of pure 31-kilodalton bovine inhibin on gonadotropin-releasing hormone (GnRH)-induced up-regulation of GnRH binding

sites in cultured rat anterior pituitary cells, *Endocrinology* 124: 363-368, 1989.

35. Tsonis, C.G., A.S. McNeilly, and D.T. Baird, Measurement of exogenous and endogenous inhibin in sheep serum using a new and extremely sensitive bioassay for inhibin based on inhibition of ovine pituitary FSH secretion *in vitro*, *J Endocrinol* 110: 341-352, 1986.

36. McLachlan, R.I., D.M. Robertson, H.G. Burger, and D.M. de Kretser, The radioimmunoassay of bovine and human follicular fluid and serum inhibin, *Mol Cell Endocrinol* 46: 175-185, 1986.

37. Hasegawa, Y., K. Miyamoto, M. Fukuda, Y. Takahashi, and M. Igarashi, Immunological study of ovarian inhibin, *Endocrinol Jpn* 33: 645-654, 1986.

38. Robertson, D.M., S. Hayward, D. Irby, J. Jacobsen, L. Clarke, R.I. McLachlan, and D.M. de Kretser, Radioimmunoassay of rat serum inhibin: changes after PMSG stimulation and gonadectomy, *Mol Cell Endocrinol* 58: 1-8, 1988.

39. Bicsak, T.A., E.M. Tucker, S. Cappel, J. Vaughan, J. Rivier, W. Vale, and A.J.W. Hsueh, Hormonal regulation of granulosa cell inhibin biosynthesis, *Endocrinology* 119: 2711-2719, 1986.

40. Rivier, C., and W. Vale, Inhibin: measurement and role in the immature female rat, *Endocrinology* 120: 1688-1690, 1987.

41. Ying, S.-Y., J. Czvik, A. Becker, N. Ling, N. Ueno, and R. Guillemin, Secretion of follicle-stimulating hormone and production of inhibin are reciprocally related, *Proc Natl Acad Sci USA* 84: 4631-4635, 1987.

42. Grootenhuis, A.J., J. Steenbergen, M.A. Timmerman, A.N.R.D. Dorsman, W.M.M. Schaaper, R.H. Meloen, and F.H. de Jong, Inhibin and activin-like activity in fluids from male and female gonads: different molecular weight forms and bioactivity/immunoactivity ratios, *J Endocrinology* 122: 293-301, 1989.

43. Bicsak, T.A., W. Vale, J. Vaughan, E.M. Tucker, S. Cappel, and A.J.W. Hsueh, Hormonal regulation of inhibin production by cultured Sertoli cells, *Mol Cell Endocrinol* 49: 211-217, 1987.

44. de Jong, F.H., and R.M. Sharpe, Evidence for inhibin-like activity in bovine follicular fluid, *Nature [Lond]* 263: 71-72, 1976.

45. Fukuda, M., K. Miyamoto, Y. Hasegawa, M. Nomura, M. Igarashi, K. Kangawa, and H. Matsuo, Isolation of bovine follicular fluid inhibin of about 32 kDa, *Mol Cell Endocrinol* 44: 55-60, 1986.

46. Robertson, D.M., F.L. de Vos, L.M. Foulds, R.I. McLachlan, H.G. Burger, F.J. Morgan, M.T.W. Hearn, and D.M. de Kretser, Isolation of a 31 kDa form of inhibin from bovine follicular fluid, *Mol Cell Endocrinol* 44: 271-277, 1986.

47. Ling, N., S.-Y. Ying, N. Ueno, S. Shimasaki, F. Esch, M. Hotta, and R. Guillemin, Pituitary FSH is released by a heterodimer of the β-subunits from the two forms of inhibin, *Nature [Lond]* 321: 779-782, 1986.

48. Miyamoto, K., Y. Hasegawa, M. Fukuda, M. Nomura, M. Igarashi, K. Kangawa, and H. Matsuo, Isolation of porcine follicular fluid inhibin of 32K daltons, *Biochem Biophys Res Commun* 129: 396-403, 1985.

49. Rivier, J., J. Spiess, R. McClintock, J. Vaughan, and W. Vale, Purification and partial characterization of inhibin from porcine follicular fluid, *Biochem Biophys Res Commun* 133: 120-127, 1985.

50. Robertson, D.M., L.M. Foulds, L. Leversha, F.J. Morgan, M.T.W. Hearn, H.G. Burger, R.E.H. Wettenhall, and D.M. de Kretser, Isolation of inhibin from bovine follicular fluid, *Biochem Biophys Res Commun* 126: 220-226, 1985.

51. Leversha, L.J., D.M. Robertson, F.L. de Vos, F.J. Morgan, M.T.W. Hearn, R.E.H. Wettenhall, J.K. Findlay, H.G. Burger, and D.M. de Kretser, Isolation of inhibin from ovine follicular fluid, *J Endocrinol* 113: 213-221, 1987.

52. Mason, A.J., J.S. Hayflick, N. Ling, F. Esch, N. Ueno, S.-Y. Ying, R. Guillemin, H. Niall, and P.H. Seeburg, Complementary DNA sequences of ovarian follicular fluid inhibin show precursor structure and homology with transforming growth factor-β, *Nature [Lond]* 318: 659-663, 1985.

53. Mayo, K.E., G.M. Cerelli, J. Spiess, J. Rivier, M.G. Rosenfeld, R.M. Evans, and W. Vale, Inhibin A-subunit cDNAs from porcine ovary and human placenta, *Proc Natl Acad Sci USA* 83: 5849-5853, 1986.

54. Forage, R.G., J.M. Ring, R.W. Brown, B.V. McInerney, G.S. Cobon, R.P. Gregson, D.M. Robertson, F.J. Morgan, M.T.W. Hearn, J.K. Findlay, R.E.H. Wettenhall, H.G. Burger, and D.M. de Kretser, Cloning and sequence analysis of cDNA species coding for the two subunits of inhibin from bovine

follicular fluid, *Proc Natl Acad Sci USA* 83: 3091-3095, 1986.

55. Esch, F.S., S. Shimasaki, K. Cooksey, M. Mercado, A.J. Mason, S.-Y. Ying, N. Ueno, and N. Ling, Complementary deoxyribonucleic acid (cDNA) cloning and DNA sequence analysis of rat ovarian inhibins, *Mol Endocrinol* 1: 388-396, 1987.

56. Woodruff, T.K., H. Meunier, P.B.C. Jones, A.J.W. Hsueh, and K.E. Mayo, Rat inhibin: molecular cloning of α- and β-subunit complementary deoxyribonucleic acids and expression in the ovary, *Mol Endocrinol* 1: 561-568, 1987.

57. Mason, A.J., H.D. Niall, and P.H. Seeburg, Structure of two human ovarian inhibins, *Biochem Biophys Res Commun* 135: 957-964, 1986.

58. Stewart, A.G., H.M. Milborrow, J.M. Ring, C.E. Crowther, and R.G. Forage, Human inhibin genes. Genomic characterization and sequencing, *FEBS Lett* 206: 329-334, 1986.

59. Jansen, E.H., J.M. Steenbergen, F.H. de Jong, and H.J. van der Molen, The use of affinity matrices in the purification of inhibin from bovine follicular fluid, *Mol Cell Endocrinol* 21: 109-117, 1981.

60. Lecomte-Yerna, M.J., M.T. Hazee-Hagelstein, C. Charlet-Renard, and P. Franchimont, Effect of neuraminidase on inhibin activity in vivo and in vitro, *Horm Res* 20: 277-284, 1984.

61. Miyamoto, K., Y. Hasegawa, M. Fukuda, and M. Igarashi, Demonstration of high molecular weight forms of inhibin in bovine follicular (bFF) by using monoclonal antibodies to bFF 32K inhibin, *Biochem Biophys Res Commun* 136: 1103-1109, 1986.

62. Bardin, C.W., P.L. Morris, C.-L. Chen, C. Shaha, J. Voglmayr, J. Rivier, J. Spiess, and W.W. Vale, Testicular inhibin: structure and regulation by FSH, androgens and EGF, In H.G. Burger, J.K. Findlay, D.M. de Kretser, and M. Igarashi (eds) *Serono Symposium: Inhibin-Non-Steroidal Regulation of Follicle Stimulating Hormone Secretion*, Raven Press, New York, pp. 179-190, 1987.

63. Grootenhuis, A.J., M.A. Timmerman, P. Hordijk, and F.H. de Jong, Inhibin in immature rat Sertoli cell conditioned medium: A 32 kDa $\alpha\beta$-B-dimer, Submitted.

64. Ying, S.-Y., A. Becker, A. Baird, N. Ling, N. Ueno, F. Esch, and R. Guillemin, Type beta transforming growth factor (TGF-β) is a potent stimulator of the basal secretion of follicle stimulating hormone (FSH) in a pituitary monolayer system, *Biochem Biophys res Commun* 135: 950-956, 1986.

65. Ling, N., S.-Y. Ying, N. Ueno, S. Shimasaki, F. Esch, M. Hotta, and R. Gullemin, A homodimer of the β-subunits of inhibin A stimulates the secretion of pituitary follicle stimulating hormone, *Biochem Biophys Res Commun* 138: 1129-1137, 1986.

66. Vale, W., J. Rivier, J. Vaughan, R. McClintock, A. Corrigan, W. Woo, D. Karr, and J. Spiess, Purification and characterization of an FSH releasing protein from porcine ovarian follicular fluid, *Nature [Lond]* 321: 776-779, 1986.

67. Eto, Y., T. Tsuji, M. Takezawa, S. Takano, Y. Yokogawa, and H. Shibai, Purification and characterization of erythroid differentiation factor (EDF) isolated from human leukemia cell line THP-1, *Biochem Biophys Res Commun* 142: 1095-1103, 1987.

68. Kitaoka, M., N. Yamashita, Y. Eto, H. Shibai, and E. Ogata, Stimulation of FSH release by erythroid differentiation factor (EDF), *Biochem Biophys Res Commmun* 146: 1382-1385, 1987.

69. Sugino, K., T. Nakamura, K. Takio, K. Titani, K. Miyamoto, Y. Hasegawa, M. Igarashi, and H. Sugino, Inhibin alpha-subunit monomer is present in bovine follicular fluid, *Biochem Biophys Res Commun* 159: 1323-1329, 1989.

70. Knight, P.G., A.J. Beard, J.H.M. Wrathall, and R.J. Castillo, Evidence that the bovine ovary secretes large amounts of monomeric inhibin α subunit and its isolation from bovine follicular fluid, *J Mol Endocrinol* 2: 189-200, 1989.

71. de Jong, F.H., A.J. Grootenhuis, I.A. Klaij, J.M.W. Toebosch, A.M. Ultee-van Gessel, S. Shimasaki, and J.A. Grootegoed, Regulation of inhibin production in rat Sertoli cells, In M. Serio (ed) *Perspectives in Andrology*, Raven Press, New York, pp. 235-242, 1989.

72. Cate, R.L., R.J. Mattaliano, C. Hession, R. Tizard, N.M. Farber, A. Cheung, E.G. Ninfa, A.Z. Frey, D.J. Gash, E.P. Chow, R.A. Fisher, J.M. Bertonis, G. Torres, B.P. Wallner, K.L. Ramachandran, R.C. Ragin, T.F. Manganaro, D.T. MacLaughlin, and P.K. Donahoe, Isolation of the bovine and human genes for Müllerian inhibiting substance and expression of the human gene in animal cells, *Cell* 45: 685-698, 1986.

73. Wozney, J.M., V. Rosen, A.J. Celeste, L.M. Mitsock, M.J. Whitters, R.W. Kriz, R.M. Hewick, and E.A.

Wang, Novel regulators of bone formation: molecular clones and activities, *Science* 242: 1528-1534, 1988.

74. Padgett, R.W., R.D. St. Johnston, and W.M. Gelbart, A transcript from a *drosophila* pattern gene predicts a protein homologous to the transforming growth factor-β family, *Nature [Lond]* 325: 81-84, 1987.

75. Weeks, D.L., and D.A. Melton, A maternal mRNA localized to the vegetal hemisphere is Xenopus eggs codes for a growth factor related to TGF-β, *Cell* 51: 861-867, 1987.

76. McCullagh, D.R., Dual endocrine activity of the testes, *Science* 76: 19-20, 1932.

77. Steinberger, A., and E. Steinberger, Secretion of an FSH-inhibiting factor by cultured Sertoli cells, *Endocrinology* 99: 918-921, 1976.

78. de Jong, F.H., and R.M. Sharpe, Gonadotrophins, testosterone and spermatogenesis in neonatally irradiated male rats: evidence for a role of the Sertoli cell in follicle-stimulating hormone feedback, *J Endocrinol* 75: 209-219, 1977.

79. Lee, W., A.J., Mason, R. Schwall, E. Szonyi, and J.P. Mather, Secretion of activin by interstitial cells in the testis, *Science* 243: 396-398, 1989.

80. Risbridger, G.P., J. Clements, D.M. Robertson, A.E. Drummond, J. Muir, H.G. Burger, and D.M. de Kretser, Immuno- and bioactive inhibin and inhibin α-subunit expression in rat Leydig cell culture, *Mol Cell Endocrinol* 66: 119-122, 1989.

81. Rodgers, R.J., S.J. Stuchbery, and J.K. Findlay, Inhibin mRNAs in ovine and bovine ovarian follicles and corpora lutea throughout the estrous cycle and gestation, *Mol Cell Endocrinol* 62: 95-101, 1989.

82. Torney, A.H., Y.M. Hodgson, R. Forage, and D.M. de Kretser, Cellular localization of inhibin mRNA in the bovine ovary by in-situ hybridization, *J Reprod Fertil* 86: 391-399, 1989.

83. McLachlan, R.I., D.L. Healy, D.M. Robertson, H.G. Burger, and D.M. de Kretser, The human placenta: a novel source of inhibin, *Biochem Biophys Res Commun* 140: 485-490, 1986.

84. Toebosch, A.M.W., D.M. Robertson, J. Trapman, P. Klaassen, R.A. de Paus, F.H. de Jong, and J.A. Grootegoed, Effects of FSH and IGF-I on immature rat Sertoli cells: inhibin α- and β-subunit mRNA levels and inhibin secretion, *Mol Cell Endocrinol* 55: 101-105, 1988.

85. Toebosch, A.M.W., D.M. Robertson, I.A. Klaij, F.H. de Jong, and J.A. Grootegoed, Effects of FSH and testosterone on highly purified rat Sertoli cells: inhibin α-subunit mRNA expression and inhibin secretion are enhanced by FSH but not by testosterone, *J Endocrinol* 122: 757-762, 1989.

86. Klaij, I.A., A.M.W. Toebosch, A.P.N. Themmen, S. Shimasaki, F.H. de Jong, and J.A. Grootegoed, Regulation of inhibin α- and β_B-subunit mRNA levels in rat Sertoli cells, *Mol Cell Endocrinol*, In Press.

87. Gonzales, G.F., G.P. Risbridger, and D.M. de Kretser, In vitro synthesis and release of inhibin in response to FSH stimulation by isolated segments of seminiferous tubules from normal adult male rats, *Mol Cell Endocrinol* 59: 179-185, 1988.

88. Handelsman, D.J., J.A. Spaliviero, E. Kidston, and D.M. Robertson, Highly polarized secretion of inhibin by Sertoli cells in vitro, *Endocrinology* 125: 721-729, 1989.

89. Ultee-van Gessel, A.M., F.G. Leemborg, F.H. de Jong, and H.J. van der Molen, In-vitro secretion of inhibin-like activity by Sertoli cells from normal and prenatally irradiated immature rats, *J Endocrinol* 109: 411-418, 1986.

90. Verhoeven, G., and P. Franchimont, Regulation of inhibin secretion by Sertoli cell-enriched cultures, *Acta Endocrinol [Copenh]* 102: 136-143, 1983.

91. Steinberger, A., Regulation of inhibin secretion in the testis, In P. Franchimont and C.P. Channing (eds) *Intragonadal Regulation of Reproduction*, Academic Press, London, pp. 283-298, 1981.

92. Rivarola, M.A., P. Sanchez, and J.M. Saez, Inhibition of RNA and DNA synthesis in Sertoli cells by co-culture with spermatogenic cells, *Int J Androl* 9: 424-434, 1986.

93. Le Magueresse, B., and B. Jégou, Possible involvement of germ cells in the regulation of oestradiol-17β and ABP secretion by immature rat Sertoli cells (in vitro studies), *Biochem Biophys Res Commun* 141: 861-869, 1986.

94. Bhasin, S., L.A. Krummen, R.S. Swerdloff, B.S. Morelos, W.H. Kim, G.S. DiZerega, N. Ling, F. Esch, S. Shimasaki, and J. Toppari, Stage dependent expression of inhibin α and β-B subunits during the cycle of the rat seminiferous epithelium, *Endocrinology* 124: 987-991, 1989.

95. Au, C.L., D.M. Robertson, and D.M. de Kretser, Effects of hypophysectomy and subsequent FSH and

testosterone treatment on inhibin production by adult rat testes, *J Endocrinol* 105: 1-6, 1985.

96. Ultee-van Gessel, A.M., M.A. Timmerman, and F.H. de Jong, Effects of treatment of neonatal rats with highly purified FSH alone and in combination with LH on testicular function and endogenous hormone levels at various ages, *J Endocrinol* 116: 413-420, 1988.

97. McLachlan, R.I., A.M. Matsumoto, H.G. Burger, D.M. de Kretser, and W.J. Bremner, Relative roles of follicle-stimulating hormone and luteinizing hormone in the control of inhibin secretion in normal men, *J Clin Invest* 82: 880-884, 1988.

98. Drummond, A.E., G.P. Risbridger, and D.M. de Kretser, The involvement of Leydig cells in the regulation of inhibin secretion by the testis, *Endocrinology* 125: 510-515, 1989.

99. Hermans, W.P., E.C.M. van Leeuwen, M.H.M. Debets, and F.H. de Jong, Involvement of inhibin in the regulation of follicle-stimulating hormone concentrations in prepubertal and adult, male and female rats, *J Endocrinol* 86: 79-92, 1980.

100. Rivier, C., S. Cajander, J. Vaughan, A.J.W. Hsueh, and W. Vale, Age-dependent changes in physiological action, content, and immunostaining of inhibin in male rats, *Endocrinology* 123: 120-126, 1988.

101. Burger, H., R.I. McLachlan, M. Bangah, H. Quigg, J.K. Findlay, D.M. Robertson, D.M. de Kretser, G.L. Warne, G.A. Werther, I.L. Hudson, J.J. Cook, R. Fiedler, S. Greco, A.B.W. Yong, and P. Smith, Serum inhibin concentrations rise throughout normal male and female puberty, *J Clin Endocrinol Metab* 67: 689-694, 1988.

102. Erickson, G.F., and A.J.W. Hsueh, Secretion of "inhibin" by rat granulosa cells in vitro, *Endocrinology* 103: 1960-1963, 1978.

103. Zhiwen, Z., R.S. Carson, A.C. Herington, V.W.K. Lee, and H.G. Burger, Follicle-stimulating hormone and somatomedin-C stimulate inhibin production by rat granulosa cells *in vitro*, *Endocrinology* 120: 1633-1638, 1987.

104. Henderson, K.M., and P. Franchimont, Regulation of inhibin production by ovine ovarian cells *in vitro*, *J Reprod Fertil* 63: 431-442, 1981.

105. Henderson, K.M., and P. Franchimont, Inhibin production by bovine ovarian tissues *in vitro* and its regulation by androgens, *J Reprod Fertil* 67: 291-298, 1983.

106. Channing, C.P., K. Tanabe, M. Chacon, and J.T. Tildon, Stimulatory effects of follicle-stimulating hormone and luteinizing hormone upon secretion of progesterone and inhibin activity by cultured infant human ovarian granulosa cells, *Fertil Steril* 42: 598-605, 1984.

107. Henderson, K.M., P. Franchimont, C. Charlet-Renard, and K.P. McNatty, Effect of follicular atresia on inhibin production by bovine granulosa cells *in vitro* and inhibin concentrations in the follicular fluid, *J Reprod Fertil* 72: 1-8, 1984.

108. Sander, H.J., E.C.M. van Leeuwen, and F.H. de Jong, Inhibin-like activity in media from cultured rat granulosa cells collected throughout the oestrous cycle, *J Endocrinol* 103: 77-84, 1984.

109. Tsonis, C.G., S.G. Hillier, and D.T. Baird, Production of inhibin bioactivity by human granulosa-lutein cells: stimulation by LH and testosterone in vitro, *J Endocrinol* 112: R11-R14, 1987.

110. Lee, V.W.K., J. McMaster, H. Quigg, J. Findlay, and L. Leversha, Ovarian and peripheral blood inhibin concentrations increase with gonadotropin treatment in immature rats, *Endocrinology* 108: 2403-2405, 1981.

111. Lee, V.W.K., J. McMaster, H. Quigg, and L. Leversha, Ovarian and circulating inhibin levels in immature female rats treated with gonadotropin and after castration, *Endocrinology* 111: 1849-1854, 1982.

112. Lee, V.W.K., and W.R. Gibson, Ovulation rate and inhibin levels in gonadotrophin-treated mice, *Aust J Biol Sci* 38: 115-120, 1985.

113. Davis, S.R., F. Dench, I. Nikolaidis, J.A. Clements, R.G. Forage, Z. Krozowski, and H.G. Burger, Inhibin A-subunit gene expression in the ovaries of immature female rats is stimulated by pregnant mare serum gonadotrophin, *Biochem Biophys Res Commun* 138: 1191-1195, 1986.

114. Fujii, T., D.J. Hoover, and C.P. Channing, Changes in inhibin activity, and progesterone, oestrogen and androstenedione concentrations, in rat follicular fluid throughout the oestrous cycle, *J Reprod Fertil* 69: 307-314, 1983.

115. DePaolo, L.V., D. Shander, P.M. Wise, C.A. Barraclough, and C.P. Channing, Identification of inhibin-like activity in ovarian venous plasma of rats during the estrous cycle, *Endocrinology* 105: 647-654, 1979.

116. Channing, C.P., P. Gagliano, D.J. Hoover, K. Tanabe, S.K. Batta, J. Sulewski, and P. Lebech, Relationship between human follicular fluid inhibin F activity and steroid content, *J Clin Endocrinol Metab* 52: 1193-1198, 1981.

117. Chappel, S.C., J.A. Holt, and H.G. Spies, Inhibin: differences in bioactivity within human follicular fluid in the follicular and luteal stages of the menstrual cycle, *Proc Soc Exp Biol Med* 163: 310-314, 1980.

118. Tsonis, C.G., H. Quigg, V.W.K. Lee, L. Leversha, A.O. Trounson, and J.K. Findlay, Inhibin in individual ovine follicles in relation to diameter and atresia, *J Reprod Fertil* 67: 83-90, 1983.

119. van de Wiel, D.F.M., S. Bar-Ami, A. Tsafriri, and F.H. de Jong, Oocyte maturation inhibitor, inhibin and steroid concentrations in porcine follicular fluid at various stages of the oestrous cycle, *J Reprod Fertil* 68: 247-252, 1983.

120. McLachlan, R.I., D.M. Robertson, D.L. Healy, H.G. Burger, and D.M. de Kretser, Circulating immunoreactive inhibin levels during the normal human menstrual cycle, *J Clin Endocrinol Metab* 65: 954-961, 1987.

121. Hasegawa, Y., K. Miyamoto, S. Iwamura, and M. Igarashi, Changes in serum concentrations of inhibin in cyclic pigs, *J Endocrinol* 118: 211-219, 1988.

122. McLachlan, R.I., D.L. Healy, D.M. Robertson, H.G. Burger, and D.M. de Kretser, Circulating immunoactive inhibin in the luteal phase and early gestation of women undergoing ovulation induction, *Fertil Steril* 48: 1001-1005, 1987.

123. Robertson, D.M., C.G., Tsonis, R.I. McLachlan, D.J. Handelsman, R. Leask, D.T. Baird, A.S. McNeilly, S. Hayward, D.L. Healy, J.K. Findlay, H.G. Burger, and D.M. de Kretser, Comparison of inhibin immunological and *in vitro* biological activities in human serum, *J Clin Endocrinol Metab* 67: 438-443, 1988.

124. McLachlan, R.I., D.M. Robertson, D.L. Healy, D.M. de Kretser, and H.G. Burger, Plasma inhibin levels during gonadotropin-induced ovarian hyperstimulation for IVF: a new index of follicular function? *Lancet* 1: 1233-1234, 1986.

125. Rivier, C., and W. Vale, Immunoneutralization of endogenous inhibin modifies hormone secretion and ovulation rate in the rat, *Endocrinology* 125: 152-157, 1989.

126. Henderson, K.M., P. Franchimont, M.J. Lecomte-Yerna, N. Hudson, and K. Ball, Increase in ovulation rate after active immunization of sheep with inhibin partially purified from bovine follicular fluid, *J Endocrinol* 102: 305-309, 1984.

127. Forage, R.G., R.W. Brown, K.J. Oliver, B.T. Atrache, P.L. Devine, G.C. Hudson, N.H. Goss, K.C. Bertram, P. Tolstoshev, D.M. Robertson, D.M. de Kretser, B. Doughton, H.G. Burger, and J.K. Findlay, Immunization against an inhibin subunit produced by recombinant DNA techniques results in increased ovulation rate in sheep, *J Endocrinol* 114: R1-R4, 1987.

128. de Greef, W.J., F.H. de Jong, J. de Koning, J. Steenbergen, and P.D.M. van der Vaart, Studies on the mechanism of the selective suppression of plasma levels of follicle-stimulating hormone in the female rat after administration of steroid-free bovine follicular fluid, *J Endocrinol* 97: 327-338, 1983.

129. Clarke, I.J., J.K. Findlay, J.T. Cummins, and W.J. Ewens, Effects of ovine follicular fluid on plasma LH and FSH secretion in ovariectomized ewes to indicate the site of action of inhibin, *J Reprod Fertil* 77: 575-585, 1986.

130. Lumpkin, M.D., L.V. DePaolo, and A. Negro-Vilar, Pulsatile release of follicle-stimulating hormone in ovariectomized rats is inhibited by porcine follicular fluid (inhibin), *Endocrinology* 114: 201-206, 1984.

131. Ireland, J.J., A.D. Curato, and J. Wilson, Effect of charcoal-treated bovine follicular fluid on secretion of LH and FSH in ovariectomized heifers, *J Anim Sci* 57: 1512-1516, 1983.

132. Condon, T.P., R.E. Leipheimer, and J.J. Curry, Preliminary evidence for a CNS site of action for ovarian inhibin, *Life Sci* 32: 1691-1698, 1983.

133. Lumpkin, M., A. Negro-Vilar, P. Franchimont, and S. McCann, Evidence for a hypothalamic site of action of inhibin to suppress FSH release, *Endocrinology* 108: 1101-1104, 1981.

134. Sawchenko, P.E., P.M. Plotsky, S.W. Pfeiffer, E.T. Cunningham, Jr., J. Vaughan, J. Rivier, and W. Vale, Inhibin β in central neural pathways involved in the control of oxytocin secretion, *Nature [Lond]* 334: 615-617, 1988.

135. Franchimont, P., F. Croze, A. Demoulin, R. Bologne, and J. Hustin, Effect of inhibin on rat testicular desoxyribonucleic acid (DNA) synthesis in vivo and in vitro, *Acta Endocrinol [Copenh]* 98: 312-320, 1981.

136. van Dissel-Emiliani, F.M.F., A.J. Grootenhuis, F.H. de Jong and D.G. de Rooij, Inhibin reduces spermatogenesis spermatogonial numbers in testes of adult mice and Chinese hamsters, *Endocrinology* 125: 1898-1903, 1989.

137. Chari, S., G. Aumüller, E. Daume, G. Sturm, and C. Hopkinson, The effects of human follicular fluid inhibin on the morphology of the ovary of the immature rat, *Arch Gynecol* 230: 239-249, 1981.

138. Ying, S.-Y., A. Becker, N. Ling, N. Ueno, and R. Guillemin, Inhibin and beta type transforming growth factor (TGFβ) have opposite modulating effects on the follicle stimulating hormone (FSH)-induced aromatase activity of cultured rat granulosa cells, *Biochem Biophys Res Commun* 136: 969-975, 1986.

139. Hutchinson, L.A., J.K. Findlay, F.L. de Vos, and D.M. Robertson, Effects of bovine inhibin, transforming growth factor-β and bovine activin-A on granulosa cell differentiation, *Biochem Biophys Res Commun* 146: 1405-1412, 1987.

140. Sugino, H., T. Nakamura, Y. Hasegawa, K. Miyamoto, M. Igarashi, Y. Eto, H. Shibai, and K. Titani, Identification of a specific receptor for erythroid differentiation factor on follicular granulosa cell, *J Biol Chem* 263: 15249-15252, 1988.

141. Hsueh, A.J.W., K.D. Dahl, J. Vaughan, E. Tucker, J. Rivier, C.W. Bardin, and W. Vale, Heterodimers and homodimers of inhibin subunits have different paracrine action in the modulation of luteinizing hormone-stimulated androgen biosynthesis, *Proc Natl Acad Sci USA* 84: 5082-5086, 1987.

142. Grootenhuis, A.J., R. Melsert, M.A. Timmerman, J.W. Hoogerbrugge, F.F.G. Rommerts, and F.H. de Jong, Short-term stimulatory effect of Sertoli cell conditioned medium on Leydig cell steroidogenesis is not mediated by inhibin, Submitted.

143. Bapat, B.V., T.D. Nandedkar, and A.R. Sheth, Reversal of the anti-implantation effect of inhibin with progesterone in the hamster, *Int J Fertil* 31: 71-76, 1986.

144. Petraglia, F., P. Sawchenko, A.T.W. Lim, J. Rivier, and W. Vale, Localization, secretion, and action of inhibin in human placenta, *Science* 237: 187-189, 1987.

145. Petraglia, F., J. Vaughan, and W. Vale, Inhibin and activin modulate the release of gonadotropin-releasing hormone, human chorionic gonadotropin, and progesterone from cultured human placental cells, *Proc Natl Acad Sci USA* 86: 5114-5117, 1989.

146. Findlay, J.K., C.G. Tsonis, B. Doughton, R.W. Brown, K.C. Bertram, G.H. Braid, G.C. Hudson, M.L. Tierney, N.H. Goss, and R.G. Forage, Immunization against the amino-terminal peptide (α_N) of the alpha$_{43}$ subunit of inhibin impairs fertility in sheep, *Endocrinology* 124: 3122-3124, 1989.

147. Meunier, H., C. Rivier, R.M. Evans, and W. Vale, Gonadal and extragonadal expression of inhibin α, βA, and βB subunits in various tissues predicts diverse functions, *Proc Natl Acad Sci USA* 85: 247-251, 1988.

148. Kondo, S., M. Hashimoto, Y. Etoh, M. Murata, H. Shibai, and M. Muramatsu, Identification of the two types of specific receptor for activin/EDF expressed on Friend leukemia and embryonal carcinoma cells, *Biochem Biophys Res Commun* 161: 1267-1272, 1989.

149. Campen, C., and W. Vale, Characterization of activin A binding sites on the human leukemia cell line K562, *Biochem Biophys Res Commun* 157: 844-849, 1988.

150. Totsuka, Y., M. Tabuchi, I. Kojima, H. Shibai, and E. Ogata, A novel action of activin A: stimulation of insulin secretion in rat pancreatic islets, *Biochem Biophys Res Commun* 156: 335-339, 1988.

151. Mine, T., I. Kojima, and E. Ogata, Stimulation of glucose production by activin-A in isolated rat hepatocytes, *Endocrinology* 125: 586-591, 1989.

152. Merchenthaler, I., M.D. Culler, P. Petrusz, and A. Negro-Vilar, Immunocytochemical localization of inhibin in rat and human reproductive tissues, *Mol Cell Endocrinol* 54: 239-243, 1987.

153. Cuevas, P., S.-Y. Ying, N. Ling, N. Ueno, F. Esch, and R. Guillemin, Immunohistochemical detection of inhibin in the gonad, *Biochem Biophys Res Commun* 142: 23-30, 1987.

154. Meunier, H., S.B. Cajander, V.J. Roberts, C. Rivier, P.E. Sawchenko, A.J.W. Hsueh, and W. Vale, Rapid changes in the expression of inhibin α-, βA-, and βB-subunits in ovarian cell types during the rat estrous cycle, *Mol Endocrinol* 2: 1352-1363, 1988.

155. Rivier, C., V. Roberts, and W. Vale, Possible role of luteinizing hormone and follicle-stimulating hormone in modulating inhibin secretion and expression during the estrous cycle of the rat, *Endocrinology* 125: 876-882, 1989.

156. Davis, S.R., Z. Krozowski, R.I. McLachlan, and H.G. Burger, Inhibin gene expression in the human corpus luteum, *J Endocrinol* 115: R21-R23, 1987.
157. Woodruff, T.K., J. D'Agostino, N.B. Schwartz, and K.E. Mayo, Dynamic changes in inhibin messenger RNAs in rat ovarian follicles during the reproductive cycle, *Science* 239: 1296-1299, 1988.
158. Woodruff, T.K., J. D'Agostino, N.B. Schwartz, and K.E. Mayo, Decreased inhibin gene expression in preovulatory follicles requires primary gonadotropin surges, *Endocrinology* 124: 2193-2199, 1989.
159. Crawford, R.J., V.E. Hammond, B.A. Evans, J.P. Coghlan, J. Haralambidis, B. Hudson, J.D. Penschow, R.I. Richards, and G.W. Tregear, α-Inhibin gene expression occurs in the ovine adrenal cortex, and is regulated by adrenocorticotropin, *Mol Endocrinol* 1: 699-706, 1987.
160. Roberts, V., H. Meunier, J. Vaughan, J. Rivier, C. Rivier, W. Vale, and P. Sawchenko, Production and regulation of inhibin subunits in pituitary gonadotropes, *Endocrinology* 124: 552-554, 1989.
161. Sugino, H., T. Nakamura, Y. Hasegawa, K. Miyamoto, Y. Abe, M. Igarashi, Y. Eto, H. Shibai, and K. Titani, Erythroid differentiation factor can modulate follicular granulosa cell functions, *Biochem Biophys Res Commun* 153: 281-288, 1988.
162. Hasegawa, Y., K. Miyamoto, Y. Abe, T. Nakamura, H. Sugino, Y. Eto, H. Shibai, and M. Igarashi, Induction of follicle stimulating hormone receptor by erythroid differentiation factor on rat granulosa cell, *Biochem Biophys Res Commun* 156: 668-674, 1988.
163. O, W.-S., D.M. Robertson, and D.M. de Kretser, Inhibin as an oocyte meiotic inhibitor, *Mol Cell Endocrinol* 62: 307-311, 1989.
164. Yu, J., L. Shao, V. Lemas, A.L. Yu, J. Vaughan, J. Rivier, and W. Vale, Importance of FSH-releasing protein and inhibin in erythrodifferentiation, *Nature [Lond]* 330: 765-767, 1987.
165. Broxmeyer, H.E., L. Lu, S. Cooper, R.H. Schwall, A.J. Mason, and K. Nikolics, Selective and indirect modulation of human multipotential and erythroid hematopoietic progenitor cell proliferation by recombinant human activin and inhibin, *Proc Natl Acad Sci USA* 85: 9052-9056, 1988.
166. Murata, M., Y. Eto, H. Shibai, M. Sakai, and M. Muramatsu, Erythroid differentiation factor is encoded by the same mRNA as that of the inhibin β_A chain, *Proc Natl Acad Sci USA* 85: 2434-2438, 1988.
167. Hedger, M.P., A.E. Drummond, D.M. Robertson, G.P. Risbridger, and D.M. de Kretser, Inhibin and activin regulate [^{3}H]thymidine uptake by rat thymocytes and 3T3 cells in vitro, *Mol Cell Endocrinol* 61: 133-138, 1989.
168. Kojima, I., and E. Ogata, Dual effect of activin A on cell growth in Balb/C 3T3 cells, *Biochem Biophys Res Commun* 159: 1107-1113, 1989.

ROLE OF ENDOTOXIN AND INTERLEUKIN-1 IN MODULATING ACTH, LH AND SEX STEROID SECRETION

C. Rivier

The Clayton Foundation Laboratories for Peptide Biology
The Salk Institute
La Jolla, CA 92037

Introduction

There is increasing evidence that the immune and the endocrine axis are functionally connected and influence each other's activity {review in (1)}. While the effects of hormones such as corticosteroids and endogenous opiates on immune functions have been abundantly described, the reciprocal arm of this bidirectional pathway, *i.e.* the influence exerted by signals from the immune system on endocrine functions, has only recently started to be investigated. It is presently believed that products of activated immune cells, such as the lymphokines, are central to this bilateral communication (2). During the early part of immune activation, in particular, stimulated macrophages produce interleukin-1 (IL-1), which can enter the circulation (3) and alter the activity of the hypothalamic-pituitary-adrenal (HPA) and/or hypothalamic-pituitary-gonadal (HPG) axis.

A number of experimental models have been proposed, which allow studies of the effect of immune activation on the endocrine axis. For example, Besedovsky *et al.* have shown that in rats injected with sheep red blood cells, the peak of the immune response corresponded to a marked increase in corticosterone secretion (4). Similarly, Dunn and colleagues have reported an increase in plasma corticosterone levels in mice injected with Newcastle Disease virus (5). Finally, lipopolysaccharides (LPS) extracted from the walls of gram-negative bacteria, which are often used as a means of mimicking some of the early events which occur after activation of the immune system, have long been known to stimulate the HPA axis (6,7).

Over the past few years, work in our laboratory has been devoted to investigate the effects of LPS on the HPA and the HPG axis, and to elucidate some of the mechanisms, including an increased secretion of endogenous lymphokines, which might mediate these effects. This chapter describes some of these studies.

Materials and Methods

Experiments

Adults male Sprague-Dawley rats (230-250 g body weight) or Balb/c mice (23-26 g body weight) were used throughout the experiments. The animals were kept under standard light and feeding regimens. Castration, when mentioned, was performed under ether anesthesia. Intravenous and/or intracerebroventricular cannulas were implanted under metofane anesthesia as previously described (8,9). Intraperitoneal injections were

Circulating Regulatory Factors and Neuroendocrine Function
Edited by J. C. Porter and D. Ježová
Plenum Press, New York, 1990

295

done in awake animals. Blood samples were collected on ice, plasmas were separated and kept frozen until the appropriate RIAs were carried out (8,10,11). Statistical analysis used the multiple range test of Duncan.

Drugs

A lipopolysaccharide (LPS) extracted from the walls of E. Coli was used to stimulate some of the events characterizing the early part of the immune response, including the production of endogenous lymphokines (12). LPS was first dissolved in 2N acetic acid, then in phosphate buffer, and injected either peripherally or into the lateral ventricle of the brain. IL-1α was a gift of Dr. Peter Lomedico, Hoffman-LaRoche (Nutley, NJ) and IL-1β a gift from Dr. Steven Gillis, Immunex (Seattle, WA). Both lymphokines were dissolved in phosphate buffer containing 0.1% BSA and 0.01% ascorbic acid. The anti-interleukin-1 receptor serum which was used in these studies was generously provided by Dr. Richard Chizzonite, Hoffman-Laroche (Nutley, NJ). These antibodies, which block IL-1 binding and bioactivity in a number of mouse cells, were injected subcutaneously 18 hours before IL-1 or LPS.

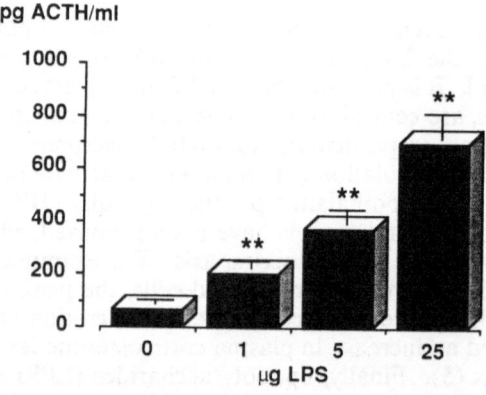

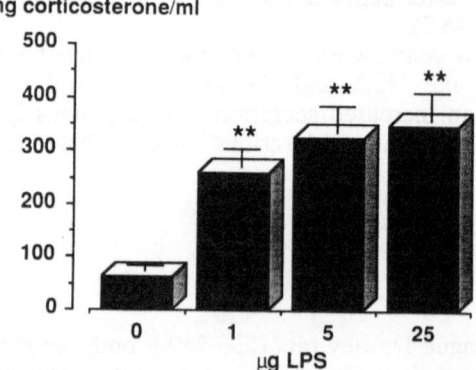

Figure 1.   Dose-related effect of endotoxin (a lipopolysaccharide, LPS), injected iv, on ACTH and corticosterone secretion in rats. Blood samples were obtained 45 min after treatment. Each bar represents the mean ± SE of 5-6 animals. ** denotes $P < 0.01$.

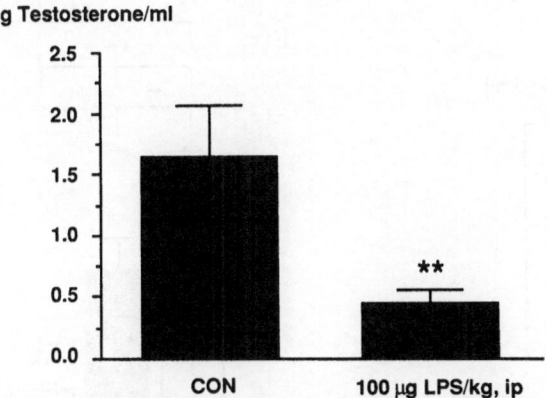

Figure 2. Effect of endotoxin, injected ip, on plasma testosterone levels in intact adult male rats. Blood samples were obtained 20 hr after treatment. Each bar represents the mean ± SE of 5 animals. ** denotes $P < 0.01$.

Results and Discussion

Effect of LPS on the HPA and HPG Axis

The injection of LPS, which stimulates the production of endogenous lymphokines (13,14), causes a marked activation of the HPA axis (6,7,15), as well as a significant inhibition of the HPG axis (16). With regard to the first axis, this effect is illustrated by the dose-related increases in plasma ACTH and corticosterone levels measured in rats or mice administered LPS. While in rats, the minimal effective dose of LPS necessary to stimulate both ACTH and corticosterone secretion was below one 8 μg per animal (Figure 1), the doses of LPS which released ACTH in mice (30 μg) were significantly higher than

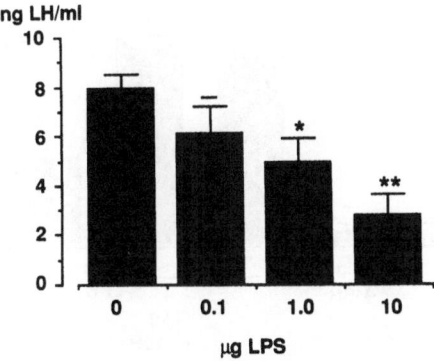

Figure 3. Dose-related effect of endotoxin, injected into the lateral ventricle of the brain, on LH secretion in castrated male rats. Blood samples were obtained 3 hr after treatment. Each bar represents the mean ± SE of 6 animals. — denotes $P > 0.05$; * denotes $P < 0.05$; ** denotes $P < 0.01$.

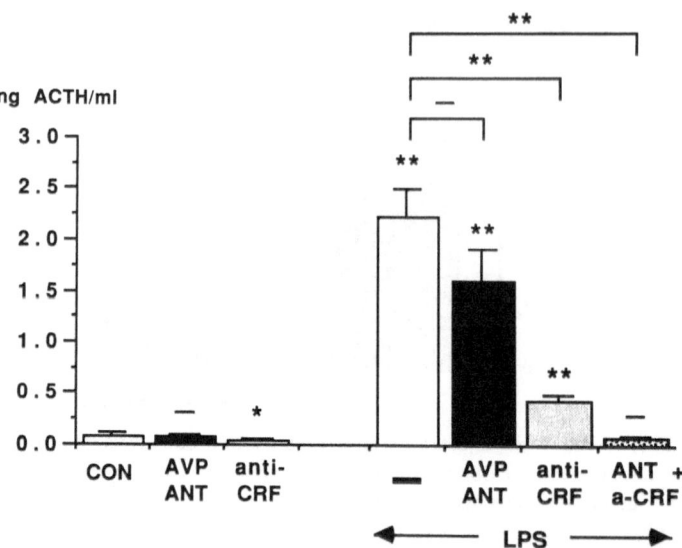

Figure 4. Interaction between LPS (25 µg), a CRF antiserum (anti-CRF, 0.2 ml) and the vasopressin antagonist {1-deaminopenicilamine-2-(O-methyl)tyrosine} arginine-vasopressin (AVPant, 50 µg) on ACTH secretion in rats. All treatments were injected iv. Blood samples were obtained 45 min later. Each bar represents the mean ± SE of 5 animals. — denotes $P > 0.05$; ∗∗ denotes $P < 0.01$.

those (3 µg) which increased corticosterone levels (Rivier and Vale, submitted). Whether the discrepancy between ACTH and corticoid secretion is due to the documented dissociation between the release of these two hormones which can happen under some experimental circumstances (17), or are indicative of a direct adrenal effect of the endotoxin (18,19), is presently unknown. In intact male rats, the peripheral administration of LPS caused a marked decrease in the circulating levels of testosterone (Figure 2),

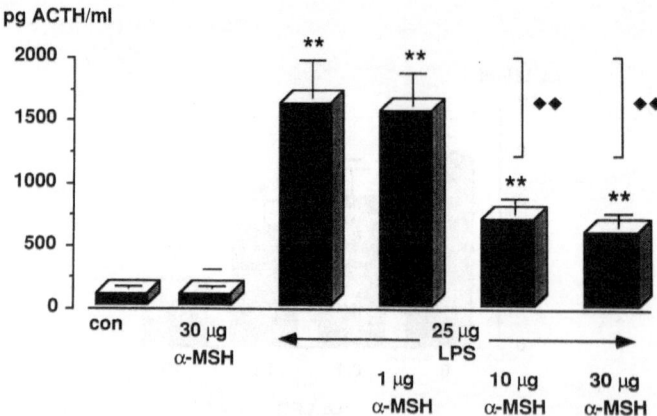

Figure 5. Interaction between LPS and α-MSH on ACTH secretion in mice. LPS was injected ip and α-MSH was administered sc. Blood samples were obtained 6 hr later. Each bar represents the mean ± SE of 8 animals. — denotes $P > 0.05$; ∗∗ denotes $P < 0.01$ from control; *closed diamonds* denote $P < 0.01$ from LPS alone.

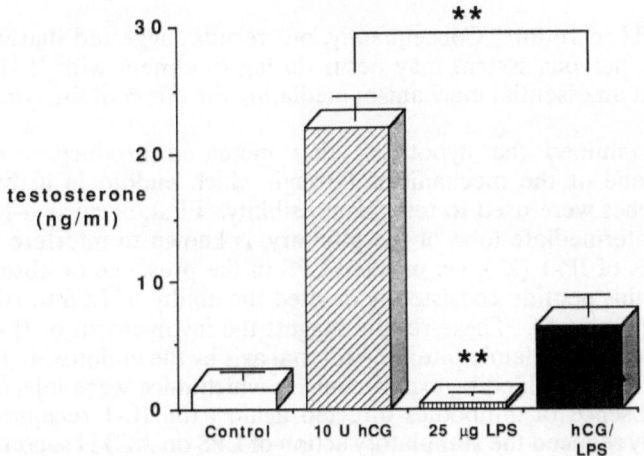

Figure 6. Interaction between LPS and hCG on testosterone secretion in intact adult male rats. LPS was injected ip at time 0, and hCG was injected sc 20 hr later. Blood samples were obtained 1 hr after the gonadotropin. Each bar represents the mean ± SE of 6 rats. ** denotes $P < 0.01$.

which was not accompanied by any measurable changes in LH or FSH release (data not shown). In contrast, LPS injected intracerebroventricularly was very effective in lowering plasma LH levels of castrated rats (Figure 3).

The next step was to investigate some of the mechanisms which might mediate the endocrine changes caused by LPS, and in particular to examine the hypothesis that LPS-induced production of endogenous lymphokines represented one such mechanism. Two experimental approaches were chosen: first, we studied the ability of IL-1 itself to alter ACTH, LH and/or sex steroid secretion; second, we explored the physiological role of endogenous IL-1 in mediating the endocrine action of LPS.

Mechanisms Mediating the Effects of LPS or IL-1 on the HPA Axis

We had observed earlier that a single peripheral injection of IL-1α augmented the portal blood concentrations of CRF and vasopressin, and that the administration of a CRF antiserum blocked the ability of IL-1 to release ACTH (20). We subsequently showed that immunoneutralization of endogenous CRF blunted LPS-induced ACTH secretion, and that the concomitant administration of the CRF antiserum and a vasopressin antagonist completely abolished the stimulatory effect of LPS (Figure 4). These results suggested that interleukins act within the brain to stimulate CRF- and vasopressin-secreting neurons, and that this might represent one of the mechanisms mediating the effect of LPS.

Several mechanisms could mediate the ability of IL-1 to activate the HPA axis, including an increased production of catecholamines. Dunn had demonstrated that the peripheral injection of IL-1 augmented catecholamine turn-over in the brain, suggesting increased release (21). We subsequently observed that both the intravenous and the intracerebroventricular administration of IL-1α to intact rats caused a marked increase in plasma epinephrine and norepinephrine levels, which was not due to changes in cardiovascular parameters or glucose levels (Rivier and Vale, submitted). Blockade of α- and β- adrenergic receptors, at doses of prazosin and propanolol which totally obliterated the effect of phenylephrine and isoproterenol on the corticotrophs, was without effect on

IL-1-induced ACTH secretion. Consequently, our results suggested that while activation of the sympathetic nervous system may occur during treatment with IL-1, this does not appear to represent an essential mechanism mediating the effect of the lymphokine on the HPA axis.

We also examined the hypothesis that increased production of endogenous lymphokines was one of the mechanisms through which endotoxin activated the HPA axis. Two approaches were used to test this possibility. First, because α-MSH, a peptide produced by the intermediate lobe of the pituitary, is known to interfere with a number of biological effects of IL-1 (22), we injected LPS in the presence or absence of α-MSH. We observed that this peptide consistently blunted the ability of LPS to stimulate ACTH secretion in mice (Figure 5). These results suggest the involvement of IL-1 in mediating the activation of the hypothalamic-pituitary-adrenal axis by the endotoxin. Further support for this hypothesis was provided by experiments in which mice were injected with LPS in the presence or absence of antibodies directed against the IL-1 receptor. Indeed, the antiserum markedly reduced the stimulatory action of LPS on ACTH secretion (Rivier and Vale, submitted). Taken together, these results indicate that at least part of the ability of the endotoxin to stimulate the HPA axis appears to be dependent upon activation of IL-1 receptors.

Mechanisms Mediating the Effect of LPS or IL-1 on the HPG Axis

In the intact male rat, the peripheral injection of LPS partially interfered with hCG-stimulated testosterone secretion (Figure 6). Similarly, we have observed that IL-1 blunted the effect of pregnant mare serum gonadotropin (PMSG) on sex steroids (23). In addition, when administered into the brain of castrated rats, both LPS (see above) and IL-1 (23) markedly lowered plasma LH levels. These results suggest that during circumstances of increased IL-1 production, such as some infectious diseases (24), at least two mechanisms might mediate the resulting impairment of reproductive functions: at the level of the brain, IL-1 might interfere with gonadotropin secretion, possibly through CRF-induced inhibition of GnRH secretion. In addition, IL-1 might act directly at the level of the gonads to block the effect of LH on sex steroid production.

Summary

We have shown that the endotoxin LPS acted both at the level of the brain and the gonads to stimulate the hypothalamic-pituitary-adrenal, and inhibit the hypothalamic-pituitary-gonadal, axis. Exogenously administered IL-1 mimics most of the effects of LPS on pituitary activity. In addition, antibodies against IL-1 receptors can interfere with LPS-induced ACTH secretion. These results suggest that at least part of the ability of LPS to alter endocrine functions appears to depend upon endogenous interleukin-1.

Acknowledgements

The author is grateful to Rosalia Chavarin, Leatrice Gandara and David Hutchinson for excellent technical help, to Dr. R. Chizzonite, Hoffman-Laroche, for generously providing the anti-interleukin receptor serum, and to Dr. J. Rivier for α-MSH. Research was supported by NIH grant No. HD13527 and conducted in part by the Clayton Foundation for Research, California Division. C.R. is a Clayton Foundation investigator.

References

1. Bateman, A., A. Singh, T. Kral, and S. Solomon, The immune-hypothalamic-pituitary-adrenal axis, *Endocrine Rev* 10: 92-112, 1989.
2. Dinarello, C.A., Interleukin-1 and its biologically related cytokines, *Adv Immuol* 44: 153-205, 1989.
3. Dinarello, C.A., and J.W. Mier, Lymphokines, *N Engl J Med* 317: 940-945, 1987.
4. Besedovsky, H., E. Sorkin, M. Keller, and J. Muller Changes in blood hormone levels during the immune response, *Proc Soc Exp Biol Med* 150: 466-470, 1975.
5. Dunn, A.J., M.L. Powell, W.V. Moreshead, J.M. Gaskikn, and N.R. Hall, Effects of Newcastle disease virus administration to mice on the metabolism of cerebral biogenic amine, plasma corticosterone, and lymphocyte proliferation, *Brain Behav Immunol* 1: 216-230, 1987.
6. Yasuda, N., and M.A. Greer, Evidence that the hypothalamus mediates endotoxin stimulation of adrenocorticotropic hormone secretion, *Endocrinology* 102: 947-953, 1978.
7. Makara, G.B., E. Stark, and T. Meszaros, Corticotrophin release induced by *E. coli* endotoxin after removal of the medial hypothalamus, *Endocrinology* 88: 412-414, 1971.
8. Rivier, C., M. Brownstein, J. Spiess, J. Rivier, and W. Vale, *In vivo* CRF-induced secretion of ACTH, β-endorphin and corticosterone, *Endocrinology* 110: 272-278, 1982.
9. Rivier, C., and W. Vale, Influence of corticotropin-releasing factor (CRF) on reproductive functions in the rat, *Endocrinology* 114: 914-921, 1984.
10. Orth, D.N., Adrenocorticotropic Hormone (ACTH), In B.M. Jaffe and H.R. Behrman (eds), *Methods of Hormone Radioimmunoassay*, Academic Press, New York, pp. 245-284, 1979.
11. Rivier, C., J. Rivier, and W. Vale, Chronic effects of [D-Trp^6, Pro^9-NEt]-LRF on reproductive processes in the male rat, *Endocrinology* 105: 1191-1201, 1979.
12. Kushner, I., The phenomenon of the acute phase response, *Ann NY Acad Sci* 389: 39-48, 1982.
13. Dinarello, C.A., Interleukin-1, *Rev Infect Dis* 6: 51-95, 1984.
14. Zuckerman, S.H., J. Shellhaas, and L.D. Butler, Differential regulation of lipopolysaccharide-induced interleukin 1 and tumor necrosis factor synthesis: effects of endogenous and exogenous glucocorticoids and the role of the pituitary-adrenal axis, *Eur J Immunol* 19: 301-305, 1989.
15. Moberg, G.P., Site of action of endotoxins on hypothalamic-pituitary-adrenal axis, *Am J Physiol* 220: 397-400, 1971.
16. Rivier, C., Stress: Definition and neuroendocrine evaluation, *Proceedings of International Workshop on Stress and Digestive Motility*, John Libbey Eurotext, Ltd., France, Mont Gabriel, Canada, In Press, 1989.
17. Wood, C.E., J. Shinsako, L.C. Keil, D.J. Ramsay, and M.F. Dallman, Apparent dissociation of adrenocorticotropin and corticosteroid responses to 15 ml/kg hemorrhage in conscious dogs, *Endocrinology* 110: 1416-1421, 1982.
18. Suzuki, S., C. Oh, and K. Nakano, Pituitary-dependent and -independent secretion of CS caused by bacterial endotoxin in rats, *Am J Physiol* 250: E470-E474, 1986.
19. Suzuki, S. and K. Nakano, Suppression of endotoxin-induced corticosterone secretion in rats by H_1-antihistamine, *Am J Physiol* 248: E26-E30, 1985.
20. Sapolsky, R., C. Rivier, G. Yamamoto, P. Plotsky, and W. Vale, Interleukin-1 stimulates the secretion of hypothalamic corticotropin-releasing factor, *Science* 238: 522-524, 1987.
21. Dunn, A.J., Systemic interleukin-1 administration stimulates hypothalamic norepinephrine metabolism paralleling the increased plasma corticosterone, *Life Sci* 43: 429-435, 1988.
22. Robertson, B.A., L.C. Gahring, and R.A. Daynes, Neuropeptide regulation of interleukin-1 activities, *Inflammation* 10: 371-385, 1986.
23. Rivier, C., and W. Vale, In the rat, interleukin-1α acts at the level of the brain and the gonads to interfere with gonadotropin and sex steroid secretion, *Endocrinology* 124: 2105-2109, 1989.
24. Ensoli, B., S. Nakamura, S.Z. Salahuddin, P. Biberfeld, L. Larsson, B. Beaver, F. Wong-Staal, and R.C. Gallo, AIDS-Kaposi's sarcoma-derived cells express cytokines with autocrine and paracrine growth effects, *Science* 243: 223-226, 1989.

NEUROENDOCRINOLOGY OF INTERLEUKIN-1

Frank Berkenbosch, Roel de Rijk, Adriana Del Rey,[1]
and Hugo Besedovsky[1]

Department of Pharmacology
Medical Faculty, Free University
1081 BT Amsterdam, The Netherlands

Introduction

Inflammatory processes and infectious diseases induce a constellation of host responses referred to as acute phase response (1,2). These responses include changes in immunologic, metabolic, neurologic, and endocrinologic functions. Although many of its components are far from being understood, it is generally believed that the acute phase response serves to regain normal homeostasis.

One of the earliest events of the acute phase response is the production and release of a mediator, presently identified as interleukin-1 (IL-1) (1). It should be noted that the identity of IL-1 has been a controversial issue for many years. Initial descriptions were mostly based on a variety of biological activities, and IL-1 has been described as endogenous pyrogen (EP, mediator of fever), proteolysis inducing factor (PIF), serum amyloid A inducer (SAA), leucocyte endogenous mediator (LEM), and others (3).

Application of molecular biology has revealed the existence of two distinct genes in various species including humans, capable of expressing IL-1 activity (4,5,6). Nowadays, the products of these genes are known as IL-1α and IL-1β. Although these two polypeptides share only 26-30% structural homology, they bind to the same receptor in a variety of biological models and have been found to exert similar biological activities (7-11). Recent research has indicated that IL-1α is associated with the outer membrane of monocytes, whereas IL-1β is actively secreted and therefore may be classified as hormone (12).

The primary source of IL-1α and β are blood monocytes, phagocytic lining cells of the liver and spleen and various other mononuclear cells related to macrophages (3). Interestingly, astrocytes and microglia cells in the central nervous system can also produce IL-1 activity, probably exerting primary effects within the brain (13-15).

Since exogenously administered IL-1 induces both laboratory and clinical features of the acute phase response, it has been proposed that IL-1 may coordinate several of the components of the acute phase response (1). IL-1 induces synthesis of acute phase proteins and of prostaglandins, stimulates bone resorption and cartilage breakdown, promotes proliferation of various cell types, alters levels of iron and zinc in the blood and induces fever (3), biological responses associated with microbial, chemical, and traumatic insults.

Recently, we identified another type of activity of IL-1: *viz.*, the capacity to stimulate the pituitary-adrenal axis (16). This finding is of particular interest since the

---

[1]Section of Neurobiology, Research Department, University of Cantonal Hospital, Basel, Switzerland

Circulating Regulatory Factors and Neuroendocrine Function
Edited by J. C. Porter and D. Ježová
Plenum Press, New York, 1990

303

pituitary-adrenal system is involved in regulation of a variety of biological functions such as host metabolic (17,18) as well as immune functions (19,20). In fact, we have presented evidence for the existence of an immunoregulatory feedback loop involving glucocorticoids and IL-1. In the present paper, we review advances that led to the discovery of this immunoregulatory feedback loop.

Pituitary-Adrenal Activity During Immune Challenges

Evidence is mounting that the pituitary-adrenal axis is activated during infectious diseases (21) and after injection of a variety of infective agents (22-25). For instance, lipopolysaccharide (LPS), well known for its induction of the production and release of IL-1 from macrophages *in vitro* (26,27), markedly increases plasma concentrations of adrenocorticotropin (ACTH) and glucocorticoids (23).

In rats, the effects of LPS are dose-dependent with an ED_{50} of approximately 1 μg per 100 g body weight. At doses of 3-5 μg per 100 g body weight, maximal responses can be obtained that last up to 6 hours after intraperitoneal injection of LPS. Also Poly I.C., an activator of monocytes, macrophages, and natural killer cells, is a potent agent to increase plasma ACTH and corticosterone concentrations in the rat. Doses as low as 30 μg per 100 g body weight injected intraperitoneally induces a full-blown pituitary-adrenal response that lasts for several hours (unpublished observations).

Viral particles are also known for their capacity to increase circulating levels of ACTH and corticosterone (22,28). In mice, intraperitoneal injection of the chicken virus, Newcastle disease virus (NDV), causes a rapid increase of pituitary-adrenal activity (29), a response that lasts up to 7 hours after NDV inoculation. Wistar rats are somewhat insensitive to NDV (unpublished observations). These responses are not dependent on the virulency of the virus; activated as well as inactivated virus particles are equally potent.

All of the mentioned agents activate pituitary-adrenal activity within 2 hours after their injection. T-cell dependent antigens, like sheep red blood cells, also raise plasma glucocorticoids. (30,31). However, this response is delayed for 4-5 days after immunization and seems to follow the course of the antibody response. It is of interest that animals with low antibody responses do not show increased plasma glucocorticoid concentrations, indicating that the immuno-messengers should reach certain threshold concentrations to elicit this response.

Involvement of Monokines?

The increased plasma concentrations of ACTH and glucocorticoids during immunological challenges could be due to a direct effect of the causal agent. However, most of the evidence points to a mediating role of factors released by activated immunological and/or accessory cells. It has been suggested that histamine secreted by mast cells and basophils mediates in part the pituitary-adrenal activation in response to intraperitoneal injection of LPS (23). No data are available on other factors mediating the LPS induced response.

As LPS, NDV has been shown to induce ACTH and endorphin production in lymphocytes and monocytes (32,33). It has been suggested that the ACTH produced by immune cells is capable of directly stimulating the adrenal cortex (22). However, other investigators have failed to obtain supporting evidence for the existence of a *lymphoid-adrenal axis* (28). In fact, the total amount of cells that produce ACTH/endorphin in response to a stimulus is a small fraction of the total immune cells (34). Recently, Besedovsky *et al.* (16,35) proposed that IL-1 serves as an important messenger between

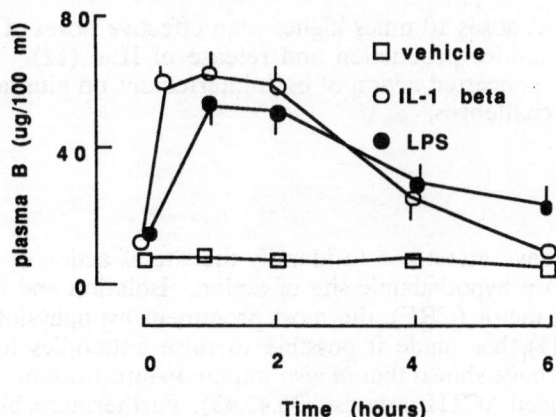

Figure 1. Time-course of the plasma corticosterone (B) response to intraperitoneal injection of vehicle (PBS containing 0.01% BSA), human recombinant interleukin-1β (IL-1β) or lipopolysaccharide (LPS) in male Wistar rats. Blood was obtained at various times after vehicle, IL-1β (1 μg per animal; specific activity of IL-1 preparation was 4×10^7 LAF units/mg protein) or LPS (3 μg per animal) injection *via* a chronic cannula inserted in the jugular vein and assayed for corticosterone using a radioimmunoassay. Data represent mean and SE (n = 10) and are evaluated by one factor ANOVA for repeated measures. This test revealed significant differences between the different treatments.

the immune system and the hypothalamo-pituitary-adrenal axis during viral infections. The evidence that supports this hypothesis derives from three observations:

a) When injected into rats or mice, supernatant fluid from homogenates of cultured DNV-infected mouse spleen cells or human peripheral blood leucocytes stimulates pituitary-adrenal activity.

b) The pituitary-adrenal response is prevented by immunoneutralization of the supernatant fluid with an antibody to IL-1.

c) The supernate-induced, pituitary-adrenal response is mimicked by intraperitoneal injection of purified IL-1β or recombinant rat or human IL-1β in doses as small as 0.5 μg per rat. After IL-1 injection, the plasma ACTH and corticosterone responses peak between 1 and 2 hours and return to baseline values at 4 hours (Figure 1).

IL-1α also stimulates pituitary-adrenal activity (Figure 2). As yet, no detailed comparisons of the relative potencies of the α and β forms of IL-1 are available. Although IL-1α and IL-1β interact with the same receptor, differences in binding affinities have been reported. IL-1α reportedly binds to T-cells more avidly than IL-1β, but less avidly to B-cells (8). In fact, there are reports showing a complete dissociation between some of the biological activities of the α and β forms (36-38), indicating the additional presence of distinct receptors for both polypeptides or their fragments.

Injection of mice or rats with other monokines or lymphocyte derived factors such as tumor necrosis factor (TNF), interleukin-2 (IL-2) or γ-interferon (γ-INF) in doses severalfold higher than the effective doses of IL-1, calculated on a molecular weight basis, were ineffective in stimulating pituitary-adrenal activity (16,35). However, interleukin-6 (IL-6), a glycoprotein produced by many different cells, has been shown to increase plasma

ACTH concentrations at doses 10 times higher than effective doses of IL-1 (39). It is of interest that IL-1 can induce production and release of IL-6 (12). No information is available for a possible concerted action of both interleukins on pituitary-adrenal activity during immunological challenges.

Site of Action

We and others have attempted to identify the site of action of IL-1. Most of the evidence now points to a hypothalamic site of action. Isolation and characterization of corticotropin releasing factor (CRF), the most prominent hypophysiotropic regulator of ACTH secretion (40,41), has made it possible to raise antibodies to CRF. Different groups, including ours, have shown that *in vivo* immunoneutralization of CRF completely abrogates the IL-1 induced ACTH response (35,42,43). Furthermore, blood samples taken from the hypothalamic-hypophysial portal system showed increased CRF concentrations after intravenous injection of IL-1 (42). Moreover, we showed that the CRF concentration in nerve terminals of the median eminence declines in response to IL-1 after blockade of axonal transport to prevent replenishment with newly synthesized CRF (35), further supporting the notion that CRF secretion from the hypothalamus mediates the pituitary-adrenal response.

By using the latter approach, we were able to study the involvement of arginine vasopressin (AVP), another hypothalamic regulator of ACTH secretion (40,41). In contrast to enhanced CRF turnover, no change in turnover of AVP in the zona externa of the median eminence was observed (44), indicating that IL-1 selectively stimulates CRF secretion without AVP secretion. This observation is especially intriguing since insulin-induced hypoglycemia concomitantly affected turnover of CRF and AVP (45). Our data are compatible with observations from Whitnall *et al.* (46,47), and Whitnall (48), indicating that approximately 50% of the nerve terminals in the median eminence that contain CRF also store AVP in the same secretory granules, whereas few terminals were found to contain only AVP. These data indicate that the selective increase in CRF turnover induced by IL-1 reflects specific activation of the population of nerve terminals that contain CRF but not AVP.

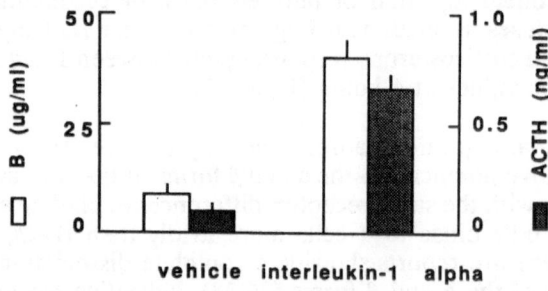

Figure 2. Effect of human recombinant interleukin-1α (IL-1α) on plasma ACTH and corticosterone (B) concentrations in male Wistar rats. Rats were decapitated 2 hr after intraperitoneal administration of IL-1 (1 μg per animal; specific activity of the preparation was 10^8 D10 units/mg protein). Plasma was assayed for corticosterone by using a fluorimetric assay and for ACTH using a radioimmunoassay. Data represent mean and SE (n = 6) and are evaluated by Scheffe-F test. $P < 0.05$ for vehicle *vs.* IL-1α injected groups.

The latter implies a functional difference between the two subtypes of CRF containing neurons. Such functional difference may be explained by differences in receptor populations of the subtypes. Alternatively, subtypes of CRF neurons may exist that are differentially activated by afferent pathways entering the paraventricular nucleus.

The studies cited earlier support the hypothesis that IL-1 stimulates ACTH release by secretion of CRF from the hypothalamus. Several groups of investigators have attempted to determine whether IL-1 stimulates ACTH release directly from pituitary cells. Using primary anterior pituitary cells in culture, we could not detect any short term (3 hours) ACTH releasing effect of human recombinant IL-1β at concentrations of 0.01-10 mM (35). Also, in the presence of submaximal concentrations of CRF, IL-1β appeared to be ineffective, excluding a possible synergism between IL-1 and CRF. Results of others also using primary cultures of anterior pituitary cells are in accord with our observations (42,49,50).

However, contradictory results have been reported. Bernton et al. (51) showed that IL-1α and IL-1β stimulate in a dose-dependent manner the secretion of ACTH, luteinizing hormone (LH), growth hormone (GH) and thyroid stimulating hormone (TSH) but inhibit prolactin secretion from anterior pituitary cells in culture. IL-1 receptors have been identified in the pituitary gland (52) and may support a direct pituitary site of IL-1 action.

It is worth noting that anterior pituitary cells contain folliculostellate and endothelial cells, which have been shown to secrete IL-6 (53). Production and secretion of IL-6 is regulated by IL-1, implying localization of IL-1 receptors on these cells. As yet, no obvious reason for the discrepancy on IL-1's site of action exists, although unsatisfactory explanations that the differences are due to the use of different forms (α and β) or batches of IL-1, the use of different sexes of rats, and/or differences in preparation of the primary cell cultures have been suggested (54).

More consistent are observations that prolonged incubation (up to 24 hours) of primary anterior pituitary cell cultures with IL-1 causes ACTH release (55,56). Such observations are intriguing since under certain inflammatory conditions, IL-1 may be present in the circulation for an extended period of time. However, more detailed work should be conducted in order to examine whether the ACTH releasing effect of prolonged exposure to IL-1 is an artifact due to the well known mitogenic (57) and/or cytotoxic effects (36) of this monokine.

Mechanism of Action

Although experiments, discussed earlier, point to an activation of CRF secretion in response to IL-1, they do not prove that IL-1 acts directly on the hypothalamus. In fact, by its various peripheral actions, IL-1 could induce reflex signals that induce CRF secretion via afferent pathways. Evidence that IL-1 acts centrally has been provided by data showing that intracerebroventricular injection of IL-1 induces ACTH secretion (38). As shown in Figure 3, doses as low as 10 ng, administered intracerebroventricularly, elicit a strong elevation of circulating ACTH concentrations in the rat that lasts for several hours. Although this observation points to a central action of IL-1, they are not conclusive for a direct action of IL-1 at the CRF neuron itself. IL-1 exerts various effects on the brain, including induction of fever (1), slow wave sleep (58), regulation of opioid receptor binding (59), and changes in norepinephrine metabolism (60,61).

Since most polypeptides such as IL-1 do not cross the blood-brain barrier, it is an intriguing question how peripherally induced IL-1 causes its central actions. IL-1 or its fragments may enter the brain at areas where the blood-brain barrier is absent. There areas include the median eminence, organum vasculosum of the laminae terminalis (OVLT), subfornical organ, and others, e.g., OVLT (62,63). The cells of these regions are reported to project to key endocrine sites such as the paraventricular nucleus, which contains CRF producing neurons, and to autonomic cell groups that control the central components of the acute phase reaction.

By the use of *in situ* hybridization with a 30 mer oligonucleotide to IL-1β, we recently found neuronal cells expressing mRNA coding for IL-1 in the rat and the mouse brain (64). These cells were located in the mediobasal hypothalamus and preoptic area. The latter cells are in close vicinity to the OVLT. Based on the localization of the IL-1 immunoreactive neurons, Breder *et al.* (63) proposed that IL-1 may serve as its own messenger. IL-1 producing neurons in the preoptic area may respond to circulating IL-1 that crosses the OVLT. These neurons may then use IL-1 as a neuromodulator, which serves as the central component of the acute phase response. Although this is an interesting hypothesis, it lacks substantial support. The release of IL-1 in response to depolarizing stimuli or the localization of IL-1 in synaptic vesicles has not been substantiated. In fact, it is unlikely that IL-1β exists in secretory granules since its precursor protein lacks a signal peptide (5). Therefore, the status of IL-1 as a neuromodulator seems questionable.

Studies conducted under *in vitro* conditions have demonstrated that IL-1 bioactivity can be produced and secreted by astrocytes and/or microglial cells (13-15). Microglial cells are phagocytic cells that are related to macrophages. It is possible that glial cells are somehow involved in the transfer of peripheral IL-1 signals to the brain. However, more likely, secretion of IL-1 by astrocytes and microglia may serve as an internal defense and as a repair mechanism of the brain (15,65) and/or may be involved in developmental (66) and pathological processes (64).

Since the median eminence lacks a blood-brain barrier, IL-1 may have privileged access to CRF neurons by entering the brain at this site. Since IL-1 at doses up to 10 nM

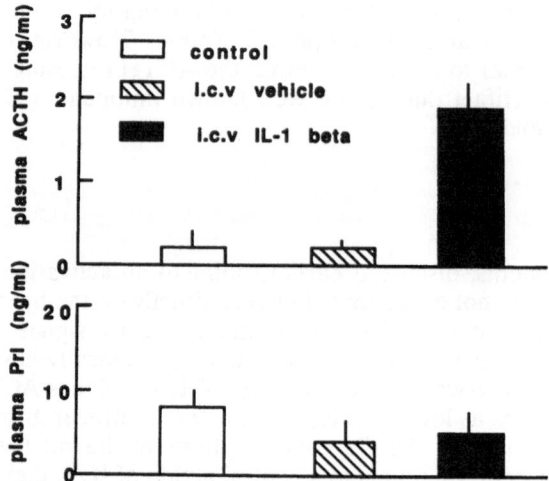

Figure 3. Effect of intracerebroventricular (icv) injection of human interleukin-1β (IL-1β) on plasma ACTH and prolactin (Prl) concentrations in male Wistar rats. Vehicle (PBS and 0.01% BSA) or 10 ng per 10 μl interleukin-1 (IL-1) (for specific activity, see legend of Fig. 1) was injected icv *via* a chronic implanted cannula. Rats were decapitated 2 hr after icv injections and plasma assayed for ACTH and Prl (reagents supplied by NIADDK). Data represent mean and SE (n = 6) and are evaluated by Scheffe-F test. $P < 0.05$ for ACTH response of vehicle *vs.* IL-1β injected rats. No differences were noted in the plasma prolactin concentrations.

were ineffective in causing CRF release from median eminence preparations *in vitro* (unpublished observations), we have excluded a direct effect of IL-1 at the level of the median eminence. However, recently IL-1α and IL-1β were shown to stimulate the release of CRF from hypothalamic framents *in vitro* (50). After a 20-minute pre-incubation period, both forms of IL-1 produce a dose-dependent increase in the release of CRF in a dose range of 1-100 U/ml with a plateau in responsiveness at 100 U/ml. Although this observation indicates that the rat hypothalamus releases CRF in direct response to low concentrations of IL-1, it does not prove that IL-1 activates CRF neurons directly. IL-1 has been shown to stimulate noradrenaline turnover in the hypothalamus (60,61). Inasmuch as noradrenergic neurons play a stimulatory role in CRF release (67), a mediating function for noradrenaline may be possible. Further studies should be conducted to elucidate the extent of the contribution of noradrenaline in IL-1 induced CRF response.

It has been shown that the central pyrogenic action of IL-1 involves entry of IL-1 into the brain through the OVLT (68,69). In this respect, it is of interest that the central pyrogenic actions completely dissociate from the pituitary-adrenal effects of IL-1. For example, doses of IL-1 that causes a full-blown activation of the pituitary-adrenal axis do not elicit a rise in colon temperature (16,42,70). In accord with the current data, we propose that peripherally induced IL-1 may enter the OVLT and median eminence and have a lower threshold for pituitary-adrenal activation than for pyrogenic action.

Other Neuroendocrine Effects of IL-1

Recently, we investigated the effects of peripheral administration of IL-1β on hormonal parameters other than pituitary-adrenal activation (44). IL-1, injected intraperitoneally in a dose that maximally activates pituitary-adrenal activity, did not affect plasma concentrations of the anterior pituitary hormones, LH, FSH, GH, and prolactin, within the time-interval of its ACTH releasing capacity. A lack of effect of IL-1β on plasma prolactin concentrations has also been reported by Uehara *et al.* (71). Moreover, intracerebroventricular administration of IL-1β does not change plasma prolactin concentrations (Figure 2) while increasing plasma ACTH concentrations.

IL-1β has been shown to reduce TSH values in the rat in doses ranging between 0.125-12.5 μg per animal (72). Maximal reductions were found at 5 hours after subcutaneous injection of IL-1, and the response lasted up to 12 hours. Although no data were presented, the researchers suggested that inhibition of TSH release was caused by reduced release of thyroid releasing hormone (TRH) from the hypothalamus. Thus, the effects of IL-1 on hormone secretion from the anterior pituitary gland appears limited to ACTH and TSH release and involves increased CRF and probably decreased TRH secretion from the hypothalamus.

Melanocyte stimulating hormone (MSH), a secretory fragment of proopiomelanocortin (POMC) produced in the intermediate lobe (41), does not respond to intraperitoneal injection of IL-1 (44). Apparently, IL-1 does not interfere with POMC secretion from the intermediate lobe either by a direct effect or by affecting regulatory mechanisms. We have shown that catecholamines secreted from the sympatho-adrenomedullary system stimulate POMC secretion from the intermediate lobe during emotional stress (73,74,41). Although IL-1 injection at a dose of 1 μg/animal elicits an increase of plasma noradrenaline and adrenaline concentrations, the responses are small and short-lived and do not follow the time-course of the ACTH response to IL-1 (44). The increased plasma adrenaline and noradrenaline concentrations at this dose of IL-1, *viz.*, 1 μg/animal, do not reach the threshold concentration necessary to elicit an MSH response. It is tempting to speculate whether higher doses of IL-1 could cause a stronger increase in plasma adrenaline and noradrenaline concentrations, thereby exceeding the threshold for stimulation of MSH secretion from the intermediate lobe. The latter mechanism may be of interest since MSH has been shown to prevent many of the biological actions of IL-1 (75,76).

The site at which IL-1 acts to elevate plasma catecholamines has not been elucidated. Presynaptic activation of sympathetic nerve within immune organs may account for this response. The latter is in accord with findings of increased turnover of noradrenaline during an immunological challenge (77) and with the close association of macrophages and noradrenergic nerve fibers in the spleen (78). As a sound alternative, central CRF projections that are known to be involved in regulation of sympathetic outflow (79) may mediate the sympathetic effect of IL-1. The latter is an interesting possibility since central administration of IL-1β has been shown to induce autonomic responses *via* mediation of CRF (80). Information whether peripheral injection of IL-1 also induces these autonomic responses is lacking.

Functional Significance

Host responses to microbial, traumatic, and chemical insults take the form of the acute phase response, in which IL-1 plays a central role. The acute phase response includes catabolic changes for the mobilization of metabolic substrates, the production of specific immune substances by the liver, induction of fever, and others. In addition, IL-1 is the first mediator to activate T-cell and B-cell functions by promoting IL-2 production, and enhances natural killer activity directed against tumor cells.

Once activated, the components of the acute phase response can cause immense damage to the body. Therefore, it seems of utmost importance for the body to have mechanisms by which the ongoing aggressive process can ultimately be regulated and reduced. One of the most important mechanisms to control the acute phase response may be provided by the capacity of IL-1 to stimulate efficiently the pituitary-adrenal axis, leading to enhanced levels of glucocorticoids.

Reiterating, IL-1 elevates circulating glucocorticoids by releasing CRF from the hypothalamus. Since CRF has multiple central actions, which modify mood learning and other behaviors, CRF secretion induced by IL-1 may also contribute to emotional and behavioral changes induced by infections, inflammations and immune responses. When the acute phase response progresses, an additional effect of IL-1 at the level of the pituitary gland may contribute to the pituitary-adrenal activation. Glucocorticoids inhibit various aspects of the acute phase response by direct effects on the organs involved as well as by inhibiting IL-1 and IL-2 production, thereby preventing clonal expansion of the immune system (19,20). This regulatory loop may have an essential role in preventing a vigorous ongoing acute phase response as explained earlier by Besedovsky *et al.* (16).

A second level of constraint of the acute phase response may be provided by the capacity of IL-1 to reduce thyroid hormones in the circulation, thereby shutting down the metabolic needs for the progression of the response. Finally, enhanced MSH secretion by mediation of catecholamines released by the sympathetic-adrenomedullary system may contribute to the regulation of a vigorous acute phase response by antagonizing various effects of IL-1.

Acknowledgements

This work was supported by the Royal Dutch Academy of Sciences and Arts, European Training Program for Brain and Behaviour, and the Swiss National Science

Foundation. We thank Dr. Alan Shaw and Dr. P. Lomedico for their gifts of interleukin-1 β and α, respectively; and Dr. J.R. Roberts and Dr. M. Blum (Fishberg Center for Neurobiology, Mount Sinai Medical School, New York) for their support in the conduct of *in situ* hybridization experiments.

References

1. Dinarello, C.A., Interleukin-1 and the pathogenesis of the acute phase response, *New Eng J Med* 311: 1413-1418, 1984.
2. Frayn, K.N. Hormonal control of metabolism in trauma and sepsis, *Clin Endocrinol* 24: 577-599, 1986.
3. Oppenheim, J.J., E.J. Kovács, K. Matsushima, and S.K. Durum, There is more than one interleukin-1, *Immunol Today* 7: 45-56, 1986.
4. Lomedico, P.T., U. Gubler, C.P. Hellman, M. Dukovich, J.C. Giri, Y.C.E. Pan, K. Collier, R. Semionow, A.O. Chua, and S.B. Mizel, Cloning and expression of murine interleukin-1 cDNA in Eschericia Coli, *Nature* 312: 458-462, 1984.
5. March, C.J., B. Mosley, A. Larsen, D.P. Ceretti, G. Breadt, V. Price, S. Gillis, C.S. Henney, S.R. Knonheim, K. Grabstein, P.J. Conlon, T.P. Hopp, and D. Kosman, Cloning, sequence and expression of two distinct human interleukin-1 complementary DNAs, *Nature* 315: 641-647, 1985.
6. Nishida, T., N. Nishino, K. Mizuno, Y. Sekiguch, M. Takano, K. Kawai, S. Nakai, and Y. Hirai, Cloning of the cDNAs for rat interleukin-1 alpha and beta, In *Monokines and Other Non-Lymphocytic Cytokines*, A.R. Liss, New York, pp.73-78, 1988.
7. Bird, T.A., and J. Saklatvala, Identification of a common class of high affinity receptors for both types of porcine interleukin-1 on connective tissue cells, *Nature* 324: 263-266, 1986.
8. Dower, S.K., S.M. Call, S. Gillis, and D.L. Urdal, Similarity between the interleukin-1 receptor on a murine T-lymphoma cell line and on a murine fibroblast cell line, *Proc Natl Acad Sci USA* 83: 1060-1064, 1986.
9. Dower, S.K., S.R. Krohnheim, T.P. Hopp, M. Cantrell, M. Deeley, S. Gillis, C.S. Henney, and D.L. Urdal, The cell surface receptors for interleukin-1α and interleukin-1β are identical, *Nature* 324: 266-268, 1986.
10. Kilian, P.L., K.L. Kaffka, A.S. Stein, D. Woehle, W.R. Benjamin, T.M. Decheria, U. Gabler, J.J. Farrar, S.B. Mizel, and P.T. Lomedico, Interleukin-1 alpha and interleukin-1 beta bind to the same receptor on T-cells, *J Immunol* 136: 4509-4514, 1986.
11. Sims, J.E., C.J. March, D. Cosman, M.B. Widmer, R.H. MacDonald, C.J. McMahan, C.E. Grubin, J.M. Wignall, J.L. Jackson, S.M. Call, D. Frien, A.R. Alpert, S. Gillis, D.L. Urdal, and S.K. Dower, cDNA expression cloning of the interleukin-1 receptor, a member of the immunoglobulin superfamily, *Science* 241: 585-589, 1988.
12. Dinarello, C.A., Biology of interleukin-1, *FASEB* 2: 108-115, 1988.
13. Fontana, A., F. Kristensen, R. Dubi, D. Gemsa, and E. Weber, Production of prostaglandin E and an interleukin-1 like factor by cultured astrocytes and C6 glioma cells, *J Immunol* 129: 2413-2419, 1982.
14. Fontana, A., E. Weber, and J.M. Dayer, Synthesis of interleukin-1/endogenous pyrogen in the brain of endotoxin-related mice: a step in fever induction, *J Immunol* 133: 1696-1698, 1984.
15. Giulian, D., T.J. Baker, L-C.N. Shih, and L.B. Lachman, Interleukin-1 of the central nervous system is produced by ameloid microglia, *J Exp Med* 164: 594-604, 1986.
16. Besedovsky, H., A Del Rey, E. Sorkin, and C.A. Dinarello, Immunoregulatory feedback between interleukin-1 and glucocorticoid hormones, *Science* 233: 652-654, 1986.
17. Urquhart, J., Physiological actions of adrenocorticotropic hormone, In S.R. Geiger (ed) *Handbook of Physiology, Endocrinology*, American Physiological Society, Washington, D.C., pp. 133, 1974.
18. DeWied, D., Pituitary-adrenal system hormones and behaviour, In F.O. Schmitt and F.G. Worden (eds) *The Neurosciences*, MIT Press, Cambridge, pp. 653, 1974.
19. Munck, A., P.M. Guyre, and N.J. Holbrook, Physiological functions of glucocorticoids in stress and their relation to pharmacological actions, *Endocrine Rev* 5: 25-44, 1984.

20. Bateman, A., A. Singh, T. Kral, and S. Solomon, The immune-hypothalamic pituitary-adrenal axis, *Endocrine Rev* 10: 92-112, 1989.

21. Beisel, W.R. and M.I. Rapoport, Interrelations between adrenocortical functions and infectious illness, *N Engl J Med* 280: 541-546, 1969.

22. Smith, E.M., W.J. Meyer, and J.E. Blalock, Virus induced corticosterone in hypophysectomized mice: a possible lymphoid-adrenal axis, *Science* 218: 1311-1312, 1982.

23. Nakano, K., S. Suzuki, and C. Oh, Significance of increased secretion of glucocorticoids in mice and rats injected with bacterial endotoxin, *Brain Behav Immun* 1: 159-172, 1987.

24. Leshin, L.S., and P.V. Malven, Bacteremia-induced changes in pituitary-hormone release and effect of naloxone, *Am J Physiol* 247: E585-E591, 1984.

25. Wolf, S.M., Biological effects of bacterial endotoxin in man, *J Infect Dis* 128: S259-S264, 1974.

26. Lachman, L.B., Interleukin-1 release from LPS-stimulated mononuclear phagocytes, In A. Nowotny (ed) *Beneficial Effect of Endotoxin*, Plenum Press, New York, pp. 283, 1983.

27. Okusawa, S., C.A. Dinarello, K.B. Yancey, S. Endies, T.J. Lawley, M.M Frank, J.F. Burke, and J.A. Gelfand, G5a induction of human interleukin-1: synergistic effects with endotoxin or interferon-gamma, *J Immunol* 139: 2635-2640, 1987.

28. Dunn, A. J., and M.L. Powell, Virus-induced increases in plasma corticosterone, Science 238: 1423-1424, 1987.

29. Berkenbosch, F., A. Del Rey, J.W.A. Van Oers, F.J.H. Tilders, and H. Besedovsky, Feedback circuit involving the immuno-hormone interleukin-1 and the pituitary-adrenal system, In R. Kvetnansky and G. Van Loon (eds) *Catecholamines and Other Neurotransmitters in Stress*, Gordon and Breach, New York, In Press, 1989.

30. Besedovsky, H.O., E. Sorkin, M. Keller, and J. Muller, Changes in blood hormone levels during the immune response, *Proc Soc Exp Med* 150: 466-470, 1975.

31. Besedovsky, H., A. Del Rey, E. Sorkin, W. Lotz, and U. Schuwela, Lymphoid cells produce an immunoregulatory glucocorticoid increasing factor (GIF) acting through the pituitary gland, *Clin Exp Immunol* 59: 622-628, 1985.

32. Harbour-McMenemamin, D., E.M. Smith, and J.E. Blalock, Bacteria lipopolysaccharide induction of leukocyte-derived corticotropin and endorphins, *Infect Immunol* 48: 813-817, 1985.

33. Smith, E. K., and E.J. Blalock, Human lymphocyte production of corticotropin and endorphin like substances: association with leukocyte interferon, *Proc Natl Acad Sci USA* 78: 7530-7534, 1981.

34. Kavelaars, A., R.E. Ballieux, and C.J. Heijnen, The role of IL-1 in the corticotropin releasing factor and arginine-vasopressin induced secretion of immunoreactive beta-endorphin by human peripheral blood mononuclear cells, *J Immunol*, In Press, 1989.

35. Besedovsky, H., and A. Del Rey, Neuroendocrine and metabolic responses induced by interleukin-1, *J Neurosci Res* 18: 172-178, 1987.

36. Bendtzen, K., T. Mandrup-Poulson, J. Nerup, J.H. Nielsen, C.A. Dinarello, and M. Svenson, Cytoxicity of human pI 7 interleukin-1 for pancreatic islets of langerhans, *Science* 232: 1545-1547, 1986.

37. Ferreira, S.H., B.B. Lorenzetti, A.F. Bristow, and S. Poole, Interleukin-beta as a potent hyperalgesic agent antagonized by a tripeptide analogue, *Nature* 334: 698-703, 1988.

38. Katsuura, G., P.E. Gottschall, R.R. Dahl, and A. Arimura, Adrenocorticotropin release induced by intracerebroventricular injection of recombinant interleukin-1 in rats: possible involvement of prostaglandins, *Endocrinology* 122: 1773-1779, 1988.

39. Naitoh, Y., J. Fukata, T. Tominaga, Y. Nakai, S. Tami, K. Mori, and H. Imura, Interleukin-6 stimulates the secretion of adrenocorticotropic hormone in conscious freely moving rats, *Biochem Biophys Res Commun* 155: 1459-1463, 1988.

40. Antoni, F.A., Hypothalamic control of adrenocorticotropin secretion: advances since the discovery of 41-residue corticotropin releasing factor, *Endocrine Rev* 7: 351-378, 1986.

41. Tilders, F.J.H., F. Berkenbosch, and P.G. Smelik, Control of secretion of peptides related to adrenocorticotropin, melanocyte stimulating hormone and endorphin, *Front Horm Res* 14: 161-196, 1985.

42. Sapolsky, R., C. Rivier, G. Yamamoto, P. Plotsky, and W. Vale, Interleukin-1 stimulates the secretion of hypothalamic corticotropin releasing factor, *Science* 238: 522-524, 1987.

43. Uehara, A., P.E. Gottschall, R.R. Dahl, and A. Arimura, Interleukin-1 stimulates ACTH release by an indirect action which requires endogenous corticotropin releasing factor, *Endocrinology* 121: 1580-1582, 1987.

44. Berkenbosch, F., D. De Goeij, A. Del Rey, and H. Besedovsky, Neuroendocrine, sympathetic and metabolic responses induced by interleukin-1, *Neuroendocrinology*, In Press, 1989.

45. Berkenbosch, F., D. De Goeij, and F.J.H. Tilders, Hypoglycemia enhances turnover of corticotropin releasing factor and of vasopressin in the zona externa of the median eminence, *Endocrinology*, In Press, 1989.

46. Whitnall, M.H., E. Mezey, and H. Gainer, Colocalization of corticotropin releasing factor and vasopressin in the median eminence neurosecretory vesicles, *Nature* 317: 248-250, 1985.

47. Whitnall, M.H., D. Smyth, and H. Gainer, Vasopressin coexists in half of the corticotropin releasing factor axons in the external zone of the median eminence, *Neuroendocrinology* 45: 420-424, 1987.

48. Whitnall, M.H., Distribution of provasopressin expressing and provasopressin deficient CRF neurons in the paraventricular hypothalamic nucleus of colchicine treated normal and adrenalectomized rats, *Comp Neurol*, In Press, 1989.

49. Uehara, A., S. Gillis, and A. Arimura, Effects of interleukin-1 on hormone release from normal rat pituitary cells in primary culture, *Neuroendocrinology* 45: 343-347, 1987.

50. Tsagarakis, S., G. Gillies, L.H. Rees, M. Besser, and A. Grossman, Interleukin-1 directly stimulates the release of corticotrophin releasing factor from the rat hypothalamus, *Neuroendocrinology* 49: 98-101, 1989.

51. Bernton, E.W., J.E. Beach, J.W. Holaday, R.C. Smallridge, and H.G. Fein, Release of multiple hormones by a direct action of interleukin-1 on pituitary cells, *Science* 238: 519-521, 1987.

52. Tracey, D.E., and E.B. De Souza, Identification of interleukin-1 receptors in mouse pituitary cell membranes and AtT-20 pituitary tumor cells, Society for Neuroscience, (Abstract #422.11), 1988.

53. Vankelecom, H., P. Carmeleit, J. Van Damme, A. Billiau, and C. Denef, Production of interleukin-6 by folliculo-stellate cells of the anterior pituitary gland in a histiotypic cell aggregate culture system, *Neuroendocrinology* 49: 102-106, 1989.

54. Lumpkin, M.D., The regulation of ACTH secretion by interleukin-1, *Science* 238: 452-454, 1987.

55. Kehrer, P., D. Turnhill, J.M. Dayer, A.F. Muller, and R.C. Gaillard, Human recombinant interleukin-1 beta and alpha, but not recombinant tumor necrosis factor alpha stimulate ACTH release from the rat anterior pituitary cells in vitro in a prostaglandin E2 and cAMP independent manner, *Neuroendocrinology* 48: 160-166, 1988.

56. Suda, T., F. Tozawa, T. Ushiyama, N. Tomori, T. Sumitomo, Y. Nakagamai, M. Yamada, H. Demura, and K. Shizume, Effects of protein kinase C related adrenocorticotropin secretagogues and interleukin-1 on proopiomelanocortin gene expression in rat anterior pituitary cells, *Endocrinology* 124: 1444-1449, 1989.

57. Pike, R.L., and G.J.V. Nossal, Interleukin-1 can act as a beta-cell growth and differentiation factor, *Proc Natl Acad Sci USA* 82: 8153-8157, 1985.

58. Krueger, J.M., J. Walter, C.A. Dinarello, S.M. Wolf, and L. Cheded, Sleep-promoting effects of endogenous pyrogen (interleukin-1), *Am J Physiol* 246: R994-R999, 1984.

59. Ahmed, M.S., Q.J. Llanos, C.A. Dinarello, and C.M. Blatteis, Interleukin-1 reduces opioid binding in guinea pig brain, *Peptides* 6: 1149-1154, 1985.

60. Dunn, A.M., Systematic interleukin-1 administration stimulates norepinephrine metabolism paralleling the increased plasma corticosterone, *Science* 43: 429-435, 1988.

61. Kabiersch, A., A. Del Rey, C.G. Honegger, and H. Besedovsky, Interleukin-1 induces changes in norepinephrine metabolism in the rat brain, *Brain Behav Immunol*, In Press, 1989.

62. Partridge, W.M., Neuropeptides and the blood-brain barrier, *Ann Rev Physiol* 45: 73-82, 1983.

63. Breder, C., C.A. Dinarello, and C.B. Saper, Interleukin-1 immunoreactive innervation of the human hypothalamus, *Science* 240: 321-324, 1988.

64. Berkenbosch, F., D. Caspers, R. Hellendall, V. Friedrich, L. Refolo, D. Lahiri, D. Blum, and N. Robakis, Roles for interleukin-1 and nerve growth factor in amyloid formation in Alzheimer's disease, Society for Neuroscience, (Abstract), In Press, 1989.

65. Giulian D., J. Woodward, D.G. Young, J.F. Krebs, and L.B. Lachman, Interleukin-1 injected into

mammalian brain stimulates astrogliosis and neurovascularization, *J Neurosci* 8: 2485-2490, 1988.

66. Giulian, D., D.G. Young, J. Woodward, D.C. Brown, and L.W. Lachman, Interleukin-1 as an astroglial growth factor in the developing brain, *J Neurosci* 8: 709-714, 1988.

67. Plotsky, P.M., Facilitation of immunoreactive corticotropin releasing factor secretion into hypophysial-portal circulation after activation of catecholaminergic pathways or central norepinephrine injection, *Endocrinology* 121: 924-930, 1987.

68. Morimoto, A., M. Murakami, T. Nakamori, and T. Watanabe, Multiple control of fever production in the central nervous system of rabbits, *J Phyiol (Lond)* 397: 269-280, 1988.

69. Stitt, J.T., Evidence for the involvement of the organum vasculosum laminae terminalis in the febrile response of rabbits and rats, *J Physiol (Lond)* 368: 501-511, 1985.

70. Berkenbosch, F., J. Van Oers, A. Del Rey, F. Tilders, and H. Besedovsky, Corticotropin releasing factor-producing neurons in the rat activated by interleukin-1, *Science* 238: 524-526, 1987.

71. Uehara A., P.E. Gottschall, R.R. Dahl, and A. Arimura, Stimulation of ACTH release by human interleukin-1 beta but not by interleukin-1 alpha in conscious freely moving rats, *Biochem Biophys Res Commun* 146: 1286-1290, 1987.

72. Dubius, J.M., J.M. Dayer, C.A. Siegrist-Kaiser, and A.G. Burger, Human recombinant interleukin-1 beta decreases plasma thyroid hormone and thyroid stimulating hormone levels in rats, *Endocrinology* 123: 2175-2181, 1988.

73. Berkenbosch, F., I. Vermes, and F.J.H. Tilders, The beta-adrenoceptor blocking drug propranolol prevents secretion of immunoreactive β-endorphin and α-melanocyte stimulating hormone in response to certain stress stimuli, *Endocrinology* 115: 1051-1059, 1984.

74. Berkenbosch, F., F.J.H. Tilders, and I. Vermes, Beta-adrenoceptor activation mediates stress-induced secretion of beta-endorphin related peptides from the intermediate but not from the anterior pituitary, *Nature* 305: 237-329, 1983.

75. Cannon, J.G., J.B. Tatro, S. Reichlin, and C.A. Dinarello, Alpha-melanocyte-stimulating hormone inhibits immunostimulatory and inflammatory actions of interleukin-1, *J Immunol* 137: 2232-2236, 1986.

76. Daynes, R.A., B.E. Robertson, B. Cho, D.K. Burnham, and R. Newton, Alpha-melanocyte stimulating hormone exhibits target cell selectivity in its capacity to affect interleukin-1 inducible responses in vivo and in vitro, *J Immunol* 139: 103-109, 1987.

77. Del Ray, A., H.O. Besedovsky, E. Sorkin, M. Da Prada, and S. Arrenbrecht, Immunoregulation mediated by the sympathetic nervous system II, *Cell Immunol* 63: 329-334, 1981.

78. Felten, D.L., S.Y. Felten, S.L. Carlson, J.A. Olschowka, and S. Livnat, Noradrenergic and peptidergic innervation of lymphoid tissue, *J Immunol* 135: 755s-765s , 1985.

79. Brown, M.R., L.A. Fisher, J. Durer, J. Spiess, C. Rivier, and W. Vale, Corticotropin releasing factor: effects on the sympathetic nervous system and oxygen consumption, *Life Sci* 30: 207-210, 1982.

80. Rothwell, N.J., CRF is involved in the pyrogenic and thermogenic effects of interleukin-1 beta in the rat, *Am J Physiol* 256: E111-E115, 1989.

ROLE OF MONOKINES IN CONTROL OF ANTERIOR PITUITARY HORMONE RELEASE

S.M. McCann, V. Rettori, L. Milenkovic,[1]
J. Jurčovičová,[2] and M.C. González

Department of Physiology, Neuropeptide Division
University of Texas Southwestern Medical Center at Dallas
Dallas, TX 75235-9040

Introduction

It has been known as far back as 1936 from the pioneering work of Selye (1) that noxious stimuli of one type or another, activate the release of ACTH which in turn releases adrenal cortical steroids. These then bring about a series of reactions in the body which consist of thymic involution, decrease in size of lymph nodes, lymphopenia, and eosinopenia (1). The role of the nervous system in these phenomena was not established at that time; however, in the early '50s it became apparent that hypothalamic lesions, particularly in the median eminence of the tuber cinereum would abolish the release of ACTH and adrenal steroids resulting from stress (2). Conversely, stimulation of the hypothalamus could evoke release of ACTH followed by release of adrenal steroids (3).

Early studies with adrenal steroids and ACTH indicated that these steroids had predominantly an immunosuppressive effect. This has been amply confirmed by later work now that the immune system is much better understood (4). Although hypercorticalism leads to suppression of immune responses, levels of the steroids which would be found under resting conditions may actually promote certain immune responses (5). The response to stress extends beyond the activation of ACTH secretion to the activation of prolactin (PRL) and growth hormone (GH) release in man (6). PRL and GH release is augmented by nearly all stresses in lower forms; however, the rat is an exception in that stress, instead of stimulating, inhibits GH release (7).

The hypophysectomized animal in general shows deficient immune responses which can be corrected by the administration of either GH (8) or PRL (9). Since these peptides are related chemically, it may be that the receptors which mediate these responses respond to portions of these two peptides that are similar in three dimensional structure. Thus, ACTH and adrenocortical hormones are immunosuppressive, whereas GH and PRL augment immune responses.

The hypothalamus may also modulate immune responses by the stimulation of the sympathoadrenal system. Beta receptors have been found on most immune responsive cells and in general appear to have a suppressive action on immune responses (10). Since stress activates the sympathoadrenal system, this system would appear to participate with

[1]Present affiliation: Institute of Molecular Biology-090, Boris Kidric Institute, P.O. Box 522, 11001 Beograd, Yugoslavia.

[2]Present affiliation: Institute for Experimental Endocrinology, Vlárska 3, 83306 Bratislava, Czechoslovakia

Circulating Regulatory Factors and Neuroendocrine Function
Edited by J. C. Porter and D. Ježová
Plenum Press, New York, 1990

and augment the immunosuppression induced by the pituitary-adrenocortical system.

With the discovery that ACTH is produced as a proopiomelanocortin molecule, which is processed differentially in different tissues, it appeared of interest to see whether β-endorphin was also released by the pituitary as well as ACTH. This has turned out to be correct (11), and β-endorphin appears to have the capability to modify immune responses (12). If this can be shown to be true at levels of the peptide which are found in the circulation in stress, then this may be another pathway for neuroimmunomodulation. Alternatively, recent evidence suggests that ACTH and endorphins (13) can actually be produced in lymphocytes where they may modulate these responses.

Following infection, the introduction of bacterial endotoxins, or most immunization procedures, a stress-like response of the hypothalamic-pituitary unit occurs (14). Therefore, in the acute phase one would have the stimulation of ACTH and adrenal corticoids tending to suppress the immune response, which would be counterbalanced by the stimulatory effects of GH (except in the rat) and PRL released during stress. Although there is abundant evidence for adrenal steroid receptors on immune cells (15), this has not been shown clearly for PRL and GH, and their effects may be mediated particularly by actions on the thymic cells (4). Multiple pathways mediate the activation of the hypothalamic-pituitary unit that takes place during stress (16). This can occur via emotional stimuli, painful stimuli, damage to tissue that produces pain and products of the immune cells themselves may evoke the response.

For example, exogenous pyrogens such as typhoid vaccine evoke the release of endogenous pyrogens that circulate through the blood to the hypothalamus to evoke fever by affecting the temperature regulating centers. Chowers *et al.* (17) demonstrated in dogs that injection of bacterial pyrogen evoked fever after a delay. The fever was presumably induced by endogenous pyrogen and was correlated in time with an elevation of plasma cortisol indicative of an activation of the pituitary-adrenal system by endogenous pyrogen.

Interleukin-1

The principal endogenous pyrogen is now known to be interleukin-1 (IL-1), a peptide having a molecular weight of about 17,000. It has become available for study, and it is apparent that peripheral administration of the peptide activates ACTH secretion (18,19). Intravenous administration of IL-1 increases plasma ACTH in the rat presumably by evoking release of corticotrophin releasing factor (CRF). This supposition is supported by recent experiments which showed that following intravenous administration of IL-1, there was an increase in CRF in portal blood and a borderline increase in vasopressin as well (19). Both of these peptides are capable of directly stimulating a release of ACTH from the pituitary gland, and vasopressin potentiates the action of CRF to release ACTH (20). In other studies, antisera directed against CRF have been shown to block the response to systemic administration of IL-1, again indicating that the response may be mediated by release of CRF (18,19).

A possible action of IL-1 to affect ACTH release directly is controversial. In AT-10 tumor cells, which consist of corticotrophs, IL-1 stimulates ACTH release *in vitro* (21). IL-1 has been found in three studies to have no effect on release of ACTH from normal pituitary cells *in vitro* (18,19,22), and furthermore it failed to alter the response to CRF (18). However, in one study a dose-related release of ACTH was found in monolayer cultured pituitary cells (23). The reason of these discrepant results is not apparent.

We have studied the effects of IL-1 on the release of other anterior pituitary hormones. Natural IL-1 was injected into the third ventricle of conscious male rats. Control animals were injected with the solvent vehicle [0.9% NaCl (saline)]. There was

Table 1

Effect of Intraventricular Injection of Saline and
of IL-1 at Doses of 5 and 25 ng on Rectal
Temperature (°C) in Male Rats*

Treatment	Time (min)		
	0	120	240
Saline	37.1 ± 0.4	34.9 ± 0.05†	34.6 ± 0.8‡
IL-1, 5 ng	36.8 ± 0.3	36.3 ± 0.9	35.8 ± 1.1
IL-1, 25 ng	37.3 ± 0.1	38.4 ± 0.2†	38.0 ± 0.3

*Time zero was approximately 30 min prior to iv injection. All values are mean ± SE (8 rats/group).
†$P < 0.005$ vs. time 0 value.
‡$P < 0.02$ vs. time 0 value.

a decline in body temperature in the control rats at 120 and 240 minutes after initiation of blood sampling (Table 1). Following the lowest dose of IL-1 [5 ng (0.3 pmol) in 2.5 μl], rectal temperature was significantly higher than in the saline-injected group at 120 minutes but was no longer significantly elevated at 240 minutes. Thus, this dose of IL-1 produced a borderline elevation of body temperature. The higher dose of 25 ng (1.5 pmol in 2.5 μl) of IL-1 evoked a highly significant elevation in body temperature at 120 minutes following initiation of blood sampling ($P < 0.005$). Temperature remained elevated at 240 minutes. Thus, the higher dose produced a frank fever.

After injection of saline into the third ventricle, plasma GH levels declined significantly at 15 minutes and remained low for the duration of the experiment; whereas following intraventricular injection of 5 ng of IL-1, plasma GH increased at 15 minutes and remained elevated for the duration of the experiment (Figure 1). These values were significantly above those in the saline-injected controls, and the area under the GH release curve was significantly elevated above that in the saline-injected group. Surprisingly, this effect was not evident following the injection of 25 ng of IL-1, and in this instance the results were almost superimposable on those of the animals that received intraventricular saline.

Plasma thyrotropin (TSH) levels did not significantly change after intraventricular injection of saline; however, after injection of 5 ng of IL-1, there was a decline in plasma TSH beginning at 15 minutes that achieved statistical significance by 30 and 60 minutes whether compared with plasma TSH in the saline-injected animals or the starting values. Similarly, the area under the plasma TSH curve was significantly less in the rats that received this dose of IL-1 than in the saline-treated rats. Although TSH values were lower following 25 ng of IL-1, these changes were no longer significant on comparison with saline-injected control values. They were also not significantly different from the values obtained with the lower dose of IL-1. The results with this higher dose were not significant when the area under the release curve was calculated.

Intraventricular injection of saline was followed by a slight but not significant decline in plasma PRL at 15 minutes. Values tended to return toward baseline by 120 minutes. After the lower dose of IL-1 (5 ng) was injected intraventricularly, an elevation of plasma PRL was observed at 60 minutes and thereafter. However, the variance was large, so these values were not significantly different from starting values or those of the saline-injected groups. This tendency to elevation was not observed following IL-1 at the dose of 25 ng. Similarly, when the area under the plasma PRL release curve was calculated, although the values were higher in the animals injected with 5 ng of IL-1, this change was not found to be significant and there was no alteration in the animals receiving the higher dose compared to the saline-injected controls.

When we compared the values at various times following injection of saline in order to calculate the maximal post-treatment increase in plasma PRL values for each animal, we found that the mean maximum increase in the saline-injected animals was 3.2 ± 2.0 ng/ml (mean ± SE), indicating no significant increase in PRL at any time following the injection of saline. However, in the animals injected with 5 ng of IL-1, the maximum mean increase in plasma PRL following the injection was 10.9 ± 3.5 ng/ml, a significant increase ($P = 0.02$). On the other hand, the maximum increase following the higher dose of IL-1 of 4.8 ± 2.5 ng/ml was not statistically significant (24).

We expected to find a stress-like pattern of hormonal release following the intraventricular injection of IL-1. Instead, we found a different picture. In general, stress decreases GH release in the rat (7); yet, we found that although there was an apparent slight stress effect of intraventricular injection of the saline diluent, as reflected in significant declines in plasma GH, the effect of IL-1 at the 5 ng (0.3 pmol) dose was to elevate plasma GH significantly above the values in the saline-injected controls. This dose, although not producing a febrile effect, prevented the decline in body temperature that was seen in the control animals following intraventricular injection of saline and multiple blood sampling. We do not know the cause of this small decline in temperature, but it may be related to the blood sampling and the maintenance of the animals in single

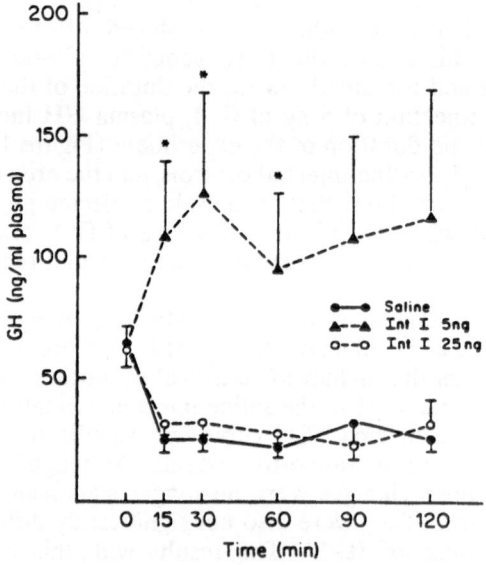

Figure 1. The effect of injection of the saline diluent or different doses of interleukin 1 (Int 1) in 2 μl of saline (0.9% NaCl) on plasma growth hormone in conscious male rats. Values in this and subsequent figures are mean ± 1 SE. *$P < 0.05$ vs. saline-injected controls.

cages so that heat loss was increased. The failure of temperature to decline in the interleukin-injected animals indicates that we had reached a dose that was biologically effective in raising body temperature.

It was very surprising that when the dose of IL-1 was increased to 25 ng (1.5 pmol), injected intraventricularly, which was associated with a clear febrile response, the response of GH was muted and no longer significantly different from values in the saline-injected animals. We have no explanation for this effect; however, it would appear that when frank elevation of temperature occurs, the response appears to be reversed.

The mechanism of the interleukin-induced elevation in GH is not known but may reflect either a decrease in release of somatostatin or an increase in the release of GH-releasing factor, or both changes may be operative. In previous studies of the effect of IL-1 on pituitary cells incubated *in vitro*, no effect on the release of GH was demonstrable in one case (22) whereas in the other experiment, a dose-related stimulation of IL-1 on GH release was obtained. Consequently, in our studies in which the hormone was injected intraventricularly, it is likely that the effects were mediated within the hypothalamus; however, we cannot rule out the possibility that sufficient amounts of the protein reached the anterior pituitary to stimulate GH release directly. Arguing against this possibility is the fact that the stimulation of GH vanished at higher doses which should have had a greater effect at the pituitary level. Therefore, it is quite probable that the effects observed were indeed due to hypothalamic action of IL-1.

Stress, in general, results in an inhibition of TSH release (7), so we expected to see a decrease in TSH in the interleukin-injected animals; and indeed this was found with significant effects at the 5 ng dose. On the other hand, there was no change in plasma TSH in the animals receiving saline. Again, as in the case of GH, there was no effect from the higher dose (25 ng) of IL-1. We speculate that frank pyrexia somehow interferes with the effects of the peptide. Presumably, the effect observed with the 5 ng dose is due to decreased TRH release from the hypothalamus.

We were expecting an increase in PRL following administration of IL-1, since this is a usual pattern with stress (7). However, there was great variability in the response, and significance was achieved only if one considered the maximal increment in PRL following injection of the 5 ng dose. Again, the response to the 25 ng dose was less and did not quite achieve statistical significance. Our results extend the pattern of hormonal response to IL-1 to include a stimulation of GH and PRL release and an inhibition of TSH release. They also show that as frank pyrexia develops, the hormonal responses are muted.

It will be necessary to do further experimentation with incubation of hemipituitaries and dispersed pituitary cells *in vitro* to distinguish clearly between the probable hypothalamic effect that we have seen and possible direct effects if IL-1 on the anterior pituitary.

Tumor Necrosis Factor (Cachectin)

An exciting recent development has been the discovery of the macrophage hormone, cachectin, which has recently been shown to be identical to tumor necrosis factor (25). It has a molecular weight of 17,000 daltons which is similar to that IL-1. Cachectin is released in response to infection, and the most effective promoter of its release is bacterial endotoxin, which is a lipopolysaccharide. Exposure of macrophages *in vitro* to endotoxin results in the synthesis of mRNA that initiates the synthesis of cachectin. This is released from the macrophages and has many effects in the organism. Low concentrations can activate phagocytosis and may play a beneficial role in combating infection; however, overwhelming infection results in massive stimulation by endotoxin and a discharge of large quantities of the hormone into the circulation where it has many adverse effects (27). It acts on endothelial cells to release IL-1 (27). IL-1 as well as cachectin act on the temperature regulating centers to induce fever. There is increased coagulability of the blood which can result in thrombosis. High concentrations alter endothelial permeability with resultant extravasation of blood. Permeability of cell membranes is altered with resultant changes in muscle membrane potential and uptake of water by cells. There is a blockade of lipoprotein lipase resulting in a failure to clear triglycerides and hypertriglyceridemia. Many other metabolic changes occur that represent manifestations of the so-called toxic shock syndrome produced by bacterial lipopolysaccharides. In fact evidence is accruing, based on passive immunization against cachectin, that the toxic shock syndrome is caused by release of massive quantities of this hormone into the circulation (26).

In dogs, hormonal changes have been demonstrated as evidenced by increases in circulating epinephrine, norepinephrine, cortisol, and glucagon following injection of cachectin (26). Therefore, it is apparent that cachectin can stimulate ACTH secretion; however, there have been no studies to determine the mechanism by which this effect takes place. It could be a result of afferent stimuli impinging on the hypothalamus from the multitudinous peripheral effects of the hormone, or a result of effects on the temperature regulating centers that in turn cause release of CRF and/or vasopressin as appears to be the case for IL-1. There may also be a direct effect on the pituitary.

We have carried out an extensive study of the effects of cachectin on hypothalamic-pituitary function in the rat. As in the case of IL-1, rectal temperature rose at the first measurement 1 hour after the intraventricular injection of tumor necrosis factor (TNF). There was no dose-response relationship since the responses to the 5 ng (0.3 pmol) dose and 100 ng (6.0 pmol) dose were similar.

In this case there was no significant decline of plasma GH following intraventricular injection of the saline diluent, but there was a significant increase in plasma GH in animals injected with the highest dose of 100 ng of TNF, which was significant only at 1 hour after injection.

Plasma PRL was not significantly altered by intraventricular injection of saline; however, following intraventricular injection of TNF at either the lower 5 ng or higher 100 ng dose, plasma PRL was elevated slightly, and the maximal increases from basal values were significant for both doses with a greater significance ($P < 0.025$) for the 100 ng dose.

There was no significant alteration of plasma TSH in control animals; however, TSH became significantly lower by 3 hours following both doses of TNF. This decrease persisted until 4 hours in the case of the lower dose, but TSH levels continued to decline and remained significantly lower at 6 hours following the higher dose of the peptide.

Plasma ACTH was slightly but significantly elevated by the higher dose of TNF but was unaltered by intraventricular injection of the diluent.

To determine the possible role of prostaglandins in these responses, the animals received a microinjection of indomethacin into the third ventricle, at a dose which had previously been found to block the release of LH in ovariectomized rats. Indomethacin had a significant suppressive effect on the response of PRL and TSH to cachectin but failed to alter significantly the response of ACTH to the peptide.

To determine whether or not the effects found with intraventricular injections were mediated by alterations in the secretion of releasing factors or by a direct action on the pituitary itself, the effects of several concentrations of TNF were tested for their ability to alter anterior pituitary hormone release from overnight cultures of dispersed anterior pituitary cells or of hemipituitaries in static incubations.

Preliminary experiments were carried out with concentrations of cachectin of 10^{-10} and 10^{-9} M to determine the optimal time for measurement of hormone release from dispersed pituitary cells of male rats. Cachectin failed to alter either TSH or GH release

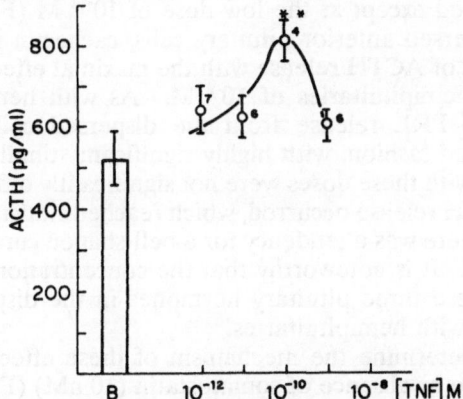

Figure 2. Effect of various doses of cachectin (TNF) on the release of ACTH from hemipituitaries incubated *in vitro* for 2 hrs. B represents the basal release. Numbers beside the mean indicate the number of flasks incubated. **$P < 0.025$ *vs.* basal release. Reproduced from Milenkovic *et al.* (28) with permission of the authors.

at 30 or 60 minutes but significantly increased the release of both hormones at both doses following 120 minutes of incubation. Consequently, this latter period was chosen for further study.

Cachectin significantly stimulated release of ACTH with a bell-shaped dose-response curve; the maximal significant response occurred at 10^{-10} M (Figure 2). This contrasted with a failure of the peptide to stimulate consistently prolactin release from the same hemipituitaries. There was stimulation in two of four experiments. The peptide also significantly stimulated GH release from these same hemipituitaries with a bell-shaped dose-response curve as in the case of ACTH (Figure 3). In this case the maximal effect occurred at a very low concentration of 10^{-12} M. Similarly, there was a dose-related release of TSH except that the peak was sharper than in the case of GH,

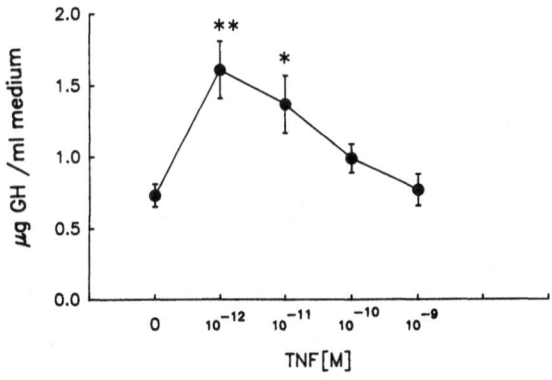

Figure 3. The effect of cachectin (TNF) on GH secretion from hemipituitaries incubated *in vitro*. **$P < 0.01$ *vs.* control; *$P > 0.05$ *vs.* control.

and no stimulation occurred except at the low dose of 10^{-12} M (Figure 4).

In the case of dispersed anterior pituitary cells, cachectin induced a dose-related and significant stimulation of ACTH release with the maximal effect occurring at a higher concentration than with hemipituitaries of 10^{-9} M. As with hemipituitaries there was inconstant stimulation of PRL release from the dispersed cells. TSH release was stimulated in a dose-related fashion, with highly significant stimulation at both 10^{-10} and 10^{-9} M. The stimulations with these doses were not significantly different from each other. Similarly, stimulation of GH release occurred, which reached statistical significance at 10^{-10} M; however, in this case there was a tendency for a bell-shaped curve and stimulation with 10^{-9} M was not significant. It is noteworthy that the concentration required to stimulate the release of each of these three pituitary hormones in the dispersed cell system was greater than that needed with hemipituitaries.

In an attempt to determine the mechanism of these effects, the responses were determined in the presence or absence of somatostatin (10 nM) (Figure 5). Somatostatin suppressed basal GH release and blocked the stimulatory effect of 10^{-10} cachectin on GH release. In the case of PRL there was no significant stimulatory effect of this dose of cachectin on prolactin release. In the presence of somatostatin basal prolactin release was not significantly reduced, and a highly significant stimulatory effect occurred with the addition of cachectin. In the case of TSH, somatostatin failed to alter the basal or stimulated release by cachectin.

To evaluate the role of cyclic AMP in the mechanism of cachectin stimulation, cachectin was incubated with the cells, and cyclic AMP in the cells was measured at the end of the experiment along with the various hormones. Cachectin lowered cell cyclic AMP concentrations in a dose-related fashion with the first significant lowering occurring at 10^{-9} M TNF (Figure 6). In the presence of somatostatin, cyclic AMP levels were lowered but now instead of lowering them further, TNF elevated the levels significantly (Figure 5). This was associated with the changes in hormone release mentioned before, *viz.*, that in the presence of somatostatin the GH response to cachectin was blocked but that of TSH was unaltered, but in the presence of somatostatin the release of PRL was dramatically increased from a nonsignificant increase to a highly significant one.

Lastly, to evaluate the role of prostaglandins in the *in vitro* effect of cachectin on pituitary hormone release, indomethacin was present at a concentration of 10 μM during both preincubation and incubation periods. Indomethacin did not alter basal release of the various hormones but completely blocked the stimulatory effect of 10^{-10} M cachectin

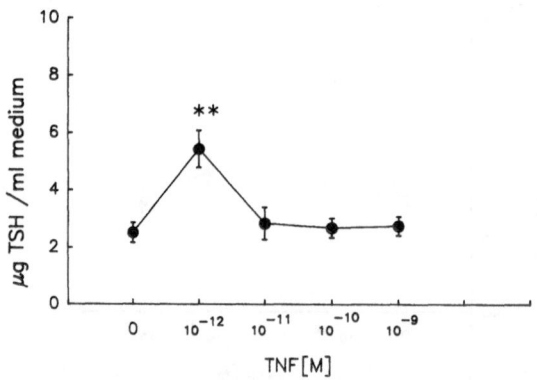

Figure 4. The effect of cachectin (TNF) on TSH secretion from hemipituitaries incubated *in vitro*. **$P < 0.01$ *vs.* control.

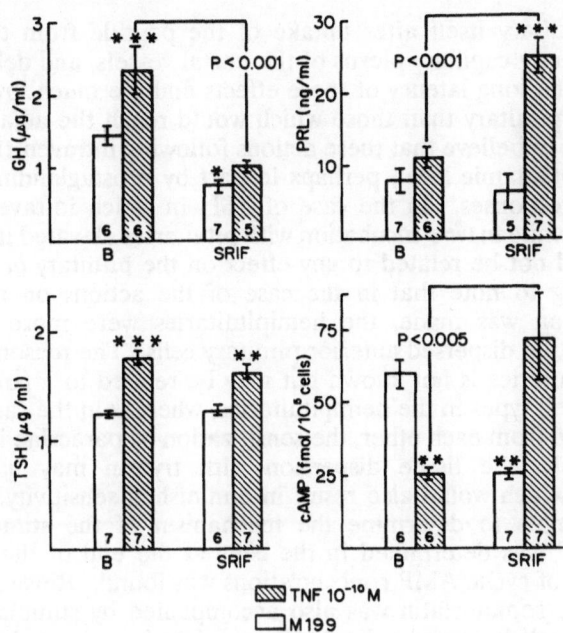

Figure 5. Effect of cachectin (TNF) (10^{-10} M) on the release of hormones from dispersed pituitary cells in the presence or absence of somatostatin (SRIF) (10 nM). Medium 199 (M199) served as the control. The effects of these treatments on the cyclic AMP content of the cells at the end of the incubation is also shown (*lower right panel*). *P < 0.05 *vs.* control; **P < 0.025 *vs.* control; ***P < 0.005 *vs.* control. Reproduced from Milenkovic *et al.* (28) with permission of the authors.

on GH and TSH release, but did not significantly decrease the ACTH-releasing action of the peptide (Figure 7) (28).

After the intraventricular injection of cachectin, results were similar to those obtained with IL-1 (0.3 pmol) with pyrexia. There was no significant further elevation of body temperature following the higher dose.

The effects of cachectin on pituitary hormone release following its intraventricular injection were similar but slightly less pronounced than those obtained with IL-1. We obtained only a slight increase in GH release following the higher dose (6 pmol) and have not yet had a satisfactory experiment with the lower dose. The elevation of PRL was modest. The most striking effects were obtained in the case of TSH, which was significantly lowered by both doses with a prolonged lowering obtained with the higher dose (6 pmol). There was also a significant stimulatory effect on ACTH release at the higher dose. The effects on GH, PRL, TSH, which followed intraventricular injection of cachectin, may have been due in part to the release of prostaglandins since the cyclooxygenase inhibitor, indomethacin, which should inhibit prostaglandin synthesis, attenuated the effects except in the case of ACTH. In view of the relatively long period before the induction of any of these responses *in vivo*, they could be mediated either by cachectin itself or by IL-1 released from astrocytes (25).

In the case of GH, ACTH, and PRL the actions of the peptide on the release of pituitary hormones *in vitro* had the same sign as those occurring following intraventricular injection. Thus, it is possible that the *in vivo* effects may have been due to actions on the pituitary itself after uptake of the peptide from the third ventricle, diffusion into the primary capillary plexus of the portal vessels, and delivery to the gland. However, in view of the long latency of these effects and the much lower concentrations that would reach the pituitary than those which would reach the adjacent hypothalamic tissue, we are inclined to believe that these actions following intraventricular injection are mediated at the hypothalamic level, perhaps in part by prostaglandins, which may also induce the pituitary responses. In the case of TSH in which intraventricular injection lowered plasma TSH while *in vitro* incubation with pituitaries elevated it, the hypothalamic action obviously could not be related to any effect on the pituitary *in vitro*.

It is interesting to note that in the case of the actions on pituitary cells that wherever a comparison was made, the hemipituitaries were more sensitive to TNF stimulation than were the dispersed anterior pituitary cells. The reason for this enhanced sensitivity of hemipituitaries is not known but may be related to a paracrine interaction between the various cell types in the hemipituitaries, whereas in the case of the dispersed cells which are isolated from each other, the sensitization by paracrine interaction was not present. Alternatively, the tissue dispersion with trypsin may have damaged cell membrane receptors which would also result in diminished sensitivity.

In initial attempts to determine the mechanism of the stimulatory actions of cachectin, cyclic AMP was determined in the cells at the end of the incubation and a dose-related lowering of cyclic AMP concentrations was found. Reversal of this lowering action of cachectin by somatostatin was also accompanied by stimulation of release of PRL. Somatostatin itself lowered cyclic AMP, as might be expected from prior results. It acts on a variety of cells in the pituitary gland to lower cyclic AMP, and the lowering is accompanied by decreased release not only of GH but also, at least at high doses, of PRL and TSH (29). The stimulatory action of cachectin on PRL release in the presence of somatostatin may be brought about by an elevation of cyclic AMP, which may have occurred in the lactotrophs, but only in the presence of somatostatin plus cachectin. Therefore, elevation of cyclic AMP may be a factor in promoting PRL release in response to cachectin. Indeed, exogenous cyclic AMP can stimulate PRL release.

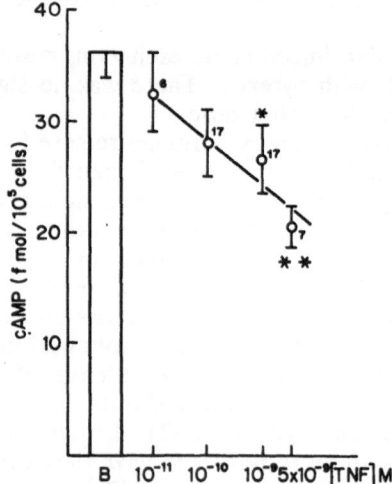

Figure 6. Pituitary cell cyclic AMP concentrations after incubation with various doses of cachectin (TNF). $*P < 0.05$ *vs.* basal; $**P < 0.025$ *vs.* basal. Reproduced from Milenkovic *et al.* (28) with permission of the authors.

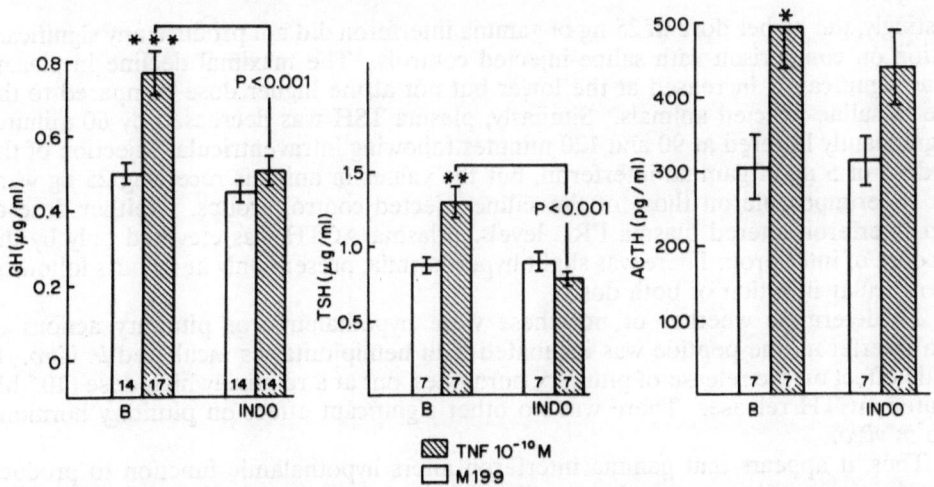

Figure 7. Effect of indomethacin on the release of pituitary hormones by cachectin (TNF) $(10^{-10}$ M). Indomethacin (INDO) was present in both preincubation and incubation periods at 10 μM. *P < 0.05 vs. basal; **P < 0.025 vs. basal; ***P < 0.005 vs. basal. Reproduced from Milenkovic et al. (28) with permission of the authors.

Although indomethacin did not alter basal hormone release, it blocked the releasing action of cachectin on both GH and TSH release in vitro. These results suggest that cachectin interaction with receptors on pituitary cells leads to activation of prostaglandin synthetase, which generates the release of prostaglandins that in turn stimulate release of GH and TSH. Earlier experiments have shown that prostaglandins can indeed release these various pituitary hormones by direct action in vitro. Thus, the results support the hypothesis that the releasing action of cachectin is mediated at least in part by prostaglandins. Recently, experiments have been reported which indicate that cachectin can stimulate endogenous production of prostaglandin E_2 by macrophages and that this action is blocked by the cyclooxygenase inhibitor indomethacin, which suppresses the metabolic activation of macrophages. It appears that the action of cachectin on pituitary cells may be similar to that on macrophages. The failure of indomethacin to block the ACTH-releasing action of cachectin is puzzling and suggests a role for other intracellular mediators in this action. The proposed interactions at the hypothalamic and pituitary level of IL-1 and TNF are illustrated in Figure 8.

Gamma Interferon

The number of monokines whose structure has been determined and synthetic versions made has increased by leaps and bounds in the last several years. Another available, important monokine is gamma interferon, and we have begun to study the possible effects of this monokine on hypothalamic-pituitary function as well.

Injection of human recombinant gamma interferon into the third ventricle of conscious male rats at a dose of 5 ng lowered plasma GH levels by 15 minutes. They were significantly lowered by 30 minutes and remained low for the 2 hour duration of the experiment on comparison with plasma GH levels in the saline-injected controls.

Interestingly, the higher dose of 25 ng of gamma interferon did not produce any significant alteration on comparison with saline-injected controls. The maximal decline in plasma GH was significantly increased at the lower but not at the higher dose compared to the decline in saline-injected animals. Similarly, plasma TSH was decreased by 60 minutes and significantly lowered at 90 and 120 minutes following intraventricular injection of the lower dose of 5 ng of gamma interferon, but the values in animals receiving 25 ng were almost superimposable on those of the saline-injected control groups. Neither dose of gamma interferon altered plasma PRL levels. Plasma ACTH was elevated only by the higher dose of interferon. There was slight hyperthermia, present only at 4 hours following intraventricular injection of both doses.

To determine whether or not these were hypothalamic or pituitary actions of gamma interferon, the peptide was incubated with hemipituitaries incubated *in vitro*. It had little effect on the release of pituitary hormones, but at a relatively high dose (10^{-8} M) stimulated ACTH release. There was no other significant effect on pituitary hormone release *in vitro*.

Thus, it appears that gamma interferon alters hypothalamic function to produce changes in pituitary hormone release. The results are unique for this monokine and different from those obtained with either IL-1 or cachetin. The principal action of gamma interferon appears to be on structures near the third ventricle to alter pituitary hormone release (30).

It will be important to examine the actions of other monokines as they become available. The picture is much more complicated than earlier envisioned, and it is probable that the response of the hypothalamic-pituitary axis to infection is brought about

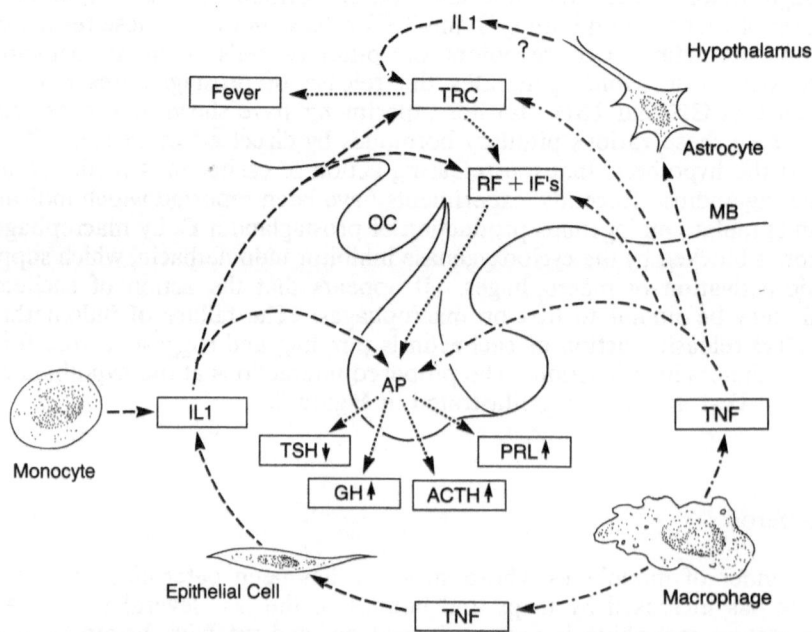

Figure 8. Summary of the proposed interactions of IL-1 and cachetin (TNF) on the hypothalamus and anterior pituitary gland. Abbreviations: TRC: temperature regulating centers; RF: releasing factor; IF: inhibiting factors; MB: mammallary bodies; OC: optic chiasm; AP: anterior pituitary. Reproduced from McCann *et al.* (31) with permission of Alan R. Liss, Inc.

by complex interactions of monokines acting both on the hypothalamus and pituitary directly.

Summary

It has long been known that endogenous pyrogen, released as a result of injection of typhoid vaccine or in response to infection, produces fever and increases ACTH secretion. Recent studies have indicated that endogenous pyrogen is, at least in part, IL-1. This monokine has now been shown to activate the release of ACTH by a hypothalamic mechanism with release of CRF and possibly vasopressin, which stimulates the corticotrophs. There may also be a pituitary action to stimulate the release of ACTH directly. In our experiments we showed that IL-1 at low but not higher doses appears to act intrahypothalamically to stimulate GH and PRL release and to inhibit TSH release.

In the meantime, another monokine, cachectin, was isolated and its structure determined. We have found that this monokine can act following its third ventricular injection to stimulate ACTH, PRL, and GH release and to inhibit TSH release, at least in part, by release of prostaglandins since indomethacin, an inhibitor of prostaglandin synthesis, produced a blockade of the responses except for those of ACTH. This peptide also has highly potent effects to alter pituitary hormone release by direct action on the pituitary to stimulate ACTH, GH, and TSH and to a slight extent PRL release. These actions appear to involve prostaglandins since indomethacin blocks all of the effects except for the effect on ACTH secretion. This monokine also produces a dose-related lowering of anterior pituitary cyclic AMP levels. When the monokine was incubated along with somatostatin, the lowering of cyclic AMP was reversed, and a potent PRL-releasing effect of the monokine was visible.

We have begun studies with a third monokine, gamma interferon, which indicate that it stimulates ACTH release but suppresses plasma GH and TSH levels by a hypothalamic action.

It is apparent that these various monokines have powerful effects to alter hypothalamic-pituitary function and that they probably mediate most of the effects of infections on the release of anterior pituitary hormones.

References

1. Selye, H., A syndrome produced by diverse noxious agents, *Nature* 138: 32, 1936.
2. McCann, S.M., Effect of hypothalamic lesions on the adrenal cortical responses to stress in the rat, *Am J Physiol* 175: 13-20, 1953.
3. DeGroot, J., and G.W. Harris, Hypothalamic control of the anterior pituitary gland and blood lymphocytes, *J Physiol* 111: 335-346, 1950.
4. Berczi, I., The immune system and its function, In I. Berczi (ed) *Pituitary Function and Immunity*, CRC Press, Inc., Boca Raton, pp. 1-25, 1986.
5. Ambrose, C.T., The essential role of corticosteroids in the induction of the immune response *in vitro*, In G.E. Wolsteinholme and J. Knight (eds) *Hormones and the Immune Response CIBA Study Group*, Churchill Livingston, London, p. 100-125, 1970.
6. Daughaday, W.H., The anterior pituitary, In J. Wilson and D. Foster (eds) *Textbook of Endocrinology*, W.B. Saunders Co., Philadelphia, pp. 568-613, 1986.
7. Krulich, L., E. Hefco, P. Illner, and C.B. Reed, The effects of acute stress on the secretion of LH, FSH, prolactin and GH in the normal male rat with comments on their statistical evaluation, *Neuroendocrinology* 16: 293-311, 1974.
8. Berczi, I., The influence of pituitary-adrenal axis on the immune system, In I. Berczi (ed) *Pituitary*

Function and Immunity, CRC Press, Inc., Boca Raton, pp. 49-132, 1986.

9. Berczi, I., and E. Nagy, Prolactin and other lactogenic hormones, In I. Berczi (ed) *Pituitary Function and Immunity*, CRC Press, Inc., Boca Raton, pp. 161-183, 1986.

10. Sanders, V.M., and A.E. Munson, Norepinephrine and the antibody response, *Pharmacol Rev* 37: 229-248, 1985.

11. Rivier, C., and W. Vale, Effect of corticotropin-releasing factor, neurohypophysial peptides and catecholamines on pituitary function, *Fed Proc* 44: 189-195, 1985.

12. Shavit, Y., J.W. Lewis, G.W. Turman, C.J. Zanes, R.P. Gale, and J.C. Lieboskine, Apparent role of opioid peptides in mediating stress-induced immunosuppression, In B.D. Jankovic, B.M. Markovic, and N.H. Spector (eds) *Proceedings of the 1st International Workshop on Neuroimmunomodulation, International Working Group on Neuroimmunomodulation*, Bethesda, pp. 95-98, 1984.

13. Smith, E.M., and J.E. Blaylock, Lymphocyte production of neurally active pituitary hormone-like molecules, In B.D. Jankovic, B.M. Markovic, and N.H. Spector (eds) *Proceedings of the 1st International Workshop on Neuroimmunomodulation, International Working Group on Neuroimmunomodulation*, Bethesda, pp. 65-68, 1984.

14. Besedovsky, H.O., A. Del Rey, and E. Sorkin, Lymphokine-containing supernatants from Con A-stimulated cells increase corticosterone blood levels, *J Immunol* 126: 385-387, 1981.

15. Homo-Delarche, F., Glucocorticoid receptors and steroid sensitivity in normal and neoplastic human lymphoid tissues - a review, *Cancer Res* 44: 431-437, 1984.

16. Yates, F.E., and J.W. Maran, Stimulation and inhibition of adrenocorticotropin, In R.O. Greep and E.B. Aswood (eds) *Handbook of Physiology* Chapter 36, Section 7, Endocrinology Volume 4, Part 2, American Physiological Society, Washington, D.C., pp. 367-404, 1974.

17. Chowers, I., H.T. Hammel, J. Eisenman, R.M. Abrams, and S.M. McCann, Comparison of effect of environmental and preoptic heating and pyrogen on plasma cortisol, *Am J Physiol* 210: 606-610, 1966.

18. Berkenbosch, F., J. Van Oers, A. Del Rey, F. Tilders, and H. Besedovsky, Corticotropin-releasing factor-producing neurons in the rat activated by interleukin 1, *Science* 238: 524-524, 1987.

19. Sapolsky, R., C. Rivier, G. Yamamoto, P. Plotsky, and W. Vale, Interleukin 1 stimulates the secretion of hypothalamic corticotropin-releasing factor, *Science* 238: 522-524, 1987.

20. McCann, S.M., M.D. Lumpkin, and W.K. Samson, The role of vasopressin and oxytocin in control of anterior pituitary hormone secretion, In A.J. Baertschi and J.J. Dreifuss (eds) *Neuroendocrinology of Vasopressin, Corticoliberin and Opiomelanocortins*, Academic Press, London, pp. 319-330, 1982.

21. Woloski, B.M.R.N.J., E.M. Smith, W.J. Meyer, III, G.M. Fuller, and J.E. Blaylock, Corticotropin-releasing activity of monokines, *Science* 230: 1035-1037, 1985.

22. Uehara, A., S. Gillis, and A. Arimura, Effects of interleukin 1 on hormone release from normal rat pituitary cells in primary culture, *Neuroendocrinology* 45: 343-347, 1987.

23. Bernton, E.W., J.E. Beach, J.W. Holaday, R.C. Smallridge, and H.G. Fein, Release of multiple hormones by a direct action of interleukin 1 on pituitary cells, *Science* 238: 519-521, 1987.

24. Rettori, V., J. Jurčovičová, and S.M. McCann, Central Action of interleukin 1 in altering the release of TSH, growth hormone, and prolactin in the male rat, *J Neurosci Res* 18: 179-183, 1987.

25. Beutler, B., N. Krochin, I.W. Milsark, C. Luedke, and A. Cerami, Control of cachectin (tumor necrosis factor) synthesis: mechanisms of endotoxin resistance, *Science* 232: 977-980, 1986.

26. Tracey, K.J., S.F. Lowry, T.J. Fahey, III, J.D. Albert, Y. Fong, D. Hesse, B. Beutler, K.R. Manogue, S. Calvano, H. Wei, A. Serami, and G.T. Shires, Cachectin/tumor necrosis factor induces lethal shock and stress hormone responses in the dog, *Surg Gyn Obstet* 164: 415-422, 1987.

27. Nawroth, P.P., I. Bank, D. Handley, J. Cassimeris, L. Chess, and D. Stern, Tumor necrosis factor/cachectin interacts with endothelial cell receptors to induce release of interleukin 1, *J Exp Med* 163: 1363-1375, 1986.

28. Milenkovic, L., V. Rettori, G.D. Snyder, B. Beutler, and S.M. McCann, Cachectin alters anterior pituitary hormone release by a direct action *in vitro*, *Proc Natl Acad Sci USA* 86: 2418-2422, 1989.

29. McCann, S.M., Physiology and pharmacology of LHRH and somatostatin, *Ann Rev Pharmacol Toxicol* 22: 491-515, 1982.

30. González, M.C., and M. Riedel, Effects of human recombinant γ-interferon on the release of growth

hormone (GH), thyroid stimulating hormone (TSH) and prolactin in the male rat, *71st Annual Meeting of The Endocrine Society*, Abstract #164, Seattle, WA, June 21-24, pp. 63, 1989.

31. Lakoski, J.M., J. Region Perez-Polo, D.K. Rassin, S.M. McCann, V. Rettori, L. Milenkovic, J. Jurčovičová, G. Snyder, and B. Beutler, Role of interleukin-1 and cachectin in control of anterior pituitary hormone release, In *Neural Control of Reproductive Function*, Alan R. Liss, Inc., New York, pp. 333-349, 1989.

INTERLEUKINS, SIGNAL TRANSDUCTION, AND THE IMMUNE SYSTEM-MEDIATED STRESS RESPONSE

R.L. Eskay, M. Grino,[1] and H.T. Chen

Laboratory of Clinical Studies
National Institute on Alcohol Abuse and Alcoholism
Bethesda, MD 20892

Introduction

The interleukins (ILs) comprise a group of small proteins or polypeptides with molecular weights in the 15-25 kD range. These polypeptides possess pleiotropic biological activities on immune as well as non-immune cells and can be produced by either cell type. At the present time, recombinant forms for at least IL-1α and β through IL-7 are available for investigative use, and these ILs possess certain overlapping biological activities. The use of the term interleukin is somewhat archaic and limiting since a variety of non-leukocytic cell types produce these molecules or others with similar biological activities. Designations for the ILs, based on cell type of origin, include monokines, lymphokines, and cytokines if they are derived from monocytes, lymphocytes, and non-lymphoid cells, respectively. Finally, there is a trend to refer to all of the above mentioned molecules as cytokines, and an appreciation of the multitude of biological effects induced by the cytokines can be found in several recent reviews (1,2,3,4).

Clearly, the immune system recognizes infective organisms and foreign antigens, and mounts an appropriate host response in which the endocrine system plays an important role. On the basis of experiments performed over the last decade, the concept of bidirectional communication between the immune and neuroendocrine systems has emerged (5). Simply stated, neuroendocrine systems, particularly the hypothalamic-pituitary-adrenal (HPA) axis, can modulate immune system responses, and immune system-derived cytokines can modulate neuroendocrine system activities. A corollary to this basic concept further suggests that lymphoid cells can express polypeptides that appear identical to certain pituitary hormones, which may serve as an extra-pituitary source of classical regulatory hormones, particularly proopiomelanocortin (POMC)-derived peptides such as corticotropin (ACTH) and β-endorphin.

The observed effects of cytokines on the HPA axis are perhaps the best example of immune modulation of a neuroendocrine system, and are the focus of this presentation. Although the precise physiological actions of cytokines on the HPA axis are unclear, *in vitro* and *in vivo* findings indicate cytokine modulation at the level of the brain, pituitary, and adrenal gland. Some recent findings germane to understanding the physiological role of immune system-derived cytokines on the HPA axis will be integrated with the current body of literature on this subject. The effect of prolonged cytokine (IL-1α and β and IL-6) exposure of corticotrophs in primary culture and of corticotroph-derived, clonal cells (AtT-20) on β-endorphin release will be presented, along with some preliminary

---

[1]Present affiliation: Laboratoire de Neuroendocrinologie Expérimentale, Faculté de Médecine Nord, Marseille, France

Circulating Regulatory Factors and Neuroendocrine Function
Edited by J. C. Porter and D. Ježová
Plenum Press, New York, 1990

331

observations (6) on cytokine modulation of peptide/catecholamine accumulation in primary cultures of bovine medullary chromaffin cells.

Early Studies Linking Immune System and Stress Axis Activation

Evidence that immune system activation could alter neuroendocrine parameters began to emerge over a decade ago. In a couple of intuitive, landmark papers, Besedovsky and co-workers (7,8) through the administration of either a particulate or soluble antigen to rats, generated a primary immune response that was accompanied by increased corticosterone levels three days later and remained elevated for as long as 10 days (7). Using identical antigen challenges, they found the firing rate of ventromedial hypothalamic neurons was increased in intact and adrenalectomized rats (8). After consideration of the response lag time following antigen administrations, it was reasoned that antigen-responsive cells of the immune system generated products that led to an enhanced hypothalamic electrical activity and elevated plasma corticosterone levels. The ability of intraperitoneally injected supernatant fluid, obtained from concanavalin A-stimulated immunocompetent cells in $vitro$, to enhance corticosterone release in $vivo$ strengthened the notion that lymphoid cell-derived cytokines could activate the HPA axis (9). From these early studies have sprung a plethora of reports aimed at delineating the responsible HPA-activating cytokines, their sites of action [neuronal $vs.$ non-neuronal (pituitary, adrenal)], and their mechanism of action. Since IL-1, tumor necrosis factor (TNF), and IL-6 are prominent mediators of the systemic $acute$ $phase$ or primary immune response, these cytokines in particular have received much attention.

CNS-Mediated Cytokine Stimulation of the HPA Axis

The results of an elegant study by Besedovsky et $al.$ (10) demonstrated for the first time that IL-1, a polypeptide produced primarily by stimulated macrophages and monocytes, could be responsible for the intuitive and scientific observations derived in earlier studies (7-9,11) that immune cell products could activate the HPA axis. Supernatant fluids from cultures of human leukocytes challenged by Newcastle disease virus contained an HPA axis-activating substance, which could be completely neutralized by antiserum generated against IL-1. Injection of purified, monocyte-derived human IL-1 or recombinant (r) human (h) IL-1β significantly enhanced plasma levels of corticosterone and ACTH in rodents. Other potentially active cytokines such as TNF, IL-2, or γ-interferon were inactive when tested at comparable or higher doses (10).

With the convincing demonstration that IL-1 stimulates the HPA axis, numerous investigators have focused on determining the primary site(s) of IL-1 action, which following a single dose or short-term exposure to rIL-1 appears to be the central nervous system or hypothalamus. Simultaneous reports by Sapolsky et $al.$ (12) and Berkenbosch et $al.$ (13) clearly established that IL-1α or β affects pituitary-adrenal activity through enhanced corticotropin-releasing hormone (CRH) secretion. Immunoneutralization of CRH with CRH antiserum markedly reduced IL-1 enhanced ACTH release in rats. Furthermore, the concentration of CRH in hypophysial portal blood (12) and the calculated median eminence-release rate (13) for CRH were enhanced by rhIL-1 treatment. Finally, neither group was able to demonstrate a direct, acute (3-4 hours) effect on corticotrophs in culture (12,13) or AtT-20 cells (13). Additional investigations are in agreement that the primary site of the acute effects of IL-1 on the HPA axis is the brain and requires CRH secretion (14). It was further indicated that IL-1β was substantially more potent that IL-1α in stimulating ACTH release following either intravenous or intracerebroventricular administration (15,16). Furthermore, both IL-1α and IL-1β produced a dose-related, enhanced release of CRH from incubated rat mediobasal hypothalami (17).

Although there is ample evidence that peripheral or central administration of IL-1 activates the HPA axis through altered CRH secretion, these effects may not be the result of a direct action on CRH-containing neurons. Complete neural deafferentation of the mediobasal hypothalamus (MBH) prevented IL-1 induced ACTH release, suggesting that extra-hypothalamic neural inputs to the hypothalamus are essential for enhanced pituitary-adrenocortical activity (18). Consistent with the conclusions of the MBH deafferentation study, earlier studies demonstrated that both particulate (12,20) and cytokine-enriched fluids (19) activated noradrenergic neurons in the brainstem and hypothalamus, which were independent of corticosterone changes (20). Clearly, IL-1 is a primary activator of central noradrenergic activity since the intraperitoneal administration of recombinant IL-1 to rats and mice activates norepinephrine metabolism in the brain stem and hypothalamus, including the paraventricular nucleus (21,22). It is possible that the observed IL-1 induced release of CRH is secondary to activation of noradrenergic cell bodies in the brainstem that send projections to the hypothalamus, which are known to regulate CRH secretion (23-25).

Consistent evidence indicates that the primary site of acute pituitary-adrenal activation by IL-1 is the brain, but numerous pivotal questions remain to be answered. The accessibility of peripheral cytokines to areas of the CNS other than the circumventricular organs remains an open question. The precise interaction between circulating hormonal cytokines, IL-1 binding sites in the brain (26,27), and an endogenous IL-1β neuronal network in the hypothalamus (28) and other areas of the CNS remains largely unexplored. Furthermore, as a result of the recent cloning and expression of rat IL-1α cDNA (29), IL-1α mRNA was found not to be expressed in rat brain of normal or lipopolysaccharide (LPS) challenged animals but was present in LPS-stimulated peritoneal macrophages. In contrast, LPS challenge did reveal the presence of IL-1β mRNA in the rat brain, which is consistent with an earlier report (26). The availability of rat recombinant IL-1α and β will make it possible to do species specific neuroendocrine challenges, binding studies in potential target tissues, generation of rat IL-1α and β antiserum for central mapping studies, quantification of circulating IL-1 levels, and immunoneutralization experiments.

In addition to the widely studied action of IL-1, other cytokines have been implicated as modulators of the HPA axis. TNF-α, like IL-1, is produced primarily by activated monocytes but can be synthesized by non-immune cells as well. Since TNF-α harbors many of the same biological activities as IL-1 and responds to many of the same immune challenges as IL-1, TNF-α may act in concert with IL-1 to activate the HPA axis. Recently, intravenous administration of human rTNF-α was shown to enhance ACTH release with a potency comparable to an equivalent dose of human rIL-1β (30,31). Furthermore, neither 2-hour (31) nor 24-hour (32) exposure of cultured pituitary cells to TNF-α enhanced ACTH release. Although IL-1β and TNF-α are potent acute ACTH-releasing compounds *in vivo*, their apparent extra-pituitary site of action may differ (31). A third monokine, IL-6, which shares many overlapping biological activities with IL-1 and TNF-α, also activates the HPA axis (33) in a dose-related manner following intravenous administration. Pre-treatment with CRH antiserum completely blocks the effects of IL-6, suggesting an extra-pituitary sites of action for IL-6 with ultimately an enhanced release of CRH or other substances that require the presence of CRH. Another potential immunohormone is IL-2, which can be synthesized by T lymphocytes following a variety of antigenic challenges. Purified natural human IL-2 (34,36) or rhIL-2 (36) when administered to humans results in elevated ACTH and cortisol. Whether or not the large doses of IL-2 administered in these studies represents a non-specific stress response needs to be clarified, since at least in one report (37) the observed ACTH response coincided with fever and chills. In addition, neither purified natural IL-2 or rIL-2 was found to activate the HPA axis in rodents (10). Support for the notion that IL-2 can directly activate corticotrophs was demonstrated using the reverse hemolytic plaque assay to monitor corticotroph ACTH secretion (38) from dispersed rat pituitary cells. The fact that the acute (2-3 hours) IL-2 effect was equipotent with a maximal stimulatory concentration of CRH represents a unique observation, which echoes an earlier finding by this group (39) that a 6-hour exposure to rIL-2 was more potent than or equipotent with CRH in increasing POMC mRNA levels in dispersed rat pituitary cells or in AtT-20 cells. In an early study, Farrar observed for the first time that prolonged exposure (24 or 48 hours) of AtT-20 cells to rhIL-2 or purified IL-1 was needed to demonstrate enhanced POMC-peptide release (40). Further exploration by other investigators of the observed direct effects of IL-2 on corticotrophs *in vitro* is needed.

Pituitary-Mediated Cytokine Stimulation of the HPA Axis

From the earliest studies, consistent findings have indicated the CNS to be the primary site of the acute activation of the HPA axis by cytokine-enriched fluids or recombinant forms of the ILs, particularly IL-1. This contrasts with the disagreement between investigators as to the importance of direct, acute effects of IL-1 on corticotrophs. A triad of reports (12,13,41) perhaps best typifies the confusion with regard to the ability of IL-1 to stimulate ACTH release from corticotrophs. In spite of reports to the contrary from one group (41,42), a consensus (12,13,32,43-45) has emerged that indicates that short-term exposure of cultured corticotrophs to IL-1 does not enhance ACTH release. Continuous short-term exposure of perfused anterior pituitary fragments to IL-1 also does not alter basal ACTH release (43).

In our studies, we have been unable to detect a consistent stimulation of β-endorphin release from primary cultures of rat anterior pituitary cells, unless cells were exposed to rhIL-1α or β for 12 hours or longer. Twenty-four hour treatment of cultured corticotrophs with IL-1 (Figure 1) results in a meager 50-70% accumulation of β-endorphin over basal, as compared to the 7-fold change observed with a maximal stimulatory concentration of CRH. In contrast to IL-1 exposure, IL-6 treatment did not alter basal β-endorphin release. The observed IL-1 effect was cell-density dependent because IL-1 was inactive in the presence of 0.04×10^6 cells per well, as compared to 0.25×10^6 cells per well. Cell density-dependent effects of IL-1 enhance β-endorphin release contrast with the cell density independent effects of CRH, a known direct stimulator of corticotrophs. However, reducing the number of cultured cells may be limiting the concentration of endogenous substances that exert a paracrine action on corticotrophs, thus enabling the manifestations of direct IL-1 effects on corticotrophs. The mechanism through which chronic exposure of pituitary cells to IL-1 increases POMC-derived peptide accumulation in the medium is for the most part unknown. Whether or not the modest changes in basal β-endorphin release following chronic IL-1 exposure represent Ca^{2+} independent, constitutive release, as appears to be the case in AtT-20 cells, remains an open question. Recent observations (32) indicate that the signal transduction mechanism mediating the observed IL-1 actions does not depend on PGE_2 or cyclic AMP. Furthermore, in contrast to CRH-induced ACTH release, dexamethasone treatment only partially blocks IL-1 induced ACTH release from corticotrophs. Different cellular mechanisms appear to mediate IL-1-induced POMC-derived peptide release than cellular mechanisms associated with other direct ACTH-releasing substances.

Finally, one group has demonstrated that IL-1 is a more potent stimulator of POMC mRNA synthesis in dispersed pituitary cells than CRH, following a short-term

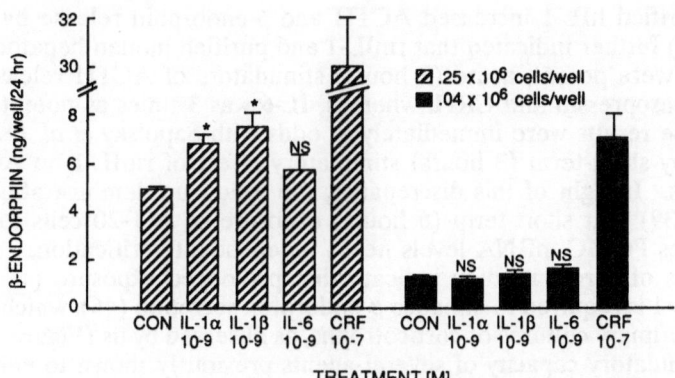

Figure 1. Prolonged exposure of rat anterior pituitary cells in culture to rhIL-1α or β enhances β-endorphin release. Anterior pituitary (AP) cells were dispersed as described previously (59) and cultured in Gibco DMEM: Nutrient Mixture F12 (1:1, v:v) containing 10% heat-inactivated fetal bovine serum (FBS). AP cells were plated into 24-well Costar plates at two different densities as indicated and cultured for 3 days. At this time the medium was removed and replace with 1 ml of fresh 2% FBS, DMEM: F12 medium with or without the test agents at the concentrations indicated. The medium was collected 24 hrs later and β-endorphin levels were determined by radioimmunoassay (60). The data represents the mean ± SE (n = 6) obtained from 1 of 3 similar experiments. Statistical comparisons were performed using Student's t test. * denotes $P < 0.05$ as compared to control (CON); NS, not significantly different from CON; Abbreviation: IL-1α, recombinant (r), human (h), interleukin (IL)-alpha (P. Lomedico, Hoffman-LaRoche, Inc.); IL-1β, rhIL-1 beta (Genzyme); IL-6, rhIL-6 (Genzyme); CRH, corticotropin-releasing factor.

(6 hours) incubation (39). This finding suggested a rapid, profound direct effect of IL-1 on the readiness of corticotrophs to release POMC-derived peptides; however, the significance of this finding must be tempered by the more recent demonstration that the direct effects of IL-1α or β on POMC gene transcription after 3 or 15 hours exposure, is minimal or non-existent (44).

In spite of the established synergistic effects of IL-1 with a variety of agents on immunocompetent (2) and other cells (46), this issue has not been adequately addressed with regard to corticotrophs at this time. Uehara *et al.* (45) noted a synergistic action of high doses of CRH with subeffective doses of IL-1β on ACTH release, but noted the interaction was not striking. Textual comments of other investigators (32,43) indicate that CRH and possibly other ACTH-releasing factors do not synergize with IL-1; however, data were not presented. In view of well known synergistic action of multiple secretogogues with CRH on corticotroph activation, a thorough evaluation of the possible potentiating effects if IL-1 on the spectrum of known ACTH-releasing compounds would be informative.

Attempts to define mechanisms of action or signal transduction pathways that may mediate a cell-specific response (*e.g.*, POMC peptide-derived synthesis in and secretion from corticotrophs) are not easy if the cell type to be studied is a minor cell type (*e.g.*, corticotrophs) in a heterogeneous population of cells (*e.g.*, dispersed anterior pituitary cells). Immortalized tumor cell lines, such as the mouse corticotroph-derived tumor cell line, AtT-20s, are often studied instead of primary cultured cells in order to obviate such problems. In an initial report, Farrar (40) indicated that 1-2 day exposure of AtT-20 cells to rhIL-2 and purified hIL-1 increased ACTH and β-endorphin release by 2- to 3-fold. Waloski *et al.* (47) further indicated that rmIL-1 and purified human hepatocyte-stimulating factor (IL-6) were potent, acute (2 hours) stimulators of ACTH release. IL-1 was equipotent with vasopressin and CRH, whereas, IL-6 was 3 times as potent as the other treatments. These results were immediately at odds with Sapolsky *et al.* (12), who were unable to find any short-term (3 hours) stimulatory effect of rmIL-1 on ACTH release from AtT-20 cells. In light of this discrepancy, the follow-up demonstration by Blalock and co-workers (39) that short-term (6 hours) exposure of AtT-20 cells to IL-1 and -2 potently stimulates POMC mRNA levels needs independent verification.

The results of a recent study indicate that prolonged exposure (> 12 hours) of AtT-20 cells to IL-1 is required to enhance β-endorphin secretion (46), which appears also to be the case in primary cultures of corticotrophs, as observed by us (Figure 1) and others (32,44). The stimulatory capacity of several agents previously shown to enhance ACTH or β-endorphin release from AtT-20 cells (48) such as CRH, forskolin, and phorbol 12-0-tetradecanoate 13-acetate (TPA), was potentiated by 24-hour pre-treatment with IL-1. The ability of IL-1 pre-treatment to amplify various secretogogue effects on β-endorphin release by CRH and forskolin, whose peptide stimulatory effects are known to be mediated by generation of cyclic AMP, was found to be independent of cAMP generation. Chronic treatment of AtT-20 cells with phorbol ester for 10 hours, which down regulates or depletes protein kinase (PK) C through an apparent increased rate of PK C degradation (49), eliminated the secretion of β-endorphin induced by phorbol ester, as well as the potentiation induced by IL-1 without altering the IL-1 induced β-endorphin secretion. These observations suggest that the IL-1 potentiation of TPA-induced β-endorphin release requires the presence of PK C and that the observed enhancement of basal β-endorphin secretion by IL-1 appears to be independent of PK C mediation.

In studies with Dr. Torda (IEE, Bratislava, Czechoslovakia), AtT-20 cells were pre-treated with IL-1 for 24 hours, washed and then incubated for 1 hour in the presence of varying concentrations of Ca^{2+} (2.5, 2.0, 1.5, 1.0, 0.5, 0 mM Ca^{2+} or 0 mM Ca^{2+} plus 0.5 mM EGTA) in otherwise normal medium. The increase over basal β-endorphin release induced by IL-1, even in the absence of Ca^{2+}, was constant (40-60%) and, therefore, independent of extracellular Ca^{2+} concentration. This suggests that prolonged IL-1 exposure enhances basal β-endorphin release from AtT-20 cells through a constitutive, Ca^{2+}-independent pathway, which differs from the observed potentiation of β-endorphin release by PK A(CRH, forskolin)- or PK C(TPA)-mediated secretogogues. Further studies revealed that down regulation of PK C through chronic exposure to TPA (Figure 2) did not abolish the IL-1 amplification of β-endorphin release induced by PK A-mediated secretogogues (forskolin, 8-bromo cAMP or CRH). This suggests that the potentiating action of IL-1 can occur independent of PK C and at a point beyond the activation of adenylate cyclase and GTP binding proteins, perhaps *via* protein kinase A or other cytosolic protein kinases or their products. Generation of intracellular messengers such as diacyglycerol (mimicked by TPA) and cAMP (mimicked by 8-bromo cAMP) are potentiated by IL-1 treatment. However, increased intracellular Ca^{2+} alone (Figure 3), another intracellular messenger in many systems, is an insufficient signal by itself to potentiate β-endorphin release following prior IL-1 pre-treatment. On the other hand, even though A23187-enhanced Ca^{2+} entry into IL-1 pre-treated AtT-20 cells did not result in amplification of β-endorphin release, this does not rule out that cytosolic Ca^{2+} changes are acting in concert with other signals to trigger potentiation of β-endorphin release.

At the present time, the mechanism through which IL-1 is able to potentiate either PK A- or PK C-signaling pathways to release β-endorphin from AtT-20 cells is unknown; however, IL-1 pre-treatment may increase AP-1 transcription factor activities (50) in perhaps a non-phosphorylated form. The subsequent exposure of AtT-20 cells to either PK A- or PK C-signal pathway secretogogues could result in rapid phosphorylation of an

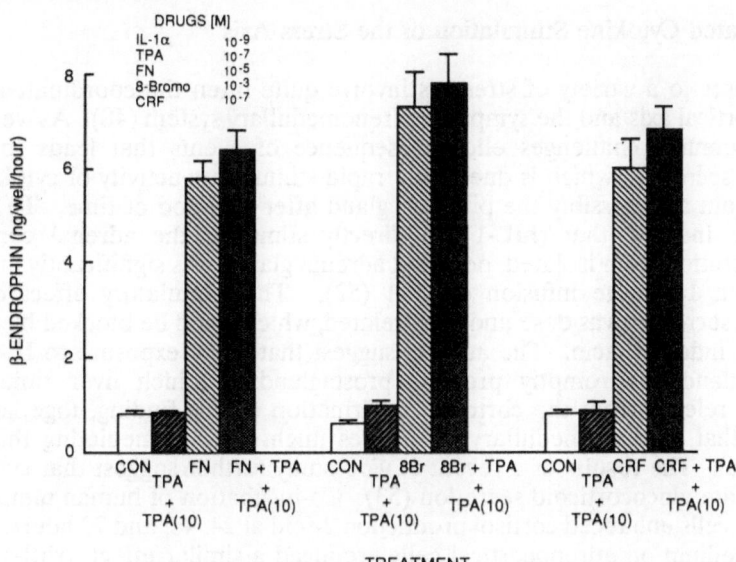

Figure 2. Down regulation of protein kinase C does not prevent IL-1α pre-treatment from potentiating protein kinase A-mediated β-endorphin secretion from AtT-20 cells. AtT-20 cells were subcultured in 24-well Costar plates in 10% FBS-DMEM medium and used 5-7 days later at 80-90% confluency. At the onset of each experiment, old medium was aspirated from the cells and replaced with fresh 10% FBS-DMEM with or without IL-1 for 24 hours. At this time cells were washed twice in release medium [DMEM containing 0.2% bovine serum albumin (BSA)]. Immediately thereafter, release medium with or without IL-1 or IL-1 with TPA was added and the incubations continued for 10 hours. Cells were again washed twice and IL-1 alone *(open bars)*, IL-1 plus TPA *(dark cross-hatched bars)*, or IL-1 plus FN,8 BR, or CRH *(light cross-hatched bars)*, or IL-1 plus TPA plus FN, 8 BR, or CRH *(stippled bars)*, and β-endorphin release was determined after 1 hour incubation. TPA (10) represents the cells which were treated with TPA plus IL-1 for 10 hours prior to the 1 hour release experiment. β-endorphin release is presented after subtracting the amount of β-endorphin released by untreated cells. Data are the mean ± SE of six observations from one of several similar experiments. The molar concentration of test agents used is indicated in the inset. Abbreviations: IL-1α, recombinant human interleukin-1 alpha (P. Lomedico, Hoffman-LaRoche, Inc.); TPA, phorbol 12-0-tetradecanoate 13-acetate; FN, forskolin; 8-bromo (8BR), 8-bromo cyclic AMP; CRH, corticotropin-releasing hormone; CON, control.

AP-1 related protein leading to the rapid transcription of POMC gene and amplification of β-endorphin release. Finally, the ability of IL-1 pre-treatment to potentiate multiple secretogogue-induced β-endorphin release from AtT-20 cells has no apparent comparable effect on primary cultures of corticotrophs. This raises the caveat that responses of transformed cells to exogenous regulators may not mimic the responses of non-transformed cells of the same lineage. Another example of the differential response of transformed corticotrophs, as compared to normal corticotrophs, to IL-1 treatment was observed when IL-1 elicited an acute (3 hour exposure), 3-fold increase in ACTH release from cultured corticotroph adenoma tissue obtained from patients with Cushing's disease (51).

Adrenal-Mediated Cytokine Stimulation of the Stress Axis

Responses to a variety of stressors involve quite often the coordinated activation of the HPA-cortical axis and the sympathoadrenomedullary system (48). As we have seen, a variety of immune challenges elicits a sequence of events that leads to enhanced glucocorticoid secretion, which is due to the rapid stimulatory activity of cytokines at the level of the brain and possibly the pituitary gland after a period of time. The results of several studies indicate that rhIL-1 can directly stimulate the adrenal cortex. Corticosterone secretion from isolated, perfused adrenal glands was significantly enhanced 90 minutes after a 1 minute infusion of IL-1 (52). The stimulatory effect of IL-1 on corticosterone secretion was dose and time related, which could be blocked by continuous perfusion with indomethacin. The authors suggest that acute exposure to IL-1 activates the adrenal gland to promptly produce prostaglandins, which over time stimulate corticosterone release from the cortex. Confirmation of this finding, together with the possible role that adrenal medullary substances might play in mediating this response would be useful. The results of a recent *in vitro* study further suggest that cytokines can directly modulate glucocorticoid secretion (53). Co-incubation of human monocytes with adrenocortical cells enhanced cortisol production 2-fold at 24, 48, and 72 hours. Monocyte conditioned medium on adrenocortical cells produced a similar effect. rhIl-1 alone was able to account for only a fraction of the observed response, suggesting that IL-1 in concert with other monokines in the conditioned medium is capable of directly stimulating cortisol production after prolonged exposure.

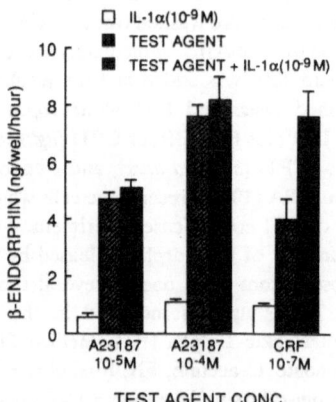

Figure 3. β-endorphin release induced by an increase in intracellular Ca^{2+} concentration alone is not potentiated by IL-1 pre-treatment. AtT-20 cells were pre-treated with or without 1 nM IL-1 for 24 hours. Next, cells were washed and untreated cells were incubated for an additional hour either with vehicle or with test agents (A23187 or CRH, *cross-hatched bars*). The IL-1 pre-treated cells were incubated with IL-1 alone *(open bars)* or with IL-1 plus test agents (A23187 or CRH, *stippled bars*) also for 1 hour. β-Endorphin released into the medium was determined by radioimmunoassay. The amount of β-endorphin released by untreated cells was subtracted from each treatment group. The *solid portion* of the *stippled* CRH plus IL-1 *bar* is representative of the ability of IL-1 pre-treatment to potentiate β-endorphin secretion induced by multiple secretogogues (46) and was included as a positive control. The data represents the mean ± SE (n = 6) obtained from 1 of several experiments. Abbreviations: IL-1α, rhIL-1 alpha (P. Lomedico, Hoffman-LaRoche, Inc.); A23187, calcium ionophore; CRH, corticotropin-releasing hormone.

To our knowledge, even though numerous stressors activate the adrenal medulla to secrete catecholamines and peptides, the possible direct effects of cytokines on the adrenal medulla have not been explored. In collaboration with Dr. Eiden (LCB, NIMH), a series of studies have been initiated to explore the possible direct effects of ILs on adrenal medullary-derived cells (6). Bovine adrenomedullary chromaffin cells in culture synthesize and secrete catecholamines and a variety of neuropeptides, amongst which are methionine enkephalin and vasoactive intestinal polypeptide (VIP) (54). The biosynthesis and release of these peptides were chosen for study because they have previously been shown to be differentially regulated by PK C (phorbol ester)- and PK A (forskolin)-mediated secretogogues (55). The results of initial experiments indicated that continuous exposure of chromaffin cells to rhIL-1α for 6-48 hours enhanced basal epinephrine accumulation in the medium from 50-200% without any discernible effect on cellular epinephrine content. This stimulatory effect was dose related in the presence of either IL-1α or β, although IL-1α was more potent than IL-1β. Twenty-four to 72 hour exposure of chromaffin cells to 1 nM rhIL-1α increased medium and cellular content of VIP, which indicates modulation of both synthesis and secretion of VIP by IL-1. Neither synthesis nor release of methionine enkephalin was augmented by exposure to rhIL-1α or β. In contrast to the stimulatory effects of IL-1, rhIL-6 (0.1 nM) treatment for 24 hours did not enhance the accumulation of catecholamines or peptides. The possibility that co-incubation of chromaffin cell secretogogues with IL-1s might potentiate peptide or catecholamine synthesis and secretion remains to be determined. The importance of these findings to possible *in vivo* activation of the sympathoadrenomedullary system by endogenous cytokines, also remains for future study. However, in one report the intravenous administration of *E. coli* endotoxin to human subjects resulted in enhanced plasma epinephrine levels at 2 hours that remained elevated even at 6 hours (37). The latency period of this response probably represents the activation time for macrophages to synthesize and release various cytokines, such as TNF-α or IL-1.

Summary and Conclusions

Overwhelming evidence indicates that the administration of cytokines such as IL-1α and β, IL-6, and TNF-1α stimulates one or more components of the HPA axis. The hypothesis driving this research is that host infection and tissue injury trigger the synthesis and release of several cytokines that act locally at sites of trauma and distally upon entering the circulation. Available evidence suggests that the primary source of HPA axis-acting or circulating cytokines is activated monocytes or macrophages; therefore, a direct relationship should exist between the appearance of monokines in plasma and the subsequent appearance of pituitary-adrenocortical hormones in plasma as well. Clarification of the physiological role of monokines as mediators of the host stress response will come from *in vivo* studies in which the type, sequence of appearance, duration of elevation, and quantification of each monokine is monitored along with ACTH and glucocorticoids, following an appropriate immune challenge.

In several recent reports, investigators have administered bacterial-derived endotoxin or LPS to stimulate the physiological events associated with infection or injury and chronicled plasma levels of IL-1, IL-6, and TNF-α (37,56,57). In human subjects, endotoxin challenge enhanced plasma TNF-α levels by 1 hour, which returned to basal levels by 4 hours (37), whereas, IL-6 plasma activity increased at 2 hours post-challenge and returned to baseline by 6 hours (56). Thus, both of these monokines are implicated as possible acute activators of the HPA axis. In perhaps the most revealing study to date, LPS challenge of mice indicated both a differential appearance and disappearance rate in serum for TNF-α and IL-1 and a differential regulation of these monokines by glucocorticoid feedback (57). Serum TNF was detected 45 minutes post-LPS, peaked by 1 hour, and returned to control levels by 3 hours. Serum corticosterone concentrations rose rapidly over a time course similar to that of TNF. Even after serum TNF concentration had returned to basal conditions, corticosterone levels remained maximally elevated, and serum corticocosterone was still significantly above basal levels 24-hour post-LPS. The rapid return of circulating TNF to pre-LPS challenge levels appeared to be regulated by negative glucocorticoid feedback, because TNF remained maximally elevated for at least 6 hours in adrenalectomized or hypophysectomized mice. LPS-induced levels of IL-1 were delayed as compared to serum TNF, peaked at 4 hours, and remained elevated even at 24 hours. These findings suggest a scenario in which TNF plays a primary role in the acute activation of the HPA axis followed by the overlapping stimulatory influence of IL-1. The prolonged elevation of IL-1 could enable certain novel actions of IL-1 such as enhancing the responsiveness of corticotrophs to β-adrenergic stimulation through unknown mechanisms or through an increase in the number of β-adrenergic receptors (43,58). Furthermore, IL-1 could directly increase the secretion of CRH or other corticotroph regulators from the median eminence. An appreciation of the sequence and magnitude of changes in circulating IL-1, TNF, IL-6, and IL-2 following an immune challenge is essential to understanding the physiological importance of the bidirectional communication between the immune system and HPA axis.

References

1. Bendtzen, K., Interleukin 1, interleukin 6 and tumor neurosis factor in infaction, inflammation and immunity, *Immun Lett* 19: 183-191, 1988.
2. Dinarello, C.A., Interleukin-1 and its biologically related cytokines, In F.J. Dixon, (ed) *Advances in Immunology* Academis Press, San Diego, Volume 44: 153-205, 1989.
3. Sipe, J.D., The molecular biology of interleukin-1 and the acute phase response, In G.H. Stollerman, W.J. Harrington, J.T. LaMont, J.J. Leonard, and M.D. Siperstein (eds) *Advances in Internal Medicine*, Year Book Medical Publishers, Inc. Chicago, Volume 34: 1-20, 1989.
4. Strober, W., and S.P. James, The interleukins, *Pediatr Res* 24: 549-557, 1988.
5. Blalock, J.E., A molecular basis for bidirectional communication between the immune and neuroendocrine systems, *Physiol Rev* 69: 1-32, 1989.
6. Eskay, R.L., A. Thigarajan, and L. Eiden, Interleukin-1 enhances the accumulation of epinephrine and vasoactive-intestinal polypeptide in cultured adrenal chromaffin cells, *Program of the 19th Annual Meeting of the Society for Neuroscience*, Phoenix, AZ, Abstract 156.2, p. 380, 1989.
7. Besedovsky, H., E. Sorkin, D. Felix, and H. Haas, Hypothalamic changes during the immune response, *Eur J Immunol* 7: 323-325, 1977.
8. Besedovsky, H., E. Sorkin, M. Keller, and J. Müller, Changes in blood hormone levels during the immune response, *Proc Soc Exp Biol Med* 150: 466-470, 1975.
9. Besedovsky, H.O., A. del Rey, and E. Sorkin, Lymphokine-containing supernatants from Con A-stimulated cells increase corticosterone blood levels, *J Immunol* 126: 385-387, 1981.
10. Besedovsky, H.O., A. del Rey, E. Sorkin, and C.A. Dinarello, Immunoregulatory feedback between interleukin-1 and glucocorticoid hormones, *Science* 233: 652-654, 1986.
11. Besedovsky, H.O., A. del Rey, and E. Sorkin, Immun-neuroendocrine interactions, *J Immunol* 135: 750s-754s, 1985.
12. Sapolsky, R., C. Rivier, G. Yamamoto, P. Plotsky, and W. Vale, Interleukin-1 stimulates the secretion of hypothalamic corticotropin-releasing factor, *Science* 238: 522-524, 1987.
13. Berkenbosch, F., J. van Oers, A. del Rey, F. Tilders, and H. Besedovsky, Cortocotropin-releasing factor-producing neurons in the rat activated by interleukin-1, *Science* 238: 524-526, 1987.
14. Uehara, A., P.E. Gottschall, R.R. Dahl, and A. Arimura, Interleukin 1 stimulates ACTH release by an indirect action which requires endogenous corticotropin releasing factor, *Endocrinology* 121: 1580-1582, 1987.

15. Katsuura, G., P.E. Gottschall, R.R. Dahl, and A. Arimura, Adrenocorticotropin release induced by intracerebroventricular injection of recombinant human interleukin-1 in rats: possible involvement of prostaglandin, *Endocrinology* 122: 1773-1779, 1988.

16. Uehara, A., P.E. Gottschall, R.R. Dahl, and A. Arimura, Stimulation of ACTH release by human interleukin-1β, but not by interleukin-1α, in conscious, freely-moving rats, *Biochem Biophys Res Commun* 146: 1286-1290, 1987.

17. Tsagarakis, S., G. Gillies, L.H. Rees, M. Besser, and A. Grossman, Interleukin-1 directly stimulates the release of corticotrophin releasing factor from rat hypothalamus, *Neuroendocrinology* 49: 98-101, 1989.

18. Ovadia, H., O. Abramsky, V. Barak, N. Conforti, D. Saphier, and J. Weidenfeld, Effect of interleukin-1 on adrenocortical activity in intact and hypothalamic deafferentated male rats, *Exp Brain Res* 76: 246-249, 1989.

19. Besedovsky, H., A. del Rey, E. Sorkin, M. da Prada, R. Burri, and C. Honegger, The immune response evokes changes in brain noradrenergic neurons, *Science* 221: 564-566, 1983.

20. Dunn, A.J., M.L. Powell, W.V. Moreshead, J.M. Gaskin, and N.R. Hall, Effects of Newcastle disease virus administration to mice on the metabolism of cerebral biogenic amines, plasma corticosterone, and lymphocyte proliferation, *Brain Behav and Immunity* 1: 216-230, 1987.

21. Dunn, A.J., Systemic interleukin-1 administration stimulates hypothalamic norepinephrine metabolism paralleling the increased plasma corticosterone, *Life Sci* 43: 429-435, 1988.

22. Kabiersch, A., A. del Rey, C.G. Honegger, and H.O. Besedovsky, Interleukin-1 induces changes in norepinephrine metabolism in the rat brain, *Brain Behav and Immunity* 2: 267-274, 1988.

23. Plotsky, P.M., Facilitation of immunoreactive corticotropin-releasing factor secretion into hypophysial-portal circulation after activation of catecholaminergic pathways or central norepinephrine injection, *Endocrinology* 121: 924-930, 1987.

24. Swanson, L.W., P.E. Sawchenko, and R.W. Lind, Regulation of multiple peptides in CRF parvocellular neurosecretory neurons: implications for the stress response, In T. Hökfelt, K. Fuxe, and B. Pernow (eds) *Progress in Brain Research*, Elsevier Science Publishers, New York, Volume 68: 169-190, 1986.

25. Szafarczyk, A., F. Malaval, A. Laurent, R. Gibaud, and I. Assenmacher, Further evidence for a central stimultory action of catecholamines on adrenocorticotropin release in the rat, *Endocrinology* 121: 883-892, 1987.

26. Farrar, W.L., J.M. Hill, A. Harel-Bellan, and M. Vinocour, The immune logical brain, *Immunol Rev* 100: 361-378, 1987.

27. Katsuura, G., P.E. Gottschall, and A. Arimura, Identification of a high-affinity receptor for interleukin-1 beta in the rat brain, *Biochem Biophys Res Commun* 156: 61-67, 1988.

28. Breder, C.D., C.A. Dinarella, and C.B. Saper, Interleukin-1 immunoreactive innervation of the human hypothalamus, *Science* 240: 321-324, 1988.

29. Nishida, T., N. Nishino, M. Takano, Y. Sekiguchi, K. Kawai, K. Mizuno, S. Nakai, Y. Masui, and Y. Hirai, Molecular cloning and expression of rat interleukin-1 alpha cDNA, *J Biochem* 105: 351-357, 1989.

30. Nakamura, H., S. Motoyoshi, and T. Kadokawa, Anti-inflammatory action of interleukin 1 through the pituitary-adrenal axis in rats, *Eur J Pharmacol* 151: 67-73, 1988.

31. Sharp, B.M., S.G. Matta, P.K. Peterson, R. Newton, C. Chao, and K. McAllen, Tumor necrosis factor-α is a potent ACTH secretogogue: comparison to interleukin-1β, *Endocrinology* 124: 3131-3133, 1989.

32. Kehrer, P., D. Turnill, J.M. Dayer, A.F. Muller, and R.C. Gaillard, Human recombinant interleukin-1 beta and -alpha, but not recombinant tumor necrosis factor alpha stimulate ACTH release from rat anterior pituitary cells *in vitro* in a prostaglandin E_2 and cAMP independent manner, *Neuroendocrinology* 48: 160-166, 1988.

33. Naitoh, Y., J. Fukata, T. Tominaga, Y. Nakai, S. Tamai, K. Mori, and H. Imura, Interleukin-6 stimulates the secretion of adrenocorticotropic hormone in conscious, freely-moving rats, *Biochem Biophys Res Commun* 155: 1459-1463, 1988.

34. Bindon, C., M. Czerniecki, P. Ruell, A. Edwards, W. McCarthy, R. Harris, and P. Hersey, Clearance rates and systemic effects of intravenously administered interleukin 2 (IL-2) containing preparations in human subjects, *Br J Cancer* 47: 123-133, 1983.

35. Lotze, M.T., L.W. Frana, S.O. Sharrow, R.J. Robb, and S.A. Rosenberg, *In vivo* administration of purified human interleukin 2. I. Half-like and immunologic effects of the Jurkat cell line-derived interleukin 2, *J Immunol* 134: 157-166, 1985.

36. Lotze, M.T., Y.L. Matory, S.E. Ettinghausen, A.A. Rayner, S.O. Sharrow, C.A. Seipp, M.C. Custer, and S.A. Rosenberg, *In vivo* administration of purified human interleukin-2. II. Half-life, immunologic effects and expansion of peripheral lymphoid cells *in vivo* with recombinant IL-2, *J Immunol* 135: 2865-2875, 1985.

37. Michie, H.R., K.R. Manogue, D.R. Spriggs, A. Revhaug, S. O'Dwyer, C.A. Dinarello, A. Cerami, S.M. Wolff, and D.W. Wilmore, Detection of circulating tumor necrosis factor after endotoxin administration, *N Engl J Med* 318: 1481-1486, 1988.

38. Smith, L.R., S.L. Brown, and J.E. Blalock, Interleukin-2 induction of ACTH secretion: pressure of an interleukin-2 receptor α chain-like molecule on pituitary cells, *J Neuroimmunol* 21: 249-254, 1989.

39. Brown, S.L., L.R. Smith, and J.E. Blalock, Interleukin 1 and interleukin 2 enhance proopiomelanocortin gene expression in pituitary cells, *J Immunol* 139: 3181-3183, 1987.

40. Farrar, W.L., Endorphin modulation of lymphokine activity, In F. Fraioli, A. Isidori, and M. Mazzetti (eds) *Opioid Peptides in the Periphery*, Elsevier, Amsterdam, pp. 159-165, 1984.

41. Bernton, E.W., J.E. Beach, J.W. Holaday, R.C. Smallridge, and H.G. Fein, Release of multiple hormones by direct action of interleukin-1 on pituitary cells, *Science* 238: 519-521, 1987.

42. Beach, J.E., R.C. Smallridge, C.A. Kinzer, E.W. Bernton, J.W. Holaday, and H.G. Fein, Rapid release of multiple hormones from rat pituitaries perfused with recombinant interleukin-1, *Life Sci* 44: 1-7, 1989.

43. Boyle, M., G. Yamamoto, M. Chen, J. Rivier, and W. Vale, Interleukin 1 prevents loss of corticotropic responsiveness to β-adrenergic stimulation *in vitro*, *Proc Natl Acad Sci USA* 85: 5556-5560, 1988.

44. Suda, T., F. Tozawa, T. Ushiyama, N. Tomori, T. Sumitomo, Y. Nakagami, M. Yamada, H. Demura, and K. Shizume, Effects of protein kinase-C-related adrenocorticotropin secretagogues and interleukin-1 on proopiomelanocortin gene expression in rat anterior pituitary cells, *Endocrinology* 124: 1444-1449, 1989.

45. Uehara, A., S. Gillis, and A. Arimura, Effects of interleukin-1 on hormone release from normal rat pituitary cells in primary culture, *Neuroendocrinology* 45: 343-347, 1987.

46. Fagarasan, M.O., R. Eskay, and J. Axelrod, Interleukin 1 potentiates the secretion of β-endorphin induced by secretagogues in a mouse pituitary cell line (AtT-20), *Proc Natl Acad Sci USA* 86: 2070-2073, 1989.

47. Woloski, B.M., E.M. Smith, W.J. Meyer, G.M. Fuller, and J.E. Blalock, Corticotropin-releasing activity of monokines, *Science* 230: 1035-1037, 1985.

48. Axelrod, J., and T. Reisine, Stress hormones: their interaction and regulation, *Science* 224: 452-459, 1984.

49. Young, S., P.J. Parker, A. Ullrich, and S. Stabel, Down-regulation of protein kinase-C is due to an increased rat of degradation, *Biochem J* 244: 775-779, 1987,

50. Mitchell, P.J., and R. Tjian, Transcriptional regulation in mammalian cells by sequence-specific DNA binding proteins, *Science* 245: 371-378, 1989.

51. Malarkey, W.B., and B.J. Zvara, Interleukin-1β and other cytokines stimulate adrenocorticotropin release from cultured pituitary cells of patients with Cushing's disease, *J Clin Endocrinol Metab* 69: 196-199, 1989.

52. Roh, M.S., M.O. Kathleen A. Drazenovich, J.J. Barbose, C.A. Dinarello, and C.F. Cobb, Direct stimulation of the adrenal cortex by interleukin-1, *Surgery* 102: 140-146, 1987.

53. Whitcomb, R.W., W.M. Linehan, L.M. Wahl, and R.A. Knazek, Monocytes stimulate cortisol production by cultured human adrenocortical cells, *J Clin Endocrinol Metab* 66: 33-38, 1988.

54. Eiden, L.E., R.L. Eskay, J. Scott, H. Pollard, and A.J. Hotchkiss, Primary cultures of bovine chromaffin cells synthesize and secrete vasoactive intestinal polypeptide (VIP), *Life Sci* 33: 687-693, 1983.

55. Pruss, R.M., J.R. Moskal, L.E. Eiden, and M.C. Beinfeld, Specific regulation of vasoactive intestinal polypeptide biosynthesis by phorbal ester in bovine chromaffin cells, *Endocrinology* 117: 1020-1026, 1985.

56. Fong, Y., L. Moldawer, M. Marano, H. Wei, S.B. Tatter, R.H. Clarick, U. Santhanam, D. Sherris, L.T. May, P.B. Sehgal, and S.F. Lowry, Endotoxemia elicits increased circulating β_2-IFN/IL-6 in man, *J Immunol* 142: 2321-2324, 1989.

57. Zuckerman, S.H., J. Shellhaas, and L.D. Butler, Differential regulation of lipopolysaccharide-induced interleukin 1 and tumor necrosis factor synthesis: effects of endogenous and exogenous glucocorticoids and the role of the pituitary-adrenal axis, *Eur J Immunol* 19: 301-305, 1989.

58. Stern,L., and G. Kunos, Synergistic regulation of pulmonary β-adrenergic receptors by glucocorticoids and interleukin-1, *J Biol Chem* 263: 15876-15879, 1988.

59. Vale, W., J. Vaughan, M. Smith, G. Yamamoto, J. Rivier, and C. Rivier, Effects of synthetic ovine corticotropin-releasing factor, glucocorticoids, catecholamines, neurohypophyseal peptides and other substances on cultured corticotropin cells, *Endocrinology* 113: 1121-1131, 1983.

60. Dave, J.R. N. Rubenstein, and R.L. Eskay, Evidence that β-endorphin binds to specific receptors in rat peripheral tissues and stimulates the adenylate cyclase-adenosine $3',5'$-monophosphate system, *Endocrinology* 117: 1389-1396, 1985.

IMMUNONEUROLOGY: A SERUM PROTEIN AFFERENT LIMB TO THE CNS

Gerald P. Kozlowski, Gajanan Nilaver,[1] and Berislav V. Zlokovič[2]

Department of Physiology
University of Texas Southwestern Medical Center
Dallas, Texas 75235-9040

Introduction

The burgeoning field of neuroimmunology currently emphasizes a brain-immune link represented by hormones, neurotransmitters, neuropeptides, cytokines (1) and other substances of the CNS which are shared in common with the immune system and can act either directly or indirectly on components of the immune system or their target tissues. This concept essentially depicts the CNS as having an efferent outflow of active principles to a variety of peripheral structures. Another concept of potential importance in the maintenance of a homeostatic brain-immune interaction is that of immunoneurology, whereby principles derived from peripheral sources feed back onto the CNS. In this sense, immune cells, their products and other related substances can be transported from their source *via* the blood stream to the CNS wherein they may access the internal milieu and can, therefore, be considered afferent to the CNS. This review describes studies on transport of immunogammaglobulins (IgG) and albumins into the CNS of rat, rabbit and guinea pig using several techniques: immunocytochemistry (ICC), colchicine administration, active immunization, induced experimental allergic encephalomyelitis (EAE), radioisotope labeling, enzyme-linked immunosorbent assay (ELISA), rocket immunoelectrophoresis and an *in situ* vascular brain perfusion method. Results from these studies reveal that serum albumin and globulin can enter the brain *via* several routes including that of active neuronal uptake by nerve endings, and subsequent transport to their cell bodies of origin *via* retrograde transport. These compounds may also enter the brain at circumventricular regions wherein functional leaks occur. Finally, there appears to be a saturable transport mechanism for IgG that is used to traverse the blood-brain barrier (BBB) and gain direct entry into the CNS. These, as well as other immunoneurologic mechanisms may be significant in establishing a pre-disease state (2,3) or may induce neuronal impairments as seen in autoimmune diseases of the CNS.

The Blood-Brain Barrier (BBB) and Blood-Cerebrospinal Fluid (CSF) Barrier

Historically, the CNS was generally considered to be an *immunologically privileged*

[1]Present affiliation: Departments of Neurology and Cell Biology and Anatomy, Oregon Health Sciences University, Portland, OR 97201

[2]Present affiliation: Department of Neurological Surgery, University of Southern California School of Medicine, Los Angeles, CA 90033

Circulating Regulatory Factors and Neuroendocrine Function
Edited by J. C. Porter and D. Ježová
Plenum Press, New York, 1990

345

site, incapable of producing its own antibodies, and normally protected from entry of systemic antibodies and immunocompetent cells by the physiological confines of the BBB (4,5). Proof for the existence of such a barrier was documented by the exclusion of systemically administered dyes such as Evan's blue or trypan blue (4,5) and the observation that organisms such as the yellow fever virus can proliferate unperturbed in the nervous system of mice (5). Some of the dye studies are now known to have been based on the formation of complexes with serum albumin, and their exclusion from the CNS reflected the inability of the dye-albumin complex to cross the BBB. Anatomically (Figure 1), the BBB is represented by the tight junctions between brain capillary endothelial cells which preclude the passage of substances along the intercellular cleft (Figure 1). Functionally, however, such a simplified view of the BBB is not tenable (6-9,29). A variety of biochemical mechanisms allow selective transport of specific substances across the BBB while excluding any *undesirable* substances either due to a lack

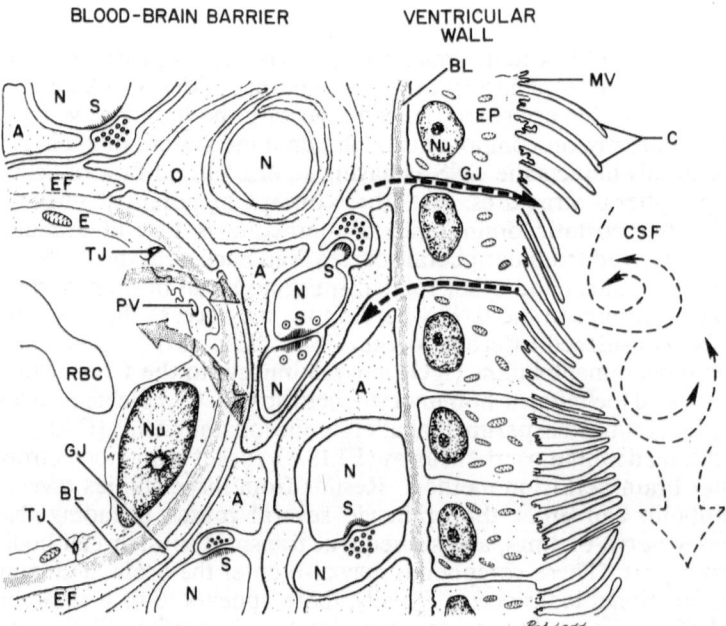

Figure 1. A region of the BBB juxtaposed to the ventricular wall. The endothelial (E) cells of a closed capillary have tight junctions (TJ) that exclude large molecular weight substances. The *large arrows* depict the flow of direction for substances taken up by pinocytotic vesicles (PV). Most of these substances enter lysosomes for eventual degradation. Historically, the flow was thought to be bidirectional (*other large arrow*), but recent studies show that experimental marker materials do not traverse from the abluminal to the luminal side. Other structures sometimes included as part of the BBB are the basal lamina (BL) and astroglial end-feet (EF). The glial cells (A=astrocytes; O=oligodendroglia) surround neuronal (N) processes and encapsulate areas of synaptic (S=synapse) activity. Ependymal (EP) cells (Nu=nucleus) lining the ventricle have microvilli (MV) and prominent cilia (C) on their apical surface for local transport of CSF (*thin broken arrows*). The *thick broken arrows* show the bidirectional flow of materials through the gap junctions (GP) of adjacent ependymal cells. Reprinted from Kozlowski (135), with permission of W. B. Saunders Co.

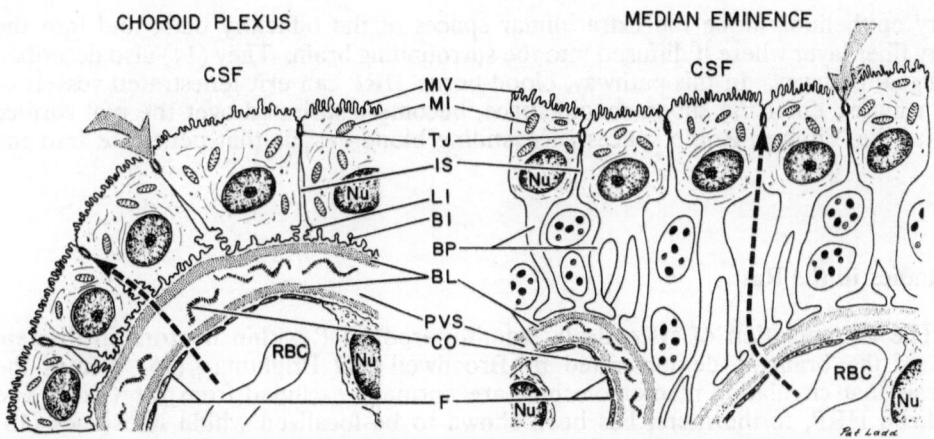

Figure 2. The choroid plexus and median eminence are two examples of circumventricular organs having, respectively, either unmodified or modified ependymal cells. As shown for the median eminence, the modified ependymal cells called tanycytes have long basal processes (135,136). The tight junction (Tj) represents the CSF-brain barrier to substances attempting to cross from the CSF side (*large arrow*) to the vascular side; and also a brain-CSF barrier for substances attempting to cross (*broken arrows*) from the vascular side to the CSF side. Some of these Tj may be of the *leaky* type. MV, microvilli; MI, mitochondria; IS, intercellular space; LI, lateral interdigitations; BI, basal interdigitations, BP, basal processes; BL, basal laminae; PVS, perivascular space; CO, collagen; F, fenestrae of capillary endothelium; RBC, red blood cell; NU, nucleus. Reprinted from Kozlowski (135), with permission of W. B. Saunders Co.

of a transport system or by virtue of being shuttled into the lysosome pathway for degradation. For example, there are degradative enzymes in the capillary wall which prevent entry of nonessential lipid soluble molecules into the CNS (10).

Recent studies have indicated that the brain is not totally isolated from the immune system by the BBB. First, there are certain circumventricular areas wherein this barrier is deficient: the median eminence and choroid plexus as illustrated in Figure 2, pars nervosa of the pituitary, area postrema, pineal, subcommissural organ, subfornical organ, and the organum vasculosum of the lamina terminalis (OVLT). These areas of the circumventricular organs (CVO) have fenestrated vessels as a common feature that easily allow substances to pass into the perivascular space and into the interependymal space but are prevented from entering the ventricle by tight junctions (Figure 2). Here, the tight junctions represent the blood-CSF barrier. Broadwell *et al.* (11) observed the passage of horseradish peroxidase (HRP) between some ependymal cells of the median eminence indicating that the tight junctions may be of the *leaky* type. Furthermore, Brightman and Reese (12) described so-called *functional leaks* at regions of attachment of CVO to the brain whereby substances could cross from the perivascular space into surrounding areas. For example, in the median eminence (Figures 2,11,12) IgG can pass out of the fenestrated vessels, travel along the perivascular spaces and diffuse into the area of the arcuate nucleus.

Blatteis and colleagues have shown that the OVLT is an important site whereby blood-borne cytokines interact with the brain since electrolytic ablation of the OVLT prevents the febrile response to circulating pyrogens (11,13).

Recently, Balin *et al.* (14) described a nose-brain pathway whereby intranasal administration of HRP could be followed as it passed through intercellular clefts of the olfactory epithelium, along the extracellular spaces of the olfactory bulb, and into the olfactory fiber layer where it diffused into the surrounding brain. They (14) also described a meningeal pathway. In this pathway, blood-borne HRP can exit fenestrated vessels of the dura mater, enter the subarachnoid space, become distributed over the pial surface and enter the Virchow-Robin spaces surrounding blood vessels that penetrate into the brain.

BBB Studies in the Rat

The accumulation of systemically administered HRP within neurons in discrete regions of the brain as demonstrated by Broadwell and Brightman (15) refutes the hypothesis that circulating macromolecules are normally excluded from the CNS. The internalized HRP, furthermore, has been shown to be localized within lysosomes (16) implying that retrograde transport mechanisms may play a role in the uptake of these compounds. Such mechanisms also mediate the uptake of macromolecules which are not present in high concentrations in the systemic circulation. This is evidenced by the ICC localization of albumin (Figure 3-7) and globulin within discrete neuronal populations of rat brain (17-21). The use of antibodies to rat albumin demonstrated its localization within supraoptic (Figures 4 and 5) and paraventricular nuclei of rat hypothalamus (17), tanycytes of the third ventricle (17,19), the medial habenular nucleus, the lateral nuclear complex of the thalamus, the dorsal hippocampus (Figures 6 and 7), the area postrema

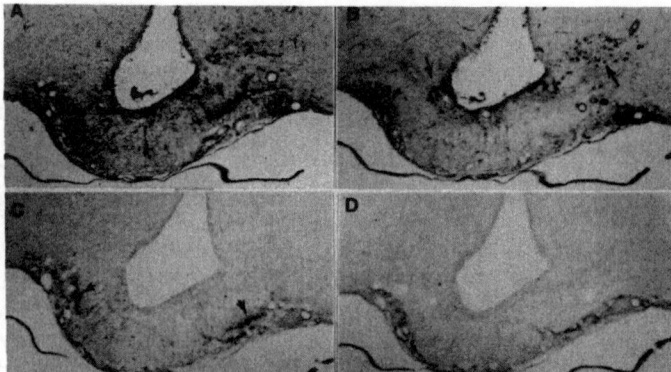

Figure 3.   Adjacent sections of the rat median eminence immunostained with various treatments of antisera to the decapeptide luteinizing hormone-releasing hormone (LHRH). This antisera was generated against a LHRH-bovine serum albumin (BSA) conjugate. (A) the *arrowheads* indicate staining for LHRH fibers which disappears when the antisera is absorbed with LHRH (B), but staining for albumins in the arcuate nucleus (*arrows* of A and B) remains undisturbed. However, if the antisera is absorbed with BSA (C), the staining in arcuate nucleus neurons disappears (A, *arrows*) while that of LHRH fibers (C, *arrowheads*) remains intense. If the antisera is absorbed with both LHRH and BSA, all staining disappears (D). Neurons of the arcuate nucleus normally contain an albumin which could be misconstrued for LHRH staining with this antisera. Original magnification ×80. Reprinted from Kozlowski and Dees (19) with permission of The Histochemical Society, Inc.

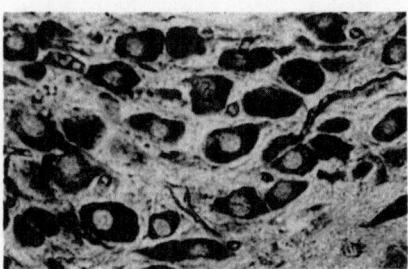

Figure 4. Rat supraoptic nucleus neurons immunostained with rabbit anti-rat albumin serum showing immunopositive albumin in the cytoplasm and proximal fiber segments (17). Original magnification ×100. Reprinted from Kozlowski and Nilaver (37), with permission of Alan R. Liss, Inc.

and nucleus tractus solitarius of the medulla (17). Albumin-labeling of neurons in the habenular nucleus, thalamus, hippocampus and nucleus tractus solitarius (which do not project to circumventricular brain regions with a deficient BBB) probably represent trans-synaptic transport of the macromolecule (22,23). Furthermore, absence of staining in these regions of colchicine-treated rats (intracerebroventricular, 100 μg; 24 hrs), points to the importance of retrograde transport mechanisms. In a similar study Meeker *et al.* (20) employing monoclonal antibodies, demonstrated the presence of circulating endogenous immunoglobulins within magnocellular neurons of the rat supraoptic and paraventricular nuclei. Accumulation of immunoreactive product was also blocked by colchicine administration. Similarly, incorporation of intravenously administered [^{125}I]-labeled rabbit or rat immunoglobulins, and the localization of the macromolecules within lysosome-like organelles of the magnocellular neurons by immunoelectron microscopy, also indicate that circulating proteins and immunoglobulins enter the CNS *via* this route. Furthermore, since viruses have been shown to be capable of entering the CNS by retrograde trans-neuronal transport mechanisms (24,25), the entry of these organisms and immunoglobulins into the brain may play an important role in pathogenesis of certain autoimmune brain diseases.

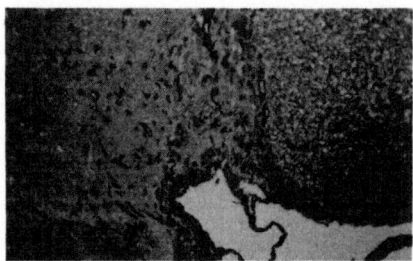

Figure 5. Supraoptic nucleus from a rat administered colchicine intracerebroventricularly prior to sacrifice and processed by immunocytochemistry with rabbit anti-rat albumin. The cells and fibers are immunonegative for albumin (17). Original magnification ×40. Reprinted from Kozlowski and Nilaver (37), with permission of Alan R. Liss, Inc.

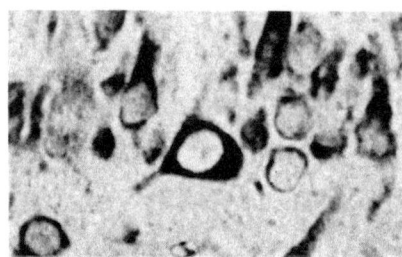

Figure 6.   Pyramidal cells of the rat hippocampus immunostained with rabbit anti-rat albumin antisera using the peroxidase-antiperoxidase technique. The neuronal cytoplasm and proximal segments of the fibers are immunopositive for albumin (17). Original magnification ×100. Reprinted from Kozlowski and Nilaver (37) with permission of Alan R. Liss, Inc.

BBB Studies in the Rabbit

Introduction

It is useful to study the pattern of distribution of brain IgG under conditions in which the BBB has been disrupted (Figures 8-10) and compare it to that seen in control animals with an intact BBB (Figures 11 and 12). Freund's adjuvant is a powerful agent that opens the BBB (26) and, most probably, the blood-CSF barrier. It was administered to our rabbits for up to two years during the process of generating antibodies to a variety of peptides.

Materials and Methods

Control (non-immunized) rabbits and those used for antibody generation were anesthetized, perfused-fixed with Zamboni's fixative, and processed for ICC as previously described (27). Their brains were sectioned serially in the coronal plane (50 μm) using a vibrating microtome, and nearly adjacent sections selected for immunostaining. In the

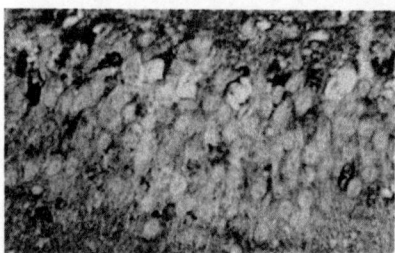

Figure 7.   Rat hippocampus from an animal administered colchicine intracerebroventricularly prior to sacrifice and processed for immunocytochemistry using rabbit anti-rat albumin. The cells and fibers are immunonegative for albumin (17). Original magnification ×100. Reprinted from Kozlowski and Nilaver (37) with permission of Alan R. Liss, Inc.

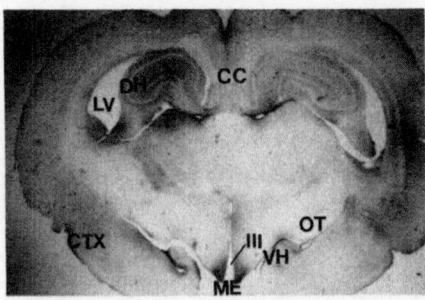

Figure 8. Transverse section of brain from an immunized rabbit used to generate antisera against the neuropeptide LHRH. Brain was stained with sheep anti-rabbit IgG (Shanti) using the PAP technique. Areas especially rich in IgG are the: dorsal and ventral hippocampus (DH and VH), thalamus (T), cerebral cortex (CTX), median eminence (ME) and periventricular areas of the third (III) and lateral ventricles (LV). Areas of white matter such as the optic tract (OC), corpus callosum (CC) and the pyramidal layer of the hippocampus (*white line* in DH) are unstained for IgG.

3-step staining procedure, employing the peroxidase-antiperoxidase (PAP) technique, the sections were incubated with primary antisera (either anti-VP or anti-LHRH), followed by a sheep anti-rabbit antibody as the secondary reagent, and then the rabbit PAP complex. For the 2-step staining procedure, one of the following reagents was used in the first step: sheep anti-rabbit IgG (Shanti), goat anti-rabbit IgG (Ganti) or mouse anti-rabbit IgG (Manti—produced as a polyclonal antibody from ascites tumor). This was followed by rabbit PAP as the secondary reagent. Antibody concentrations and incubation times were optimized to produce maximal intensity of staining. Negative controls consisted of using Shanti, Ganti or Manti immunoadsorbed with rabbit IgG, or using rabbit PAP only.

Results and Discussion

When the 3-step procedure was used, staining for LHRH or VP was detected in appropriate areas, but it was obvious that the *background* staining (due to secondary

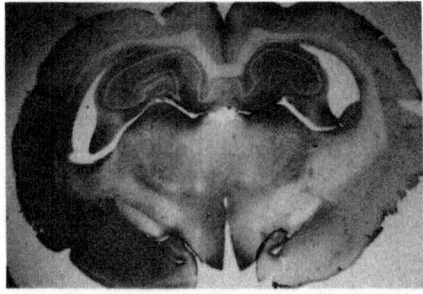

Figure 9. Transverse section of brain shown in Figure 8 except stained with goat anti-rabbit IgG (Ganti) and rabbit PAP. Ganti is less intense than Shanti but the pattern of distribution for IgG is similar for both sections.

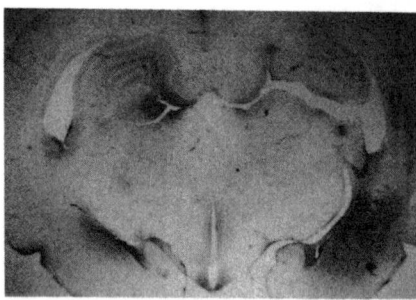

Figure 10. Transverse section of brain shown in Figure 8 except stained with mouse anti-rabbit IgG (Manti) from ascites fluid. Although the stain is lighter and fewer areas stain, the pattern of distribution is similar to that seen with Ganti (Figure 8) and Shanti (Figure 9).

antibody) was high. The primary antisera against the neuropeptides were therefore deleted from the procedure, and the 2-step protocol was used. Maximal intensity of IgG staining was noted with Ganti followed by Shanti, and least with Manti. As shown in Figure 8, staining was preferentially confined to grey matter regions, such as the cerebral cortex and hippocampus, with no reactivity being observed in white matter tracts, *e.g.*, the corpus callosum and optic tract. These observations indicate that the *background* staining represents authentic intra-neuronal labeling of endogenous IgG, rather than non-specific binding of the bridging antibodies to the tissue sections. However, unlike the albumin staining noted in rat hippocampus, not all the rabbit pyramidal cells were reactive, the IgG-labeling being confined to only some pyramidal neurons (Figures 13 and 14).

Long-term disruption of the BBB results in high concentrations of CSF IgG which are probably responsible for the preferential staining of the cortical surfaces and periventricular regions noted in these animals. The periventricular areas of human and rabbit brain are rich in Fc receptors for IgG, suggesting a protective role for these areas in the removal of IgG or IgG-antigen complexes from the CSF (28).

The staining of CVO regions such as the median eminence-arcuate nucleus areas seen in control animals (Figures 11 and 12), probably reflects the presence of functional leaks in the BBB, first described by Brightman and Reese (12). Staining was also seen along heavily vascularized areas of the subarachnoid membranes in the hippocampus and

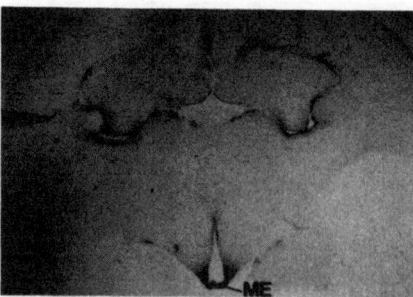

Figure 11. Transverse section of brain from a control rabbit, *i.e.*, one not used for immunization or other experimental purposes. Sections stained with Shanti and showing IgG present on the surface of thalamus and the area of the median eminence (ME).

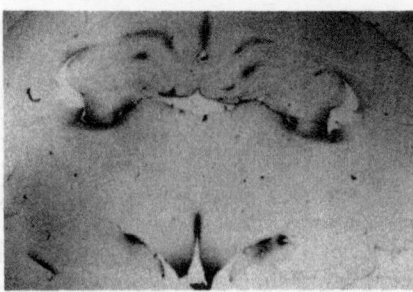

Figure 12. Transverse section of brain shown in Figure 11 stained with Ganti. Periventricular areas of the III and lateral ventricles, as well as heavily vascularized areas of the dorsal and ventral hippocampus have IgG.

along the ventral brain surface, probably due to leakage of IgG from fenestrated vessels of the meninges.

BBB Studies in the Guinea Pig

Introduction

The classical view of a cerebral capillary wall acting solely as a fixed cellular tube is no longer tenable and the dynamic nature of endothelia, long familiar to immunologists, has now been gradually integrated into our current concept of the BBB (29). Considerations that apply to blood-brain transport of the plant enzyme, HRP and micro HRP (12,30) do not necessarily apply to BBB transport of other studied peptides and proteins. For example, recent *in vitro* work has revealed that a number of biologically important proteins and large peptides may bind to capillary endothelial cells, followed by subsequent internalization into endosomes (31) and abluminal exocytosis of the engulfed molecules, giving rise to the concept of receptor- and/or carrier-mediated transcytosis

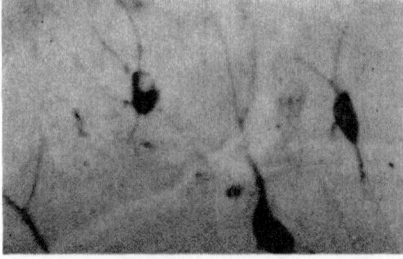

Figure 13. Pyramidal neurons of the rabbit hippocampus immunostained using Shanti. This rabbit was used to generate antisera against LHRH. Analysis of results using both serum and technical controls demonstrate that these neurons contain globulin which freely accessed the CNS due to the breakdown of the BBB by Freund's adjuvant during the immunization process. Original magnification ×100. Reprinted from Kozlowski and Nilaver (37) with permission of Alan R. Liss, Inc.

(32,33). Specific by subsequent internalization into endosomes (31) and abluminal exocytosis of the engulfed molecules, giving rise to the concept of receptor- and/or carrier-mediated transcytosis (32,33). Specific blood-to-brain transport systems for a variety of brain peptides have been also shown at the BBB under *in vivo* experimental conditions (34-36).

Although there are a number of potential pathways for serum proteins to reach the interior milieu of the CNS (37), the BBB interface is the most likely candidate given that it has the largest surface area and the shortest path length between the blood and the brain (35). One approach to studying a serum protein afferent limb to the CNS under *in vivo* conditions was our use of a vascular brain perfusion method (38) to quantitate transport of blood-borne homologous IgG across the BBB of normal guinea-pigs (39) and guinea-pigs with EAE (40). IgG were selected as a model for studying immune protein interactions with the BBB because these molecules could potentially be used as both therapeutic agents for brain disease (41) and neurodiagnostic imaging agents (42).

Materials and Methods

VASCULAR BRAIN PERFUSION TECHNIQUE. Adult Hartley guinea-pigs of both sexes, weighing 300-350 g, were anesthetized with thiopentone-sodium (30-35 mg/Kg) intraperitoneally. Perfusion of the ipsilateral forebrain was performed through the right common carotid artery using an extracorporeal perfusion circuit as previously described (38). The perfusion medium consisted of 20% sheep red cells (oxygen carrier) and artificial plasma salts (38) containing dextran (molecular weight 70,000; 48 g/liter). Immediately before the start of the perfusion, the contralateral carotid artery was ligated and both jugular veins were cut to allow drainage of the perfusate. Perfusion pressure (13 to 16 kPa) and pCO_2 (5-5.5 kPa) were standardized to maintain cerebral blood flow in the ipsilateral forebrain at about 1 ml min^{-1} g^{-1} brain tissue. The effective perfusion pressure was maintained 1-3 kPa above the animal's own arterial blood pressure to ensure the functional separation between the artificial and vertebral circulation (35,38).

Isotopically labelled [^{125}I]-IgG solutions at concentrations of 2.5 μg/ml administered either in the presence or absence of unlabelled guinea pig IgG (Sigma, St. Louis, MO), were introduced into the perfusion circuit by a slow-drive syringe at a known rate and concentration for times ranging from 1 to 20 minutes. Perfusion was terminated by cutting

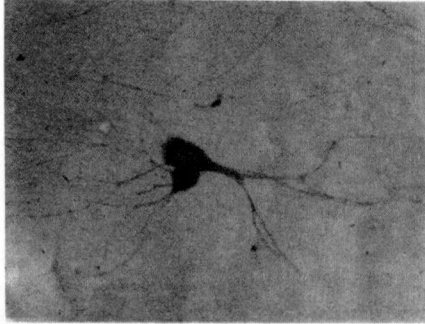

Figure 14. Pyramidal neurons of the rabbit hippocampus immunostained using Shanti. This rabbit was a control rabbit, *i.e.*, not used for immunization or other experimental studies. Original magnification ×100. Reprinted from Kozlowski and Nilaver (37) with permission of Alan R. Liss, Inc.

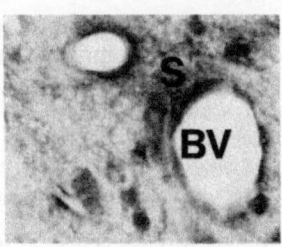

Figure 15. Stain (S) for IgG using the avidin-biotin-peroxidase reaction seen around blood vessels (BV) from brain of guinea pig that was first perfused for 10 min with artificial blood having no IgG, followed by perfusion with 4 mg/ml guinea pig IgG added to the perfusion medium. Original magnification ×400.

the right common carotid artery and decapitating the animal. The brain was removed, dissected into different regions, and prepared for scintillation counting. In a separate series of experiments, brains were perfused with unlabelled homologous IgG for 10 minutes followed by ICC for visualizing its distribution.

EXPERIMENTAL ALLERGIC ENCEPHALITIS (EAE) IN THE GUINEA PIG. Experiments were carried out in outbred adult Hartley guinea pigs weighing 300-350 g. As described by Colover (43), EAE (also called Experimental Autoimmune Encephalitis) was induced by administrating homologous myelin basic protein (MBP) after treatment with ovalbumin (OA) and muramyl dipeptide (MDP) using the following scheme: *(a)* on day 1, animals were injected subcutaneously with a water-in-oil emulsion of 15 μg of MBP, 200 μg of OA and Freund's complete adjuvant (FCA); *(b)* on day 28, they were injected intraperitoneally with 10 mg of OA in saline under a Piriton cover; and *(c)* on day 35 the survivors were injected with 50 μg of MBP and 0.1 ml of Freund's incomplete adjuvant (FIA). Following MBP injection, animals were assigned a clinical score daily for up to 45 days and then sacrificed. Brains and spinal cord, contralateral to the vascular perfusion side, were fixed in formal-saline, and paraffin embedded sections were stained with hematoxylin and eosin for routine histology, and with gallocyanin-Darrow red stain for myelin. Radioisotope studies with [^{125}I]-labelled IgG as well as ICC analysis of the distribution of blood-borne IgG following the perfusion were performed in these animals in a manner identical to that of the normal non-treated guinea pig.

SCINTILLATION COUNTING. The ipsilateral forebrain was divided into the parietal cortex, caudate nucleus, and hippocampus. Brain samples of approximately 100 mg wet weight were dispensed into pre-weighed scintillation vial inserts and plasma samples of 25 μl were taken for counting. For gamma counting, samples and standards were counted directly on an LKB spectrometer. In a separate series of experiments, [^{3}H]-dextran (molecular weight 70,000, Amersham) was perfused in the absence of radiolabelled [^{125}I]-IgG, samples were solubilized overnight in 0.5 ml of Soluene (Packard), and treated with 4 ml of scintillation cocktail before counting. Samples for β-counting were measured on an LKB Spectral β-scintillation spectrometer in the low energy channel using an internal quench curve program.

ISOTOPICALLY LABELLED IGG. IgG was labelled with ^{125}I by the micro-chloramine-T method (137) modified by adding a 10-fold excess of chloramine-T as well as phosphate-buffered saline (PBS, pH 7.4). Unbound iodide was removed by gel filtration on Sephadex G-200 and homogeneity was verified by thin-layer chromatography.

The specific activity of the immune protein enabled the use of highly radioactive perfusion fluids at concentrations about three orders of magnitude lower than that normally found in guinea pig blood.

MEASUREMENT OF IGG AND PLASMA ALBUMIN LEVELS IN CSF. Samples of CSF (50-100 μl) and plasma from anesthetized EAE and normal guinea pigs were obtained from the cisterna magna and jugular vein, respectively, prior to commencement of the vascular perfusion. IgG was determined by the ELISA method, while albumins were measured by rocket immunoelectrophoresis.

IMMUNOCYTOCHEMICAL (ICC) STUDIES. Brains were vascularly perfused with either 4 mg/ml unlabelled IgG or only perfusion medium for 10 minutes, followed by perfusion with fixative (2% periodate-lysine monochloride-paraformaldehyde) for 15 minutes. The brain was removed and processed for ICC. The avidin-biotin-peroxidase method was used as previously described (40). Sections were sequentially incubated with: *a)* goat anti-guinea pig IgG, *b)* biotinylated anti-goat IgG, *c)* avidin-peroxidase, and *d)* diaminobenzidine and H_2O_2. The sections were lightly counterstained with hematoxylin and eosin.

CALCULATION OF THE UNIDIRECTIONAL BLOOD-TO-BRAIN TRANSFER CONSTANT, K_{in}. As previously reported (38,39), the unidirectional blood-to-brain transfer constant-K_{in} for $[^{125}I]IgG$ both in normal and EAE animals, either in the presence or absence of different concentrations of unlabelled IgG, was estimated from the multiple time-point/graphic analysis of the brain uptake data during the first 20 minutes of perfusion. The following equation was employed:

$$C_{BR}(T)/C_{PL}(T) = K_{in} T + V_i$$

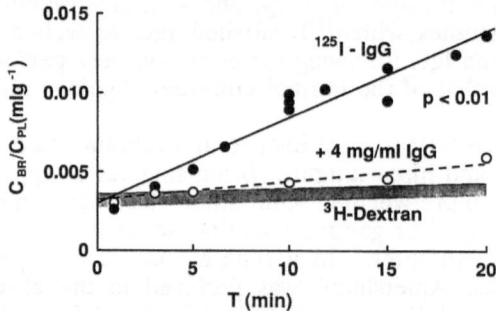

Figure 16. Kinetics of entry of homologous $[^{125}I]IgG$ (2.5 μg/ml) into the hippocampus of the perfused guinea-pig brain in the absence (*solid points* and *line*) and presence of 4 mg/ml unlabelled IgG (*open points* and *interrupted line*). $[^{125}I]$-radioactivity, cpm g^{-1} brain/cpm ml^{-1} plasma perfusate, is plotted against the perfusion time, T. Each point represents a single experiment; P difference in slopes by ANOVA. The shaded area represents $[^3H]$dextran space determined in a separate series of experiments and illustrated by its K_{in} SE and V_i SE (n = 9 brains). Adapted from Zlokovič *et al.*, (39) with permission of Experimental Neurology.

where $C_{BR}(T)$ is the radioactivity of IgG measured per unit mass of brain at time T, and $C_{PL}(T)$ is the constant radioactivity of IgG per unit mass of plasma during the perfusion; T is the time when perfusion is terminated. The radioactivity of IgG in the arterial inflow (C_{PL}) is constant under present experimental conditions. The equation defines a straight line with a slope, K_{in}, and an ordinate intercept, V_i, which includes the possibility of initial distribution of the solute in a rapidly reversible compartment(s).

Results

Figure 16 illustrates the time-dependent brain uptake of [^{125}I]-IgG as measured in the perfused hippocampus of the normal guinea pig in the presence and absence of unlabelled IgG in concentrations as high as 4 mg/ml. It can be seen that the slope for [^{125}I]-IgG in the absence of unlabelled protein is significantly higher when compared to either the slope for [^{125}I]-IgG obtained in the presence of unlabelled IgG, or the slope of [^{3}H]-dextran obtained in a separate series of experiments. These results indicate that permeability of the BBB to IgG is significantly higher than for the inert polar molecule, dextran (molecular weight 70,000), due to the presence of a carrier-mediated uptake mechanism for the immune protein located at the BBB. ICC analysis of the brain tissue following perfusion with unlabelled IgG (Figure 15) demonstrated that IgG penetrated the BBB rather than staying bound to the luminal side of the endothelial cells. Reaction product for IgG was found both in the endothelial cells of microvessels, as well as in the surrounding perivascular tissue (Figure 15) indicating that IgG was transferred across the endothelial cell wall into the surrounding brain tissue. Under the same staining conditions, when control animals were perfused with artificial blood free of IgG, the avidin-biotin peroxidase reaction was negative. The proposed existence of a saturable mechanism for transport of proteins across the BBB is in agreement with previous *in vitro* work with insulin (44), insulin-like growth factors (45), the cationized form of albumin (46), transferrin (47), and more recently cationized form of IgG (48). This concept is also supported by recent findings of saturable transport mechanisms for peptides, such as delta-sleep inducing peptide and leucine-enkephalin (36,49-51), both at the BBB and blood-CSF barrier (34,49,52). These findings may lead to the conclusion that some naturally-occurring peptides and proteins including IgG, may in fact be transported across the BBB by a carrier-mediated process. However, brain perfusion data with IgG suggests that the transport system for the immune protein located at the luminal side of the BBB may be saturated at normal physiological plasma levels of IgG (4 mg/ml in the guinea pig), indicating that the carrier-mediated process is of limited capacity.

Regional unidirectional transfer constants, K_{in}, estimated by multiple time-point/graphic analysis for [^{125}I]-IgG and [^{3}H]-dextran in the parietal cortex of the guinea pig are given in Table 1. For comparison, the K_{in} value for [^{125}I]-IgG estimated in EAE animals 20 days following MBP injection are also presented. In comparison to the control group, the EAE animals had a significant increase in uptake of circulating IgG accompanied by varying degrees of increased BBB permeability to the inert polar molecule, mannitol, depending on the particular brain region (40) studied. Results obtained in EAE animals were also confirmed by ICC analysis of the distribution of blood-borne IgG which were strongly reactive. Seven days after administering MBP, there was an increased uptake of IgG into the vascular endothelium and perivascular areas of the brain. Since the extracellular space marker, D-[^{3}H]mannitol, consistently penetrated much less, then the data suggests that greater specific transport of IgG into the CNS occurs prior to the onset of demyelination and cellular proliferation. Studies conducted during earlier time points in this EAE model showed that the earliest demyelination and cellular infiltration occurs 16 days after MBP administration (53,54). Thus, considering a possible chain of events leading to demyelination, there appears to be a gap between

Table 1
Unidirectional Transfer Constant K_{in} for Homologous
$[^{125}I]IgG$ (2.5 $\mu g/ml$) During Vascular Perfusion
of the Guinea-Pig Parietal Cortex in Normal and EAE Animals

	K_{in} ($\mu l/min \times g$)	P (nm/ms)
$[^{125}I]IgG$	0.58 ± 0.05	0.97 ± 0.08
+ 4 mg/ml Unlabelled IgG	0.13 ± 0.02	0.22 ± 0.03
EAE, 20 Days	2.45 ± 0.17	4.08 ± 0.28
$[^3H]Dextran$	0.054 ± 0.01	0.09 ± 0.02

Values for $[^3H]$dextran in normal animals are given for comparison. Values are means ± SE, from 4 to 12 animals. Cerebrovascular permeability constant, P, is calculated on the assumption of capillary surface area of 100 cm^2/g brain. nm denotes nanometers. Data are from Zlokovič *et al.* (39,40) with permission of Experimental Neurology.

early increased permeability to IgG and onset of demyelination and cellular infiltration. The abnormal BBB permeability to IgG in EAE animals reverts back to normal again at about 30 days after MBP injection. In the CSF, the albumin quotient rises steadily up to 42 days, while the rise in IgG quotient is transient and diminished (40), which may suggest that these two serum proteins access the CSF in different ways during development of EAE.

Autoimmune Diseases

Introduction

Studying immunoneurological mechanisms is important since it provides a sound framework for understanding the pathophysiology of autoimmune diseases of the CNS. Results from many of the basic studies using endogenous and exogenous markers described so far, have their counterpart in clinically relevant observations, both as to passage of immune cells and/or material through the BBB and the nature of antigens expressed by the CNS. For example, activated T cells have been shown to be capable of penetrating the BBB (55); and endothelial cells, pericytes and astrocytes have been shown to be capable of expressing HLA-DR or Ia (the murine equivalent of HLA-DR) antigens (56-59) even though the BBB remains intact. Brain capillary permeability can also be increased in certain inflammatory disease states (probably secondary to the secretion of chemical mediators by sensitized immune cells). The entry of systemic immunocompetent cells and/or viruses into the CNS could initiate a series of events including secretion of lymphokinins by T lymphocytes, antibody production by plasma cells, and complement binding, eventually leading to cell death (60). Thus, gamma interferon has been shown to induce expression of Ia antigen in murine brain (61), and cultured astrocytes following lymphokinin (61) or viral (62) stimulation. All of these observations may have relevance to disease states since exposure of the CNS to the immune system during a temporary breach in the BBB could result in antigenic stimulation by brain areas that were not exposed to the immune system during the *self* recognition developmental stages, and now regarded as *non-self* entities. Such immunoneurological mechanisms may play a role in predisposing or inducing certain pathological processes within the CNS. Autoantibodies to axonal neurofilaments have been documented in patients with Kuru and Creutzfelt-Jacob disease (63), Ia antigen expression has been demonstrated in guinea pig (64) and rat brain (65) following induction of EAE, and HLA-DR expression has been documented in endothelial cells and astrocytes in multiple sclerosis.

Multiple Sclerosis (MS)

A salient pathological feature of MS is the perivenular and periventricular distribution of the demyelinating sclerotic lesions. The paucity of peri-arteriolar lesions, and the observation that oligoclonal bands in the CSF have a different clonal origin from those found in the serum (66), have led to speculation that the triggering factor in the autoimmune response is derived from within the CNS. Several studies have attempted to characterize the nature of CNS antigen in MS (67-73). Although MS serum has been shown to be capable of inducing axonal demyelination (74-76), specific binding of IgG to oligodendrocytes or intact myelin remains to be demonstrated (69). EAE induced in laboratory animals by immunization with MBP, produces demyelinating lesions similar to those seen in MS. This has led several investigators to consider that a cell-mediated immune response to MBP is present in the pathogenesis of MS (74,76-79).

Homology in a six amino acid sequence in the *encephalitogenic* domain of MBP with six consecutive amino acids of the hepatitis B viral polymerase (80) has renewed speculation on the viral etiology of MS. While there is no evidence linking the hepatitis B virus with MS, several investigators have considered a role for the measles virus in its pathogenesis (81-83). Although the measles virus has not been identified in MS tissue, serum or CSF, the observation that gamma globulin *spikes* in the serum of MS patients are comparable to those in the CSF of patients suffering from subacute sclerosing panencephalitis (SSPE), a documented measles infection, has fueled speculation that the measles virus may be responsible for the pathogenesis of this disease. The fact that CSF immunoglobulins in SSPE are directed to a nucleocapsid of the measles virus, also points to the possibility that this virus may be a factor in the pathogenesis of MS as well (74).

Recent advances in the field of molecular biology have rekindled speculation regarding the possible viral etiology of MS. Both the availability of cDNA probes and development of the polymerase chain reaction (PCR) has provided a means for amplifying rare, low copy viral genomic sequences (84,85), thereby facilitating their detection when conventional techniques have failed (86,87). Thus, the early penetration of the BBB by the HIV virus (88), and the observation that antibodies to the HIV group of viruses can be detected in the serum and CSF of MS patients (89) prompted investigators to search for viral nucleotides in this disease as corroborative evidence. Interestingly, nucleotide sequences homologous to HTLV-I were documented in systemic mononuclear cells of MS patients employing primer pairs to the gag and env regions in PCR amplification (90) despite the inability to document significant HTLV-I antibody titers in the same group. A subsequent study further demonstrated homologous sequences to human T-cell leukemia/lymphoma virus type 1 in mononuclear cells of MS patients employing similar gene amplification techniques (91), again unaccompanied by a rise in antibody titre. However, the DNA sequences in these patients did not correspond to known endogenous human retroviral sequences by Southern analysis. It was felt that the homologous sequences represented a new human retrovirus related to, but not identical with, HTLV-I, which in turn could account for the failure to demonstrate an increase in circulating antibodies.

Since the BBB is not generally compromised in MS (60), the putative viral pathogen probably gained access to the CNS *via* retrograde transneuronal transport mechanisms (24,25) or during an early stage of the inflammatory process when the barrier was relatively permeable. Once in the CNS, with the BBB re-established, the virus could proliferate without being challenged by systemic immunocompetent cells. The CNS lesions could then result from direct viral infection, or be secondary to a cell-mediated response to the virus by the rarely-found lymphocytes which encounter viruses in the CNS. The pathogen, however, would have to be a *slow* or latent virus in order to meet these criteria. The chronic relapsing nature of the demyelinating process in MS and the persistence of lymphocytes in the CSF for several years (70) have led to speculation about latent virus infection in MS, or at least a role for a virus-initiated autoimmune response (75,81,92,93). The fact that epidemiological studies implicate environmentally-related exposure factor with prolonged latency periods (52,63) also tend to support a slow or latent virus hypothesis in the pathogenesis of MS.

While direct evidence for antibody-specificity or cell mediated autoimmunity to identifiable CNS antigens is lacking in MS (76), several lines of indirect evidence suggests that autoimmune mechanisms may induce this disease. Thus MS patients subjected to plasmapheresis and immunosuppressive therapy demonstrated a greater degree of improvement or stabilization when compared to patients on a drug regimen alone (94-96); whereas no significant improvement was documented when only lymphocytopheresis was employed (97-99). In another study, a less severe progression of the disease was noted in patients subjected to total lymphoid irradiation (100).

Alternatively, the autoimmune response could be mounted against an antigenic determinant shared by the pathogen and a structural element in the CNS. During the process of eradicating this invading systemic pathogen (which has not entered the CNS) memory cells could be generated having antibody specificity for the common antigenic epitope. Those memory cells could then lie dormant in the circulation unaccompanied by clinical symptomatology. The entry of these memory cells into the CNS, secondary to a breach in the BBB at some future point (possibly secondary to a measles virus infection), and their subsequent activation by the shared CNS antigen could then trigger antibody propagation against this antigen and initiate the demyelinating process.

Yet another hypothesis involves over-expression of HLAA-3 and HLAA-7 markers in MS patients (74). The location of the HLA group of genes in close proximity to the immune response gene (ir) has led to speculation that certain HLA markers could signal genetic defects in the immune response locus. Such genetic predisposition to autoimmunity in MS patients could result from their inability to recognize certain antigens as *self*, resulting in a gradual immune response to these antigens. Such a response could eventually lead to the progressive relapsing demyelinating lesions of MS. Natural killer cells have also been implicated in the disease process. The demonstration that a monoclonal antibody generated to natural killer cells also reacts with oligodendroglia, myelin sheaths, neurons and astroglial cells (101) has led to speculation that a naturally occurring mechanism for suppressing natural killer cells could be rendered overactive by genetic predisposition or viral infection. The stimulated production of antibodies against natural killer cells, in turn could result in their binding to antigenic epitopes in myelin and oligodendrocytes, with resulting demyelination.

Paraneoplastic Cerebellar Degeneration (PCD)

PCD is manifested by progressive deterioration of cerebellar function (102-105) in cancer patients in which there is no direct involvement of the cerebellum in the neoplastic process. Onset of symptoms may precede the diagnosis of the neoplasm, occur in the course of the malignancy, or on occasions occur after a remission or cure. The neoplasms most commonly associated with this syndrome are ovarian, breast and lung carcinomas, although neuroblastomas, Hodgkin's disease, non-Hodgkin's lymphomas, and carcinomas of the uterus, stomach, colon and larynx have been implicated as well (103-106). Histological changes in the cerebellum include: cerebellar atrophy characterized by Purkinje neuronal degeneration, proliferation of Bergmann astrocytes (102,107,108), some loss of granule cells, and mononuclear infiltration of the perivascular and leptomeningeal spaces. Other regions of the CNS can also be affected as evidenced by demyelinating lesions of the spinal cord, and loss of neurons in brain stem motor nuclei and anterior horn of the spinal cord (102,104).

While the paraneoplastic cerebellar syndrome is considered to represent a remote effect of the neoplastic process, the factors leading to, and the precise mechanism of, cerebellar degeneration are not well understood. A role for nutritional deficiency secondary to the associated malignancy has been considered, as also the production of cerebellotoxic factors by the neoplasm, or the generation of Purkinje cell antibodies by immunocompetent cells in response to the tumor (105). The presence of perivascular inflammatory cell infiltrates, and the demonstration of antibodies to Purkinje neurons in the serum of these patients (103-106,109), provide compelling evidence in support of an autoimmune mechanism in the pathogenesis of PCD. It has also been shown that Purkinje cell antibodies are more commonly seen in PCD associated with breast, ovarian and uterine carcinomas, than in other PCD-associated neoplasms (104,110,111), leading to speculation that these two entities represent different pathogenic processes, with a final common clinical expression of cerebellar degeneration (104,111,112). Irrespective of the mechanisms involved, when present, antibodies to Purkinje cells provide a high index of suspicion for the existence of an underlying neoplasm, having been documented in the sera of ovarian carcinoma patients with no signs of PCD (103), and not detected in patients with non-neoplastic cerebellar degeneration (104,111,112). Surgical removal of the primary tumor associated with PCD has also been shown to result in remission of the cerebellar dysfunction (105,113), and there have been mixed reports of beneficial effects of plasmapheresis in affecting the outcome (104,114).

Studies have recently focused on characterizing the antigen involved in PCD, in an effort to determine the presence of shared epitopes with tumor cells. Sera from patients with PCD were screened for Purkinje cell antibodies by indirect immunofluorescence and immunoperoxidase techniques (106). The antibodies where found to recognize two specific groups of antigens with relative masses of 34-38 kd and 62-64 kd in immunoblots of purified human Purkinje cell preparations. No bands were detected in preparation of purified cerebral cortical neurons, or when control sera were employed. While the identity of these protein bands have not been established by N-terminal sequence analysis, the antibodies have been employed to screen and clone a gene from a human cerebellar cDNA library (115). The amino acid sequence predicted from the nucleotide sequence of a positive clone has been shown to contain 34 tandem repeats of a hexapeptide unit. Antisera against this synthetic peptide sequence has been shown to label the cytoplasm of human Purkinje cells and react with the 34 kd (but not 62-64 kd) band in immunoblots of human Purkinje cell preparations (116).

The selective vulnerability of Purkinje neurons to the paraneoplastic degenerative process also needs to be addressed. It has recently been shown that Purkinje neurons can selectively extract certain molecules (117) and immunoglobulins (118) from the CSF. Intraventricular injection of propidium iodide, granular blue, and wheat germ agglutinin-horse radish peroxidase complex in rats resulted in retrograde labeling of cerebellar Purkinje neurons (117). Accumulation of propidium iodide in Purkinje cells, furthermore, was associated with features of cerebellar dysfunction. The ability of intraventricularly-administered colchicine and ouabain to block the retrograde labeling of

Purkinje cells and prevent the development of cerebellar symptoms in these animals implies active microtubular transport of these compounds in the induction of cerebellar dysfunction. In a related study, Jaeckle *et al.* (118) demonstrated IgG labeling of Purkinje neurons in rat, following intraventricular injections of Purkinje cell antibodies. Similar injections of control IgG did not show labeling. It would thus appear Purkinje cell antibodies are responsible for PCD, and that Purkinje neurons expedite their own degeneration by selectively facilitating the retrograde transport of these antibodies to their neuronal perikarya. The factors that induce the production of these antibodies in the first place, and their path of entry into the CNS, however, remain to be determined.

Alzheimer's disease (AD)

AD is the most prevalent form of presenile dementia, accounting for about 50% of all dementias. Approximately 1.5 million Americans suffer from this degenerative condition. Mental deterioration, incomprehensible speech, and total loss of cognitive functions render the patients incapable of caring for themselves and pose major economical and health care problems to the immediate family, the medical community, and society in general. Based on pharmacological and behavioral studies in both humans and animals, a consensus has emerged that the neurotransmitter, acetylcholine, is markedly impaired in AD. The disease is histologically characterized by marked neuronal loss, glial proliferation, and the presence of neurofibrillary tangles and neuritic plaques, the latter two being considered hallmarks of this disease. Amyloid (119), immunoglobulins (120), and complement proteins (121,122) are also deposited in these lesions. These observations, and the focal accumulation of scavenger cells (118,123) have suggested parallels between AD pathology and systemic immune response. It has also been speculated that focal lesions at the site of immunoglobulin and complement deposition could provide antigenic stimuli (124-126). The expression of HLA-DR by glial scavenger cells has recently been demonstrated at lesion sites in AD brain, as well as apposition between putative glial and T cells (127). Since these findings provide all elements of antigen presentation, Rogers *et al.* (127) proposed that the glial proliferation of AD may not represent reactive gliosis (secondary to neuronal loss), but rather occur as an immune response and thereby contribute directly to the AD pathology.

Discussion and Conclusions

The crucial question regarding the regulatory mechanisms by which the BBB selectively excludes certain macromolecules awaits elucidation. The studies discussed above challenge the concept that the BBB totally excludes serum proteins such as albumin and immunoglobulin, and viruses from the CNS. Nevertheless, the BBB normally operates efficiently to exclude a majority of other proteins. Simplistically, the functions of the BBB can be viewed from the perspective of two different levels of activity. To begin with, bulk amounts of serum proteins are excluded from the CNS resulting in low protein levels normally found in the CSF, which also reflect the enormous amount of effort exerted by the brain in protecting itself from blood-borne substances. The orchestrated efforts of various components of the barrier (endothelia, basal lamina and glial cells) acting in concert with enzymes, transport molecules, and other biochemical mechanisms are directed towards maintaining the most protected environment possible for brain function. Factors such as trauma and disease can cause serious breaches in the BBB at this level of function, resulting in severe impairment of brain activity. The resulting initiation of autoimmune mechanisms in the CNS could also trigger destructive cellular mechanisms leading to clinical disease states such as MS, PCD and perhaps AD.

There is also a second level of function for the BBB that is less well understood. This involves mechanisms such as the nose-brain and meningeal pathways, as well as transcytosis of IgG through closed capillaries, resulting in direct entry of substances into the CNS. It is also at this level, that molecules can enter the brain *via* more indirect routes. Thus neuroendocytosis of substances at the sites of neurohemal contacts outside the BBB, can lead to their retrograde transport to neuronal perikarya followed by subsequent transynaptic transport to connecting neurons. Similarly, materials could exit areas of the CVO *via* functional leaks. The factors regulating the amount of serum proteins that enter the CNS *via* these routes, as well as their eventual fate are important issues that await elucidation. Questions also remain regarding modification of these substances during their transcytotic- or transynaptic-transport, and the mode of their eventual degradation.

Whether the BBB normally allows some accessibility to the CNS, or if this represents altered physiology, will ultimately be resolved by identifying the functional significance in the brain of the very substances that appear to *preferentially* circumvent the BBB. In a sense, these perspectives on the BBB are analogous to the apparent paradoxical function of the choroid plexus which appears to both secrete and absorb CSF. The degree of accessibility of circulating factors and macromolecules to the CNS could be selectively regulated by the BBB at various set points as a mechanism for allowing specific systemic signals to reach the brain in coordinating central and systemic activity. Circulating cholecystokinin could thus modulate satiety by acting as a systemic hormone while also acting centrally to mediate feeding (128). Similar mechanisms involving circulating angiotensin (129) and atrial natriuretic factor (130) could coordinate central and systemic control of blood pressure and volume. Finally, since stress plays an important role in the immune response, it could be argued that immunocompetent cells are given access to the CNS in order to relay immunological information critical for secretion of neurohormones in modulating the hypothalamo-pituitary-adrenal stress axis. In this context, it is interesting to note that fibers immunoreactive for interleukin-1 (IL-1) have been found to innervate endocrine and autonomic centers in human brain that control the central component of the acute phase reaction (131). IL-1 can act centrally to alter the release of several pituitary hormones (138). However, sites of action of IL-1 on the release of pituitary hormones is controversial. IL-1 may (132) or may not (139,140) act directly on the pituitary to induce the release of, for example, ACTH. Serum albumin is a known carrier for many drugs and hormones (133,134), and it probably enters the CNS to transport these compounds to relevant brain sites. The selective permeability of the BBB to these agents may therefore play a critical role in mediating autonomic and endocrine adjustments in immunoneurologic mechanisms as involved in acute phase reactions, and in maintaining CNS homeostasis.

Acknowledgements

This work was supported in part by: USPHS grants AA-06014 (GPK) and DK-37205 (GN); NSF grant BNS-8820600 (GN); and generous grants to BVZ from the Wellcome Trust, the British Council, and the Republicki Zavod za Medjunarodnu Saradnju SR Srbije. We thank Drs. Albert Sidney Whiting Jr. (Portland, OR) and Krzysztof Lyson (Dallas, TX) for their help in preparing this manuscript.

References

1. Dinarello, C.A., and J.W. Mier, Current concepts - lymphokines, *N Engl J Med* 317: 940-945, 1987.

2. Nandy, K., Immune reactions in aging brain and senile dementia, In K. Nandy and I. Sherwin (eds) *The Aging Brain and Senile Dementia, Advances in Behavioral Biology, Volume 23*, Plenum Press, New York, pp. 181-196, 1977.

3. Nandy, K., Brain-reactive antibodies in aging and senile dementia, In R. Katzman, R. Terry, K. Bick (eds) *Alzheimer's Disease-Senile Dementia and Related Disorder*, Raven Press, New York, pp. 503-512, 1978.

4. Neuwelt, E.A., and W.K. Clark, Unique aspects of the immunology of the central nervous system, In *Clinical Aspects of Neuroimmunology*, Waverly Press, Baltimore, pp. 39-72, 1978.

5. Leibowitz, S., and R.A.C. Hughes, Immunology and the blood-brain barrier, In *Immunology of the Nervous System*, Edward Arnold, Ltd., London, pp. 1-19, 1983.

6. Bradbury, M.W., Transport across the blood-brain barrier, In E.A. Neuwelt (ed) *Implication of the Blood-Brain Barrier and Its Manipulation, Volume I*, Plenum Medical Book Company, New York, pp. 119-136, 1989.

7. Broadwell, R.D., and M. Salcman, Expanding the definition of the blood-brain barrier to protein, *Proc Natl Acad Sci USA* 78: 7820-7824, 1981.

8. Broadwell, R.D., B.J. Balin, and M. Salcman, Transcytotic pathway for blood-borne protein through the blood-brain barrier, *Proc Natl Acad Sci USA* 85: 632-636, 1988.

9. Broadwell, R.D., B.J. Balin, M. Salcman, and R.S. Kaplan, Brain-blood barrier? Yes and no, *Proc Natl Acad Sci USA* 80: 7352-7356, 1983.

10. Oldendorf, W.H., Permeability of the blood-brain barrier, In D.B. Tower, R.O. Brady, D.P. Purpura, C.D. Clemente, W.M. Landau, and S.E. Mayer SE (eds) *The Nervous System, Volume 1*, Raven Press, New York, pp. 279-89, 1975.

11. Blatteis, C.M., N. Quan, and R.D. Howell, The organum vasculosum laminae terminalis (OVLT) is critical for fever induced in guinea pigs by blood-borne cytokines, *Soc Neurosci Abstr* 15: 718, 1989.

12. Brightman, M.W., and T.S. Reese, Junctions between intimately apposed cell membranes in the vertebrate brain, *J Cell Biol* 40: 648-677, 1969.

13. Blatteis, C.M., W.S. Hunter, J.M. Wright, R.A. Ahokas, Q.J. Llanos, and T.A. Mashburn, Jr., Thermoregulatory responses of guinea pigs with anteroventral third ventricle lesions, *Can J Physiol Pharmacol* 65: 1261-1266, 1987.

14. Balin, B.J., R.D. Broadwell, M. Salcman, and M. El-Kalliny, Avenues for entry of peripherally administered protein to the central nervous system in mouse, rat, and squirrel monkey, *J Comp Neurol* 251: 260-280, 1986.

15. Broadwell, R.D., and M.W. Brightman, Entry of peroxidase into neurons of the central and peripheral nervous systems from extracerebral and cerebral blood, *J Comp Neurol* 166: 257-283, 1976.

16. Broadwell, R.D., and M.W. Brightman, Cytochemistry of undamaged neurons transporting exogenous protein *in vivo*, *J Comp Neurol* 185: 31-73, 1979.

17. Nilaver, G., H. Brem, and E.A. Zimmerman, Immunocytochemical localization of albumin in rat brain, *Neurology* 32: A107, 1982.

18. Sparrow, J.R., Immunocytochemical localization of plasma protein in neuronal perikarya, *Brain Res* 212: 159-163, 1981.

19. Kozlowski, G.P., and W.L. Dees, Immunocytochemistry for LHRH neurons in the arcuate nucleus of the rat: fact or artifact? *J Histochem Cytochem* 32:83-91, 1984.

20. Meeker, M.L., R.B. Meeker, and J.N. Hayward, Accumulation of circulating endogenous and exogenous immunoglobulins by hypothalamic magnocellular neurons, *Brain Res* 423: 45-55, 1987.

21. Fabian, R.H., and G. Petroff, Intraneuronal IgG in the central nervous system: update by retrograde axonal transport, *Neurology* 37: 1780-1784, 1987.

22. Baker, H., and R.F. Spencer, Transneuronal transport of peroxidase-conjugated wheat germ agglutinin (WGA-HRP) from the olfactory epithelium to the brain of the adult rat, *Exp Brain Res* 63: 461-473, 1986.

23. Harrison, P.J., H. Hultborn, E. Jankowska, R. Katz, B. Storai, and D. Zytnicki, Labelling of interneurones by retrograde transsynaptic transport of horseradish peroxidase from motoneurones in rats and cats, *Neurosci Lett* 45: 15-19, 1984.

24. Ugolini, G., H.G.J.M. Kuypers, and A. Simmons, Retrograde transneuronal transfer of Herpes simplex virus type 1 (HSV 1) from motorneurons, *Brain Res* 422: 242-256, 1987.

25. Jones, E.G., and B.K. Hartman, Recent advances in neuroanatomical methodology, *Ann Rev Neurosci* 1: 215-296, 1978.

26. Reiber, H., A.J. Suckling, and M.G. Rumsby, The effect of Freund's adjuvants on blood-cerebrospinal fluid barrier permeability, *J Neurol Sci* 63: 55-61, 1984.

27. Kozlowski, G.P., and G. Nilaver, Immunoelectron microscopy of neuropeptides: theoretical and technical considerations, In J.L. Barker and J.F. McKelvy (eds) *Current Methods in Cellular Neurobiology*, John Wiley and Sons, Inc., New York, pp. 133-174, 1983.

28. Peress, N.S., J. Siegelman, and H.B. Fleit, High avidity periventricular IgG-Fc receptor activity in human and rabbit brain, *Clin Immunol Immunopath* 42: 229-238, 1987.

29. Segal, M.B., and B.V. Zlokovič, *The Blood-Brain Barrier, Amino Acids and Peptides*, Kluwer Academic Publishers, London, pp. 1-11, 1989.

30. Reese, T.S., and M.J. Karnovsky, Fine structural localization of a blood-brain barrier to exogenous peroxidase, *J Cell Biol* 34: 207-217, 1967.

31. Broadwell, R., A. Wolf, and M. Tangoren, Transcytosis of blood-borne ferrotransferrin and insulin through the blood-brain barrier, *Soc Neurosci Abstr* 15: 821, 1989.

32. Pardridge, W.M., Receptor-mediated peptide transport through the blood-brain barrier, *Endocr Review* 7: 314-330, 1986.

33. Pardridge, W.M., Recent advances in blood-brain barrier transport, *Ann Rev Pharmacol Toxicol* 28: 25-39, 1988.

34. Zlokovič, B.V., V.T. Susic, H. Davson, D.J. Begley, R.M. Jankov, D.M. Mitrovič, and M.N. Lipovac, Saturable mechanism for delta-sleep-inducing peptide (DSIP) at the blood-brain barrier of the vascularly perfused guinea-pig brain, *Peptides* 10: 249-254, 1989.

35. Zlokovič, B.V., *In vivo* approaches for studying peptide interactions at the blood-brain barrier, *J Control Rel*, In Press.

36. Zlokovič, B.V., J.B. Mackič, B. Djuricič, and H. Davson, Kinetic analysis of leucine-enkephalin cellular uptake at the luminal side of the blood-brain barrier of an *in situ* perfused guinea-pig brain, *J Neurochem*, 53: 1333-1340, 1989.

37. Kozlowski, G.P., and G. Nilaver, Structural and functional relationships between the immune and central nervous systems in Alzheimer's Disease, *Drug Dev Res* 15: 129-142, 1988.

38. Zlokovič, B.V., D.J. Begley, B.M. Djuricič, and D.M. Mitrovič, Measurement of solute transport across the blood-brain barrier in the perfused guinea-pig brain: method and application of N-methyl-alpha-aminoisobutric acid, *J Neurochem* 46: 1444-1451, 1986.

39. Zlokovič, B.V., D.S. Skundric, M.B. Segal, M.N. Lipovac, J.B. Mackič, and H. Davson, A saturable mechanism for transport of immunoglobulin G across the blood-brain barrier of the guinea-pig, *Exp Neurol*, In Press.

40. Zlokovič, B.V., D.S. Skundric, M.B. Segal, J. Colover, R.M. Jankov, N. Pejnovič, V. Lackovič, J.B. Mackič, M.N. Lipovac, H. Davson, E. Kasp, D. Dummonde, and L.J. Lackovič, Blood-brain barrier permeability changes during acute allergic encephalomyelitis induced in the guinea-pig, *Metab Brain Dis* 4: 33-40, 1989.

41. Erlendsson, K., T. Swartz, and J.M. Dwyer, Successful reversal of echo virus encephalitis in x-linked hypogammaglobulinemia by intraventricular administration of immunoglobulin, *N Engl J Med* 312: 351-353, 1985.

42. Keenan, A.M., J.C. Harbert, and S.M. Larson, Monoclonal antibodies in nuclear medicine, *J Nucl Med* 26: 531-537, 1985.

43. Colover, J., Acute demyelination in EAE after pretreatment with foreign protein and muramyl dipeptide (MDP), In A.R. Liss (ed) *Experimental Allergic Encephalomyelitis: A Useful Model for Multiple Sclerosis*, Alan R. Liss, New York, pp. 37-41, 1984.

44. Pardridge, W.M., J. Eisenberg, and J. Yang, Human blood-brain barrier insulin receptor, *J Neurochem* 44: 1771-1778, 1985.

45. Frank, H.J.L., W.M. Pardridge, W.L. Morris, R.G. Rosenfeld, and T.B. Choi, Binding and internalization of insulin and insulin-like growth factors by isolated brain microvessels, *Diabetes* 35: 654-661, 1986.

46. Kumagai, A.K., J.B. Eisenberg, and W.M. Pardridge, Absorptive-mediated endocytosis of cationized albumin and a β-endorphin-cationized albumin chimeric peptides by isolated brain capillaries. Model

system of blood-brain transport, *J Biol Chem* 262: 15214-15219, 1987.

47. Pardridge, W.M., J. Eisenberg, and J. Yang, Human blood-brain barrier transferrin receptor, *Metabolism Clin Exper* 36: 892-895, 1987.

48. Triguero, D., J.B. Buciak, J. Yang, and W.M. Pardridge, Blood-brain barrier transport of cationized immunoglobulin G: Enhanced delivery compared to native protein, *Proc Natl Acad Sci USA* 86, 4761-4765, 1989.

49. Zlokovič, B.V., D.J. Begley, M.B. Segal, H. Davson, L.J. Rakič, M.N. Lipovac, D.M. Mitrovič, and R.M. Jankov, Neuropeptide transport mechanisms in the central nervous system, In L.J. Rakič, D.J. Begley, H. Davson, and B.V. Zlokovič (eds) *Peptide and Amino Acid Transport Mechanisms in the Central Nervous System*, Macmillan Press Ltd, London, pp. 3-21, 1988.

50. Zlokovič, B.V., M.N. Lipovac, D.J. Begley, H. Davson, and L. Rakič, Transport of leucin-enkephalin across the blood-brain barrier in the perfused guinea-pig brain, *J Neurochem* 49: 310-315, 1987.

51. Zlokovič, B.V., M.B. Segal, H. Davson, and D.M. Mihovič, Unidirectional uptake of enkephalins at the blood-tissue interface of the blood-cerebrospinal fluid barrier: a saturable mechanism, *Regul Peptides* 20: 33-44, 1988.

52. Zlokovič, B.V., M.B. Segal, H. Davson, and R.M. Jankov, Passage of delta-sleep-inducing peptide (DSIP) across the blood-cerebrospinal fluid barrier, *Peptides* 9: 533-538, 1988.

53. Colover, J., A new pattern of spinal-cord demyelination in guinea pigs with acute experimental allergic encephalomyelitis mimicking multiple sclerosis, *Br J Exp Biol Pathol* 61: 390-400, 1980.

54. Colover, J., and A. Qureshi Myelin debris in demyelinating process: a disease activity marker? In C. Confavreux, G. Aimard, and M. Devič (eds) *Trends in European Multiple Sclerosis Research*, Elsevier, London, pp. 229-234, 1988.

55. Wekerle, H., C. Linnington, H. Lassmann, and R. Meyerman, Cellular immune reactivity within the CNS, *Trends in Neurosci* 9: 271-277, 1986.

56. Fontana, A., Astrocytes and lymphocytes: intercellular communication by growth factors, *J Neurosci Res* 8: 443-451, 1982.

57. Lampson, L.A., and W.F. Hickey, Monoclonal antibody analysis of MHC expression in human brain biopsies: tissues ranging from "histologically normal" to that showing different levels of glial tumor involvement, *J Immunol* 136: 4054-4062, 1986.

58. Luber-Narod, J., and J. Rogers, Immune system associated antigens expressed by cells of the human central nervous system, *Neurosci Lett* 94: 17-22, 1988.

59. McCarron, R.M., O. Kempski, M. Spatz, and D.E. McFarlin, Presentation of myelin basic protein by murine cerebral vascular endothelial cells, *J Immunol* 134: 3100-3103, 1985.

60. Leibowitz, S., and R.A.C. Hughes, The immune response in the central nervous system, In *Immunology of the Nervous System*, Edward Arnold, Ltd., London, pp. 20-40, 1983.

61. Wong, G.H.W., P.F. Bartlett, I. Clark-Lewis, J.L. McKimm-Breschkin, and J.W. Schrader, Interferon-λ induces the expression of H-2 and Ia antigens on brain cells, *J Neuroimmunol* 7: 255-278, 1985.

62. Massa, P.T., R. Dorries, ter Muelen, Viral particles induce Ia antigen expression on astrocytes, *Nature* 320: 543-546, 1986.

63. Sotelo, J., C.J. Gibbs, Jr., and D.C. Gajdusek, Autoantibodies against axonal neurofilaments in patients with kuru and Creuetzfeld-Jacob disease, *Science* 210: 190-193, 1980.

64. Antoniou, A.V., H. El-Sady, C. Butler, and J.L. Turk, The modulation of class II histocompatibility antigens and activated macrophage determinants in the spinal cord during the development of chronic relapsing experimental allergic encephalomyelitis (CREAE) in the guinea pig - relevance to the induction of remission, *J Neuroimmunol* 15: 57-71, 1987.

65. Matsumoto, Y., N. Hara, R. Tanaka, M. Fugiwara, Immunohistochemical analysis of rat central nervous system during experimental allergic encephalomyelitis, with special reference to Ia-positive cells with dendritic morphology, *J Immunol* 136: 3668-3676, 1986.

66. Neuwelt, E.A., and W.K. Clark, The immunology of demyelinating disease, In *Clinical Aspects of Neuroimmunology*, Waverly Press, Baltimore, pp. 233-72, 1978.

67. Chou, C.H.J., F.C.H. Chou, W.W. Tourtellotte, and R.F. Kibler, Search for a multiple sclerosis specific brain antigen, *Neurology* 33: 1300-1304, 1983.

68. Gannuszkina, I.W., H. Weinrauder, and I.G. Zirnowa, Use of the immunofluorescence and

immunodiffusion methods for the detection of anti-brain antibodies in sera of patients with central nervous system diseases, *Neurpath Pol* 21: 183-190, 1983.

69. Johnson, A.B., and M.B. Bornstein, Myelin binding antibodies *in vitro*: immunoperoxidase studies with experimental allergic encephalomyelitis, anti-galactocerebroside and multiple sclerosis sera, *Brain Res* 159: 173-182, 1978.

70. Lisak, R.P., R.G. Heinze, G.A. Falk, and M.W. Keis, Search for antiencephalitogen antibodies in human demyelinating diseases, *Neurology* 18: 122-128, 1968.

71. Paterson, P.Y., Autoimmune neurologic disease: experimental animal systems and implications for multiple sclerosis, In N. Talal (ed) *Autoimmunity. Genetic Immunologic Virologic and Clinical Aspects*, Academic Press, New York, pp. 643-707, 1977.

72. Rostrom, B., Specificity of antibodies in oligoclonal bands in patients with multiple sclerosis and cerebrovascular disease, *Acta Neurol Scand* 63 [Suppl 86]: 10-84, 1981.

73. Wisniewski, H.M., Immunopathology of demyelination and autoimmune disease and virus infections, *Brit Med Bull* 33: 54-59, 1977.

74. Arnason, B.G.W., Clinical immunology, In D.B.Tower, T.N. Chase, A.L. Sahs, P.R. Dodge, H.L. Heyl, R.J. Joynt, E.S. Goldenshon, R. DeJong, and A. Pope (eds) *The Nervous System, Volume 2*, Raven Press, New York, pp. 61-66, 1975.

75. Dean, G., The multiple sclerosis problem, *Sci Amer* 223: 40-46, 1970.

76. Lisak, R.P., B. Zweiman, J.B. Burns, A. Rostami, and D.H. Silberberg, Immune responses to myelin antigens in multiple sclerosis, *Ann NY Acad Sci* 436: 221-232, 1984.

77. Lisak, R.P., and B. Zweiman, *In vitro* cell-mediated immunity of cerebrospinal fluid lymphocytes to myelin basic protein in primary demyelinating diseases, *New Eng J Med* 297: 850-853, 1977.

78. Lisak, R.P., B. Zweiman, D. Waters, H. Koprowski, and D.E. Pleasure, Cell-mediated immunity to measles, myelin basic protein and central nervous system extract in multiple sclerosis: a longitudinal study employing direct buffy coat migration assays, *Neurology* 28: 798-803, 1978.

79. Lisak, R.P., B. Zweiman, and J.N. Whitaker, Spinal fluid basic protein immunoreactive material and spinal fluid lymphocyte reactivity to basic protein, *Neurology* 31: 180-182, 1981.

80. Fujinami, R.S., and M.B.A. Oldstone, Amino acid homology between the encephalitogenic site of myelin basic protein and virus: mechanisms for autoimmunity, *Science* 230: 1043-1045, 1985.

81. Fraser, K.B., Population serology, In A.N. Davison, J.H. Humphrey, A.L. Liversedge, W.I. McDonald, and J.S. Porterfield, (eds) *Multiple Sclerosis Research*, Elsevier, New York, pp. 53-79, 1975.

82. Petz, L.D., G.C. Sharp, N.R. Cooper, and W.S. Irvin, Serum and cerebrospinal fluid complement and serum autoantibodies in systemic lupus erythematosus, *Medicine* (Baltimore) 50: 259-275, 1971.

83. Rostagi, S.C., J. Clausen, H. Offner, G. Konat, and T. Fog, Partial purification of MS brain-specific antigen, *Acta Neurol Scand* 59: 281-296, 1979.

84. Mulles, K.B., and F.A. Faloona, Specific synthesis of DNA in vitro via a polymerase catalyzed chain reaction, *Meth Enzymol* 155: 335, 1987.

85. Saikai, R.K., D.H. Gelfand, S. Stoffel, S.J. Scharf, R. Higuchi, G.T. Horn, K.B. Mullis, and H.A. Erlich, Primer-directed enzymatic amplification of DNA with a thermostable DNA polymerase, *Science* 239: 487-491, 1988.

86. Duggan, D.B., G.D. Ehrlich, F.P. Davey, S. Kwok, J. Sninsky, J. Goldberg, L. Baltrucki, and B.J. Poiesz, HTLV-1 induced lymphoma mimicking Hodgkin's disease. Diagnosis by polymerase chain reaction amplification of specific HTLV-1 sequences in tumor DNA, *Blood* 71: 1027-1032, 1988.

87. Manzari, V., A. Gismondi, G. Barellari, S. Morrone, A. Modesti, L. Albonici, L. de Marchis, V. Fazio, A. Gradilone, M. Zani, L. Frati, and A. Santoni, HTLV-V: a new human retrovirus isolated in a tac-negative T-cell lymphoma/leukemia, *Science* 238: 1581-1583, 1987.

88. Resnick, L., J.R. Berger, P. Shapshak, and W.W. Tourtellotte, Early penetration of the blood-brain-barrier by HIV, *Neurology* 38: 9-14, 1988.

89. Koprowski, H., E.C. de Freitas, M.E. Harper, M. Sandberg-Wollheim, W.A. Sheremata, M. Robert-Guroff, C.W. Saxinger, M.B. Feinberg, F. Wong-Staal, and R.C. Gallo, Multiple sclerosis and human T-cell lymphotropic retroviruses, *Nature* 318: 154-160, 1985.

90. Reddy, E.P., M. Sandberg-Woelheim, R.V. Meltus, P.E. Ray, E. deFreitas, and H. Koprowski, Amplification and molecular cloning of HTLV-I sequences from DNA of multiple sclerosis patients,

Science 243: 529-533, 1989.

91. Greenberg, S.J., G.D. Ehrlich, M.A. Abbott, B.J. Hurwitz, T.A. Waldmann, and B.J. Poiesz, Detection of sequences homologous to human retroviral DNA in multiple sclerosis by gene amplification, *Proc Natl Acad Sci USA* 86: 2878-2882, 1989.

92. Fraser, K.B., Multiple Sclerosis, a viral disease? *Brit Med Bull* 33: 34-39, 1977.

93. Johnson, R.T., Virological data supporting the viral hypothesis in multiple sclerosis, In A.N. Davison, J.H. Humphrey, A.L. Liversedge, W.I. McDonald, and J.S. Porterfield, (eds) *Multiple Sclerosis Research*, Elsevier, New York, pp. 155-183, 1975.

94. Höcker, P., V. Stellamor, K. Summer, and M. Mann, Plasma exchange (PE) and lymphocytapheresis (LCA) in multiple sclerosis (MS), *Int J Artif Organs* 7: 39-42, 1984.

95. Khatri, B.O., S.M. Koethe, and M.P. McQuillen, Plasmapheresis with immunosuppressive drug therapy in progressive multiple sclerosis: a pilot study, *Arch Neurol* 41: 734-738, 1984.

96. Khatri, B.O., M.P. McQuillen, G.J. Harrington, D. Schmoll, and R.G. Hoffman, Chronic progressive multiple sclerosis: double-blind controlled study of plasmapheresis in patients taking immunosuppressive drugs, *Neurology* 35: 312-319, 1985.

97. Ghezzi, A., M. Zaffaroni, D. Caputo, R. Guaschino, P. Gasco, R. Montanini, and C.L. Cazzullo, Lymphocytoplasmapheresis in multiple sclerosis: preliminary laboratory findings, *Ital J Neurol* Sci 6: 85-87, 1985.

98. Hauser, S.L., M. Fosberg, S.V. Kevy, and H.L. Weiner, Lymphocytopheresis in chronic progressive multiple sclerosis: immunologic and clinical effects, *Neurology* 34: 922-926, 1984.

99. Tindall, R.S.A., J.E. Walker, A.L. Ehle, L. Near, J. Rollins, and D. Becker, Plasmapheresis in multiple sclerosis: prospective trial of pheresis and immunosuppression versus immunosuppression alone, *Neurology* 32: 739-743, 1982.

100. Cook, S.D., C. Devereux, R. Troiano, M.P. Hafstein, G. Zito, E. Hernandy, M. Larvenhar, R. Vidaver, and P.C. Dowling, Effect of total lymphoid irradiation in chronic progressive multiple sclerosis, *Lancet* 1: 1405-1409, 1986.

101. Gebhart, W., S. Schuller-Petrovic, H. Lassmann, and D. Kraft, NK cells and the nervous system, *Wein Klin Wochenschr* 95: 828-831, 1983.

102. Brain, L., and M. Wilkinson, Subacute cerebellar degeneration associated with neoplasms, *Brain* 88: 465-478, 1965.

103. Greenlee, J.E., and H.R. Brashear, Antibodies to cerebellar Purkinje cells in patients with paraneoplastic cerebellar degeneration and ovarian carcinoma, *Ann Neurol* 14: 609-613, 1983.

104. Jaeckle, K.A., F. Graus, A. Houghton, C. Cordon-Cardo, S.L. Nielsen, and J.B. Posner, Autoimmune response of patients with paraneoplastic cerebellar degeneration to a Purkinje cell cytoplasmic antigen, *Ann Neurol* 18: 592-600, 1985.

105. Kearsley, J.H., P. Johnson, and M. Halmagyi, Paraneoplastic cerebellar disease: remission with excision of the primary tumor, *Arch Neurol* 42: 1208-1210, 1985.

106. Cunningham, J., F. Graus, N. Anderson, and J.B. Posner, Partial characterization of Purkinje cell antigens in paraneoplastic cerebellar degeneration, *Neurology* 36: 1163-1168, 1986.

107. Brain, W.R., P.M. Daniel, and J.G. Greenfield, Subacute cortical cerebellar degeneration and its relation to carcinoma, *J Neurol Neurosurg Psych* 14: 59-75, 1951.

108. Henson, R.A., and H. Urich, Remote effects of malignant disease: certain intracranial disorders, In P.J. Vinken, G.W. Bruyn, and H.L. Klawans (eds) *Handbook of Clinical Neurology, Volume 38*, North-Holland Publishing, Amsterdam, pp. 625-668, 1979.

109. Hammack, J.E., D.W. Kimmel, B.P. O'Neill, and V.A. Lennon, Comparison of patients with paraneoplastic cerebellar degeneration (PCD) who are seropositive and seronegative for Purkinje cell cytoplasmic antibodies (PCAb), *Neurology* 38 [Suppl 1]: 127, 1988.

110. Jaeckle, K.A., and J.E. Greenlee, Immunohistochemical patterns of antibody response in paraneoplastic neurological syndromes correlate with specific syndromes and with tumor types, *Ann Neurol* 24: 121, 1988.

111. Smith, J.L., J.C. Finley, and V.A. Lennon, Autoantibodies in paraneoplastic cerebellar degeneration bind to cytoplasmic antigens of Purkinje cells in humans, rats, and mice and are of multiple immunoglobulin classes, *J Neuroimmunol* 18: 37-48, 1988.

112. Anderson, N., M.K. Rosenbaum, and J.B. Posner, Paraneoplastic cerebellar degeneration: clinical-immunological correlations, *Ann Neurol* 24: 559-567, 1988.

113. Paone, J.F., and K. Jeyasingham, Remission of cerebellar dysfunction after pneumonectomy for bronchogenic carcinoma, *New Eng J Med* 302: 156, 1980.

114. Cocconi, G., G. Leci, G. Juvarra, M.R. Monopoli, T. Cocchi, F. Fraccadori, A. Lechi, and P. Boni, Successful treatment of subacute cerebellar degeneration in ovarian carcinoma with plasmapheresis: a case report, *Cancer* 56: 2318-2320, 1985.

115. Dropcho, E.J., Y.T. Chen, J.B. Posner, and L.J. Old, Cloning of a brain protein identified by autoantibodies from a patient with paraneoplastic cerebellar degeneration, *Proc Natl Acad Sci USA* 84: 4552-4556, 1987.

116. Furneaux, H.M., E.J. Dropcho, D. Barbut, Y.T. Chen, M.K. Rosenblum, L.J. Old, and J.B. Posner, Characterization of a cDNA encoding a 34-kDa Purkinje neuron protein recognized by sera from patients with paraneoplastic cerebellar degeneration, *Proc Natl Acad Sci USA* 86: 2873-2877, 1989.

117. Borges, L.F., P.J. Elliot, R. Gill, S.D. Iversen, and L.L. Iversen, Selective extraction of small and large molecules from the cerebrospinal fluid by Purkinje neurons, *Science* 228: 346-348, 1985.

118. Jaeckle, K.A., W.G. Stroop, J.E. Greenlee, M. Price, and Q.S. Deng, Intraventricular injection of paraneoplastic anti-Purkinje cell antibody in a rat model, *Neurology* 36: 332, 1986.

119. Wisniewski, H.M., and R.D. Terry, Neuropathology of the aging brain, In R.D. Terry and S. Gershon (eds) *Neurobiology of Aging*, Raven Press, New York, pp. 265-280, 1976.

120. Ishii, T., S. Haga, and F. Shimizu, Identification of components of immunoglobulins in senile plaques by means of fluorescent technique, *Acta Neuropath* 32: 157-162, 1975.

121. Eikelenboom, P., and F.C. Stam, Immunoglobulins and complement factors in senile plaques, *Acta Neuropathol* 57: 239-242, 1982.

122. Ishii, T., and S. Haga, Immunoelectron microscopic localization of complements in amyloid fibrils of senile plaques, *Acta Neuropathol* 63: 296-300, 1984.

123. Schechter, R., S.H.C. Yen, and R.D. Terry, Fibrous astrocytes in senile dementia of Alzheimer type, *J Neuropath Exp Neurol* 40: 95-101, 1981.

124. Hildemann, W.H., *Essentials of Immunology*, Elsevier, New York, pp. 114, 1984.

125. Elovaara, I., A. Icen, J. Palo, and T. Erkinjuntti, CSF in Alzheimer's disease. Studies on blood-brain barrier function and intrathecal protein synthesis, *J Neurol Sci* 70: 73-80, 1985.

126. Elovaara, I., J. Palo, T. Erkinjuntti, and R. Sulkava, Serum and cerebrospinal fluid proteins and the blood-brain barrier in Alzheimer's disease and multi-infarct dementia, *Eur Neurol* 26: 229-234, 1987.

127. Rogers, J., J. Luber-Narod, S.D. Styren, and W.H. Civin, Expression of immune system-associated antigens by cells of the human nervous system: relationship to pathology of Alzheimer's disease, *Neurobiol Aging* 9: 339-349, 1988.

128. Della-Fera, M.A., and C.S. Baile, Cholecystokinin octapeptide: continuous picomole injections into the cerebral ventricles of sheep suppress feeding, *Science* 206: 471-473, 1979.

129. Lind, R.W., L.W. Swanson, and D. Ganten, Organization of angiotensin II immunoreactive cells and fibers in the rat central nervous system. An immunohistochemical study, *Neuroendocrinology* 40: 2-24, 1985.

130. Saper, C.B., D.G. Standaert, M.G. Currie, D. Schwartz, D.M. Geller, and P. Needleman, Atriopeptin-immunoreactive neurons in the brain: presence in cardiovascular regulatory areas, *Science* 227: 1047-1049, 1985.

131. Breder, C.D., C.A. Dinarello, and C.B. Saper, Interleukin-1 immunoreactive innervation of the human hypothalamus, *Science* 240: 321-234, 1988.

132. Bernton, E.W., J.E. Beach, J.W. Holaday, R.C. Smallridge, and H.G. Fein, Release of multiple hormones by a direct action of interleukin-1 on pituitary cells, *Science* 238: 519-521, 1987.

133. Pardridge, W.M., Carrier-mediated transport of thyroid hormone through the rat blood-brain-barrier. Primary role of albumin-bound hormones, *Endocrinology* 105: 605-612, 1979.

134. Pardridge, W.M., and L.J. Meitus, Transport of steroid hormone through the rat blood-brain-barrier: primary role of albumin-bound hormones, *J Clin Invest* 64: 145-154, 1979.

135. Kozlowski, G.P., Hormone pathways in cerebrospinal fluid, In E.A. Zimmerman and G. Abrams (eds) *Neurologic Clinics: Neuroendocrinology and Brain Peptides*, W.B. Saunders, Philadelphia, pp. 907-917, 1986.

136. Kozlowski, G.P., Ventricular route hypothesis and peptide-containing structures of the cerebroventricular system, In T.B. van Wimersma Greidanus (ed) *Frontiers of Hormone Research, Volume 9*, S. Karger, Basel, pp. 105-118, 1982.

137. Hunter, W.M., and F.C. Greenwood, Preparation of 131 labelled human growth hormone of high specific activity, *Nature* 194: 495-496, 1962.

138. Rettori, V., J. Jurčovičová, and S.M. McCann, Central action of interleukin-1 in altering the release of TSH, growth hormone, and prolactin in the male rat, *J Neurosci Res* 18: 179-183, 1987.

139. Uehara, A., S. Gillis, and A. Arimura, Effects of interleukin-1 on hormone release from normal rat pituitary cells in primary culture, *Neuroendocrinology* 45: 343-347, 1987.

140. Katsuura, G., and A. Arimura, Interleukin-1β, Messenger of the immune signal to the neuroendocrine system, may act primarily on the hypothalamus: presence of its receptor in the hypothalamus, but not in the pituitary, of the rat, *Enocrinology* [Suppl] 122: 132, 1988.

NEUROENDOCRINE MECHANISMS IN THE THERMOGENIC RESPONSES TO DIET, INFECTION, AND TRAUMA

Nancy J. Rothwell

University of Manchester
Manchester, United Kingdom

Thermogenesis—Basic Concepts and Biological Value

The term thermogenesis can be applied to any component of heat production including basal metabolic rate, physical activity and the energy costs of growth, but is now considered to represent regulatory or adaptive forms of heat production. Of these, non-shivering thermogenesis (NST), which is activated in response to cold, is perhaps the most extensively studied, but it is now recognized that many other forms of thermogenesis share common effector mechanisms. For example, diet-induced thermogenesis (DIT) is stimulated by hyperphagia or nutritional deficiency such as protein deficient diets, and forms an important component of energy balance regulation particularly in small mammals (1,2). Fever is generated by increases in heat production as well as by suppression of heat loss, and the thermogenic responses to fever share many features with NST (3-5). In addition, other pathological stimuli such as injury, cancer and even psychological stress may result in increased rates of heat production that are not necessarily associated with fever but have an important impact on energy balance (2).

Effector Mechanisms of Thermogenesis

It was the results of studies on the mechanism of NST that first highlighted the importance of the sympathetic nervous system and brown adipose tissue (BAT) as effectors of thermogenesis (6). BAT is small and highly specialized tissue, representing only 1-2% of body weight in most mammals. The primary function of BAT is heat production. The remarkable thermogenic capacity of this tissue (aerobic capacity of 500 W/kg for BAT, compared to approximately 60 W/kg for skeletal muscle) means that it can result in a twofold increase in total metabolic rate of small mammals (6). Even in adult humans where BAT is much less abundant, calculations indicate that as little as 50 g of BAT can produce a 20% increase in metabolic rate (2).

Sympathetic activation of thermogenesis is now recognized as a primary mediator of DIT and of the hypermetabolic responses to infection and injury (1,5,7,8). In small mammals the effector tissue for these processes appears to be BAT (6,9), although there is some doubt about the contribution of BAT to diet-induced thermogenesis in adult men, and skeletal muscle has been proposed as an alternative (10).

Central Control of Thermogenesis

Research into the central control of thermogenesis is quite recent and hence

Circulating Regulatory Factors and Neuroendocrine Function
Edited by J. C. Porter and D. Ježová
Plenum Press, New York, 1990

somewhat limited. However, information and insight into these processes have been gained from studies aimed at investigating the central control of body temperature, food intake or fever. As a result of these and a few experiments that have followed directly the effects on thermogenesis (i.e., metabolic rate or brown fat activity), a number of brain regions or pathways have now been identified. Several hypothalamic nuclei exert marked effects on thermogenesis, particularly the preoptic/anterior hypothalamus (POAH), best known for its involvement in thermoregulation, and the ventromedial hypothalamus (VMH) which has been strongly implicated in appetite control and body weight regulation (11,12). Other hypothalamic areas that may affect thermogenesis include the paraventricular (PVN), supraoptic and suprachiasmatic nuclei (See Table 1) (13,14). Noradrenergic, dopaminergic, gabaminergic, and serotonergic pathways have all been shown either directly or indirectly to affect thermogenesis (14-17). For example, stimulation of serotonergic pathways by 5HT uptake inhibitors or releasers such as fenfluramine or fluoxetine inhibits feeding (16) has been used as a means of treating obesity (18). Recent data suggest that the potent effects of these compounds on thermogenesis may be of greater importance than their effects on body weight since peripheral or central injections of 5HT or 5HT uptake inhibitors cause dose-dependent increases in metabolic rate and brown fat activity in experimental animals (19,20).

Hypothalamic-Pituitary-Adrenal Axis

The importance of the hypothalamic-pituitary-adrenal axis in the control of thermogenesis was first recognized from studies on genetically obese rats and mice. These mutants deposit excess fat from very early in life as a result of both hyperphagia and impaired thermogenesis, and both defects appear to result from some abnormality in hypothalamic function (21-23). Surgical adrenalectomy or hypophysectomy normalize many of the defects in obese animals including food intake, thermogenesis and body weight gain (24-26). The normalizing effects of adrenalectomy or hypophysectomy are reversed when the animals are treated with either ACTH or corticosterone (27,28). In fact, genetically obese rats are particularly sensitive to the inhibitory effects of glucocorticoids, requiring very low doses to suppress thermogenesis after adrenalectomy (28). This increased sensitivity may explain the dramatic effects of adrenalectomy in the obese mutants when compared to their lean counterparts, since circulating concentrations of corticosterone are normal or only slightly elevated in the obese animals (28). Adrenalectomy also attenuates the development of obesity in aging animals and restores thermogenesis to the levels seen in normal young animals (29).

A likely explanation for the effects of adrenalectomy on thermogenesis is an increase in the release or turnover of corticotropin releasing factor (CRF) due to removal of the negative feedback from glucocorticoids. CRF concentrations within the hypothalamus are elevated following adrenalectomy or hypophysectomy (30), and CRF is a potent thermogenic agent (31).

In addition to its well-established actions on pituitary release of ACTH, CRF also has a number of actions within the brain including behavioral and cardiovascular actions, suppression of food intake, activation of the sympathetic nervous system thermogenesis, and stimulation of BAT activity (31-33). These effects are distinct from the actions of CRF on the pituitary, and probably involve different receptors (34).

Genetically obese rodents exhibit normal thermogenic responses to central injections of CRF (35) although it is possible that concentrations of the peptide may be altered in specific hypothalamic regions, and release of CRF in response to other stimuli may be impaired (see later in this review). In contrast, we have observed (unpublished data) that the thermogenic responses to CRF are impaired in aging rats although this may

Table 1
Changes in Brown Adipose Tissue (BAT) Temperature in
Response to Electrical Stimulation of
Hypothalamic Nuclei in the Rat

	Increase in BAT temperature ($^\circ$C)
Positive responses	
Ventromedial nucleus	0.50 ± 0.05
Preoptic anterior	0.50 ± 0.04
Suprachiasmatic nucleus	0.33 ± 0.04
Paraventricular nucleus	0.51 ± 0.03
No responses	
Dorsomedial nucleus	
Ventral premamillary nucleus	
Dorsal premamillary nucleus	
Supraoptic nucleus	
Posterior hypothalamus	
Paraqueductal grey	
Raphé nucleus	

Mean values ± SE (n = 12-20). Measurements were made in anesthetized rats. (Unpublished data of LeFeuvre and Rothwell.)

be related to a reduced capacity of thermogenic effector mechanisms such as brown fat.

Unlike their action on ACTH release, circulating glucocorticoid concentrations do not influence the behavioral or thermogenic responses to CRF, which are unaltered by adrenalectomy or treatment with dexamethasone (Table 2) (36), and the discrepancy emphasizes the dissociation between the hypothalamic and pituitary responses. However, glucocorticoids probably influence CRF release (or turnover) and hence modify thermogenesis. Central or peripheral injection of a glucocorticoid antagonist (RU486) stimulates thermogenesis in the rat, and this effect is prevented by pretreatment of the animals with a CRF receptor antagonist (37).

The involvement of CRF in responses to stress or trauma is well established, and it now seems likely that this peptide also mediates thermogenic responses to other stimuli. For example, angiotensin and interleukin-1 (see below) both stimulate thermogenesis, and their effects are attenuated by a CRF receptor antagonist (α helical CRF 9-41) administered intracerebroventricularly (Table 3) (38). Our recent data also indicate that angiotensin is involved in the hypermetabolic response to cerebral ischemia in the rat (39).

It seems likely, though not proven, that other hypothalamic pathways exert their effects on thermogenesis by modifying CRF release. Serotonergic and noradrenergic neurons act on cells within the PVN (the major site of CRF synthesis) to affect food intake, and these actions are modified by changes in circulating glucocorticoids (40). Stimulation of α_2 adrenoceptors in the PVN by localized injection of α_2-adrenoceptor agonists inhibits food intake (17,41), and we have observed that the α_2 receptor agonist, chlonidine, is a potent stimulator of thermogenesis in the rat (unpublished data). As yet

Table 2
Effect of Glucocorticoid Status on the Thermogenic
Responses to CRF in the Rat

	Thermogenic responses (% increase in $\dot{V}_{O_2}$)
CRF (1 nmol)	10 ± 2
(2 nmol)	15 ± 1
CRF (2 nmol) + dexamethasone (50 µg/kg)	19 ± 2
CRF + glucocorticoid antagonist (RU486, 10 mg/kg)	17 ± 2

Mean values ± SE (n = 8). Rat CRF-41 was injected intracerebralventricularly into in conscious rats. Dexamethasone and RU486 were injected intraperitoneally. $\dot{V}_{O_2}$ was measured at 24°C for 2 hr before and 2 hr after CRF.

it is not know if, or how, these actions influence CRF release, but it has been reported that electrolytic lesions that destroy the paraventricular nucleus result in the development of obesity (17,42).

The effects of GABA on thermogenesis are complex since agonists of the $GABA_A$ receptor (*e.g.*, muscimol) inhibit thermogenesis, whereas $GABA_B$ receptor agonists (*e.g.*, baclofen) stimulate metabolic rate and brown fat activity (43). $GABA_A$ receptors inhibit the release of CRF (44), and this may be the reason why sodium valproate, a GABA agonist and antiepileptic drug, inhibits thermogenesis and results in obesity in experimental animals (45) and human subjects (46). The peripheral stimuli that may affect GABA release and action on thermogenesis are unknown.

Dietary Related Signals for Thermogenesis

The search for afferent signals relating to nutrient availability or the size of body energy stores is relevant to the control of appetite and energy balance as well as to the central control of diet-induced thermogenesis. Numerous hypotheses, dating back almost forty years, have been proposed to explain how the CNS senses or assesses energy balance and nutrient availability (47-51). While none of these provide an adequate explanation, all may be of some value. For example, it has been suggested that signals relating directly to the size of body fat stores modify the central control of food intake and hence energy balance, and candidates for such signals include some factor that is partitioned between aqueous and lipid phases (52). This hypothesis could well be extended now to include the control of thermogenesis. No specific factor relating to body fat stores has yet been shown to influence energy balance although the recently identified substance, adipsin, may have such actions (53).

Specific nutrients appear to have varied effects on thermogenesis. Carbohydrate ingestion or overfeeding with carbohydrate rich diets stimulates DIT, and glucose

Table 3
Thermogenic Effects of Angiotensin in the Rat

	% Increase in Resting $\dot{V}_{O_2}$
<u>Intraperitoneal Injection</u>	
Angiotensin I (10 μg/kg)	15 ± 2
Angiotensin I (20 μg/kg)	21 ± 2
<u>Intracerebroventricular Injection</u>	
Angiotensin I (0.4 μg/kg)	11 ± 2
Angiotensin I (0.4 μg/kg) + CRF antagonist (α helical CRF 9-41, 0.1 mg/kg)	3 ± 1*

Mean values ± SE (n = 8). All injections were made in conscious Sprague Dawley rats. Oxygen consumption ($\dot{V}_{O_2}$) was measured at 24°C for 2 hr before and 2 hr after treatments. (Unpublished data of Balment, Cooper & Rothwell.) Peak responses occurred 60-90 min after injections.
*P < 0.05 vs. angiotensin alone.

availability within the CNS may modify sympathetic outflow. Glucopaenia, induced by injections of the glucose analogue, 2-deoxyglucose, suppresses metabolic rate (54); whereas central injections of small amounts of glucose increase the firing rate of sympathetic nerves supplying brown fat in the rat (55) and stimulate oxygen consumption (56). Intracerebroventricular infusion of nutrients such as glucose or ketone bodies suppresses body weight gain in normal rats (57) and thus cannot simply be related to a reduction in food intake, suggesting that thermogenesis is also increased. Insulin has also been proposed as a centrally acting satiety signal that may influence directly or modify thermogenic responses (49,58).

The most potent dietary stimulus to thermogenesis appears to be protein deficiency, although the mechanism of this is unknown (59). Protein deficiency results in an increase in the ratio of tryptophan to other large neutral amino acids in circulation (47,60). These amino acids compete for uptake into brain so an increased uptake of tryptophan will occur. This in turn may result within the brain in increased synthesis of 5HT, which would suppress appetite, particularly for carbohydrate (47,60). Since 5HT also stimulates thermogenesis (see above), the same mechanisms could be involved in the thermogenic responses to protein deficiency, and 5HT antagonists do indeed suppress the increased metabolic rates of protein deficient animals (unpublished data).

Fever

Fever, a controlled rise in body temperature, is activated in response to viral or bacterial infection and is mediated by increases in metabolic rate and/or reductions in heat loss. Fever is activated by release of endogenous pyrogen(s) which act either directly or indirectly on hypothalamic neurons to alter the set point for temperature regulation. This seemingly simple and straightforward explanation of fever masks some major problems and controversies in our current understanding. For example, most active

Table 4
Effects of IL-1α and IL-1β in Genetically
Obese Zucker Rats

	Lean	Obese
IL-1α		
Rise in $\dot{V}_{O_2}$ (%)	17 ± 3	16 ± 2*
Rise in body temperature (°C)	0.1 ± 0.1	0.3 ± 0.2
IL-1β		
Rise in $\dot{V}_{O_2}$ (%)	38 ± 3	15 ± 4†
Rise in body temperature (°C)	1.5 ± 0.2	0.5 ± 0.2†

Mean values ± SE of 6-8 male lean (+/?) and obese (*fa/fa*) Zucker rats, age 8-10 weeks. $\dot{V}_{O_2}$ and colonic temperatures were measured before and after intracerebroventricular injection of human recombinant IL-1α (10 ng) or human recombinant IL-1β (5 ng)–doses which are maximal in normal rats. Unpublished data of Busbridge, Dascombe and Rothwell.
*P < 0.05.
†P < 0.001 *vs.* lean, Student's t test.

endogenous pyrogens are large molecules (>15 kD) interleukin-1 (IL-1) and tumor necrosis factor (α TNF)] and therefore fail to cross the blood-brain barrier (61). In febrile conditions endogenous pyrogens are often undetectable or present only in very low concentrations in peripheral circulation. We have observed no significant increase in circulatory IL-1 in rats injected with endotoxin (unpublished data) or in normal human volunteers injected with typhoid vaccine (62) in spite of marked fevers.

There are receptors for IL-1 in the brain and IL-1β, and their messenger RNAs have been detected in the CNS (63,64). Thus, it seems likely that IL-1 is synthesized within the brain and exerts its actions there. Intracerebroventricular injections of nanogram doses of IL-1 induce fever and stimulate thermogenesis in rats and mice (65). These actions are dependent on eicosanoid production and on activation of sympathetic outflow to brown adipose tissue (65). IL-1 is also a potent activator of the hypothalamic-pituitary-adrenal axis and causes release of CRF from the hypothalamus under *in vivo* (66) and *in vitro* conditions (67). The involvement of CRF in the pituitary actions of IL-1 has now been extended since it has been demonstrated that CRF also mediates the central effects of IL-1β in fever and thermogenesis in the rat (38). Glucocorticoid inhibition of the actions of IL-1 on fever has previously been attributed to suppression of prostaglandin synthesis but may also be related to inhibition of CRF release or action.

Genetically obese (*fa/fa*) Zucker rats, which exhibit impaired DIT, show markedly attenuated responses to IL-1β (68) but respond normally to IL-1α (Table 4). The defective response to IL-1β may be due to increased circulating concentrations or sensitivity to glucocorticoids since surgical adrenalectomy completely restores the responses to IL-1 in obese mutants but has little effect on their lean littermates (69).

Other endogenous pyrogens such as TNF and interleukin-6 (IL-6) also act centrally to stimulate fever and thermogenesis (69). IL-6 does stimulate release of CRF (70) although it is not known whether this is essential for its pyrogenic action or whether it is effective in genetically obese rodents. IL-6 is a strong candidate for a circulating endogenous pyrogen since its concentration in plasma is markedly increased in febrile patients and animals (71,72). IL-6 is a potent agonist that also suppresses food intake (69). Thus, it is possible that IL-6 could provide the afferent signal for fever causing release of IL-1 from within the CNS. However, the mechanisms underlying the effect of IL-6 is unknown. In view of its size, it seems unlikely that IL-6 crosses the blood-brain barrier although it may act at sites such as the VMH where the barrier is "leaky."

Thermogenic Responses to Injury

Hypermetabolic responses to injury often occur in the absence of bacterial infection and are not always associated with fever. Burns and head injuries are particularly noted for their rapid and potent effects on metabolic rate, and both appear to involve activation of the sympathetic nervous system (8,73) although peripheral ischemia also stimulates the sympathetic nervous system (9). Cytokines such as IL-6 and IL-1 are released by local tissue damage and IL-6 concentrations are elevated in burn patients (74), but afferent and central pathways involved in these hypermetabolic responses are largely unknown. Our studies on the hypermetabolic response to cerebral ischemia in the rat indicate an involvement of CRF, although unlike bacterial infection, thermogenesis is not modified by suppressing eicosanoid production (75).

Chronic increases in metabolic rate following injury or during sepsis and cancer may exert severe and detrimental effects on body weight and composition resulting in cachexia and impaired survival and recovery. Understanding of the afferent signals and central neuroendocrine mechanisms responsible for these effects is therefore of potential therapeutic benefit as well as scientific interest.

Conclusions

Research into the central control of thermogenesis and circulating factors relating physiological and pathological stimuli to activation of thermogenesis is in its infancy. It may, therefore, be considered premature to attempt any conclusions about the results obtained. However, several common features are already emerging from the seemingly diverse studies discussed above. For example, CRF appears to play an important role in activation of thermogenesis in response to a number of physiological and pathological stimuli. It is also clear that valuable information can be gained from studies of related function such as food intake and temperature regulation, which emphasizes the value of an integrated approach. Finally, observations on neuroendocrine mechanisms involved in the control of thermogenesis have proven relevant to other neuroendocrine processes. An example of this is the interaction between IL-1 and CRF and the defective responses seen in genetically obese mutants.

References

1. Rothwell, N.J., and M.J. Stock, Diet-induced thermogenesis, In L. Girardier and M.J. Stock (eds), *Mammalian Thermogenesis*, Chapman & Hall, London, pp. 208-233, 1983.
2. Rothwell, N.J., and M.J. Stock, Whither brown fat? *Biosci Rep* 6: 3-18, 1986.
3. Harris, W.H., D.O. Foster, and B.E. Nadeau, Evidence for a contribution by brown adipose tissue to the development of fever in the young rabbit, *J Physiol Pharmacol* 63: 595-598, 1985.
4. Blatteis, C.M., Fever: exchange of shivering by non-shivering pyrogenesis in cold-acclimated guinea pigs,

J Appl Physiol 40: 29-34, 1976.

5. Jepson, M.M., D.J. Millward, N.J. Rothwell, and M.J. Stock, Involvement of sympathetic nervous system and brown fat in endotoxin induced fever in rats, *Am J Physiol* 255: E617-E620, 1988.

6. Girardier, L., Brown fat: an energy dissipating tissue, In L. Girardier and M.J. Stock (eds) *Mammalian Thermogenesis*, Chapman & Hall, London, pp. 598, 1983.

7. Landsberg, L., and J.B. Young, Autonomic regulation of thermogenesis, In L. Girardier and M.J. Stock (eds), *Mammalian Thermogenesis*, Chapman & Hall, London, pp. 99-140, 1983.

8. Wilmore, D.W., J.M. Long, A.D. Mason, R.W. Skreen, and B.A. Pruitt, Catecholamines: mediator of the hypermetabolic response to thermal injury, *Ann Surg* 870: 653-670, 1974.

9. Young, J.B., S. Fish, and L. Landsberg, Sympathetic nervous system and adrenal medullary responses to ischemic injury in mice, *Am J Physiol* 245: E67-E73, 1983.

10. Astrup, A., J. Bulow, J. Madsen, and N.J. Christensen, Contribution of BAT and skeletal muscle to thermogenesis induced by ephedrine in man, *Am J Physiol* 248: E507-E515, 1985.

11. Stellar, E., The physiology of motivation, *Psychol Rev* 61: 5-22, 1954.

12. Han, P.W., Hypothalamic obesity in rats without hyperphagia, *Trans NY Acad Sci* 30: 229-243, 1967.

13. Holt, S.J., H.V. Wheal, and D.A. York, Hypothalamic control of brown adipose tissue in Zucker lean and obese rats: effects of electrical stimulation of the ventromedial nucleus and other hypothalamic centres, *Brain Res* 405: 227-233, 1987.

14. Rothwell, N.J., Central control of brown adipose tissue, *Proc Nutr Soc* 48: 241-250, 1989.

15. Sawchenko, P.E., and L.W. Swanson, Central noradrenergic pathways for the integration of hypothalamic neuroendocrine and autonomic responses, *Science* 214: 685-687, 1981.

16. Liebowitz, S.F., Neurochemical systems of the hypothalamic control of feeding and drinking behavior and water electrolyte excretion, In P.J. Morgane and J. Panksepp (eds), *Handbook of the Hypothalamus*, Vol. 3, Dekker, New York, pp. 299-243, 1980.

17. Liebowitz, S.F., Brain monoamines and peptides. Role in the control of eating behavior, *Fed Proc* 45: 1396-1403, 1986.

18. Bray, G.A., and M. Cairella (ed), Drugs influencing food intake and energy balance, *Int J Obes* 11: 1987.

19. Rothwell, N.J., and M.J. Stock, Effect of diet and fenfluramine on thermogenesis in the rat: possible involvement of serotonergic mechanisms, *Int J Obes* 11: 319-324, 1987.

20. Latham, A., Ros A. LeFeuvre, and N.J. Rothwell, Enhanced thermogenic effect of D-fenfluramine following neurotoxic lesions of central 5-hydroxytryptamine pathways in the rat, *Proc Nutr Soc* 48: 38A, 1989.

21. Bray, G.A., and D.A. York, Hypothalamic and genetic obesity in experimental animals: an autonomic and endocrine hypothesis, *Physiol Rev* 59: 719-809, 1979.

22. Trayhurn, P., and W.P.T. James, Thermogenesis and obesity, In L. Girardier and M.J. Stock (eds) *Mammalian Thermogenesis*, Chapman & Hall, London, pp. 234-258, 1983.

23. York, D.A., Neural activity in hypothalamic and genetic obesity, *Proc Nutr Soc* 46: 105-117, 1987.

24. Holt, S., and D.A. York, The effect of adrenalectomy on GDP binding to brown adipose tissue mitochondria of obese rats, *Biochem J* 208: 819-822, 1982.

25. Marchington, D., N.J. Rothwell, M.J. Stock, and D.A. York, Energy balance, diet-induced thermogenesis and brown adipose tissue in lean and obese (*fa/fa*) Zucker rats after adrenalectomy, *J Nutr* 113: 1395-1402, 1983.

26. Rothwell, N.J., M.J. Stock, and D.A. York, Effects of adrenalectomy on energy balance diet-induced thermogenesis and brown adipose tissue in adult cafeteria-fed rats, *Comp Biochem Physiol* 78A: 565-569, 1984.

27. Rothwell, N.J., and M.J. Stock, Thermogenesis and BAT activity in hypophysectomized rats with and without corticotrophin replacement, *Am J Physiol* 249: E333-E336, 1985.

28. York, D.A., S.J. Holt, J. Allars, and J. Payne, Glucocorticoids and the central sympathetic activity in the obese *fa/fa* rat, In P. Bjorntorp and S. Rossner (eds), *Obesity in Europe 88*, John Libbey, London, pp. 247-252, 1989.

29. Rothwell, N.J., and M.J. Stock, Influence of adrenalectomy on age-related changes in energy balance, thermogenesis and brown fat activity in the rat, *Comp Biochem Physiol* 89A: 265-269, 1988.

30. Kovacs, K., J.Z. Kiss, and G.B. Markara, Glucocorticoid implants around the hypothalamic paraventricular nucleus prevent the increase of corticotrophin releasing factor and arginine vasopressin immunostaining induced adrenalectomy, *Neuroendocrinology* 44: 229-234, 1986.

31. LeFeuvre, R.A., N.J. Rothwell, and M.J. Stock, Activation of brown fat thermogenesis in response to central injection of corticotropin releasing hormone in the rat, *Neuropharmacology* 26: 1217-1221, 1987.

32. Brown, M.R., and L.A. Fisher, Corticotrophin releasing factor: effects on the autonomic system and visceral systems, *Fed Proc* 44: 243-248, 1985.

33. Siggins, G.R., D. Gruol, J. Aldenoff, and Q. Pittman, Electrophysiological actions of corticotrophin-releasing factor in the central nervous system, *Fed Proc* 44: 237-242, 1985.

34. Grigoriadis, D.E., and E.G. de Souza, The brain corticotrophin-releasing factor (CRF) receptor is of lower apparent molecular weight than the CRF receptor in anterior pituitary, *J Biol Chem* 263: 10927-10931, 1988.

35. Carnie, J.A., R.A. LeFeuvre, E.A. Linton, H.D. McCarthy, and N.J. Rothwell, Thermogenic effects of corticotrophin releasing factor in genetically obese Zucker rats, *Proc Nutr Soc* 47: 165A, 1988.

36. Britton, K.T., G. Lee, R. Dana, S.C. Risch, and G.F. Koob, Activity and anxiogenic effects of corticotrophin releasing factor are not inhibited by blockade of the pituitary-adrenal system with dexamethasone, *Life Sci* 39: 1281-1286, 1986.

37. Hardwick, A.J., E.A. Linton, and N.J. Rothwell, Thermogenic effects of the antiglucocorticoid RU-486 in the rat: involvement of corticotropin-releasing factor and sympathetic activation of brown adipose tissue, *Endocrinology* 124: 1684-1688, 1989.

38. Rothwell, N.J., CRF is involved in the pyrogenic and thermogenic effects of interleukin-1β in the rat, *Am J Physiol* 256: E111-E115, 1989.

39. Cooper, A.L., H.D. McCarthy, C.T. O'Shaughnessy, and N.J. Rothwell, Involvement of angiotensin in hypermetabolic responses to injury and interleukin-1β in the rat, *Neurosci Lett*, In Press, 1989.

40. Jhanwar-Uniyal, M., C.R. Roland, and S.F. Liebowitz, Diurnal rhythm of α_2-noradrenergic receptors in the paraventricular nucleus and other brain areas: relation to circulating corticocosterone and feeding behavior, *Life Sci* 38: 473-482, 1986.

41. Liebowitz, S.F., and C. Rossakis, Mapping study of brain dopamine and epinephrine sites which cause feeding suppression in the rat, *Brain Res* 12: 107-113, 1979.

42. Weingarten, H.P., P. Cheng, and T.J. MacDonald, Comparison of the metabolic and behavioral disturbances following paraventricular and ventromedial hypothalamic lesions, *Brain Res Bull* 14: 551-559, 1985.

43. Horton, R.W., R.A. LeFeuvre, N.J. Rothwell, and M.J. Stock, Opposing effects of activation of central $GABA_A$ and $GABA_B$ receptors on brown fat thermogenesis in the rat, *Neuropharmacology* 27: 363-366, 1988.

44. Fehm, H.L., K.L. Voigt, R.E. Lang, and E.R. Pfeiffer, Effects of neurotransmitter on the release of corticotropin releasing hormone (CRH) by rat hypothalamic tissue in vitro, *Exper Brain Res* 39: 229-234, 1980.

45. Astrup, A., and N.J. Rothwell, Sodium valproate promotes obesity by reducing energy expenditure, (Abstract) *Int J Obes* 13 Suppl 1: 125, 1989.

46. Dinesen, H., L. Gram, T. Anderson, and M. Dam, Weight gain during treatment with valproate, *Acta Neurol Scand* 70: 65-69, 1984.

47. Fernstrom, J.D., Acute and chronic effects of protein and carbohydrate ingestion on brain tryptophan levels and serotonin synthesis, *Nutr Rev Suppl* 5: 25-36, 1986.

48. Oomura, Y., Regulation of feeding by neural responses to endogenous factors, *Trends Physiol Sci* 2: 199-203, 1987.

49. Rothwell, N.J., and M.J. Stock, Insulin and thermogenesis, *Int J Obes* 12: 93-102, 1988.

50. Steffens, A.B., The evidence that humoral factors are involved in the regulation of food intake and body weight, *Brain Res Bull* 5: 13-16, 1980.

51. Williamson, D.H., Brain substrates and the effects of nutrition, *Proc Nutr Soc* 46: 81-87, 1987.

52. Rothwell, N.J., and M.J. Stock, Regulation of energy balance, *Ann Rev Nutr* 1: 235-256, 1981.

53. Lavau, M., I. Dugail, A. Quignard-Boulangé, R. Bazin, and M. Guere-Milo, Adipsin in different obesity

models, In P. Bjorntorp and S. Rossner (eds) *Obesity in Europe 88*, John Libbey, London, pp. 167-172, 1989.

54. Rothwell, N.J., M.E. Saville, and M.J. Stock, Metabolic responses to food, atropine and 2 deoxy-D-glucose in Zucker rats, *Proc Nutr Soc* 41: 37A, 1982.

55. York, D.A., Corticosteroid inhibition of thermogenesis in obese animals, *Proc Nutr Soc*, In Press, 1989.

56. LeFeuvre, R.A., N.J. Rothwell, M.J. Stock, and A.J. Woods, Central actions of nutrients on thermogenesis, (Abstract) *Circulating Regulatory Factors and Neuroendocrine Function*, pp. 34, 1989.

57. Brief, D.J., and J.D. Davies, Reduction of food intake and body weight by chronic intraventricular insulin infusion, *Brain Res Bull* 12: 571-575, 1984.

58. Posner, B.I., Insulin interaction with the central nervous system: nature and possible significance, *Proc Nutr Soc* 46: 97-103, 1987.

59. Rothwell, N.J., and M.J. Stock, Influence of carbohydrate and fat intake on diet-induced thermogenesis and brown fat activity in rats fed low protein diets, *J Nutr* 117: 1721-1726, 1987.

60. Leathwood, P.D., Tryptophan availability and serotonin synthesis, *Proc Nutr Soc* 46: 143-156, 1987.

61. Blatteis, C.M., C.A. Dinarello, M. Shibata, Q.J. Llanos, and J. Quan, Does circulating interleukin-1 (IL-1) enter the brain? *Proc Int Cong Physiol Sci*, Helsinki, In Press, 1989.

62. Horan, M.A., L. Gibbons, S.J. Hopkins, A. Cooper, P. Strijbos, N.J. Rothwell, and R.A. Little, Changes in plasma interleukin-6 during experimentally-induced fever in humans, (Abstract) *Lymphokine Res*, In Press, 1989.

63. Breder, C.D., C.A. Dinarello, and C.B. Saper, Interleukin-1 immunoreactive innervation of the human hypothalamus, *Science* 240: 321-324, 1988.

64. Farrar, W.L., J.M. Hill, A. Havel-Bellen, and M. Vinocour, The immune logical brain, *Immunol Rev* 100: 361-378, 1987.

65. Dascombe, M.J., N.J. Rothwell, B.O. Sagay, and M.J. Stock, Pyrogenic and thermogenic effects of interleukin-1β in the rat, *Am J Physiol* 256: E7-E11, 1989.

66. Berkenbosch, F., J. van Oers, A. del Rey, F. Tilders, and H. Besedovsky, Corticotrophin-releasing factor-producing neurons in the rat activated by interleukin-1, *Science* 238: 524-526, 1987.

67. Tsagarakis, S., G. Gillies, L.H. Rees, M. Besser, and A. Grossman, Interleukin-1 directly stimulates the release of corticotrophin-releasing factor from rat hypothalamus, *Neuroendocrinology* 49: 98-101, 1989.

68. Dascombe, M.J., A.J. Hardwick, R.A. LeFeuvre, and N.J. Rothwell, Impaired effects of interleukin-1β on fever, thermogenesis and brown fat in genetically obese rats, *Int J Obes*, In Press, 1989.

69. Busbridge, N.J., M.J. Dascombe, S. Hopkins, and N.J. Rothwell, Acute central effects of interleukin-6 on body temperature, thermogenesis and food intake in the rat, *Proc Nutr Soc* 48: 48A, 1989.

70. Naitoh, Y., J. Fukata, T. Tominaga, Y. Nakai, S. Tamai, K. Mori, and H. Imura, Interleukin-6 stimulates the secretion of adrenocorticotropic hormone in conscious freely-moving rats, *Biochem Biophys Res Comm* 155: 1459-1463, 1988.

71. Houssiau, F.A., J.P. Devoyelaer, J. van Dumme, C. Nagant, and J. van Snick, Interleukin-6 in synovial fluid and serum of patients with rheumatoid arthritis and other inflammatory arthritides, *Arthritis Rheum* 31: 784-788, 1988.

72. Nijsten, M.W.N., E.R. de Groot, H.J. ten Duis, H.J. Klasen, C.E. Hack, and L.A. Aarden, Serum levels of interleukin-6 and acute phase responses, *Lancet* 11: 921, 1987.

73. Little, R.A., Metabolic rate and thermoregulation after injury, In I. Ledingham (ed) *Recent Advances in Critical Care Medicine*, Churchill, London, Vol. 3, pp. 159-172, 1988.

74. Childs, C., R.J. Ratcliff, I. Holt, and S.J. Hopkins, Relationship between interleukin 1, interleukin 6 and pyrexia in burned children, (Abstract) *Lymphokine Res*, In Press, 1989.

75. McCarthy, H.D., C.T. O'Shaughnessy, and N.J. Rothwell, Endogenous mediators of cerebral ischaemia-induced hypermetabolism in the rat, *Br J Pharmacol*, In Press, 1989.

THE BLOOD-BRAIN BARRIER IN DIABETES MELLITUS

M. Lorenzi

Eye Research Institute and Departments of Ophthalmology
and Medicine
Harvard Medical School
Boston, MA

Introduction

The blood-brain barrier (BBB) is the function whereby the cerebral microvasculature selectively shields the brain from the rapidly changing milieu of the systemic circulation. Its strict anatomical counterpart is the tight junctions between capillary endothelial cells, but if the concept of *barrier* is broadened to mean not only partition and exclusion, but also exchanges and transport, then its morphologic counterpart becomes the whole endothelial cells, with its carrier molecules, membrane receptors, and biosynthetic repertory.

In either case, the question of whether the BBB is altered in diabetes is a highly pertinent one, since long-term diabetes leads to a degenerative vasculopathy, most characteristically affecting microvessels and their endothelial lining. Despite the relevance that studies of the BBB would have toward a comprehensive description of diabetic microangiopathy and, potentially, the understanding of some manifestations of this multisystem disorder, the investigations are few and their results often contradictory. My report on these studies will thus attempt to provide a perspective for their paucity and some discussion of the inconsistencies.

Central Nervous System (CNS) Abnormalities in Diabetes

The most compelling reason to study the BBB in diabetes would be to ascertain whether altered barrier characteristics cause, or contribute to, diabetes-induced disturbances in CNS function. A class of CNS abnormalities well known to occur in diabetic patients are those consequent to occasional hypoglycemia provoked by excess insulin administration: the subtle symptomatology of neuroglycopenia, or frank seizures and coma. However, once these abnormalities of known etiology are excluded, are there CNS disturbances that can be attributed to the diabetic state and, especially, to its long-term microvascular sequelae? These are best known to affect the eyes, chiefly the retina, and the kidneys, and to contribute to peripheral nerve dysfunction. The degenerative process in these organs and tissues results in florid clinical syndromes (retinopathy, nephropathy, neuropathy) that constitute the well known chronic complications of diabetes. Whichever lesions diabetes may induce in the CNS, no syndrome with a specific clinical configuration appears to arise from them.

There are, nonetheless, suggestions that complex or specific CNS functions might be altered by long-term diabetes. I consider the evidence suggestive rather than firmly

Circulating Regulatory Factors and Neuroendocrine Function
Edited by J. C. Porter and D. Ježová
Plenum Press, New York, 1990

indicting, since for many of these abnormalities pathogenetic elements other than the long-term effects of diabetes may be invoked. The subject of CNS abnormalities in diabetes was recently and comprehensively reviewed by Mooradian (1), and I shall therefore focus my discussion on representative findings especially pertinent to possible BBB dysfunction.

Cognitive Function

It has been reported that individuals who developed insulin-dependent diabetes (IDDM) early in life—before the age of 5 years—earn significantly poorer neuro-psychological test scores than either diabetic subjects who developed the disease after 5 years of age, or demographically similar non-diabetic control subjects (2). Although in the particular cohort under study the poorer performance could not be correlated with previous history of severe hypoglycemic episodes, the likelihood of an association remains high, since children who experience multiple hypoglycemic seizures early in life are prone to showing significant CNS disturbances (reviewed in Ref. 2), and the frequency of severe hypoglycemia is greater in children who develop IDDM before the age of 3 years (3). However, recurrent hypoglycemia is unlikely to play a role in the cognitive dysfunction reported in aging, non-insulin-dependent diabetic patients (4). In these patients the impairment in memory retrieval did not correlate with known duration of diabetes, but it was greater in patients with poor metabolic control and peripheral neuropathy (4). Although these correlations suggest a possible connection with the chronic sequelae of diabetes, the lack of comparison with control group affected by another well-defined chronic disorder precludes a definitive attribution of the abnormalities to unique alterations induced by diabetes.

EEG Changes

The EEG changes reported in children and adolescents with IDDM can generally be attributed to repeated episodes of severe hypoglycemia (5), or to the metabolic disturbances characteristic of hyperglycemic emergencies, in particular to hyperosmolality (6). The fact, however, that a positive correlation was found between EEG changes and vascular changes in the retina (5), and that in several children EEG abnormalities persisted long past the acute hyperglycemic episode (6), suggest the possibility that changes in vascular or neural structures may sustain the abnormality.

Nociception and Neuroendocrine Abnormalities

Patients with poorly controlled diabetes have been reported to be hyperalgesic when compared with normal subjects (7). This may relate to the elevated glucose levels insofar as normal individuals made hyperglycemic with glucose infusion did manifest a decrease in both the threshold level of pain and the maximal pain tolerated (7). In experimentally diabetic mice, the antinociceptive potency of morphine was significantly decreased (8), and, again, this could be mimicked in non-diabetic mice by infusion of hypertonic glucose or fructose, but not of a non-metabolizable analog (8). A most interesting observation made by the same authors was that the antinociceptive potencies of phenazocine and levorphanol were altered similarly to that of morphine in experimentally diabetic and transiently hyperglycemic mice, but the potencies of methadone, propoxyphene, and meperidine were not altered by changes in the serum glucose levels (9). This has suggested the hypothesis that, in some way, high glucose may alter the conformation of the opiate receptor, so that molecules with a rigid structure and high affinity binding (such as morphine) would be impaired in the binding, while molecules with a less rigid structure (such as methadone) would not face such impairment.

The possibility that the level of circulating glucose may modulate receptor binding of molecules with specific structures is intriguing, and should be further pursued experimentally. It must be noted that it is unknown whether decreased sensitivity to narcotic analgesics also occurs in poorly controlled human diabetes.

With regard to endocrine function, a host of pituitary-hypothalamic abnormalities have been described in experimental diabetes (1,10,11), but they have not always been confirmed (1). In human diabetes a mild elevation of fasting serum prolactin levels has been reported in male patients aged 26-74 years (12), but no correlation was found with the presence of impotence or retinopathy (12). In our studies of IDDM patients, prolactin levels were similar or lower than in control individuals (13,14) and normally responsive to dopamine (13) and apomorphine (14) suppression.

The best established and metabolically important (15) endocrine abnormality characteristically occurring in IDDM patients is the increased circadian level of circulating growth hormone (16,17), and the exaggerated or paradoxical growth hormone response to a variety of stimuli. Thus, the response to arginine (18,19), insulin-induced hypoglycemia (19), and physical exercise (20) is greater in IDDM patients than in age- and sex-matched controls, and stimuli that do not consistently elicit growth hormone secretion in normal subjects (glucagon, TRH, small doses of peripherally administered dopamine) are effective secretagogues in IDDM patients (13,18,21,22). Since growth hormone hyperresponsiveness is especially evident during poor metabolic control (16,20), which also entails lower circulating somatomedin levels (23), and since both somatomedin and insulin suppress basal and stimulated secretion of pituitary somatotrophs (24), it is conceivable that the growth hormone hyperresponsiveness encountered in poorly controlled IDDM patients might reflect lack of restraining influences on the pituitary. However, in our study with systemic dopamine administration (13) we were able to exclude pituitary-hypothalamic hypersensitivity as a mechanism for the paradoxical growth hormone response in IDDM patients, and we thus proposed the hypothesis that subtle changes in BBB permeability might account for the phenomenon. This study is further discussed in the permeability section.

Abnormalities of the Cerebral Microvasculature in Diabetes

The strongest clinical evidence that cerebral vessels are involved in the degenerative angiopathy characteristic of long-term diabetes is the increased incidence of cerebrovascular accidents (CVA) in diabetic patients. In the Framingham study the relative risk for CVA, adjusted for other risk factors, was 2.2 in both men and women with diabetes (25). Whether the increased risk of brain infarction is solely attributable to accelerated atherosclerosis, and thus to large vessel disease, or also to compromised patency of microvessels is not established.

Notwithstanding this clinical consideration, there is an analogical reason to investigate whether and how the cerebral microvasculature is affected by diabetes. The reason stems from the morphologic, functional, and embryologic similarities that brain capillaries share with retinal capillaries (26), which in diabetes develop severe lesions leading to retinopathy and blindness. The well-described abnormalities of retinal capillaries in diabetes are increased permeability, increased thickness of basement membranes, loss of pericytes and endothelial cells, and, ultimately, non-perfusion. These abnormalities can be used as a frame for comparing cerebral to retina capillaries in diabetes.

Permeability

It has been known for many years that in long-term diabetes there is a generalized increased permeability of microvessels (27). In advanced background retinopathy the fundus shows hard exudates and dot-hemorrhages, and fluorescein angiography shows widespread leakage of retinal capillaries (28). More recently, employing the technique of vitreous fluorophotometry which measures low concentrations of fluorescein in the vitreous body, several investigators have identified fluorescein accumulation in the vitreous at very early stages of diabetes, in the absence of other signs of retinopathy (29,30). It is currently uncertain whether such accumulation can indeed be attributed to leaky retinal vessels for two related reasons. First, the blood-aqueous barrier (the ciliary process and the iris) appears more permeable at these early stages of diabetes (29,30), and it is possible that fluorescein flows from the ciliary body posteriorly to the vitreous (31). Second, increased vitreous fluorescence has been detected in the streptozotocin-treated, hyperglycemic guinea pig (31), an animal known to have no retinal vasculature.

Relevant studies of the blood-retinal barrier in experimental diabetes are summarized in Table 1. The data on fluorescein leakage must be interpreted within the caveat just discussed; otherwise there appears to be good agreement that in animals with short duration of diabetes the blood-retinal barrier is not readily permeable to large molecular weight substances such as horseradish peroxidase (HRP). However, after several years of diabetes there is clear histologic evidence of alterations in the blood-retinal barrier. In humans, the wall of microaneurysms is markedly permeable to all elements of the blood (35), and in dogs with 5 years of diabetes HRP is seen to fill the interendothelial clefts, basement membranes, and intercellular spaces of perivascular tissue (34). Since the tracer was not transported through the endothelial cytoplasm, the mechanism likely to account for the increased permeability is loss of fusion of the outer leaflets of plasma membranes of adjacent endothelial cells, *i.e.*, compromise of tight junctions.

Do similar findings pertain to the BBB? As summarized in Table 2, there is again good agreement that at early stages of diabetes the barrier is generally competent. However, when we allowed a small molecular weight substance (inulin, molecular weight ≈ 5,000) to circulate for 15 minutes (instead of 5 minutes) prior to sacrifice, we identified, in rats with 4 weeks of diabetes, discrete areas of leakage (38). Of 12 brain regions examined, only the medio-basal and medio-dorsal hypothalamus and the periaqueductal gray exhibited increased permeability. The selectivity might have been more apparent than real, if the more rapid decline in circulating inulin radioactivity in the diabetic rats (neither diabetic nor control rats were nephrectomized) had resulted in insufficient exposure of several brain regions. Nonetheless, to explain why increased permeability had been readily demonstrable only in those specific regions, we considered that the hypothalamic periventricular regions (and posterior hypothalamic area) host the descending pathway for the drinking response, which ends in the mesencephalic periaqueductal gray. It could be speculated that enhanced activity of the drinking pathway in the diabetic animals, with attendant increased blood flow to the structures involved, might have been responsible for uncovering a slowly developing defect in BBB competence. A direct relationship between increase in blood flow and increase in barrier permeability has been demonstrated in other experimental situations (40).

In agreement with Jakobsen (36) and Malmgren (whose studies were in galactosemic rats; 37) we did not detect increased permeability to larger molecular weight substances. In a single diabetic rat we observed occasional focal areas of HRP leakage; the significance of this leakage remains unclear since it occurred in only one of seven animals. These negative results obtained with large molecular weight tracers must be reconciled with the observation reported by Stauber *et al.* (39) of extravasation of

Table 1
Blood-Retinal Permeability in
Experimental Diabetes

Author	Ref #	Species	Duration of DM	Tracer	Permeability
Kirber	32	Rat	4 weeks	Fluorescein (microscopy)	-
				HRP	-
Tso	33	Rat	1 week	Fluorescein (FPM)	+
				HRP	-
Klein	31	Guinea Pig	4 weeks	Fluorescein (FPM)	+
				HRP	-
Wallow	34	Dog	5 years	HRP	+

DM: diabetes mellitus.
HRP: horseradish peroxidase, M.W. 40,000.
FPM: fluorophotometry.

endogenous albumin—but not of IgG or complement—in the cerebral cortex of rats with only 2 weeks of diabetes. It is tempting to speculate that such abnormal finding may have been consequent, not to mechanical incompetence of the barrier but, rather to altered interactive properties of the endogenous albumin molecule with vascular endothelium, caused by non-enzymatic glycosylation. Excessive glycosylation of serum albumin is well known to occur in diabetes (41), and the overly glycosylated molecule is, at least *in vitro*, avidly and preferentially pinocytosed by microvessels (42).

There is an astounding paucity of studies on the permeability of the BBB in human diabetes. A study we performed a few years ago (13) offered, through indirect evidence, the suggestion that in diabetic subjects there might be increased barrier permeability at the level of hypothalamic regions. In 12 patients aged 21-32 years and with a 2-29 year history of insulin-dependent diabetes, we observed a substantial growth hormone response to the systemic administration of 1.5 μg/kg/minute of dopamine, a dose that was consistently ineffective in non-diabetic control subjects. The results could not be accounted for by spontaneous fluctuations in growth hormone secretion, monitored during infusion of saline solution. Metabolic and cardiovascular responses to the infusion were similar in the diabetic and control subjects, militating against the possibility that higher circulating levels of dopamine might have been achieved in the diabetic group. Other growth hormone stimuli (apomorphine in decreasing amounts, glucagon, graded physical exercise) failed to indicate that pituitary or hypothalamic hypersensitivity could account for the consistent growth hormone rise. Effects of dopamine known to be exerted at the pituitary level, such as suppression of prolactin, were of the same magnitude in diabetic and control subjects. We interpreted this constellation of data to indicate that the ventral hypothalamus, the locus of the signal for growth hormone secretion that has transitional characteristics *vis-a-vis* the BBB, might be reached in diabetic patients by greater

Table 2
Blood-Brain Barrier Permeability
in Experimental Diabetes

Author	Ref #		Duration of DM (weeks)	Tracer	Permeability
Jakobsen	36	Rat	4-20	Evans-blue albumin	-
				HRP	-
				Cytochrome C	-
Malmgren	37	Rat	4-22 Galactosemia	HRP	-
Lorenzi	38	Rat	2	Inulin	-
				Sucrose	-
			4	Inulin (15 min)*	+
				HRP	-
Stauber	39	Rat	2	Albumin†	+
				IgG†	-
				C3†	-

DM: diabetes mellitus.
HRP: horseradish peroxidase; C3: complement.
*Inulin was allowed to circulate for 15 min before sacrifice. Experiments in which inulin circulated for only 5 min failed to show leakage.
†Endogenous

dopamine concentrations as a result of augmented permeability. These studies, and those performed with TRH (21,22) and glucagon (18), by demonstrating that a systemic perturbation without consequences in normal subjects resulted instead in a paradoxical hormonal response in diabetic patients, propose the possibility that, in these patients, neuroendocrine regulation might occasionally be disrupted by circulating factors.

To my knowledge, studies of tight junctions similar to those performed in retinal capillaries (34) are not available for diabetic cerebral capillaries.

Transport

Because neural tissue utilizes glucose as its main energy source, and because glucose uptake by neural tissue is independent of insulin, the rate of glucose delivery from the circulation is a critical feature of neural tissue homeostasis. Thus, glucose transport across retinal and brain capillaries has been quite extensively studied in both normal and pathologic conditions. There is good agreement that D-glucose moves through the blood-retina and the blood-brain barriers by a saturable, stereospecific, carrier-mediated mechanism. Using a modification of the single-injection technique developed by Oldendorf, Ennis and colleagues (43) have demonstrated in the rat retina two transport systems for D-glucose, a high affinity (Km of 0.24 mM) and a lower affinity one (Km of 7.8 mM). The Km values of both transport systems were significantly increased in rats with 2 months of streptozotocin-induced diabetes.

The effect of diabetes on blood-to-brain glucose transport appears currently somewhat controversial, but the discrepancies may not be irreconcilable. Two studies (44,45) concurred in finding a decreased BBB glucose transport capacity in rats with chronic hyperglycemia, and this was detectable over a range of glycemic concentrations at the time of testing (44). Harik and La Manna (46) have recently challenged these results, but their study used historical data for the control rats, and reported nonetheless that the carrier-mediated influx of D-glucose in the hyperglycemic animals was 80% of that found in controls (46). A study performed in insulin-dependent diabetic patients using [^{11}C]3-0-methylglucose and positron emission tomography has shown that, in patients rendered acutely normoglycemic by clamp technique, the brain [^{11}C] MeG extraction fractions were similar to those found in normal controls (47). The decreased extraction found at hyperglycemic levels could be explained entirely in terms of increased competition between the tracer and circulating glucose. The normal extraction found in these patients may appear at variance with the results obtained in experimental animals in whom decreased transport was observed irrespective of acute changes in glycemic levels (44), or was normalized only after a relatively prolonged period of normoglycemia (45). It is, however, possible that the diabetic rats might experience a degree of transport abnormalities much more severe, and thus less readily correctable, than human patients because their degree of hyperglycemia is so much greater (25-40 mM, vs. 10-15 mM in customarily treated diabetic patients).

The finding of decreased BBB glucose transport in severely diabetic animals has two potential implications: a homeostatic one, whereby the decreased transport might decrease the osmotic load (net glucose flux) entering the brain in the presence of hyperglycemia, and a potentially harmful one, involving enhanced susceptibility to hypoglycemia because of decreased availability of glucose to the brain when glycemia declines (44,45). Because this latter implication may have great clinical relevance, further studies should be devoted to the subject in humans.

Other alterations in BBB transport systems may occur in diabetes. Extraction of β-hydroxybutyrate was increased in diabetic rats (45), but this was concomitant with increased plasma levels of the ketone body. The BBB permeability to sodium ions was found decreased by 25% in the cerebral cortex of rats with 2 weeks of streptozotocin diabetes, in contrast to unchanged permeability to sucrose and chloride (48). Whether this abnormality indicates that the activity of the Na-K-ATPase pump is decreased in cerebral microvessels as it is in diabetic peripheral nerve (49), or that Na^+ shares in the same abnormality affecting glucose because of co-transport of the two molecules, remains to be established.

Histologic Abnormalities

The fundamental structural lesion of the small blood vessels in diabetic patients is increased thickness of the basement membranes (50), which has been noted in practically all microvascular districts examined, including cerebral capillaries (51,52). Although the disturbed metabolic environment appears to be critical for the development of this abnormality (reviewed in Ref. 53), it is unclear whether specific elements of the diabetic milieu play a preeminent causative role and, eventually, how they produce their effects. Recent studies in my laboratory have shown that high glucose is a perturbation sufficient to increase the expression of basement membrane components in cultured human endothelial cells (53), and we are currently investigating the underlying mechanisms.

The second best documented histologic abnormality of diabetic capillaries is loss of pericytes and endothelial cells. Noted in lower extremities (54,55), renal glomeruli (56), and retina (57,58), it is in the latter that the consequences of the phenomenon have been best established: capillary closure, ischemia, neovascularization, and attendant risk of hemorrhage and retinal detachment. The trypsin digest method that has enabled the elegant studies of retinal capillary cellularity cannot, unfortunately, be applied to the brain (57). Employing a modification of the technique, Ashton failed to detect abnormalities of cerebral capillaries in 26 diabetic patients with severe retinopathy (57). Pericyte degeneration, considered an early event in diabetic retinopathy, was not detected in the brain of patients exhibiting the lesion in their retinal vessels (58,59). Although this information may suggest a sparing of cerebral microvascular cells and thus a better preservation of capillary patency in the brain than in the retina, other studies indicate that, at least in experimental animals, long-term diabetes results in marked shortening of the length of the cortical capillary network (60).

It is obvious that more studies on the subject are needed. If indeed cerebral capillaries are less affected by diabetes than retinal capillaries, this should be positively and precisely known, since the information may foster an understanding of whether and how local factors (topography, tissue architecture, blood flow, etc.) modulate the impact of the metabolic abnormalities on vascular integrity.

References

1. Mooradian, A.D., Diabetic complications of the central nervous system, *Endocrine Rev* 9: 346-356, 1988.
2. Ryan, C., A. Vega, and A. Drash, Cognitive deficits in adolescents who develop diabetes early in life, *Pediatrics* 75: 921-927, 1985.
3. Ternand, C., V.L.W. Go, J.E. Gerich, and M.W. Haymond, Endocrine pancreatic response of children with onset of insulin-requiring diabetes before age 3 and after age 5, *J Pediatrics* 101: 36-39, 1982.
4. Perlmuter, L.C., M.K. Hakami, C. Hodgson-Harrington, J. Ginsberg, J. Katz, D.E. Singer, and D.M. Nathan, Decreased cognitive function in aging non-insulin-dependent diabetic patients, *Am J Med* 77: 1043-1048, 1984.
5. Haumont, D., H. Dorchy, and S. Pelc, EEG abnormalities in diabetic children. Influence of hypoglycemia and vascular complications, *Clin Pediatr* 18: 750-753, 1979.
6. Tsalikian, E., D.J. Becker, P.K. Crumrine, D. Daneman, and A. Drash, Electroencephalographic changes in diabetic ketosis in children with newly and previously diagnosed insulin-dependent diabetes mellitus, *J Pediatr* 98: 355-359, 1981.
7. Morley, G.K., A.D. Mooradian, A.S. Levine, and J.E. Morley, Mechanism of pain in diabetic peripheral neuropathy. Effect of glucose on pain perception in humans, *Am J Med* 77: 79-82, 1984.
8. Simon, G.S., and W.L. Dewey, Narcotics and diabetes. I. The effects of streptozotocin-induced diabetes on the antinociceptive potency of morphine, *J Pharmacol Exper Ther* 218: 318-323, 1981.
9. Simon, G.S., J. Borzelleca, and W.L. Dewey, Narcotics and diabetes. II. Streptozotocin-induced diabetes selectively alters the potency of certain narcotic analgesics. Mechanism of diabetes: morphine interaction, *J Pharmacol Exper Ther* 218: 324-329, 1981.
10. Bestetti, G., V. Locatelli, F. Tirone, G.L. Rossi, and E.E. Müller, One month of streptozotocin-diabetes induces different neuroendocrine and morphological alterations in the hypothalamo-pituitary axis of male and female rats, *Endocrinology* 117: 208-216, 1985.
11. Bestetti, G.E., C.E. Boujon, M.J. Reymond, and G.L. Rossi, Functional and morphological changes in mediobasal hypothalamus of streptozotocin-induced diabetic rats, *Diabetes* 38: 471-476, 1989.
12. Mooradian, A.D., J.E. Morley, C.J. Billington, M.F. Slag, M.K. Elson, and R.B. Shafer, Hyperprolactinemia in male diabetics, *Postgrad Med J* 61: 11-14, 1985.
13. Lorenzi, M., J.H. Karam, M.B. McIlroy, and P.H. Forsham, Increased growth hormone response to dopamine infusion in insulin-dependent diabetic subjects. Indication of possible blood-brain barrier

abnormalities, *J Clin Invest* 65: 146-153, 1980.

14. Lorenzi, M., E. Tsalikian, N.V. Bohannon, J.E. Gerich, J.H. Karam, and P.H. Forsham, Differential effects of L-Dopa and apomorphine on glucagon secretion in man: evidence against central dopaminergic stimulation of glucagon, *J Clin Endocrinol Metab* 45: 1154-1158, 1977.

15. Gerich, J.E., Role of growth hormone in diabetes mellitus, *N Engl J Med* 310: 848-850, 1984.

16. Johansen, K., and A.P. Hansen, Diurnal serum growth hormone levels in poorly and well-controlled juvenile diabetics, *Diabetes* 20: 239-245, 1971.

17. Zadik, Z., R. Kayne, M. Kappy, L.P. Plotnick, and A.A. Kowarski, Increased integrated concentration of norepinephrine, epinephrine, aldosterone, and growth hormone in patients with uncontrolled juvenile diabetes mellitus, *Diabetes* 29: 655-658, 1980.

18. Drash, A., J.B. Field, L.Y. Garces, F.M. Kenny, D. Mintz, and A.M. Vazquez, Endogenous insulin and growth hormone response in children with newly diagnosed diabetes mellitus, *Pediatr Res* 2: 94-102, 1968.

19. Burday, S.Z., P.H. Fine, and D.S. Schalch, Growth hormone secretion in response to arginine infusion in normal and diabetic subjects: relationship to blood glucose levels, *J Lab Clin Med* 71: 897-911, 1968.

20. Hansen, A.P., Abnormal serum growth hormone response to exercise in juvenile diabetics, *J Clin Invest* 49: 1467-1478. 1970.

21. Dasmahapatra, A., E. Urdanivia, and M.P. Cohen, Growth hormone response to thyrotropin-releasing hormone in diabetes, *J Clin Endocrinol Metab* 52: 859-862, 1981.

22. Ceda, G.P., G. Speroni, E. Dall'Aglio, G. Valenti, and U. Butturini, Nonspecific growth hormone responses to thyrotropin-releasing hormone in insulin-dependent diabetes: sex- and age-related pituitary responsiveness, *J Clin Endocrinol Metab* 55: 170-174, 1982.

23. Winter, R.J., L.S. Phillips, M.N. Klein, H.S. Traisman, and O.C. Green, Somatomedin activity and diabetic control in children with insulin-dependent diabetes, *Diabetes* 28: 952-954, 1979.

24. Yamashita, S., and S. Melmed, Effects of insulin on rat anterior pituitary cells. Inhibition of growth hormone secretion and mRNA levels, *Diabetes* 35: 440-447, 1986.

25. Kannel, W.B., and D.L. McGee, Diabetes and glucose tolerance as risk factors for cardiovascular disease: the Framingham study, *Diabetes Care* 2: 120-126, 1979.

26. Rapoport, S.I. (ed), *Blood-Brain Barrier in Physiology and Medicine*, Raven Press, New York, pp. 63 and 108, 1976.

27. Parving, H.H., Increased microvascular permeability to plasma proteins in short- and long-term juvenile diabetics, *Diabetes [Suppl 2]* 25: 884-889, 1976.

28. Palmberg, P.F., Diabetic retinopathy, *Diabetes* 26: 703-711, 1977.

29. Cunha-Vaz, J.G., J.R. Gray, R.C. Zeimer, M.C. Mota, B.M. Ishimoto, and E. Leite, Characterization of the early stages of diabetic retinopathy by vitreous fluorophotometry, *Diabetes* 34: 53-59, 1985.

30. White, N.H., S.R. Waltman, T. Krupin, and J.V. Santiago, Reversal of abnormalities in ocular fluorophotometry in insulin-dependent diabetes after five to nine months of improved metabolic control, *Diabetes* 31: 80-85, 1982.

31. Klein, R., R.L. Engerman, and J.T. Ernest, Fluorophotometry II. Streptozotocin-treated guinea pigs, *Arch Ophthalmol* 98: 2233-2234, 1980.

32. Kirber, K.M., C.W. Nichols, P.A. Grimes, A.I. Winegrad, and A.M. Laties, A permeability defect of the retinal pigment epithelium. Occurrence in early streptozotocin diabetes, *Arch Ophthalmol* 98: 725-728, 1980.

33. Tso, M.O.M., J.G. Cunha-Vaz, C.Y. Shih, and C.W. Jones, Clinicopathologic study of blood-retinal barrier in experimental diabetes mellitus, *Arch Ophthalmol* 98: 2032-2040, 1980.

34. Wallow, I.H.L., and R.L. Engerman, Permeability and patency of retinal blood vessels in experimental diabetes, *Invest Ophthalmol Vis Sci* 16: 447-461, 1977.

35. Bloodworth, J.M.B., and D.L. Molitor, Ultrastructural aspects of human and canine diabetic retinopathy, *Invest Ophthalmol Vis Sci* 4: 1037-1048, 1965.

36. Jakobsen, J., L. Malmgren, and Y. Olsson, Permeability of the blood-nerve barrier in the strep-tozotocin-diabetic rat, *Exper Neurol* 60: 277-285, 1978.

37. Malmgren, L.T., J. Jakobsen, and Y. Olsson, Permeability of blood-nerve barrier in galactose-fed rats, *Exper Neurol* 66: 758-770, 1979.

38. Lorenzi, M., D.P. Healy, R. Hawkins, J.M. Printz, and M.P. Printz, Studies on the permeability of the blood-brain barrier in experimental diabetes, *Diabetologia* 29: 58-62, 1986.

39. Stauber, W.T., S.H. Ong, R.S. McCuskey, Selective extravascular escape of albumin into the cerebral cortex of the diabetic rat, *Diabetes* 30: 500-503, 1981.

40. Heistad, D.D., and M.L. Marcus, Effect of sympathetic stimulation on the permeability of the blood-brain barrier to albumin during acute hypertension in cats, *Circ Res* 45: 331-338, 1979.

41. Guthrow, C.E., M.A. Morris, J.F. Day, S.R. Thorpe, and J.W. Baynes, Enhanced nonenzymatic glycosylation of human serum albumin in diabetes mellitus, *Proc Natl Acad Sci USA* 76: 4258-4261, 1979.

42. Williams, S.K., J.J. Devenny, and M.W. Bitensky, Micropinocytic ingestion of glycosylated albumin by isolated microvessels: possible role in pathogenesis of diabetic microangiopathy, *Proc Natl Acad Sci USA* 78: 2393-2397, 1981.

43. Ennis, S.R., J.E. Johnson, and E.L. Pautler, In situ kinetics of glucose transport across the blood-retinal barrier in normal rats and rats with streptozotocin-induced diabetes, *Invest Ophthalmol Vis Sci* 23: 447-456, 1982.

44. Gjedde, A., and C. Crone, Blood-brain glucose transfer: repression in chronic hyperglycemia, *Science* 214: 456-457, 1981.

45. McCall, A.L., W.R. Millington, and R.J. Wurtman, Metabolic fuel and aminoacid transport into the brain in experimental diabetes mellitus, *Proc Natl Acad Sci USA* 79: 5406-5410, 1982.

46. Harik, S.I, J.C. LaManna, Vascular perfusion and blood-brain glucose transport in acute and chronic hyperglycemia, *J Neurochem* 51: 1924-1929, 1988.

47. Brooks, D.J., J.S.R. Gibbs, P. Sharp, S. Herold, D.R. Turton, S.K. Luthra, E.M. Kohner, S.R. Bloom, and T. Jones, Regional cerebral glucose transport in insulin-dependent diabetic patients studied using $[^{11}C]$3-0-methyl-D-glucose and positron emission tomography, *J Cereb Blood Flow Metab* 6: 240-244, 1986.

48. Knudsen, G.M., J. Jakobsen, M. Juhler, and O.B. Paulson, Decreased blood-brain barrier permeability to sodium in early experimental diabetes, *Diabetes* 35: 1371-1373, 1986.

49. Greene, D.A., S.A. Lattimer, and A.A.F. Sima, Sorbitol, phosphoinositides, and sodium-potassium-ATPase in the pathogenesis of diabetic complications, *N Engl J Med* 316: 599-606, 1987.

50. Østerby, R., Basement membrane morphology in diabetes mellitus In M. Ellenberg and H. Rifkin (eds) *Diabetes Mellitus: Theory and Practice*, Medical Examination Publishing Company, New Hyde Park, NY, pp. 323-341, 1983.

51. Junker, U., C. Jaggi, G. Bestetti, and G.L. Rossi, Basement membrane of hypothalamus and cortex capillaries from normotensive and spontaneously hypertensive rats with streptozotocin-induced diabetes, *Acta Neuropathol* 65: 202-208, 1985.

52. Johnson, P.C., K. Brendel, and E. Meezan, Thickened cerebral cortical capillary basement membranes in diabetes, *Arch Pathol Lab Med* 106: 214-217, 1982.

53. Cagliero, E., M. Maiello, D. Boeri, S. Roy, and M. Lorenzi, Increased expression of basement membrane components in human endothelial cells cultured in high glucose, *J Clin Invest* 82: 735-738, 1988.

54. Vracko, R., R.E. Pecoraro, and W.B. Carter, Basal lamina of epidermis, muscle fibers, muscle capillaries, and renal tubules: changes with aging and in diabetes mellitus, *Ultrastruct Pathol* 1: 559-574, 1980.

55. Tilton, R.G, A.M. Faller, J.K. Brukhardt, P.L. Hoffman, C. Kilo, and J.R. Williamson, Pericyte degeneration and acellular capillaries are increased in the feet of human diabetic patients, *Diabetologia* 28: 895-900, 1985.

56. Bloodworth, J.M.B., A re-evaluation of diabetic glomerulo-sclerosis 50 years after the discovery of insulin, *Human Pathol* 9: 439-453, 1978.

57. Ashton, N., Studies of the retinal capillaries in relation to diabetic and other retinopathies, *Br J Ophthalmol* 47: 521-538, 1963.

58. De Oliveira, F., Pericytes in diabetic retinopathy, *Br J Ophthalmol* 50: 134-143, 1966.

59. Addison, D.J., A. Garner, and N. Ashton, Degeneration of intramural pericytes in diabetic retinopathy, *Br Med J* 1: 264-266, 1970.

60. Jakobsen, J., P. Sidenius, H.J.G. Gundersen, and R. Østerby, Quantitative changes of cerebral neocortical structure in insulin-treated long-term streptozotocin-induced diabetes in rats, *Diabetes* 36: 597-601, 1987.

CIRCULATING BLOOD GLUCOSE AND HYPOTHALAMIC-PITUITARY SECRETION

M. Grino, V. Guillaume, A. Caraty,[1] B. Conte-Devolx,
P. Joanny, F. Boudouresque, G. Pesce, J. Steinberg,
G. Peyre, A. Dutour, P. Giraud, and C. Oliver

Laboratoire de Neuroendocrinologie Expérimentale
INSERM U 297
Faculté de Médecine Nord
Bd P. Dramard
13326 Marseille Cédex 15, France

Introduction

Normally glucose accounts for more than 90% of the metabolic fuel of the brain (1). A constant supply and utilization of glucose is essential for normal cerebral metabolism since brain carbohydrate stores are very small. Hypoglycemia as well as diabetes mellitus result in a variety of neuroendocrine alterations probably through changes in the rate of glucose uptake and metabolism in the hypothalamus. At the hypothalamic-pituitary level, two metabolic conditions can be realized: either increased glucose disposal after administration of exogenous glucose, or glucopenia due to hypoglycemia or diabetes. In the latter condition, blood glucose increased, but cannot be utilized in brain cells because of the lack of insulin secretion. Changes in pituitary hormone release under these acute or chronic alterations in blood glucose levels have been well characterized. However, the role of the hypothalamus in driving these variations in pituitary function is still controversial due to difficulties in measuring the secretion of the hypophysiotropic factors. Therefore, the mechanisms by which hyper- or hypoglycemia exerts an influence on hypothalamic neurons remain to be determined. In this report, we will briefly review this tropic including some results from our laboratory, keeping in mind the potential clinical applications of animal studies.

Effect of Acute Hyperglycemia on Hypothalamic-Pituitary Secretion

The effect of acute hyperglycemia on the release of pituitary hormones has been investigated mainly in human and only in a limited series of animal studies (Table 1). Hyperglycemia does not change basal and stimulated luteinizing hormone (LH), follicle stimulating hormone (FSH), and prolactin (PRL) secretion. Indeed, in a group of insulin-dependent diabetic patients, increments in serum LH, FSH, thyroid stimulating hormone (TSH), and PRL after luteinizing hormone-releasing hormone (LHRH) and thyrotopin-releasing hormone (TRH) injection are similar under euglycemic and hyperglycemic (14-18 mmol/liter) steady states clamped by feedback control with an

---

[1]Laboratoire de Neuroendocrinologie, Physiologie de la Reproduction, INRA, 37380 Nouzilly, France

Circulating Regulatory Factors and Neuroendocrine Function
Edited by J. C. Porter and D. Ježová
Plenum Press, New York, 1990

Table 1
Effects of Acute Hyperglycemia on the
Secretion of Pituitary Hormones

Hormone	Human	Rat
LH	↔	?
FSH	↔	?
PRL	↔	?
TSH	↓	?
GH	↓	↔
ACTH	↓	?
AVP	?	?

↓ Decrease
↑ Increase
↔ No change
? Not tested

artificial pancreas device. In the same subjects, the growth hormone (GH) response to arginine is suppressed during the hyperglycemic clamp. Since the hypothalamus is thought to be the site of action of arginine, these data suggest that elevated blood glucose concentrations exert a modulatory influence on the hypothalamus (2). In several studies, it has been shown that basal and growth hormone-releasing hormone (GHRH)-stimulated GH secretion are suppressed by intravenous or oral glucose administration. The availability of more sensitive assay methods for TSH has made it possible to demonstrate that plasma TSH is also lowered under hyperglycemia. Since TSH release in inhibited by somatostatin, the latter data suggest that glucose may stimulate the secretion of hypothalamic somatostatin in human with a subsequent inhibiting action on GH secretion. However, there is no direct evidence for this hypothesis (3).

The ingestion of 100 g glucose induces a significant fall in plasma cortisol and suppresses the enhancement of cortisol response to the combination of physical exercise and propranolol infusion. This effect is probably mediated by an action of glucose on adrenocorticotropic hormone (ACTH) secretion (4). However, one does not know if glucose acts directly at the pituitary level or indirectly through the hypothalamus, since no information is available as yet on the influence of hyperglycemia on corticotropin releasing factor (CRF) and arginine vasopressin (AVP) secretion.

Effect of Diabetes Mellitus on Hypothalamic-Pituitary Secretion

Most animal studies have been performed in streptozotocin (STZ)-induced diabetic rats. Only a few data are available in the spontaneously diabetic *(BB/W)* rat. Hormonal investigations have also been conducted in type 1 human diabetes. However, one must be cautious in the interpretation of neuroendocrine alterations found in STZ-diabetes rats and in the applications of these results to human diabetes. Indeed, metabolic changes other than an increase in blood glucose levels are present in STZ-diabetic rats, including increased ketonic bodies and changes in plasma lipids and ions. These changes may alter hypothalamic functions. Reduced food intake and body weight are also present and may

Table 2
Effects of Diabetes on the
Secretion of Pituitary Hormones

Hormone	Human	Rat
LH	↔ or ↓	↓
FSH	↔ or ↓	↓
PRL	↔ or ↑	↓
TSH	↔	↓
GH	↑	↓
ACTH	?	↑
AVP	↑	↑

↓ Decrease
↑ Increase
↔ No change
? Not tested

play a similar role. Furthermore, a decrease in the levels of peripheral hormones (thyroxine and gonadal steroids) are found in these animals and may indirectly generate neuroendocrine alterations. Besides, the general condition of STZ-diabetic rats is very bad in general and cannot be compared with the condition of type 1 diabetes, even poorly equilibrated. The data available in the literature are summarized in Table 2. In the rat, diabetes influences the release of most pituitary hormones. In contrast, the alterations in the secretion of these hormones are found only in a limited number of type 1 diabetic patients. Whenthey are present, they are less marked than in the rat. Some discrepancies have even been noticed in the results of experiments performed in human and in rat.

TSH secretion is unaltered in humans. In STZ-diabetic rats, there is a marked reduced secretion of T3, T4, and TSH as well as a blunted TSH response to TRH (5). Morphological changes in hypothalamic TRH neurons have been detected suggesting a central origin for the hypothyroidism of STZ-diabetic rats (6). The TRH binding capacity of pituitary plasma membranes is unchanged (7).

In STZ-diabetic rats, basal PRL secretion is reduced as well as its response to TRH or estradiol injection (8). These rats display an enhanced activity of the tuberoinfundibular dopaminergic (TIDA) neurons (9). The subsequent increased secretion of hypothalamic dopamine may be responsible for the diminution of plasma PRL levels. Similar changes have been found in hypothyroid rats, but Reymond and Lemarchand-Béraud (10) have provided evidence that the hyperactivity of the TIDA neurons in the STZ-diabetic rats was not solely caused by hypothyroidism. In human diabetes, plasma PRL levels are generally normal although a mild hyperprolactinemia is found in about 18% of the patients (11). There is no explanation for the discrepancy between human and rat studies.

The data available on the hypothalamic-hypophysial-adrenal axis are more limited. An increased adrenal activity has been reported in a few studies (12). Increased basal levels of corticosterone have been found in long-term chemical (alloxan-treated and STZ-treated rats) and genetic (ob/ob mice) diabetes (13,14). STZ-treated rats have also been shown to have chronic hypersecretion of ACTH accompanied by defective suppression of ACTH by glucocorticoids (15). In our laboratory, we found no difference in circulating ACTH and corticosterone levels determined during surgical stress in 5- and 9-STZ diabetic rats. However, a moderate but significant reduction in the *in vitro* release of hypothalamic CRF has been demonstrated 5 days but not 9 days after the injection of STZ (P. Johanny *et al.*, unpublished observations). Such measurements have not been performed at later stages after STZ injection. A marked depletion of β-endorphin stores in the hypothalamus and neurointermediate lobe, but not in the anterior pituitary has been reported after one month in STZ-diabetes rats. Depletion of β-endorphin stores was returned to normal with insulin treatment (16).

In several studies, it has been shown that plasma AVP increases in diabetic rats. It is not clear from these studies whether the increase in AVP concentrations results from osmotic or hemodynamic disturbances. Spontaneously diabetic *BB/W* rats display a much higher increase in plasma AVP than STZ-diabetic rats. In these animals, a similar increase in plasma glucose and osmolarity was observed. However, a significant reduction in arterial blood pressure was found only in *BB/W* rats. This change in arterial blood pressure may account for the stronger stimulation of AVP release in genetic diabetes mellitus. This suggests that both osmotic and hemodynamic changes are involved in the enhancement of AVP secretion during diabetes. Since plasma sodium variations are moderate in these animals, the increase in plasma glucose is likely the main component of the osmotic stimulus (17). In type 1 diabetic patients, basal AVP levels are similar to those measured in the control subjects; however, the AVP response to insulin-induced hypoglycemia is twofold to threefold higher than in controls (18).

Diabetes mellitus has often been associated with reproductive disorders and alterations of gonadal function both in humans and in animals. Most experiments in animals have been performed in one-month STZ-diabetic rats. In these animals, plasma gonadal steroids are significantly lowered and there is now evidence for a defect in the hypothalamic control of their pituitary-gonadal axis. In male rats, the LHRH content of the median eminence and the plasma LH response to exogenous LHRH are unaltered. Plasma LH is decreased under basal conditions and after treatment by naloxone, a drug which acts through stimulation of LHRH release. In female rats, the median eminence LHRH content and the plasma LH response to exogenous LHRH or naloxone are reduced. In this study, Bestetti *et al.* (19) have shown that these alterations are diabetes- and not nutrition-dependent. When hypothalami from diabetic animals were examined morphologically, degenerated axons, mainly of the LHRH type, were found in the arcuate nucleus and the median eminence. These hypothalamic lesions were more pronounced in male than in female rats. Functional and morphological changes in LHRH have recently been re-evaluated by Bestetti *et al.* (20) again in one-month STZ-induced diabetes. By light microscopy, the dilated axon cross-sections are more numerous in the basal arcuate nucleus and in the median eminence of diabetic rats. By electron microscopy, the ratio of exocytosis to neurosecretory granules observed in the median eminence axon cross-sections is lower in the diabetic group. The total LHRH immunoreactivity, the number of labeled axons, and the amount of positive material in the axons are reduced as compared to controls. In diabetic rats, basal *in vitro* LHRH release from the mediobasal hypothalamus does not change. However, it is markedly reduced, as compared to controls, after incubation in K^+-enriched medium.

In humans, several studies have been performed for a better understanding of the pituitary gonadal disorders in diabetes mellitus. Plasma testosterone levels are normal in most male patients (21). Similarly, there is no alteration in the menstrual cycle of most diabetic women. However, amenorrhea is present in some of these patients. Diabetic women with functional amenorrhea have low basal LH levels, decreased LH response to LHRH despite low estrogen levels, and low basal and metoclopramide-stimulated PRL levels (22). These hormonal changes may be due in part to an increased hypothalamic dopaminergic activity leading to a depression of pituitary-driven ovulatory mechanisms.

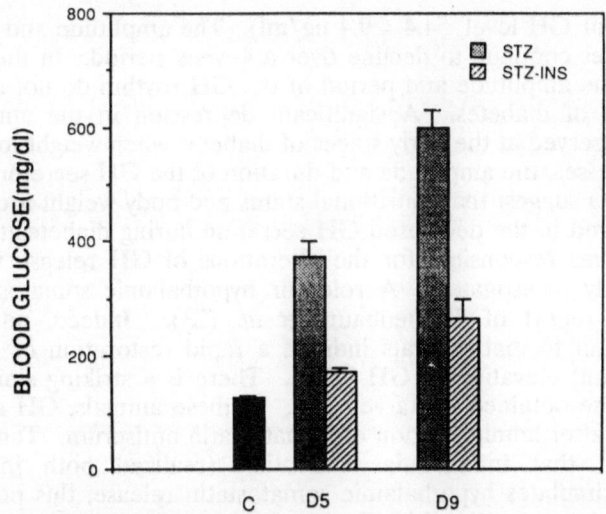

Figure 1. Blood glucose levels in controls (C), streptozotocin-diabetic (STZ) rats, and streptozotocin-diabetic rats treated with insulin (STZ-INS). Blood glucose was determined 5 and 9 days after streptozotocin injection (65 mg/kg BW, ip). Values are mean ± SE.

The secretion pattern of GH during diabetes has been extensively investigated in rats and in humans. Tannenbaum (23) has studied the effects of STZ-induced diabetes on the dynamics of GH secretion in chronically cannulated rats. Normal rats display a typical ultradian rhythm of GH secretion with most peak GH values greater than 500 ng/ml (mean 6 hour GH level, 125.3 ± 7.2 ng/ml). In diabetic rats, a significant depression in the amplitude of GH pulses is observed as early as 18 hours after STZ

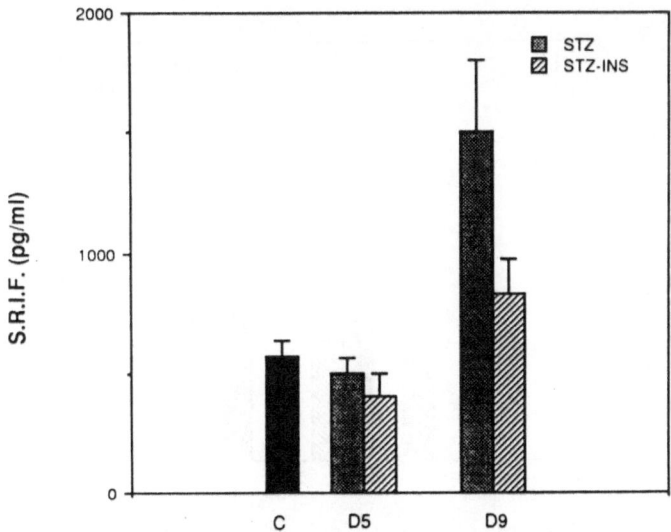

Figure 2. Somatostatin concentrations in hypophysial portal blood from controls (C), streptozotocin-diabetic rats (STZ), and streptozotocin-diabetic rats treated with insulin (STZ-INS), 5 and 9 days after streptozotocin injection. Values are mean ± SE.

injection (mean 6 hour GH level, 54.4 ± 9.4 ng/ml). The amplitude and duration of the GH secretory episodes continue to decline over a 4-week period. In the spontaneously diabetic *BB/W* rat, the amplitude and period of the GH rhythm do not change within 5 days after the onset of diabetes. A significant depression in the amplitude of GH secretory bursts is observed at the early stages of diabetes when weight loss is important. As the disease progresses, the amplitude and duration of the GH secretory peaks decline markedly. These data suggest that nutritional status and body weight are, in addition to hyperglycemia, involved in the decreased GH secretion during diabetes (24).

The mechanisms responsible for the alterations of GH release in diabetic rats have been extensively investigated. A role for hypothalamic somatostatin has been suspected after the report of Tannenbaum *et al.* (23). Indeed, administration of somatostatin antiserum to diabetic rats induces a rapid restoration of high amplitude pulses and a significant elevation of GH levels. There is a striking similarity between these results and those obtained in starved rats. In these animals, GH secretion is low and can be restored after administration of somatostatin antiserum. Tannenbaum *et al.* (23) have proposed that intracellular starvation (realized both in diabetic and food-deprived rat) stimulates hypothalamic somatostatin release, this peptide being an important sensor of changes in nutritional status.

In our laboratory, we have directly investigated the secretion of hypothalamic somatostatin in STZ-diabetic rats. These studies have been performed both *in vivo* and *in vitro* using the following paradigms:

—control rats (with normal blood glucose levels);
—5- and 9-day STZ-diabetic rats;
—5- and 9-day STZ-diabetic rats receiving daily replacement insulin therapy.

Blood glucose levels in these groups of animals are shown in Figure 1.

Somatostatin levels in hypophysial portal blood (HPB) were measured after urethan-γ-hydroxybutyrate anesthesia, which does not suppress the normal pulsatility of

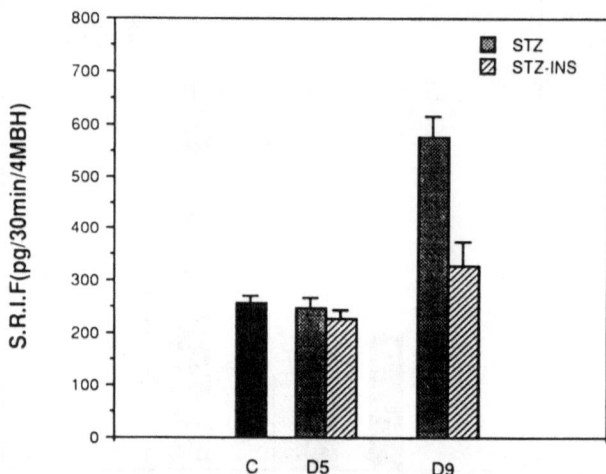

Figure 3. Somatostatin secretion during *in vitro* incubation from controls (C), streptozotocin-diabetic rats (STZ), and streptozotocin-diabetic rats treated with insulin (STZ-INS), 5 and 9 days after streptozotocin injection. Values (expressed in pg/4 hypothalamic fragments/30 min) are mean ± SE.

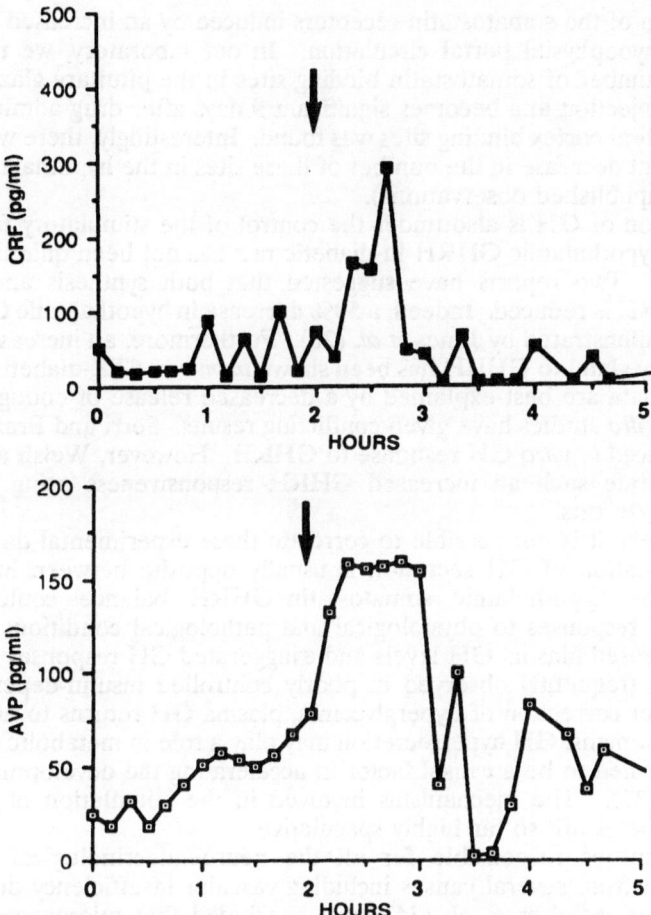

Figure 4. CRF and AVP secretion into the hypophysial portal blood of ram #557 during the first 5 hours (1-5) of the experiment. The *arrow* indicates the timing of insulin injection (0.2 IU/kg BW, iv).

GH secretion. Five days after STZ injection, somatostatin release was unaltered. Nine days after STZ injection, a threefold increase in portal somatostatin was found. This increase was significantly reduced after substitutive insulin therapy (Figure 2). Similar data have been obtained by measuring the *in vitro* somatostatin release from hypothalamic fragments (Figure 3) (25). The 250% enhancement of hypothalamic somatostatin mRNA levels found by Jones *et al.* (26) in STZ-diabetic rats is in good agreement with our findings. Therefore, both synthesis and secretion of hypothalamic somatostatin appears to increase in STZ-diabetic rats.

The action of somatostatin on GH release from the pituitary gland is decreased in diabetic rats. Indeed, Welsh and Szabo (27) have shown recently an impaired somatostatin suppression of stimulated GH release in cultured adenohypophysial cells of spontaneously diabetic *BB/W* rats. Prolonged exposure to somatostatin of a similar preparation from control rats *in vitro* produces a similar resistance. This suggests that the reduction in pituitary sensitivity to somatostatin found in diabetic rats may be consecutive to a desensitization of the somatostatin receptors induced by an increased release of this peptide into the hypophysial portal circulation. In our laboratory, we found that the reduction in the number of somatostatin binding sites in the pituitary gland is moderate 5 days after STZ injection and becomes significant 9 days after drug administration. No change in the cerebral cortex binding sites was found. Interestingly, there was in the same animals a significant decrease in the number of these sites in the hypothalamus (G. Pesce and colleagues, unpublished observations).

The secretion of GH is also under the control of the stimulatory factor, GHRH. The secretion of hypothalamic GHRH in diabetic rats has not been quantified as yet for technical reasons. Two reports have suggested that both synthesis and secretion of hypothalamic GHRH is reduced. Indeed, a 50% decrease in hypothalamic GHRH mRNA levels has been demonstrated by Jones *et al.* (26). Furthermore, an increased responsiveness of the pituitary gland to GHRH has been shown *in vivo* in STZ-diabetic and in *BB/W* rats (28). These data are best explained by a decreased release of endogenous GHRH into the HPB. *In vitro* studies have given conflicting results. Serri and Brazeau (29) have reported an enhanced *in vitro* GH response to GHRH. However, Welsh and Szabo (27) failed to demonstrate such an increased GHRH responsiveness using pituitary cells prepared from *BB/W* rats.

Unfortunately, it is not possible to correlate these experimental data with human studies. The regulation of GH secretion is usually opposite between human and rat. Differences in the hypothalamic somatostatin/GHRH balance could explain the differences in GH responses to physiological and pathological conditions between both species (30). Increased plasma GH levels and exaggerated GH responses to provocative stimuli have been frequently observed in poorly controlled insulin-dependent diabetic patients (31). After correction of hyperglycemia, plasma GH returns to normal within a few days (32). In humans, GH hypersecretion may play a role in metabolic derangements, and it is also suggested to be a causal factor in accelerating the development of diabetic microangiopathy (33). The mechanisms involved in the stimulation of GH secretion during human diabetes are so far highly speculative.

The mechanisms responsible for all the neuroendocrinological alterations in diabetes can result from several causes including vascular insufficiency due to microangiopathy. However, Affolter *et al.* (34) have concluded that microangiopathy plays a relatively minor role in diabetic lesion of the hypothalamus and CNS. The blood-brain barrier for several substrates (glucose, β-hydroxybutyrate, choline, sodium) is altered during diabetes and may participate in neuronal dysfunction (12).

The general cause of diabetic neuropathy is unknown although it has been extensively investigated at the level of peripheral nerves. Currently it is thought that early abnormalities in axon function, such as delayed axonal flow and slowed nerve conduction times are due to metabolic rather than to structural abnormalities. In diabetes, hyperglycemia lowers the myoinositol content of Schwann cells and axons, probably by increasing sorbitol-fructose content of the nerve via the polyol pathway. The sequence appears to be (35):

> *hyperglycemia* → *increased sorbitol-fructose* →
> *decrease myoinositol in Schwann cells and axons* →
> *decreased phosphoinositol turnover* → *decreased*
> $(Na^+ + K^+)$-*ATPase activity* → *abnormal energy*
> *metabolism, nerve dysfunction, structural damage.*

Diabetes may induce a *hypothalamic neuropathy* similar to the well established peripheral neuropathy. A variety of other neurobiochemical alterations have been identified in the CNS of diabetic animals. Glucose utilization and glycogen content as well

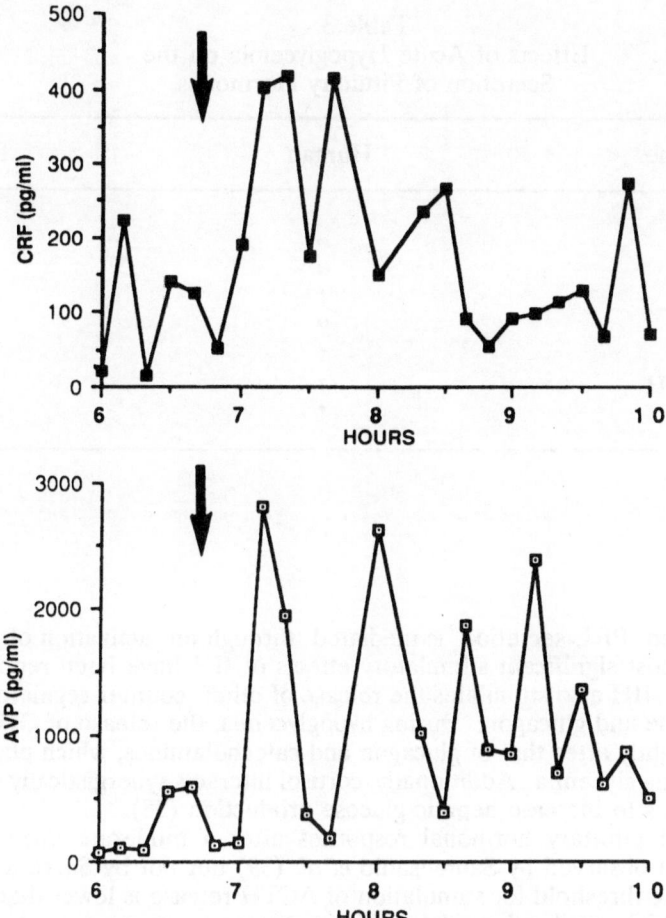

Figure 5. CRF and AVP secretion into the hypophysial portal blood of ram #557 during the last 5 hours (6-10) of the experiment. The *arrow* indicates the timing of insulin injection (2 IU/kg BW, iv).

as the level of aromatic amino acids are decreased while the levels of branched-chain amino acids are increased. Changes in brain neurotransmitters have also been documented. Diabetic animals have a reduced rate of both serotonin and dopamine synthesis within the brain. The concentrations of norepinephrine in the hypothalamus and cortex are increased. Such alterations in the level of brain monoamines may play a role in the genesis of hypothalamic dysfunction (12).

Effect of Hypoglycemia on Hypothalamic-Pituitary Secretion

Insulin-induced hypoglycemia (IIH) alters the secretion of several pituitary hormones (Table 3) and is frequently used as a test in the clinical investigation of pituitary disorders. No change in plasma LH and FSH has been detected (36). No variation in plasma TSH has been reported. A moderate but inconsistent increase in plasma PRL has been described in human (37). There is indirect evidence suggesting that hypo-

Table 3
Effects of Acute Hypoglycemia on the
Secretion of Pituitary Hormones

Hormone	Human	Rat
LH	↔	?
FSH	↔	?
PRL	↑	?
TSH	?	?
GH	↑	↓
ACTH	↑	↑
AVP	↑	↑

↓ Decrease
↑ Increase
↔ No change
? Not tested

glycemia-stimulated PRL secretion is mediated through an activation of hypothalamic serotonin. The most significant stimulatory effects of IIH have been reported for GH, ACTH, and AVP. IIH also stimulates the release of other counter-regulatory hormones, such as epinephrine and glucagon. During hypoglycemia, the release of GH, ACTH, and cortisol occurs slightly after that of glucagon and catecholamines, which play major roles in correcting plasma glycemia. Additionally, cortisol interacts synergistically with glucagon and catecholamines to increase hepatic glucose production (38).

Changes in pituitary hormonal responses after a moderate decrease in blood glucose have been observed by Santeusanio *et al.* (39) but not by Sacca and colleagues (40). The glycemic threshold for stimulation of ACTH release is lower than that for GH or epinephrine, which implies that a lower blood glucose is necessary to stimulate ACTH release than other pituitary hormones (41).

Acute and pronounced IIH is a potent activator of the hypothalamic-pituitary-adrenal axis in several species including humans (42), dogs (43), and rats (44). The mechanisms of this phenomenon and the factors responsible for it are not fully elucidated. They can involve peripheral and/or central pathways. Initially, the hypothalamus was thought to play a dominant role. However, in one report it was proposed that IIH stimulates ACTH release by a mechanism in which peripheral catecholamines act directly on the anterior pituitary gland (45). In contrast Ježová *et al.* (46) have shown that the rise in plasma ACTH during IIH was reduced in rats with lesions of the mediobasal hypothalamus and that propranolol pre-treatment was ineffective both in sham-operated and lesioned animals, suggesting that the hypoglycemia-induced ACTH release is mediated mostly by hypothalamic factors. Later on, the participation of the hypothalamus in the stimulation of ACTH release during hypoglycemia has been demonstrated by the measurement of CRF and AVP secretion in rat HPB. Nevertheless, a discrepancy was noticed between two studies. Plotsky *et al.* (47) have reported that IIH in rats results in no change in HPB CRF levels, but causes significant increase in AVP concentrations. Since they found that immunoneutralization of endogenous CRF can suppress plasma ACTH levels, they suggested that CRF plays a permissive role and that AVP is the mediator hypoglycemia-induced ACTH secretion in the rat. In contrast, Guillaume *et al.* (48) have observed, under comparable experimental conditions, that CRF concentrations in HPB increase significantly after hypoglycemia. The discrepancy between both investigations may be explained by the methodological approach used for collecting HPB in the rat. This method requires major surgery under deep anesthesia with consequent strong activation of pituitary-adrenal hormones secretion. Furthermore, sample volumes in rats are restricted because of the size of the animal and the IIH must be kept constant for at least 60 minutes in order to allow CRF and AVP determinations in HPB. Therefore, we studied the effect of IIH on CRF and AVP release into HPB using a more physiological animal model, namely unrestrained, conscious sheep (49). This study (50) has been performed in castrated rams injected at 4 hour intervals with two different doses of insulin:

—*First*, the low dose of insulin (0.2 U/kg body weight intraperitoneally) induced a moderate and short-lasting hypoglycemia, reproducing the test used in clinical investigation of pituitary function. Under that condition, a moderate increase of ACTH and cortisol was found. CRF and AVP release into HPB were similarly activated and the molar portal CRF/AVP ratio remained stable;

—*Second*, the high dose of insulin (2 U/kg body weight intraperitoneally) induced a deeper and long-lasting hypoglycemia, reproducing clinical states of severe hypoglycemia or hypoglycemic coma. Under this condition, a further stimulation of CRF and AVP secretion in HPB was observed. However, the increase in AVP levels was much more pronounced leading to a significant decrease in the CRF/AVP ratio.

Table 4 shows the mean integrated hormonal pattern after the injection of the low and high doses of insulin in 7 animals. Figures 4 and 5 depict the pattern of CRF and AVP secretion into hypophysial portal blood in one representative animal under the same experimental conditions.

These results indicate that hypoglycemia acts through CRF and AVP to stimulate ACTH secretion. They do not exclude the possible role of peripheral factors. Taken together with other observations, these data suggest that an increased CRF secretion is responsible for ACTH stimulation, whereas the increased AVP release potentiates the stimulatory effect of CRF.

The afferent pathways responsible for the sensation of altered plasma glucose concentrations and the subsequent stimulation of CRF, AVP, and ACTH secretion have not yet been described. The response to hypoglycemia may be mediated by hypothalamic and/or extra-hypothalamic glucoreceptors which affect the firing of hypothalamic CRF neurons. Indeed, glucose-sensitive neurons have been located in the ventromedial and lateral hypothalamic nuclei (51). However, reversal of hypothalamic glucopenia by intracarotid glucose infusion, which prevents stimulation of hypothalamic glucose receptors by hypoglycemia, reduces but does not eliminate the ACTH response to insulin injection (43). We and others have recently demonstrated that central catecholamines have a stimulatory effect on CRF release into HPB (52-54). Interestingly, hypothalamic norepinephrine activity increases during IIH in rats, and this increase parallels the variations in plasma corticosterone (55). Thus, an increase in central catecholamine activity may be responsible for the increased secretion of hypothalamic CRF (and AVP) following IIH.

In human, IIH also stimulates the secretion of GH. The mechanisms of the activation of somatotrophs is unknown (increased GHRH secretion or decreased somatostatin secretion?). The regulation of GH secretion in rat is in most cases in opposition with that observed in human. Thus, plasma GH levels decrease after IIH in

Table 4
Hormonal Secretory Pattern During
Insulin-Induced Hypoglycemia (IIH)

	Base	First Hour after IIH	Second Hour after IIH
ACTH	4.98 ± 1.31	10.10 ± 1.97	8.90 ± 1.22
Cortisol	0.83 ± 0.01	1.65 ± 0.78	2.12 ± 0.70
CRF	3.57 ± 0.78	10.23 ± 1.55	6.78 ± 1.55
AVP	1.64 ± 0.63	4.55 ± 1.46	4.30 ± 1.60
CRF/AVP	0.50 ± 0.18	0.52 ± 0.13	0.37 ± 0.10
ACTH	5.60 ± 1.15	35.28 ± 13.42	41.58 ± 15.28
Cortisol	1.03 ± 0.28	2.86 ± 0.63	3.61 ± 0.61
CRF	4.83 ± 0.88	23.05 ± 5.18	30.80 ± 8.75
AVP	2.68 ± 0.90	134.98 ± 34.38	137.42 ± 39.67
CRF/AVP	0.42 ± 0.13	0.039 ± 0.020	0.052 ± 0.015

The *upper panel* shows the effects of the low dose of insulin (0.2 U/kg BW). The *lower panel* shows the effects of the high dose of insulin (2 U/kg BW). Each value (pg/ml/min for ACTH, CRF, AVP; ng/ml/min for cortisol; molar ratio for CRF/AVP) is the mean ± SE for 7 animals.

the rat (56). This phenomenon has been attributed to an increased secretion of hypothalamic somatostatin. Indeed, during *in vitro* incubation of hypothalamic fragments, medium somatostatin was significantly higher when glucose concentrations in the medium were lowered from 20 to 2.5 mmol/liter (25,57). The stimulation of hypothalamic somatostatin release found in the absence of medium glucose can be replicated, in the presence of glucose, by agents that interfere with glucose uptake (3-0-methyl-D-glucose, phlorizin, cytochalasin-B) or with glucose metabolism (2-deoxy-D-glucose). Glucose influence on *in vitro* somatostatin secretion displays a regional specificity since cerebral cortical somatostatin is unaffected by perturbations of glucose concentration, uptake, or metabolism. Increased somatostatin release when glucose medium is low provides an explanation for the observation of decreased serum GH levels during IIH (58).

Any attempt to discuss glucose exclusively as a regulator or mediator of neuronal activity is limited. Glucose may directly affect neurons in the hypothalamus and the CNS. However, its effects may be modified by other factors (nutrients, metabolites, hormones, neurotransmitters) which vary during hypoglycemia. In addition to a direct effect of glucopenia on hypothalamic and brain cells, there are other possibilities for hypoglycemia to alter hypothalamic functions.

Changes in blood glucose levels may affect peripheral sites where glucose-responsive nerve elements have been neurophysiologically identified: liver, intestine, pancreas, etc. Nerve inputs from these viscera are carried by the vagus and glossopharyngeal nerve to the rostral part of the nucleus tractus solitarius. Then, information can proceed rostrally to the hypothalamus, either directly or through the parabrachial nucleus.

The glucose-responsive cells residing on the higher region of the brainstem, in the lateral hypothalamic area and in the ventromedial nucleus may also obtain systemic glucose information from the blood or the third ventricle (51).

Conclusions

The effects of changes in blood glucose upon the secretion of pituitary hormones have been well documented for several years. However, the involvement of hypothalamic neurons in driving such changes has been demonstrated only in recent years. Further investigations are clearly needed to:

—completely determine the role of the hypothalamus in all variations in pituitary function;

—trace the pathways through which changes in blood glucose influence hypothalamic neurons;

—link the putative metabolic changes in neuronal glucose metabolism and the neuronal secretory pathway of hypothalamic neuropeptides.

References

1. Pardridge, W.B., Brain metabolism. A perspective from the blood-brain barrier, *Physiol Rev* 63: 1481-1535, 1983.
2. Vierhapper, H., B. Grubeck-Loebenstein, P. Bratush-Marrain, S. Panzer, and W. Waldhausl, The impact of euglycemia and hyperglycemia on stimulated pituitary hormone release in insulin-dependent diabetics, *J Clin Endocrinol Metab* 52: 1230-1234, 1981.
3. Shibasaki, T., A. Masuda, M. Hotta, N. Yamauchi, N. Hizuka, K. Takano, H. Demura, and K. Shizume, Effects of ingestion of glucose on GH and TSH secretion: evidence for stimulation of somatostatin release from the hypothalamus by acute hyperglycemia in normal man and its impairment in acromegalic patients, *Life Sci* 44: 431-438, 1989.
4. Ježová-Repčeková, D., M. Vigaš, and I. Klimeš, Decreased plasma cortisol response to pharmacological stimuli after glucose load in man, *Endocrinol Exp* 14: 113-120, 1980
5. Bestetti, G.E., M.J. Reymond, I.V. Perrin, P.C. Kniel, T. Lemarchand-Béraud, and G.L. Rossi, Thyroid and pituitary secretory disorders in streptozotocin-diabetic rats are associated with severe structural changes of these glands, *Virchows Arch B Cell Pathol* 53: 69-78, 1987.
6. Bestetti, G.E., H.P. Jacob, C.E. Boujon, M.J. Reymond, and G.L. Rossi, Le diabète induit par la streptozotocine altère le fonctionnement et la morphologie de l' hypothalamus médiobasal du rat mâle: étude des axones à TRH dans l' éminence médiane à l' aide d' un modèle *in vitro, Congrès de la Société de Neuroendocrinologie Expérimentale, Anales d' Endocrinologie*, Rennes, September, 1988, p. 15 N.
7. Wilber, J.F., A. Banergi, C. Prasa, and M. Mori, Alterations in hypothalamic pituitary-thyroid regulation produced by diabetes mellitus, *Life Sci* 28: 1757-1763, 1981.
8. Tesone, M., R.G. Ladenheim, and E.H. Charreau, Alterations in the prolactin secretion in streptozotocin-induced diabetic rats. Correlation with pituitary and hypothalamus estradiol receptors, *Mol Cell Endocrinol* 43: 135-140, 1985.
9. Reymond, M.J., and T. Lemarchand-Béraud, Hyperactivity of the hypothalamic dopaminergic neurons and hyposecretion of prolactin in diabetic rats: influence of the thyroid status, *8th International Congress of Endocrinology*, July 17-23, Kyoto, Japan, Abstract 08-19-045, 1988.
10. Reymond, M.J., and T. Lemarchand-Béraud, Effects of thyroid hormones on the hypothalamic dopaminergic neurons, In J.C. Porter and D. Ježová (eds) *Circulating Regulatory Factors and Neuroendocrine Function*, Plenum Press, New York, pp.257-270, 1990.

11. Mooradian, A.D., J.E. Morlay, C.J. Billington, M.F. Slag, M.K. Elson, and R.R. Shafer, Hyperprolactinoemia in male diabetics, *Postgraduate Med J* 61: 11-14, 1985.

12. Mooradian, A.D., Diabetic complications of the central nervous system, *Endocrine Rev* 9: 346-356, 1988.

13. L'Age, M., J. Langholz, W. Fechner, and H. Salzman, Disturbances of the hypothalamo-hypophysial-adrenocortical system in the alloxan diabetic rat, *Endocrinology* 95: 760-765,, 1974.

14. De Nicola, A.F., O. Fridman, E.J. Del Castillo, and V.G. Foglia, The influence of streptozotocin diabetes on adrenal function in male rats, *Horm Metab Res* 8: 388-392, 1976.

15. De Nicola, A.F., O. Fridman, E.J. Del Castillo, and V.G. Foglia, Abnormal regulation of adrenal function in rats with streptozotocin diabetes, *Horm Metab Res* 9: 469-473, 1977.

16. Locatelli, V., F. Petraglia, N. Tirloni, and E.E. Müller, Beta-endorphin concentrations in the hypothalamus, pituitary and plasma of streptozotocin-diabetic rats with and without insulin substitution therapy, *Life Sci* 38: 379-386, 1986.

17. Brooks, D.P., D.F. Nutting, J.T. Crofton, and L. Share, Vasopressin in rats with genetic and streptozotocin-induced diabetes, *Diabetes* 38: 54-57, 1989.

18. Thompson, C.J., J. Thow, I.R. Jones, and P.H. Baylis, Vasopressin secretion during insulin-induced hypoglycemia: exaggerated response in people with type 1 diabetes, *Diabetic Med* 6: 158-163, 1989.

19. Bestetti, G.E., V. Locatelli, F. Tirone, G.L. Rossi, and E.E. Müller, One month of streptozotocin-diabetes induces different neuroendocrine and morphological alterations in the hypothalamo-pituitary axis of male and female rats, *Endocrinology* 117: 208-216, 1985.

20. Bestetti, G.E., C.E. Boujon, M.J. Reymond, and G.L. Rossi, Functional and morphological changes in mediobasal hypothalamus of streptozotocin-induced diabetic rats: *in vitro* study of LHRH release, *Diabetes* 38: 471-476, 1989.

21. Conte-Devolx, B., C. Oliver, and J.L. Codaccioni, Diabète et fonction Leydigienne, *Progrès en Andrologie* 2: 39-45, 1989.

22. Djursing H., C. Hagen, H.C. Hyholm, L. Carstensen, and A.N. Andersen, Gonadotropin responses to gonadotropin-releasing hormone and prolactin responses to thyrotropin-releasing hormone and metoclopramide in women and amenorrhea and insulin-treated diabetes mellitus, *J Clin Endocrinol Metab* 56: 1016-1021, 1983.

23. Tannenbaum, G.S., Growth hormone secretion dynamics in streptozotocin diabetes: evidence of a role for endogenous-circulating somatostatin, *Endocrinology* 108: 76-82, 1981.

24. Tannenbaum, G.S., E. Colle, W. Gurd, and L. Wanamaker, Dynamic time-course studies of the spontaneously diabetic BB Wistar rat. I. Longitudinal profiles of plasma growth hormone, insulin, and glucose, *Endocrinology* 109: 1872-1879, 1981.

25. Joanny, P., G. Peyre, J. Steinberg, B. Conte-Devolx, and C. Oliver, Secrétion hypothalamique de somatostatine chez les rats diabétiques, *Diabète et Métabolisme* (Abstract 76) 14: 168, 1988.

26. Jones, P.M., J.M. Burrin, Y. Yiangou, and S.R. Bloom, Altered synthesis of hypothalamic somatostatin and growth hormone releasing factor may explain growth hormone abnormalities in the streptozotocin diabetic rat, *Diabetic Med 5 [Suppl P20]*, 1988.

27. Welsh, J.B., and M. Szabo, Impaired suppression of growth hormone release by somatostatin in cultured adenohypophyseal cells of spontaneously diabetic BB/W rats, *Endocrinology* 123: 2230-2234, 1988.

28. Locatelli, V., S. Rovati, H. Miyoshi, and E.E. Müller, Growth hormone hyperresponsiveness to human pancreatic growth hormone releasing hormone in streptozotocin-diabetic rats, *Horm Metab Res* 16: 507, 1984.

29. Serri, O., and P. Brazeau, Growth hormone responsiveness *in vivo* and *in vitro* to growth hormone releasing factor in the spontaneously diabetic BB Wistar rat, *Neuroendocrinology* 46: 162-166, 1987.

30. Müller, E.E., Neural control of somatotropic function, *Physiolog Rev* 67: 962-1053, 1987.

31. Hansen, A.P., and K. Johansen, Diurnal pattern of blood glucose, serum free fatty acids, insulin, glucagon and growth hormone in normals and juvenile diabetics, *Diabetologia* 6: 27-33, 1970.

32. Press, M., Tamborlane, W.V., and R.S. Sherwin, Importance of raised growth hormone levels in mediating the metabolic derangements of diabetes, *N Engl J Med* 310: 810-815, 1984.

33. Holly, J.M.P., S.A. Amiel, R. Sandhu, L.H. Rees, and J.A.H. Wass, The role of growth hormone in diabetes mellitus, *J Endocrinology* 118: 353-364, 1988.

34. Affolter, V., P. Boujon, G. Bestetti, and G.L. Rossi, Hypothalamic and cortical neurons of normotensive and spontaneously hypertensive rats are differently affected by streptozotocin diabetes, *Acta Neuropathol* 70: 135-141, 1986.

35. Finegold, D., S.A. Lattimer, S. Nolle, M. Berstein, and D.A. Green, Polyol pathway activity and myo-inositol metabolism. A suggested relationship in the pathogenesis of diabetic neuropathy, *Diabetes* 32: 988-992, 1983.

36. Cumming, D.C., M.E. Quigley, and S.S.C. Yen, Acute suppression of circulating testosterone levels by cortisol in men, *J Clin Endocrinol Metab* 57: 671-673, 1983.

37. Whitaker, M.D., B. Corenblum, P.J. Taylor, and P.H. Harasym, Control of the hypoglycemia release of prolactin, In M. L' Hermite and S.J. Judd (eds) *Progress in Reproductive Biology: Advances in Prolactin, Volume 6*, Karger, Basel, pp. 77-82, 1980.

38. Cryer, P.E., Glucose counterregulation in man, *Diabetes* 30: 261-264, 1981.

39. Santeusanio, F., G. Bolli, M. Massi-Benedetti, P. Defeo, G. Angeletti, P. Compagnucci, G. Calabrese, and P. Brunetti, Counterregulatory hormones during moderate, insulin-induced, blood glucose decrements in man, *J Clin Endocrinol Metab* 52: 477-482, 1981.

40. Sacca, L., R. Sherwin, R. Hendler, and P. Felig, Influence of continuous physiologic hyperinsulinemia on glucose kinetics and counterregulatory hormones in normal and diabetic man, *J Clin Invest* 63: 849-857, 1979.

41. Schwartz, N.S., W.E. Clutter, S.D. Shah, and P.E. Cryer, Glycemic thresholds for activation of glucose counterregulatory systems are higher than the threshold for symptoms, *J Clin Invest* 79: 777-781, 1987.

42. Watabe, T., K. Tanaka, M. Kumagae, S. Itoh, F. Takeda, K. Morio, M. Hasegawa, T. Horiuchi, S. Miyabe, and N. Shimizu, Hormonal responses to insulin-induced hypoglycemia in man, *J Clin Endocrinol Metab* 65: 1187-1191, 1987.

43. Keller-Wood, M.E., C.E. Wade, J. Shinsako, L.C. Keil, G.R. Van Loon, and M.F. Dallman, Insulin-induced hypoglycemia in conscious dogs: effect of maintaining carotid arterial glucose levels on the adrencorticotropin, epinephrine and vasopressin responses, *Endocrinology* 112: 624-632, 1982.

44. Karteszi, M., M.F. Dallman, G.B. Markara, and E. Stark, Regulation of the adenocortical response to insulin-induced hypoglycemia, *Endocrinology* 111: 535-541, 1982.

45. Mezey, E., T.D. Reisine, M.J. Brownstein, M. Palkovits, and J. Axelrod, β-adrenergic mechanism of insulin-induced adrenocorticotropin release from the anterior pituitary, *Science* 226: 1085-1087, 1984.

46. Ježová, D., R. Kvetňanský, K. Kovács, Z. Opršalová, M. Vigaš, and G.B. Makara, Insulin-induced hypoglycemia activates the release of adrenocorticotropin predominantly via central and propranolol insensitive mechanisms, *Endocrinology* 120: 409-415, 1987.

47. Plotsky, P.M., T.O. Bruhn, and W. Vale, Hypophysiotropic regulation of adrenocorticotropin secretion in response to insulin-induced hypoglycemia, *Endocrinology* 117: 323-329, 1985.

48. Guillaume, V., M. Grino, B. Conte-Devolx, F. Boudouresque, and C. Oliver, Corticotropin-releasing factor secretion increases in rat hypophysial portal blood during insulin-induced hypoglycemia, *Neuroendocrinology* 49: 676-679, 1989.

49. Caraty, A., M. Grino, A. Locatelli, and C. Oliver, Secretion of corticotropin releasing factor (CRF) and vasopressin (AVP) into hypophysial portal blood of conscious, unrestrained rams, *Biochem Biophys Res Comm* 155: 841-849, 1988.

50. Caraty, A., M. Grino, A. Locatelli, V. Guillaume, F. Boudouresque, B. Conte-Devolx, and C. Oliver, Effect of insulin-induced hypoglycemia on corticotropin-releasing factor (CRF) and arginine vasopressin (AVP) secretion into hypophysial portal blood of conscious, unrestrained rams, *J Clin Invest* Submitted.

51. Oomura, Y., Glucose as a regulator of neuronal activity, *Adv Metab Disorders* 10: 31-65, 1983.

52. Guillaume, V., B. Conte-Devolx, A. Szafarczyk, F. Malaval, N. Pares-Herbute, M. Grino, G. Alonso, I. Assenmacher, and C. Oliver, The corticotropin-releasing factor release in rat hypophysial portal blood is mediated by brain catecholamines, *Neuroendocrinology* 46: 143-146, 1987.

53. Szafarczyk, A., F. Malaval, A. Laurent, R. Gibaud, and I. Assenmacher, Further evidence for a central stimulatory action of catecholamines on adrenocorticotropin release in the rat, *Endocrinology* 121: 883-892, 1987.

54. Plotsky, P.M., Facilitation of immunoreactive corticotropin-releasing factor secretion into the hypophysial

portal circulation after activation of catecholaminergic pathways or central norepinephrine injection, *Endocrinology* 121: 924-930, 1987.

55. Smythe, G.A., J.E. Bradshaw, M.V. Nicholson, H.S. Grunstein, and L.H. Storelien, Rapid bidirectional effects of insulin on hypothalamic noradrenergic and serotoninergic neuronal activity in the rat: role of glucose homeostasis, *Endocrinology* 117: 1590-1597, 1985.

56. Takahashi, K., W.H. Daughaday, and D.M. Kipnis, Regulation of immunoreactive growth hormone secretion in male rats, *Endocrinology* 88: 909-917, 1971.

57. Berelowitz, M., D. Dudlak, and L.A. Frohman, Release of somatostatin-like immunoreactivity from incubated rat hypothalamus and cerebral cortex, *J Clin Invest* 69: 1293-1301, 1982.

58. Berelowitz, M., N.C. Ting, and L. Murray, Glucopenia-mediated release of somatostatin from incubated rat hypothalamus: monosaccharide specificity and role of glycolytic intermediates, *Endocrinology* 124: 826-830, 1989.

NUTRITIONAL AND HEMODYNAMIC FACTORS INFLUENCING ADENOPITUITARY FUNCTION IN MAN

M. Vigaš, P. Tartár, D. Ježová, J. Jurčovičová,
R. Kvetňanský, J. Malatinsky,[1] and R. Tigranyan[2]

Institute of Experimental Endocrinology, Centre of Physiological Sciences
Slovak Academy of Sciences
Bratislava, Czechoslovakia

Hypothalamic-pituitary functions are characterized by great variations in their activity. These include not only regular programmed alternations according to biological rhythms (lunar, circadian, ultradian, as well as the sleep-wake cycle) but also irregular ones induced by the given requirements for maintaining the steady state in different physiological situations.

Hypothalamic-pituitary activity is predominantly under nervous control effected by means of neurotransmitters. Feedback mechanisms play important regulatory roles, and are particularly pronounced for tropic functions that directly control target glands but are less marked for hormones having direct effects on tissues [*e.g.*, prolactin (PRL)]. In addition to major control mechanisms, hypothalamic-pituitary functions are affected by factors that are usually of minor importance but may under certain conditions become important. The most prominent of these factors are *a)* extrahypothalamic releasing hormones and statins, *b)* peripheral hormones of certain target glands, *c)* nutritional factors, and *d)* hemodynamic changes in the central nervous system and in endocrine tissues.

Cells of the immune system and gastrointestinal tract as well as some tumor cells produce extrahypothalamic releasing hormones and statins. The involvement of these substances in the physiological control of the neuroendocrine system has not yet been fully elucidated.

Peripheral hormones produced by tissues other than the target gland of the corresponding hypothalamic-pituitary function exert specific effects upon the pituitary to stimulate its growth, stimulate the biosynthesis and secretion of pituitary hormones (as in the case of estrogens), or exert actions that include a broad range of effects that influence proteosynthesis (as in the case of thyroid hormones). The effects of pathologically increased hormone concentrations in blood and pharmacological effects of a hormone after its exogenous administration (*e.g.*, a corticosteroid) do not fall within this group. Circulating hormones with neurotransmitter actions can act as neuroendocrine regulators only upon opening of the blood-brain barrier (BBB), which otherwise prevents their passage from blood to cerebral tissue. Nutritional factors, such as an excess or a deficiency of glucose, certain amino acids, and non-esterified fatty acids, exert their effect at different levels of neuroendocrine regulation.

---

[1]Present affiliation: Department of Anesthesiology, Derer's Hospital, Bratislava, Czechoslovakia

[2]Present affiliation: Institute of Biomedical Problems, Moscow, Russia

Circulating Regulatory Factors and Neuroendocrine Function
Edited by J. C. Porter and D. Ježová
Plenum Press, New York, 1990

407

Hemodynamic factors have an indirect effect on neuroendocrine regulation through diminished perfusion of the brain or endocrine tissues. Changes in the systemic circulation alter the biosynthesis, secretion, transport and degradation of hormones. Thus, in interpreting a change in the concentration of a circulating hormone or assessing the causes and mechanisms involved in the change, hemodynamic factors must be considered.

In this presentation, we shall discuss the influence of nutritional factors upon neuroendocrine systems, and assess changes in the concentration of circulating hormones induced by hemodynamic factors.

General Conditions of the Investigation

Mostly healthy volunteers and outpatients were involved in the present study. The subjects presented for the examination on an empty stomach at 0730. After insertion of an indwelling catheter into the cubital vein, each subject waited in a sitting position for at least 30 minutes prior to withdrawal of the control blood sample. Blood for analysis of hormones and metabolic substrates was collected in cooled test tubes, and the plasma was stored at -20°C until assayed. The samples of one investigation were analyzed in one assay. Growth hormone (GH) and insulin were assayed by radioimmunoassay using commercial kits (METRONEX, Poland); PRL was assayed using reagents from CEA-IRE-SORIN. ACTH was determined as previously described (1). Blood glucose and free fatty acids were quantified using commercial kits (LACHEMA, Czechoslovakia). Fructose was measured by the method of Heyrovsky (2) using indoleacetic acid. The results in the

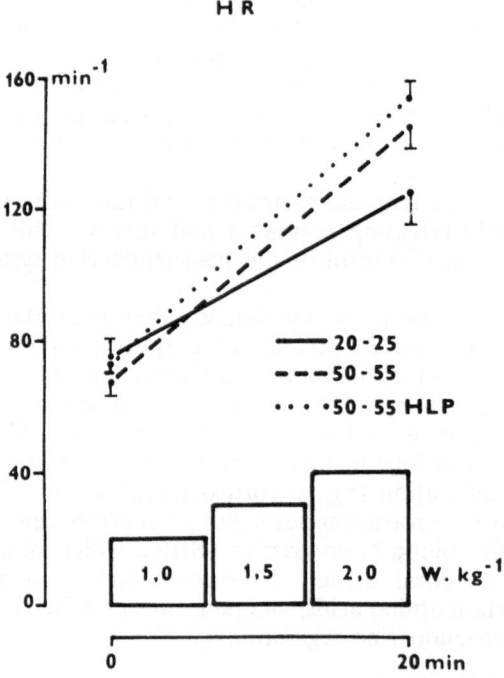

Figure 1.   Heart rate (HR) during graded exercise of seven hyperlipidemic patients *(dotted line)*, seven age-matched healthy controls *(broken line)*, and seven young healthy controls *(solid line)*.

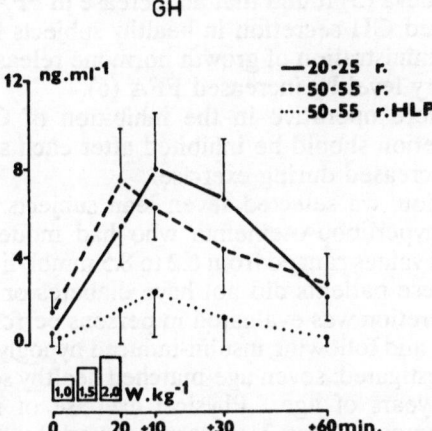

Figure 2. Plasma GH before and after graded exercise of seven hyperlipidemic patients *(dotted line)*, seven age-matched healthy controls *(broken line)*, and seven young healthy controls *(solid line)*. Mean and SE.

figures are given as mean ± SE. The data were analyzed for statistical signficance using paired and unpaired t tests.

Effect of Hyperlipidemia on Growth Hormone Secretion

It is known that in patients with hyperlipidemia there is no increase in GH secretion after exercise and certain other stimuli (3). A feedback relationship has been postulated to exist between GH and free fatty acids (FFA). Pharmacological reduction of circulating FFA results in GH release, whereas FFA elevations reduce or block GH

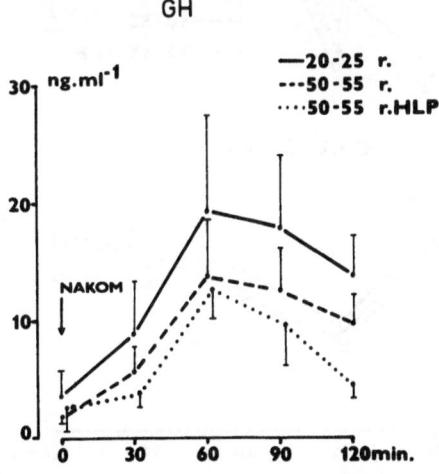

Figure 3. Plasma GH, before and after oral administration of 500 mg of L-dopa (Nakom, GALENIKA) of seven hyperlipidemic patients *(dotted line)*, seven age-matched healthy controls *(broken line)*, and seven young healthy controls *(solid line)*. Mean and SE.

secretion in man (4). Casanueva (5) found that an increase in FFA induced by an infusion of lipid and heparin, inhibited GH secretion in healthy subjects following exercise. GH release in response to an administration of growth hormone releasing hormone was found to be blocked at the pituitary level by increased FFA (6).

If this mechanism were operative in the inhibition of GH secretion in hyperlipidemic patients, GH secretion should be inhibited after each stimulus, and circulating FFA should be markedly increased during exercise.

To test this assumption, we selected seven lean subjects, 50 to 55 years of age, with type IIb or type IV hyperlipoproteinemia who had moderately elevated plasma cholesterol levels (individual values ranged from 6.2 to 8.5 mmol/liter) and/or triglycerides (1.9 to 5.6 mmol/liter). These patients did not have diabetes or any other metabolic or endocrine disorder. GH secretion was evaluated in persons performing physical exercise, after L-dopa administration, and following insulin-induced hypoglycemia. Simultaneously, two control groups were investigated: seven age-matched healthy subjects and seven young healthy students, 20 to 25 years of age. Physical exercise of moderate intensity was performed on a bicycle ergometer, using 3 consecutive work loads of graded intensity of 1.0, 1.5, and 2.0 W/kg body weight. Each exercise at a given work load lasted 6 minutes followed by a pause of 1 minute, so that the entire exercise period was of 20 minutes duration (Figure 1).

The heart rates of the patients and those of the age-matched control subjects were

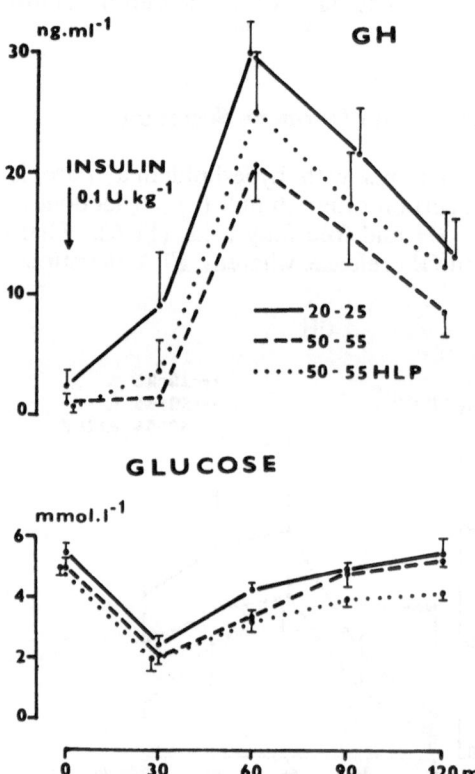

Figure 4. Plasma GH and glucose, before and after iv administration of insulin (0.1 U/kg BW), of seven hyperlipidemic patients *(dotted line)*, seven age-matched healthy controls *(broken line)*, and seven young healthy controls *(solid line)*. Mean and SE.

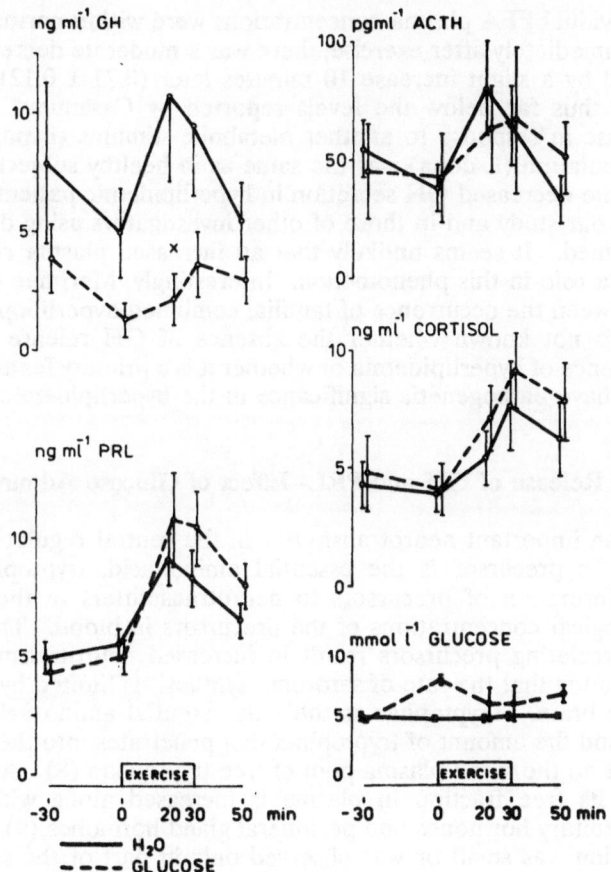

Figure 5. Plasma growth hormone (GH), prolactin (PRL), ACTH, cortisol, and glucose of eight healthy volunteers before and after graded exercise. The exercise was begun 30 min after oral administration of water (400 ml) or glucose solution (100 g glucose in 400 ml water). The difference in GH values in water-treated subjects *vs.* glucose-treated subjects is significant, $P < 0.01$. Mean and SE.

comparable during the exercise, suggesting comparable physical fitnesses (Figure 1). Yet, the post-exercise increase in GH secretion, which reached the same value in the two control groups, did not occur in the patients with hyperlipidemia (Figure 2).

The L-dopa test, which consisted of oral administration of 2 tablets of NAKOM (Galenika, Yugoslavia) containing 500 mg of L-dopa and 500 mg carbidopa (a peripheral inhibitor of L-dopa decarboxylation that does not penetrate the BBB), led to the same increase of plasma GH in the hyperlipidemic patients as in the two control groups (Figure 3).

Plasma GH after intravenous insulin administration (0.1 U/kg body weight of Pur Inzulin, SPOFA, Czechoslovakia) was increased within normal limits in hyperlipidemic patients and in the control subject (Figure 4).

In healthy subjects infused with lipid and heparin to increase the plasma concentration of FFA, Casanueva *et al.* (5) observed an inhibition of the post-exercise release of GH. The suppressive threshold level of FFA was estimated to be 3 mmol/liter.

In our study the individual FFA plasma concentrations were within normal range (0.67 ± 0.16 mmol/liter). Immediately after exercise, there was a moderate decrease (0.57 ± 0.06 mmol/liter) followed by a slight increase 10 minutes later (0.71 ± 0.12). Plasma FFA concentrations were thus far below the levels reported by Casanueva and associates. Moreover, GH release in response to another metabolic stimulus (hypoglycemia) or to pharmacological stimulation (L-dopa) was the same as in healthy subjects.

The basis of the decreased GH secretion in hyperlipidemic patients after physical exercise observed in our study and in those of other investigators using different stimuli, remains to be explained. It seems unlikely that an increased plasma concentration of circulating FFA has a role in this phenomenon. Interestingly, Merimee (7) reported an independent link between the occurrence of familial combined hyperlipoproteinemia and GH deficiency. It is not known whether the absence of GH release in response to exercise is a consequence of hyperlipidemia or whether it is a primary feature of endocrine regulation that may have pathogenetic significance in the hyperlipidemic state.

Tryptophan-Induced Release of GH and PRL: Effect of Glucose Administration

Serotonin is an important neurotransmitter in the central regulation of neuroendocrine functions. Its precursor is the essential amino acid, tryptophan. Enzymes catalyzing the transformation of precursors to neurotransmitters in the brain are not saturated at physiological concentrations of the precursors in blood. Therefore, higher concentrations of circulating precursors result in increased neurotransmitter synthesis. There are data indicating that the rate of serotonin synthesis is limited by the availability of tryptophan in the brain. Tryptophan is the only essential amino acid that binds to albumin in plasma, and the amount of tryptophan that penetrates into the brain seems to be closely dependent on the small plasma pool of free tryptophan (8). After administration of tryptophan, its free fraction in plasma is increased along with an increased secretion of adenopituitary hormones and peripheral gland hormones (9). In some cases the hormonal elevation was small or was observed only in part of the subjects studied. However, the transport of tryptophan into the brain not only depends on its free fraction in plasma but also on the ratio to other amino acids competing for the same transport mechanism through the BBB (10). These include valine, isoleucine, leucine, and phenylalanine. Insulin, which reduces plasma amino acids by increasing their uptake into peripheral tissues, should enhance the transport of tryptophan into brain tissue.

We investigated the effect of tryptophan in the presence and absence of exogenous glucose on the release of GH and PRL. Tryptophan was administered orally (150 mg/kg body weight) in 300 ml of water or in 300 ml of water containing 75 g glucose. The study was conducted on seven healthy women, 19 to 37 years of age, and was performed between days 12 and 17 of their ovarian cycle. Each subject was examined twice (tryptophan without glucose and tryptophan with glucose) within a period of 3 to 5 days in random order.

This amount of tryptophan caused nausea in all seven of the subjects, and vomiting in some. Administration of tryptophan without glucose did not result in a significant rise in plasma GH. Of the seven subjects, only two responded with incremental increases of 8.2 and 6.5 ng/ml. The PRL increase was also not significant with only three subjects responding with elevated plasma PRL concentrations. Administration of tryptophan with glucose resulted in a twelvefold increase in the concentration of insulin (IRI) in plasma. Yet, plasma GH concentrations were reduced in all subjects treated in this manner, and only one subject responded with a mild rise in plasma PRL concentration (Table 1).

In our study, oral administration of a large dose of tryptophan failed to exert a stimulatory effect on the release of GH and PRL. In all subjects, including those who

Table 1
Plasma GH, PRL, Insulin, and Glucose after Treatment
with Tryptophan or Tryptophan with Glucose

Tryptophan (150 mg/kg BW)	---------------------------------Time (min)---------------------------------					
	0	30	60	75	90	120
GH (ng/ml)						
Without Glucose†	4.3 ± 1.0*	2.3 ± 0.6	9.8 ± 1.3	3.3 ± 1.4	2.6 ± 1.2	1.9 ± 0.5
With Glucose‡	1.6 ± 0.5	0.6 ± 0.2	0.7 ± 0.5	0.4 ± 0.2	0.4 ± 0.2	1.4 ± 0.8
PRL (ng/ml)						
Without Glucose	7.0 ± 1.4	10.3 ± 3.4	39.3 ± 19.8	23.9 ± 9.3	17.7 ± 6.3	12.0 ± 3.4
With Glucose	6.9 ± 1.4	10.1 ± 2.4	8.9 ± 2.0	9.7 ± 1.4	8.9 ± 1.7	8.2 ± 1.8
IRI (μU/ml)						
Without Glucose	5.8 ± 0.6	9.0 ± 1.3	8.5 ± 1.2	10.1 ± 1.3	8.3 ± 2.2	8.0 ± 1.6
With Glucose	4.8 ± 0.7	55.3 ± 5.2	52.5 ± 6.4	60.5 ± 7.2	56.8 ± 7.6	49.3 ± 7.9
Glucose (mmol/liter)						
Without Glucose	4.6 ± 0.2	4.8 ± 0.2	5.2 ± 0.3	5.1 ± 0.3	5.0 ± 0.3	5.1 ± 0.3
With Glucose	4.6 ± 0.2	6.3 ± 0.2	6.2 ± 0.4	6.3 ± 0.5	6.2 ± 0.5	5.6 ± 0.4

*Mean ± SE.
†Tryptophan in 300 ml water.
‡Tryptophan in 300 ml water containing 75 g glucose.

responded to tryptophan alone, the glucose-induced rise in circulating insulin inhibited rather than enhanced the secretion of GH and PRL. The direct neuromodulatory effect of glucose, which diminished neuroendocrine responsiveness, prevailed over the insulin-induced improvement of tryptophan transport into the brain (11,12).

Effect of Glucose on Exercise-Induced Release of GH, ACTH, and PRL

Glucose deficiency stimulates the release of GH, ACTH, PRL as well as other hormones. Conversely, its administration prevents or inhibits GH secretion after exercise (13), surgical trauma (14), hyperthermia (12), and after pharmacological stimulation with L-dopa (15) or apomorphine (16). We have extended these studies by investigating the effect of glucose on exercise-induced secretion of ACTH and PRL. A group of eight volunteers was examined during exercise on a bicycle ergometer with a graded load (1.5, 2.0, or 2.5 W/kg body weight) as in the previous studies. Two examinations were performed within a period of one week. Thirty minutes before exercise, the subjects received in random order, 400 ml water only in one examination and 400 ml water containing 100 g glucose in the other one.

Table 2
Plasma GH (ng/ml) Before and After Exercise.
Water or Hexose (1 g/kg BW) was Given
Orally 30 min before Exercise

Treatment	-30	0	20	+10	+30	+60 min
Water	$0.4 \pm 0.1^*$	0.3 ± 0.1	11.4 ± 2.5	12.2 ± 2.6	6.8 ± 1.7	2.0 ± 0.9
Glucose	0.2 ± 0.1	0.2 ± 0.1	3.1 ± 1.0	5.2 ± 1.4	1.7 ± 0.5	0.6 ± 0.1
Fructose	0.3 ± 0.1	0.3 ± 0.1	8.3 ± 1.7	10.0 ± 1.9	5.1 ± 1.2	1.4 ± 0.3

*Mean ± SE.
The post-exercise increase of plasma GH after glucose administration is significantly ($P < 0.01$) less than that after fructose or water.

In the investigation using water only, exercise induced a significant increase in circulating GH, ACTH, PRL, and cortisol. Administration of glucose, however, inhibited the secretion of GH but did not inhibit ACTH, PRL, or cortisol release (Figure 5). Thus, during short-term exercise, glucose does not have a modulatory effect on ACTH and PRL release from the adenohypophysis. Inhibition of the neuroendocrine response in animals (17) and humans (18), observed during prolonged muscular work after glucose administration, presumably operates by a different mechanism: glucose becomes here an important source of energy during a period of depleted carbohydrate reserves.

During the last decade or so, much effort has been devoted to the elucidation of mechanisms controlling energy metabolism during physical work. As yet, however, the mechanisms determining the ratio of utilization of different substrates by the working muscle cell and the mobilization of energy sources from extramuscular tissues are poorly understood. The neuroendocrine response is known to play an important role, but the origin of the triggering stimulus is unknown. In attempting to fill this gap, two main concepts have been developed (19): *a)* the *central command hypothesis* in which it is proposed that hypothalamic neurons controlling pituitary hormone secretion are activated by impulses originating in the motor cortex and *b)* the *peripheral control hypothesis* in which it is proposed that neurohormonal responses to exercise are elicited by afferent nervous impulses from receptors of working muscles. The muscle receptors are thought to be activated by metabolic byproducts or other chemical changes in the interstitial muscular milieu.

Prevention of exercise-induced GH release by glucose administration can be accounted for either by an inhibitory effect of glucose on the central command in the CNS or by nutritional inhibition of the signal from the muscle cell. To clarify the mechanism and identify the site of action of the inhibitory effect of glucose on GH secretion during exercise, we exploited the differences in the distribution of glucose and fructose in the body. Both hexoses are readily metabolized in all cells of the body, but fructose is not transported *via* the BBB into the CNS (20). Thus, fructose can inhibit exercise-induced GH release only if the signal came from receptors of the working muscles.

Six healthy male volunteers, 22 to 23 years of age, received glucose or fructose (1 g/kg body weight) in 400 ml water 30 minutes before graded exercise. There was a 5

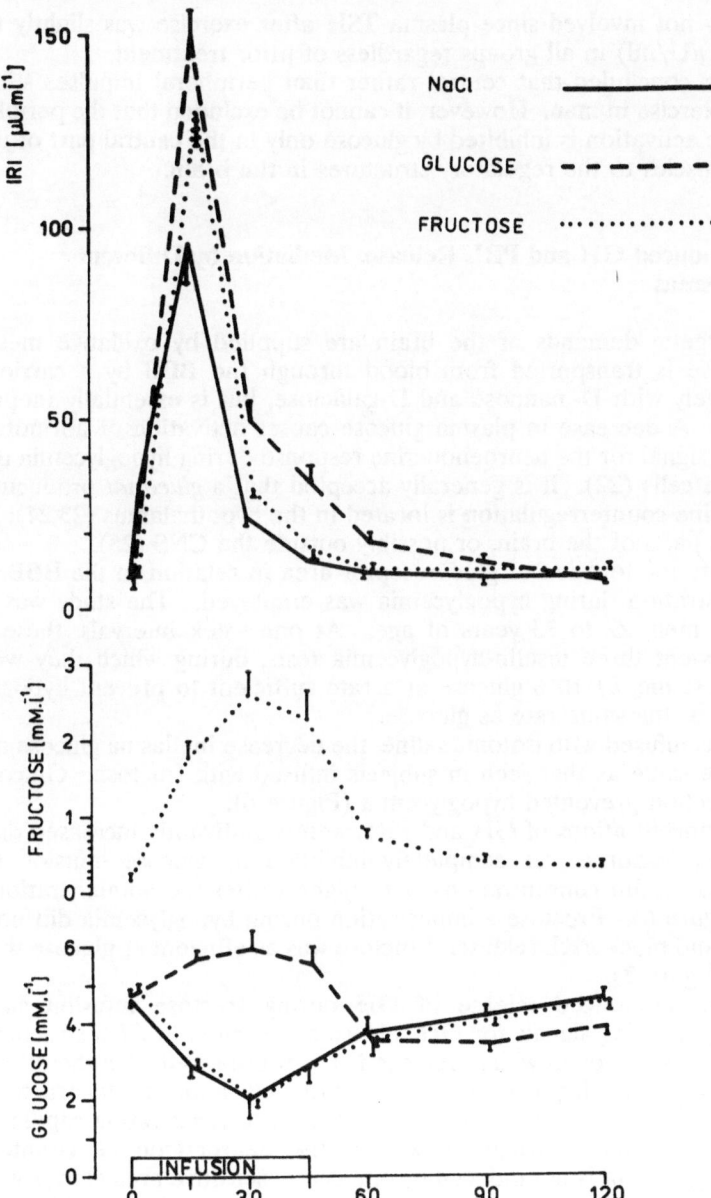

Figure 6. Plasma insulin (IRI), fructose and glucose concentrations after iv injection of insulin followed by a 40 min infusion of isotonic saline, glucose, or fructose. Mean and SE.

to 8-day span between the examinations. The post-exercise GH increase after fructose administration was not statistically different from the results seen after water administration. GH release was inhibited only after glucose administration ($P < 0.01$) (Table 2).

These results show that, during exercise, glucose exerts its inhibitory effect on GH release on brain structures protected by the BBB. The glucose-induced somatostatin release, which would inhibit GH secretion independently of the nature of the stimulus

(21), is probably not involved since plasma TSH after exercise was slightly but equally increased ($+0.2$ $\mu U/ml$) in all groups regardless of prior treatment.

It may be concluded that central rather than peripheral impulses stimulate GH release during exercise in man. However, it cannot be excluded that the peripheral signal for somatotropic activation is inhibited by glucose only in the central part of its pathway from working muscles to the regulatory structures in the brain.

Hypoglycemia-Induced GH and PRL Release: Mediation by Different Glucoreceptor Areas

The energetic demands of the brain are supplied by oxidative metabolism of glucose. Glucose is transported from blood through the BBB by a carrier that also operates effectively with D-mannose and D-galactose, but is essentially inoperable with D-fructose (20). A decrease in plasma glucose causes activation of hormonal counter-regulation. The signal for the neuroendocrine response during hypoglycemia is generated by glucosensitive cells (22). It is generally accepted that a *glucostat* producing impulses for neuroendocrine counterregulation is located in the hypothalamus (23,24); however, it may be in other parts of the brain, or possibly outside the CNS (25).

To identify the location of glucoreceptor area in relation to the BBB, glucose or fructose administration during hypoglycemia was employed. The study was performed on nine normal men, 22 to 23 years of age. At one week intervals, these volunteers randomly underwent three insulin-hypoglycemia tests, during which they were infused with *a)* isotonic saline, *b)* 10% glucose at a rate sufficient to prevent hypoglycemia, or *c)* 10% fructose at the same rate as glucose.

In subjects infused with isotonic saline, the decrease in plasma glucose after insulin injection was the same as that seen in subjects infused with fructose. Glucose infusion after insulin injection prevented hypoglycemia (Figure 6).

Plasma concentrations of GH and PRL were significantly increased during saline infusion, but the response was completely inhibited by glucose infusion (Figure 7). However, plasma insulin concentrations were highest after the administration of insulin and glucose (Figure 6). Fructose administration during hypoglycemia did not influence GH release but did block PRL release. Fructose was as efficient as glucose in preventing PRL increase (Figure 7).

Hypoglycemia-induced release of GH during fructose infusion suggests that glucoreceptors generating signals for GH release are located in CNS structures that are protected by the BBB through which fructose is not transported. On the other hand, the secretion of PRL during hypoglycemia was completely inhibited by fructose infusion, localizing the responsible glucoreceptors to an area with fenestrating capillaries.

In conclusion, these results show that the neuroendocrine counterregulatory response to hypoglycemia is not induced by one chemosensitive area since signals for GH and PRL secretion in man are generated by glucoreceptors in different locations.

Plasma GH, PRL, and Cortisol after Transient Cerebral Ischemia

Under basal conditions, approximately 800 ml of blood per minute flows through the human brain, representing 15% of the stroke volume (26). Reduction of brain blood flow by half results in loss of consciousness in young healthy persons. Diminished cerebral perfusion causes hypoxia and glucopenia, even at normal blood oxygen and glucose levels. When a reduction in cerebral blood flow is so severe that the normal function, metabolism, or structure of the brain cannot be maintained, ischemia occurs (27). The

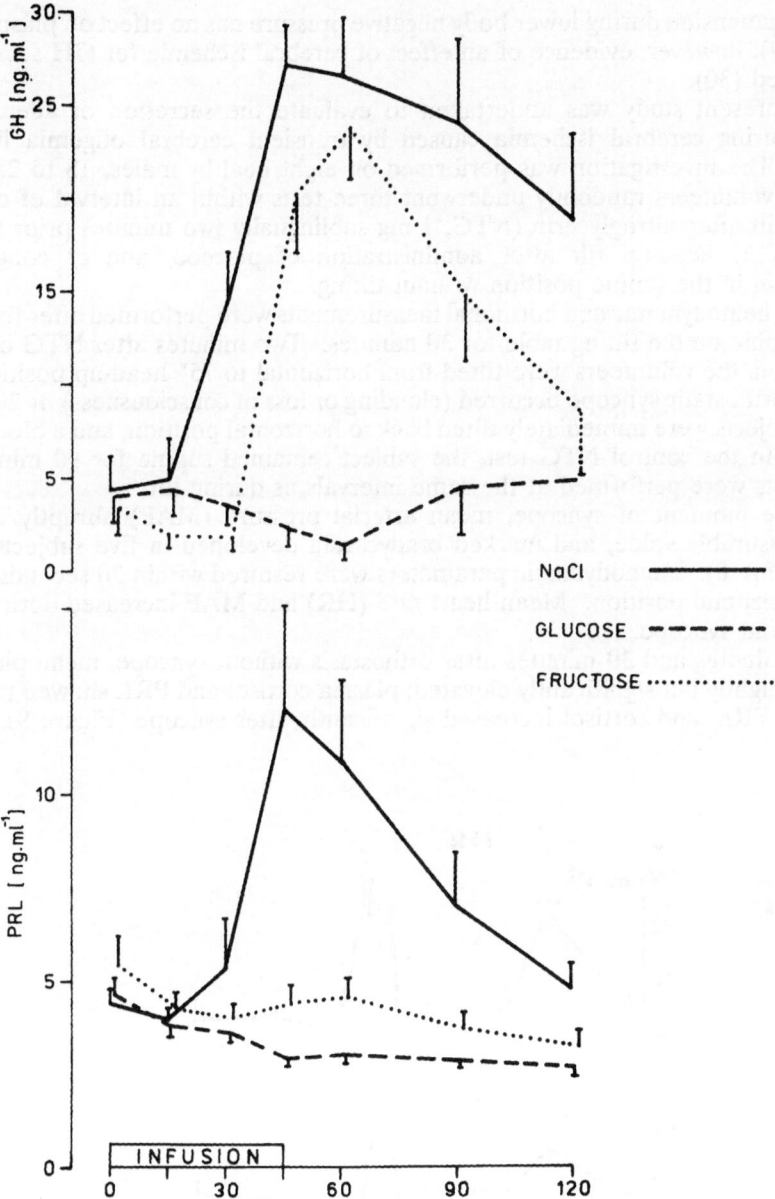

Figure 7. Plasma GH and PRL after iv administration of insulin and saline, insulin and glucose, or insulin and fructose. Fructose infusion did not prevent GH increase during hypoglycemia; however, it completely blocked PRL release. Mean and SE.

objective manifestation of transient cerebral ischemia is syncope. The regulation of the anterior pituitary is mediated *via* the blood flowing from the brain into the pituitary through the hypophysial portal system, which is rather vulnerable to circulatory derangements (28). Under such conditions, any acute change in the blood supply to the brain and pituitary may influence neuroendocrine functions. It has been found that moderate hypotension during lower body negative pressure has no effect on plasma ACTH and PRL (29); however, evidence of an effect of cerebral ischemia for GH secretion has been observed (30).

The present study was undertaken to evaluate the secretion of adenopituitary hormones during cerebral ischemia caused by transient cerebral oligemia in healthy volunteers. The investigation was performed on eight healthy males, 18 to 22 years of age. These volunteers randomly underwent three tests within an interval of one week: a) head-up tilt after nitroglycerin (NTG, 1 mg sublingually two minutes prior to change of position), b) head-up tilt after administration of placebo, and c) control NTG administration in the supine position without tilting.

Basal hemodynamic and hormonal measurements were performed after the subjects had been supine on the tilting table for 30 minutes. Two minutes after NTG or placebo administration, the volunteers were tilted from horizontal to 75° head-up position, which lasted until orthostatic syncope occurred (clouding or loss of consciousness) or 20 minutes. Then, the subjects were immediately tilted back to horizontal position, and a blood sample was taken. In the control NTG test, the subject remained supine for 90 minutes, and measurements were performed at the same intervals as during tilt.

At the moment of syncope, mean arterial pressure (MAP) abruptly decreased below a measurable value, and marked bradycardia developed in five subjects prior to syncope (Figure 8). Hemodynamic parameters were restored within 30 seconds after the return to horizontal position. Mean heart rate (HR) and MAP increased during upright posture without syncope.

Ten minutes and 30 minutes after orthostasis without syncope, mean plasma GH levels were slightly but significantly elevated; plasma cortisol and PRL showed no change. Plasma GH, PRL, and cortisol increased significantly after syncope (Figure 9).

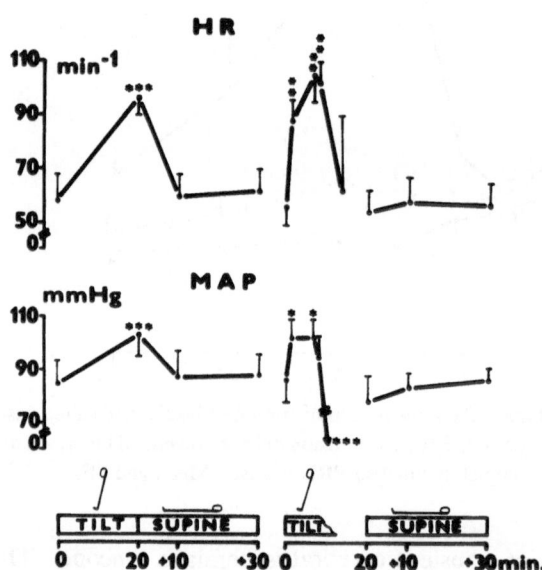

Figure 8. Mean heart rate (HR) and mean arterial blood pressure (MAP) during orthostatic syncope (right panel) and during orthostasis without syncope (left panel). Mean and SE. Asterisks signifies statistical significance of difference from basal values: * denotes $P < 0.025$, ** denotes $P < 0.01$, *** denotes $P < 0.001$.

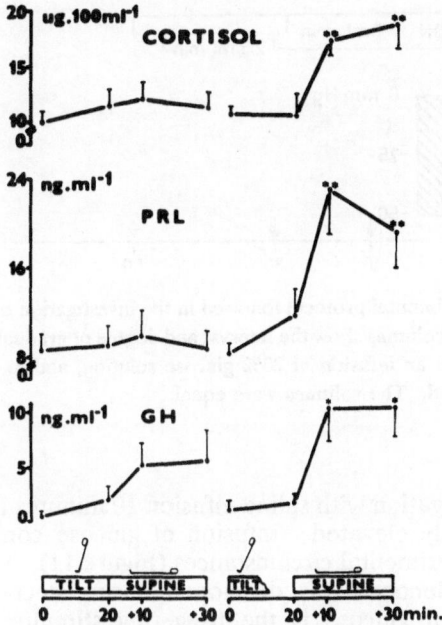

Figure 9. Plasma cortisol, PRL, and GH concentrations during orthostatic syncope (*right panel*) and during orthostatis without syncope (*left panel*). Mean and SE. Significant difference from basal values: * denotes $P < 0.025$, ** denotes $P < 0.05$.

The increased release of GH, PRL, and cortisol in response to syncope may result from the action of specific activating mechanisms or it may be a common non-specific response to the loss of consciousness. The relative cerebral glucopenia resulting from cerebral hypoperfusion may indicate the existence of a central triggering mechanism. To evaluate this hypothesis, the decrease of cerebral blood perfusion, with and without glucose supplementation, was induced in a model of lower body negative pressure.

The application of lower body negative pressure (LBNP) resulted in a sudden redistribution of cardiopulmonary or central blood volume to the lower extremities. These changes in the distribution of blood volume are similar, as seen during a sudden rise from the horizontal position. If LBNP is greater than -20 mm Hg, some presyncopal or syncopal symptoms may as a consequence of the vasovagal reaction with a precipitous fall in blood pressure, a result of peripheral vasodilation.

Ten healthy male volunteers, 25 to 40 years of age, who were experienced with the procedure, participated in the study. After the subjects were placed in the LBNP box, catheters were inserted into both cubital veins and the electrocardiogram, rheoencephalogram, and blood pressure were continuously recorded. One hour later, the first blood sample was taken, and an infusion of 20% glucose was begun. Ten minutes later, graded negative pressure was applied until -50 mm Hg was attained, and this pressure was maintained for 5 minutes. In some cases where symptoms of presyncope or syncope were observed, the application of negative pressure was immediately interrupted. After 30 minutes, the rate of glucose infusion was decreased to 0.5 g per minute for 30 minutes, and then stopped. One week later the same investigation was performed on the same volunteers with the exception that glucose was replaced by isotonic saline infusion (Figure 10).

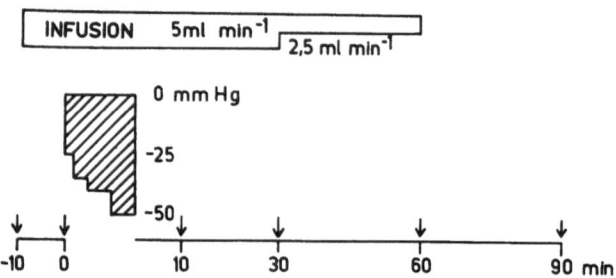

Figure 10. Scheme of experimental protocol followed in the investigation of endocrine responses to LBNP. Striated columns show the interval and degree of gradually applied LBNP. In the first investigation an infusion of 20% glucose solution, and in the second one isotonic saline was applied. The volumes were equal.

In the control investigation with saline infusion 10 minutes after LBNP application, plasma GH was significantly elevated. Infusion of glucose completely prevented GH release under the same experimental circumstances (Figure 11). This finding supports the hypothesis that cerebral glucopenia is developed during decreased perfusion of the cerebral tissue, and lack of glucose is the triggering stimulus for GH release (30). However, the extent of the decrease in blood flow under LBNP is not sufficient for the stimulation of ACTH and PRL release (Figure 12).

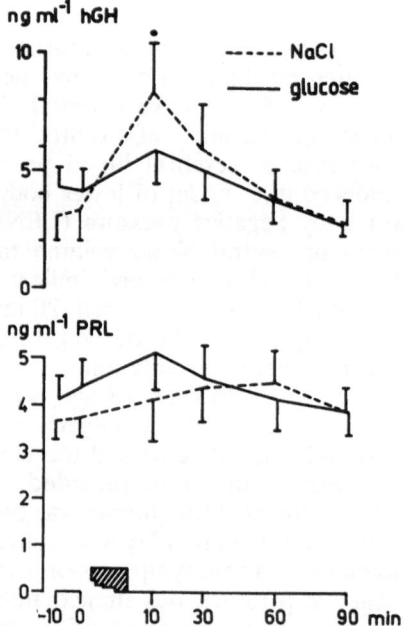

Figure 11. Plasma GH and PRL, before and after the application of LBNP, during infusion of glucose or isotonic saline. Mean and SE. Plasma GH after saline infusion is significantly increased ($P < 0.05$).

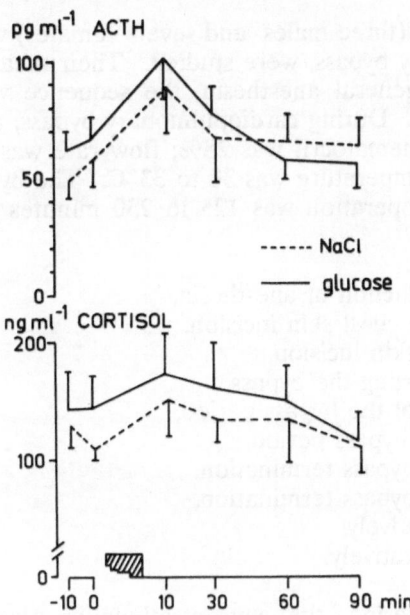

Figure 12. Plasma ACTH and cortisol concentrations before and after the application of LBNP. Mean and SE.

The preventative effect of glucose during LBNP on the release of GH cannot be considered as definite proof of a glucogenic mechanism that initiates neuroendocrine activation during transient cerebral ischemia. It is known that glucose inhibits GH release in situations where cerebral glucopenia does not act as a stimulus for GH release (exercise, hyperthermia). Moreover, mechanisms independent of hypothalamic regulatory processes could contribute to the increase in plasma hormone concentrations. After cessation of hormone outflow during temporary ischemia, the restoration of perfusion may mechanically wash out hormones from the pituitary. This mechanism can manifest itself more intensely after lengthy hypoperfusion of the brain and endocrine glands.

Effects of Open Heart Surgery on Growth Hormone, Cortisol, and Insulin Concentrations

The cardiopulmonary bypass operation represents a complicated stress situation in which the plasma concentration of certain hormones is affected. These involve emotional stress, effect of anesthetic drugs, injury, hypothermia with consecutive rewarming, and other factors such as heparinization, hemodilution, destruction of hormones by the heart-lung machine, as well as the effect of exogenous glucose in the perfusate on the secretion of some hormones. An important situation for hormonal secretion is the drop in blood pressure in the systemic circulation during the bypass period with diminished blood flow through the brain and endocrine tissues. In contrast to the complete failure of perfusion in syncope, which may last only a second or so, the blood pressure during bypass operation is maintained at the lower limit of the physiologic range, and hypoperfusion may last more than an hour.

Ten cardiac patients (three males and seven females) who underwent cardiac surgery with cardiopulmonary bypass, were studied. Their mean age was 28 years with a range of 8 to 44. For general anesthesia, the sequence with the diazepam-ketamine-nitrous oxide was used. During cardiopulmonary bypass, approximately 25,000 ml of solution was used. The hematocrit was 28%; flow rate was 2.5 minute/m^2 of body surface. Their total body temperature was 30 to 33°C. The bypass period lasted 21 to 67 minutes, and the entire operation was 125 to 230 minutes long. The sequence of sampling was as follows:

1)–shortly before induction of anesthesia,
2)–shortly before the chest skin incision,
3)–30 minutes after skin incision,
4)–shortly before starting the bypass,
5)–after 15 minutes of the bypass period,
6)–at the end of the bypass period,
7)–30 minutes after bypass termination,
8)–60 minutes after bypass termination,
9)–2 hours postoperatively,
10)–24 hours postoperatively.

During the bypass period, the concentrations of plasma hormones may be influenced by the hemodilution as well as the hypothermia. Cold is known to have no effect on basal secretion of ACTH, but has a mild inhibiting effect on GH release (31). Glucose in the perfusate could induce hyperglycemia, thereby inhibiting GH release (14). The attenuated tissue perfusion during the bypass period is characterized by a non-pulsatile blood flow. In this period, the blood flow through the secondary complex of the hypothalamic-pituitary portal circulation is probably reduced due to the low perfusion pressure in spite of diminished viscosity of the diluted blood.

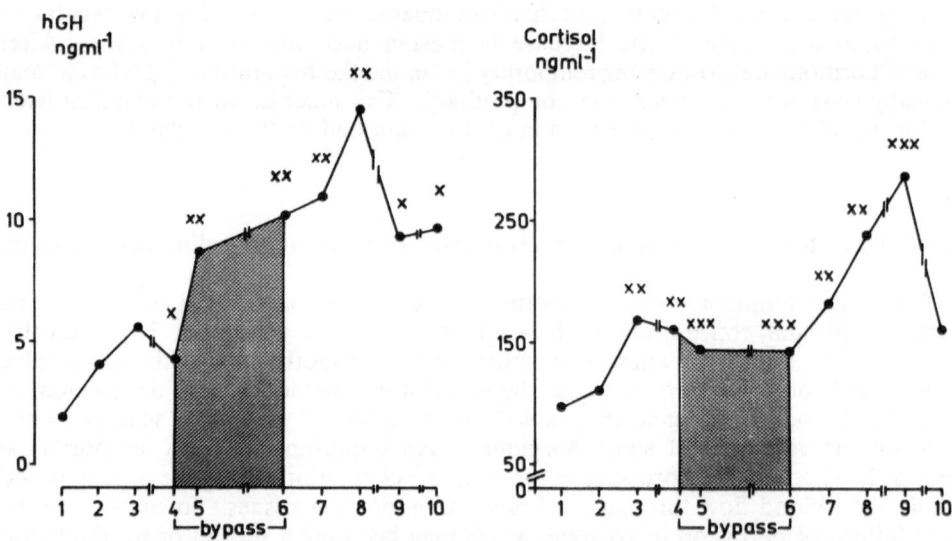

Figure 13. Plasma GH and cortisol concentrations during cardiopulmonary bypass surgery. See text for explanation. × denotes $P < 0.5$, ×× denotes $P < 0.01$, ××× denotes $P < 0.001$.

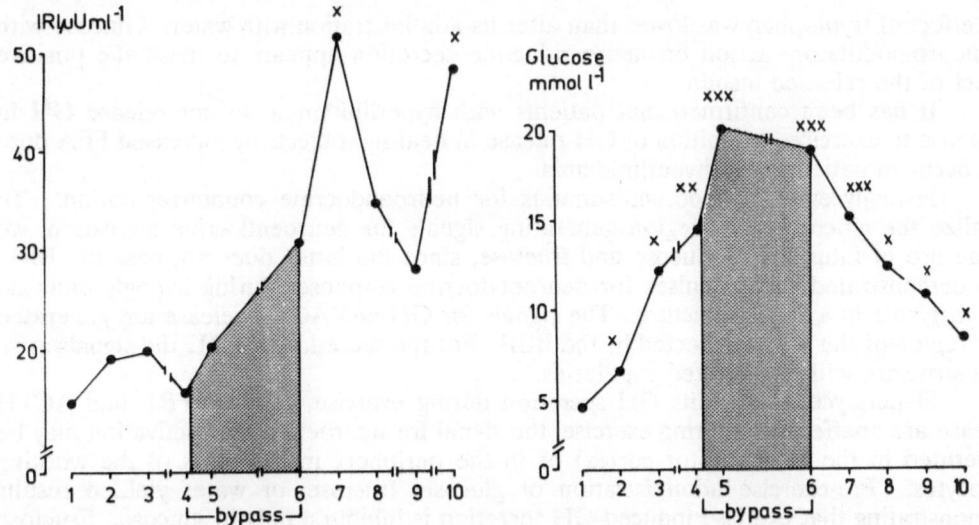

Figure 14. Plasma insulin (IRI) and glucose concentrations before and during cardiopulmonary bypass surgery. See text for explanation. Statistical significance: × denotes $P < 0.05$, ×× denotes $P < 0.01$, ××× denotes $P < 0.00$

In our study, the effect of injury on cortisol release was significant after the skin incision (Figure 13). However, the marked hyperglucemia (Figure 14) was probably responsible for the reduction in GH release. A significant increase of GH release occurred following the start of bypass in spite of hemodilution. Hemodilution after the start of bypass, resulted in a sustained moderate decrease in cortisol levels. The correction of blood hormone values for dilution revealed an increase of almost 50% in cortisol levels. It was found that synthetic ACTH administered during cardiac bypass stimulated the adrenal cortex, suggesting a primary failure of hormone production at the hypothalamic-pituitary level (32). Passage of blood through the bubble oxygenator led to approximately 30% degradation of GH and reduction of cortisol (33).

Upon rewarming and restoring the spontaneous circulation, the concentration of cortisol and GH after surgery began to increase, attaining their maxima at 1 hour and 2 hours, respectively. After termination of the bypass, plasma glucose began to decrease (Figure 14). Plasma insulin levels, inappropriately low for the corresponding plasma glucose levels until termination of cardiac bypass, started to rise after restoration of the spontaneous circulation despite the 21% decrease of plasma glucose concentration (Figure 14). Similar observations for blood glucose and insulin levels have been reported by others (34).

The paradoxical increases in plasma hormone levels at the end of surgery (after restoration of spontaneous circulation) suggest that after the cessation of hormone outflow during the decreased perfusion of bypass, restoration of blood flow mechanically washes hormones out of the endocrine glands.

Conclusions

The aim of this study was to evaluate the hypothesis that elevated circulating insulin potentiates the effect of exogenous tryptophan on the secretion of GH and PRL. After co-administration of glucose and tryptophan, the insulin concentration increased; yet, the effect of tryptophan was lower than after its administration with water. Glucose with its neuromodulatory action on neuroendocrine secretion appears to mask the positive effect of the released insulin.

It has been confirmed that patients with hyperlipidemia do not release GH in response to exercise. Inhibition of GH release in healthy subjects by increased FFA does not occur in patients with hyperlipidemia.

Hypoglycemia is a potent stimulus for neuroendocrine counterregulation. To localize the glucoreceptor region generating signals for neuroendocrine activation, we made use of infusions of glucose and fructose, since the latter does not pass the BBB. We demonstrated that impulses for neuroendocrine responses during hypoglycemia do not originate in a single structure. The signals for GH and ACTH release are generated in a region of the brain protected by the BBB. For the secretion of PRL, the signals arise in a structure with fenestrated capillaries.

Hyperglycemia inhibits GH secretion during exercise, whereas PRL and ACTH release are unaffected. During exercise, the signal for neuroendocrine activation may be generated in the CNS (motor cortex) or in the periphery in receptors of the working myocytes. Pre-exercise administration of glucose, fructose, or water yielded results demonstrating that exercise induced GH secretion is inhibited only by glucose. Fructose was as ineffective as water, revealing the CNS rather than peripheral tissues as the origin of impulses triggering GH secretion during exercise.

During transient cerebral ischemia (syncope) induced by orthostasis, the secretion of GH, PRL, and ACTH are stimulated. To analyze the mechanism of neuroendocrine activation, the LBNP model with glucose infusion was used. However, only GH secretion, which was inhibited by glucose, was found to be elevated during LBNP. The evidence indicating that the neuroendocrine system is activated by cerebral glucopenia can not be considered conclusive.

In heart surgery with bypass, which is associated with diminished perfusion of the tissues, a sudden increase in the secretion of GH, cortisol, and insulin occurred after restoration of the spontaneous circulation. The improved perfusion of the endocrine glands after restored circulation is assumed to account for this finding.

References

1. Ježová, D., R. Kvetňanský, K. Kovacs, Z. Opršalová, M. Vigaš, and G.B. Makara, Insulin induced hypoglycemia activates the release of a adrenocorticotropin predominantly via central and propranolol insensitive mechanism, *Endocrinology* 120: 409-415, 1987.
2. Heyrovský, A., Kolorimetriké stanoveni fruktosy kyselinou indolyloctovou, *Chemicke listy* 50: 1593-1597, 1956.
3. Schollberg, K., W. Jaross, E. Seker, A. Haschke, W. Wilke, G. Schmidt, R. Hentschel, B. Assmus, and G. Schirmer, Pituitary and adrenal hormones in patients after myocardial infarction under ergometric load, *Atherosclerosis* 49: 163-170, 1983.
4. Daughaday, W.H., The anterior pituitary gland, In J.D. Wilson and D.W. Foster (eds) *William's Textbook of Endocrinology*, W.B. Saunders, Philadelphia, pp. 568-613, 1985.
5. Casanueva, F., L. Villanueva, A. Penalva, T. Vila, and J. Cabezas-Cerrato, Free fatty acid inhibition of exercise-induced growth hormone secretion, *Horm Metab Res* 13: 348-350, 1981.
6. Casanueva, F.F., L. Villanueva, C. Dieguez, Y. Diaz, J.A. Cabranes, B. Szoke, M.F. Scanlon, A.V. Schally, and A. Fernandez-Cruz, Free fatty acid block growth hormone (GH) releasing hormone-stimulated GH secretion in man directly at the pituitary, *J Clin Endocrinol Metab* 65: 634-642, 1987.
7. Merimee, T.J., Familial Combined hyperlipoproteinemia. Evidence for a role of growth hormone deficiency in effecting its manifestation, *J Clin Invest* 65: 829-835, 1980.

8. Collu, R., Neuroendocrine control of pituitary hormone secretion, In R. Collu (ed) *Pediatric Endocrinology*, Raven Press, New York, pp. 1-36, 1989.

9. Tuomisto, J., and P. Männistä, Neurotransmitter regulation of anterior pituitary hormones, *Pharmacol Rev* 37: 249-332, 1985.

10. Pardridge, W.M., Regulation of amino acid availability to the brain, In R.J. Wurtman and J.J. Wurtman (eds) *Nutrition and the Brain*, Raven Press, New York, pp. 141-204, 1977.

11. Ježová-Repceková, D., M. Vigaš, and I. Klimeš, Decreased plasma cortisol response to pharmacological stimuli after glucose load in man, *Endocrinol Exper* 14: 113-120, 1980.

12. Vigaš, M., Neuroendocrine reaction of man in stress, Veda, Bratislava, 1985 (in Slovak).

13. Hunter, W.M., C.C. Fonseka, and R. Passmore, The rôle of growth hormone in the mobilization of fuel for muscular exercise, *Quart J Exper Physiol* 50: 406-416, 1965.

14. Vigaš, M., J. Malatinský, S. Németh, and J. Jurčovičová, Alpha-adrenergic control of growth hormone release during surgical stress in man, *Metabolism Clin Exper* 26: 399-402, 1977.

15. Mims, R.B., C.L. Scott, O.M. Modebe, and J.E. Bethune, Prevention of L-dopa-induced growth hormone stimulation by hyperglycemia, *J Clin Endocrinol Metab* 37: 660-663, 1973.

16. Ettigi, P., S. Lal, J.B. Martin, and H.G. Friesen, Effect of sex, oral contraception, and glucose loading on apomorphine-induced growth hormone secretion, *J Clin Endocrinol Metab* 40: 1094-1098, 1975.

17. Nazar, K., Adrenocortical activation during long-term exercise in dogs: evidence for a glucostatic mechanism, *Pflug Arch* 329: 156-166, 1971.

18. Galbo, H., N.J. Christensen, and J.J. Holst, Glucose induced decrease in glucagon epinephrine responses to exercise in men, *J Appl Physiol* 42: 525-530, 1977.

19. Galbo, H., Hormonal and metabolic adaptation to exercise, George Thieme Verlag, Stuttgart, 1983.

20. Oldendorf, W.H., Brain uptake of radiolabeled amino acids, amines and hexoses after arterial injection, *Am J Physiol* 221: 1629-1639, 1971.

21. Masuda, A., T. Shibajaki, M. Nakahara, T. Imaki, Y. Kiosawa, K. Jibiki, H. Demura, K. Shizume, and N. Ling, The effect of glucose on growth hormone (GH) - releasing hormone - mediated GH secretion in man, *J Clin Endocrinol Metab* 60: 523-526, 1985.

22. Oomura, Y., and H. Yoshimatsu, Neural network of glucose monitoring system, *J Auto Nervous Sys* 10: 359-372, 1984.

23. Himsworth, R.L., P.W. Carmel, and A.G. Frantz, The location of the chemoreceptor controlling growth hormone secretion during hypoglycemia in primates, *Endocrinology* 91: 217-226, 1972.

24. Aizawa, T., N.N. Yasuda, and M.A. Greer, Hypoglycemia stimulates ACTH secretion through a direct effect on the basal hypothalamus, *Metabolism* 30: 996-1000, 1981.

25. Cane, P., R. Artal, and R.N. Bergman, Putative hypothalamic glucoreceptors play no essential role in the response to moderate hypoglycemia, *Diabetes* 35: 268-277, 1986.

26. Sokolof, L., G.G. Fitzgerald, and E.E. Kaufman, Cerebral nutrition and energy metabolism, In R.J. Wurtman and J.J. Wurtman (eds) *Nutrition and the Brain, Volume 1*, Raven Press, New York, pp. 87-139, 1977.

27. Siesjö, B.K., *Brain Energy Metabolism*, John Willey and Sons, Chichester, pp. 453-526, 1978.

28. Kovach, A.G.B., and P. Sandor, Cerebral Blood flow and brain function during hypotension and shock, *Ann Rev Physiol* 38: 571-596, 1976.

29. Mills, D.E., and D. Robertshaw, Plasma prolactin responses to acute changes in central blood volume in man, *Horm Res* 18: 153-159, 1983.

30. Kellerová, E., and M. Vigaš, Cerebral hypoperfusion as a stimulus for growth hormone release in man, *Horm Res* 12: 260-265, 1980.

31. Vigaš, M., E. Martino, M. Bukovská, and P. Langer, Effect of acute cold exposure and insulin hypoglycemia on plasma thyrotropin levels by IRMA in healthy young males, *Endocrinol Exper* 22: 229-234, 1988.

32. Taylor, K.M., J.V. Jones, and M.S. Walker The cortisol response during heart-lung bypass, *Circulation* 54: 20-25, 1976.

33. Malatinský, J., M. Vigaš, D. Vršanký, R. Kvetňanský, J. Jurčovičová, and D. Ježová, In vitro study of hormone degradation by heart-lung machine with bubble oxygenator, *Resuscitation* 11: 69-77, 1984.

34. Yakota, H., Y. Kawashima, S. Hashimoto, H. Manabe, T. Onishi, T. Aono, and K. Matsumoto, Plasma cortisol, luteinizing hormone (LH), and prolactin secretory response to cardiopulmonary bypass, *J Surg Res* 23: 196-200, 1977.

DIFFERENCES IN THE EFFECTS OF ACUTE AND CHRONIC ADMINISTRATION OF DEXFENFLURAMINE ON CORTISOL AND PROLACTIN SECRETION

C. Oliver, D. Ježová,[1] M. Grino, V. Guillaume,
F. Boudouresque, B. Conte-Devolx, G. Pesce,
A. Dutour, and D. Becquet

*Laboratoire de Neuroendocrinologie Expérimentale
INSERM U 297, Faculté de Médecine Nord
Bd P. Dramard
13326 Marseille Cédex 15, France*

Introduction

The involvement of serotonin on neuroendocrine regulation has been substantiated by much experimental data. Indeed, serotonin has been shown to play a role in the regulation of pituitary hormones that display a secretory rhythmicity and/or changes after exposure by to various types of stress. These results are supported by the existence of a serotoninergic innervation of hypothalamic regions that control hormonal rhythmical secretion of hormones (suprachiasmatic nucleus) and that respond to stress (paraventricular nucleus) (1).

Most studies on the pituitary actions of serotonin have been based on the acute administration of serotonin agonists or precursors. However, chronic studies with such agents are lacking. Recently, we were able to investigate both acute and chronic effects of a serotoninergic agonist, dexfenfluramine (d-F), on some hypothalamic-pituitary secretions.

d-F is the dextro stereoisomer of fenfluramine (d,1-F), the racemic mixture of the dextro(+) and levo(-) steroisomers, that has been used for treatment of human obesity for 25 years (2). The dextroisomer has been shown to be twice as effective as the racemic mixture in reducing food intake. Therefore, d-F was introduced into clinical treatment of obesity, giving good results with few side effects (3).

The anorectic effect of d-F was shown to be due to increased serotoninergic tone (4). The decrease in food intake after administration of other serotonin agonists appears to be the result of an action that occurs mostly at the level of the paraventricular nucleus (PVN) (5). Interestingly, this nucleus has been shown to play an important role in the control of the secretion of several pituitary hormones including adrenocorticotropin hormone (ACTH), thyrotropin stimulating hormone (TSH), and prolactin (PRL) (1,6).

In our investigation, we have tested the effect of d-F on the regulation of two pituitary hormones that are predominantly (ACTH) or partially (PRL) controlled by the PVN. In the first study, we have evaluated the effect of acute and chronic administration of d-F on the secretion of hypothalamic CRF, ACTH, and corticosterone. In the second study, we have investigated the effects of acute and chronic administration of d-F on

---

[1]Present affiliation: Institute of Experimental Endocrinology, Slovak Academy of Sciences, Vlárska 3, 83306 Bratislave, Czechoslovakia

Circulating Regulatory Factors and Neuroendocrine Function
Edited by J. C. Porter and D. Ježová
Plenum Press, New York, 1990

cortisol (an index of ACTH secretion) and PRL plasma levels in normal volunteers and obese women.

Effect of Dexfenfluramine on the Hypothalamic-Pituitary-Adrenal Axis in the Rat

Serotonin (5-HT) and serotonin agonists can stimulate the hypothalamic-pituitary adrenal (HPA) axis of the rat as indicated by measurements of plasma corticosterone (7). It has been reported that activation of either 5-HT_2 or 5-HT_{1A} receptors results in an increase in plasma corticosterone levels (8). Serotonin fibers have been identified at the level of CRF neurons in the PVN (9), and it has been proposed that PVN serotonin mediates some of the neurally-stimulated adrenocortical secretions (10). As noted earlier, this nucleus may be the main site of action of d-F in reducing food intake. Acute injection of d,1-F stimulates corticosterone secretion, an effect that may be mediated through the release of hypothalamic CRF. However, *in vivo* pharmacological manipulations suggest an additional peripheral site of action of d,1-F on serum corticosterone levels, which may include direct stimulation of pituitary corticotrophs or adrenal cells (11,12). Therefore, the first aim of our study was to investigate the release of CRF into the hypophysial portal blood (HPB) after acute d-F administration. Since little information is available on the changes of the HPA axis during chronic administration of this drug, CRF, ACTH, and corticosterone secretions were determined under this condition.

Materials and Methods

Animals

Adult male Sprague-Dawley rats (300-350 g body weight) were used in all experiments. The rats were housed under controlled conditions (temperature 23°C; lights on between 0700 and 1900 hours). They were fed commercial rat chow and had free access to tap water.

Collection of HPB

All collections were performed between 1100 and 1700 hours on sodium thiopental (40 mg/kg body weight, intraperitoneally) anesthetized rats. The infundibular region was exposed and the collection of HPB was performed according to a method adapted from Porter and Smith (13). A postsurgical stabilization period of 60 minutes was incorporated before initiation of HPB sampling. Then, the pituitary stalk was cut, and every 10 seconds during succeeding 60 to 90 minutes, HPB was collected by aspiration with a polystyrene pipette.

Cannulation of the Tail Artery

The day before peripheral blood sampling, the rats were anesthetized with methohexital (Brietal, Lilly, 45-55 mg/kg, intraperitoneally) and an indwelling cannula was inserted into the tail artery as previously described (14).

Acute d-F Treatment

In one group of rats, the infundibular region was exposed. Then, the pituitary stalk

Table 1
Experimental Procedure During Chronic Dexfenfluramine
(d-F) Treatment for 14 Days

Day	Procedure
1	—Body weight measurement —Insertion or sham-insertion of osmotic minipumps filled with d-F
12	—Tail artery cannulation

13	—Blood sampling for ACTH in conscious animals:
	    Basal plus novelty stress at 0800
	    Basal at 1800

14	—Body weight measurment —Surgery required for HPB collection anesthesia —Peripheral blood sampling under anesthesia —HPB collection for 60 min —Rapid hemorrhage —HPB collection for 30 min

HPB denotes hypophysial portal blood.

was sectioned, and HPB collection started 15 minutes after intraperitoneal injection of saline or d-F (15 mg/kg). In another groups of rats, the infundibular region was exposed. The pituitary gland was left intact. Saline or d-F was injected intraperitoneally and a systemic blood sample was withdrawn 15 minutes later for ACTH determination.

Chronic d-F Treatment

An osmotic minipump (Alzet, model 2002, 0.49 μl/hr) was inserted subcutaneously under light ether anesthesia. The minipumps were preloaded with d-F at a concentration of 95 mg/ml in normal saline, which resulted in the delivery of 3 mg/kg/day of the drug for 14 days (15). Control rats were treated the same way but no pump was inserted. The incision was closed with a wound clip, and the rats were returned to their cages.

The order of the further experimental procedures is schematically presented in Table 1. Body weights were recorded before the last day of d-F treatment. To evaluate the activity of the HPA axis in conscious, nonstressed animals, blood samples for ACTH assay were taken *via* the tail artery catheter at 0800 hours and 1800 hours on day 13 of d-F administration. Immediately after the morning blood sample was taken, the rats were transferred, while still in their cages, to a novel environment (novelty stress) and peripheral blood was drawn 10 minutes later. Then, the rats were returned to the animal room. The next day (day 14 of d-F treatment) the rats were anesthetized with thiopental and the surgery required for HPB collection was performed. At the end of surgery, but before hypophysectomy, a blood sample was obtained from the femoral artery for ACTH determination. After a postsurgery stabilization period, HPB was collected during 60 minutes. This was followed by a rapid hemorrhage (10% of estimated blood volume within 1 minute, *via* femoral artery) and HPB was collected for the next 30 mintues.

Hormone Measurements

CRF was determined in acetone-extracts prepared from hypophysial portal plasma, using a radioimmunoassay as previously described (16). ACTH and corticosterone were measured by radioimmunoassays previously described, after apropriate extraction of peripheral plasma (17).

Statistical Analysis

Statistical analysis of the CRF and body weight data was performed using Student's *t* test. Since the ACTH values did not have a normal distribution on a linear scale, they were subjected to logarithmic transformation and then evaluated by one-way analysis of variance followed by multiple range tests of Dunnett and Dunn, as appropriate.

Results

Acute d-F Treatment

The single injection of d-F induced a significant increase in the release of CRF into HPB as compared to the saline-treated group. Plasma ACTH levels were already high in the control animals, probably due to the anesthesia and surgical stress. The increase

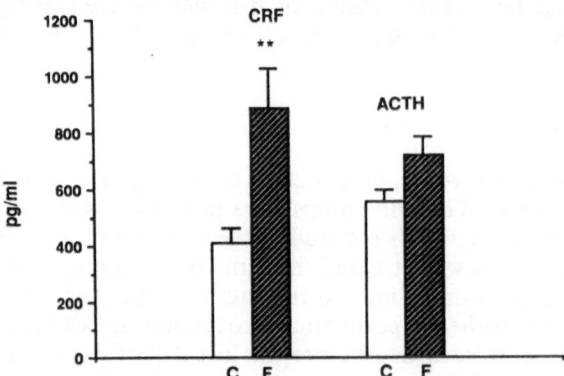

Figure 1. Effect of acute dexfenfluramine (d-F) injection (15 mg/kg, ip) on CRF release into HPB and on plasma ACTH levels in anesthetized rats. Means of 7 values ± SE. C denotes controls, F denotes, d-F. Statistical significant *vs.* controls: ** denotes $P < 0.01$.

Table 2
Change in Body Weight of Rats Chronically Treated With
Dexfenfluramine (d-F) (3 mg/kg/day for 14 days)

| Group | --------------------Body Weight-------------------- (g) | |
	Before Treatment	After Treatment
Control	$375 \pm 9^*$	413 ± 7
d-F	374 ± 7	$381 \pm 10\dagger$

*Mean ± SE; (n = 8).
†Statistical significance vs. control group ($P < 0.01$).

in ACTH levels following acute intraperitoneal injection of d-F did not reach significance
(Figure 1).

Chronic d-F Treatment

In contrast to the normal rise in the control group, the final body weight of the
animals treated with d-F did not change in comparison with the pretreatment values. The
mean body weight after 14 days of d-F treatment was significantly lower as compared to
that of the control animals (Table 2).

Chronic treatment with d-F resulted in an alteration of the circadian variation of
plasma ACTH levels, as estimated in conscious unstressed animals. In the control group,
no difference was found between morning and evening ACTH levels, while in d-F treated
animals, ACTH concentrations were significantly lower at 1800 hours as compared to
those at 0800 hours. Exposure to a novel environment in the morning—a stress stimulus
of moderate intensity—resulted in a significant increase in ACTH levels in both groups,
which was somewhat less pronounced in d-F treated animals. However, the difference in
stress-induced ACTH levels between control and d-F treated rats did not reach
significance (Figure 2).

The ACTH responses to a major stress stimulus are shown in Figure 3. The
hormone levels were very high, as the blood samples were taken at the end of the surgical
exposure of the hypothalamic-hypophysial region. The ACTH concentrations were higher
in rats chronically treated with d-F, but this change did not reach significance.

The CRF release into HPB was significantly increased in response to rapid
hemorrhage. Neither initial nor post-hemorrhage CRF secretion was modified by chronic
d-F administration (Figure 4).

Discussion

Our data demonstrate that acute d-F administration induces a significant increase
in the release of CRF into HPB. However, prolonged treatment with d-F has few, if any,
effects on the activity of the HPA axis.

The present data clearly show that the effect of acute d-F administration on pituitary-adrenocortical function is mediated, at least in part, *via* hypothalamic CRF release. No other data are available on the effects of d-F or d,1-F on CRF release *in vivo*. Among other serotoninergic drugs, only the effect of the serotonin re-uptake inhibitor, fluoxetine, has been investigated and this drug was found to increase CRF concentrations in HPB (18). Our data are consistent with the results of Holmes *et al.* (19) showing d,1-F-induced stimulation of bioactive CRF release from the hypothalamus *in vitro*.

The ability of serotonin agonists, including d,1-F, to increase corticosterone release was initially considered to support the hypothesis that brain serotonin neurons have a stimulatory influence on HPA function (7). Indeed, acute administration of d-F causes an immediate increase of serotonin in the synaptic cleft (20). It has been shown recently that acute intraperitoneal injection of d-F (doses of 3 mg and 10 mg/kg) increases extracellular serotonin in the perifornical lateral hypothalamus of unanesthetized rats (21). However, it is now appears that other mechanisms may be involved in d-F or d,1-F-induced corticosterone secretion (11,12). The major metabolite of d-F, d-norfenfluramine, has a high affinity for serotonin receptors (20) and, as a consequence, the mediation of d-F effects *via* d-norfenfluramine and postsynaptic receptors cannot be excluded. Besides, a peripheral site of action to initiate the effect of d,1-F on corticosterone secretion, including the possibility of a direct action at the pituitary level, has been proposed by several investigators on the basis that intraventricular injection of d,1-F does not increase corticosterone levels (7,11,22). A similar hypothesis has been raised for serotonin and some of its agonists and may be extended to d-F. Van de Kar *et al.* (11) have reported that the serotonin-releasing agent p-chloroamphetamine (PCA) increases corticosterone secretion in rats with ablation of central 5-HT pathways (23). Administration of p-chlorophenylalanine (PCPA), a central and peripheral serotonin synthesis inhibitor, abolishes the corticosterone response to PCA (24). These data suggest that a peripheral action of serotonin is possible and that most of the serotonin agonists may act at the level of the adrenal gland. Indeed, at least two-thirds of the epinephrine-containing

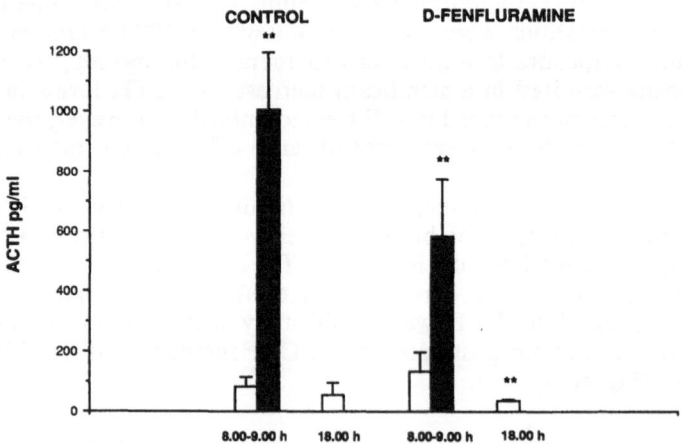

Figure 2. Plasma ACTH levels after chronic treatment with dexfenfluramine (d-F) (3 mg/kg/day sc for 14 days) in conscious rats bearing indwelling cannulas. The geometric mean ± SE of 8 values are given. *Open columns* correspond to basal (resting) levels, *full columns* correspond to novelty stress (10 min). Statistical significance *vs.* appropriate basal levels at 0800: ** denotes $P < 0.01$.

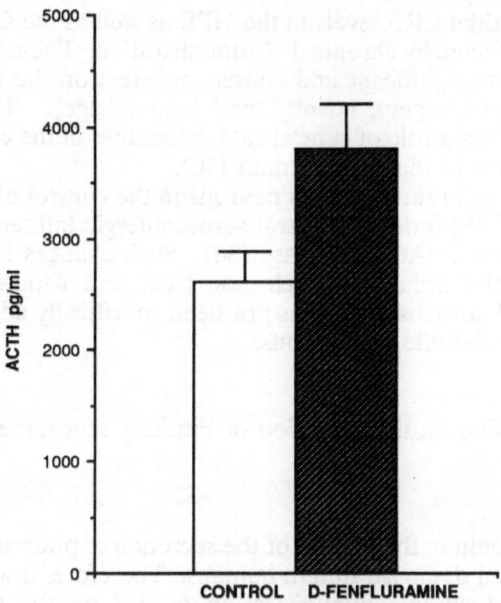

Figure 3. Plasma ACTH levels during surgical stress in controls and in dexfenfluramine- (d-F) treated (3 mg/kg/day sc for 14 days) anesthetized rats. A blood sample was collected just before HPB collection whose results are shown in Figure 7. The geometric mean ± SE of 8 values is given.

cells adrenal medulla contain serotonin (25). Under *in vitro* conditions, serotonin stimulates the synthesis and release of both corticosterone and aldosterone and may act as a paracrine agent on 5-HT$_2$ receptors in the adrenal cortex (26). However, most of these conclusions are based on the measurement of plasma corticosterone, which is not an absolutely adequate indicator of ACTH release (14).

Because of the lack of information on the effects of chronic d-F treatment on HPA function, a complex approach was chosen in this study, employing the advantage of repeated blood sampling in rats bearing indwelling cannulas. Chronic administration of d-F was effected using osmotic minipumps. According to Rowland (15), the dose of d-F (3 mg/kg/day) used in our study mimics the clinically effective daily dose in humans. To our knowledge, there are only two reports on the effects of chronic d-F treatment on HPA function. Daily intraperitoneal injection of d-F for 14 days did not affect circulating corticosterone levels (27). In another study chronic treatment with d-F (10 mg/kg/day) for 12 days decreases the duration, but not the peak, of the corticosterone response to the stress of acute ingestion of a meal of fructose, a test that is not currently used in neuroendocrinology (28).

Brain serotoninergic system interferes with ACTH response to stress. Earlier reports have suggested an inhibitory effect of serotonin on stress-induced secretion of ACTH (1) whereas more recent studies favor a stimulatory role for central serotonin in the control of ACTH release during stress (10,29). In the present experiments, ACTH response to a stress stimulus of moderate intensity (novelty stress) was slightly reduced in rats chronically treated with d-F. On the other hand, ACTH levels in response to a major stress stimulus (surgical stress) were somewhat higher in the same animals as compared to untreated controls. Initial CRF levels in the HPB as well as the CRF response to rapid hemorrhage were not affected by chronic d-F administration. Then, in our studies, chronic d-F administration has no significant and consistent effect on the stimulation of ACTH release by these stresses (surgery, novelty, and hemorrhage). These results are best explained by the predonimant role of central catecholamines in the control of CRF release during stress as compared to that of serotonin (16).

The involvement of brain serotonin neurons in the control of hormonal rhythms is well established (1) and disruption of central serotoninergic influence has been found to abolish circadian variation in ACTH release (30). Such changes have been observed in our study. However, the influence of chronic treatment with d-F on the circadian variations of ACTH and corticosterone has not been specifically addressed, and our data are too limited to allow definite conclusions.

Effects of Dexfenfluramine on the Secretion of Pituitary Hormones in Human

Effect in Human

The role of serotonin in the control of the secretion of pituitary hormones including ACTH and PRL has been demonstrated in humans. Therefore, it was of interest to study the effect of acute and chronic administration of d-F on the secretion of pituitary hormones. Furthermore, it is known that in obese patients, the secretion of hypophysial hormones is altered. In particular, it has been reported that PRL secretion is often reduced in massively obese patients (31), an effect that has been linked to a reduction in serotoninergic tone (32). An influence of obesity on pituitary adrenal function has been demonstrated in several studies and will be discussed below. In the human, cortisol and PRL were measured after acute and chronic administration of d-F, throughout the day

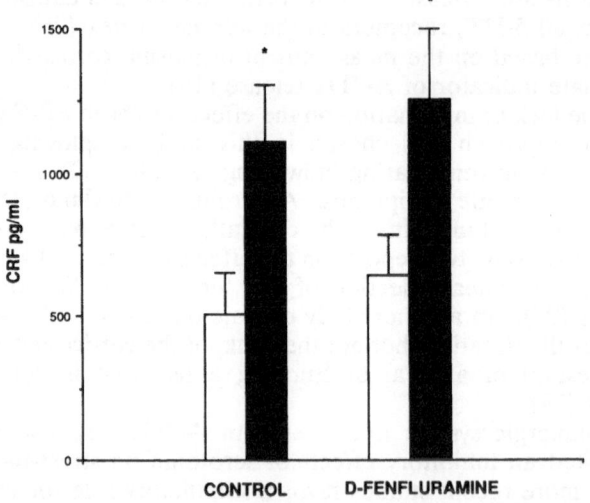

Figure 4. CRF release into the HPB before *(open columns)* and after *(full columns)* rapid hemorrhage in control and d-F treated anesthetized rats. The geometric mean ± SE of 8 values is given. Statistical significance *vs.* appropriate prehemorrhage values: * denotes $P < 0.05$.

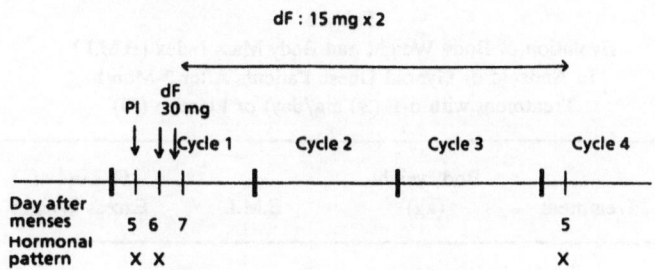

Figure 5. Outline of the clinical trial.

and during the post-prandial period. Studies on plasma TSH, GH, β-endorphin, insulin, C-peptide, free fatty acids, and blood glucose are in progress.

Materials and Methods

Subjects

The eight normal volunteer women (100-110% index body weight) were 27.9 ± 2.8 years old (mean ± SE). The android obese women (130-140% index body weight) were 31.4 ± 1.8 years old. The acute administration of d-F was performed in 14 android patients and its chronic administration in six of them. The gynoid obese women (130-140% index body weight) were 27.8 ± 2.0 years old. Twelve gynoid patients were studied during the acute administration of d-F and eight of them during its chronic administration. The menstrual cycle of all subjects was normal.

Protocols

In normal volunteers and obese patients, the circadian and post-prandial hormonal pattern was assessed during the early follicular phase (day 5 to 7 of the menstrual cycle).

The acute investigation was performed in normal volunteers in both groups of obese patients. Placebo or d-F (30 mg) was given per os at 0800 hours.

In android and gynoid patients, blood was also collected after chronic d-F treatment, i.e., after daily administration of d-F (15 mg × 2) for three menstrual cycles, and the same hormonal investigations were repeated.

In all studies (after placebo and after acute or chronic administration of d-F), a catheter was inserted at 0700 hours into an antecubital vein, and blood samples were collected at 0800, 1200, 1600, 2000, 2400, 0400 hours and every 30 minutes between 1200 hours and 1600 hours. The outline of the trial is shown in Figure 5.

During the studies, the normal volunteers as well as the obese patients were given a caloric intake of 1800 Kcal/day, consisting of breakfast (400 Kcal) at 0800 hours, lunch (800 Kcal) at 1200 hours, and supper (about 600 Kcal) at 1900 hours. During the 3-month treatment with d-F, the obese patients were allowed daily food intake ranging between 1200 and 1400 Kcal.

Determination of Hormones

Plasma ACTH and cortisol were determined by radioimmunoassay methods

Table 3
Evolution of Body Weight and Body Mass Index (B.M.I.)
In Android or Gynoid Obese Patients After 3-Month
Treatment with d-F (30 mg/day) or Placebo (Pl)

Group	Treatment	Bodyweight (kg)	B.M.I.	% Loss of Excess Weight	P
Android	Pl	100.9 ± 6.4	36.9 ± 1.6	16.0 ± 2.5	.001
	d-F	91.0 ± 6.0	33.3 ± 1.6		
Gynoid	Pl	83.9 ± 5.5	32.6 ± 2.2	10.5 ± 2.0	.001
	d-F	76.1 ± 5.4	29.6 ± 2.2		

Mean ± SE.

developed in our laboratory and previously described (33). Plasma PRL was measured with commercially available radioimmunoassay kit (Immunotech, Maresille, France).

Statistical Analysis

The results are expressed as the mean ± SE. A two-way or three-way analysis of variance was used.

Results

As shown in Table 3, both android and gynoid obese patients showed a significant reduction in body weight after chronic d-F treatment. No side effect was observed during the 3 months of treatment. The percent loss of excess weight was higher in android (16.0 ± 2.5) than in gynoid (10.5 ± 2.0) patients. The higher pre-treatment body mass index in the group of android patients may account for the difference.

After acute d-F administration, plasma cortisol levels were moderately but significantly increased in the gynoid obese patients. No modification of cortisol secretion was observed in normal volunteers and android obese patients (Table 4). The post-prandial cortisol pattern was not significantly modified in android obese patients. A slight but significant increase in cortisol levels was observed during the post-prandial period in volunteers and gynoid obese patients (data not shown).

The acute administration of d-F was followed by a moderate increase in plasma PRL levels in normal volunteers as well as in android and gynoid obese patients. During the post-prandial period, PRL levels were significantly higher after d-F than after placebo in the three subgroups (data not shown). The most striking increase in plasma PRL was observed at 1600 hours and 2000 hours. No further enhancement of the physiological increase in plasma PRL occuring at 0400 hours was induced by the acute d-F

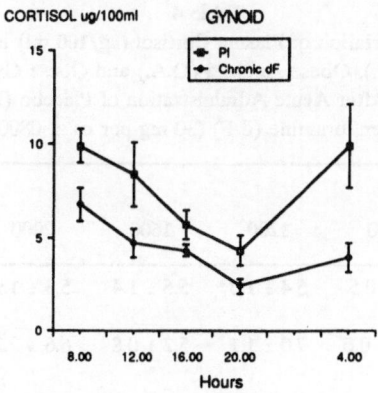

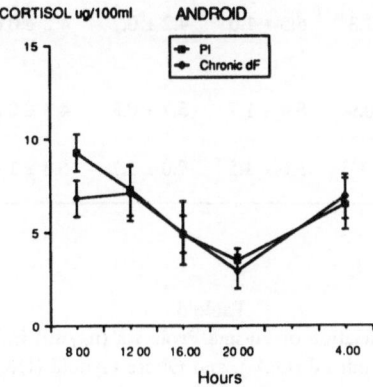

Figure 6. Circadian variation of plasma cortisol (μg/100 ml) after chronic treatment with dexfenfluramine (d-F) (30 mg per os per 24 hrs) of android and gynoid patients *(closed circles)*. Plasma cortisol values under placebo, before the onset of d-F treatment are shown *(open squares)*.

administration. Interestingly, plasma PRL levels after the placebo were not significantly different in control and obese women (Table 5).

After chronic d-F administration, plasma cortisol values at 0800 hours were lower than pre-treatment values. In the group of gynoid patients, plasma cortisol levels remained lower throughout the day in treated patients. However, in android patients plasma cortisol levels before and after treatment were similar during the rest of the day (Figure 6). A trend to lower post-prandial cortisol levels was observed after d-F treatment, without reaching significance (data not shown). After chronic treatment with d-F, PRL levels returned to pre-treatment values, determined under placebo in both groups of obese patients (Figure 7). The post-prandial PRL pattern was similar to the pre-treatment pattern.

Discussion

Administration of serotoninergic agents in humans stimulates ACTH and cortisol release in most cases, but not always. Oral administration of 5-hydroxytryptophan, the

Table 4

Circadian Variation of Plasma Cortisol (μg/100 ml) in Normal
Volunteers (N.V.), Obese Android (O.A.) and Obese Gynoid (O.G.)
Patients After Acute Administration of Placebo (Pl) or
Dexfenfluramine (d-F) (30 mg per os at 0800)

Group	Treatment	0800	1200	1600	2000	0400	Profile P
N.V.	Pl	9.3 ± 0.5	5.4 ± 1.0	5.5 ± 1.4	5.3 ± 1.9	5.3 ± 1.0	NS
	d-F	11.6 ± 0.6	7.0 ± 1.1	5.2 ± 0.8	6.6 ± 2.2	5.5 ± 1.4	
O.A.	Pl	9.3 ± 1.0	7.3 ± 1.4	4.9 ± 1.0	3.5 ± 0.6	6.5 ± 1.4	NS
	d-F	9.3 ± 1.3	6.5 ± 1.0	4.2 ± 0.7	4.2 ± 0.6	6.1 ± 0.7	
O.G.	Pl	9.9 ± 0.9	8.4 ± 1.7	5.7 ± 0.8	4.3 ± 0.8	9.9 ± 2.2	.05
	d-F	10.3 ± 1.3	8.6 ± 1.5	7.0 ± 1.1	6.8 ± 1.4*	7.2 ± 1.1*	

Mean ± SE.

Table 5

Circadian Variation of Plasma Prolactin (ng/ml) in Normal
(N.V.), Obese Android (O.A.), and Obese Gynoid (O.G.) Patients
After Acute Administration of Placebo (Pl) or
Dexfenfluramine (d-F) (30 mg per os at 0800)

Group	Treatment	0800	1200	1600	2000	0400	Profile P
N.V.	Pl	15.3 ± 0.9	7.2 ± 1.3	10.0 ± 1.5	13.0 ± 4.0	21.1 ± 3.1	NS
	d-F	16.6 ± 1.6	10.8 ± 1.4	13.3 ± 1.9	23.9 ± 5.3	23.4 ± 2.3	
O.A.	Pl	9.2 ± 1.1	7.8 ± 1.2	10.2 ± 1.3	10.9 ± 1.6	18.4 ± 2.1	NS
	d-F	10.2 ± 1.2	9.5 ± 1.6	14.6 ± 2.0	15.5 ± 2.4	21.5 ± 2.5	
O.G.	Pl	10.3 ± 1.1	7.9 ± 1.4	9.7 ± 1.0	11.6 ± 1.2	20.7 ± 2.7	.01
	d-F	12.4 ± 0.9	8.7 ± 1.3	15.2 ± 2.4*	17.1 ± 1.5*	19.5 ± 2.3	

Mean ± SE.

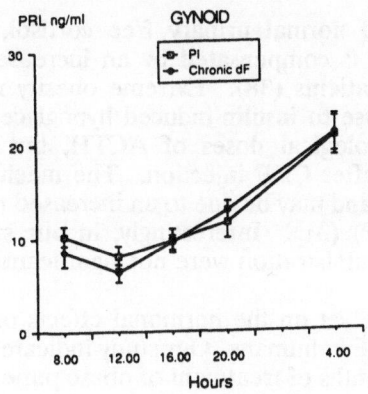

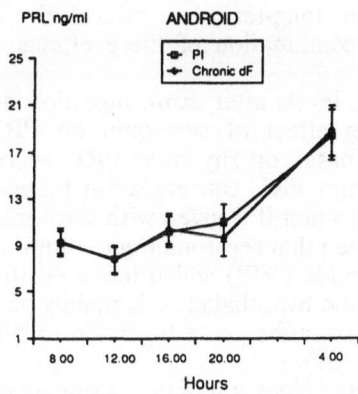

Figure 7. Circadian variation of plasma prolactin (ng/ml) after chronic treatment with dex-fenfluramine (d-F) (30 mg per os per 24 hrs) of android and gynoid patients *(closed circles)*. Plasma prolactin values under placebo, before the onset of d-F treatment are shown *(open squares)*.

direct metabolic precursor of serotonin, increases plasma ACTH and cortisol levels in normal men (34). Modlinger and colleagues (35) found that ACTH and cortisol levels rise after oral tryptophan load. Conversely, Woolf and Lee (36) reported that tryptophan causes a significant decrease in both basal and hypoglycemia-stimulated ACTH release. Lewis and Sherman (37) reported a dose-dependent stimulatory effect of d,1-F on both ACTH and cortisol levels. After the highest dose of d,1-F (1.5 mg/kg), mean ACTH and cortisol levels increased from 20.8 pg/ml and 7.3 μg/dl to 35.5 pg/ml and 15.1 μg/dl, respectively. The drug was administered orally at 1700 hours, and the ACTH and cortisol increase was observed 2 to 4 hours later. The absence of side effect as well as the blunting of ACTH and cortisol responses to d,1-F after cyproheptadine pretreatment suggest that the action of d,1-F on ACTH release is specific. This result is in good agreement with the present demonstration of a stimulating effect of acute d-F on cortisol release in man.

The clinical investigations on ACTH and cortisol secretion in obese patients did not give a clear picture. Obese humans have normal circulating plasma cortisol with a normal circadian rhythm and normal urinary free cortisol, but have an accelerated degradation of cortisol which is compensated by an increased cortisol production rate especially in android obese patients (38). Extreme obesity may be characterized by a normal plasma control response to insulin-induced hypoglycemia, a normal response to adrenal stimulation by physiological doses of ACTH, but an impaired response to pituitary-adrenal stimulation after CRF injection. The mechanisms responsible for the latter observation are unclear and may be due to an increased release of endogenous CRF or arginine vasopressin (AVP) (31). Interestingly, in our study the cortisol circadian rhythms following placebo administration were not significantly different in obese and in controls.

No data are avilable as yet on the hormonal effects of chronic administration of serotonin agonists including d-F in humans. Our study indicates that plasma cortisol levels are lowered slightly after 3 months of treatment of obese patients with d-F. However, the mechanisms responsible for the decreased cortisol levels are unclear. Our studies were performed in obesity during a weight-reducing program. One cannot eliminate a role for either decreased body weight or long-term low calorie diet in reducing plasma cortisol levels in our patients or the combination of these effects with a direct effect of the compound.

The stimulation of PRL levels after acute ingestion of d-F was expected in view of the well-known stimulating effect of serotonin on PRL secretion. Indeed, the involvement of serotonin in basal or rhythmic PRL secretion seems to be clearly demonstrated. In the rat, serum PRL concentration increases after administration of 5-hydroxytryptophan, especially when it is given with serotonin agonists or releasers such as d,1-F. It has been hypothesized that serotoninergic stimuli act by enhancing the release of vasoactive intestinal polypeptide (VIP) which has a positive effect on PRL secretion by the pituitary gland. VIP in the hypothalamus is mainly produced in the PVN (1). An acute stimulation of PRL release after acute ingestion of 5-hydroxytryptophan has also been demonstrated in humans (39).

The PRL response to acute ingestion of d-F is similar in obese and control women. Similarly, it has been reported that basal plasma PRL, 24-hour integrated PRL, and PRL responses to thyrotropin-releasing hormone (TRH) are normal in obesity. However, in some subjects there is no PRL response to insulin-induced hypoglycemia (31,32). In the same subjects the PRL release after metoclopramide (a dopamine antagonist that crosses the blood-brain barrier) is significantly lowered while PRL response to domperidone (a dopamine antagonist which does not cross the blood-brain barrier) is unchanged. These results suggest that the changes in PRL secretion seen in some obese patients may occur at the hypothalamic level and may involve changes in serotoninergic and/or dopaminergic pathways. Indeed, the PRL response to insulin-induced hypoglycemia is, at least in part, serotonin-mediated (40).

In our study, a tolerance to the stimulatory effect of d-F on PRL secretion developed in both groups of obese patients after chronic d-F administration. Similar data have been reported in the rat. In the rat, Serri and Rasio (27) have shown that acute administration of d-F stimulates PRL secretion. After 2 weeks of treatment, PRL levels returned to the pre-treatment values. In these animals, the number of striatal dopaminergic receptors was significantly reduced after chronic administration of the drug for 14 days.

Conclusion

In the rat, the acute injection of d-F stimulates pituitary-adrenocortical function, an effect that is at least partly mediated *via* an increased release of CRF into the HPB.

On the other hand, chronic d-F treatment in rats, which mimics the mode of the drug administration to humans, has only minor if any influence on the activity of HPA axis.

In the human, the effects of d-F on the pituitary-adrenal axis are more discrete. Acute d-F administration causes a slight increase in plasma cortisol levels whereas chronic treatment with d-F has no effect or only a slight lowering effect on plasma cortisol values. PRL secretion is stimulated during the acute administration of d-F, but it remains unaffected after chronic treatment with d-F.

These results have obvious clinical implications. They also raise questions about the mechanisms responsible for the differential effects of acute and chronic d-F administration: a) down regulation of d-F on serotonin receptors, and b) predominant role of factors other than serotonin in the regulation of cortisol and PRL.

Acknowledgements

The authors than Ms. R. Quérat for her excellent secretarial assistance. They also thank Drs. B. Guardiola-Lemaitre, C. Nathan, O. Arnaud, and R. Lavielle for helpful comments and discussion. The financial support of Institut Recherches Internationales Servier (Neuilly sur Sein, France) is gratefully acknowledged.

References

1. Montange, M., and A. Calas, Serotonin and endocrinology, the pituitary, In N.N. Osborn and M. Hamon (eds) *Neuronal Serotonin*, Wiley & Sons, New York, pp. 271-303, 1988.

2. Kyriakides, M., and T. Silverstone, Comparison of the effects of d-amphetamine and fenfluramine on hunger and food intake in man, *Neuropharmacology* 18: 1007-1008, 1979.

3. Wurtman, J.J., and R.J. Wurtman, D-fenfluramine selectively decreases carbohydrate but not protein intake in obese subjects, *Int J Obes* 8: 79-84, 1984.

4. Garattini, S., W. Buczo, A. Jori, and R. Samanin, The mechanism of action of fenfluramine, *Postgrad Med J* 15: 27-35, 1975.

5. Leibowitz, S.F., G.F. Weiss, and G. Shor-Posner, Hypothalamic serotonin: pharmacological, biochemical, and behavioral analyses of its feed-suppressive action, *Clin Neuropharmacol* 11: S51-S71, 1988.

6. Kiss, J.Z., Dynamism of chemoarchitecture in the hypothalamic paraventricular nucleus, *Brain Res Bull* 20: 699-708, 1988.

7. Fuller, R.W., and H.D. Snoddy, Effect of serotonin-releasing drugs on serum corticosterone in rats, *Neuroendocrinology* 31: 96-100, 1980.

8. Koenig, J.I., G.A. Gudelsky, and H.Y. Meltzer, Stimulation of corticosterone and β-endorphin secretion in the rat by selective 5-HT receptor subtype activation, *Eur J Pharmacol* 137: 1-8, 1987.

9. Lipolists, Zs., C. Phelix, and W.K. Paull, Synaptic interaction of serotonergic axons and corticotropin-releasing factor (CRF) synthesizing neurons in the hypothalamic paraventricular nucleus of the rat, *Histochemistry* 86: 541-549, 1987.

10. Feldman, S., N. Conforti, and E. Melamed, Paraventricular nucleus serotonin mediates neurally stimulated adrenocortical secretion, *Brain Res Bull* 18: 165-168, 1987.

11. Van de Kar, L.D., J.H. Urban, K.D. Richardson, and C.L. Bethea, Pharmacological studies on the serotoninergic and nonserotonin-mediated stimulation of prolactin and corticosterone secretion by fenfluramine. Effects of pretreatment with fluoxetine, indalpine, PCPA, and L-tryptophan, *Neuroendocrinology* 41: 283-288, 1985.

12. McElroy, J.F., J.M. Miller, and J.S. Meyer, Fenfluramine, p-chloroamphetamine and p-fluoroamphetamine stimulation of pituitary-adrenocortical activity in rat: evidence for differences in site and mechanism of action, *J Pharmacol Exper Ther* 228: 593-599, 1984.

13. Porter, J.C., and K.R. Smith, Collection of hypophysial stalk blood in rat, *Endocriology* 81: 1182-1185, 1967.

14. Ježová, D., R. Kvetňanský, K. Kovács, Z. Opršalová, M. Vigaš, and G.B. Makara, Insulin-induced hypoglycemia activates the release of adrenocorticotropin predominantly via central and propranolol insensitive mechanisms, *Endocrinology* 120: 409-415, 1987.

15. Rowland, N.E., Effect of continuous infusions of dexfenfluramine on food intake, body weight and brain amines in rats, *Life Sci* 39: 2581-2586, 1986.

16. Guillaume, V., B. Conte-Devolx, A. Szafarczyk, F. Malaval, N. Pares-Herbut, M. Grino, G. Alonso, I. Assenmacher, and C. Oliver, The corticotropin-releasing factor release in rat hypophysial portal blood is mediated by brain catecholamines, *Neuroendocrinology* 46: 143-146, 1987.

17. Conte-Devolx, B., C. Oliver, P. Giraud, E. Castanas, F. Boudouresque, P. Gillioz, and Y. Millet, Adrenocorticotropin, β-endorphin and corticosterone secretion in Brattleboro rats, *Endocrinology* 110: 2097-2100, 1982.

18. Gibbs, D.M., and W. Vale, Effect of the serotonin reuptake inhibitor fluoxetine on corticotropin-releasing factor and vasopressin secretion into hypophysial portal blood, *Brain Res* 280: 176-179, 1983.

19. Holmes, M.C., G. Di Renzo, U. Beckford, B. Gillham, and M.T. Jones, Role of serotonin in the control of secretion of corticotrophin releasing factor, *J Endocrinol* 93: 151-160, 1982.

20. Garattini S., S. Caccia, T. Mennini, R. Samanin, S. Consolo, and H. Ladinsky, Biochemical pharmacology of the anorectic drug fenfluramine: a review, *Curr Med Res Opin* 6: 15-27, 1979.

21. Schwartz, D., L. Hernandez, and B.G. Hoebel, Fenfluramine administered systematically or locally increases extracellular serotonin in lateral hypothalamus as measured by microdialysis, *Brain Res* 482: 261-270, 1989.

22. Heybach, J.P., and J. Vernikos-Danellis, Effects of fenfluramine administration on activity of the pituitary-adrenal system in the rat, *West Pharmacol Soc Proc* 21: 19-25, 1978.

23. Van de Kar, L.D., C.W. Wilkinson, Y. Skrobik, M.S. Brownfield, and W.F. Ganong, Evidence that serotonergic neurons in the dorsal raphe nucleus exert a stimulatory effect on the secretion of renin but not of corticosterone, *Brain Res* 235: 233-243, 1982.

24. Van de Kar, L.D., C.W. Wilkinson, and W.F. Ganong, Pharmacological evidence for a role of brain serotonin in the maintenance of plasma renin activity in unanesthetized rats, *J Pharmacol Exper Ther* 219: 85-90, 1981.

25. Verhofstad, A.A.J., and G. Jonsson, Immunohistochemical and neurochemical evidence for the presence of serotonin in the adrenal medulla of the rat, *Neuroscience* 10: 1443-1453, 1983.

26. Rácz, K., I. Wolf, R. Kiss, G. Lada, S. Vida, and E. Gláz, Corticosteroidogenesis by insolated human adrenal cells: effect of serotonin and serotonin antagonists, *Experientia* 35: 1532-1535, 1979.

27. Serri, O, and E. Rasio, Effect of chronic administration of dextrofenfluramine (d-F) on prolactin (PRL) and corticosterone secretion and on central dopaminergic receptors, *70th Annual Meeting of the Endocrine Society*, New Orleans, LA, June, Abstract 1177, p. 315, 1988.

28. Brindley, D.N. Metabolic and hormonal effects of dexfenfluramine on stress stiuations, *Clin Neuropharmacol [Suppl 11]* 1: S86-S89, 1988.

29. Yehuda, R., and J.S. Meyer, A role for serotonin in the hypothalamic-pituitary-adrenal response to insulin stress, *Neuroendocrinology* 38: 25-32, 1984.

30. Szafarczyk, A., G. Ixart, F. Malaval, J. Nouguier-Soulé, and I. Assenmacher, Effects of lesions of the suprachiasmatic nuclei and of p-chlorophenylalanine on the circadian rhythms of adrenocorticotrophic hormone and corticosterone in the plasma and on locomotor activity of rats, *J Endocrinol* 83: 1-16, 1979.

31. Kopelman, P.G., Neuroendocrine function in obesity, *Clin Endocrinol* 28: 675-689, 1988.

32. Bernini, G.P., G.F. Argenio, M.S. Vivaldi, C. Del Corso, M. Sgro, F. Franchi, and M. Luisi, Effects of fenfluramine and ritanserin on prolactin response to insulin-induced hypoglycemia in obese patients: evidence for failure of the serotoninergic system, *Horm Res* 31: 133-137, 1989.

33. Giaufre, E., B. Conte-Devolx, G. Morisson-Lacombe, F. Boudouresque, M. Grino, B. Rousset-Rouvière, V. Guillaume, and C. Oliver, Anesthésie péridurale par voie caudal chez l'enfant. Etude des variations endocriniennes, *La Presse Médicale* 14: 201-203, 1985.

34. Imura, H., Y. Nakai, and T. Yoshimi, Effect of 5-hydroxytryptophan (5-HTP) on growth hormone and

ACTH release in man, *J Clin Endocrinol Metab* 36: 204-206, 1973.

35. Modlinger, R.S, J.M. Schonmuller, and S.P. Arora, Adrenocorticotropin release by tryptophan in man, *J Clin Endocrinol Metab* 50: 360-363, 1980.

36. Woolf, P.D., and L. Lee, Effects of the serotonin precursor, tryptophan, on pituitary hormone secretion, *J Clin Endocrinol Metab* 45: 123-133, 1977.

37. Lewis, D.A., and B.M. Sherman, Serotonergic stimulation of adrenocorticotropin secretion in man, *J Clin Endocrinol Metab* 58: 458-462, 1984.

38. Vague, J., P.H. Vague, J. Boyer, and M.C. Cloix, Antropometry of obesity. Diabetes, adrenal and beta-cell functions, In R.R. Rodriguez and J. Vallance-Owen (eds) *Diabetes*, Excerpta Medica, Amsterdam, pp. 517-525, 1971.

39. Mashchak, C.A., O.A. Kletzky, C. Spencer, and R. Artal, Transient effect of L-5-hydroxytryptophan on pituitary function in men and women, *J Clin Endocrinol Metab* 56: 170-176, 1983.

40. Masala, A., G. Delitala, and L. Devilla, Enhancement of insulin-induced prolactin secretion by fluoxetine in man, *J Clin Endocrinol Metab* 49: 350-352, 1979.

Participants

Aguilera, G., National Institute of Child Health and Human Development, Bethesda, Maryland

Atrens, D., University of Sydney, Sydney, Australia

Bakoš, P., Institute of Experimental Endocrinology, Slovak Academy of Sciences, Bratislava, Czechoslovakia

Banks, W.A., Veterans Administration Medical Center, New Orleans, Louisiana, U.S.A.

Bartha, L., Institute of Experimental Medicine, Budapest, Hungary

Berkenbosch, F., Free University, Amsterdam, The Netherlands

Brtko, J., Institute of Experimental Endocrinology, Slovak Academy of Sciences, Bratislava, Czechoslovakia

Bugajski, J., Polish Academy of Sciences, Krakow, Poland

Burkov, A., Research Institute of Animal Production, Nitra, Czechoslovakia

Chesnokova, V., Siberian Branch of the USSR Academy of Science, Novosibirsk, U.S.S.R.

Ciccarelli, E., Molinette Hospital, Torino, Italy

De Boer, S.F., University of Utrecht, Utrecht, The Netherlands

Danišová, A., Institute of Experimental Endocrinology, Slovak Academy of Sciences, Bratislava, Czechoslovakia

De Jong, F.H., Erasmus University Rotterdam, Rotterdam, The Netherlands

Dobrakovová, M., Institute of Experimental Endocrinology, Slovak Academy of Sciences, Bratislava, Czechoslovakia

Ermisch, A., Karl Marx University, Leipzig, German Democratic Republic

Eskay, R.L., National Institute on Alcohol Abuse and Alcoholism, Bethesda, Maryland, U.S.A.

Farah, J.M., G.D. Searle and Company, St. Louis, Missouri, U.S.A.

Ferguson, A.V., Queen's University, Kingston, Ontario, Canada

Fodor, M., Semmelweis University Medical School, Budapest, Hungary

Gadek-Michalska, A., Polish Academy of Sciences, Krakow, Poland

Götz, F., Humboldt University Medical School, Berlin, German Democratic Republic

Grässler, J., Medical Academy of Dresden, Dresden, German Democratic Republic

Grossman, A., Medical College of St. Bartholomew's Hospital, London, Great Britain

Héber, S., Markusovszky Teaching Hospital, Szombathely, Hungary

Inoué, S., Tokyo Medical and Dental University, Tokyo, Japan

Jeske, W., Centre for Postgraduate Medical Education, Warszawa, Poland

Ježová, D., Institute of Experimental Endocrinology, Slovak Academy of Sciences, Bratislava, Czechoslovakia

Johansson, B.B., University of Lund, Lund, Sweden

Jurčovičová, J., Institute of Experimental Endocrinology, Slovak Academy of Sciences, Bratislava, Czechoslovakia

Kaiser, K., ČTK, Bratislava, Czechoslovakia

Kiem, T.D., Institute of Experimental Medicine, Budapest, Hungary

Kiss, A., Comenius University Medical School, Bratislava, Czechoslovakia

Korányi, L., Postgraduate Medical School, Budapest, Hungary

Kouřilová, M., Comenius University, Bratislava, Czechoslovakia

Kozlowski, G.P., University of Texas Southwestern Medical Center, Dallas, Texas, U.S.A.

Kvetňanský, R., Institute of Experimental Endocrinology, Slovak Academy of Sciences, Bratislava, Czechoslovakia

Lackovič, V., Institute of Virology, Slovak Academy of Sciences, Bratislava, Czechoslovakia

Larina, I.M., Institute of Biomedical Problems, Moscow, U.S.S.R.

LeFeuvre, R., University of Manchester, Manchester, Great Britain

Lichardus, B., Institute of Experimental Endocrinology, Slovak Academy of Sciences, Bratislava, Czechoslovakia

Linton, E.A., University of Reading Whiteknights, Reading, Great Britain

Lorenzi, M., Harvard Medical School, Boston, Massachusetts, U.S.A.

Macho, L., Institute of Experimental Endocrinology, Slovak Academy of Sciences, Bratislava, Czechoslovakia

Marek, J., Third Department of Medicine, Praha, Czechoslovakia

Maslova, L.N., Siberian Branch of the Academy of Sciences, Novosibirsk, U.S.S.R.

Mazza, E., University of Torino, Torino, Italy

McCann, S.M., University of Texas Southwestern Medical Center, Dallas, Texas, U.S.A.

Medvedev, O., All-Union Cardiology Research Center, Moscow, U.S.S.R.

Meyerhoff, J.L., Walter Reed Army Institute of Research, Washington, District of Columbia, U.S.A.

Micic, D., Institute for Endocrinology, University Cinical Center, Beograd, Yugoslavia

Montini, M., General Hospital of Bergamo, Bergamo, Italy

Murgaš, K., Institute of Experimental Endocrinology, Slovak Academy of Sciences, Bratislava, Czechoslovakia

Naumenko, E., Siberian Branch of the Academy of Sciences, Novosibirsk, U.S.S.R.

Nieber, K., Research Institute of Lung Disease and Tuberculosis, Berlin, German Democratic Republic

Nishizuka, M., Juntendo University School of Medicine, Tokyo, Japan

Oliver, C., Faculté de Médecine Nord, Marseille, France

Opršalová, Z., Institute of Experimental Endocrinology, Slovak Academy of Sciences, Bratislava, Czechoslovakia

Pagani, G., OORR Bergamo Largo, Bergamo, Italy

Pammer, C., Semmelweis University Medical School, Budapest, Hungary

Pavlik, A., Faculty of Medicine, Safat, Kuwait

Ponec, J., Institute of Experimental Endocrinology, Slovak Academy of Sciences, Bratislava, Czechoslovakia

Popovic, V., University Clinical Center, Beograd, Yugoslavia

Porta, S., Institute of Functional Pathology, Graz, Austria

Porter, J.C., University of Texas Southwestern Medical Center, Dallas, Texas, U.S.A.

Printz, M., University of California-San Diego, La Jolla, California, U.S.A.

Rešetková, E., Institute of Experimental Endocrinology, Slovak Academy of Sciences, Bratislava, Czechoslovakia

Rettori, V., University of Texas Southwestern Medical Center, Dallas, Texas, U.S.A.

Reul, J.M.H.M., Rudolf Magnus Institute, Utrecht, The Netherlands

Reymond, M.J., C.H.U.V.-EH 19, Lausanne, Switzerland

Reznikov, A., Research Institute of Endocrinology and Metabolism, Kiev, U.S.S.R.

Rinner, I., Institute of Functional Pathology, Graz, Austria

Rivier, C., The Salk Institute, San Diego, California, U.S.A.

Rolla, M., Centre of Adolescents, University of Pisa, Pisa, Italy

Roske, L, Institute of Drug Research of the Academy of Sciences of the German Democratic Republic, Berlin, German Democratic Republic

Rothwell, N.J., University of Manchester, Manchester, Great Britain

Rovenský, J., State Institute for Control of Drugs, Bratislava, Czechoslovakia

Rühle, H., Karl-Marx University, Leipzig, German Democratic Republic

Saavedra, J.M., National Institute of Mental Health, Bethesda, Maryland, U.S.A.

Samson, W.K., University of Missouri School of Medicine, Columbia, Missouri, U.S.A.

Segal, M., St. Thomas's Hospital, London, Great Britain

Shin, S., Queen's University, Kingston, Ontario, Canada

Sikora, J., Charles University School of Medicine, Praha, Czechoslovakia

Sirotkin, A., Research Institute of Animal Production, Nitra, Czechoslovakia

Smirnova, T., Activity of Tartu State University, Tartu, U.S.S.R.

Soszynski, P., Medical Center for Postgraduate Education, Bielanski Hospital, Warsaw, Poland

Štrbák, V., Institute of Experimental Endocrinology, Slovak Academy of Sciences, Bratislava, Czechoslovakia

Sweep, F.C.G.J., Rudolf Magnus Institute of Pharmacology, Utrecht, The Netherlands

Tilders, F.J.H., Free University, Amsterdam, The Netherlands

Van Oers, J.W.A.M., Free University, Amsterdam, The Netherlands

Vietor, L, Institute of Experimental Endocrinology, Slovak Academy of Sciences, Bratislava, Czechoslovakia

Vigaš, M., Institute of Experimental Endocrinology, Slovak Academy of Sciences,Bratislava, Czechoslovakia

Watkins, P., University of Auckland, Auckland, New Zealand

Watkins, W.B., University of Auckland, Auckland, New Zealand

Žižkovský, V., Charles University School of Medicine, Praha, Czechoslovakia

Zlokovič, B.V., UCS School of Medicine, Los Angeles, California, U.S.A.

Index

A23187 336, 338
A3V 214-217
A3V and renal sodium excretion
  anterior commissurae 214
  BBB 214
  effect of an AV3V lesion on
    plasma osmolality 217
  effect of hypertonic saline load
    215
  effect of isotonic saline load 216
  glomerular filtration rate 215
  inadequate water intake 217
  median preoptic nucleus 214
  natriuresis following an isotonic
    saline load 217
  natriuresis induced by ECFV
    expansion 217
  osmotically free water 215
  OVLT 214
  periventricular organs 214
  plasma sodium 215
  potassium concentrations 215
  potassium excretion 215
  preoptic periventricular tissue
    214
  push-pull technique 214
  renal blood flow 215
  sodium excretion 217
  sodium (osmotic?) sensors 214
  urine excretion 215
  VIP secretion 217
Abdominal pain 166
Ablated A3V region 214-216
Abluminal membrane 29, 78
Abortion 166
Access to the brain 136
Accumulation of leucine 85
ACE 30, 191
ACE inhibitor 193
Acetic acid 296
Acetoacetate 28
Acetylcholine 64, 170, 175, 362
Acromegaly 171-174
ACTH 41, 42, 44, 45, 47-58, 91, 93, 95,
  96, 98-105, 107-110, 115,
  124, 126, 128, 133,
  137-141, 144, 145, 147,
  149, 151, 152, 154-156,
  159-164, 172, 173, 175,
  178, 180, 189, 201, 205,
  206, 229, 231-235, 240,
  251, 255, 256, 295,
  296-301, 304-309,
  313-316, 320-323,
  325-327, 331-337,
  339-342, 363, 372, 373,
  392-394, 400-402, 408,
  411, 413, 414, 418, 420-
  425, 427-435, 437, 439,
  440, 443
ACTH-secreting tumor 172
α_1-adrenoceptor 172
β-adrenoceptor pathway 172
ANF 172
AVP 172
basophilic adenoma 172
bromocriptine 172
cholinergic tone 172
corticotrophs 173
CRF 172, 173
Cushing's disease 172, 173
cyproheptadine 172
dopamine agonist 172
GABA 172
GABA aminotransferase inhibitor 172
hypercortisolemia 172
hypothalamic control 172
hypothalamic-pituitary axis 172
α-MSH 172
Nelson's Syndrome 172
pituitary-adrenal axis 173
release 172
responsiveness to bromocriptine 172
secretion 172, 173
sodium valproate 172
somatostatin analogues 172
transsphenoidal surgery 173
ACTH and cortisol levels 45, 48, 439
ACTH and testicular steroidogenesis 183
ACTH assay 429

ACTH infusion 231
ACTH release 44, 45, 48-55, 57, 103, 108,
                147, 154, 159, 172, 180,
                205, 307, 313, 314, 316,
                322, 323, 326, 327,
                332-337, 340, 341, 373,
                400, 424, 433, 434, 439,
                443
ACTH releasing activity 159
ACTH releasing capacity 309
ACTH response 47-50, 101, 104, 172,
                306, 308, 309, 334, 401,
                433
ACTH response to serotonin 47
ACTH secretion 41, 44, 49, 51, 53, 55-57,
                100, 101, 108, 133, 137,
                138, 140, 144, 152, 159,
                162, 172, 173, 175, 189,
                206, 251, 256, 298-300,
                306, 307, 313, 315, 316,
                320, 327, 334, 342, 400,
                401, 425, 428
ACTH-induced corticosterone secretion
                140
ACTH-secreting tumor 172
Activin 271, 275, 276, 278-280, 282-285,
                287, 289, 292, 293
Activity of TH 4-9, 11-19, 21, 258, 263
Addiction 59, 65, 66
Adenohypophysis 49, 173, 258, 260-264,
                414
Adenosine 31
Adenylate cyclase 81, 189, 336, 343
Adipsin 374, 379
Adrenal cortex 103, 140, 172, 173, 234,
                240, 293, 304, 338, 342,
                423, 433
Adrenal gland 147, 181, 240, 241, 277,
                331, 338, 432
Adrenal glomerulosa cell function during
                sodium restriction 228
Adrenal medullectomy 49, 234, 242
Adrenal sensitivity to AII 229, 230, 232,
                238, 239
        adrenal AII receptors 230
        ANF 230
        dopaminergic system 230
        enzymes of the aldosterone
                biosynthetic pathway
                230
        modulation of adrenal sensitivity

        to AII 230
        sensitivity of adrenal glomerulosa cell to
                AII 229
        sodium intake 229
        sodium loading 230
        sodium restriction 229
        somatostatin 230
        trophic effects of AII 230
Adrenalectomy 102, 107, 109, 143, 163, 194, 200,
                204, 206, 209, 243, 247, 249,
                254-256, 313, 332, 340, 372, 373,
                376, 378, 379
Adrenaline 31, 37, 56, 309, 310
β-Adrenergic pathways 170
α-Adrenergic receptors 31, 49, 340, 343
β-Adrenergic receptors 31, 49, 340, 343
Adrenocorticotropin, see ACTH
Affinity-purified CRF 158
Aged brain 7, 8, 17, 22
Aging 1, 22, 59, 63-66, 68, 250, 255, 364, 369, 372,
                382, 388, 390
AI 93, 103, 104, 106, 115, 126, 224, 375
^{125}I[Sar^1,Ile^8]AII 236
AII 30, 41, 56, 79, 80, 88, 98, 102-104, 109, 110,
                113, 117, 124-126, 130, 132, 133,
                178, 179, 188, 190-209, 223, 224,
                226-233, 235, 236, 238-242, 369
        cerebellum 79
        cortex 79
AII binding density
        area postrema 195
        enalapril-treated Wistar-Kyoto rat 195
        inferior olive 195, 199
        median preoptic nucleus 195, 199
        nucleus of the solitary tract 195, 199
        organum vasculosum of the lamina ter-
                minalis 199
        paraventricular nucleus 195, 199
        spontaneously hypertensive rat 195
        SFO 195, 199
        suprachiasmatic nucleus 195
AII dependent steroidogenesis 232
AII infusion 124, 125, 240
        epinephrine 125
        blood pressure 125
        norepinephrine 125
AII receptor 192, 196-199, 201-203, 205, 206
AII receptor concentration
        area postrema 198
        median preoptic nucleus 198
        nucleus of solitary tract 198

suprachiasmatic 198
AII receptor distribution 192
AII binding density
    area postrema 199
AII receptor concentration
    olfactory bulb 198
Albumin 34, 38, 62, 64, 69, 154, 218, 241,
    337, 345, 346, 348-350,
    352, 356-358, 362-365,
    369, 385, 386, 390, 412
Aldosterone 103, 109, 110, 114, 115, 126,
    127, 129, 133, 178, 181,
    187, 191, 201, 206, 209,
    227-235, 238, 239, 240,
    241, 244, 246, 247,
    251-254, 389, 433
Aldosterone and AII
    ACTH 229
    ACTH decreases aldosterone
        secretion 229
    ACTH levels during sodium defi-
        ciency 229
    adrenal AII receptors 228-231
    adrenal glomerulosa AII recep-
        tors 229
    adrenal glomerulosa cell function
        during sodium restriction
        228
    adrenal responses to sodium
        restriction 230
    AII during sodium deficiency
        229
    AII receptors 229
    aldosterone and adrenal glomeru-
        losa cells 227
    aldosterone and sodium intake
        227
    aldosterone responses 228
    aldosterone responses to sodium
        restriction 229
    aldosterone secretion 227
    ANF 232
    AII infusion on adrenal
        glomerulosa function
        230
    anterior pituitary gland 228
    basal activity of enzymes in al-
        dosterone biosynthetic
        pathway 228
    CEI on adrenal glomerulosa
        function 230

    decrease in AII receptors 228
    effects of AII infusion on blood
        aldosterone 231
    effects of sodium restriction 230
    effects on the adrenal glomerulosa cell
        229
    enhanced sensitivity of adrenal during
        sodium restriction 229
    extracellular potassium 229
    glomerulosa cell function 228
    18-hydroxylase activity 228, 229, 230, 231
    hypophysectomy 228
    increases in aldosterone secretion 229
    increases in AII receptors 229
    infusion of AII 229
    peptide levels during sodium restriction
        229
    physiological control of aldosterone
        secretion 228
    plasma aldosterone 229, 230
    potassium intake 227
    production and secretion 178
    receptor 244
    regulation 232, 240, 241
    secretion 103, 187, 227-234, 239-241
    sensitivity of adrenal glomerulosa cell to
        AII 232
    side chain cleavage enzyme 230
    sodium intake 227, 231
    sodium loading 228
    sodium restriction 230
    sodium restriction plus CEI 230
    somatostatin 232
    stimulator of aldosterone secretion 229
Aldosterone secretion after hypophysectomy 232
Alkaline phosphatase 29
Allergic encephalitis 345, 354-359, 365
Alloxan-treatment 393
Alterations in the BBB permeability to amino
        acids
    leucine 83
    lysine 83
    phenylalanine 83
Altering the BBB 34
    hyperosmolar solutions 34
    hypertension 34
    long-term effects of an altered brain en-
        vironment 34
    permanent neuronal damage 34
    transient dysfunction of the BBB 34
Aluminum 65, 66, 68, 69

Alzheimer's disease 66, 68, 313, 362, 364, 365, 369
Amenorrhea 165, 394, 404
Amines 26, 30, 36, 56, 131, 341, 425, 442
γ-Amino butyric acid, see GABA
Amino acid 22, 36, 63, 68, 78, 82-84, 87, 88, 167, 169, 172, 177, 186, 187, 268, 271-274, 276, 359, 361, 366, 367, 412, 425
Ammon's horn 245, 248
Amyloid 303, 313, 362, 369
Amyotrophic lateral sclerosis 66, 68
Analgesia 59, 64, 66, 68, 75, 87
Anatomical basis of the BBB 26
Androgens 279, 281, 285, 288, 290
Android 435-440
ANF 88, 89, 110, 172, 178-180, 182, 183, 186-190, 207-210, 222, 226, 228-230, 232, 239, 363
Angiotensin 30, 31, 41, 56, 79, 80, 88, 93, 98, 102-104, 106, 109, 110, 113-115, 117, 124-127, 130, 132, 133, 178, 179, 187, 188, 190-193, 202-205, 207-209, 223, 224, 226, 227, 230, 233, 237, 240-242, 363, 369, 373, 375, 379
Angiotensin converting enzyme, see ACE
Angiotensin I, see AI
Angiotensin II, see AII
Anorectic effect 427
Anorexia 165
ANP 77-81, 88, 91, 93, 96-99, 103, 110, 177-192, 195-197, 199-201, 203-210, 220-227
    cortex 79
    hippocampus 79
    striatum 79
    Wistar rat 81
ANP antagonists 185
ANP binding density
    area postrema 197
    choroid plexus 197
    nucleus of the solitary tract 197
    paraventricular nucleus 197
    SFO 197
ANP binding

binding affinity 201
binding capacity 201
Brattleboro rat 201
microvessels 81
ANP receptor 177, 203, 207
α_2-Antagonists 170
Anterior bregma 199
Anterior commissurae 214
Anterior pituitary 9, 11, 52, 56, 58, 81, 91, 101-103, 108, 110, 132, 139, 141, 145, 152, 164, 165, 174, 175, 179, 186, 189, 200, 205, 206, 210, 228, 233, 240, 242, 244, 246, 251, 285, 287, 307, 309, 313-316, 319, 320, 322, 324, 326-329, 334, 335, 341, 342, 379, 389, 394, 400, 405, 417, 424, 425
Anterior pituitary grafts 9, 11
Anterior third ventricle, see A3V
Anterior ventricle third ventricle, see AV3V
Anterograde transport 143, 145
Anti-androgen 279
Anti-ANP serum 185
Anti-cholinesterase drug 170
Anti-interleukin-1 receptor serum 296
Antibodies to neuropeptides
    anterograde transport 143
    AVP 143, 144
    CRF 143
    CRF antibodies 144
    CRF antibody and CRF producing neurons 144
    CRF receptor antagonist 143
    IL-1β 143
    internalization of antibodies 143
    neuropeptide Y 144
    oxytocin 144
    retrograde transport 143
    substance P 143
    substance P neurons 143
    TH immunoreactive neurons 144
Antibodies to substance P 143
Antibody and CRF producing neurons 144
Antibody to β-endorphin 99
Antibody to IL-1 305
Antibody-cell interaction concept 141
    antibodies to neuropeptides 142
    epitopes 141
    Fc receptors 141
    neuronal uptake of antibodies to enzymes 141

receptors 141
Antibody-cell interaction transport
    antibody-induced neuronal
        lesions 142
Antibody-induced neuronal lesions
    complement mediated antibody
        interaction 142
    cytoplasmic and particle bound
        enzymes 142
    cytoplasmic free calcium 142
    endocytotic processes 142
    intracellular potassium 142
    membrane depolarization 142
    passage of immunoglobulins
        across cell membrane
        142
    release of primary secretory
        product 142
    sympathetic degeneration 142
Antibody-ligand binding characteristics
    137
Antibody-peptide interaction concept 135
    access to the brain 136
    affinity and biological activity
        139
    antibody-ligand binding
        characteristics 137
    arcuate nucleus 136
    BBB 136
    binding kinetics and biological
        activity 139
    circulating immunoglobulins 136
    circumventricular organs 136
    distribution volume of
        immunoglobulins 136
    epitopes and biological activity
        137
    ME 136
    passive immunization studies 137
    ventromedial hypothalamic area
        136
    ventromedial hypothalamic
        nucleus 136
Antigen-producing cells 185
Antihypertensive agents 191, 203
Antinociceptive potencies 382
Antisera against CRF 316
Antiserum against ANP 185
Antiserum against rat prolactin 4, 7
Apomorphine 42, 45, 48, 57, 383, 385,
    389, 413, 425

Apomorphine-induced ACTH release 45, 48
Arabinose 34
Arcuate nucleus 15, 22, 23, 114, 136, 184, 245,
    268, 347, 348, 352, 364, 394
Area postrema 31, 38, 128, 129, 131, 133, 193,
    195, 197-199, 201, 205, 239, 347,
    348
Arginine vasopressin, see AVP
Arginine vasotocin 62, 63
Aromatase activity 283, 292
Aromatic L-amino acid decarboxylase 22, 268
Arterial baroreflex 129, 131
Arterial pressure 43, 53, 113, 124-127, 129, 131,
    208, 221, 225, 354, 394, 418
Aspartate 28
Astrocytes 26, 35, 36, 267, 303, 308, 311, 323, 346,
    358, 359, 361, 366, 369
Astroglial end-feet 346
AT-10 tumor cells 316
Atrial appendectomy 218, 219, 225
Atrial appendectomy and natriuresis
    ANP 221
    atrial natriuretic system 218, 221
    auricular ablations 221
    CSF sodium concentration 221
    hypertonic saline 219, 220
    isotonic saline 220
    plasma ANP 220
    plasma osmolality 218, 221
    plasma sodium concentration 218
    sodium excretion 218, 219
    urine 220
    venous pressure 220
Atrial natriuretic factor, see ANF
Atrial natriuretic peptide, see ANP
Atrial natriuretic system 218, 219, 221, 225
AtT-20 cells 332, 334-338
Auricular ablations 221
Autoimmune diseases
    acetylcholine 362
    Alzheimer's disease 362
    amyloid 362
    astrocytes 358
    astroglial cells 360
    BBB 359, 360
    Bergmann astrocytes 361
    carcinomas 361
    Creutzfelt-Jacob disease 359
    dementia 362
    endothelial cells 358
    glial proliferation 362

hepatitis B virus 359
HIV virus 359
Hodgkin's disease 361
immunoglobulins 362
γ-interferon 358
killer cells 360
Kuru 359
lymphokinin 358
lymphomas 361
measles virus 359
multiple sclerosis 359
myelin sheaths 360
neuritic plaques 362
neurofibrillary tangles 362
neurons 360
oligodendrocytes 359
oligodendroglia 360
paraneoplastic cerebellar
        degeneration 360
pericytes 358
Purkinje cell antibodies 361
Purkinje neuronal degeneration
        361
sclerosing panencephalitis 359
sclerotic lesions 359
T cells 358
AV3V lesion on plasma osmolality 217
AVP 31, 41, 57, 60, 62, 63, 65, 67, 68,
        71-89, 91, 93, 94, 96-104,
        106-110, 113-122, 124,
        126-133, 137, 143, 144,
        145, 146, 154, 163, 172,
        178, 179, 185, 187, 188,
        190, 191, 194, 203,
        205-211, 217, 224, 225,
        227-229, 298, 299, 306,
        312, 313, 316, 320, 327,
        328, 336, 379, 392-394,
        397, 399-402, 404, 405,
        440, 442
    Brattleboro rat 81
    cortex 79
    hippocampus 79
    striatum 79
AVP binding to microvessels 81
AVP and performance 83
AVP binding sites at the BBB 77, 81
AVP infusion 101
AVP receptor 83
AVP releases ACTH 101
AVP responses to various stressors 99

AVP-induced information processing 76
AVP-positive cells 102
AVPergic fibers 73
Basement membranes 383, 384, 387, 390
Basophilic adenoma 172
BAT 371-373, 378
BBB 25-39, 41-45, 47-62, 64-69, 71, 72, 74, 76-78,
        80-83, 88, 89, 101, 103, 104, 191,
        192, 195, 197, 201, 203, 214, 238,
        307, 308, 313, 345-350, 352-354,
        357-360, 362-366, 369, 376, 377,
        381-390, 398, 403, 407, 411, 412,
        414, 415, 416, 424, 440
    allergic encephalitis 355
    anti-AVP 351
    anti-LHRH 351
    anti-rabbit IgG 351
    area postrema 348
    AVP 351
    BBB permeability 358
    BBB transfer constant 356
    BBB transport 353
    blood-CSF barrier 350
    brain uptake of $[^{125}I]$-IgG 357
    capillary surface area 358
    cationized form of albumin 357
    cellular infiltration 358
    CNS 348, 354
    colchicine 349, 350
    CSF IgG 352
    demyelination 358
    dextran 358
    DSIP 357
    endosomes 354
    Fc receptors 352
    habenular nucleus 348
    hippocampus 348, 350, 354
    hypothalamus 348
    IGF 357
    IgG 350, 352, 354, 358
    immunoglobulins 349
    insulin 357
    leu-enkephalin 357
    LHRH 351
    mannitol 357
    permeability to IgG 358
    retrograde transport 349
    supraoptic nucleus 349
    tanycytes 348
    thalamus 348
    tractus solitarius 349

transfer constants 357, 358
transferrin 357
vascular brain perfusion 354
BBB and CSF
    antibodies 346
    astroglial end-feet 346
    BBB 346, 347
    blood-CSF barrier 347
    choroid plexus 347
    cilia 346
    circumventricular areas 347
    CSF 346
    cytokines 347
    dura mater 348
    ependymal cells 346
    fenestrated vessels 347
    gap junctions 346
    glial cells 346
    horseradish peroxidase 347, 348
    IgG 347
    ME 347
    microvilli 346
    olfactory bulb 348
    OVLT 347
    pyrogens 347
    subarachnoid space 348
    Virchow-Robin spaces 348
BBB as a target for vasopressin 76
    accumulation of leucine 85
    alterations in the BBB
        permeability to amino
        acids 83
    AII 80
    ANP binding to microvessels 81
    AVP 77, 82, 84
    AVP binding sites 77, 83
    AVP binding to microvessels 81
    AVP receptors 83
    binding sites on isolated cerebral
        microvessels 79
    B_{max} 80
    Brattleboro rat 80, 82
    carrier-mediated amino acid
        transport 82
    β-casomorphin 81
    cerebrovascular permeability 78
    dehydrated animals 80
    endothelial cells 77
    hippocampal microvessels 80
    intracellular calcium storage 81
    K_D 80

    K_m 82
    leucine 82, 84
    leucine and phenylalanine transport from
        blood to brain 84
    Long-Evans rat 82
    lysine 82
    neocortex 80
    passive permeation 78
    peptide extraction 78
    phenylalanine 82, 84, 85
    receptor-mediated endocytosis 78
    receptors resembling the V_1 subtype 81
    specific binding sites 83
    striatum 80
    substance P 77
    tight junctions 77
    transcytosis of macromolecules by
        endothelial cells 78
    transport across the BBB 77
    V_1 subtype 83
    V_{max} 82, 84
    Wistar rats 80
    zonulae occludentes 76
BBB concept 25
BBB glucose transport 387
BBB: a modified tight epithelium 27
    brain endothelium 28
    electrical resistance 27, 28
    frog mesenteric endothelium 28
    frog muscle endothelium 28
    frog skin 28
    ion permeability 27
    necturus gallbladder 28
    necturus proximal tubule 28
    potassium permeability 28
    rabbit collection tubule 28
    rat proximal kidney tubule 28
    sodium permeability 28
    toad urinary bladder 28
Bergmann astrocytes 361
Binding kinetics and biological activity
    ACTH-induced corticosterone secretion
        140
    antibody to rCRF 140
    corticosterone 140
    off-rate constant 139
    on-rate constant 139
Binding protein 148, 153, 157-159, 162, 164, 268,
    283
Binding sites on isolated cerebral microvessels
    AII 79

ANP 79
AVP 79
insulin 79
oxytocin 79
Bioactivity studies 155
Biochemical responses 92
β-Blockers 43
Blood-brain barrier, see BBB
Blood flow 30, 37, 56, 57, 67, 68, 82, 88, 108, 178, 215, 354, 384, 388, 390, 416, 420-423, 425
Blood pressure 34, 43, 50, 51, 53, 55, 74, 101, 110, 117, 124-126, 130-133, 178, 179, 185, 187, 188, 191, 192, 197, 205, 206, 211, 221, 246, 255, 354, 363, 394, 418, 419, 421
Blood supply of the intermediate pituitary 234
Blood volume 99, 106, 226, 419, 425, 430
Blood-aqueous barrier 384
Blood-CSF barrier 347, 350, 357
Blood-retinal barrier 384, 389, 390
Blood-retinal permeability 385
Blood-to-brain transport 63, 354
B_{max} 77, 79-81, 201, 202, 247, 249
BMP 276
BNP 185, 190
Body fat stores 374
Body mass 436
Body temperature 75, 317-319, 323, 372, 375, 376, 380, 422
Bone marrow 277, 284, 285
Bone morphogenetic proteins, see BMP
Brain capillaries 25-29, 35, 37, 50, 67, 79, 85, 88, 346, 358, 365, 383, 386
Brain endothelial cells 25-27, 29, 30, 36, 65, 77
Brain endothelium 28
Brain extracellular fluid 26, 32
Brain natriuretic peptide, see BNP
Brain neurotransmitters 399
Brain trauma 34
Brain uptake index 29, 31
Brain uptake index of β-hydroxybutyric acid 31
Brain-to-blood transport 63, 65
Brainstem 130, 197, 199, 201, 203, 333, 402
Brattleboro rat 132, 133, 209
Brightness-discrimination reaction 75
Bromocriptine 2, 3, 165-175, 231, 241
Brown adipose tissue, see BAT
Brown fat activity 372, 374, 378, 380
Burns and head injuries 377
Butyric acid 172
Cabergoline 167, 168, 174
Calcium 81, 142, 334, 336, 338
cAMP 64, 109, 189, 279, 281, 283, 313, 336, 341
Cancer 328, 341, 360, 369, 371, 377
Capillary closure 388
Capillary surface area 358
Captopril 103, 110, 224, 229
Carbidopa 44, 48, 411
Carbohydrate ingestion 374, 379
Carcinoma cells 284, 292
Carcinomas 361
Cardiac hormones in the CNS
    ACTH 178
    actions on testicular steroidogenesis 183
    adrenal gland 181
    aldosterone 181
    AII 178, 179, 185
    ANP 179-181, 183-185
    ANP receptors 180
    anterior pituitary gland 179
    antiserum to ANP 185
    arcuate nucleus 184
    ANF 180
    AVP 178, 185
    blood pressure 185
    BNP 185
    brain 181
    cGMP 179
    circumventricular organs 178
    CNS 178
    CRF 180
    diuresis 181
    domperidone 179, 181
    dopamine 180
    dopaminergic D-2 receptor blocker 181
    electrolyte homeostasis 178, 185
    GH 184, 185
    GH releasing factor secretion 184
    hypophysial portal vessels 181
    kidney 181
    LH 182, 183
    LH secretion 184
    LHRH 184

ME 180
α-methyltyrosine 180, 181
natriuresis 181
natriuretic hormones 185
opiate antagonist naloxone 184
pituitary hormone secretion 180
pituitary hormone synthesis 180
plasma ANP 185
portal blood 184
prolactin 179-181, 184
somatostatin 184
TH 180
VIP 179
volume expansion 182, 185
water intake 178, 185
Cardiac hormones
    aldosterone production and
        secretion 178
    ANP 177
    ANP receptor 177
    β-receptor 177
    cGMP 177
    corticosteroid synthesis 178
    electrolyte homeostasis 177
    G-protein 177
    glomerular filtration rate 177
    guanylate cyclase 177
    half-life 177
    inositol triphosphate 177
    osmotic pressures 177
    prepro-ANP 177
    renin 178
Cardiac-derived ANP 185
Cardiac hormones in the CNS
    ANP 182
Cardiopulmonary bypass operation 421
Cardiovascular regulation 75, 87, 131,
        133, 205
Carrier-mediated amino acid transport
        82
Carrier-mediated transport systems 27
β-Casomorphin 62, 63, 78, 81
Castration 267, 281, 290, 295
Catechol-O-methyltransferase 30
Catecholamine turn-over 299
Catecholamines 41, 46, 49-51, 55-57, 102,
        104, 105, 108-110, 113,
        117, 120, 124, 130-133,
        164, 205, 206, 209, 228,
        241, 242, 266, 268, 299,
        309, 310, 312, 328, 339,

        341, 343, 378, 400, 401,
        405, 434, 442
Cationization 32, 33
    BBB 33
    diabetes 33
    glycosylation 33
Cationized albumin 34, 357, 365
CBS abnormalities
    diabetic capillaries 388
CEI 230
Cell polarity 29
    abluminal membrane 29
    alkaline phosphatase 29
    λ-glutamyl transpeptidase 29
    luminal cell membrane 29
    Na,K-ATPase 29
    5′-nucleotidase 29
    polarity of receptors 29
    transcellular transport 29
Cell size 258
Cellular infiltration 357, 358
Central command hypothesis 414
Central control of thermogenesis 371, 377
Central nervous system, see CNS
Central receptors 85, 238
Cerebral blood flow 37, 56, 57, 82, 354, 416, 425
Cerebral capillaries 88, 386-388
Cerebral glucopenia 419-421, 424
Cerebral hypoperfusion 419, 425
Cerebral ischemia 373, 377, 416-418, 421, 424
    ACTH 418, 420, 421
    arterial pressure 418
    blood volume 419
    bradycardia 418
    cerebral blood flow 416
    cerebral glucopenia 419, 420
    cerebral hypoperfusion 419
    cerebral oligemia 418
    cortisol 418, 419, 421
    GH 419, 420
    glucopenia 416
    glucose 420, 421
    heart rate 418
    hypophysial portal system 417
    hypotension 418
    hypoxia 416
    isotonic 420
    nitroglycerin 418
    prolactin 418-420
    syncope 417-419
    transient cerebral ischemia 421

Cerebral natriuretic system 214, 222
Cerebral oligemia 418
Cerebral origin of a natriuretic hormone
          211
Cerebrospinal fluid, see CSF
Cerebrovascular permeability 37, 56, 78,
          358
cGMP 64, 110, 177, 179, 186, 189, 190,
          203
Changes in hypothalamic dopaminergic
          tone 237
Chimeric peptides 34, 38, 365
Chloride 28, 37, 387
Chlorpromazine 165
Choleratoxin 279
Choline 27, 146, 398
Cholinergic tone 172
Choroid plexus 31, 65, 68, 196, 197, 201,
          347, 363
Chromaffin cells 141, 142, 232, 332, 339,
          340, 342
Chromosome 8 149, 163
Chronic dexfenfluramine treatment 429,
          431, 433, 435, 436, 441
Chronic renal failure 165, 186
Chorionic gonadotropin, see CG
Cilia 346
Ciliary body 384
Ciliary process 384
Circadian variation 31, 43, 56, 194, 249,
          251, 431, 434, 437-439
Circadian variation in the permeability of
          the BBB 31
Circulating AVP 113
Circulating immunoglobulins 136
Circulating mineralocorticoids 238
Circulatory homeostasis 227, 232, 234,
          239
Circumventricular organ 201
Classical GH deficiency 171
CNS 26, 35, 37-39, 41, 50, 55, 59, 60,
          62-64, 66, 67, 71, 87, 88,
          117, 130, 141, 146, 178,
          184, 185, 188, 189, 201,
          207, 226, 234, 239, 240,
          243, 246, 248, 252, 256,
          257, 291, 303, 311, 314,
          332-334, 345-349, 353,
          354, 357, 358-364, 366,
          367, 369, 374-377,
          379-382, 388, 398, 402,

          404, 407, 414, 416, 424
CNS abnormalities
          β-hydroxybutyrate 387
          antinociceptive potencies 382
          apomorphine 385
          basement membranes 383, 387
          BBB 381, 383, 384, 386
          BBB glucose transport 387
          blood flow 384
          blood-aqueous barrier 384
          blood-retina 386
          blood-retinal barrier 384
          blood-retinal permeability 385
          brain capillaries 383, 386
          capillary closure 388
          cerebral capillaries 387, 388
          chloride 387
          ciliary body 384
          ciliary process 384
          CNS function 381
          cognitive function 382
          D-glucose 387
          degenerative angiopathy 383
          diabetes 383-388
          dopamine 385
          EEG changes 382
          endothelial cells 383, 387, 388
          eyes 381
          fluorescein 385
          fluorescein angiography 384
          GH 383, 385
          glucagon 383, 385, 386
          glycosylation 385
          hemorrhage 388
          histologic abnormalities 387
          horseradish peroxidase 384, 385
          hyperglycemic 387
          hypoglycemia 382
          hypothalamic regions 385
          hypothalamus 384
          IgG 385
          insulin 383, 386
          insulin-dependent diabetes 382
          inulin 384
          ischemia 388
          kidneys 381
          levorphanol 382
          meperidine 382
          methadone 382
          methylglucose 387
          microaneurysms 384

microvasculature in diabetes 383
microvessels 384
morphine 382
Na-K-ATPase pump 387
neovascularization 388
nephropathy 381
neuroglycopenia 381
neuropathy 381
nociception and neuroendocrine
   abnormalities 382
periaqueductal gray 384
pericytes 383, 388
permeability 383, 384
permeability of the BBB 385
phenazocine 382
prolactin 385
propoxyphene 382
renal glomeruli 388
retina 381, 382, 388
retinal capillaries 383, 384, 386
retinal detachment 388
retinopathy 381, 384, 388
somatomedin 383
STZ 384
STZ-induced diabetes 387
sucrose 387
tomography 387
transport 386
TRH 383, 386
CNS function 59, 381
Codeine 33
Cognitive function 382, 388
Colchicine 313, 345, 349, 350, 361
Collagenase 47-55, 58
Collagenase to induce BBB dysfunction
   47-49, 53
Constipation 166
Control of appetite 374
Converting enzyme inhibition, see CEI
Corpus luteum 21, 277, 278, 281-283, 289,
   293
Cortex 3-5, 7, 8, 76, 79, 83, 89, 103, 140,
   172, 173, 234, 239, 240,
   245, 248, 268, 277, 293,
   304, 338, 342, 351, 352,
   355, 357, 358, 385, 387,
   390, 398, 399, 406, 414,
   423, 424, 433
Corticoid receptors
   Ammon's horn 248
   cortex 248

   corticosteroid receptor 246
   dentate gyrus 248
   glucocorticoid receptors 246, 247
   hippocampus 248
   hypothalamus 248
   lateral septum 248
   mineralocorticoid receptors 246, 247
   thalamic regions 248
Corticosteroid receptor sites
   forebrain 245
Corticosteroid receptor 243-246, 248-250, 254, 255
Corticosteroid receptor gene expression 249, 254
   glucocorticoid receptor mRNA 248, 249
   mineralocorticoid receptor mRNA 248,
      249
   receptor mRNA 248
Corticosteroid receptor sites 244-246
   glucocorticoid receptor cDNA 245
   glucocorticoid receptor mRNA 245
   neuronal and glial cells 244
   Northern blot 245
   receptor gene expression 245
   septo-hippocampal complex 244
   solution hybridization/S_1 nuclease
      analysis 245
Corticosteroid receptor types
   aldosterone 244
   aldosterone-binding receptor 244
   [^{3}H]-corticosterone 244
   corticosterone 244
   corticosterone-binding receptor 244
   [^{3}H]-dexamethasone 244
   glucocorticoid receptors 244
   hippocampal formation 243
   intracellular binding site 243
   limbic brain regions 243
   mineralocorticoid receptors 244
   motor neurons 243
Corticosterone 44, 46, 48, 50-52, 55, 57, 101, 105,
   108, 109, 115, 126, 128, 133,
   138-140, 194, 200, 204, 206, 233,
   235, 240, 242-244, 246-256, 265,
   268, 295-298, 301, 304-306, 312,
   313, 328, 332, 333, 338, 340, 341,
   372, 393, 394, 401, 427, 428, 430,
   432-434, 441, 442
[^{3}H]-Corticosterone 243, 244, 253
Corticosterone binding globulin 246
Corticosterone-binding receptor 244, 246, 252
Corticotropin, see ACTH
Corticotropin releasing factor, see CRF, CRH

Corticotropin releasing hormone, see
CRH, CRF
Cortisol 43-45, 48-50, 57, 58, 91, 93,
96-98, 100-102, 105, 109,
132, 149, 152, 153, 159,
161-164, 174, 246, 247,
255, 301, 311, 316, 320,
328, 333, 338, 342, 392,
400-403, 405, 411, 414,
416, 418, 419, 421-428,
434, 435, 436-442
late pregnancy 159
third trimester 152
Creutzfelt-Jacob disease 359
CRF 41, 49, 57, 100-104, 107, 108, 110,
131, 138, 140, 143, 144,
146-163, 172, 173, 176,
180, 185, 189, 201, 205,
206, 234, 298-301,
306-310, 312-314, 316,
320, 327, 340, 341,
372-377, 379, 392, 394,
397, 399-402, 405, 427,
428, 430-432, 434, 440,
441
ACTH releasing activity 159
detected in fetal blood 149
fetal plasma 160
gestation 147
hypothalamic and placental 149
maternal circulation 149
men 147
non-pregnant women 147
placental origin 148
pre-term labor 150
pregnancy 147
pregnancy-induced hypertension
149
premature rupture of the
membranes 150
suppressive effect of
glucocorticoids 160
CRF antibodies 144
CRF antiserum 157, 298, 299
CRF in the maternal circulation 149, 150,
153
CRF neurons in hypothalamus 185
CRF receptor antagonist 143, 373
CRF half-life 148
CRF-positive axons 100, 102
CRF-producing neurons 101

CRF release into the portal vasculature 102, 316
CRH 101, 103, 107, 113, 114, 161, 265, 266,
332-338, 340, 379
CRH in hypophysial portal blood 332
CSF 8-12, 21, 28, 31, 32, 36-38, 59-62, 67, 68, 106,
161, 212-214, 221, 222, 225,
345-347, 350, 352, 356-363,
365-367, 369
CSF and ECFV
cerebral natriuretic system 214
CSF concentration of sodium 212, 213
ECFV expansion 212, 213
hypertonic artificial CSF 212
intracerebroventricular infusions 214
mannitol 214
renal sodium excretion 212, 213
sodium salts in CSF 213, 214
urine 212, 213
CSF barrier 31
area postrema 31
brain extracellular fluid 32
choroid plexus 31
CSF 32
ME 31
OVLT 31
pineal gland 31
pituitary neural lobe 31
retrograde axonal transport mechanisms
32
subcommissural organ 32
SFO 32
Cushing's disease 172, 173, 175, 176, 337, 342
CV 205-502 168, 169, 174
structure 169
CycloLeu-Gly 63
Cyclosporin 27, 36
Cyproheptadine 64, 172, 173, 175, 176, 439
Cytokine stimulation
A23187 336, 338
ACTH 334-336
adenylate cyclase 336
adrenal medulla 339
AtT-20 cells 334
AVP 336
cAMP 334, 336, 337
catecholamines 339
chromaffin cells 339
corticosterone 338
corticotrophs 334
CRH 334-336, 338
cytokines 338, 339

dexamethasone 334
β-endorphin 334-338
endotoxin 339
enkephalin 339
epinephrine 339
extracellular sodium 336
forskolin 336, 337, 339
glucocorticoid secretion 338
GTP binding proteins 336
HPA axis 334, 338
human interleukin-1 336
indomethacin 338
IL-1 334, 338, 339
IL-1α 335, 337
IL-1β 335
IL-6 334
intracellular calcium 338
PGE$_2$ 334
phorbol ester 336, 339
POMC gene 337
POMC gene transcription 335
POMC mRNA 336
POMC-derived peptide 334, 335
protein kinase A 336, 337, 339
protein kinase C 336, 337, 339
sodium 334
sympathoadrenomedullary system 338
TPA 336, 337
VIP 339
Cytokine stimulation of the HPA axis 332, 334
Cytoplasmic and particle bound enzymes 142
Cytoplasmic volume 26, 27
Cytotrophoblast 148, 277
D-fructose 416
D-galactose 416
D-glucose 27, 29, 30, 380, 386, 387, 390, 402
D-mannose 416
DAMME 170, 313
DBH antibody uptake 142
dDAVP 117, 123, 124, 126, 129, 130, 132
epinephrine 123
norepinephrine 123
1-Deamino-8-D-arginine-vasopressin, see dDAVP
Decarboxylase inhibitor 30, 235
Decrease in AII receptors 228, 233
Decreased transport 65, 387

Defervescence 75
Degenerative angiopathy 383
Dehydrated animal 80
Dehydration 98, 188, 194, 199, 205-207, 210, 223, 225, 234, 241
Delta sleep-inducing peptide, see DSIP
Dementia 59, 66, 68, 69, 362, 364, 369
Demyelination 357-360, 365-367
Dentate gyrus 87, 243-245, 248
Deoxycorticosterone acetate, see DOCA
2-Deoxy-D-glucose 27, 402
2-Deoxyglucose 375
Depression 20, 108, 166, 172, 255, 394-396
Deprivation of food and water 66
Dexamethasone 101, 106, 108, 140, 159, 206, 209, 244, 246-250, 252-256, 334, 373, 374, 379
[^{3}H]-Dexamethasone 244, 253
Dexfenfluramine 427-442
ACTH 430-434, 437, 439, 440
aldosterone 433
android 436, 437
AVP 440
BBB 440
body mass 436
body weight 431, 436
chronic administration 434, 437
chronic treatment 431, 436
circadian variation 431, 437-439
circadian rhythms 440
corticosterone 432, 433
cortisol 436-440
CRF 428, 430, 434, 440
CRF neurons 428
CRF release 434
cyproheptadine 439
gynoid 436, 437
HPA axis 428
5-hydroxytryptophan 439, 440
hypoglycemia 440
hypoglycemia-stimulated ACTH release 439
hypophysial hormones 434
obese android 438
obese gynoid 438
pituitary adrenal function 434
prolactin 434, 436-439
serotonin 434, 439
serotoninergic agents 437
TRH 440
VIP 440

obesity 434
hypophysectomy 430
hypophysial portal blood 428, 430, 434
hypothalamic CRF release 432
obese humans 440
obese patients 434, 436
p-chloroamphetamine 432
paraventricular nucleus 428
paraventricular nucleus and serotonin 428
pituitary hormones in human 434
pituitary-adrenocortical function 432
5-HT$_{1A}$ receptors 428
5-HT$_2$ receptors 428
release of CRF 430
serotonin 428, 433
stress 433
stress-induced ACTH 431
thiopental 430
Dextran 354, 355, 357
Diabetes 33, 36, 88, 89, 114, 118, 120, 121, 129, 132, 133, 262, 264, 265, 269, 365, 381-396, 398, 403-405, 410, 425, 443
ACTH 393, 394
alloxan 393
amenorrhea 394
anterior pituitary 394
arcuate nucleus 394
aromatic amino acids 398
AVP 393, 394
BBB 398
body weight 396
brain neurotransmitters 399
choline 398
corticosterone 393
CRF 394
diabetic rats 397
dopamine 399
β-endorphin 394
fructose 398
FSH 393
genetic (*ob/ob* mice) diabetes 393
GH 393, 395-398
GHRH 398
GHRH mRNA 398

β-hydroxybutyrate 398
glucocorticoids 394
glucose 395, 398
glucose utilization 398
glycogen content 398
gonadal steroids 393
human diabetes 393
hypophysial portal blood 395
hypophysial portal circulation 398
hypothalamic dopamine 393
hypothalamic dopaminergic activity 394
hypothalamic neuropathy 398
hypothalamic somatostatin 397
hypothalamic-hypophysial-adrenal axis 393
hypothalamus 394, 399
insulin 394
LH 393, 394
LHRH 394
metoclopramide 394
microangiopathy 398
myoinositol 398
neurointermediate lobe 394
norepinephrine 399
nutritional status 396
pituitary gland 398
pituitary-gonadal axis 394
portal blood somatostatin 397
prolactin 393
Schwann cells 398
secretion of pituitary hormones 393
serotonin 399
sodium 398
somatostatin 395-398
somatostatin antiserum 396
somatostatin binding sites 398
somatostatin levels in hypophysial portal blood 396
somatostatin mRNA 397
somatostatin on GH release 397
somatostatin receptors 398
sorbitol 398
spontaneously diabetic (BB/W) rat 392, 394
STZ diabetic rats 392, 393
testosterone 394
thyroxine 393
TIDA neurons 393
TRH 393
TRH binding capacity 393
TRH neurons 393

TSH 393
Diabetes insipidus 88, 114, 118, 120, 121, 129, 132, 133
Diabetic BB/W rat 396
Diabetic capillaries 388
Diabetic rat 263-265, 268, 269, 384, 387, 388, 392-398, 403, 404
Diabetic retinopathy 171, 388-390
Diet-induced thermogenesis 371, 374, 377, 378, 380
Dietary related signals 374
Differentiation 243, 246, 257, 258, 265, 266, 275, 276, 280, 284, 285, 288, 292, 293, 313
Dihydroxyphenyalanine, see DOPA
Distribution volume of immunoglobulins 136
Diuresis 99, 106, 181, 188, 191, 226
Dizziness 165
DOCA 193, 194, 197-199, 203-206, 208, 209
Domperidone 42, 45, 48, 179, 181, 440
DOPA 3, 4, 9, 11, 15, 28, 30, 44, 48, 57, 170, 235, 258, 259, 264, 268, 389, 409-413, 425
DOPA accumulation 258
DOPA decarboxylase inhibitor 235
Dopamine 2-5, 8, 11, 13-15, 17, 18, 20-22, 43-45, 48, 55, 57, 58, 64, 106, 108, 110, 117, 120, 132, 141, 145, 146, 165-169, 171-174, 179-181, 211, 228-239, 241, 242, 259, 260, 263, 267, 268, 379, 383, 385, 386, 388, 393, 399, 440
Dopamine agonist 2, 166-169, 172, 174, 239
Dopamine agonists on corticosterone concentrations 44
Dopamine and AII
    ACTH infusion 231
    adrenal AII receptors 231
    aldosterone responses to AII infusion 231
    aldosterone responses to graded AII infusion 231
    AII during high sodium intake 230
    AII elevated plasma aldosterone 231

AII on plasma aldosterone 233
AII receptors 231
bromocriptine 231
decreased aldosterone response 231
dopamine agonists 231
dopamine antagonist 230
dopamine infusion 231
dopaminergic inhibition of endogenous AII 231
effects of dopamine agonists during low sodium intake 231
furosemide 231
high sodium intake 231
18-hydroxylase activity 231
interaction 230
metoclopramide 230, 231, 233
metoclopramide stimulates aldosterone secretion in vivo 230
modulate the adrenal actions of AII 230
modulatory action 231
plasma aldosterone 231
renin-angiotensin system 230
sensitization of adrenal to AII 231
sodium restriction 231
Dopamine and sodium intake
    adrenal medullectomy 234
    aldosterone 238
    AII 238
    AII receptors in the circumventricular organs 239
    arcuate nucleus 236
    area postrema 239
    BBB 238
    blood supply of the intermediate pituitary 234
    central receptors 238
    changes in hypothalamic dopaminergic tone 237
    choroidal plexus 239
    circulating mineralocorticoids 238
    circumventricular organs 238
    control of adrenal sensitivity to AII 238
    cortex 234
    DOPA decarboxylase inhibitor 235
    dopamine in plasma 234
    dopamine in renal tissue 234
    dopamine in urine 234
    dopamine turnover in the neurointermediate pituitary lobe 235-238
    dopaminergic activity 237

dopaminergic innervation of the
   pituitary 236, 238
dopaminergic input from the
   hypothalamus 234
dopaminergic input from the
   posterior pituitary 234
dopaminergic neurons 234
dorsal thalamic nuclei 239
ganglionic blockade 234
hypothalamic dopaminergic tone
   238
intermediate pituitary 234
intracerebroventricular
   administration of AII
   238
kidney denervated rats 234
lateral septum 239
ME 236, 239
α-methyltyrosine 235
mineralocorticoid antagonist 238
negative feedback by dopamine
   from the neural lobe
   237
neural lobe 236
non-neuronal dopaminergic
   secreting cells 234
nucleus of the solitary tract 239
OVLT 239
periventricular nucleus 236
pituitary regulation 234
sodium intake can affect
   hypothalamic
   dopaminergic activity
   237
sodium loading 234
sodium restriction 234
spironolactone 238
steroid hormones 238
SFO 239
Dopamine antagonist 230, 231, 233, 440
Dopamine-β-hydroxylase, see DBH
Dopamine during changes in sodium
   intake 234
Dopamine inhibits aldosterone secretion
   232
Dopamine in plasma 234
Dopamine in the cortex 234
Dopamine receptor blocker 45, 179
Dopamine receptors 44, 48, 171, 232, 241
Dopamine stimulates GH secretion 171
Dopamine to pituitary regulation 234

Dopamine turnover in neurointermediate pituitary
   lobe 237
Dopaminergic D-2 receptor blocker 181
Dopaminergic innervation of the pituitary 236,
   238
Dopaminergic input from the hypothalamus 234
Dopaminergic mechanisms and AII 230
Dopaminergic neurons 1, 3, 21-23, 234, 242, 257,
   267-269, 403
Dopaminergic regulation of aldosterone 231, 239,
   240
   ACTH 232, 235
   adrenal function and sodium diet 233
   adrenal glomerulosa 233
   adrenal sensitivity to AII 232
   aldosterone stimulating factor 232
   AII dependent steroidogenesis 232
   AII receptors 233, 235
   AII receptors in the intermediate pituitary
      233
   AII stimulated aldosterone 232
   AII stimulated aldosterone secretion 232
   attenuated aldosterone responses
      following hypophysectomy 232
   binding of ^{125}I[Sar^1,Ile^8]AII 236
   chromaffin cells 232
   circulatory homeostasis 232, 234
   corticosterone 235
   dopamine inhibits aldosterone secretion
      232
   dopamine levels 231
   dopamine receptors 232
   effect on plasma aldosterone 233
   β-endorphin stimulates aldosterone
      secretion 234
   high sodium diet 235
   high sodium intake 232
   hypertonic saline modifies α-MSH 234
   hypophysectomized animals 232, 233, 235
   hypothalamic CRF mRNA 234
   intermediate lobe of the pituitary 233
   intermediate pituitary AII receptors and
      sodium diet 233
   α-MSH increases aldosterone secretion
      234
   β-MSH increases aldosterone secretion
      234
   γ-MSH mediates natriuresis 234
   γ-MSH stimulates adrenal steroidogenesis
      234
   metoclopramide 232, 233, 235

neurointermediate pituitary 233
neurointermediate pituitary
          membranes 236
non-ACTH POMC peptides 232
normal sodium diet 235
paracrine regulators in the
          adrenal 232
plasma of ACTH 233
plasma aldosterone 233, 235
plasma aldosterone in
          hypophysectomized rats
          233
plasma corticosterone 233
plasma prolactin 233
plasma prolactin and sodium
          intake 233
PRA 233, 235
POMC mRNA 234
prolactin 235
sodium-loaded rats 232
somatostatin 232
Dopaminergic system in ACTH
          regulation 44
Dopaminergic regulation of aldosterone
          non-ACTH POMC peptides 234
DSIP 61, 62, 64, 65, 67-69, 365, 366
Dura mater 348
Dynamics of BBB functions
          circadian variation in the
                    permeability of the BBB
                    31
          rhythmic variations in BBB
                    function 31
Dynorphin 62, 63, 186
Dysfunction of the BBB 34, 47-49, 53-55,
          382
E_2 13, 313, 325, 341
ECFV 211, 213
EDLA 211
EEG changes 382
Effects on glomerulosa cell function 228
Effects on the adrenal glomerulosa cell
          229
Electrical resistance 27, 28, 34, 36, 38
Electrolyte homeostasis 177, 178, 185
Emotional stress 91, 99, 106, 107, 109,
          309, 421
Enalapril 193, 195, 197, 203, 208
Endocrine disorders 165
Endocrine manipulation
          adrenalectomy 200, 204

AII receptor 204
ANP binding sites 200, 204
anterior pituitary 204
corticosterone 200, 204
DOCA 204
hypophysectomy 200, 204
magnocellular 200
magnocellular paraventricular nucleus
          204
parvocellular paraventricular nucleus 200,
          204
pituitary gland 200
SFO 201, 204
supraoptic nucleus 200, 204
Endogenous CRF 143, 153, 155, 157, 299, 400,
          440
Endogenous digoxin-like activity, see EDLA
Endogenous pyrogens 316, 376
β-Endorphin 3, 48, 57, 66, 69, 78, 91, 93, 96, 99-
          101, 103-106, 108, 113, 161, 234,
          241, 301, 314, 316, 331, 334-338,
          342, 343, 365, 394, 435, 441, 442
β-Endorphin stimulates aldosterone secretion 234
Endothelial cell 26, 29, 78, 80, 81, 328, 357
Endothelial cell membrane 78
Endothelial mitochondria 65
Endotoxin 295-298, 300, 301, 311, 312, 319, 328,
          339, 342, 376, 378
Energy balance 87, 371, 374, 378, 379
Energy stores 374
Enkephalin 3, 30, 59, 60, 62, 63, 67-69, 170, 339,
          357, 365, 366
Enkephalin analogue 170
Enzymatic barriers 29
          aminopeptidase A degrading AII 30
          CEI 30
          aromatic L-amino acid decarboxylase 29
          catechol-O-methyltransferase 30
          decarboxylase inhibitor 30
          DOPA 30
          endothelial cell membrane 29
          enkephalin degrading aminopeptidase 30
          enzymes present in brain endothelial cells
                    30
          5HTP 30
          monoamine oxidase 30
Enzymes in brain endothelial cells 30
Ependymal cells 143, 346, 347
Epinephrine 49, 64, 98, 103-105, 108, 110,
          113-120, 123-125, 130, 132, 145,
          205, 209, 299, 320, 339, 340, 379,

389, 400, 405,
    425, 432
Epitopes 137, 138, 141, 144, 360, 361
Epitopes and biological activity
    ACTH 138
    anti-idiotype immunoglobulin
        138
    AVP antisera 137
    biological activity of a
        monoclonal antibody to
        ACTH 138
    CRF 138
    GH 138
    polyuria 137
Ergoline 166-168
Ergot alkaloid derivatives 165
Estradiol 12-15, 17-21, 23, 242, 279-282,
    285, 393, 403
Estrous cycles 3, 5
Ethanol 64-66, 69
Ethanol addiction 65, 66
Ethanol withdrawal 65, 69
Ethers 26
Exercise 46, 49, 50, 101, 102, 105, 106,
    383, 385, 389, 392,
    408-416, 421, 424, 425
Exercise-induced GH release 414
Experimental diabetes 36, 264, 383-386,
    389, 390
Extracellular calcium 336
Extracellular fluid volume, see ECFV
Extracellular potassium 229, 239
Extraction of peptides by rat brain 78
    ANP 78
    AVP 78
    β-casomorphin-5 78
    β-endorphin 78
    oxytocin 78
    substance P 78
Extrahypothalamic limbic brain regions
    243
Eyes 381
Fatigue 166
Fc receptor 365
Febrile body temperature 75
Feeding behavior 75, 256, 379
Fenestrated vessels 347, 348, 353
Fenfluramine 372, 378, 427, 428, 432,
    439-442
Fetal glucocorticoids 160
Fever 74, 75, 86, 87, 146, 303, 307, 310,

311, 314, 316, 317, 320,
    327, 334, 346, 364, 371,
    372, 375-378, 380
    BAT 376
    BBB 376, 377
    CNS 377
    CRF 376
    food intake 377
    genetically obese *(fa/fa)* Zucker rats 376
    HPA axis 376
    hypothalamic neurons 375
    IL-1 376
    IL-1α 376
    IL-1β 376
    IL-6 376, 377
    metabolic rate 375
    pyrogen 377
    pyrogens 375-377
    sympathetic outflow 376
    thermogenesis 376
    TNF 376
    viral or bacterial infection 375
Fibroblasts 280
Fluid intake 185
Fluid regulation 201, 203, 207
Fluorescein 384, 385
Fluorescein angiography 384
Fluoxetine 46, 52, 372, 432, 441-443
FMRF 59
Follicle stimulating hormone, see FSH
Follicle stimulating hormone releasing hormone, see
FSHRH
Follicular fluid 273-276, 278, 281-283, 285-288,
    290-292
Follicular fluid inhibin 273, 274, 287, 291, 292
Food intake 372-375, 377-380, 392, 427, 428, 435,
    441, 442
Forebrain 87, 193, 196, 199, 201, 203, 205, 225,
    245, 246, 267, 354, 355
Forskolin 189, 279, 281, 336, 337, 339
Free fatty acids 171, 404, 408-412, 424, 435
Frog mesenteric endothelium 28
Frog muscle endothelium 28
Frog skin 28
Fructose 382, 398, 408, 414-417, 424, 433
FSH 271, 272, 275, 276, 278-293, 299, 309, 327,
    391-393, 399, 400
FSHRH 271
Functions of the BBB 65, 362
Furosemide 231
G-protein 177

GABA 64, 142, 172, 176, 374
GABA aminotransferase inhibitor 172
Galactorrhea 2, 165
Ganglionic blockade 234
Gap junctions 346
General capillary 26
Genetic diabetes 393
Gestation 23, 147-149, 152, 155, 164, 283, 289, 291
GH 50, 56, 138, 145, 165, 168-170, 173-175, 184, 189, 269, 307, 315, 318, 328, 329, 370, 383, 385, 388, 389, 392, 404, 406, 408-411, 421, 424, 425, 442
   acetylcholine 170
   acromegaly 171, 172
   α_2-adrenoceptor agonist 170, 171
   β-adrenergic pathways 170
   β-adrenoceptor antagonist 170
   α_2-antagonists 170
   anti-cholinesterase drug 170
   bromocriptine 171, 172
   classical GH deficiency 171
   clonidine 170, 171
   diabetic retinopathy 171
   dopamine stimulates GH secretion 171
   dopaminergic stimulation with drugs 170
   enkephalin analogue, DAMME 170
   free fatty acid 171
   GH secretion 170
   GH-deficient children 171
   GHRH 168-171
   GHRH(1-29)-NH2 169
   glucose 171
   growth velocity 171
   idozoxan 170
   IGF-1 171
   ME 168, 169
   octreotide 172
   opiate and serotoninergic pathways 171
   pirenzepine 170
   pituitary 168
   prolactinomas 172
   propranolol 170
   pulsatile GH secretion 171
   pyridostigmine 170, 171

   release 168, 169, 171
   response to GHRH 170
   secretion 168, 170, 171
   somatostatin 168, 169, 171
   somatostatin analogue 172
   tall stature 171
   tumor shrinkage 172
   yohimbine 170
GH release 168-171, 184, 185, 315, 317-323, 327, 396, 397, 409, 410, 412, 414-416, 420-424
GH secretion 168, 170, 171, 174, 185, 321, 392, 395-398, 401, 409-414, 418, 424, 425
GH-deficient children 171
GH-releasing factor 319
GHRH mRNA 398
Glial cells 26, 36, 244, 245, 252, 308, 346, 362
Glial foot processes 26
Glial proliferation 362
Glomerular filtration rate 177, 187, 215, 221, 224
Glucagon 320, 383, 385, 386, 389, 400, 404, 425
Glucocorticoid receptor cDNA 254
Glucocorticoid receptor mRNA 245, 248-250, 252, 254
Glucocorticoid receptors 243, 244, 246-252, 254, 255, 328
Glucopenia 375
Glucoreceptor area 416
Glucose 27-30, 36, 57, 60, 81, 82, 89, 97, 98, 170, 171, 194, 269, 292, 299, 374, 375, 380, 382, 383, 386-392, 394-396, 398, 400-408, 410-417, 419-425, 435
   ACTH 411, 414
   apomorphine 413
   BBB 412, 414, 415
   central command hypothesis 414
   CNS 414
   cortisol 411, 414
   exercise 414, 416
   exercise-induced GH release 414
   fructose 414
   GH 411-416
   glucose deficiency 413
   hyperthermia 413
   insulin 412, 413
   isoleucine 412
   L-dopa 413
   leucine 412
   neuroendocrine response 414

peripheral control hypothesis
        414
phenylalanine 412
prolactin 411-414
serotonin 412
somatostatin 415
somatotropic activation 416
surgical trauma 413
tryptophan 412, 413
TSH 416
valine 412
water 414
Glucose deficiency 413
Glucose transporter 27, 28, 36
Glucose utilization 398
Glucose-sensitive neurons 401
Glucosensitive cells 416
Glucostat 416
Glutamate 28, 106, 142, 146
λ-Glutamyl transpeptidase 29, 36
Glycogen content 398
Glycogenolysis 280, 284
Glycosylation 33, 273, 385, 390
GnRH 165, 271, 272, 280, 282-284, 286,
        300
GnRH superagonists 165
Gonadotropin 2, 3, 22, 184, 189, 190, 271,
        272, 283, 285, 286,
        290-293, 299-301, 404
Gonadotropin releasing hormone, see
        GnRH
Granulosa cells 277, 280, 281, 283, 285,
        290, 292
Growth hormone, see GH
Growth hormone releasing hormone, see
        GHRH
Growth velocity 171
GTP binding proteins 336
Guanine triphosphate, see GTP
Guanylate cyclase 177, 186, 190
Gynoid 435-439
Habenular nucleus 348, 349
Half-life 148, 150, 177, 186, 249, 255, 342
Hallucination 166
Haloperidol 45, 48, 57, 165, 267
hCG 272, 279-281, 284, 299, 300
hCRF 140, 147, 148, 151, 154-157, 159
Headache 166
Heart rate 92, 94-97, 101, 113, 124, 133,
        134, 408, 418
Heart-lung machine 421, 425

Heat loss 319, 371, 375
Heat production 371
Hematopoietic cells 280, 284
Hemodilution 421-423
Hemorrhage 86, 99, 100, 103, 106, 122, 124, 130,
        132, 133, 188, 301, 388, 429-431,
        434
Heparin 410, 411
Heparinization 421
Hepatitis B virus 359
Heroin 33
Heterozygous 80-82, 88, 194, 200
High sodium intake 229-234, 236, 239
Hippocampal formation 243, 245, 247
Hippocampal microvessels 80, 81, 83, 86
Hippocampus 73-77, 79, 83, 87-89, 243, 245-249,
        251-254, 348-357
Histamine 31, 58, 64, 65, 172, 304
Histamine stimulated PTS 64
Histologic abnormalities 387
History of the BBB concept 25
HIV virus 359
HMG 279
Hodgkin's disease 361, 367
Homovanillic acid 28
Homozygous 80-82, 88, 194, 200, 225
Horseradish peroxidase 100, 107, 347, 364,
        384-386
HPA axis 251, 295, 297, 299, 300, 331-334, 339,
        340, 428, 429, 431, 441
        ACTH 332, 333
        ACTH release 332
        AtT-20 cells 332, 334
        brain stem 333
        circumventricular organs 333
        corticosterone 332
        CRH 332-334
        CRH antiserum 333
        CRH in hypophysial portal blood 332
        hypothalamus 333
        IL-1 333
        γ-interferon 332
        IL-1 334
        IL-1 binding sites 333
        IL-1α 333
        IL-1α mRNA 333
        IL-1β 332, 333
        IL-1β mRNA 333
        IL-2 332
        IL-6 333
        Newcastle disease virus 332

noradrenergic neurons 333
paraventricular nucleus 333
pituitary-adrenal activity 332, 333
POMC mRNA 334
rIL-2 334
TNF 332
TNF-α 333
HPA axis 49, 102, 107, 300, 301, 372, 376, 428
HPG axis 295, 297, 300
5HT 372, 375
5HTP 30
Human diabetes 383, 385, 392, 393, 398
Human IL-1 336
Human immunodeficiency virus, see HIV
Human menopausal gonadotropins, see
    HMG
Hydroxybenzylhydrazine, see NSD 1015
β-Hydroxybutyrate 28, 36, 387, 396, 398
β-Hydroxybutyric acid 31
6-Hydroxydopamine 43, 56
5-Hydroxyindoleacetic acid 28
18-Hydroxylase activity 228, 229, 231
11β-Hydroxysteroid dehydrogenase 246,
    252, 254
L-5-Hydroxytryptophan 30, 443
Hypercortisolemia 172
Hyperglycemia 28, 36, 265, 387, 390-392,
    396, 398, 403, 422, 424,
    425
    ACTH 392
    AVP 392
    cortisol 392
    CRF 392
    FSH 391, 392
    GH 392
    GHRH 392
    LH 391, 392
    LHRH 391
    prolactin 391, 392
    propranolol 392
    release of pituitary hormones
        391, 392
    somatostatin 392
    TRH 391
    TSH 391, 392
Hyperlipidemia 409, 411, 412, 424
    BBB 411
    carbidopa 411
    exercise 408, 409, 411, 412
    free fatty acids 409-412
    GH 409-412

GHRH 410
glucose 410
heart rate 408, 410
heparin 410, 411
hyperlipidemic patient 408-410
hyperlipoproteinemia 410, 412
hypoglycemia 412
insulin 410, 411
insulin-induced hypoglycemia 410
L-dopa 409, 411, 412
lipid 411
plasma cholesterol 410
post-exercise release of GH 411
Hyperlipidemia and growth hormone secretion
    409
Hyperlipoproteinemia 410, 412, 424
Hypernatremia 217
Hyperosmolar solutions 34, 38
Hyperosmolarity 34, 217
Hyperphagia 371, 372, 378
Hyperprolactinemia 2-9, 14, 20, 22, 165-168, 388,
    393
Hypertension 31, 34, 37, 38, 43, 50, 56, 57, 109,
    131, 133, 150, 152, 186-188, 191,
    193, 197, 203, 205, 207, 208, 209,
    225, 240, 242, 254, 390
    AII binding density 195
    AII receptor 197, 199
    AII receptor concentration 198
    ANP binding density 197
    blood pressure 197
    DOCA 197, 199
    enalapril 197
    spontaneously hypertensive rat 197
    Wistar-Kyoto rat 197
Hypertension-induced opening of the BBB 43
Hypertensive rat 131, 208
Hyperthermia 326, 413, 421
Hypertonic saline load 215, 217, 219
Hypertonic saline modifies α-MSH 234
Hypoadrenal 173
Hypoglycemia 28, 57, 98, 99, 107, 132, 265, 306,
    313, 381-383, 387, 388, 391, 394,
    399-402, 404, 405, 410, 412, 416,
    417, 424, 425, 439, 440, 442
    ACTH 400, 402
    AVP 397, 400-402
    AVP secretion into hypophysial portal
        blood 401
    BBB 416
    cortisol 400, 402

CRF 397, 400-402
CRF/AVP 402
D-fructose 416
D-galactose 416
D-mannose 416
fructose 415, 416
FSH 399, 400
GH 400, 401, 416, 417
glucagon 400
glucoreceptor area 416
glucoreceptors 401
glucose 402, 415-417
glucose-sensitive neurons 401
glucosensitive cells 416
glucostat 416
hypoglycemia on the secretion of
        pituitary hormones 400
hypoglycemia-induced release of
        GH 416
hypophysial portal blood 397,
        401
insulin 401, 415-417
insulin-induced hypoglycemia
        399, 400, 402
LH 399, 400
prolactin 399, 400, 416, 417
saline 417
serotonin 400
somatostatin 402
somatotrophs 401
TSH 400
Hypoglycemia-induced release of GH
        416
Hypophysectomy 10, 11, 18-21, 106, 194,
        200, 206, 209, 221-224,
        228, 232, 233, 235, 241,
        268, 312, 315, 340, 378
Hypophysial hormones 434
Hypophysial portal blood 3-5, 11, 13, 15,
        17, 21, 22, 107, 163, 265,
        268, 269, 332, 395-397,
        399, 401, 405, 428, 429,
        442
Hypophysial portal system 306, 417
Hypotension 165, 167, 171, 418, 425
Hypothalamic and placental CRF 149
Hypothalamic control 107, 172, 184, 190,
        312, 327, 378, 394
Hypothalamic CRF mRNA 234
Hypothalamic CRF release 432
Hypothalamic function 326, 372

Hypothalamic neuropathy 398
Hypothalamic peptides 113, 147, 165, 168
Hypothalamic somatostatin 170, 184, 265, 269,
        392, 396-398, 402, 404
Hypothalamic-dopaminergic neurons 257
Hypothalamic-hypophysial portal system 306
Hypothalamic-hypophysial-adrenal axis 393
Hypothalamic-pituitary axis 49, 102, 105, 107, 108,
        149, 161, 165, 172, 173, 250, 266,
        269, 295, 300, 301, 316, 320,
        325-327, 331, 372, 376, 391, 392,
        399, 407, 422, 423, 427, 428, 442
Hypothalamic-pituitary-adrenal, see HPA
Hypothalamus 3, 15, 20, 23, 49, 71-73, 76,
        83, 100, 102, 107-109,
        113, 114, 130-132, 134,
        139, 143, 147, 161, 165,
        171, 174, 175, 185, 186,
        209, 232, 234, 238, 241,
        245, 246, 248, 250, 252,
        257, 258, 264-267, 269,
        278, 280, 282, 283, 285,
        301, 306-310, 313, 315,
        316, 319, 320, 326, 327,
        332, 333, 341, 348, 369,
        370, 372, 373, 376, 378,
        380, 384, 385, 388,
        390-392, 394, 398-400,
        402-404, 406, 416, 425,
        432, 440, 442
Hypothyroidism 165, 264, 265, 267, 269, 393
Hypoxia 99, 416
Idozoxan 170
IGF 171, 281, 289
IGF-1 171
IgG 138, 143, 345, 347, 350-359, 362-365, 385, 386
Immune cells 295, 304, 316, 331, 333, 345, 358
Immune responses 310, 315, 316, 367
Immune system
        corticosterone 332
        cytokines 332
        HPA axis 332
        hypothalamic neurons 332
        immune response 332
        IL-1 332
        IL-6 332
        TNF 332
Immunological activity 273
Immunological aspects 31
        BBB 31
        lymphatic drainage 31

T-cells 31
Immunoneutralization 7, 9, 14, 15, 20, 21, 185, 272, 274, 286, 291, 299, 305, 306, 332, 333, 400
Immunoneutralization of CRF 306
Immunoneutralized inhibin 281
Impotence 165, 383
In situ activity of TH 4-9, 11, 14-19, 21
In situ molar activity of TH 4-7, 9, 11-16, 18, 19, 21, 263
Interleukin, see IL
In vitro labeling of peptide receptors
   AII receptors 194, 195
   ANP receptors 195
Increases in AVP levels in the hypophysial-portal circulation of conscious sheep 100
Increasing lipid solubility 32
   cholinesterase 33
   codeine 33
   heroin 33
   morphine 33
Indomethacin 320, 322, 323, 325, 327, 338
Infected mouse spleen cells 305
Infertility 2, 166, 286
Inhibin 271-293
   FSH 272
   GnRH 272
   GnRH receptors 272
   hCG 272
   immunological activity 273
   inhibin activity 271, 273
   inhibin secretion 273
   LH 272
   pituitary cells 272
Inhibin activity 271, 272, 276, 278, 281, 282, 286, 288, 290
Inhibin and activin
   aromatase activity 283
   carcinoma cells 284
   corpora lutea 283
   fibroblasts 280
   follicular fluid 282
   FSH 282
   glycogenolysis 284
   GnRH 282-284
   hCG 284
   hematopoietic cells 280
   hypothalamus 280

   inositol phosphatidyl metabolism 284
   Leydig cells 283
   liver 280
   oocytes 280
   ovary 280
   oxytocin 283
   pancreas 280
   pancreatic islands 284
   pituitary stalk blood 282
   placenta 280
   progesterone 283, 284
   protein kinase C 284
   receptors for activin 284
   rete testis 283
   secretion of insulin 284
   Sertoli cell 283
   spermatogonia 283
   testis 280
   TGF-β 283
   theca cells 283
Inhibin bioactivity 273, 275, 290
Inhibin production by granulosa cells 281
Inhibin secretion 273, 279, 281, 289, 290, 293
Inhibin subunits 273, 275, 278, 282, 292, 293
Inhibin-β-subunits 275
Inhibition of prolactin 189
Inhibitory effect of CS on the hypothalamic synthesis of CRF 102
Injuries
   burns and head 377
   cancer 377
   CRF 377
   IL-1 377
   IL-6 377
   peripheral ischemia 377
   sepsis 377
   sympathetic nervous system 377
Injury 255, 339, 371, 377-380, 421, 422
Inositol phosphatidyl metabolism 284
Inositol triphosphate 177
Insulin 34, 57, 79, 80, 88, 89, 107, 132, 145, 171, 280, 281, 284, 292, 306, 357, 365, 375, 379-383, 385-391, 394-399, 401-406, 408, 410-413, 415-417, 421, 423-425, 435, 440, 442, 443
Insulin-dependent diabetes 382, 385, 388, 389
Insulin-induced hypoglycemia 57, 107, 132, 306, 383, 394, 399, 402, 404, 405, 410, 440, 442
Insulin-like growth factor, see IGF
IL-1

ACTH releasing capacity 309
antisera against CRF 316
AT-10 tumor cells 316
AVP 316
CRF 316
CRF in hypophysial portal blood
316
FSH 309
GH 309, 317-319
GH-releasing factor 319
LH 309
molecular weight 316
prolactin 309, 317-319
rectal temperature 317
TSH 317, 319
IL-1 as a neuromodulator 308
IL-1 binding sites 333
IL-1 bioactivity 308
IL-1 mechanism of action
ACTH 308
adrenaline 309
BBB 308
brain 307
catecholamines 310
CRF 309
CRF neurons 308
induction of fever 307
IL-1 as a neuromodulator 308
IL-1 bioactivity 308
intermediate lobe 309
ME 309
mediobasal hypothalamus 308
mRNA coding 308
MSH 309, 310
noradrenaline 309
norepinephrine metabolism 307
opioid receptor binding 307
OVLT 309
POMC 309
preoptic area 308
prolactin 308
slow wave sleep 307
TSH 309
IL-1 site of action
ACTH 307
CRF 307
GH 307
hypothalamus 307
IL-1α 307
IL-1β 307
LH 307

prolactin 307
TSH 307
IL-1β 143, 301, 305, 306, 308, 311, 341, 342, 370,
379, 380
IL-1α mRNA 333
IL-2 305, 310, 332-334, 340-342
IL-6 305-307, 312, 313, 331-336, 339-342, 376, 377,
380
IL-7 331
γ-Interferon 305, 325-328, 332, 358
ACTH 326
GH 325, 326
hyperthermia 326
prolactin 326
TSH 326
Intermediate lobe 85, 110, 172, 175, 189, 233, 236,
237, 242, 300, 309, 310
Internalization of antibodies 143, 144
Interview 91, 94-99, 101-104
Intracellular binding sites for corticosteroids 243
Intracellular calcium 81
Intracellular potassium 142
Intracerebral pituitary grafts 9
Intracerebroventricular infusion 181, 184, 214,
254, 375
Intravascular pressure 34, 37
Inulin 26, 36, 62, 78, 82, 384, 386
Ion permeability 27
Ischemia 34, 373, 377, 388, 416-418, 421, 424
Isotonic saline load 216, 217
Ka 139, 140, 158, 201
Kainic acid 64
K_D 77, 79-81, 139, 202, 250, 273, 274, 279, 283,
331, 361, 376
Ketone bodies 28, 36, 375
Kidney 3, 5, 6, 8, 9, 28, 106, 120, 132, 133, 177,
181, 187, 189, 209, 228, 234, 237,
242, 244, 246, 253, 254, 277
Killer cells 304, 360
K_m 65, 82, 84, 387
Kuru 359, 366
Kyotorphin 59
L-dopa 57, 170, 268, 389, 409-413, 425
Lamina terminalis 31, 196, 199, 214, 239, 246, 347
Large neutral amino acids 28, 82, 85, 88, 375
Lateral septum 239, 244, 245, 247, 248
Lateral ventricle 8, 10, 11, 60, 196, 212, 296, 297
Learning performance and AVP 76
Leucine 62-65, 68, 82-85, 88, 357, 365, 412
Leucine enkephalin 59, 63
Levorphanol 382

Leydig cells 190, 277-279, 290
LH 2, 3, 49, 54-56, 58, 108, 141, 144,
182-185, 189, 271, 272,
280-283, 285, 286,
290-292, 295, 297, 299,
300, 307, 309, 320, 327,
348, 391-394, 399, 400,
426
LH concentrations 49
LHRH 54, 62, 63, 68, 144, 154, 156, 183,
184, 265, 271, 285, 286,
328, 348, 351, 353, 364,
391, 394, 404
Libido 165
Lighting 59, 64-66, 68
Limbic brain regions 243
Limbic structures 75, 85, 253
Lipid 26, 27, 29, 32, 33, 35, 36, 61-63,
347, 374, 410, 411
Lipid solubility 26, 29, 32, 33, 35, 36,
61-63
  amines 26
  amino acids 29
  brain uptake index 29
  cyclosporin 27
  D-glucose 29
  esters 26
  ethers 26
  lipophilicity 27
  partition coefficient 26, 29
  transport mechanisms 29
Lipid solubility and molecular weight of
  iodinated peptides 61
Lipophilicity 27, 67
Lipopolysaccharide, see LPS
Liposome 32, 33, 38
Liposome entrapment
  brain trauma 34
  ischemia 34
  liposomal superoxide dismutase,
    a free radical scavenger
    34
  permeability of the brain to
    liposomes 33
β-Lipotrophic hormone, see β-LPH
Lisuride 167, 168, 174
Liver 8-11, 33, 245, 246, 254, 280, 284,
303, 310, 402
Liver grafts 9, 11
Localization of inhibin 277, 278, 289, 292
  adrenal 277

bone marrow 277
CNS 277
corpus luteum 277
cortex 277
cytotrophoblast 277
granulosa cells 277
hypothalamus 278
inhibin activity 278
inhibin-subunits 277
kidney 277
leydig cells 277, 278
ovarian follicular fluid 278
pituitary gland 277, 278
placenta 278
Sertoli cells 277, 278
spleen 277, 278
theca cells 277
Long-term effects of an altered brain environment
34
β-LPH 91, 93, 96
LPS 295-300, 301, 304, 305, 312, 319 333, 339,
340, 342
Luminal cell membrane 29
Luteinizing hormone, see LH
Luteinizing hormone releasing hormone, see LHRH
Lymphatic drainage 31
Lymphocytes 31, 105, 141, 304, 316, 327, 331, 333,
358, 360, 366, 367
Lymphokines 295-297, 299-301, 331, 363
Lymphokinin 358
Lymphoma 311, 359, 367
Lysine 62, 63, 82, 83, 117, 356
Lysine vasopressin 63
Macrophages 295, 303, 304, 308, 310, 319, 325,
332, 333, 339
Magnocellular neurosecretory system 100, 102,
133
Magnocellular paraventricular nucleus 100, 200,
202
Mannitol 34, 213, 214, 357
MAP 113, 124, 418
Mass of TH 4-6, 8-11, 15, 16, 18, 21, 263
Mast cells 304
Maternal plasma 147, 149-157, 159-164
Maturation 26, 258, 265-268, 280, 291
ME 3-19, 21-23, 31, 56, 57, 75, 100-102, 107, 108,
110, 111, 114, 132, 136, 143, 163,
165, 168, 169, 180, 192, 193, 196,
201, 236, 239, 258, 259, 261-264,
266-268, 285, 306-309, 313, 315,
331, 332, 340, 347, 348, 351, 352,

391, 394, 403,
                404, 427, 442
Mean arterial pressure, see MAP
Measles virus 359, 360
Mechanism of action 58, 135, 136, 144,
                284, 307, 332, 441
Medial basal hypothalamus 114-116, 118,
                120, 333
Medial bregma 199
Medial parvocellular paraventricular
                nucleus 100-102
Median eminence, see ME
Median preoptic nucleus 193, 195-199,
                203, 205, 214
Mediobasal hypothalamus 23, 49, 265,
                269, 308, 333, 388, 394,
                400, 404
Melanocyte stimulating hormone, see
                MSH
Membrane depolarization 142
Meperidine 22, 382
Mesotocin 62
Mesulergine 167, 174
Met-enkephalin 60, 63, 339
Metabolic rate 371, 372, 374, 375, 377,
                380
Methadone 382, 383
1-Methyl-4-phenyl-tetrahydropyridine
3-O-Methylglucose 27
Methylglucose 27, 387
α-Methyltyrosine 180, 181, 235
Metoclopramide 165, 174, 230-233, 235,
                236, 239, 241, 394, 404,
                440
Microaneurysms 384
Microangiopathy 381, 390, 398
Microcirculation 72
Microglia cells 303
Microvasculature in diabetes 383
Microvilli 346, 347
Mineralocorticoid antagonist 238, 252
Mineralocorticoid deficiency 194
Mineralocorticoid receptor mRNA 245,
                246, 249, 252
Mineralocorticoid receptors 206, 243-254
                corticosterone binding globulin
                246
                11β-hydroxysteroid
                        dehydrogenase 246
                mineralocorticoid receptor
                        mRNA 246

salt appetite 246
MIS 276
Mitochondrial content 26, 36
Modification of transport
                acetylcholine 64
                aged choroid plexus 65
                aging 64, 65
                albumin 64
                aluminum 65, 66
                Alzheimer's disease 66
                amino acids 64
                amyotrophic lateral sclerosis 66
                analgesia 64, 66
                AVP 65
                BBB 64
                brain endothelial cells 65
                cAMP 64
                cGMP 64
                CNS 64
                cyproheptadine 64
                decreased transport 65
                deprivation of food and water 66
                dopamine 64
                DSIP 65
                endothelial mitochondria 65
                enkephalin 66
                epinephrine 64
                ethanol 66
                ethanol addiction 65, 66
                ethanol consumption 64
                ethanol withdrawal 65
                functions of the BBB 65
                GABA 64
                histamine 65
                histamine stimulated PTS 64
                5-hydroxytrytophan 64
                increased transport 65
                K_m for brain-to-blood transport 65
                kainic acid 64
                leucine 64, 65
                lighting 64, 65
                methionine enkephalinamide 65
                monoamines 64
                morphine 65, 66
                naltrexone 65, 66
                opiates 64
                Parkinsonism-dementia syndrome of
                        Guam 66
                peptide BBB interactions 64
                peptides across the BBB 65
                restraint 64

serotonin 64
serotonin agonists 65
serotonin antagonist 65
stress 64, 65
stresses selectively affect BBB
    transport 64
transmembrane diffusion and
    saturable transport of
    peptides 63
transport of AVP 65
Tyr-DSIP 64
V_{max} for the brain-to-blood
    transport 65
water loading 65
Modulation of adrenal sensitivity to AII
    230
Modulation of renal sodium 227
Modulatory effect of plasma potassium
    227
Molecular weight 6, 27, 61, 62, 153, 154,
    163, 177, 273, 274, 287,
    288, 305, 316, 319, 346,
    354, 355, 357, 379, 384
Monoamine oxidase 30
Monoamines 22, 30, 35, 37, 41, 44, 50,
    51, 58, 59, 64, 66, 378,
    399
Monocarboxylic acid 28
Monocarboxylic acid transporter 28
Monocytes 303, 304, 331-333, 338, 339,
    342
Monokines
    ACTH 304, 306
    antibody to IL-1 305
    basophils 304
    corticosterone 305, 306
    endorphin 304
    glucocorticoids 304
    histamine 304
    immune cells 304
    immunoneutralization 305
    infected mouse spleen cells 305
    IL-1 306
    IL-1α 305, 306
    IL-1β 305
    IL-2 305
    IL-6 305, 306
    γ-interferon 305
    leucocytes 305
    LPS 305
    lymphocytes 304
    mast cells 304
    monocytes 304
    pituitary-adrenal activity 305
    TNF 305
Mood change 166
Morphiceptin 59
Morphine 3, 22, 33, 65, 66, 382, 383, 388
Morphology of brain endothelial cells 25
    anatomical basis of the BBB 26
    brain capillary 26
    cytoplasmic volume 26, 27
    general capillary 26
    mitochondrial content 26
    perivascular space 25
    pinocytotic vesicles 25
    tight junctions 25
Motor neurons 243
Motor neurons of the spinal cord 243
MPTP 1, 2, 22
mRNA for IL-1 308
mRNA for inhibin 278, 279
α-MSH 62, 67, 86, 172, 175, 229, 234, 241, 242,
    298, 300
α-MSH increases aldosterone secretion 234
β-MSH increases aldosterone secretion 234
γ-MSH mediates natriuresis 234
γ-MSH stimulates adrenal steroidogenesis 234
MSH 57, 62, 67, 86, 172, 175, 229, 234, 241, 242,
    298, 300, 309, 310
Müllerian inhibiting substance, see MIS
Multiple sclerosis 359, 365-368
Myelin sheaths 360
Na,K-ATPase 29, 211
NA,K-ATPase dependent efflux 29
Na-K-ATPase pump 387
Naloxone 99, 105, 184, 312, 394
Naltrexone 65, 66, 106
Nasal congestion 166
Natriuresis 181, 187, 211, 213, 214, 217-219,
    221-226, 234, 241, 242
Natriuresis induced by ECFV expansion 217
Nausea 165, 412
Necturus gallbladder 28
Necturus proximal tubule 28
Negative feedback by dopamine 237
Nelson's Syndrome 172, 175, 176
Neocortex 80
Neovascularization 388
Nephropathy 381
Neural lobe 100-102, 121, 236, 237, 242
Neurite length 258

Neuritic plaques 362
Neuroendocrine function 37, 52, 55, 56,
         177, 184, 185, 269, 380,
         403, 442
Neuroendocrine response 38, 56, 414,
         416
Neurofibrillary tangles 362
Neurogenic influences 30
     adenosine-2 31
     α-adrenergic receptors 31
     β-adrenergic receptors 31
     angiotensin 31
     AVP 31
     BBB 30, 31
     blood flow 30
     calcitonin gene-related peptide
         30
     histamine 31
     neuropeptide Y 30
     noradrenergic terminals 30
     serotonin 31
     substance P 30
     VIP 30
     water permeability 30
Neuroglycopenia 381
Neuroimmunomodulation 316, 328
Neurointermediate pituitary lobe 101,
         235, 237
Neuromedin N 59
Neuronal and glial cells 244, 245, 252
Neuronal uptake of antibodies to enzymes
     antibodies to glutamate
         decarboxylase 142
     chromaffin cells 141
     DBH 141
     DBH antibody uptake 142
     GABA-neuron 142
     retrograde transport 141
     sympathetic ganglia 142
Neuropathy 381, 382, 388, 398, 405
Neuropeptide Y 30, 144
Neuropharmacological manipulation 165
Neurophysin 73, 75, 107, 143, 146
Neurotoxic substances 2
Neurotoxins 59
Neurotropic agents 2
Newcastle disease virus 295, 301, 304,
         332, 341
Nitroglycerin 418
Nociception and neuroendocrine
         abnormalities 382

Non-ACTH POMC peptides 232, 234
Non-lymphoid cells 331
Non-neuronal dopaminergic secreting cells 234
Non-shivering thermogenesis 371
Nontransportable peptide 34
Noradrenergic neurons 37, 141, 146, 309, 333,
         341, 373
Noradrenergic terminals 30
Norepinephrine 22, 48-50, 94, 98, 102-105, 110,
         113-126, 130, 132, 141, 142, 175,
         267, 299, 301, 307, 313, 314, 320,
         328, 333, 341, 389, 399, 401, 405,
         425
Norepinephrine metabolism 301, 307, 313, 333,
         341
Northern blot 245, 273, 277
NSD 1015 3, 259
Nuclear receptors 258, 263
Nucleic acid precursors 27
5′-Nucleotidase 29
Obese rats 372, 378, 380
Obesity 170, 372, 374, 378-380, 427, 434, 440, 442,
         443
Obesity in aging animals 372
Octreotide 165, 172
Off-rate constant 139
Olfactory bulb 196, 198, 199, 348
Oligodendrocytes 359, 360
Oligodendroglia 346, 360
On-rate constant 139, 140
Ontogenesis 257, 258, 265, 266
Oocytes 280
Open heart surgery 421
     ACTH 422, 423
     adrenal cortex 423
     anesthetic drugs 421
     cardiac bypass 423
     cardiopulmonary bypass surgery 421, 422,
         423
     cortisol 422, 423
     emotional stress 421
     GH 422, 423
     glucose 423
     heart-lung machine 421
     hemodilution 421, 423
     heparin 421
     hyperglycemia 422, 423
     hypothermia 421
     injury 421
     insulin 423
     syncope 421

Opening of the BBB 32, 34, 35, 43, 44, 50, 55, 82
Opiate antagonist naloxone 184
Opiates 64, 170, 171, 295
Opioid receptor binding 307
Opioid-mediated mechanisms inhibit AVP release 99
OVLT 193, 196, 214, 239, 307-309, 347, 364
Osmotic stimuli 122
Osmotically free water 212, 215, 222
Ovarian hormones 11, 13-20, 22
Ovarian hormones and TIDA neurons 11
    aged brain 17
    aged female rats 14, 15
    aged ovariectomized rats 15, 16
    arcuate nuclei 15
    DOPA 15
    dopamine 11, 13-15, 17, 18
    estradiol 12, 13, 17-19
    estrogen 14
    hyperprolactinemia 14
    hypophysial portal blood 11, 13, 15, 17
    immunoneutralization 14, 15
    in situ activity of TH 14, 16-18
    in situ molar activity of TH 18
    male rats 17
    mass of TH 15
    ME 13-15, 17-19
    ovarian hormones 12, 13, 15, 16
    periventricular region 15
    portal blood 18
    progesterone 12-14, 17-19
    prolactin 14, 15
    prolactin antiserum 14
    TH 13, 18, 19
    TH-containing neurons 15
    TIDA neurons 12-15, 17, 18
    young adult female rats 11
    young brain 17
Ovary 12, 271, 275, 277, 280, 281, 283, 285, 287-289, 292
Oxytocin 63, 71, 73-75, 78, 79, 81, 86-88, 106, 107, 110, 114, 121, 131, 133, 144, 163, 164, 211, 280, 282, 283, 291, 328
    cortex 79
    hippocampus 79
    striatum 79

p-Chloroamphetamine 432, 441
Pancreas 147, 161, 174, 280, 392, 402
Pancreatic islands 284
Paraneoplastic cerebellar degeneration 360-362, 368, 369
Paraventricular nucleus 41, 56, 74, 86, 87, 100-104, 109, 110, 114, 128, 133, 134, 143, 144, 192, 193, 195, 197, 198-209, 245, 256, 265, 269, 307, 308, 333, 372-374, 379, 427, 428, 440, 441
Parkinsonism 1, 22, 30, 66, 68
Parlodel LAR 166, 167, 174
Partition coefficient 26, 29
Parvocellular paraventricular system 100
Passage of peptides 28, 59-61, 66
    albumin 62
    arginine vasotocin 62
    CNS 62
    CSF 62
    DSIP 61
    inulin 62
    lipid solubility and molecular weight of iodinated peptides 61
    lysine vasotocin 62, 63
    α-MSH 62
    passage of peptides across the BBB 61
    peptide transport systems 62
    TRH 62
    Tyr-DSIP 62
    Tyr-MIF-1 62
Passage of peptides across the BBB 59-61, 66
Passage of substances from blood to brain 25, 32
    cationization 32
    lipid solubility 32
    liposome 32
    opening of the BBB 32
    receptor-mediated transport 32
Passive immunization studies 135, 137, 141
Passive permeation 78
Patchy protein extravasation 34
Peptide receptor and stress
    AII binding 200
    anterior pituitary 200
    optic tract 202
    stressed rat 202
Peptide receptor and water balance
    AII receptor 199
    Brattleboro rat 199
    DI rat 200
Peptide receptors and stress

AII receptor 202
AII receptors 200
B_{max} 202
K_d 202
magnocellular paraventricular
    nucleus 202
parvocellular paraventricular
    nucleus 200, 202
SFO 202
Peptide receptors and water balance
    AII Binding Density 199
    ANP binding sites 201
    ANP receptors 199
    Brattleboro rat 200
    dehydration 199
    DI rat 199
    Long-Evans rat 200
    SFO 199
Peptide-T analog 62
Peptide transport system, see PTS
Peptides across the BBB 59-61, 65, 66
Peptides and the BBB
    AVP 60
    CSF 59, 60
    FMRF 59
    gut-brain axis 59
    kyotorphin 59
    Leu-enkephalin 59
    Met-enkephalin 59, 60
    morphiceptin 59
    neuromedin N 59
    passage of peptides across the
        BBB 60
    proctolin 59
    TRH 59
    tuftsin 59
    Tyr-MIF-1 59
Pergolide 57, 167, 174
Periaqueductal gray 384
Pericytes 358, 383, 388, 390
Peripheral control hypothesis 414
Peripheral ischemia 377
Perivascular space 25, 347
Permanent neuronal damage 34
Permeability of the BBB 31, 34, 42, 43,
        45, 50, 52, 55, 81-83, 357,
        358, 363, 383, 385, 387
    arabinose 34
    BBB dysfunction 34
    intravascular pressure 34
    mannitol 34

patchy protein extravasation 34
protamine sulphate 34
urea 34
Permeability to IgG 358
PGE_2 313, 325, 334, 341
Phenazocine 382
Phenylalanine 28, 82-85, 412
Phenylketonuria 28
Phorbol ester 336, 339
Phorbol 12-0-tetradecanoate 13-acetate, see TPA
Phosphorylation of TH 263
Physiological control of aldosterone 228
Pineal gland 32
Pinocytotic vesicles 25, 346
Pirenzepine 170
Pituitary
    ANF 222
    AII 224
    ANP antagonist 224
    captopril 224
    control of ANP secretion 223
    ECFV expansion 223
    hypophysectomized rat 221
    infusion of isotonic saline 221
    plasma AII 223
    PRA 223
    sodium excretion 222, 224
    volume loading 222
    volume natriuresis 224
    water excretion 222
Pituitary adenoma 2, 3
Pituitary AII receptor 233
Pituitary gland 3-6, 9, 10, 20, 21, 54
Pituitary graft 3-7, 9, 11
Pituitary venous effluent of conscious horses 100
Pituitary-adrenal activity 144, 251, 267, 304-306,
        309, 332
    ACTH 304
    glucocorticoids 304
    infectious diseases 304
    IL-1 304
    LPS 304
    macrophages 304
Pituitary-adrenal axis 49, 102, 107, 173, 188, 207,
        300, 301, 303-305, 309, 310, 312,
        327, 341, 342, 372, 376, 428, 441
Pituitary-adrenocortical function 432, 440
Pituitary-gonadal axis 285, 394
Placenta 147-149, 159, 162, 163, 277, 278, 280,
        282, 283, 285, 287, 289, 292
Placental CRF 148-150, 153, 159, 160

Placental origin for CRF 148
Plasma β-endorphin 96, 105
Plasma ACTH 42, 47, 48, 53, 57, 95, 101, 104, 110, 115, 126, 128, 152, 159, 229, 233, 297, 304-306, 308, 309, 316, 320, 326, 400, 418, 421, 430-433, 439
Plasma AII 228
Plasma aldosterone 115, 126, 127, 129, 133, 229-233, 235, 241
Plasma ANP 97, 99, 110, 182, 185, 189, 190, 207
Plasma AVP 94, 106, 117, 122, 124, 133, 207
Plasma β-LPH 96
Plasma catecholamines 46, 57, 104, 110, 113, 124, 130, 131, 310
Plasma corticosteroid-binding globulin 152
Plasma corticosterone 128
Plasma cortisol 44, 48, 57, 97, 174, 316, 328, 392, 403, 418, 419, 425, 426, 436-438, 440, 441
Plasma CRF 147-152, 157, 160
Plasma epinephrine 114, 299, 339
Plasma noradrenaline 115, 120, 121, 309
Plasma norepinephrine 105, 113-118, 120-122, 124, 126, 130, 132
Plasma osmolality 98, 100, 106, 217-219, 221
Plasma prolactin 2, 20, 97, 107, 110, 173, 233, 308, 309, 425, 438, 439
Plasma renin activity, see PRA
Plasma sodium 103, 215, 218, 394
Plasticity of the corticosteroid receptor 248
    corticosterone 247
    dexamethasone 247
    glucocorticoid receptor 247
    mineralocorticoid receptor 247
PMSG 281, 287, 300
POAH 372
Polarity in brain endothelial cells 26
Polarity of receptors 29
Polycystic ovaries 165
Polyunsaturated fatty acids 258, 265, 267
POMC 91, 97, 105, 107, 108, 110, 163,

164, 189, 232, 234, 241, 242, 309, 313, 316, 331, 334-337, 342
POMC gene 335, 337
POMC mRNA 234, 334, 336
POMC-derived peptide 334
Portal blood 3-5, 11, 13, 15, 17, 18, 21, 22, 100, 107, 139, 152, 163, 184, 265, 268, 269, 299, 316, 332, 395, 396, 397, 399, 401, 405, 428, 429, 442
Portal somatostatin 397
Post-exercise release of GH 411
Posterior pituitary 101, 107, 108, 187, 203, 208, 234, 241
Postural hypotension 165, 167, 171
Potassium excretion 212, 215
PRA 91, 93-98, 102-104, 109, 115, 126-129, 133, 205, 206, 209, 223, 233, 235, 442
Pregnancy 22, 147-150, 152-155, 159-164, 166, 173, 282, 283
Pregnancy-induced hypertension 150, 152
Pregnant mare serum gonadotropin, see PMSG
Premature rupture of the membranes 150
Preoptic/anterior hypothalamus, see POAH
Preoptic periventricular tissue 214
Prepro-ANP 177
Proctolin 59
Progesterone 12-15, 17-21, 23, 56, 106, 130, 145, 180, 190, 240, 241, 253, 256, 266, 280, 286, 282-285, 290, 292
Prolactin 2-5, 7-12, 14, 15, 18, 20-23, 45, 48, 50, 52, 55, 56, 58, 91, 93, 96, 97, 103-105, 107-110, 138, 141, 145, 165-168, 173, 174, 179-181, 183, 184, 189, 233, 235, 240-242, 258, 260-262, 264, 267-269, 307-309, 315, 316, 317-329, 370, 383, 385, 391-394, 399, 400, 403-405, 407, 408, 411-414, 416-420, 423-428, 434, 436-443
    abdominal pain 166
    abortion rates 166
    α_2-adrenoceptor antagonist 167
    amenorrhea 165
    anorexia 165
    bromocriptine 165-168
    cabergoline 167, 168
    chlorpromazine 165
    chronic renal failure 165
    constipation 166
    CV 205-502 168

D1 agonist 168
D2 agonist 168
depression 166
dizziness 165
dopamine 165
dopamine agonist 166
ergoline 168
ergot alkaloid derivatives 165
fatigue 166
galactorrhea 165
hallucination 166
haloperidol 165
headache 166
hyperprolactinemia 165-167
hyperprolactinemic patients 167
hypothyroidism 165
impotence 165
infertility 166
libido 165
lisuride 167, 168
mesulergine 167
metoclopramide 165
mood change 166
nasal congestion 166
nausea 165
Parlodel LAR 167
Pergolide 167
polycystic ovaries 165
postural hypotension 165, 167
pregnancy rates 166
prolactin secretion 167
prolactinoma 165, 166
psychiatric disturbance 166
psychosis 166
side-effects 165
teratogenic 166
terguride 167
thyroxine 168
vomiting 165
Prolactin and fertility 2
aged rats 4
bromocriptine 3
DOPA 3
dopamine 3, 4
dopaminergic neurons 3
β-endorphin 3
enkephalin 3
estrous cycles 5
galactorrhea 2
gonadotropin 2
hyperprolactinemia 2

hypophysial portal blood 3, 4
hypothalamus 3
infertility 2
mass of TH 4
ME 4
morphine 3
NSD 1015 3
pituitary adenoma 3
pituitary gland 3
pituitary graft 3, 5
prolactin 2
serotonin 3
TIDA neurons 3
TH 3
Prolactin and TIDA neurons 4
aged brain 8
aged female rat 5
aged intact female rat 7
aged ovariectomized rat 9
anterior pituitary graft 11
antiserum against rat prolactin 4, 7
CSF 8-12
dopamine 5, 8
hyperprolactinemia 4, 6-9
hypophysectomy 10, 11
hypophysial portal blood 5
immunoneutralization 7, 9
in situ activity of TH 4, 9, 11
in situ molar activity of TH 6, 9, 12
intracerebral pituitary graft 9
lateral ventricle 10, 11
liver graft 11
mass of TH 4, 6, 8, 10
ME 4-12
pituitary gland 11
pituitary graft 4, 6, 8, 10
prolactin 4, 5, 9-12
prolactin antiserum 5, 7
TH 5, 11
TH activity 7
TIDA neuron 7, 8, 11
Prolactin antiserum 5, 7, 9, 14, 20
Prolactin release 45, 48, 50, 55, 110, 165, 174, 321,
322
Prolactin response to stressors 103
Prolactinoma 165, 166
Proopiomelanocortin, see POMC
Propoxyphene 382
Propranolol 57, 132, 170, 174, 314, 392, 400, 405,
424, 442
Prostaglandin E, see PGE_2

Prostaglandin synthetase 325
Prostaglandins 109, 211, 303, 312, 320,
     322-325, 327, 338
Protamine sulfate 42, 45, 50, 52, 55
Protein deficiency 375
Protein kinase A 336, 337
Protein kinase C 108, 284, 313, 337
proTRH 265
proTRH mRNA 265
Psychological responses 92
Psychological stress 109, 110, 371
Psychosis 166
PTS-1 62-66
PTS-2 62, 63, 65
PTS-3 62, 63, 66
PTS-4 62, 63, 66
Pulsatile GH secretion 171
Purkinje cell antibodies 361, 362
Purkinje neuronal degeneration 361
Pyridostigmine 170-172, 175
Pyrogens 316, 347, 376
Rabbit collection tubule 28
Rat proximal kidney tubule 28
rCRF mRNA 148
β-Receptor 177
Receptor gene expression 245, 248-250,
     254
Receptor occupancy 246, 250, 251
    glucocorticoid receptor 251
    mineralocorticoid receptor 251
Receptor plasticity vs. receptor 251
    glucocorticoid receptor 251
    mineralocorticoid receptor 251
Receptor-mediated endocytosis 78
Receptor-mediated transport 32, 34
    cationized albumin 34
    chimeric peptides 34
    insulin 34
    nontransportable peptide 34
    transferrin 34
    transportable peptide 34
Receptors and senescence 250
    ACTH 250
    dexamethasone 250
    glucocorticoid receptor 250
    mineralocorticoid receptor 250
Receptors for activin 284
Receptors resembling the V_1 subtype 81
Rectal temperature 317, 320
Reduced aldosterone responses 228
Regulation of inhibin

activin 279
androgens 281
anti-androgen 279
cAMP 279, 281
castration 281
choleratoxin 279
corpus luteum 282
5α-dihydrotestosterone 279
estradiol 279, 282
forskolin 279
FSH 279, 281
HCG 279, 281
HMG 279
IGF-I 281
immunoneutralized inhibin 281
inhibin production by granulosa cells 281
LH 281
mRNA for the inhibin α-subunit 279
PMSG 281
progesterone 282
secretion of inhibin 279
Sertoli cells 279
β-subunit 279
Regulation of sympathetic activity 114, 227
Regulation of the sympathetic nervous system 113
Release of CRF 147, 309, 316, 320, 327, 372, 374,
     376, 428, 430, 431, 441
Release of vasopressin
    antipyretic action 74
    BBB 74
    dorsal motor nucleus 74
    febrile body temperature 75
    fever 74
    nucleus tractus solitarius 74
    paraventricular nucleus 74
    prostaglandin E_1 74
    V_1 receptor antagonist 75
Renal blood flow 215
Renal glomeruli 388
Renal regulation of ECFV 213
Renal tissue 234, 237
Renin 91, 95, 98, 103-106, 109, 110, 114, 115,
     126-130, 132-134, 178, 187, 191,
     203, 205-209, 223, 224, 228, 230,
     233, 237, 240-242, 442
Renin-angiotensin system 103, 110, 126, 130, 132,
     191, 207, 208, 224, 230, 237
Reserpine-induced depletion of renal nerve
     catecholamines 102
Responsiveness to bromocriptine 172
Resting conditions 114, 119, 120, 123, 125, 129,

244, 315
Restraint 64-66, 99, 106, 109
Retina 381-383, 386, 388
Retinal capillaries 383, 384, 386, 388, 390
Retinal detachment 388
Retinopathy 171, 381, 383, 384, 388-390
Retrograde axonal transport mechanisms
    32
Retrograde flow 100
Retrograde transport 145, 345, 348, 349,
    362, 363
Rhythmic variations in BBB function 31
Role of the astrocytes
    glial cells 26
    glial foot processes 26
    polarity in brain endothelial cells
      26
    tight junctions 26
Sagittal section of mouse brain 32
Salt appetite 188, 191, 209, 246, 252, 254
Saturable transport 60, 63, 66, 68, 345,
    357
    arginine vasotocin 62, 63
    AVP 62
    AVP transport 63
    blood-to-brain transport 63
    brain-to-blood transport 63
    β-casomorphin 62, 63
    CNS transport 63
    cycloLeu-Gly 63
    DSIP 63
    dynorphin 62, 63
    leucine enkephalin 62, 63
    LHRH 62, 63
    lysine vasopressin 63
    lysine vasotocin 62, 63
    mesotocin 62
    methionine enkephalin 62, 63
    oxytocin 63
    peptide T 63
    peptide-T analog 62
    PTS 1-4 63
    Tyr-MIF-1 62, 63
    tyrosine 63
Scatchard analysis 158, 159, 233, 236, 244
Schwann cells 398
Sclerosing panencephalitis 359
Sclerotic lesions 359
Secretion of AVP and PRA elicited by
    the stressful interview
    103

Secretion of insulin 284
Secretion of pituitary hormones 41, 392, 393, 400,
    403, 407, 434
Sensitivity of the cells to other regulators 227
Sephacryl S-200 154, 156
Sephadex G50 153-155, 157, 158
Sepsis 311, 377
Septo-hippocampal complex 244
Septum 73-76, 83, 87, 214, 239, 243-245, 247, 248
Serotonin 3, 31, 34, 38, 46-55, 57, 58, 64, 65, 67,
    167, 174, 229, 379, 380, 399, 400,
    412, 427, 428, 432-434, 439-443
Serotonin agonists 427, 428, 432, 440
Serotonin and ACTH secretion
    ACTH 50, 53, 55
    ACTH release 51
    ACTH response 48
    ACTH response to serotonin 47
    arterial blood pressure 53
    catecholamines 55
    collagenase 50-55
    collagenase to induce BBB dysfunction
      47-49, 53, 54
    corticosterone 46, 51, 55
    Dysfunction of the BBB 54, 55
    fluoxetine 46, 52
    LH 49, 54, 55
    LHRH 54
    Permeability of the BBB 52
    pituitary-adrenocortical function 51
    prolactin 52, 55
    protamine sulfate 52, 55
    serotonin 47-55
    serotonin on ACTH release 52
    serotonin uptake inhibitor 46, 52
    serotonin-induced rise in ACTH levels
      55
    stress-induced corticosterone 52
    stress-induced secretion of ACTH 51,
      433
Serotonin antagonist 64, 65
Serotonin on ACTH release 52
Serotonin uptake inhibitor 46, 52
Serotonin-induced rise in ACTH levels 55
Sertoli cells 277-279, 285-289
Serum dopamine-β-hydroxylase activity 120, 132
SFO 32, 38, 41, 56, 103, 110, 192, 193, 195-209,
    239, 246, 307, 347
Sheep 100, 106, 107, 154, 212-217, 222, 225, 266,
    282, 286, 287, 291, 292, 295, 304,
    351, 354, 369, 401

Side chain cleavage enzyme 230
Side effect 168, 174, 436, 439
Site of action
    AVP 306
    CRF 306
    hypothalamic site 306
    hypothalamic-hypophysial portal
        system 306
    immunoneutralization of CRF
        306
    ME 306
Sodium 37, 63, 103, 126, 172, 176, 178,
        186-188, 191, 193,
        211-215, 217-241, 276,
        278, 354, 374, 379, 387,
        390, 394, 398, 428
Sodium appetite 227, 228
Sodium concentration of CSF 213
Sodium deficiency 229, 231, 240
Sodium diet 231, 233-237, 239
Sodium intake
    adrenal catecholamines 228
    ANF 228
    AII 228
    AVP 228
    baroreflex 228
    hormonal regulators 228
    levels of regulation 228
    local regulators 228
    mechanisms of regulation 228
    mineralocorticoids 228
    neural regulation 228
    paracrine factors 228
    peptide hormones 228
    renal afferent signals 228
    renal AII 228
    renal dopamine 228
    sodium and water transport 228
    somatostatin 228
    sympathetic tone 228
    vascular tone 228
    water and sodium appetite 228
Sodium loading 228, 230, 233, 234, 241
Sodium restriction 227-231, 234, 238, 240
Sodium valproate 172, 176, 374, 379
Soldier 91, 92
Solution hybridization/S_1 nuclease
        analysis 245
Somatostatin 113, 168-173, 175, 176, 184,
        227-230, 232, 239, 241,
        265, 266, 269, 270, 319,

        322-324, 327, 328, 392,
        395-398, 401-404, 406,
        415
Somatostatin analogue 172
Somatostatin binding sites 398
Somatostatin mRNA 397
Somatostatin on GH release 397
Somatostatin on the adrenal sensitivity to AII 227
Somatotropic activation 416
Specific binding sites 73, 77-80, 83, 189
Spermatogonia 280, 283
Spironolactone 238
Spleen 33, 38, 277, 278, 303, 305, 310
Spontaneously hypertensive rat 131, 203
Stereospecific facilitated diffusion 27
Streptozotocin 262, 264, 268, 269, 384, 387-390,
        392, 395, 396, 403-405
Streptozotocin-induced, see STZ
Stress 34, 45, 46, 49-53, 56-59, 64-66, 68, 72, 85,
        87, 91, 98-110, 114-120, 122-133,
        138, 140, 144, 145, 147, 150, 172,
        194, 200, 205-209, 243, 246, 248,
        250-253, 255, 256, 301, 309, 311,
        312, 314-316, 318, 319, 327, 328,
        331, 332, 333, 338, 339, 341, 342,
        363, 371, 373, 394, 421, 425, 427,
        429-434, 442
Stress conditions 51, 110, 117, 130
Stress hormones 49, 57, 342
Stress-induced ACTH 108, 138, 144, 431
Stress-induced corticosterone 46, 52, 140
Stress-induced increase in PRA 102, 103
Stress-induced increases in adrenal corticoid
        secretion 101
Stressful interview 98
    ANP 98
    AVP 98
    glucose 98
    PRA 98
Stressful social interaction 91, 101
Striatum 9, 76, 79, 80, 83
Structure in inhibin
    β-subunit 276
Structure of inhibin
    activin 275
    BMP 276
    cDNAs 273
    follicular fluid inhibin 273
    inhibin subunits 275
    inhibin-β-subunits 275
    MIS 276

molecular weight 273
N-terminal amino acid sequences 274
pituitary cells 275
rete testis fluid 274
secretion of FSH 275
α-subunit 273
β-subunit 273, 274
testis homogenates 274
STZ 392-398
STZ diabetes 387, 388, 390, 404
Subarachnoid space 348
Subcommissural organ 32, 347
Subfornical organ, see SFO
Substance P 30, 77, 78, 88, 113, 143, 146, 211
Substance P neurons 143, 146
α-Subunit 273-276, 279, 283, 284, 288, 289
β-Subunit 273-276, 279, 283, 284, 288, 289
Sucrose 26, 36, 386, 387
Suprachiasmatic nucleus 72, 114, 192, 195, 198, 199, 372, 373, 427, 442
Supraoptic nucleus 100, 103, 114, 143, 200, 204, 245, 349, 373
Surgical trauma 413
Sympathectomy 49
Sympathetic degeneration 142
Sympathetic ganglia 117, 142, 203, 209
Sympathetic nervous system 113, 117, 120, 128, 132, 133, 300, 314, 371, 372, 377, 378
Sympathetic outflow 113, 310, 375, 376
Sympathoadrenal system 113, 115, 124, 126, 315
Sympathoadrenomedullary system 338, 339
Synaptogenesis 258, 267
T cells 31, 305, 311, 358, 362
T3 258, 259, 264, 265, 267, 268, 393
T4 259-261, 263-265, 393
Tall stature 171
Tanycytes 32, 347, 348
Temperature regulating centers 316, 320, 326
Teratogenic effect 166
Terguride 167, 168, 174
Testis 271, 273, 274, 276, 277, 280, 283, 285, 286, 289, 290
Testosterone 279-281, 289, 290, 297-300, 394, 405

TGF-β 275, 276, 283, 288, 289
TH 3-23, 83, 144, 146, 180, 208, 254, 257, 258, 261-263, 266-268, 286
TH activity 7-9, 14, 17, 21, 263
TH gene expression 263
TH immunoreactive neurons 144
TH-containing neurons 15, 20
Thalamus 83, 348, 349, 351, 352
Theca cells 277, 283
Thermogenesis
BAT 371
BAT temperature 373
body temperature 372
brown fat activity 372
cancer 371
central control 371
fenfluramine 372
fever 372
fluoxetine 372
food intake 372
heat loss 371
heat production 371
5HT uptake inhibitors 372
hyperphagia 371
hypothalamic nuclei 373
hypothalamus 372
injury 371
metabolic rate 372
non-shivering thermogenesis 371
obesity 372
paraventricular nucleus 372
POAH 372
psychological stress 371
sympathetic nervous system 371
thermogenic capacity 371
Thermogenesis and diet
adipsin 374
body fat stores 374
carbohydrate ingestion 374
control of appetite 374
2-deoxyglucose 375
diet-induced thermogenesis 374
energy balance 374
energy stores 374
glucopenia 375
5HT 375
insulin 375
large neutral amino acids 375
protein deficiency 375
sympathetic nerves 375
thermogenic responses 375

tryptophan 375
Thermogenesis and HPA axis
    ACTH 372
    adrenalectomy 372
    α_2-adrenoceptor agonists 373
    α_2-adrenoceptors 373
    angiotensin 373
    BAT activity 372
    brown fat activity 374
    chlonidine 373
    corticosterone 372
    CRF 372
    CRF receptor antagonist 373
    dexamethasone 373
    GABA 374
    glucocorticoid antagonist 373
    glucocorticoid concentrations
        373
    glucocorticoids 372
    hypophysectomy 372
    hypothalamic function 372
    IL-1 373
    obese rat 372
    obesity in aging animals 372
    thermogenic effects of
        angiotensin 375
    thermogenic responses to CRF
        372-374
Thermogenic capacity 371
Thermogenic responses to CRF 372, 373
Thermogenic responses to injury 377
Thiopental 428, 430
Thyroid hormones 257, 258, 261, 263-269,
        310, 403, 407
    activity of TH 258
    adenohypophysis 258
    cell size 258
    cytosolic receptors 263
    5′-deiodinase 258
    differentiation 258
    DOPA accumulation 258
    DOPA concentration 259
    dopamine synthesis 259
    hypophysial portal plasma 258,
        260
    hypothalamic dopaminergic
        neurons 258
    in situ molar activity 263
    in situ molar activity of TH 263
    mass of TH 261
    maturation 258

    ME 258, 263
    neurite length 258
    nuclear receptors 258, 263
    ontogenesis 258
    phosphorylation of TH 263
    prolactin 258, 261
    synaptogenesis 258
    T3 258, 259
    T3 receptors 258
    T4 259, 263
    TH 262
    TH gene expression 263
    TH mass 263
    TH-containing cells 257
    TIDA neurons 258, 259, 261, 263
    TRH-containing neurons 258
    TRH-secreting neurons 258
    TSH 259
    turnover of dopamine 258
Thyroid stimulating hormone, see TSH
Thyrotropin releasing hormone, see TRH
Thyroxine 59, 168, 266, 268, 269, 393
TIDA 1, 3, 22, 257, 268, 393
TIDA neurons 3, 4, 7, 8, 11, 12, 14, 15, 17, 18, 20,
        21, 258-261, 263-265, 393
    adenohypophysis 262, 263
    CRH 265
    diabetic 264
    diabetic rat 263, 265
    DOPA 264
    dopamine synthesis 263
    experimental diabetes 264
    hypophysial portal plasma 265
    hypothyroidism 264
    LHRH-secreting neurons 265
    prolactin 262, 264
    somatostatin 265
    T3 264
    T4 264
    TRH 264
    TRH-secreting neurons 265
    TSH 264
Tight junctions 25, 26, 32, 35, 77, 78, 82, 346, 347,
        381, 384, 386
TNF 305, 319-329, 332, 333, 339, 340, 376
    ACTH 320-323
    anterior pituitary gland 326
    AVP 320
    cachectin 320
    cAMP 322-325
    cortisol 320

endotoxin 319
epinephrine 320
GH 320-323, 325
glucagon 320
hypothalamus 326
indomethacin 322
IL-1 325, 326
lipopolysaccharides 319, 320
norepinephrine 320
prolactin 320, 323-325
prostaglandin E_2 325
prostaglandin synthetase 325
prostaglandins 320, 323, 325
rectal temperature 320
release of CRF 320
somatostatin 322, 323
temperature regulating centers
        320
toxic shock syndrome 320
TSH 320-325
TNF-α 333, 339
Toad urinary bladder 28
Tomography 27, 166, 387, 390
Toxic shock syndrome 320
TPA 336, 337
Tractus solitarius 74, 86, 87, 240, 283,
        349, 402
Transcellular transport 29
Transfer constant 356, 358
Transferrin 34, 357, 366
Transforming growth factor, see TGF
Transient cerebral ischemia 416, 417,
        421, 424
Transient dysfunction of the BBB 34
Transport across the BBB 77
Transport mechanisms 27-29, 31, 32, 87,
        88, 348, 349, 357, 360,
        366
    acetic acid 28
    acetoacetate 28
    amino acid 28
    amino acids 27
    aspartate 28
    BBB 28, 29
    brain uptake index of
            β-hydroxybutyric acid 31
    butyric acid 28
    carrier-mediated transport
            systems 27
    choline 27
    choroid plexuses 28

2-deoxy-D-glucose 27
D-glucose 27
DOPA 28
glucose transporter 27
glutamate 28
homovanillic acid 28
β-hydroxybutyric acid 28
5-hydroxyindoleacetic acid 28
hyperglycemia 28
hypoglycemia 28
ketone bodies 28
lactic acid 28
3-O-methylglucose 27
monoamine metabolites 28
monocarboxylic acid transporter 28
monocarboxylic acids 27
NA,K-ATPase dependent efflux 29
nucleic acid precursors 27
phenylalanine 28
phenylketonuria 28
pyruvic acid 28
saturable carrier 28
stereospecific carrier 28
stereospecific facilitated diffusion 27
transporter 28
Transportable peptide 34
Transporter 27, 28, 30, 36, 82
Transsphenoidal surgery 173
TRH 59, 62, 113, 257, 258, 264-266, 268, 269, 309,
        319, 383, 386, 391, 393, 403, 440
TRH binding capacity 393
TRH gene expression 265
TRH levels in portal plasma 265
TRH mRNA 265
TRH neurons 265, 393
TRH secretion 258, 265, 309
TRH-containing neurons 258
TRH-secreting neurons 257, 258, 265
    corticosterone 265
    hypophysial portal circulation 265
    hypothyroidism 265
    ontogenesis 265
    paraventricular nucleus 265
    polyunsaturated fatty acids 265
    proTRH 265
    proTRH mRNA 265
    thyroid hormones 265
    TRH gene expression 265
    TRH levels in portal plasma 265
    TRH mRNA 265
    TRH secretion 265

Triiodothyronine 258, 267-269
Triiodothyronine receptors 258, 267
Tryptophan 146, 375, 379, 380, 412, 413, 423, 424, 439, 441, 443
TSH 145, 173, 259, 264, 265, 269, 307, 309, 317, 319-329, 370, 391-393, 399, 400, 403, 416, 427, 435
Tuberoinfundibular dopaminergic, see TIDA
Tubular cells of the nephron 234
Tuftsin 59
Tumor necrosis factor, see TNF
Tumor shrinkage 172
Turnover of dopamine 258
Tyrosine hydroxylase, see TH
Typhoid vaccine 316, 327, 376
Tyr-DSIP 62, 64
Tyr-MIF-1 59, 62, 63, 65, 67-69
Tyrosine 3, 15, 22, 23, 62, 63, 144, 146, 180, 181, 208, 257, 261, 262, 266-268, 298
Uptake of [^{125}I]-IgG 357
Urea 34, 153, 154
V_1 receptor 75
V_1 Receptor antagonist 75
Valine 412
Vascular brain perfusion 345, 354
Vascular changes in the retina 382
Vascular tone 34, 43, 227, 228
Vasopressin
    release 72
Vasopressin and information processing within the brain 75
        analgesia 75

AVP as a transmitter/modulator 75
AVP-induced information processing 76
brightness-discrimination 75
cardiovascular regulation 75
defervescence 75
feeding behavior 75
learning and memory processes 75
learning performance and AVP 76
limbic structures 75
neurophysin 75
Vasopressin and neuronal performance 72
Viral or bacterial infection 375
Virchow-Robin spaces 348
V_{max} 65, 82, 84
Volume expansion 126, 182, 185, 226
Volume loading 186, 222, 223
Vomiting 165, 412
Water deprivation 65, 194, 200, 205, 209
Water excretion 117, 191, 222
Water intake 106, 132, 178, 185, 188, 194, 217
Water loading 65
Water permeability 30
Water transport 227, 228
Water-deprived rat 205
Wistar rat 55, 80, 81, 114, 121, 129, 133, 193-195, 197, 209, 304-306, 308, 404
Yohimbine 106, 170
Zonulae occludentes 76
Zucker rat 376, 378-380